C0-BVE-080

Life at Extremes

Environments, Organisms and Strategies for Survival

Dedication

For my family, in particular Isaac, Kael and Freya
(who was born mere hours after this book).

Life at Extremes: Environments, Organisms and Strategies for Survival

Edited by

Elanor M. Bell

Scottish Institute for Marine Science, Dunbeg, Argyll, United Kingdom

CABI is a trading name of CAB International

CABI
Nosworthy Way
Wallingford
Oxfordshire OX10 8DE
UK

Tel: +44 (0)1491 832111
Fax: +44 (0)1491 833508
E-mail: cabi@cabi.org
Website: www.cabi.org

CABI
875 Massachusetts Avenue
7th Floor
Cambridge, MA 02139
USA

Tel: +1 617 395 4056
Fax: +1 617 354 6875
E-mail: cabi-nao@cabi.org

A catalogue record for this book is available from the British Library, London, UK.

Library of Congress Cataloging-in-Publication Data

Life at extremes: environments, organisms, and strategies for survival / Elanor M. Bell (editor).
p. cm.
Includes bibliographical references and index.
ISBN 978-1-84593-814-7 (alk. paper)
1. Biotic communities. 2. Extreme environments. 3. Organisms. 4. Adaptation (Biology). 5. Acclimatization. 6. Survival. 7. Climatic changes. 8. Seasons. 9. Pollution. I. Bell, Elanor.

QH543.L53 2012
577.8'2--dc23

2011026535

ISBN-13: 978 1 84593 814 7

Commissioning editor: Rachel Cutts
Editorial assistant: Alexandra Lainsbury and Gwenan Spearing
Production editor: Simon Hill

Typeset by SPi, Pondicherry, India.
Printed and bound by Gutenberg Press, Malta

The paper used for this book is FSC-certified and totally chlorine-free. FSC (the Forest Stewardship Council) is an international network to promote responsible management of the world's forests.

The costs of colour production were covered by Professor Richard A. Lutz, Institute of Marine and Coastal Sciences, Rutgers University, USA and the European Commission FP7 Coordination Action, CAREX. The editor is extremely grateful to both.

CAREX (*Coordination Action for Research Activities on life in Extreme Environments*) is a FP7 Coordination Action funded for three and a half years (2008 to 2011). This project tackles the issues of enhancing coordination of life in extreme environments research in Europe by providing networking and exchange of knowledge opportunities to the scientific community and by developing a strategic European research agenda in the field.

In its developments, CAREX intends to involve key European experts in the field of life in extreme environments research. CAREX's network includes 60 European and non-European partners from 24 countries.

CAREX is a truly interdisciplinary initiative as its approach to life in extreme environment research covers microbes, plants and animals evolving in various marine, polar, terrestrial extreme environments as well as outer space.

> CAREX Main Objectives

- **Establish** interactions, coordinate activities and promote a community identity.
- **Identify** the current status of life in extreme environments research within Europe.
- **Furthering** the scientific knowledge of life in extreme environments on key issues.
- **Identify** the priorities for future life in extreme environments research within Europe.
- **Identify** the environment-specific technological challenges and infrastructure necessary to support life in extreme environments research priorities.
- **Harmonise** protocols and approaches used in life in extreme environments research and promote knowledge transfer across the community.
- **Promote** the development of young career scientists.
- **Establish** an interactive information hub to support and develop a dynamic European community.

> CAREX Key Actions

Scientific Priority Setting

CAREX's high level scientific workshops addresses the issues of Model Ecosystems, Technology and Infrastructures and Model Organisms in order to define a European roadmap for research on life in extreme environments.

Exchange of Knowledge

Through the organisation of field trips, a laboratory seminar, a summer school and the award of 20 short visit grants, CAREX catalyses the transfer of knowledge and exchange of best practices among the scientific community and towards young researchers.

Information Hub

CAREX website (www.carex-eu.org) provides an interactive platform for the exchange of information, job and funding opportunities, news and events specific to life in extreme environment research.

Databases

CAREX website hosts an open searchable directory of international experts, a database of life in extreme environments research projects and a portfolio of specific infrastructures and technologies.

> CAREX Core Partners

- **NERC – British Antarctic Survey,** UK (Coordinator)
- **CNRS**, France
- **CNR – Istituto di Biologia Agro-ambientale e Forestale**, Italy
- **Deutsches Zentrum für Luft- und Raumfahrt (DLR)**, Germany
- **European Science Foundation**, France
- **Ifremer**, France
- **Institute of Botany, Academy of Sciences of the Czech Republic**
- **INTA – Centro de Astrobiología**, Spain
- **MATIS – Prokaria**, Iceland

Contents

Contributors

Anesio, Alexandre M. Bristol Glaciology Centre, School of Geographical Sciences, University of Bristol, University Road, Clifton, Bristol, BS8 1SS, UK. Tel: +44 (0)117 3314157; Fax: +44 (0)117 9287878; e-mail: a.m.anesio@bristol.ac.uk

Arora, Rajesh. Radiation Biotechnology Group, Institute of Nuclear Medicine and Allied Sciences, Defence Research and Development Organization, Brig. SK Mazumdar Road, Timarpur, Delhi-110054 and Staff Officer to Distinguished Scientist and Chief Controller Research and Development (Life Sciences), DRDO Bhawan, Room 339, Rajaji Marg, New Delhi-110105, India. Tel: +91 (0)11 23014873; Fax: +91 (0)11 23014259; e-mail: rajesharoradr@rediffmail.com

Aubrecht, Roman. Department of Geology and Palaeontology, Faculty of Natural Sciences, Comenius University, Mlynská dolina, 842 15 Bratislava 4, Slovakia. e-mail: aubrecht@fns.uniba.sk

Bagshaw, Elizabeth A. Bristol Glaciology Centre, School of Geographical Sciences, University of Bristol, University Road, Clifton, Bristol, BS8 1SS, UK. Tel: +44 (0)117 3314128; e-mail: liz.bagshaw@bristol.ac.uk

Baskar, Ramanathan. Department of Environmental Science and Engineering, Guru Jambheshwar University of Science and Technology, Hisar 125001, Haryana, India. e-mail: rbaskargjuhisar@yahoo.com

Baskar, Sushmitha. Department of Environmental Science and Engineering, Guru Jambheshwar University of Science and Technology, Hisar 125001, Haryana, India. e-mail: rbaskargjuhisar@yahoo.com

Barnes, David K.A. British Antarctic Survey, High Cross, Madingley Road, Cambridge, CB3 0ET, UK. Tel: +44 (0)1223 221400; e-mail: dkab@bas.ac.uk

Barták, Miloš. Section of Plant Physiology and Anatomy, Department of Experimental Biology, Faculty of Science, Masaryk University, Kamenice 5, 62500 Brno, Czech Republic. Tel: +420 549 49 3087; e-mail: mbartak@sci.muni.cz

Bell, Elanor M. Scottish Marine Institute, The Scottish Association for Marine Science, Dunbeg, Argyll PA37 1QY, UK. Tel: +44 (0)1631 559000; Fax: +44 (0)1631 559001; e-mail: elanor.bell@googlemail.com

Bellas, Chris. Bristol Glaciology Centre, School of Geographical Sciences, University of Bristol, University Road, Clifton, Bristol, BS8 1SS, UK. Tel: +44 (0)117 331 7317; Fax: +44 (0)117 9287878; e-mail: chris.bellas@bristol.ac.uk

Callaghan, Terry V. Department of Animal and Plant Sciences, Alfred Denny Building, University of Sheffield, Western Bank, Sheffield, S10 2TN, UK. Tel: +44 (0)114 222 6101; e-mail: t.v.callaghan@sheffield.ac.uk

Cass, James. Institute for Microbial Biotechnology and Metagenomics, University of the Western Cape, Bellville 7575, Cape Town, South Africa. Tel: +27 2 1959 2325; e-mail: cassjames@live.co.uk

Convey, Peter. British Antarctic Survey, High Cross, Madingley Road, Cambridge CB3 0ET, UK. Tel: + 44 (0)1223 221588; Fax: + 44 (0)1223 221259; e-mail: pcon@bas.ac.uk

Cowan, Don. Institute for Microbial Biotechnology and Metagenomics, University of the Western Cape, Bellville 7575, Cape Town, South Africa. Tel: +27 2 1959 2083; e-mail: dcowan@uwc.ac.za

Davidson, Andrew T. Australian Antarctic Division, 203 Channel Highway, 7050 Kingston, Tasmania, Australia. Tel: +61 (0)3 62323444; e-mail: Andrew.Davidson@aad.gov.au

Degu, Asfaw. French Associates Institute for Agriculture and Biotechnology of Drylands, Blaustein Institutes for Desert Research, Ben-Gurion University of the Negev. Tel: +972 8 656 3432/5; Fax: +972 8 659 6742; e-mail: asfaw1621@yahoo.com

Eppel, Amir. French Associates Institute for Agriculture and Biotechnology of Drylands, Blaustein Institutes for Desert Research, Ben-Gurion University of the Negev. Tel: +972 8 656 3432/5; Fax: +972 8 659 6742; e-mail: amireppel@gmail.com

Friedjung, Avital. French Associates Institute for Agriculture and Biotechnology of Drylands, Blaustein Institutes for Desert Research, Ben-Gurion University of the Negev. Tel: +972 8 656 3432/5; Fax: +972 8 659 6742; e-mail: avitush@hotmail.com

Glover, Adrian G. Zoology Department, The Natural History Museum, Cromwell Road, London SW7 5BD, UK. Tel: +44 (0)20 7942 5056; e-mail: a.glover@nhm.ac.uk

Gómez, Felipe. Centro de Astrobilogía (INTA-CSIC), Carretera de Ajalvir Km 4, Torrejón de Ardoz, Madrid 28850, Spain. Tel: +34 91 520 6461; e-mail: gomezgf@cab.inta-csic.es

Heuer, Verena B. Organic Geochemistry Group, Department of Geosciences and MARUM Center for Marine Environmental Sciences, University of Bremen, 28359 Bremen, Germany. Tel: +49 (0)421 218 65702; e-mail: vheuer@uni-bremen.de

Hill, Amber. French Associates Institute for Agriculture and Biotechnology of Drylands, Blaustein Institutes for Desert Research, Ben-Gurion University of the Negev. Tel: +972 8 656 3432/5; Fax: +972 8 659 6742; e-mail: bamerbright@gmail.com

Hogg, Ian D. Department of Biological Sciences, University of Waikato, Private Bag 3105, Hamilton 3240, New Zealand. Tel: +64 7 838 4139; e-mail: hogg@waikato.ac.nz

Hourdez, Stéphane. Station Biologique de Roscoff, CNRS – UPMC, Place G. Teissier, 29680 Roscoff, France. Tel: + 33 298 29 2340; Fax: + 33 298 29 2324; e-mail: hourdez@sb-roscoff.fr

Hughes, David. Scottish Association for Marine Science, Dunstaffnage Marine Laboratory, Dunbeg, Argyll, PA37 1QA, UK. Tel: +44 (0)1631 559355; e-mail: david.hughes@sams.ac.uk

Kovačik, Lubomir. Department of Botany, Faculty of Natural Sciences, Comenius University, Mlynská dolina, 842 15 Bratislava 4, Slovakia. e-mail: kovacik@fns.uniba.sk

Laybourn-Parry, Johanna. Bristol Glaciology Centre, School of Geographical Sciences, University of Bristol, University Road, Bristol BS8 1SS, UK. Tel: +44 (0)117 3314120; Fax: +44 (0)117 9287878; e-mail: jo.laybourn-parry@bristol.ac.uk

Lee, Natuschka M. Department of Microbiology, Technische Universität München, D-85354 Freising, Germany. Tel: +49 8161 71 5444; Fax: +49 8161 71 5475; e-mail: nlee@microbial-systems-ecology.de

Lettner, Herbert. University of Salzburg, Division of Physics and Biophysics, Department of Materials Engineering and Physics, Hellbrunnerstrasse 34, 5020 Salzburg, Austria. Tel: +43 (0)662 8044 5702; e-mail: herbert.lettner@sbg.ac.at

Liebl, Wolfgang. Department of Microbiology, Technische Universität München, D-85354 Freising, Germany. Tel: +49 8161 71 5444; Fax: +49 8161 71 5475; e-mail: wliebl@microbial-systems-ecology.de

Lomax, Barry H. University Park, The School of Biosciences, Division of Agricultural and Environmental Sciences, The University of Nottingham, Nottingham, NG7 2RD, UK. Tel: +44 (0)115 951 6258; e-mail: barry.lomax@nottingham.ac.uk

Lütz, Cornelius. University of Innsbruck, Institute of Botany, Sternwartestrasse 15, 6020 Innsbruck, Austria. Tel: +43 (0)512 57 5930; e-mail: cornelius.luetz@uibk.ac.at

Lutz, Richard A. Institute of Marine and Coastal Sciences, School of Environmental and Biological Sciences, Rutgers University, 71 Dudley Road, New Brunswick, NJ 08901-8525, USA. Tel: +1 732 9326555 ext. 200; e-mail: rlutz@marine.rutgers.edu

McGenity, Terry J. Department of Biological Sciences, University of Essex, Wivenhoe Park, Colchester, CO4 3SQ, UK. Tel: +44 (0)1206 872535; Fax: +44 (0)1206 872592; e-mail: TJMcGen@essex.ac.uk

Meisinger, Daniela B. Department of Microbiology, Technische Universität München, D-85354 Freising, Germany. Tel: +49 8161 71 5444; Fax: +49 8161 71 5475; e-mail: dmeisinger@microbial-systems-ecology.de

Mulako, Inonge. Institute for Microbial Biotechnology and Metagenomics, University of the Western Cape, Bellville 7575, Cape Town, South Africa. Tel: +27 2 1959 2424; e-mail: imulako@uwc.ac.za

Neal, Lenka. Zoology Department, The Natural History Museum, Cromwell Road, London SW7 5BD, UK. Tel: +44 (0)20 7942 5056; e-mail: l.neal@nhm.ac.uk

Newsham, Kevin K. British Antarctic Survey, High Cross, Madingley Road, Cambridge, CB3 0ET, UK. Tel: + 44 (0)1223 221400; e-mail: kne@bas.ac.uk

Oren, Aharon. Department of Plant and Environmental Sciences, The Institute of Life Sciences, and the Moshe Shilo Minerva Center for Marine Biogeochemistry, The Hebrew University of Jerusalem, 91904 Jerusalem, Israel. Tel: +972 2 658 4951; Fax: +972 2 6584425; e-mail: orena@cc.huji.ac.il

Pearce, David A. British Antarctic Survey, High Cross, Madingley Road, Cambridge, CB3 0ET, UK. Tel: +44 (0)1223 221561; e-mail: dpearce@bas.ac.uk

Porter, Megan. Department of Biological Sciences, University of Maryland, Baltimore County, 1000 Hilltop Circle, Baltimore, MD 21250, USA. Tel: +1 410 455 1634; e-mail: porter@umbc.edu

Post, Barbara. University of Innsbruck, Institute of Ecology, Technikerstrasse 25, 6020 Innsbruck, Austria. Tel: +43 (0)512 507 6124; e-mail: barbara.post@student.uibk.ac.at

Pradillon, Florence. Département Etude des Ecosystèmes Profonds/Laboratoire Environnements Profonds, Ifremer Centre de Brest, BP70, 29280 Plouzané, France. Tel: +33 2 98 22 49 57; e-mail: Florence.Pradillon@ifremer.fr

Psenner, Roland. University of Innsbruck, Institute of Ecology, Technikerstrasse 25, A-6020 Innsbruck, Austria. Tel: +43 (512) 507 6130; e-mail: Roland.Psenner@uibk.ac.at

Rachmilevitch, Shimon. French Associates Institute for Agriculture and Biotechnology of Drylands, Blaustein Institutes for Desert Research, Ben-Gurion University of the Negev. Tel: +972 8 656 3432/5; Fax: +972 8 659 6742; e-mail: rshimon@bgu.ac.il

Remias, Daniel. University of Innsbruck, Institute of Pharmacy, Innrain 52, 6020 Innsbruck, Austria. Tel.: +43 (0)512 507 5302; e-mail: daniel.remias@uibk.ac.at

Rewald, Boris. French Associates Institute for Agriculture and Biotechnology of Drylands, Blaustein Institutes for Desert Research, Ben-Gurion University of the Negev. Tel: +972 8 656 3432/5; Fax: +972 8 659 6742; e-mail: rewald@rootecology.de

Saiz-Jimenez, Cesareo. Instituto de Recursos Naturales y Agrobiologia, CSIC, Apartado 1052, 41080, Sevilla, Spain. Tel: +34 954 624909; Fax: +34 954 624002; e-mail: Saiz@cica.es

Sattler, Birgit. University of Innsbruck, Institute of Ecology, Technikerstrasse 25, 6020 Innsbruck, Austria. Tel.: +43 (0)512 507 6124; e-mail: birgit.sattler@uibk.ac.at

Shelef, Oren. French Associates Institute for Agriculture and Biotechnology of Drylands, Blaustein Institutes for Desert Research, Ben-Gurion University of the Negev. Tel: +972 8 656 3432/5; Fax: +972 8 659 6742; e-mail: shelefo@bgu.ac.il

Spijkerman, Elly. Department of Ecology and Ecosystem Modelling, Institute for Biochemistry and Biology, Potsdam University, Am Neuen Palais 10, D-14699 Potsdam, Germany. Tel: +49 (0)331 977 1907; Fax: +49 (0)331 977 1948; e-mail: spijker@uni-potsdam.de

Stibal, Marek. Bristol Glaciology Centre, School of Geographical Sciences, University of Bristol, University Road, Clifton, Bristol, BS8 1SS, UK. Tel: +44 (0)117 3314126; Fax: +44 (0)117 9287878; e-mail: marek.stibal@bristol.ac.uk

Summers, Annette E. Department of Geology and Geophysics, Department of Biological Sciences, E235 Howe-Russell Geoscience Complex, Louisiana State University, Baton Rouge, LA 70803, USA. New address: Department of Earth and Planetary Sciences, University of Tennessee, Knoxville, TN 37996, USA. Tel: +1 225 578 2469; Fax: +1 225 578 2302; e-mail: aengel@lsu.edu or asummers@utk.edu

Telling, Jon. Bristol Glaciology Centre, School of Geographical Sciences, University of Bristol, University Road, Clifton, Bristol, BS8 1SS, UK. Tel: +44 (0)117 331 4192; Fax: +44 (0)117 9287878; e-mail: jon.telling@bristol.ac.uk

Thomas, David N. School of Ocean Sciences, College of Natural Science, Bangor University, Menai Bridge, Anglesey, LL59 5AB, UK and Finnish Environment Institute (SYKE), Marine Research Centre, PO Box 140, FI-00251 Helsinki, Finland. e-mail: d.thomas@bangor.ac.uk

Tranter, Martyn. Bristol Glaciology Centre, School of Geographical Sciences, University of Bristol, University Road, Clifton, Bristol, BS8 1SS, UK. Tel: +44 (0)117 3314119; Fax: +44 (0)117 9287878; e-mail: m.tranter@bristol.ac.uk

Tuffin, Marla. Institute for Microbial Biotechnology and Metagenomics, University of the Western Cape, Bellville 7575, Cape Town, South Africa. Tel: +27 2 1959 9725; e-mail: ituffin@uwc.ac.za

Wadham, Jemma L. Bristol Glaciology Centre, School of Geographical Sciences, University of Bristol, University Road, Clifton, Bristol, BS8 1SS, UK. Tel: +44 (0)117 3314158; Fax: +44 (0)117 9287878; e-mail: j.l.wadham@bristol.ac.uk

Wall, Diana H. Department of Biology, and Natural Resource Ecology Laboratory, Colorado State University, Fort Collins, CO 80523, USA. Tel: +1 970 491 2504; e-mail: diana@nrel.colostate.edu

Weithoff, Guntram. Department of Ecology and Ecosystem Modelling, Institute for Biochemistry and Biology, Potsdam University, Am Neuen Palais 10, D-14699 Potsdam, Germany. Tel: +49 (0)331 977 1949; Fax: +49 (0)331 977 1948; e-mail: weithoff@uni-potsdam.de

1 What are Extreme Environments and What Lives in Them?

Elanor M. Bell[1] and Terry V. Callaghan[2,3]
[1]*Scottish Institute for Marine Science, Dunbeg, Argyll, UK;*
[2]*Department of Animal and Plant Sciences, Alfred Denny Building, University of Sheffield, UK;*
[3]*Royal Swedish Academy of Science, Stockholm, Sweden*

1.1 History and Definition of the Terms 'Extreme' and 'Extremophile'

The Romans were first to employ the term '*extremus*', the superlative of '*exterus*' (outside), and somewhere between AD 1425 and 1475 the word 'extreme' is thought to have entered common usage in Europe. Extreme is defined in the Oxford English Dictionary as 'reaching a high or the highest degree; very great; not usual; exceptional; very severe or serious' (Oxford Dictionaries, 2010). We now define a multitude of environments on planet Earth and beyond as extreme, and we continue to discover organisms capable not only of surviving but also thriving in many of them. These organisms Macelroy (1974) named 'extremophiles'; lovers (from the Greek, 'philos') of extreme environments. There are two basic degrees of extremophile-ness: those organisms that can *tolerate* an extreme and become dominant over others and those that really *love* the extreme environment and actually thrive there without release of competition.

The term extremophile frequently refers to prokaryotes such as bacteria and is sometimes used interchangeably with Archaea. However, extremophiles come in all shapes and sizes, and our understanding of the phylogenetic diversity of extreme habitats increases almost daily (Table 1.1). All hyperthermophiles (heat lovers) are members of the Archaea and Bacteria (Rothschild and Mancinelli, 2001), but eukaryotes are common among the psychrophiles (cold lovers), acidophiles (low pH lovers), alkaliphiles (high pH lovers), xerophiles (desiccation lovers) and halophiles (salt lovers). Multicellular organisms, such as invertebrate and vertebrate animals, can also be extremophiles; some psychrophilic examples include a small number of frogs, turtles and one snake that use freezing of extracellular water (water outside cells) as a survival strategy to protect their cells during the winter (Storey and Storey, 1992, 1996). Even more extraordinarily, the invertebrate Antarctic nematode, *Panagrolaimus davidi*, can survive intracellular freezing (freezing of the water inside their cells) and subsequently grow and reproduce (Wharton and Ferns, 1995). Alternatively, fish in the family Channichthyidae, living in the cold waters of the Antarctic and southern South America use anti-freeze proteins to protect their cells against freezing altogether (e.g. Bilyk and DeVries, 2010). *Poly-extremophiles* are also common in all extreme environments; organisms that can survive more than one extreme condition at the same time, for example desert plants that survive extreme heat, limited water availability and, often, high salinity.

Table 1.1. Extremophile definitions and examples.

Environmental parameter	Extremophile	Definition	Examples
Temperature	Hyperthermophile	Growth >80°C	*Pyrolobus fumarii*, 113°C (Blöchl *et al.*, 1997)
	Thermophile	Growth 60–80°C	*Synechococcus lividus* (cyanobacterium; Dyer and Gafford, 1961); *Echinamoeba thermarum* (amoeba; Baumgartner *et al.*, 2003)
	Mesophile	15–60°C	*Homo sapiens* (humans)
	Psychrophile	<15°C	*Psychrobacter; Diamesa* sp. (midge; Kohshima, 1984); *Panagrolaimus davidi* (nematode; Wharton and Ferns, 1995)
Desiccation	Xerophiles	Anhydrobiotic	*Artemia salina* (brine shrimp), nematodes, microbes, fungi, lichens (Yancey *et al.*, 1982)
Pressure	Barophile	Weight loving	Unknown
		Weight tolerant	*Plantago lanceolata* (ribwort plantain; Burden and Randerson, 1972)
	Piezophile	Pressure loving	*Shewanella violacea* (bacterium; Nakasone *et al.*, 1998; Yamada *et al.*, 2000)
pH	Acidophile	Low pH loving	Fish (pH 4; Gonzalez and Wilson, 2001); plants and insects (pH 2–3; Ponge, 2000); *Cyanidium caldarium* (red alga; Doemel and Brock, 1971); *Ferroplasma* sp. (Archaeon; Golyshina *et al.*, 2000); fungi (pH 0; Schleper *et al.*, 1995)
	Alkaliphile	High pH loving; pH >9	*Methanosalsus zhilinaeae* (pH 11.5–12; methanogenic Archaeon; Kevbrin *et al.*, 1997); *Spirulina* spp. (pH 10.5; cyanobacteria; Zavarzin *et al.*, 1999); *Ephydra hians* (pH 10; alkali fly; Thorp and Covich, 2001)
Salinity	Halophile	Salt loving (2–5 M NaCl)	*Halobacteriacea* (bacteria; Hartmann *et al.*, 1980); *Dunaliella salina* (alga; Oren, 2005)
	Osmophile	Osmotic aspects: turgor pressure, cellular dehydration and desiccation	*Saccharomyces* spp. (yeasts; e.g. Koh, 1975)
Oxygen (O_2)	Anaerobe	Cannot tolerate O_2	*Methanococcus jannaschii* (methanogenic bacterium; Jones *et al.*, 1983) *Loricefera* sp. (animal phylum; Danovaro *et al.*, 2010)
	Microaerophil	Tolerates some O_2	*Clostridium* spp. (bacteria; Johnson *et al.*, 2007)
	Aerobe	Requires O_2	*Homo sapiens* (humans)
Radiation		UV-B tolerant	*Phaeocystis pouchetti* (Marchant *et al.*, 1991)
		Infra-red	
		Ionizing radiation tolerant	*Deinococcus radiodurans* (Cox and Battista, 2005)

However, the terms *extreme* and *extremophile* are relatively anthropocentric, i.e. we humans judge habitats and their inhabitants based on what would be considered extreme for our existence. Many organisms, for example, consider oxygen to be poisonous and they flourish in anoxic environments. Because oxygen is a necessity for life as we know it, we call them extremophiles, but from their perspective, the way *we* live is extreme.

Similarly, extreme environments are usually an anthropogenic construct based on extremes of the physical environment around a baseline that is acceptable for human life. Thus extreme heat, cold, wet, dry, anoxic, toxic, etc. beyond what humans could tolerate are common concepts used to define extreme environments. Notable is the fact that this construct does not take into account either the temporal dynamics of an environment or the developmental cycle of an organism in which the tolerance of the same environment often differs. Highly dynamic environments may, for example, become very cold or very hot at certain times of year but be comfortable for humans the majority of the time. Nor does the construct consider extreme events such as natural disasters (Fig. 1.1). Furthermore, the anthropogenic construct is biased towards the adult human with no consideration that an environment benign to the adult might be extreme for the new-born.

Defining extreme environments should really be based on different concepts or currencies that relate to the organism's perspectives of a particular environment, rather than the human perspective. Such currencies include biomass, productivity, biodiversity (number of species and representation of higher taxa), diversity of ecosystem processes and specific adaptations to the environment. The sum total of all these currencies denotes the overall 'extremeness' of an environment: for example, in general an environment with low biomass and productivity, a dominance of archaic life forms and an absence of higher life forms, absence of photosynthesis and nitrogen fixation but dependence on other metabolic pathways, and specific physiological, metabolic, morphological and/or life cycle adaptations, denotes an extreme environment.

Fig. 1.1. An extreme event: the impact of a single day of rainfall in Abisko, Sub Arctic Sweden, in 2004. © Christer Jonasson.

1.2 Types of Extreme Environment

We traditionally define environments as extreme based on abiotic constructs such as their temperature, water availability, pressure, pH, salt content, oxygen status and ambient radiation levels. Similarly, extremophiles are described on the basis of the conditions they live under (Table 1.1).

Extremely cold environments are broadly conceived as habitats that either lie consistently, or fall with regular frequency or for protracted periods of time to below 5°C. This would include the poles, montane regions and some deep ocean habitats, and even some 'hot' deserts at night-time. Other organisms living in the cryosphere, for example, in ice matrices or in permafrost, or under permanent or periodic snow covers, must be tolerant of multiple extremes, including the cold, desiccation and high radiation levels. Conversely, *extremely hot* habitats remain consistently or periodically reach more than 40°C, for example, hot deserts, geothermal sites and deep-sea hydrothermal vents. Often associated with temperature extremes are environments where available *water is extremely limited*, whether persistently, or for regular or protracted periods of time such as hot and cold deserts, and some endolithic (inside rock) habitats.

Other environments are subject to *extreme hydrostatic pressure*, for example deep oceans and deep lakes. Alternatively, *hypobaric extremes* are experienced on mountains and in other elevated regions of the world.

In general, *extremely acidic* environments are conceived as being persistently or regularly below pH 5. Low pH, in particular, renders environments more extreme because it increases the solubility of metals (Delhaize and Ryan, 1995) and the organisms living in them need to be adapted to multiple abiotic extremes to cope. *Extremely alkaline* environments are those persistently or regularly above pH 9. Examples of extreme pH environments include highly acidic or alkaline lakes, groundwaters and soils. *Hypersaline* (high salt) environments are defined as those with salt concentrations greater than that of seawater, that is >35‰ or >3.5%. This includes hypersaline lakes and salterns. Habitats with limited (hypoxic) or without (anoxic) free *oxygen*, whether persistently or at regular intervals, are also considered extreme, for example, anoxic basins in oceans and lakes, and deeper sediment strata.

Last but not least, *radiation* – typically, *ultraviolet* (UV) or *infra-red* (IR) – can also impose extreme conditions on organisms. Extreme radiation environments are broadly defined as those exposed to abnormally high radiation or to radiation outside the normal range. Polar and high altitude environments, both terrestrial and aquatic, are good examples of these. Species exposed to UV or IR radiation are adapted in a range of ways, one example being the Antarctic marine algae *Phaeocystis pouchetii*, which produces water-soluble 'sunscreens' which strongly absorb UV-B wavelengths (280–320 nm) and protect the algal cells from the extremely high UV-B levels they experience in the upper 10 m of the water column after the break-up of the sea ice (Marchant *et al.*, 1991). Some organisms are also remarkably tolerant of *ionizing radiation*: the bacterium *Deinococcus radiodurans* can compensate for extensive DNA damage following exposure to ionizing radiation using intercellular mechanisms to limit DNA degradation and restrict diffusion of DNA fragments to preserve genetic integrity, as well as possessing extremely efficient DNA-repair proteins (Cox and Battista, 2005). Even in seemingly low or radiation-less environments, extremophilic organisms are adapted to respond to changes in ambient radiation levels. Blind Mexican cave fish, *Astyanax hubbsi*, have no superficially discernible eye structures and yet can distinguish minute differences in ambient light. They use extra-ocular photoreceptors to detect and respond to moving visual stimuli (Teyke and Schaerer, 1994).

At a more contemporary level, we now find ourselves living in a world where additional extremes are being imposed upon environments and the organisms that live in them as a result of human activities. These so-called *anthropogenically impacted* environments are globally wide ranging and extremely varied including, for instance, environments impacted by climate change, pollution, resource extraction, physical disturbance and over-exploitation, and can no longer be ignored when defining extremeness.

1.3 Life in Transition

In addition to the above-mentioned extreme environments, terrestrial ecologists have long been aware of the special nature of transition zones between two or more diverse communities or environments, such as the treeline on mountains (Fig. 1.2) or the boundary between forests and grasslands; these are termed *tension belts* or *ecotones* (Thomas *et al.*, 2008). Such ecotones also exist in the marine environment, for example the transition between ice and water represented by the sea ice edge (Thomas *et al.*, 2008). These transition zones typically exhibit greater species diversity and biomass density than the flanking communities, largely because the organisms living in them can draw upon resources from the adjacent environments, which can confer an advantage. However, these ecotones are continually shifting and dynamic regions, often exhibiting a wider range of abiotic and biotic conditions over an annual period than the adjacent environments. This could reasonably be considered 'extreme' in that it requires organisms to continually adapt their behaviour, phenology (seasonal timing) and interactions with other species. Species living on both sides of the ecotone are often more tolerant of the dynamics, whereas species whose range is confined to one side will consider the other side to be extreme. Pertinently, these transition zones are often also the first affected by changes in climate and/or disturbances both natural and anthropogenic (Pounds *et al.*, 2005, 2006; Parmesan, 2006).

1.4 One Man's Meat is Another Man's Poison

From the perspective of an individual organism, an extreme environment is one outside its centre of distribution or

Fig. 1.2. The Subarctic-subalpine treeline in Swedish Lapland. In the foreground is an experiment to modify the temperature of tree seedlings. © Oriol Grau.

environmental envelope. This environmental envelope can contain extreme biotic, as well as the aforementioned abiotic, environments. Thus, as a general example, Arctic or alpine vascular plants growing close to an area of permanent ice or snow will have an upper/northward boundary determined by what they experience as an extreme physical environment – ice and snow. In contrast, those adaptations that allow them to survive in their optimum environment (e.g. slow growth over many years, parsimonious use of resources such as nutrients, low fecundity) will render them poor competitors with their more 'aggressive' neighbours lower down the mountain or further southwards: these competitive neighbours represent an extreme biotic environment (e.g. Brooker and Callaghan, 1998; Cornelissen *et al.*, 2001; Callaway *et al.*, 2002). This example applies to animals too: animals of the north and mountain dwellers are relatively free of parasites and diseases and have fewer predators than those of the lowlands/tropics, although this situation is likely to alter as a result of climate change (Kutz *et al.*, 2004, 2005; Pounds *et al.*, 2006).

A difficult question therefore, that challenges the idea of the currencies described in Section 1.1, is what is the most extreme environment? The highly diverse and productive tropical rainforest in which most temperate and northern species would be out-competed in a biotically extreme environment, or the climatically, and thus abiotically, harsh tundra which would kill any plant not tolerant of freezing? This question shows the importance of having multiple currencies for comparing the extremeness of environments: whereas the two extreme environments of the Arctic and the tropical rainforest can both kill maladapted plants for very different reasons, the rainforest has greater biodiversity, representation of more life forms and higher productivity and is therefore less extreme than the Arctic.

1.5 Timing is Everything

Even outside ecotones, environments are dynamic: just when you think they are benign, there can be an extreme period or an extreme event. Temperate latitudes with high biodiversity and productivity are by name 'temperate' – except in winter time. Because winter comes every year and is predictable, specific adaptations to this 'extreme', such as the senescence and fall of deciduous leaves, seed production in annual plants, hibernation in mammals, are often overlooked as responses to the extremes of prolonged low temperatures, shorter day length and limited or absent food supplies (Storey, 1997, 1998; Morin and Storey, 2009). Extreme events, such as fire, flood, drought and storms, are more noticeable. Such events are less predictable but nevertheless demand specific adaptations.

Longer term changes can expose organisms to more or fewer extremes as well. Take for example global climate change (see also Psenner and Sattler, Chapter 23, this volume). Climate change is affecting natural systems by shifting the phenology of reproduction and migrations, shifting geographical boundaries and species distributions, and even altering species' body sizes (Parmesan, 2006).

Elevated sea temperatures as small as 1°C above long-term summer averages have subjected coral reefs to abiotic extremes. This has caused widespread coral bleaching (loss of coral algal symbionts) in many tropical oceans (Hoegh-Guldberg, 1999), resulting in community shifts toward high-temperature-tolerant species (Baker *et al.*, 2004) and the global extinction of some species (Hoegh-Guldberg, 1999, 2005). Further studies indicate that whole reefs are already at their thermal tolerance limits and that any further increases in temperature will push them beyond the extremes they can tolerate and toward mass extinction (Hoegh-Guldberg, 1999).

Moreover, there are increasing numbers of documented extinctions of entire species directly attributable to global warming. Mountain-restricted species appear to be particularly vulnerable. For instance, as a direct result of climate change the geographic range of cloud-forest-dependent amphibians in the Central and South American tropics has been restricted, and their exposure to epidemic diseases increased, leading to population declines and extinctions; among harlequin frogs alone, 67% of species have disappeared in the last 20 to 30 years (Pounds *et al.*, 2005, 2006).

Recent studies have revealed the impact of climate on the demographic processes of both animals and plants. Climate change is directly affecting population numbers of the yellow-bellied marmot (*Marmota flaviventris*) that live in subalpine habitats in the USA (Ozgul *et al.*, 2010; Visser, 2010). As a direct result of increasing local temperatures, the marmots now emerge from hibernation earlier in spring and also wean their young earlier. Their growing season has been extended and therefore they are heavier before they begin hibernation. This in turn enhances winter survival rates and probability of reproduction, leading to a threefold increase in marmot numbers since 2000 (Ozgul *et al.*, 2010). The marmots are now living within a less extreme abiotic environment, but will they soon be faced with more extreme biotic conditions as individuals within the expanding population start to compete for limited resources such as food and space? Examples are not restricted to the animal kingdom: in simulated environmental change experiments (e.g. Wookey *et al.*, 1993; Walker *et al.*, 2006) and in recent observations (e.g. Tape *et al.*, 2006), Arctic shrubs proliferate in response to climate warming but a higher and denser canopy together with more litter produce shading that decreases the abundance of plants such as lichens (e.g. Cornelissen *et al.*, 2001).

At the poles, climate change has led to declines in sea ice extent and duration, which alters both abiotic and biotic conditions (see also Thomas, Chapter 4, this volume). In the Antarctic, sea ice reduction has led to reduced abundances of ice algae, in turn leading to declines in krill from a region where they have historically been concentrated, enhancing the biotic pressures on the fish, seabirds and marine mammals that rely on them as a primary food source (Atkinson *et al.*, 2004; Parmesan, 2006). In the North, satellite-derived indices of photosynthesis show a general increase in plant cover and productivity that has led to a 'greening of the Arctic' (Bhatt *et al.*, 2010). Measurements on the ground show earlier plant phenology and changes in composition of species and life forms (Post *et al.*, 2009). In the animal kingdom, northern range margin extent has increased in some invertebrates and vertebrates and populations of several mammals have decreased (Post *et al.*, 2009).

1.6 A Frog is not a Frog but Merely a Transient Phase Between a Tadpole and an Egg

Not only are environments dynamic, sometimes extreme, sometimes not, but organisms are also dynamic: they have life cycles in which the environment that supports one stage is an extreme for other stages. Consider the coconut: the seed is adapted to transport by sea but the mature tree grows on land (Howe and Smallwood, 1982). In the spore-bearing vascular plants (vascular cryptogams) such as ferns, club mosses and spike mosses, the gametophyte generation can be without photosynthetic pigment, without roots and dependent on a wet environment. This contrasts with the sporophyte generation, for example of *Pteridium aquilinum* (bracken), which has rhizomes, roots, and tall green shoots that can tolerate hot, dry conditions in full sunlight (Foster and Gifford, 1988). Often the difference in anatomy, physiology and morphology between gametophyte and sporophyte generations is of the magnitude of the difference between major taxa. Closer to home, the young of many species, including humans, are intolerant of the extreme environment surrounding the adult and require a specific biotic environment during a period of post-natal care.

An illustrative example is the generalized life cycle of a long-lived plant such as a tree (Fig. 1.3; Callaghan *et al.*, 2002). At the germination/establishment phase (A), mortality is intense as the physical environment is 'extreme' for this life cycle stage (e.g. too little soil, strong winds, rain washing

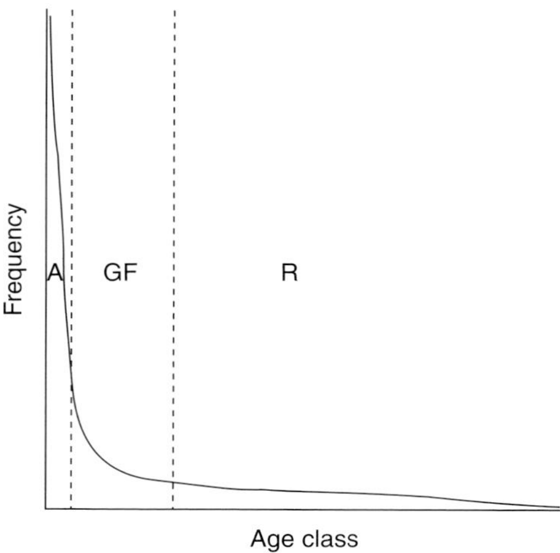

Fig. 1.3. Mortality of long-lived plants over time. At the germination/establishment phase (A), mortality is intense as the physical environment is 'extreme' for this life cycle stage; later, there is a transition phase (GF) when the seedling or sapling becomes a young tree. This encompasses a form-shift and mortality is caused by an extreme biotic environment (e.g. shade) or physical environment (e.g. wind). Once the adult tree is established (R), it survives most of the 'extremes' that the younger stages cannot and subsequently can survive and grow for hundreds of years. From Callaghan *et al.*, 2002.

seedlings away, drought before good roots are produced). Later, there is a transition phase (GF) when the seedling or sapling becomes a young tree. This encompasses a form-shift and mortality is caused by an extreme biotic environment (e.g. shade) or physical environment (e.g. wind). Once the adult tree is established (R), it survives most of the 'extremes' that the younger stages cannot and subsequently can survive and grow for hundreds of years. More often than not, the causes of mortality are biotic extremes, insect pests, diseases, competition and herbivory, although physical extreme events such as storms and fire also occur. To quantify this, a birch tree (*Betula pubescens*) in Lapland produces about 40 million seeds in its life time and only one is required to replace the parent (Callaghan *et al.*, 2002). Surely the death of 40 million seeds minus one denotes an extreme environment!

1.7 Friends and Neighbours

In the Darwinian world, neighbours are *not* friends: they are competitors for resources. Unfortunately this part-truth dominated ecology until recently. It had flourished because Darwin (and most ecologists) worked within biotically extreme environments where competition is important in determining community structure and the survival of an individual (Todes, 2009). Had Darwin worked in hot or Arctic deserts, he would have noticed that species were aggregated even though free niches on bare ground were everywhere. Consequently, the benefit of aggregation must presumably out-weigh any harmful effects of competition. A series of studies has recently proved these benefits and the processes of plant–plant help is termed *facilitation* – a new concept arising from physically extreme environments (Brooker and Callaghan, 1998; Callaway *et al.*, 2002; Bruno *et al.*, 2003; Todes, 2009; see also references in Rewald *et al.*, Chapter 11, this volume). In this theory, positive interactions play a critical, but often underappreciated, role in ecological communities by reducing physical or biotic stresses in existing habitats and by creating new habitats upon which many species depend (Brooker and Callaghan, 1998; Stachowicz, 2001; Todes, 2009). In the case of extreme environments, they may modify conditions sufficiently to make life more hospitable for others that would not otherwise be able to survive. For example, in a high salt marsh intertidal zone in New England, USA, growth of the saltmarsh rush *Juncus geradi* provides shade beneath its canopy, thus reducing the evaporation of water and the local concentration of salts and its rhizosphere increases oxygenation of the surrounding soil. The resultant lower salinity and increased oxygen levels allow the marsh elder shrub *Iva frutescens* to extend its distribution lower down the intertidal and into typically more saline, anoxic areas than it would otherwise be able to tolerate (Hacker and Bertness, 1995). Similarly, Shevtsova *et al.* (1995) demonstrated that removing the common bilberry *Vaccinium myrtillus* from Arctic shrub tundra in northern Finland led to reduced growth in the black crowberry, *Empetrum nigrum* (a dwarf shrub), because the protective, warmer microclimate provided by the larger shrub was removed. Many more examples exist in the literature, several of which are summarized in Brooker and Callaghan (1998). In such plant–plant positive interactions there also exists another life cycle/developmental dynamic in that the relationships between neighbours will change with age: a moss or shrub that protects/facilitates the survival of a tree seedling by providing shade and moisture might itself be protected by the tree once the tree seedling has grown above the ground-layer canopy.

1.8 Where to Next?

The discovery of extreme environments and flourishing communities of organisms living in them on Earth has opened up a huge range of biotechnological and biomedical opportunities and, as a result, extremophile research has expanded rapidly in the last 20 years. Now genomes are regularly sequenced, patents filed and concerted national and international programmes of research are launched on a regular basis (Arora and Bell, Chapter 25, this volume). Our continued discovery of extreme environments and understanding of the extreme conditions that life can endure has also, literally, opened up new worlds of possibilities of discovering life on other planets, including the concept of *panspermia*, the transport of life from one planet to another via, for example, meteorites (Horneck, 2006; Gómez *et al.*, Chapter 26, this volume).

However, exploring extremes in an unconventional way is not simply an intellectual academic exercise, it is a framework within which adaptation, evolution and response to environmental change – biotic or abiotic – can be better understood. Too often, the physiologist researches the adaptations to the environment of the mature individual that has huge tolerance compared with the vulnerable younger stages of the life cycle. Too often, the productive, biodiverse tropical paradise is seen as abiotically benign instead of biotically extreme. Finally, although we can currently accept exclusive communities of microbes as indicators of a physically extreme environment, do we know how genetically diverse they are, how their life cycle stages respond to various environments and what extreme micro-events they experience? We are beginning to unravel some of the answers to these questions and many of the chapters in this book will address them, but the field lies wide open for further and future research.

In summary, in this volume we discuss the range of extreme environments that exist on Earth, the organisms, often extremophiles, which live in them and their evolutionary implications. We explore contemporary issues such as the extreme changes that anthropogenically-created extremes such as climate change, pollution, habitat destruction and over-exploitation will exert upon these environments and their inhabitants. We consider the realized and potential biotechnical and biomedical applications of extremophiles in modern society and conclude with a discussion of how Earth's extreme environments are facilitating current research in astrobiology (life on other planets), and how this and the aforementioned topics feed into our understanding of how life on Earth evolved and continues to do so.

References

Atkinson, A., Siegel, V., Pakhomov, E. and Rothery, P. (2004) Long-term decline in krill stock and increase in salps within the Southern Ocean. *Nature* 432, 100–103.

Baker, A.C., Strager, C.J., McClanahan, T.R. and Glynn, P.W. (2004) Coral reefs: coral's adaptive response to climate change. *Nature* 430, 741.

Baumgartner, M., Yapi, A., Gröbner-Ferreira, R. and Stetter, K.O. (2003) Cultivation and properties of *Echinamoeba thermarum* n. sp., an extremely thermophilic amoeba thriving in hot springs. *Extremophiles* 7, 267–274.

Bhatt, U.S., Walker, D.A., Raynolds, M.K., Comiso, J.C., Epstein, H.E., Jia, G., Gens, R., Pinzon, J.E., Tucker, C.J., Tweedie, C.E. and Webber, P.J. (2010) Circumpolar Arctic tundra vegetation change is linked to sea-ice decline. *Earth Interactions* 14, 1–20.

Bilyk, K. and DeVries, A. (2010) Freezing avoidance of the Antarctic icefishes (Channichthyidae) across thermal gradients in the Southern Ocean. *Polar Biology* 33, 203–213.

Blöchl, E., Rachel, R., Burggraf, S., Hafenbradlm, D., Jannasch, H.W. and Stetter, K.O. (1997) *Pyrolobus fumarii*, gen. and sp. nov., represents a novel group of archaea, extending the upper temperature limit for life to 113 degrees C. *Extremophiles* 1, 14–21.

Brooker, R.W. and Callaghan, T.V. (1998) The balance between positive and negative plant interactions and its relationship to environmental gradient: a model. *Oikos* 81, 196–207.

Bruno, J.F., Stachowicz, J.J. and Bertness, M.D. (2003) Inclusion of facilitation into ecological theory. *Trends in Ecology & Evolution* 18, 119–125.

Burden, R.F. and Randerson, P.F. (1972) Quantitative studies of the effects of human trampling on vegetation as an aid to the management of semi-natural areas. *Journal of Applied Ecology* 9, 439–457.

Callaghan, T.V., Werkman, B.R. and Crawford, R.M.M. (2002) The Tundra–taiga interface and its dynamics: concepts and applications. *Ambio Special Issue* 12, 6–14.

Callaway, R.M., Brooker, R.W., Choler, P., Kikvidze, Z., Lortie, C.J., Michalet, R., Paolini, L., Pugnaire, F.I., Newingham, B., Aschehoug, E.T., Armas, C., Kikodze, D. and Cook, B.J. (2002) Positive interactions among alpine plants increase with stress. *Nature* 417, 844–848.

Cornelissen, J.H.C., Callaghan, T.V., Alatalo, J.M., Michelsen, A., Graglia, E., Hartley, A.E., Hik, D.S., Hobbie, S.E., Press, M.C., Robinson, C.H., Henry, G.H.R., Shaver, G.R., Phoenix, G.K., Gwynn Jones, D., Jonasson, S., Chappin III, F.S., Molau, U., Neill, C., Lee, J.A., Melillo, J.M., Sveinbjörnsson, B. and Aerts, R. (2001) Global change and arctic ecosystems: is lichen decline a function of increases in vascular plants? *Journal of Ecology* 89, 984–994.

Cox, M.M. and Battista, J.R. (2005) *Deinococcus radiodurans* – the consummate survivor. *Nature Reviews Microbiology* 3, 882–892.

Danovaro, R., Dell'Anno, A., Pusceddu, A., Gambi, C., Heiner, I. and Møbjerg Kristensen, R. (2010) The first metazoa living in permanently anoxic conditions. *Biomed Central, Biology* 8, 30.

Delhaize, E. and Ryan, P.R. (1995) Aluminium toxicity and tolerance in plants. *Plant Physiology* 107, 315–321.

Doemel, W.N. and Brock, T.D. (1971) The physiological ecology of *Cyanidium caldarium*. *Journal of General Microbiology* 67, 17–32.

Dyer, D.L. and Gafford, R.D. (1961) Some characteristics of a thermophilic blue-green alga. *Science* 134, 616.

Foster, A.S. and Gifford, E.M. (1988) *Morphology and Evolution of Vascular Plants*, 3rd edn. W.H. Freeman and Company, New York, 626 pp.

Golyshina, O.V., Pivovarova, T.A., Karavaiko, G.I., Kondratéva, T.F., Moore, E.R., Abraham, W.R., Lünsdorf, H., Timmis, K.N., Yakimov, M.M. and Golyshin, P.N. (2000) *Ferroplasma acidiphilum* gen. nov., sp. nov., an acidophilic, autotrophic, ferrous-iron-oxidizing, cell-wall-lacking, mesophilic member of the Ferroplasmaceae fam. nov., comprising a distinct lineage of the Archaea. *International Journal of Systematic and Evolutionary Microbiology* 50, 997–1006.

Gonzalez, R. and Wilson, R. (2001) Patterns of ion regulation in acidophilic fish native to the ion-poor, acidic Rio Negro. *Journal of Fish Biology* 58, 1680–1690.

Hacker, S.D. and Bertness, M.D. (1995) Morphological and physiological consequences of a positive plant interaction. *Ecology* 76, 2165–2175.

Hartmann, R., Sickinger, H.-D. and Oesterhelt, D. (1980) Anaerobic growth of halobacteria. *Proceedings of the National Academy of Sciences of the United States of America* 77, 3821–3825.

Hoegh-Guldberg, O. (1999) Climate change, coral bleaching and the future of the world's coral reefs. *Marine and Freshwater Research* 50, 839–866.

Hoegh-Guldberg, O. (2005) Marine ecosystems and climate change. In: Lovejoy, T. and Hannah, L. (eds) *Climate Change and Biodiversity*. Yale University Press, New Haven, Connecticut, pp. 256–271.

Horneck, G. (2006) Bacterial spores survive simulated meteorite impact. In: Cockell, C., Gilmour, I. and Koeberl, C. (eds) *Biological Processes Associated with Impact Events*. Springer, Berlin, Heidelberg, Germany, pp. 41–53.

Howe, H.F. and Smallwood, J. (1982) Ecology of seed dispersal. *Annual Review of Ecology and Systematics* 13, 201–228.

Johnson, E.A., Summanen, P. and Finegold, S.M. (2007) Clostridium. In: Murray, P., Baron, E., Jorgensen, J., Landry, M. and Pfaller, M. (eds) *Manual of Clinical Microbiology*, 9th edn. American Society for Microbiology Press, Washington DC, 2256 pp.

Jones, W.J., Leigh, J.A., Mayer, F., Woese, C.R. and Wolfe, R.S. (1983) *Methanococcus jannaschii* sp. nov., an extremely thermophilic methanogen from a submarine hydrothermal vent. *Archives of Microbiology* 136, 254–261.

Kevbrin, V.V., Lysenko, A.M. and Zhilina, T.N. (1997) Physiology of the alkaliphilic methanogen Z-7936, a new strain of *Methanosalsus zhilinaeae* isolated from Lake Magadi. *Microbiology* 66, 261–266.

Koh, T.Y. (1975) Studies on the 'osmophilic' yeast *Saccharomyces rouxii* and an obligate osmophilic mutant. *Journal of General Microbiology* 88, 101–114.

Kohshima, S. (1984) A novel cold-tolerant insect found in a Himalayan glacier. *Nature* 310, 225–227.

Kutz, S.J., Hoberg, E.P., Nagy, J., Polley, L. and Elkin, B. (2004) 'Emerging' parasitic infections in Arctic ungulates 1. *Integrative and Comparative Biology* 44, 109–118.

Kutz, S.J., Hoberg, E.P., Polley, L. and Jenkins, E.J. (2005) Global warming is changing the dynamics of Arctic host–parasite systems. *Proceedings of the Royal Society B: Biological Sciences* 272, 2571–2576.

Macelroy, R.D. (1974) Some comments on the evolution of extremophiles. *Biosystems* 6, 74–75.

Marchant, H.J., Davidson, A.T. and Kelly, G.J. (1991) UV-B protecting compounds in the marine alga *Phaeocystis pouchetti* from Antarctica. *Marine Biology* 109, 391–395.

Morin, P. and Storey, K.B. (2009) Mammalian hibernation: differential gene expression and novel application of epigenetic controls. *International Journal of Developmental Biology* 53, 433–442.

Nakasone, K., Ikegami, A., Kato, C., Usami, R. and Horikoshi, K. (1998) Mechanisms of gene expression controlled by pressure in deep-sea microorganisms. *Extremophiles* 2, 149–154.

Oren, A. (2005) A hundred years of *Dunaliella* research: 1905–2005. *Saline Systems* 1, doi: 10.1186/1746-1448-1-2.

Oxford Dictionaries (2010) http://oxforddictionaries.com/view/entry/m_en_gb0282940#m_en_gb0282940 (accessed 18 August 2010).

Ozgul, A., Childs, D.Z., Oli, M.K., Armitage, K.B., Blumstein, D.T., Olson, L.E., Tuljapurkar, S. and Coulson, T. (2010) Coupled dynamics of body mass and population growth in response to environmental change. *Nature* 466, 482–485.

Parmesan, C. (2006) Ecological and evolutionary responses to recent climate change. *Annual Review of Ecology, Evolution, and Systematics* 37, 637–669.

Ponge, J.-F. (2000) Acidophilic Collembola: living fossils? *Contributions from the Biological Laboratory, Kyota University* 29, 65–74.

Post, E., Forchhammer, M.C., Bret-Harte, S., Callaghan, T.V., Christensen, T.R., Elberling, B., Fox, A.D., Gilg, O., Hik, D.S., Ims, R.A., Jeppesen, E., Klein, D.R., Madsen, J., McGuire, A.D., Rysgaard, S., Schindler, D.E., Stirling, I., Tamstorf, M.P., Tyler, N.J.C., van der Wal, R., Welker, J., Wookey, P.A., Schmidt, N.M. and Aastrup, P. (2009) Ecological dynamics across the Arctic associated with recent climate change. *Science* 325, 1355–1358.

Pounds, J.A., Fogden, M.P.L. and Masters, K.L. (2005) Responses of natural communities to climate change in a highland tropical forest. In: Lovejoy, T. and Hannah, L. (eds) *Climate Change and Biodiversity*. Yale University Press, New Haven, Connecticut, pp. 70–74.

Pounds, J.A., Bustamente, M.R., Coloma, L.A., Consuegra, J.A., Fogden, M.P.L., Foster, P.N., La Marca, E., Masters, K.L., Merino-Viteri, A., Puschendorf, R., Ron, S.R., Sánchez-Azofeifa, G.A., Still, C.J. and Young, B.E. (2006) Widespread amphibian extinctions from epidemic disease driven by global warming. *Nature* 398, 611–615.

Rothschild, L.J. and Mancinelli, R.L. (2001) Life in extreme environments. *Nature* 409, 1092–1101.

Schleper, C., Piihler, G., Kuhlmorgen, B. and Zillig, W. (1995) Life at extremely low pH. *Nature* 375, 741–742.

Shevtsova, A., Ojala, A., Neuvonen, S., Vieno, M. and Haukioja, E. (1995) Growth and reproduction of dwarf shrubs in a subarctic plant community: annual variation and above-ground interactions with neighbours. *Journal of Ecology* 83, 263–275.

Stachowicz, J.J. (2001) Mutualism, facilitation, and the structure of ecological communities. *BioScience* 51, 235–246.

Storey, K.B. (1997) Metabolic regulation in mammalian hibernation: enzyme and protein adaptations. *Comparative Biochemistry and Physiology: Part A* 118, 1115–1124.

Storey, K.B. (1998) Survival under stress: molecular mechanisms of metabolic rate depression in animals. *South African Journal of Zoology* 33, 55–64.

Storey, K.B. and Storey, J.M. (1992) Natural freeze tolerance in ectothermic vertebrates. *Annual Review of Physiology* 54, 619–637.

Storey, K.B. and Storey, J.M. (1996) Natural freezing survival in animals. *Annual Review of Ecology and Systematics* 27, 365–386.

Tape K., Sturm, M. and Racine, C. (2006) The evidence for shrub expansion in Northern Alaska and the Pan-Arctic. *Global Change Biology* 12, 686–702.

Teyke, T. and Schaerer, S. (1994) Blind Mexican cave fish (*Astyanax hubbsi*) respond to moving visual stimuli. *Journal of Experimental Biology* 188, 89–101.

Thomas, D.N., Fogg, G.E., Convey, P., Fritsen, C.H., Gili, J.-M., Gradinger, R., Laybourn-Parry, J., Reid, K. and Walton, D.W.H. (2008) *The Biology of Polar Regions*. Oxford University Press Inc., New York, 394 pp.

Thorp, J.H. and Covich, A.P. (2001) An overview of freshwater habitats. In: Thorp, J.H. and Covich, A.P. (eds) *Ecology and Classification of North American Freshwater Invertebrates*. Academic Press, San Diego, California, pp. 19–42.

Todes, D. (2009) Global Darwin: contempt for competition. *Nature* 462, 36–37.

Visser, M.E. (2010) Fatter marmots on the rise. *Nature* 466, 445–446.

Walker, M.D., Wahren, H.C., Hollister, R.D., Henry, G.H.R., Ahlquist, L.E., Alatalo, J.M., Bret-Harte, M.S., Calef, M.P., Callaghan, T.V., Carroll, A.B., Copass, C., Epstein, H.E., Jonsdottír, I.S., Klein, J.A., Magnusson, B., Molau, U., Oberbauer, S.F., Rewa, S.P., Robinson, C.H., Shaver, G.R., Suding, K.N., Thompson, C.C., Tolvanen, A., Totland, Ø., Turner, P.L., Tweedie, C.E., Webber, P.J. and Wookey, P.A. (2006) Plant community responses to experimental warming across the tundra biome. *Proceedings of the National Academy of Sciences of the United States of America* 103, 1342–1346.

Wharton, D.A. and Ferns, D.J. (1995) Survival of intracellular freezing by the Antarctic nematode, *Panagrolaimus davidi*. *Journal of Experimental Biology* 198, 1381–1387.

Wookey, P.A., Parsons, A.N., Welker, J.M., Potter, J.A., Callaghan, T.V., Lee, J.A. and Press, M.C. (1993) Comparative responses of phenology and reproductive development to stimulated environmental change in sub-Arctic and high Arctic plants. *Oikos* 67, 490–502.

Yamada, M., Nakasone, K., Tamegai, H., Kato, C., Usami, R. and Horikoshi, K. (2000) Pressure regulation of soluble cytochromes c in a deep-sea piezophilic bacterium, *Shewanella violacea*. *Journal of Bacteriology* 182, 2945–2952.

Yancey, P.H., Clark, M.E., Hand, S.C., Bowlus, R.D. and Somero, G.N. (1982) Living with water stress: evolution of osmolyte systems. *Science* 217, 1214–1222.

Zavarzin, G.A., Zhilina, T.N. and Kevbrin, V.V. (1999) The alkaliphilic microbial community and its functional diversity. *Microbiology* 68, 503–521.

2 Past Extremes

Barry H. Lomax
The School of Biosciences, The University of Nottingham, UK

2.1 Introduction

> The past is a foreign country: they do things differently there.
>
> L.P. Hartley

The concept of life in extreme environments and what constitutes an extreme environment cannot be regarded as static through the vastness of geological time. At present the atmosphere has a relatively low carbon dioxide (CO_2) concentration (~390 ppm), and although it is rising rapidly it is greatly reduced when compared to concentrations that have occurred throughout the Phanerozoic (last 542 million years) (e.g. Berner and Kothavala, 2001; Berner, 2006). Oxygen (O_2) currently makes up approximately 21% of the Earth's atmosphere and again this concentration has varied widely throughout the Phanerozoic, with model predictions suggesting periods of prolonged O_2 concentration above and below current levels (e.g. Berner, 2001, 2006; Fig. 2.1). Looking further back into geological time, prior to the Great Oxidation Event 2400 million years before present (Ma), almost no free oxygen was present in the Earth's atmosphere and CO_2 concentrations were greatly elevated due to high levels of volcanic activity and the lack of a long-term CO_2 sink.

Superimposed on these long-term variations are short-term changes that disrupt this equilibrium and lead to the development of short-term (geologically speaking) extreme environments, which are often accompanied by mass extinction events or species turnover. Plants as sessile organisms must adapt to meet these new environmental challenges, thus they provide an excellent tool for tracking long-term changes in palaeoclimate and the potential to identify extreme palaeoenvironments.

This chapter uses recent advances in a wide variety of scientific disciplines to look at four key periods in the context of life in extreme past environments: (i) the Early Archaean (3500 Ma); (ii) the appearance of embryophytes (terrestrial plants) (~470 Ma); (iii) the Permian/Triassic boundary (251 Ma); and (iv) the Cretaceous/Tertiary boundary (65 Ma). These events will be discussed in terms of atmospheric O_2 and CO_2 concentrations, UV-B radiation flux and temperature.

2.2 Archaean Climate and Life (~3500 Ma)

Understanding the nature of past environments becomes increasingly fraught when investigating deep time. The identification of chemical signatures of life and climate along with identifying the microscopic

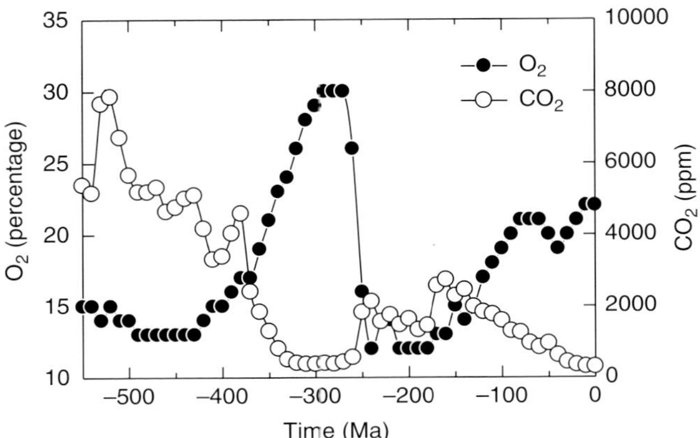

Fig. 2.1. Model prediction of long-term variation in atmospheric O_2 (%) and CO_2 (ppm). Oxygen concentrations are from Berner (2001) and CO_2 predictions are from the GEOCARB III model (Berner and Kothavala, 2001). Note that more recent formulation of the GEOCARB model family such as GEOCARBSULF (Berner, 2006; Berner, 2009) show the same overall pattern in terms of O_2 and CO_2 excursion but there is a difference in the magnitude of the excursion. Redrawn from Berner, 2001 and Berner and Kothavala, 2001 (A revised model of atmospheric CO_2 over Phanerozoic time. In: *American Journal of Science*. Reprinted with permission from the American Journal of Science).

remains of life becomes more difficult, and indeed the rock record becomes more sporadic and altered due to its antiquity. However, recent advances in technology enabling the preparation of ever smaller samples for geochemical analysis coupled with an interdisciplinary approach and the development of systems based models, is enabling progress to be made in establishing the presence of life and characterizing the environment it lived in.

Thought to be one of the long-standing rate limiting steps surrounding the development of life on Earth was the establishment of an effective stratospheric ozone (O_3) layer for the efficient screening of shortwave UV radiation (240–280 nm, UV-C and 280–315 nm UV-B). The formation of O_3 and the development of a stratospheric O_3 layer are coupled to atmospheric O_2 through the Chapman reactions:

$$O_2 + hv_1 \rightarrow O + O \quad (2.1)$$

$$O + O_2 \rightarrow O_3 \quad (2.2)$$

$$O_3 + hv_2 \rightarrow O_2 + O \quad (2.3)$$

$$O + O_3 \rightarrow O_2 + O_2 \quad (2.4)$$

where hv is radiation at a specific wavelength: $hv_1 = \lambda \leq 242$ nm and $hv_2 = \lambda \sim 1180$ nm.

This series of reactions show that prior to the development of an oxygen-rich atmosphere, the stratospheric O_3 layer would have been absent or greatly reduced implying an increase in the flux of shortwave UV radiation. The Earth's atmosphere can essentially be regarded as anoxic until 2400 million years ago (2.4 Ga). Evidence supporting this assertion comes from a variety of disciplines and includes the occurrence of detrital mineral grains that are sensitive to oxidative weathering such as uraninite (UO_2), pyrite (FeS_2) and siderite ($FeCO_3$), which would be weathered on exposure in an oxygen-rich atmosphere, plus the absence of extensive red beds produced by the oxidative weathering of iron.

One of the most compelling lines of evidence supporting an oxygen poor atmosphere in the Archaean and, thus, the absence of a stratospheric O_3 layer, is the anomalous fractionation of sulfur isotopes over this time period (Farquhar *et al.*, 2000; Farquhar and Wing, 2003). Sulfur exists as four stable isotopes (^{32}S, ^{33}S, ^{34}S and ^{36}S). Under normal biotic fractionation the ratio of S isotopes is controlled by their mass. For example, the fractionation of ^{33}S is approximately half that of ^{34}S (Fig. 2.2). This physical relationship can be used to predict the $\delta^{33}S$ isotopic value of a sample based on measurement of $\delta^{34}S$ to give $\Delta^{33}S$.

Analysis of rock samples younger than 2.4 Ga shows that predicted values of ^{33}S are similar to measured values and that fractionation is mass dependent, i.e. being driven by biotic mechanisms and $\Delta^{33}S$ is ≈ 0‰. Analysis of Archaean rocks older than 2.4 Ga reveals that this relationship no longer holds, i.e. sulfur is being fractionated in a mass independent manner (Farquar *et al.*, 2000; Farquhar and Wing, 2003; Fig. 2.3) and $\Delta^{33}S$ is ≠ 0‰. Mass independent fractionation (MIF) of sulfur isotopes has been suggested to have occurred via the photolysis of atmospheric SO_2 by high energy short wavelength UV radiation (<220 nm). Therefore a preserved MIF signature can be regarded as evidence for the absence of stratospheric O_3 (as O_3 is an effective screen of shortwave radiation) and can be used to determine the rise of O_2 within the atmosphere (Farquhar *et al.*, 2000).

There now appears to be growing evidence for the occurrence of life prior to the development of a stratospheric O_3 layer, e.g. stromatolites from the early Middle Archaean (~3.5 Ga) (Allwood *et al.*, 2006) and microbial mats from the Barberton greenstone belt of South Africa (3.47–3.33 Ga). These microbial mats are in close association with evaporite minerals (water-soluble minerals that result from and are concentrated by evaporation from an aqueous solution) and desiccation cracks that are primary evidence for microbial growth at or around the water/land interface in the littoral zone with periodic subaerial exposure. These samples are regarded as the oldest example of life in a subaerial ('under the air' or in other words on the Earth's surface) environment (Westall *et al.*, 2006). The occurrence of subaerial life ~3.47 Ga has profound implications for our understanding of how life is affected by UV radiation (see also Newsham and Davidson, Chapter 22, this volume). Reconstructed UV radiation fluxes (weighted to DNA-damaging wavelengths) at the Earth's surface at 3.5 Ga are calculated to be in the order of three times that of present (Cockell *et al.*, 2000). Archaean palaeoclimate reconstructions and model simulations (discussed below) point to an Archaean atmosphere rich in CO_2 and or methane (CH_4). Both of these gases have the capacity to absorb UV radiation. However, the sulfur isotope MIF signature confirms that the Earth's atmosphere and surface must have

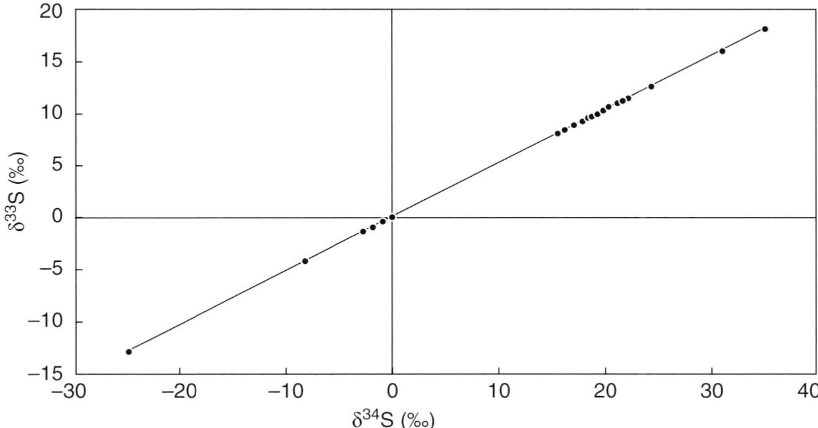

Fig. 2.2. Terrestrial mass-dependent sulfur isotope fractionation of $\delta^{33}S$ and $\delta^{34}S$. The plot is a summary plot of all mass-dependent fractionation processes with the slope of the regression indicating that the fractionation of $\delta^{33}S$ is approximately half that of $\delta^{34}S$. Taken from Farquhar and Wing, 2003.

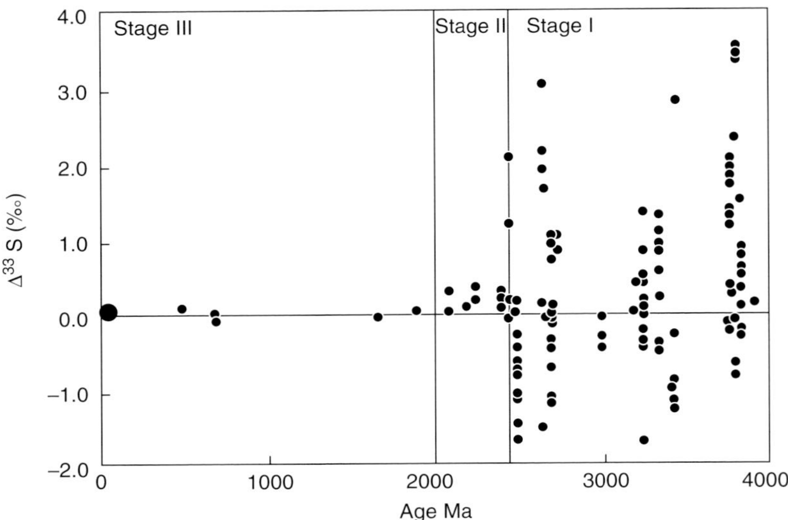

Fig. 2.3. Mass-independent fractionation (MIF) of ^{33}S plot over geological time. Data indicate that from 4000 to 2400 Ma the sulfur isotope signal was dominated by MIF processes suggesting that the Earth's atmosphere was oxygen poor and there was no stratospheric O_3 layer (stage one). From 2400 to 2000 Ma there is a reduction in the MIF signal suggesting oxygenation of the atmosphere and the initiation of a stratospheric O_3 layer (stage two). From 2000 Ma to date the MIF signal is lost and fractionation becomes dominated by mass dependency, i.e. predicted and measured δ^{33}S plot on the regression shown in Fig. 2.2. Taken from Farquhar and Wing, 2003.

been by bombarded high energy UV radiation. This in turn suggests that early life must have had the capacity to protect itself from the deleterious effects of UV radiation either via repair mechanisms or via passive UV screening through habitat preference.

2.2.1 Archaean climate

One of the most contentious and oft debated aspects of the Archaean palaeoclimates is determining the temperature of the Archaean oceans ca. 3.5 Ga. Currently two schools of thought exist: (i) ocean temperatures were high with estimates ranging between 55 and 85°C based on oxygen (Knauth and Lowe, 2003) and silicon (Robert and Chaussidon, 2006) isotope ratios found in cherts; or (ii) that ocean temperatures were similar to modern values with a range of 26–40°C based on combined oxygen and hydrogen isotope analysis of cherts (Hren *et al.*, 2009) and the oxygen isotope analysis of phosphate (Blake *et al.*, 2010). Differences between temperature predictions has been ascribed to the fact that oxygen and silicon isotope analysis of cherts primarily reflects changes in the isotopic composition of the ocean rather than changes in temperature *per se*, with the δ^{18}O of seawater changing as a result of increases in ocean depth with time (Kasting and Howard, 2006; Kasting *et al.*, 2006).

Palaeoclimate simulations by Kasting and Howard (2006) were used to determine what concentrations of CO_2 and/or CH_4 would have been needed, given a reduced solar luminosity (the sun's output at 3.3 Ga is suggested to have been at 77% of present), to achieve an ocean temperature of 70°C at 3.3 Ga as predicted by the oxygen isotope study of Knauth and Lowe (2003). Kastings and Howard (2006) suggested that a temperature of 70°C could be achieved with a CO_2 atmosphere of between 2 and 6 bar coupled with a CH_4 mixing ratio of 0–0.01. Higher CH_4 mixing ratios would have led to a reduction in CO_2 concentration.

An atmosphere with these concentrations is feasible. However, such conditions would have led to highly acidic rain water (pH 3.7) and driven intense chemical weathering (Kastings and Howard, 2006) and there is no evidence for intensive weathering over this time span and what evidence there is suggests only moderate chemical weathering (Sleep and Hessler, 2006). These data therefore support the assertion that Archaean temperatures were not excessively hot. However, estimates of Archaean temperature derived from the resurrected proteins of microbes (Gaucher *et al.*, 2008) closely match those derived from the oxygen (Knauth and Lowe, 2003) and silicon (Robert and Chaussidon, 2006) isotope analysis of chert.

2.3 Transition to the Land (~470 Ma)

The greening of the Earth can be regarded as the most significant event in the development of terrestrial ecosystems. It led to the establishment of complex feedback mechanisms within the Earth system that eventually led to a decline in atmospheric CO_2 concentrations. This culminated in a hyperoxic (elevated O_2) atmosphere, with elevated levels of O_2. These levels are thought to be responsible for the development of gigantism, notably the giant dragonfly, *Meganeura monyi*, with a wingspan of ~75 cm compared to a wingspan of ~20 cm for the largest extant dragonfly (*Megaloprepus caerulatus*).

2.3.1 The rise of the embryophytes

Embryophytes (non-vascular and vascular land plants) are thought to have evolved from the Charophyceae group of predominantly freshwater algae which contain the Coleochaetales and the Charales (reviewed in Kenrick and Crane, 1997). Recent phylogenetic DNA analysis of four genes representing three plant genomes (the plastid, the mitochondrial and nuclear) indicates that the Charales are the most likely sister group to the embryophytes (Karol *et al.*, 2001).

Cryptospores, enigmatic spores that are believed to have been produced by early embryophytes, provide the first evidence for the occurrence of embryophytes in the geological record, although not all cyrptospores are thought to have been produced by embryophytes. A recent discovery of cryptospores with definite embyrophyte affinities from the early Middle Ordovician (~470 Ma; Rubinstein *et al.*, 2010) provides evidence for unequivocal colonization of the Earth by land plants. These fossil spores are thought to be related to the non-vascular liverworts, which are the most basal extant plants in modern phylogenies (reviewed in Kenrick, 2000). These fossil spores pre-date the first macrofossil evidence for embryophytes, also thought to be liverworts, by ~20 million years (Wellman *et al.*, 2003) and the occurrence of the first vascular plants (Steemans *et al.*, 2009) by ~20 million years.

However, cryptospores are found in older sediments with specimens being retrieved from the Middle Cambrian (Strother and Beck, 2000; Strother *et al.*, 2004) and the Upper Cambrian (Taylor and Strother, 2009), although the placement of these cryptospores within the embryophytes is seen as contentious due to their structure (Steemans and Wellman, 2003; Wellman, 2003). However, the relatively diverse Argentinean early Middle Ordovician cryptospore flora (five identified genera) suggests that embryophytes may have first colonized the land in the Cambrian or Early Ordovician (Rubinstein *et al.*, 2010), lending support to the assertion of Taylor and Strother (2009) that Cambrian cryptospores may represent the origins of the embryophytes. The fact that these early Middle Ordovician cryptospores are very similar to Early Silurian genera some 35–45 million years later, indicates that initial diversification and evolution rates were slow (Rubinstein *et al.*, 2010), which could again be taken as evidence to support Taylor and Strother (2009) that the embryophytes were present in the Late Cambrian.

In terms of a transition from the aquatic to the terrestrial realm, embryophytes faced two major environmental challenges when they moved into this new extreme environment: (i) desiccation; and (ii) exposure to UV-B radiation.

In vascular plants, desiccation has been avoided by the development of a cuticle, a complex biomacromolecule composed of cutan and cutin, which provides protection both against desiccation via its physical and chemical properties, and from the harmful effects of UV-B radiation due to its chemical structure which absorbs and dissipates the UV radiation and prevents DNA damage. The cuticle, whilst protecting from desiccation, also isolates the plant from the atmosphere, retarding gas exchange which is vital for photosynthesis. Thus, specialized epidermal cells, the stomata, evolved which can open and close to regulate CO_2 uptake and H_2O (water vapour) loss. Extant non-vascular plants are represented by three lineages: the hornworts, the liverworts and the mosses. Non-vascular plants do not have a physical barrier protecting them from desiccation. Instead, desiccation is avoided either through habitat preference or via desiccation tolerance.

Atmospheric O_2 was low (~13–17%) in the Late Cambrian/Early Ordovician atmosphere (Berner, 2001, 2006) compared to the current atmospheric O_2 concentration of 21%. Since O_2 is the precursor to O_3 via the Chapman reaction series (see above) this suggests that during periods of Earth history when atmospheric O_2 concentrations were low, stratospheric O_3 formation would have been reduced resulting in an increase in the flux of UV-B radiation. Using the Cambridge 2D atmospheric chemistry model adapted to run with varying O_2 concentration, Harfoot *et al.* (2007) evaluated the impact of changes in the concentration of O_2 on the stratospheric O_3 layer over the Phanerozoic. Results indicate that although O_2 concentration have varied widely (Fig. 2.1; Berner, 2001, 2006), stratospheric O_3 concentration remained fairly stable over the Phanerozoic (Fig. 2.4) due the 'self-healing' nature of the stratospheric O_3 layer. 'Self-healing' occurred at low latitudes in the early Palaeozoic low O_2 atmospheres as a result of a reduction in the O_3 formation above 25 km, which led to an increase in the transmission of UV radiation to the lower stratosphere and drove an increase in the rate of photolysis of O_2 and, thus, the establishment of a

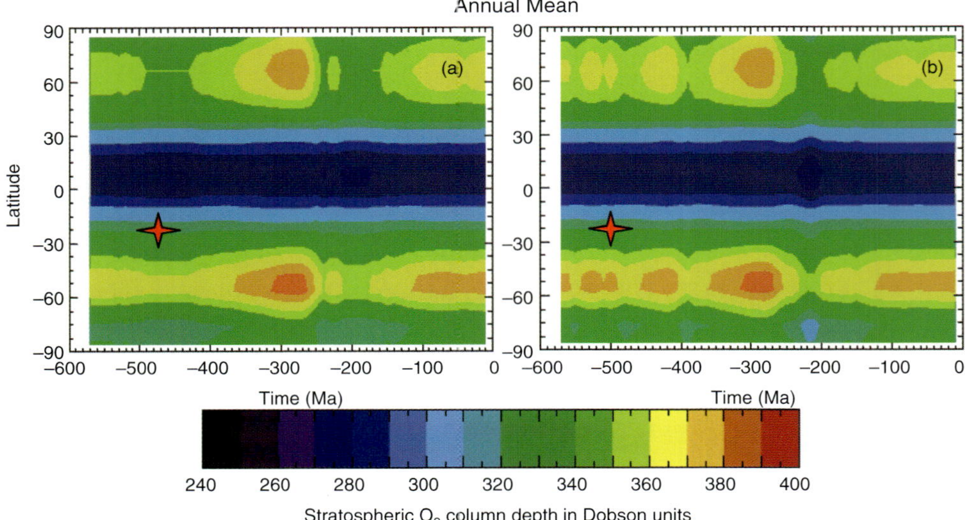

Fig. 2.4. Stratospheric ozone column depth variation with latitude over the Phanerozoic: (a) based on using the O_2 reconstruction of Berner (2001) and (b) the O_2 reconstruction of Berner (2006). The red star represents the approximate latitude and date of the first unequivocal embryophytes from Rubinstein *et al.* (2010). Redrawn from Harfoot *et al.*, 2007.

stratospheric O_3 layer in the lower stratosphere (Fig. 2.5; Harfoot *et al.*, 2007). This O_3 layer was then transported to higher latitudes via the Brewer Dobson Cell.

Palaeogeographical reconstructions suggested that the first unequivocal embryophytes were located on the southern supercontinent of Gondwana at mid latitudes (Fig. 2.6). This suggests they would have experienced overhead stratospheric O_3 concentrations of between 300 and 340 Dobson units, which are comparable to modern mid-latitude values and much higher than those experienced in the recent past under the Antarctic 'ozone hole', where O_3 depths fell to a minimum of ~110 Dobson

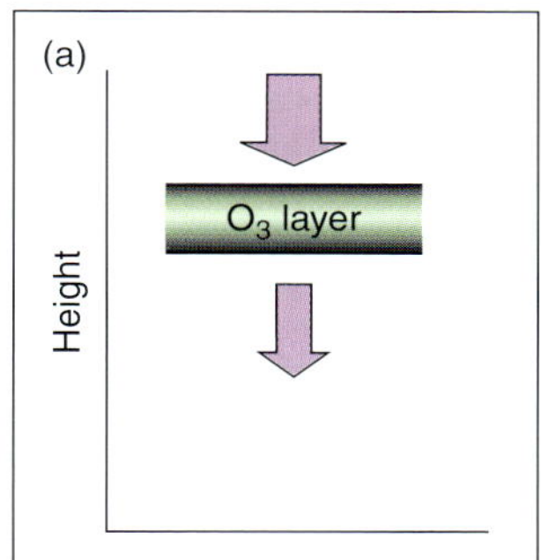

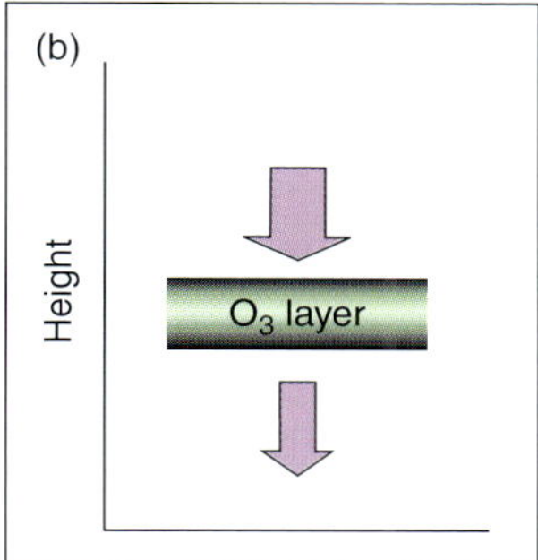

Fig. 2.5. Schematic representation of the 'self healing' of the stratospheric ozone layer: (a) the O_3 in a high ($\geq$21% O_2) atmosphere and (b) the location of the O_3 layer in a low (<21% O_2) atmosphere. In a low O_2 atmosphere more UV radiation passes through the upper stratosphere causing photolysis of O_2 lower in the stratosphere driving O_3 formation at a lower height in the stratosphere; the width of the arrow schematically represents UV radiation flux. © Barry Lomax.

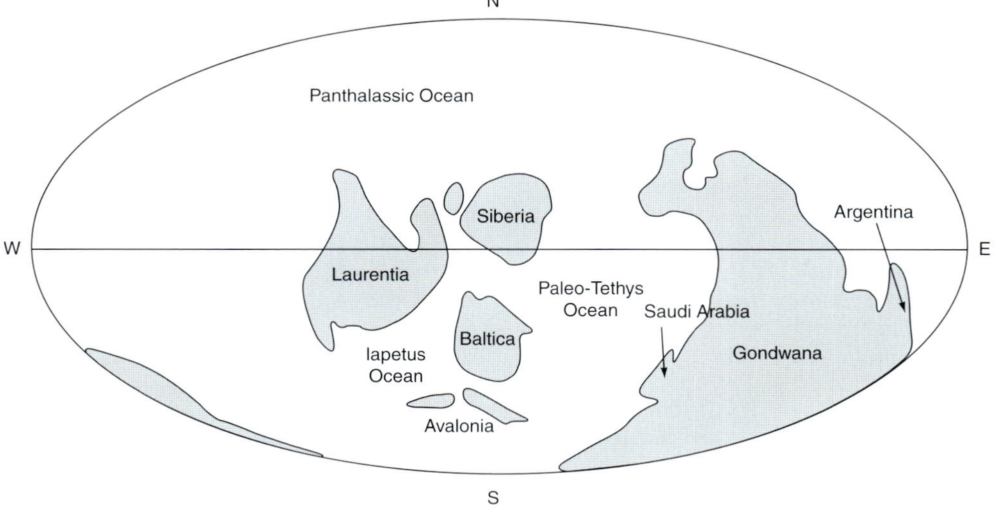

Fig. 2.6. Schematic Middle Ordovician paleogeographic map (from www.scotese.com) showing the position of Saudi Arabia on the western margin and Argentina on the eastern margin of Gondwana. The map shows the location of the earliest embyrophytes: Saudi Arabia is the location of the earliest macrofossil evidence for non-vascular plants (Wellman *et al.*, 2003) and the earliest spore record for vascular plants (Steemans *et al.*, 2009); Argentina is the location for the earliest spores of non-vascular plants. From Rubinstein *et al.*, 2010.

units. So, although atmospheric O_2 concentrations were low when compared to the present, the 'self healing' nature of the stratospheric O_3 layer as outlined above still had the capacity to shield the Earth's surface from excessive fluxes of UV-B radiation.

Traditionally the transition from the aquatic to the terrestrial habit for embryophytes and the establishment of terrestrial ecosystems was thought to have been severely limited/delayed due to a thin stratospheric ozone layer. This may have been compounded by the fact that water attenuates UV-B radiation, meaning that the 'first steps' on to the land would have been accompanied by an intense flux of UV-B. However, the modelling work of Harfoot *et al.* (2007) discussed above implies that an effective ozone shield was in place during this transition period. Furthermore, analysis of freshwater's capacity to attenuate UV-B radiation reveals that attenuation is primarily driven by the concentration of dissolved organic carbon (DOC) held within the water column, with high mountain lakes situated above the treeline that receive a diminished flux of DOC being the most UV transparent aquatic environments extant today (Sommaruga, 2001). Prior to the evolution of embyrophytes, the Earth's terrestrial surface was inhabited by microbial and algal mats which would have reduced the amount of DOC entering the watercourse and negatively impacted the capacity of freshwater to attenuate UV-B radiation. This implies that transitional species may well have been exposed to UV-B and suggests that UV-B radiation may not have played an important role in preventing the establishment of embyrophytes in terrestrial environments.

2.3.2 Embryophytes and the decline in atmospheric CO_2 and the rise of O_2

Following the colonization of the land, vascular plants diversified; one the of most intriguing aspects of this diversification is the delayed occurrence of true leaves (megaphylls). This delay is thought to be related to the high concentrations of CO_2 prevalent at the time (Beerling *et al.*, 2001). CO_2 is a greenhouse gas and as such its concentration is coupled to temperature. The long-term carbon cycle models of Berner (2006) suggest that CO_2 concentration through the Ordovician to the Silurian were greater than 3000 ppm, implying intense greenhouse warming.

CO_2 exerts selective pressures on plants and this is particularly noticeable in the relationship between stomata and atmospheric CO_2 concentration, with stomatal numbers declining as CO_2 concentration increases. This relationship was first described by Woodward (1987) and has subsequently been demonstrated across a wide range of plants (Woodward and Kelly, 1995) and over geological time (e.g. McElwain and Chaloner, 1996; McElwain *et al.*, 1999; Beerling and Royer, 2002; Beerling *et al.*, 2002).

As well as regulating CO_2 uptake, stomata control loss of water vapour and thus conductive cooling. At high CO_2 concentrations plants are predicted to have less stomata, that in turn should remain closed for longer, implying that leaf temperature would rise and conductive cooling would be minimized (Beerling *et al.*, 2001). This scenario effectively acts as a positive selective pressure for small microphyll leaves, which would absorb less solar radiation and ensure that leaf temperature does not exceed air temperature. Conversely, this scenario would have prevented the development of megaphylls (large leaves) as they would have absorbed more solar radiation, been unable to cool via transpiration, and would have overheated (Beerling *et al.*, 2001). The occurrence of large leaves and an increase in leaf size in the geological record accompanies the long term decline in atmospheric CO_2 (Osborne *et al.*, 2004), suggesting that elevated CO_2 concentrations and extreme greenhouse conditions may have delayed plant evolution. The increase in leaf size was accompanied by an increase in the rooting depth and complexity of root architecture (Algeo and Scheckler, 1998). This increase would have accelerated weathering and increased the rate of removal of CO_2 from the atmosphere (Berner, 1998).

The other key feedback mechanism associated with embyrophytes and the development of extreme environments is the increase in carbon burial. An increase in terrestrial carbon burial over the long term is linked to an increase in atmospheric O_2 because there is a decrease in decomposition which consumes O_2. Long-term models suggest that the O_2 concentration increased (Fig. 2.1) through the Carboniferous with the development of coal swamps and an increase in terrestrial carbon burial (Berner, 2006). Analysis of the charcoal over this interval shows that, as the predicted concentration of O_2 increased, the occurrence of fires within ever wet mire ecosystems increased (Scott and Glasspool, 2006), providing supportive evidence for elevated atmospheric O_2. However, recent work (Belcher *et al.*, 2010) shows that the use of charcoal abundance data to quantitatively predict high atmospheric O_2 concentrations is ill-advised due to the sigmoidal relationship between O_2 and fire probability, with a 90% likelihood of fire once O_2 is >22%. This suggests that charcoal abundance data overestimate palaeoatmospheric O_2 concentration.

2.4 Mass Extinction and Extreme Environments

Having discussed some of the long-term implications of extreme environments and their impact on life and on how life has helped fashion them, the next section of this chapter will focus on two extreme short-term (geologically speaking) mass extinction events that shaped the evolution of the Earth's biota and mark the transition from the Palaeozoic to the Mesozoic – the Permian/Triassic boundary (P/Tr boundary, ~251 Ma); and the transition from the Mesozoic to the Tertiary – the Cretaceous/Tertiary boundary (K/T boundary, ~65 Ma). Mass extinction can be defined as the extinction of a large number of species from a wide range of distinct environments occurring over a geologically short period of time (Hallam and Wignall, 1997). The Phanerozoic has been punctuated by five periods of mass extinction, which are marked by a dramatic collapse in marine biodiversity (Fig. 2.7), with this pattern being reflected in the terrestrial realm.

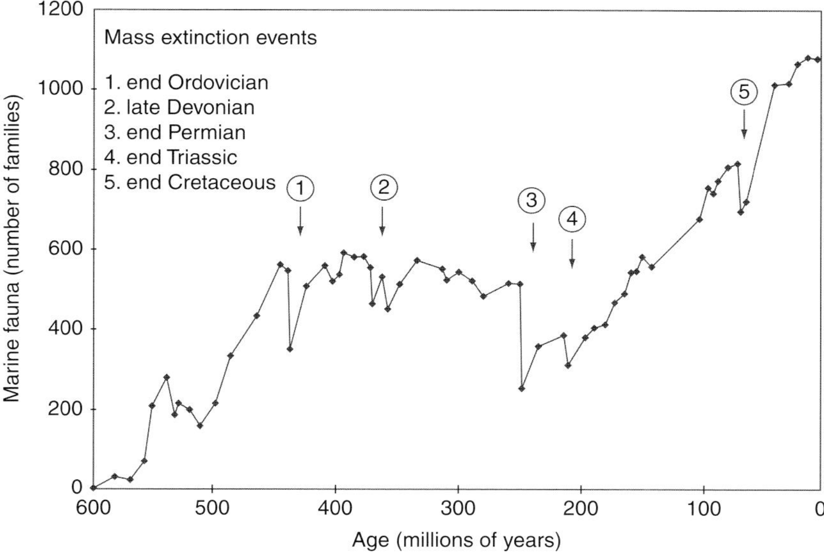

Fig. 2.7. Marine diversity curve, as reflected by the number of families over time. Major drops in diversity (mass extinction) are shown in age order. This chapter discusses the end Permian (~251 Ma, event 3) and the end Cretaceous (~65 Ma, event 5) mass extinctions. Redrawn from Raup and Sepkoski, 1982.

2.4.1 Permian/Triassic (P/Tr) boundary

The end Permian mass extinction event, occurring 251 Ma, was the largest of all the mass extinction events, resulting in the loss of approximately 95% of all marine species. In the marine realm, the extinction was marked by the final extinction of the Trilobita and, in the terrestrial realm, the extinction of the Gorgonopsidae mammal-like reptiles along with a myriad of other species. In terms of environmental change and the development of extreme environments, the most likely causal mechanism driving the environmental degradation that led to this mass extinction was the emplacement and eruption of the Siberian Traps volcanic province and the associated feedback mechanisms.

The Siberian Traps volcanic province is the largest known terrestrial igneous province, covering a total area of 1.5×10^6 km^2 although the original area was undoubtedly much larger (Wignall, 2001). Current estimates suggest that the igneous province can be spilt by volume to a rough approximation into the following constituents: 20% pyroclastic material, 50% intrusive dykes and sill complexes, and 30% basaltic lava flows (Reichow *et al.*, 2002). The eruptions are estimated to have produced between 2 and $4 \times 10^6 km^3$ of volcanic rock, although only $4 \times 10^5 km^3$ remain (Courtillot *et al.*, 1999). Radiometric dating ($^{40}Ar/^{39}Ar$) indicates that the main eruptive phase was broadly contemporaneous with the P/Tr boundary extinction event (Renne *et al.*, 1995; Reichow *et al.*, 2002), and U–Pb dating of lava flows indicates that the Siberian flood volcanism was initiated at 251.7 ± 0.4 Ma and had diminished by 251.1 ± 0.3 Ma (Kamo *et al.*, 2003). These dates establish the duration of the main eruptive phase as 0.6 Ma, with evidence for the majority being emplaced within 0.1 Ma (Kamo *et al.*, 2003). Geological evidence also indicates that the Siberian eruptions may have been unusually explosive (Campbell *et al.*, 1992), with pyroclastic deposits (rocks formed by explosive volcanic eruptions) reaching a thickness of up to 800 m (Khain, 1985).

The uniquely explosive nature of the Siberian Traps, when combined with their high latitude and the nature of the underlying regional geology, may have led to the development of extreme environments due to the collapse in the stratospheric ozone layer and increase in the flux of UV-B radiation. Volcanic eruption at high latitudes favours ozone depletion for two reasons: (i) the height of the stratosphere is lower when compared to low latitudes, implying that the explosive force needed to deliver volcanic halogens into the stratosphere is reduced; and (ii) the atmosphere at high latitudes holds less water vapour and consequently less hydrochloric acid (HCl) is washed out of volcanic plumes allowing more chlorine (Cl) to travel though the troposphere into the stratosphere. Visscher *et al.* (2004) suggested that the near global occurrence of 'mutated' fossil herbaceous lycopsids (clubmoss) spores (Fig. 2.8) in end-Permian rocks was evidence for the collapse in the Earth's ozone shield and an increase in the flux of UV-B radiation incident on the Earth's surface. Subsequently, fossil gymnosperm pollen mutations with disrupted sacci (bladder or wing-like structures that occur on many conifer pollen grains) that are indicative of a failure to form normal bisaccate grains have been reported from localities in Russia and north-west China (Foster and Afonin, 2005) and India (Banerji and Maheswari, 1975), further supporting the occurrence of a global ozone hole (Fig. 2.9). Experimental evidence from extant (modern) *Glycine max* (soybean) supports the contention that moderately enhanced fluxes of UV-B radiation can induce damage to pollen and impair germination (Koti *et al.*, 2005).

Visscher *et al.* (2004) postulated that increased surface UV-B radiation resulted not from the direct chemical effects of the massive Siberian flood basalt eruption, but by metamorphism of the surrounding rock. The geology of the region is dominated by coals (Czamanske *et al.*, 1998), hydrocarbon source rocks (Kontorovich *et al.*, 1997) and evaporites (Melnikov *et al.*, 1997), leading to the production, through metamorphism, of large quantities of organohalogens (discussed later). A series of geochemical

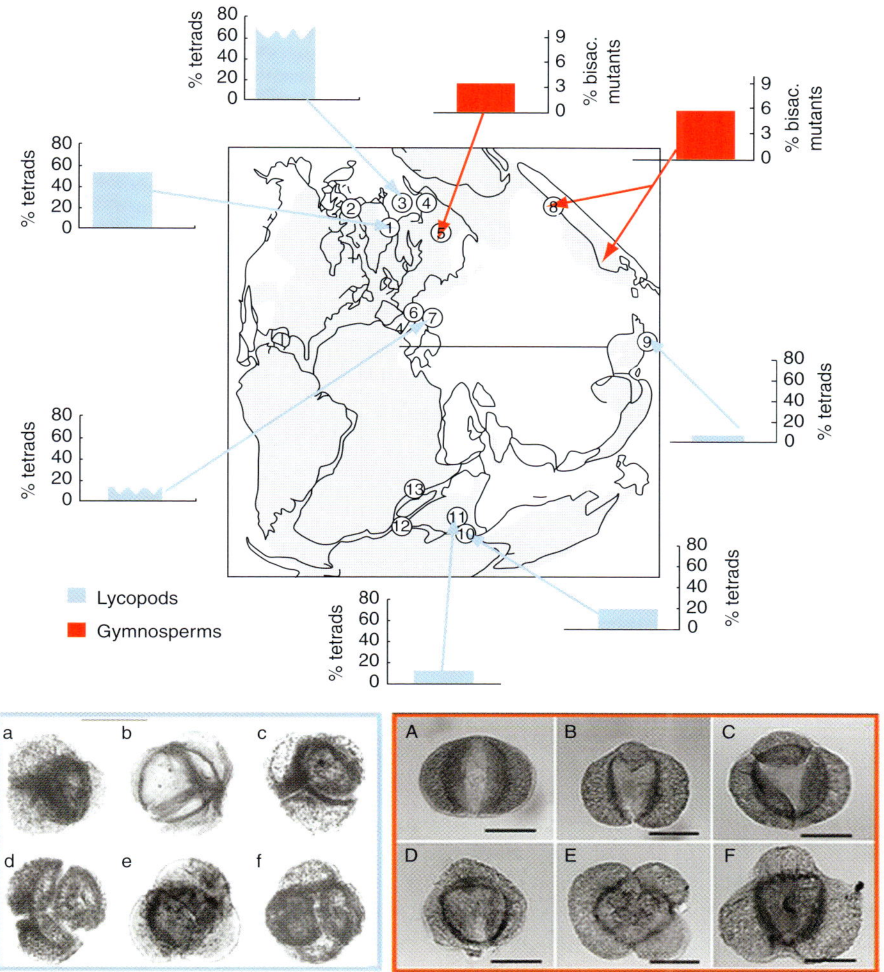

Fig. 2.8. Plot of mutated lycopod spores and gymnosperm pollen grains. Mutated permanent tetrads of lycopods are shown in blue and gymnosperm pollen in red. Note jagged bars represent maximum percentage of mutation. Base map and locations are redrawn from Vischer *et al.*, 2004, with the numbers on the base map referring to known locations of mutated spores. The percentage mutation where known were extracted from the original reference (1) East Greenland; (2) Sverdrup Basin, Arctic Canada (Utting, 1994); (3) Barents Sea (Mangerud, 1994); (4) Pechora Basin/Urals, Russia (Tuzhikova, *1985)*; (5) Russian Platform (Afonin, 2002); (6) Southern Alps, Italy (Van de Schootbrugge, 1997); (7) Transdanubian Mountains, Hungary (Haas *et al.*, 1986); (8) Jungar Basin, North China (Oujang and Norris, 1999); (9) Meishan, South China (Oujang and Utting, *1990*); (10) Raniganj Basin, India (*Tiwari and Meena,* 1988); (11) Auranga Basin, India (Banerji and Maheswari, *1975;* at this location mutated bisccate gymnosperm pollen is also found);(12) Sri Lanka (Dahanayake *et al., 1989); and (*13) Mombasa Basin, Kenya (Hankel*, 1992*). Gymnosperm pollen mutant percent data are from Foster and Afonin (2005). Pictures of mutated permanent tetrads of lycopods spore are shown edged in blue are taken from Vischer *et al.* (PNAS 101, 12952–12956. Copyright (2004) National Academy of Sciences, USA) and mutant gymnosperm pollen are edged in blue and are taken from Foster and Afonin (2005). Letters in both photographic plates refer to species, see original publication for details. Note image A of the gymnosperms is a normal bisccate pollen grain. The higher percentage of mutated grains in the northern hemisphere, close to the Siberian Traps, the proposed agent of stratospheric ozone destruction over the P/Tr boundary.

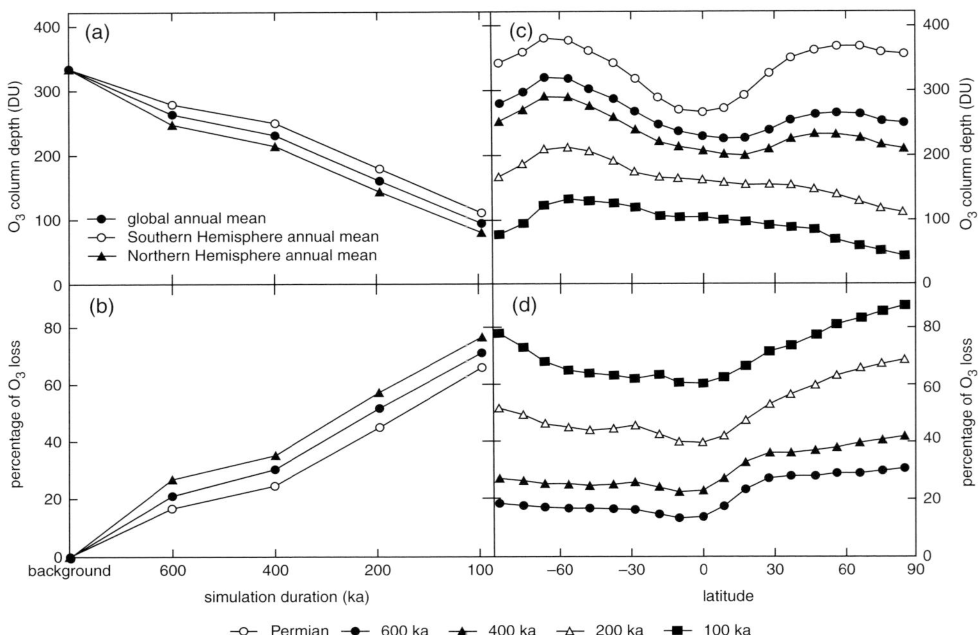

Fig. 2.9. Model simulations of the change in thickness of the stratospheric ozone column over the P/Tr boundary as a result of the eruption and emplacement of the Siberian Traps volcanic province and the release of organohalogens via metamorphism as a function of (a,b) the duration of the main eruptive phase (background, 100, 200, 400 and 600 ka) and (c,d) expressed as a function of latitude. DU, Dobson units. From Beerling *et al.*, 2007.

investigations into the generations of organohalogens by the heating of evaporates (retrieved from the Siberian Traps) to 275°C liberated both methyl chloride (CH_3Cl) and methyl bromide (CH_3Br; Svenson *et al.*, 2009), supporting the above theoretical discussion.

Using the Cambridge 2D atmospheric chemistry model, Beerling *et al.* (2007) set out to test whether the eruption and emplacement of the Siberian Traps igneous province could have led to the global destabilization of the ozone layer and a large increase in the UV-B radiation flux. Predicted concentrations of organohalgones were released at fixed flux rates to simulate emplacement/eruption over a 100, 200, 400 and 600 thousand-year (ka) period in various combinations to test the sensitivity of the O_3 layer to collapse. Combining all sources of (organo) halogens (volcanic and metamorphically derived) into one emissions scenario caused a significant reduction in the thickness of the stratospheric ozone layer, with average losses in the Northern and Southern Hemispheres of 20–30% for the 600 ka duration experiment, increasing to 60–80% depletion for the 100 ka simulation (Fig. 2.10). This simulation also resulted in the near total collapse of the natural latitudinal gradient in O_3 column thickness with losses of 70–90 and 55–80% in the high northern and southern latitudes, respectively, for emission durations of 100 and 200 ka (Beerling *et al.*, 2007). This collapse was also observed when emission rates were reduced to simulate a longer eruptive phase of either 400 or 600 ka, and resulted in the ozone column over the high northern and southern latitudes being thinned by 30–40% and 20–30%, respectively (Fig. 2.10). These data can then be used to calculate the flux of UV-B radiation incident on the Earth surface. To enable comparison to experimental work assessing plant responses to UV-B, flux values where adjusted using the generalized plant action spectrum normalized to 300 nm (Caldwell, 1971) and are reported as kJ m^{-2} day^{-1}$_{BE}$, where BE is biological equivalence. UV-B fluxes ranged from 30 to 60 (kJ m^{-2} day^{-1})$_{BE}$ throughout the growing season for the 600–400 ka simulations and 50–100 (kJ m^{-2} day^{-1})$_{BE}$ for the 100–200 ka simulations, which compare to a pre-eruption/emplacement Permian

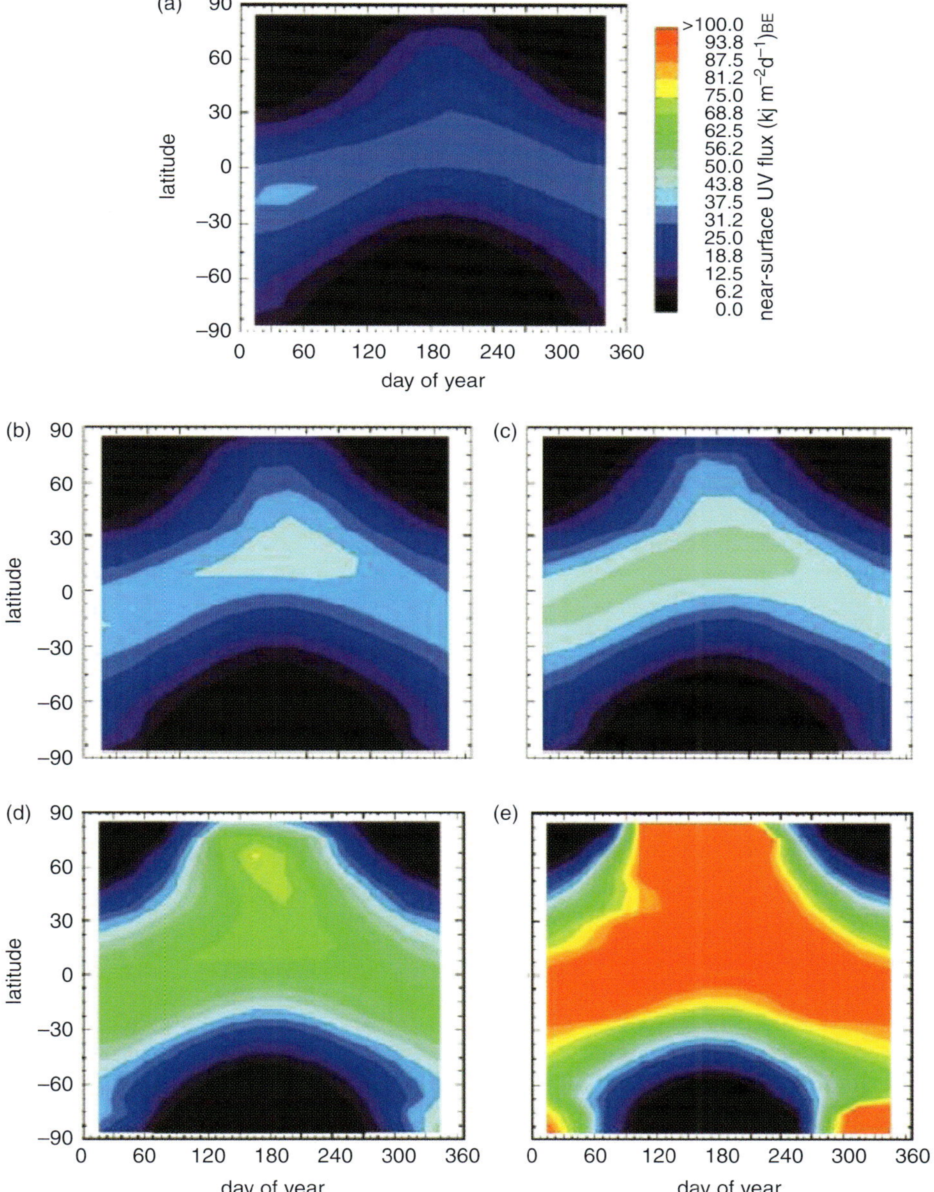

Fig. 2.10. Calculated latitudinal and seasonal daily near-surface UV-B radiation fluxes normalized at 300 nm (kJ m^{-2} day^{-1})$_{BE}$ (Caldwell, 1971) for the (a) end-Permian 'control' atmosphere and for (b) 600 ka, (c) 400 ka, (d) 200 ka and (e) 100 ka simulations. From Beerling *et al.*, 2007.

background UV-B radiation flux of 10–20 kJ m^{-2} day^{-1}_{BE} (Fig. 2.10). Earliest Triassic fluxes also far exceed those experienced at the Earth's surface today, for example, the flux at northern latitudes of between 40 and 60° is 2.3–6.6 kJ m^{-2} day^{-1}_{BE} and 18.6 kJ m^{-2} day^{-1}_{BE} over the Antarctic 'ozone hole'.

Analysis of experimental data from transgenic lines of the plants *Arabidopsis thaliana* and *Nicotiana tabacum*, grown at high UV-B flux loads (27.1 kJ m^{-2} day^{-1}_{BE}), indicates that high levels of UV-B are capable of causing homologous recombination events, which can be regarded as being analogous to mutations within a plant genome (Ries *et al.*, 2000). Intriguingly, these results also show that the effects of UV-B induced mutations persist into subsequent generations even when the plants were grown without UV-B, implying that the mutations are heritable and suggesting that there could be long-term additive effects of exposure to elevated levels of UV-B (Ries *et al.*, 2000; Beerling *et al.*, 2007). Further analysis (Boyko *et al.*, 2005) of the same transgenic lines of *A. thaliana* show that elevated temperatures do not cause widespread mutation events. These model simulations when combined with experimental data support the original palaeobotanical hypothesis that the mutated pollen and spores found within Permian/Triassic rocks were formed due to a collapse in the stratospheric ozone layer and a concomitant increase in the flux of UV-B.

A further scenario recently invoked as being a causal mechanism for the destabilization of the stratospheric O_3 layer during end Permian extinction, was the release of large volumes of H_2S from the oceans as a result of sulfide enrichment (euxinia) of the surface waters of the world's oceans (Kump *et al.*, 2005). Using a 1D model Kump *et al.* (2005) suggested that large concentrations of H_2S (2000–4000 Tg S $year^{-1}$) could have been emitted from the world's oceans due to the upward shoaling of the chemocline. This venting would result in a reduction in the concentration of OH^- radicals in the troposphere, leading to the build up of CH_4 and the eventual thinning of the stratospheric O_3 layer due to H_2S reacting with singlet O (Kump *et al.*, 2005) and a breakdown of the Chapman reactions. However, a reinvestigation of this purported mechanism using a 2D model revealed that the Earth's O_3 shield remained unaffected even when flux rates of H_2S were increased to 15,000 Tg S $year^{-1}$ (Harfoot *et al.*, 2008). This difference occurs because in the 2D model there is no collapse in the production of the OH radical in the troposphere due to its continued production via the reaction of singlet O with H_2O in the tropics, a mechanism which was not captured by the 1D simulations of Kump *et al.* (2005) (Harfoot *et al.*, 2008).

Although compelling evidence has been presented for a collapse in the stratospheric O_3 layer coincident with the end Permian extinction event, as yet there is no direct evidence for this postulated O_3 collapse. Two avenues are available for exploring this potential increase in UV-B radiation flux: (i) the examination of the S isotope record of early Triassic terrestrial sediments to look for MIF isotope signature similar to those found in the Archaean when low O_2 concentrations prevented the development of a stratospheric O_3 layer; and (ii) biogeochemical examination of fossil spores and pollen grains to determine the concentration of UV-B absorbing compounds which have been shown to correlate to ambient UV-B radiation flux (Rozema *et al.*, 2002; Lomax *et al.*, 2008; Fraser *et al.*, 2011).

2.4.2 Cretaceous/Tertiary (K/T) boundary (65 Ma)

Occurring ~65 Ma, the K/T boundary event is synonymous with the extinction of the dinosaurs and the impact of a large bolide (meteorite or comet) at Chicxulub on the Yucatan Peninsula, Mexico. The association of the K/T boundary with a large impact event was first suggested by Alvarez *et al.* (1980), with the authors showing evidence for highly elevated levels of iridium (Ir: a platinum group element) at three marine locations (Denmark, Italy and New Zealand) within the K/T boundary clay (Fig. 2.11). This was followed by the discovery of the Ir anomaly in the boundary clay in terrestrial sections of

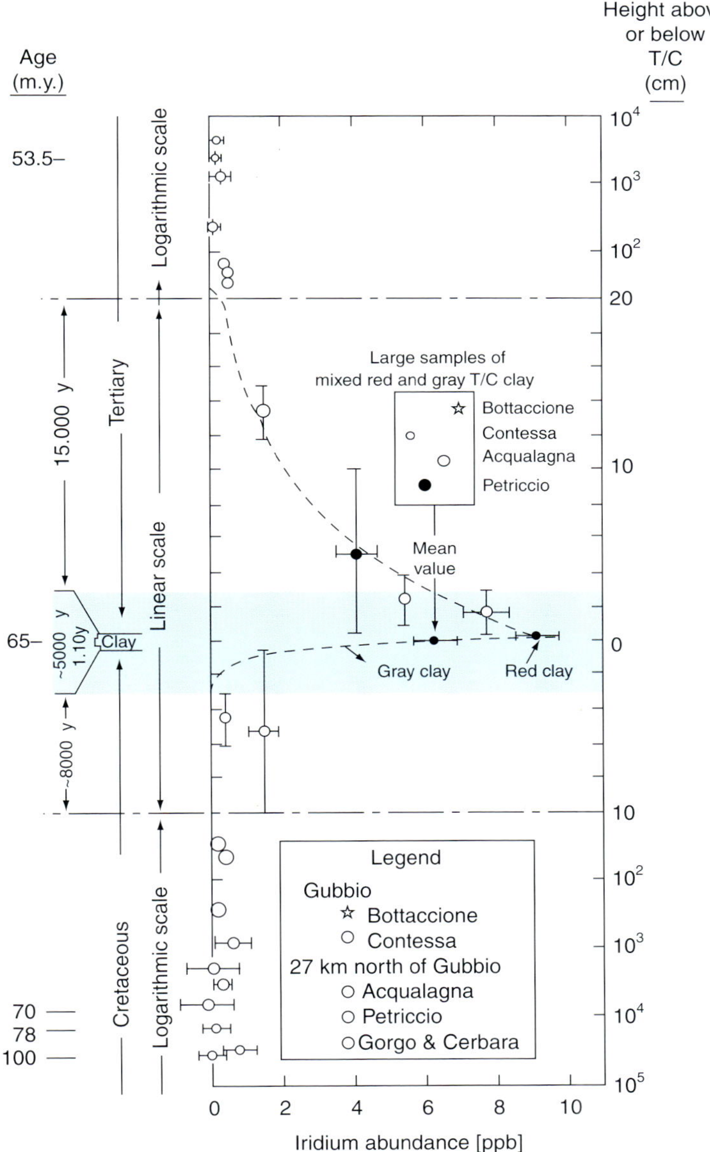

Fig. 2.11. Plot of the iridium (Ir) anomaly from the Gubbio, Italy, showing highly elevated concentrations of Ir coincident with the K/T boundary clay (marked in blue across the plot). From Alvarez *et al.*, 1980.

the Raton Basin in New Mexico, USA (Orth *et al.*, 1981). Further evidence suggesting an extraterrestrial origin for the boundary clay came in the form of shock metamorphosed quartz grains with multiple intersecting lamellae and the occurrence of microtektites, both of which are found associated with impact craters. Alvarez *et al.* (1980) hypothesized that the impact event would have caused prolonged cooling due to a global dust cloud attenuating incoming solar radiation, a concomitant reduction in photosynthesis and a collapse in net primary productivity which would have cascaded up through the food chain. This scenario became known as the 'impact winter' due its the similarities with the modelled effects of a global nuclear weapons exchange, the 'nuclear winter', reflecting the politics of the time. The Alvarez *et al.* (1980) paper led to the causal mechanisms behind the K/T extinction event being hotly debated with two disparate schools, the first focusing on extraterrestrial causes (of which the author of this chapter is a member) and the second on terrestrial mechanisms, chiefly the eruption and emplacement of the Deccan Traps volcanic province, India.

More recent work has suggested that the environmental perturbations associated with the impact were primarily driven by the nature of the target stratigraphy of the impact site. The Chicxulub site was identified and dated to the K/T boundary by Hildebrand *et al.* (1991). Subsequent analysis of drill core data revealed that the bolide had struck a carbonate ($CaCO_3$) platform which was interbedded with anhydrite ($CaSO_4$). Impact-induced vaporization of these rocks would have had profound effect on the Earth's climate inducing both short-term cooling (due to anhydrite vaporization delivering sulfate aerosols) and long-term warming (as a result of carbonate vaporization leading to a rapid release of CO_2). These changes in atmospheric composition would have been accompanied by an increase in dust loading from the bolide and the crater, as originally hypothesized by Alvarez *et al.* (1980).

The western interior of the USA has an excellent plant fossil record at both the macro- and microfossil scale, with the fossils being used to establish the level of extinction and predict the likely change in environmental conditions. The plant microfossil record shows evidence of a near instantaneous extinction accompanied by the replacement of a diverse flora with a fern assemblage dominated by a single species, called the 'fern spore spike' (Tschudy *et al.*, 1984). Investigations of the extinction event reveal strong evidence for a latitudinal gradient in extinction with low latitude sites experiencing higher levels of extinction, and extinction being selective with an increase in the mortality of evergreens when compared to deciduous species (reviewed in Upchurch *et al.*, 2007).

An early prediction of the environmental consequences of the K/T impact event was the burning of standing biomass and the development of global wildfires based on the occurrence of soot in several marine K/T boundary sections (Wolbach *et al.*, 1985). More recent work searching for evidence of biomass burning within terrestrial sections has focused on looking for an increase in the concentration and flux of charcoal over the boundary; however, evidence for both is scant (Scott *et al.*, 2000; Belcher *et al.*, 2003, 2005). A reinvestigation of the soot found in marine sediments suggested that it could have been produced by the combustion of hydrocarbons (Belcher *et al.*, 2009). These findings strongly suggest that the burning of standing biomass and the outbreak of a global wildfire did not occur. This has implications in terms of quantifying the thermal energy delivered to the Earth's atmosphere as a consequence of the impact event and trying to determine what happened to the standing biomass.

Early model simulations (Melosh *et al.*, 1990) calculated that the transfer of thermal radiation to the atmosphere as a result of ballistic re-entry of impact and crater debris to be in the region of 50 kW m^{-2} globally and >150 kW m^{-2} within 7000 km of the impact site, which in 1990 was unknown. The lack of evidence for wildfires at the boundary (Scott *et al.*, 2000; Belcher *et al.*, 2003, 2005) suggests, however, that the transfer of thermal radiation must have been much lower and cannot have exceeded 19 kW m^{-2} or been >6 kW m^{-2} for more than a few hours (Belcher *et al.*, 2005).

The P/Tr boundary is marked by the occurrence of a global 'fungal spike' (Vishcher *et al.*, 1996) suggesting that dead vegetation was rapidly broken down by fungus. However, to date, evidence for a K/T boundary 'fungal spike' is limited to one section in New Zealand (Vajda and McLoughlin, 2004). Understanding the fate of this missing carbon remains a key unanswered question in the study of the K/T boundary event and its impact on the carbon cycle (Upchurch *et al.*, 2007). The lack of a 'fungal spike' combined with a lack of evidence for an increase in biomass burning suggests an increase in microbial activity, but direct evidence for this assertion remains elusive.

One of the original predictions of the Alvarez *et al.* (1980) 'impact winter' scenario was the development of freezing conditions followed by prolonged cooling. Initial analysis of fossil plants supported both freezing and subsequent cooling. Wolfe (1991) suggested that deformations in the cuticle of fossil water lilies found at the Teapot Dome formation, Wyoming, USA, implied that the lake habitat had frozen over whilst the plants were growing. Comparisons with modern water lilies suggest that the lake/pond would have frozen in the summer, lending strong support to post-impact freezing. However, subsequent re-analysis of modern water lilies and their response to freezing failed to reproduce the deformation fabric, suggesting that the deformation observed by Wolfe (1991) was the result of preservation or processing (McIver, 1999; Upchurch *et al.*, 2007). The extinction of a high proportion of evergreens when compared to deciduous flora has also been taken as evidence for short-term cooling (Upchurch *et al.*, 2007 and references therein).

Since the bolide struck a large carbonate platform it is predicted to have resulted in the vaporization of $CaCO_3$ and the transfer of large volumes of CO_2 from the geosphere into the atmosphere (O'Keefe and Ahrens, 1989). Using stomatal analysis (as discussed above) it has been possible to semi-quantify the amount of CO_2 released over the K/T boundary. By constructing species-specific calibration curves characterizing the change in stomatal numbers to known changes in CO_2 (historical and experimental), pre- and post-impact palaeo-atmospheric CO_2 concentrations have been determined. Long-term pre- and post-impact CO_2 concentrations were estimated using stomatal counts from extant and extinct Ginkgo cuticle, and estimates of CO_2 directly above the boundary were achieved using the cuticle of *Stenochlaena* ferns (Fig. 2.12). Data indicate that pre- and post-impact atmospheric CO_2 concentrations were in the region of 500 ppm, whereas estimates directly above the boundary indicated CO_2 concentrations in excess of 2300 ppm ~10 ka after the impact event (Fig. 2.13; Beerling *et al.*, 2002). Long-term carbon cycle models enabled the prediction of atmospheric CO_2 concentration in the immediate aftermath of the impact and suggest that CO_2 concentrations rose to ~7000 ppm, enough to trigger large scale greenhouse warming with an estimated increase in temperature of ~7.5°C (Beerling *et al.*, 2002).

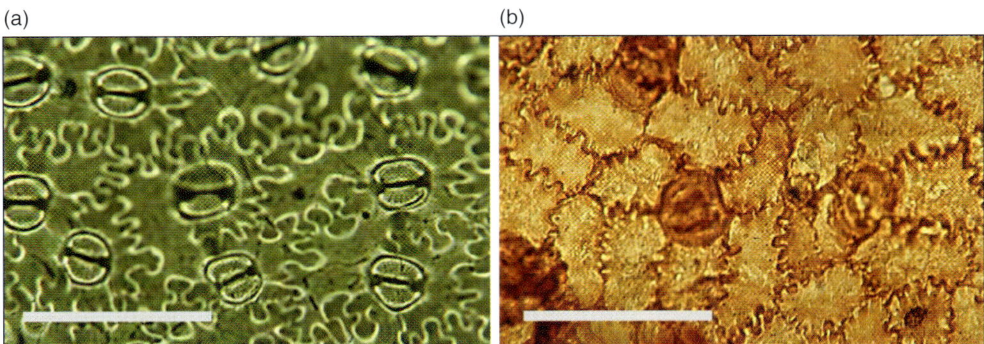

Fig. 2.12. (a) Photomicrograph of fossil leaf cuticle of the fern *Stenochlaena* from just after the Cretaceous/Tertiary (K/T) boundary; and (b) nearest living relative, *Stenochlaena palustris*. The stomatal index of the fossil cuticle is considerably lower than the extant cuticle (scale bar 100 μm). From Lomax, 2001.

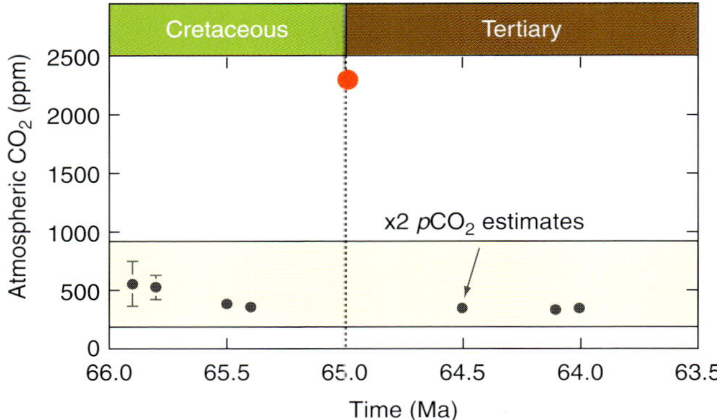

Fig. 2.13. Cretaceous/Tertiary atmospheric CO_2 reconstruction. Latest Cretaceous CO_2 values are based on the analysis of fossil *Ginkgo adiantoides* cuticle (solid circles) and the K/T estimate is derived from analysing the cuticle of *Stenochlaena* from Clear Creek South in the Raton Basin, Colorado, USA. The yellow box indicates the range in predicted long-term CO_2 based on the Geocarb III model of Berner and Kothavala (2001). Redrawn from Beerling *et al.* (2002).

2.5 Conclusion

In this chapter evidence for large scale changes in the Earth's climate over both the geological short- and long-term has been put forward. This evidence has focused on changes in the UV-B flux, temperature and changes in the atmospheric composition of O_2 and CO_2. These conditions are the most likely to create 'extreme' environments and they have also, for the most part, been underpinned by robust mechanistic proxies, which can be applied to determine past changes over the time period of interest.

However, these records become increasingly difficult to read with the passage of time. The change in ocean temperature over the Archaean remains a 'hot' topic of debate with evidence suggesting that Archaean oceans were either anomalously hot or similar to the present and indistinguishable from the Phanerozoic. What is clear is that the two scenarios are mutually exclusive and they have profound implications for early life on this planet.

Evidence discussed in this chapter suggests that although in the past UV-B fluxes have varied greatly, life seems to have found a way, with microbial mats and stromatolites becoming established within the photic zone before the presence of a stratospheric O_3 layer. Analysis of the literature also suggests that UV-B would not have limited the capacity of embryophytes to have invaded the land. The freshwater habitat of the ancestors to the first embryophytes would have been exposed to UV-B radiation and even though O_2 concentrations were reduced compared to the present, an effective stratospheric O_3 layer would have been in place due to the 'self healing' properties of O_3.

On shorter time scales, the two largest mass extinction events at the end of the Permian and the end of the Cretaceous can both be regarded as periods in which extreme environments existed, but for different reasons. Modelling and palaeobotanical evidence suggests that the end Permian event was accompanied by a global collapse in the stratospheric O_3 layer due to volcanism and associated metamorphic activity. In contrast, the end Permian event affected many different habitats, including deep-water marine environments, so UV-B cannot have been the main mechanism driving extinction (although undoubtedly if the stratospheric O_3 did collapse, the effects may have been profound as multicellular life would not have experienced such an intense UV-B radiation load prior to this). The end Cretaceous extinction event has been linked to dramatic short-term cooling followed by long-term greenhouse warming and it is this combination of fluctuating extremes that is likely to have led to the extinction.

Evidence from a diverse range of scientific disciplines indicates that humanity's use of the Earth's atmosphere as a dumping ground for the by-products of industrialization is forcing the Earth system. This has driven the rapid increase in the concentration of greenhouse gases forcing climate warming. These changes are having a profound effect on the biosphere with many commentators suggesting that we are in the grip of the sixth mass extinction event (see also Psenner and Sattler, Chapter 23 and Glover and Neal, Chapter 24, this volume).

Since the publication of Charles Lyell's seminal series of books *Principles of Geology* was published in the 1830s the theory of uniformitarianism (the present is the key to the past) as outlined by Hutton in the 1780s has influenced generations of geologists and scientists in allied disciplines. A wealth of information (some of which has been presented in this chapter) on ancient climates, extreme environments and life is retained in the rock record, the reading of which is becoming ever more refined. This suggests that it may be time to use the past as a guide to help us understand the future impact of prolonged anthropogenic climate change and perhaps time to suggest that the past may be the key to the future (Beerling, 1998).

References

Afonin, S.A. (2002) A palynological assemblage from the transitional Permian–Triassic deposits of European Russia. *Paleontological Journal* 34, 29–34.

Algeo, T.J. and Scheckler, S.E. (1998) Terrestrial-marine teleconnections in the Devonian: links between the evolution of land plants, weathering processes, and marine anoxic events. *Philosophical Transactions of the Royal Society B: Biological Sciences* 353, 113–128.

Allwood, A.C., Walter, M.R., Kamber, B.S., Marashall, C.P. and Burch, I.W. (2006) Stromatolite reef from the Early Archaean era of Australia. *Nature* 441, 714–718.

Alvarez, L.W., Alvarez, W., Asaro, F. and Michel, H.V. (1980) Extraterrestrial cause for the Cretaceous-Tertiary extinction – experimental results and theoretical interpretation. *Science* 208, 1095–1108.

Banerji, J. and Maheswari, H.K. (1975) Palynomorphs from the pancchet group exposed in the Sukri River Auranga coalfield, Bihar. *Palaeobotanist* 22, 158–170.

Beerling, D.J. (1998) The future as the key to the past for palaeobotany? *Trends in Ecology and Evolution* 13, 311–316.

Beerling, D.J. and Royer, D. (2002) Reading a CO_2 signal from fossil stomata. *New Phytologist* 153, 387–397.

Beerling, D.J., Osborne, C.P. and Chaloner, W.G. (2001) Evolution of leaf-form in land plants linked to atmospheric CO2 decline in the Late Palaeozoic era. *Nature* 410, 352–354.

Beerling, D.J., Lomax, B.H., Royer, D.L., Upchurch, G.R. and Kump, L.R. (2002) An atmospheric pCO_2 reconstruction across the Cretaceous-Tertiary boundary from leaf megafossils. *Proceedings of the National Academy of Sciences of the United States of America* 99, 7836–7840.

Beerling, D.J., Harfoot, M., Lomax, B. and Pyle, J.A. (2007) The stability of the stratospheric ozone layer during the end-Permian eruption of the Siberian Traps. *Philosophical Transactions of the Royal Society A* 365, 1843–1866.

Belcher, C.M., Collinson, M.E., Sweet, A.R., Hildebrand, A.R. and Scott, A.C. (2003) Fireball passes and nothing burns – the role of thermal radiation in the Cretaceous-Tertiary event: evidence from the charcoal record of North America. *Geology* 31, 1061–1064.

Belcher, C.M., Collinson, M.E. and Scott, A.C. (2005) Constraints on the thermal energy released from the Chicxulub impactor: new evidence from multi-method charcoal analysis. *Journal of the Geological Society* 162, 591–602.

Belcher, C.M., Finch, P., Collinson, M.E., Scott, A.C. and Grassineau, N.V. (2009) Geochemical evidence for combustion of hydrocarbons during the K-T impact event. *Proceedings of the National Academy of Sciences of the United States of America* 106, 4112–4117.

Belcher, C.M., Yearsley, J.M., Hadden R.M., McElwain, J.C. and Rein, G. (2010) Baseline intrinsic flammability of Earth's ecosystems estimated from paleoatmospheric oxygen over the past 350 million years. *Proceedings of the National Academy of Sciences of the United States of America* 107, 22448–22453.

Berner, R.A. (1998) The carbon cycle and CO_2 over Phanerozoic time: the role of land plants. *Philosophical Transactions of the Royal Society B: Biological Sciences* 353, 75–81.

Berner, R.A. (2001) Modeling atmospheric O_2 over Phanerozoic time. *Geochimica et Cosmochimica Acta* 65, 685–694.

Berner, R.A. (2006) GEOCARBSULF: a combined model for Phanerozoic atmospheric O_2 and CO_2. *Geochimica et Cosmochimica Acta* 70, 5653–5664.

Berner, R.A. (2009) Phanerozoic atmospheric oxygen: New results using the GEOCARBSULF model. *American Journal of Science* 309, 603–606.

Berner, R.A. and Kothavala, Z. (2001) GEOCARB III: a revised model of atmospheric CO_2 over Phanerozoic time. *American Journal of Science* 301, 182–204.

Blake, R.E., Chang, S.J. and Lepland, A. (2010) Phosphate oxygen isotopic evidence for a temperate and biologically active Archaean ocean. *Nature* 464, 1029–1032.

Boyko, A., Filkowski, J. and Kovalchuk, I. (2005) Homologous recombination in plants is temperature and daylength dependent. *Mutation Research* 572, 73–83.

Caldwell, M.M. (1971) Solar ultraviolet and the growth and development of higher plants. In: Giese A.C. (ed.) *Photophysiology*. Academic Press, New York, pp. 131–177.

Campbell, I.H., Czamanske, G.K., Fedorenko, V.A., Hill, R.I. and Stepanov, V. (1992) Synchronism of the Siberian Traps and the Permian-Triassic Boundary. *Science* 258, 1760–1763.

Cockell, C.S., Catling, D.C., Davis, W.L., Snook, K., Kepner, R.L., Lee, P. and McKay, C.P. (2000) The ultraviolet environment of Mars: biological implications past, present, and future. *Icarus* 146, 343–359.

Courtillot, V., Jaupert, C., Manighetti, I., Taonier, P. and Besse, J. (1999) On the causal links between flood basalts and continental break-up. *Earth Planetary Science Letters* 166, 177–195.

Czamanske, G.K., Gurevitch, A.B., Fedorenko, V. and Simonov, O. (1998) Demise of the Siberian plume: paleogeographic and paleotectonic reconstruction from the prevolcanic and volcanic record, north-central Siberia. *International Geological Review* 41, 95–115.

Dahanayake, K., Jayasena, H.A.H., Singh, B.K., Tiwari, H.K. and Tripathi, A. (1989) A Permo-Triassic(?) plant microfossil assemblage from Sri Lanka. *Review of Palaeobotany and Palynology* 58, 197–203.

Farquhar, J. and Wing, B.A. (2003) Multiple sulphur isotopes and the evolution of the atmosphere. *Earth Planetary Science Letters* 13, 1–13.

Farquhar, J., Bao, H.M. and Thiemens, M. (2000) Atmospheric influence of Earth's earliest sulphur cycle. *Science* 289, 756–758.

Foster, C.B. and Afonin, S.A. (2005) Abnormal pollen grains: an outcome of deteriorating atmospheric conditions around the Permian–Triassic boundary. *Journal of the Geological Society* 162, 653–659.

Fraser, W.T., Sephton, M.A., Watson, J.S., Self, S., Lomax, B.H., James, D.I., Wellman, C.H., Callaghan, T.V. and Beerling, D.J. (2011) UV-B Absorbing pigments in spores: biochemical responses to shade in a Swedish birch forest. *Polar Research*, in press.

Gaucher, E.A., Govindarajan, S. and Ganesh, O.K. (2008) Palaeotemperature trend for Precambrian life inferred from resurrected proteins. *Nature* 451, 704–707.

Haas, J., Góczán, F., Oravecz-Scheffer, A., Barabás-Stuhl, A., Majoros, G. and Bérczi-Makk, A. (1986) Permian–Triassic boundary in Hungary. *Società Geologica Italinana Memoire* 34, 221–241.

Hallam, A. and Wignall, P.B. (1997) *Mass Extinctions and Their Aftermath*. Oxford University Press, Oxford, UK, 328 pp.

Hankel, O. (1992) Late Permian to early Triassic microfloral assemblages from the Maji ya Chumvi formation, Kenya. *Review of Palaeobotany and Palynology* 72, 129–147.

Harfoot, M.B.J., Beerling, D.J., Lomax, B.H. and Pyle, J.A. (2007) A two-dimensional atmospheric chemistry modelling investigation of Earth's Phanerozoic O_3 and near-surface ultraviolet radiation history. *Journal of Geophysical Research* 112, DO7308.

Harfoot, M.B., Pyle, D.J. and Beerling, D.J. (2008) End-Permian ozone shield unaffected by oceanic hydrogen sulphide and methane releases. *Nature Geoscience* 1, 247–252.

Hildebrand, A.R., Penfield, G.T., Kring, D.A., Pilkington, M., Camargo, A., Jacobsen, S.B. and Boynton, W.V. (1991) Chicxulub crater – a possible Cretaceous Tertiary boundary impact crater on the Yucatan peninsula, Mexico. *Geology* 19, 867–871.

Hren, M.T., Tice, M.M. and Chamberlain, C.P. (2009) Oxygen and hydrogen isotope evidence for a temperate climate 3.42 billion years ago. *Nature* 462, 205–208.

Kamo, S.L., Czamanske, G.K., Amelin, Y., Fedorenko, V.A., Davis, D.W. and Trofimov, V.R. (2003) Rapid eruption of Siberian flood-volcanic rocks anc evidence for coincidence with the Permian–Triassic boundary and mass extinction at 251 Ma. *Earth Planetary Science Letters* 214, 75–91.

Karol, K.G., McCourt, R.M., Cimino, M.T. and Delwiche, C.F. (2001) The closest living relatives of land plants. *Science* 294, 2351–2353.

Kasting, J.F. and Howard, M.T. (2006) Atmospheric composition and climate on the early Earth. *Philosophical Transactions of the Royal Society B: Biological Sciences* 361, 1733–1741.

Kasting, J.F., Howard, M.T., Wallmann, K., Veizer, J., Shields, G. and Jaffres, J. (2006) Paleoclimates, ocean depth, and the oxygen isotopic composition of seawater. *Earth Planetary Science Letters* 252, 82–93.

Kenrick, P. (2000) The relationships of vascular plants. *Philosophical Transactions of the Royal Society B: Biological Science* 355, 847–855.

Kenrick, P. and Crane, P.R. (1997) The origin and early evolution of plants on land. *Nature* 389, 33–39.

Khain, V.E. (1985) *Geology of the USSR, Beiträge zur Regionalen Geologie der Erde*. Gebrüder Bornträger, Berlin-Stuttgart, Germany.

Knauth, L.P. and Lowe, D.R. (2003) High Archean climatic temperature inferred from oxygen isotope geochemistry of cherts in the 3.5 Ga Swaziland Supergroup, South Africa. *Geological Society of America Bulletin* 115, 566–580.

Kontorovich, A.E., Khomenko, A.V., Burshtein, L.M., Likhanov, I.I., Pavlov, A.L., Staroseltsev, V.S. and Ten, A.A. (1997) Intense basic magmatism in the Tunguska petroleum basin, eastern Siberia, Russia. *Petroleum Geoscience* 3, 359–369.

Koti, S., Reddy, K.R., Reddy, V.R., Kakani, V.G. and Zhao, D.L. (2005) Interactive effects of carbon dioxide, temperature, and ultraviolet-B radiation on soybean (*Glycine max* L.) flower and pollen morphology, pollen production, germination, and tube lengths. *Journal of Experimental Botany* 56, 725–736.

Kump, L.R., Pavlov, A. and Arthur, M.A. (2005) Massive release of hydrogen sulfide to the surface ocean and atmosphere during intervals of oceanic anoxia. *Geology* 33, 397–400.

Lomax, B.H. (2001) The use of fossil plants to detect environmental change across the Cretaceous Tertiary boundary. PhD thesis. University of Sheffield, Sheffield, UK.

Lomax, B.H., Fraser, W.T., Sephton, M.A., Callaghan, T.V., Self, S., Harfoot, M., Pyle, J.A., Wellman, C.H. and Beerling, D.J. (2008) Plant spore walls as a record of long-term changes in ultraviolet-B radiation. *Nature Geoscience* 1, 592–596.

Mangerud, G. (1994) Palynostratigraphy of the Permian and lowermost Triassic succession, Finnmark platform, Barents Sea. *Review of Palaeobotany and Palynology* 82, 317–349.

McElwain, J.C. and Chaloner, W.G. (1996) The fossil cuticle as a skeletal record of environmental change. *Palaios* 11, 376–388.

McElwain, J.C., Beerling, D.J. and Woodward, F.I. (1999) Fossil plants and global warming at the Triassic-Jurassic boundary. *Science* 285, 1386–1390.

McIver, E.E. (1999) The paleoenvironment of *Tyrannosaurus rex* from southwestern Saskatchewan, Canada. *Canadian Journal of Earth Sciences* 39, 207–221.

Melnikov, N.V., Khomenko, A.V., Kuznetsova, E.N. and Zhidkova, L.V. (1997) The effect of traps on salt redistribution in the Lower Cambrian of the western Siberian Platform. *Russian Geology and Geophysics* 38, 1378–1384.

Melosh, H.J., Schneider, N.M., Zahnle, K.J. and Latham, D. (1990) Ignition of global wildfires at the Cretaceous Tertiary boundary. *Nature* 343, 251–254.

O'Keefe, J.D. and Ahrens, T.J. (1989) Impact production of CO_2 by the Cretaceous Tertiary extinction bolide and the resultant heating of the Earth. *Nature* 338, 247–249.

Orth, C.J., Gilmore, J.S., Knight, J.D., Pillmore, C.L., Tschudy, R.H. and Fassett, J.E. (1981) An iridium abundance anomaly at the palynological Cretaceous-Tertiary boundary in northern New-Mexico. *Science* 214, 1341–1342.

Osborne, C.P., Beerling, D.J., Lomax, B.H. and Chaloner, W.G. (2004) Biophysical constraints on the origin of leaves inferred from the fossil record. *Proceedings of the National Academy of Sciences of the United States of America* 101, 10360–10362.

Ouyang, S. and Norris, G. (1999) Earliest Triassic (Induan) spores and pollen from the Junggar Basin, Xinjiang, northwestern China. *Review of Palaeobotany and Palynology* 106, 65–103.

Ouyang, S. and Utting, J. (1990) Palynology of Uer Permian and Lower Triassic rocks, Meishan, Changxing County, Zhejiang Province, China. *Review of Palaeobotany and Palynology* 66, 65–103.

Raup, D.M. and Sepkoski, J.J. (1982) Mass extinctions in the marine fossil record. *Science* 215, 1501–1503.

Reichow, M.K., Saunders, A.D., White, R.V., Pringle, M.S., Al'Mukhamedov, A.I., Medvedev, A.I. and Kirda, N.P. (2002) Ar^{40}/Ar^{39} dates from the West Siberian Basin: Siberian flood basalt province doubled. *Science* 296, 1846–1849.

Renne, P.R., Zhang, Z.C., Richards, M.A., Black, M T. and Basu, A.R. (1995) Synchrony and causal relations between Permian–Triassic Boundary crises and Siberian flood volcanism. *Science* 269, 1413–1416.

Ries, G., Heller, W., Puchta, H., Sandermann, H., Seidlitz, H.K. and Hohn, B. (2000) Elevated UV-B radiation reduces genome stability in plants. *Nature* 406, 98–101.

Robert, F. and Chaussidon, M. (2006) A palaeotemperature curve for the Precambrian oceans based on silicon isotopes in cherts. *Nature* 443, 969–972.

Rozema, J., van Geel, B., Bjorn, L.O., Lean, J. and Madronich, S. (2002) Paleoclimate: toward solving the UV puzzle. *Science* 296, 1621–1622.

Rubinstein, C.V., Gerrienne, P., de la Puente, G.S., Astini, R.A. and Steemans, P. (2010) Early Middle Ordovician evidence for land plants in Argentina (eastern Gondwana). *New Phytologist* 188, 365–369.

Scott, A.C. and Glasspool, I. (2006) The diversification of Paleozoic fire systems and fluctuations in atmospheric oxygen concentration. *Proceedings of the National Academy of Sciences of the United States of America* 103, 10861–10865.

Scott, A.C., Lomax, B.H., Collinson, M.E., Upchurch, G.R. and Beerling, D.J. (2000) Fire across the K-T boundary: initial results from the Sugarite Coal, New Mexico, USA. *Palaeogeography Palaeoclimatology Palaeoecology* 164, 381–395.

Sleep, N.H. and Hessler, A.M. (2006) Weathering of quartz as an Archaean climate indicator. *Earth Planetary Science Letters* 241, 594–602.

Sommaruga, R. (2001) The role of solar UV radiation in the ecology of alpine lakes. *Journal of Photochemistry and Photobiology B: Biology* 62, 35–42.

Steemans, P. and Wellman, C.H. (2003) Miospores and the emergence of land plants. In: Webby, B.D., Droser, M.L. and Percival, I.G. (eds) *The Great Ordovician Biodiversity Event*. Columbia University Press, New York, pp. 361–368.

Steemans, P., Le Herisse, A., Melvin, J., Miller, M.A., Paris, F., Verniers, J. and Wellman, C.H. (2009) Origin and radiation of the earliest vascular land plants. *Science* 324, 353.

Strother, P.K. and Beck, J.H. (2000) Spore-like microfossils from Middle Cambrian strata: expanding the meaning of the term cryptospore. In: Harley, M.M., Morton, C.M. and Blackmore, S. (eds) *Pollen and Spores: Morphology and Biology*. Royal Botanic Gardens, Kew, UK, pp. 413–424.

Strother, P.K., Wood, G.D., Taylor, W.A. and Beck, J.H. (2004) Middle Cambrian cryptospores and the origin of land plants. *Memoirs of the Association of Australasian Palaeontologists* 29, 99–113.

Svenson, H., Schmidbauer, N., Roscher, M., Stordal, F. and Planke, S. (2009) Contact metamorphism, halocarbons, and environmental crises of the past. *Environmental Chemistry* 6, 466–471.

Taylor, W.A. and Strother, P.K. (2009) Ultrastructure, morphology, and topology of Cambrian palynomorphs from the Lone Rock Formation, Wisconsin, USA. *Review of Palaeobotany and Palynology* 153, 296–309.

Tiwari, R.S. and Meena, K.L. (1988) Abundance of spore tetrads in Early Triassic sediments of India and their significance. *Palaeobotanist* 37, 210–214.

Tschudy, R.H., Pillmore, C.L., Orth, C.J., Gilmore, J.S. and Knight, J.D. (1984) Disruption of the terrestrial plant ecosystem at the Cretaceous-Tertiary boundary, western interior. *Science* 225, 1030–1032.

Tuzhikova, V.I. (1985) Miospores and Stratigraphy of Reference Sections in the Triassic of the Urals (Akad. Nauk SSSR, Ural'sk Nauchn. Tsentr, Sverdlovsk).

Upchurch, G.R., Lomax, B.H. and Beerling, D.J. (2007) Paleobotanical evidence for climatic change across the cretaceous-tertiary boundary, North America: twenty years after Wolfe and Upchurch. In: Jarzen, D.M., Manchester, S.R., Retallack, G.J. and Jarzen, S.A. (eds) *Advances in Angiosperm Paleobotany and Paleoclimatic Reconstruction – contributions honouring David L Dilcher and Jack A Wolfe Book Series*. Courier Forschungsinstitut Senckenberg, Germany, 258, 57–74.

Utting, J. (1994) Palynostratigraphy of Permian and Lower Triassic rocks, Sverdrup Basin, Canadian Arctic Archipelago. *Geological Survey of Canada Bulletin* 47, 1–107.

Vajda, V. and McLoughlin, S. (2004) Fungal proliferation at the Cretaceous-Tertiary boundary. *Science* 303, 1489.

Van de Schootbrugge, B. (1997) MSc thesis. Utrecht University, Utrecht, The Netherlands.

Visscher, H., Brinkhuis, H., Dilcher, D.L., Elsik, W.C., Eshet, Y., Looy, C.V., Rampino, M.R. and Traverse, A. (1996) The terminal Paleozoic fungal event: evidence of terrestrial ecosystem destabilization and collapse. *Proceedings of the National Academy of Sciences of the United States of America* 93, 2155–2158.

Visscher, H., Looy, C.V., Collinson, M.E., Brinkhuis, H., Van Cittert, J.H.A., Kurschner, W.M. and Sephton, M.A. (2004) Environmental mutagenesis during the end-Permian ecological crisis. *Proceedings of the National Academy of Sciences of the United States of America* 101, 12952–12956.

Wellman, C.H. (2003) Dating the origin of land plants. In: Donoghue, P.C.J. and Smith, M.P. (eds) *Telling the Evolutionary Time: Molecular clocks and the fossil record*. CRC Press, London, UK, pp. 119–141.

Wellman, C.H., Osterloff, P.L. and Mohiuddin, U. (2003) Fragments of the earliest land plants. *Nature* 425, 282–285.

Westall, F., de Ronde, C.E.J., Southam, G., Grassineau, N., Colas, M., Cockell, C.S. and Lammer, H. (2006) Implications of a 3.472–3.333-Gyr-old subaerial microbial mat from the Barberton greenstone belt, South Africa for the UV environmental conditions on the early Earth. *Philosophical Transactions of the Royal Society B: Biological Sciences* 361, 1857–1875.

Wignall, P.B. (2001) Large igneous provinces and mass extinctions. *Earth Science Review* 53, 1–33.

Wolbach, W.S., Lewis, R.S. and Anders, E. (1985) Cretaceous extinctions – evidence for wildfires and search for meteoritic material. *Science* 230, 167–170.

Wolfe, J.A. (1991) Palaeobotanical evidence for a marked temperature increase following the Cretaceous Tertiary boundary. *Nature* 343, 153–156.

Woodward, F.I. (1987) Stomatal numbers are sensitive to increases in CO_2 from preindustrial levels. *Nature* 327, 617–618.

Woodward, F.I. and Kelly, C.K. (1995) The influence of CO_2 concentration on stomatal density. *New Phytologist* 131, 311–327.

3 Polar Marine Ecosystems

David K.A. Barnes
British Antarctic Survey, Natural Environment Research Council, Cambridge, UK

3.1 Introduction

From the human perspective the polar regions are extreme in many ways, for example in light climate, temperature, availability of ice-free land and liquid freshwater, wind speed and wave heights, as well as isolation. The nature and context of these polar extremes is very different across the realms of the sea, the sea surface (such as sea ice), the littoral zone, land and freshwater habitats. This chapter examines the subsurface marine environment, or put another way the volume that (as elsewhere on the planet) forms 99% of living space for macroscopic life. Living at extremes poses strong challenges to organisms, some of which are long term and evolutionary in scale whilst others are short term and ecological, some are both and some could be argued to be of little challenge, despite appearing as 'hostile' to us (see Bell and Callaghan, Chapter 1, this volume). Polar seas have the most extreme light, food and disturbance climate of the Earth's seas. They are also the coldest and have the lowest calcium carbonate saturation levels (which organisms need to make skeletons), are frozen over for much of the year and have the widest and deepest continental shelves, of which a third (in Antarctica) are underneath 'permanent' ice shelves (Dayton, 1990). However, perhaps the greatest extremity affecting organism survival has been the massive expansion of ice sheets during each glaciation period (up to 90,000 of each 100,000 years for the last few million years). These ice sheets were so thick that they grounded out on the bottom of continental shelves, even as deep as 1000 m (Clarke and Johnston, 2003). The ice sheets would have bulldozed most life off the continental shelves as they advanced (Thatje *et al.*, 2005) and would have to be recolonized in the brief interglacial phases, such as the current time. In addition, they have an extreme future as the most severe levels of warming, glacier retreat and sea surface (e.g. sea ice loss) change are also occurring now and projected ahead in polar seas (IPCC, 2007; see Psenner and Sattler, Chapter 23, this volume). In contrast, polar seas have had amongst the most oceanographically constant environments (e.g. in temperature and salinity) in the last few million years, the highest oxygen levels, the fewest crushing predators (for example crabs that catch and break the shells of molluscs with their claws), the least historical anthropogenic impacts (with the exception of whaling) and are the only marine areas on the planet without established non-indigenous or invasive species. The composition of organisms in polar seas represents life's response to the complex mosaic of all these conditions across scales in time and space, thus many are polyextremophiles.

Most major animal types, which are termed phyla, evolved in the sea and many did not (or could not) colonize freshwater or the land. Only to those that evolved on land, such as the insects, is the sea such an extreme environment that they are still virtually absent from it, and are absent from polar seas. Some conditions in polar seas are too extreme for some whole animal groups to be present at all, for example it is too cold for brachyuran crabs, reptiles or hermatypic corals. Some other groups are just capable of life there but are represented by very few species which are rarely abundant, such as barnacles and cartilaginous fish (e.g. sharks and rays). Gastropods (sea snails) may be ubiquitous and locally abundant in polar seas, but few species occur there (especially in the Arctic) compared with elsewhere. About 8% of the world's known ascidians (sea squirts), amphipods (sand hoppers) and isopods (lice) occur on Antarctic continental shelves, which is about the proportion of global shelf area they occupy (Barnes and Peck, 2008). Finally, some animals have become anomalously abundant or rich in polar seas. This is perhaps most obvious in terms of being the feeding ground for most of the great whales, many seals and penguins and the home of some of the most abundant species such as krill and some copepods. On the seabed, sea spiders, polychaete worms, bryozoans and brachiopods are all very well represented compared with most other environments (Fig. 3.1). Clarke and Johnston (2003) estimated that Antarctic continental shelves are as rich as, or richer than, others elsewhere except in coral reef habitats. At the South Orkney Islands (a Southern Ocean archipelago) 24 phyla, 50 classes and 1224 species have been recorded, which exceeds (at each taxonomic level) biodiversity at Galapagos and many other tropical and temperate archipelagos of comparative size (Barnes *et al.*, 2009). One key to success is to be 'bone idle': many of the most successful species move very little and have a high proportion of skeleton to tissue (thus demand for oxygen and food is reduced). To even watch one of Antarctica's most voracious benthic predators, the proboscis worm *Parborlasia corrugatus* (Fig. 3.2), chase prey has to be filmed then played back at 20 times normal speed to see the pursuit. So it is clear that polar seas clearly have both many 'extreme' physical parameters but also considerable and diverse life that has coped

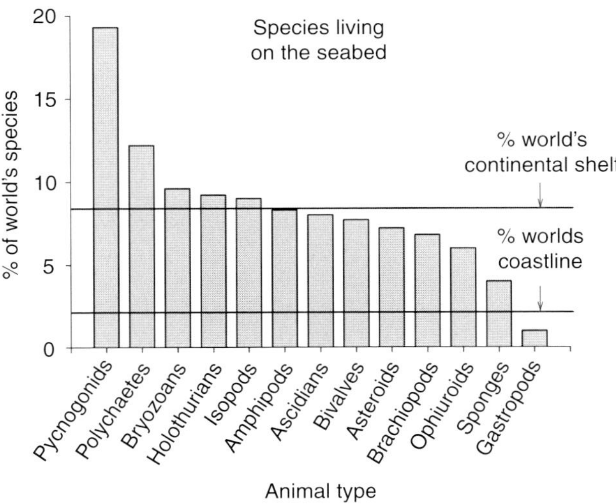

Fig. 3.1. Described species from continental shelves around Antarctica compared with continental shelf statistics. From Barnes and Peck (2008) updated from Clarke and Johnston (2003).

Fig. 3.2. Antarctica's most voracious benthic predator, the proboscis worm, *Parborlasia corrugatus*. © David K.A. Barnes.

with such extremes. There are however considerable differences between life, and life's response to extremes, in the two polar regions far beyond polar bears in the Arctic versus penguins in the Antarctic.

3.2 The Arctic Seas Versus the Southern Ocean

The Arctic seas and Southern Ocean contrast in many ways, most obviously in being water surrounded by land compared with water surrounding the land, respectively (Fig. 3.3, Table 3.1). There are many consequences to this such as typically colder and less fluctuating temperatures around Antarctica, compared with more freshening and pollution (from river input) in the Arctic. Perhaps the most relevant Arctic–Antarctic differences to the ability of life to respond to extremes are age, size and isolation. The Arctic is young and organisms are still in the process of recolonizing following expulsion by the last glacial maximum or ice age (Dunton, 1992). Size comparisons are less straightforward, although the area of sea ice peaks are higher in the Southern Ocean than in the Arctic (Table 3.1), the summer minima in the latter is double that around Antarctica (see Thomas, Chapter 4, this volume). Generally the southern polar region is considered, oceanographically and biologically, to be within the Polar Front (PF – the strongest jet of the world's most powerful major current, the Antarctic Circumpolar Current). The PF 'line in the sea' is mostly north of 60 °S. By comparison, in the Arctic marine environment conditions could be considered 'cool temperate' even at 70 °N in some places (e.g. the East Atlantic). Differences in terms of isolation are quite apparent as the Arctic has considerable land, continental shelf and slope connections with low latitude, whereas the Antarctic has none (except the isolated Kerguelen Plateau). The southern polar region is oceanographically and biologically the most isolated surface area on Earth, largely caused by the long term persistence

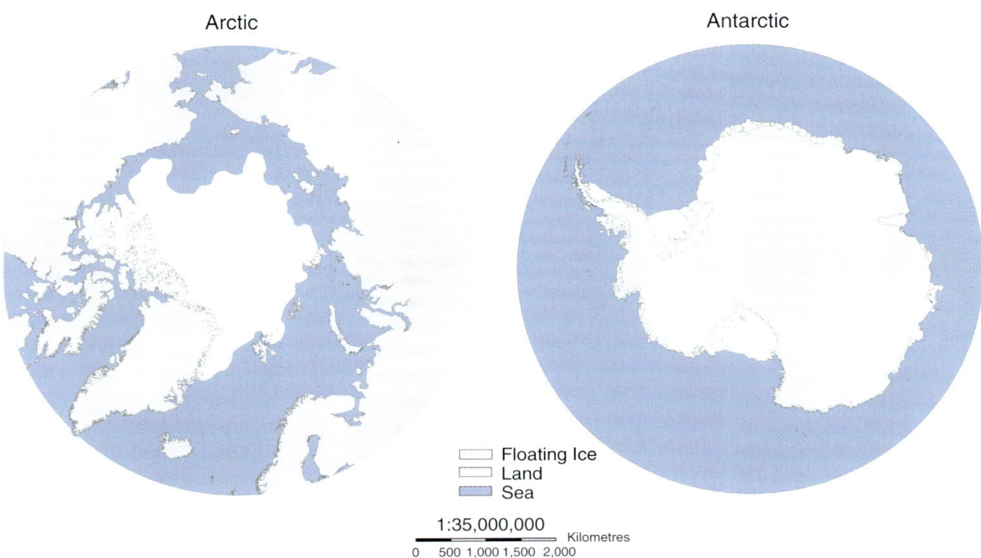

Fig. 3.3. Polar projections of the Arctic and Antarctic regions. Areas of land, sea and permanent floating ice are shown.

Table 3.1. Some key differences between the Arctic and Antarctic environments.

	Arctic	Antarctic
Gross structure	Sea surrounded by continents	Continent surrounded by sea
Ice cap age	<4 my	~15 my
Ice status	Floating ice over pole	Massive grounded ice sheet over pole
Key oceanography	Heat flow in (North Atlantic Current)	Cold outflow (Antarctic bottom water)
Ocean area	~8,000,000 km^2	~20,000,000 km^2
Mean depth	~1040 m	>4000 m
Seasonal max sea ice	~13,000,000 km^2	~18,000,000 km^2
River input	~10 major rivers	Virtually none
Continental shelf	Continuous with Americas and Eurasia	No links

of the Polar Front and other fronts. However, despite bipolar differences, the way in which life has persevered, thrived and solved some of the problems posed by living in extremes varies, perhaps mostly according to the type of ecosystem they live in. Thus, in this chapter, we consider six components separately from the shallows to the continental slope (abyssal and hadal depths are considered in Lutz, Chapter 13 and Hughes, Chapter 15, this volume). By far the largest of these six environments is the water column and in summer the top few hundred metres of polar seas are so productive that they provide the food for the largest animals that have ever lived, blue and fin whales (Fig. 3.4).

3.3 Water Column

In polar seas there is a latitudinal cline in many extremities with increasing seasonality of light, sea ice and primary production towards the poles. However, the timing and duration of these are all fairly predictable for organisms and, as with the extreme cold,

Fig. 3.4. Fin whale (*Balaenoptera physalus*). © British Antarctic Survey.

they have had millions of years to adapt to these, both physiologically and behaviourally. There are many aspects of physics in polar environments that organisms can do little to counter, such as the cold slowing enzyme action or reducing the efficiency of crushing muscles, or the higher viscosity of cold water affecting feeding or external fertilization success. Organisms have developed novel body chemistry, morphology and behaviour to cope with, and maximize the benefits from, polar environments. This is most famous in the development of glyco- and antifreeze proteins in fish. Several different types of these can be found in a single species alone, which help lower the blood freezing point, and the 'ice fish' do not even have haemoglobin so appear a ghostly semi-transparent white (Box 3.1). It is also evident in the utilization of high oxygenation levels to increase size (e.g. in arthropod groups such as sea spiders, amphipod and isopod crustaceans but also others such as ctenophores) and living low energy lifestyles, so that metabolic costs are low in winter when there is little if any primary production. A key chemical strategy for dealing with such intense seasonality of primary production is winter storage of energy reserves as oils within the body (Box 3.2).

Many endotherms (animals that regulate their body temperature, mostly mammals) migrate to the polar regions such that they are only present during the intense summer phase of summer food production. Small nekton (such as krill) can migrate within the polar regions to follow the productive seasonal ice edge but can also migrate vertically to utilize a variety of food sources. It was only recently realized that Euphausiids (krill), normally associated with the photic zone (near the surface), migrate as deep as abyssal (>3000 m) depths to forage for food (Clarke and Tyler, 2008). Similarly, new technology has enabled us to look at horizontal movements made, such as the discovery (using the robot submarine 'Autosub') that krill densities are highest in the water column a few kilometres behind (so underneath) the seasonal ice edge. Much of the extreme nature of the water column in the polar regions is associated with sea ice, and the ability of life to cope with this is covered by Thomas (Chapter 4, this volume).

Box 3.1. The Icefish – *Chaenocephalus aceratus*

A single group, the notothenioids, dominate Southern Ocean fish species. Like Galapagos finches a single group has radiated many new species across many niches within a restricted area – known as a species 'flock'. These fish all have antifreeze glycoproteins in their body fluids, such as blood. These compounds are types of polypeptides and reduce or prevent ice crystals from growing in freezing temperatures. Notothenioid fish also lack swim-bladders and live on or close to the seabed. One of the most charismatic and famous of these fish is *C. aceratus*, which only lives in the Southern Ocean but generally deeper than 50 m so are rarely seen by scuba divers. This species has a large head and mouth but is most obvious because of its lack of haemoglobin (which increases the efficiency of oxygen capture and transport around the body). Their survival using plasma rather than haemoglobin to absorb and carry oxygen, which is unique in vertebrates, is due to the ability of the cold water of the Southern ocean to hold high amounts of gas, and thus oxygen. Nevertheless, investigations of their genetics have revealed genes that would have once been involved in production of haemoglobin. Like other fish, icefish are thought to be very sensitive to even small temperature rises, probably due to an inability to deliver enough oxygen to tissues.

However, one polar extreme is associated with sea ice but does not affect the organisms in or on it. This indirect influence of sea ice on the water column below it, and the organisms in it, is its stabilization of the water column and reduction of disturbance. On average the strongest wind speeds and wave heights occur at high southern latitudes which support and suspend organisms in the (food rich) photic zone. However, when the sea surface freezes in winter the near surface volume changes from the most disturbed globally into amongst the most calm (hence the very high visibility levels measured and seen by scuba divers). This potentially poses problems for animals and phytoplankton that need to maintain positions in the upper part of the water column. The change from open water to frozen sea ice generates some of the greatest extremes in the shallows around the Antarctic continent.

Box 3.2. Krill – *Euphausia superba*

Krill are a group of shrimp-like crustaceans some of which are particularly abundant, dense and thus important as food for higher predators such as fish, seabirds, seals and whales. The Southern Ocean is the most important feeding ground for baleen whales because of krill productivity and so unsurprisingly it is also a key resource for fisheries. Nearly all krill belong to one family, the bioluminescent Euphausiidae, which includes one superabundant species *Euphausia superba*, which may swarm in many thousands of millions at densities of more than thousands per cubic metre. These reach a few centimetres in length, have prominent compound eyes, swimming abdominal appendages called pleopods and feed by filtering tiny phytoplankton from the water column. Much Southern Ocean science has focused on aspects of krill biology such as diurnal vertical migrations, spatial, seasonal and annual variability in production, and possible responses to environmental change. The success of krill is striking in many

Continued

Box 3.2. Continued

ways and seems strongly linked to its ability to convert and store energy as oil for winter reserves. Lipids form up to 4% of a live individual but can make up to nearly half the drymass of *E. superba* and likewise in other pelagic crustaceans such as amphipods and copopods. This storage can be obvious even to the naked eye and is striking when viewed under the microscope. The specific type of fat used by particular species depends on where they live and the duration of storage; for example, winter under-ice feeders use triacylglycerol. *Euphausia superba* are particularly rich in Omega-3 fatty acids and Omega-3 phospholipids, which are used by humans as natural and strong antioxidants.

3.4 Shallows

The shallows around the planet tend to be characterized by extremes of environmental conditions. On short time scales this is because of storm action and erosion, freshwater and sediment runoff from land, periodic warming, pollution and harvesting (e.g. fishing) and on longer time scales by tsunami activity and sea level rise and fall. This is exacerbated in the polar regions by one of the most destructive natural forces – ice scour.

3.4.1 Ice scour

Icebergs typically drift away from Antarctica in a westwards spiral, propelled by the combination of the highest mean wind speeds and fastest flowing major ocean current until they hit seabed shallower than their keel. The impact forces are enough to splinter rock and reshape the seabed, creating troughs and berms metres deep and kilometres long. Such impacts are locally catastrophic but some individuals of some species (e.g. *Mysella charcoti*) survive them (Smale *et al.*, 2007) and, like tree-falls or fires, these disturbance events both open up new space for colonization and modify conditions promoting wider organism diversity. To some extent there is a stochastic element to which precise area will be impacted, but some areas are much more naturally protected than others by the position of ridges and sills. Thus benthos vary from the highly depauperate that have been recently scoured through to densely packed communities of climax sponges that are hundreds of years old and their predators. The upper 50 m explorable by scuba and small remotely operated vehicles (ROVs) appear as a 'patchwork quilt', in which the density and diversity of life present is largely reflective of how recently the immediate area was scoured (Fig. 3.5). Likewise in the Arctic, the shallows tend to have an impoverished community often consisting of pioneers and mobile species that can crawl, swim or drift in (Fig. 3.6). The larger icebergs ground out in deeper water so potentially the impact force increases with depth up to the maximum depth scoured (about 800 m), but the impact frequency increases with decreasing depth because there are very many small pieces of floating ice. By 5–6 m impacts have a greater than annual frequency and ice can often be heard and seen smashing into the same intertidal zone on each high tide. The severity of ice scour disturbance in the shallows is compounded by the globally high levels of wave height (hence force) and salinity fluctuation from autumn hypersalinity (brine expulsion during sea ice formation) to strong freshening during summer sea ice melt and glacial runoff. As glaciers reach grounding lines this is increasingly coupled with sedimentation from terrestrial debris carried in melt streams. In many ways most of these impacts reach their most severe in the shallowest waters of the intertidal zone. However, the meaningful extremity of any impact on organisms should be measured with respect to their ability to recolonize, grow and reproduce to the state prior to disturbance.

(a)

(b)

Fig. 3.5. The upper 50 m explorable by scuba and small remotely operated vehicle appear as a 'patchwork quilt', in which the density and diversity of life present is largely reflective of how recently the immediate area was scoured by icebergs: (a) a 'patch' of diversity within a scoured area and (b) a clearly visible 'track' where an iceberg has scoured the seabed. Images © Kirsty Brown.

As elsewhere, polar communities comprise species with a range of strategy from pioneers to slow growing and developing climax forms. Aspects of the ecology or physiology of polar organisms measured to date show that these ranges strongly differ from those at lower latitudes. For example, within a guild such as the limpets, the fastest polar species grow quicker than the slowest of temperate or tropical equivalent species but across species, mean growth is about an order of magnitude slower. The time taken to maturity (being able to reproduce) is also about an order of magnitude

(a)

(b)

Fig. 3.6. In the Arctic the shallows tend to have an impoverished community often consisting of pioneers and mobile species that can crawl, swim or drift in: (a) serpulid worms are pioneer species that come to dominate in the shallows and (b) mobile species such as jellyfish can drift in and colonize a scoured area. Images © David K.A. Barnes.

slower as are most aspects of development (see Pearse *et al.*, 1991). Pearse and other field workers also demonstrated (as predicted by Thorson's Rule) that fewer polar species have planktotrophic larvae, potentially limiting the dispersive ability relative to those at lower latitude. Instead many have lecithotrophic larvae, which have their own food supply so are independent of food resources in the water column, giving adults more flexibility on the timing of their production. Many species brood direct developing young either because their group are obligatory direct developers such as isopods (Fig. 3.7) or because families with this reproductive strategy have been evolutionarily successful, such as many sea urchins (Fig. 3.8). Thus it would seem that for a given level of disturbance, polar benthos should take longer to recover than elsewhere. As disturbance is considered to be higher than typical environments elsewhere we should expect very much slower recovery from impacts. This is not entirely borne out by data. A number of experiments have evaluated recolonization of hard surfaces in the polar shallows by using artificial substrates such as perspex or slate panels (such as in Canada and Svalbard in the Arctic, the South Orkney, South Shetland and Adelaide islands, Ross Sea and Davis Sea in Antarctica). These studies did reveal slow colonization rates but also high between-site variability, which showed that polar levels of recolonization/recovery were not necessarily slower despite high disturbance and slow development tempo. However, it is apparent that many or most subtidal locations around the globe have a more developed megafauna (and large macrofauna) than the shallows (e.g. 10 m) around Antarctica. Amongst the most noticeable Antarctic megafauna of the shallows (Fig. 3.9) include the common sea urchin (*Sterechinus neumayeri*), sea star (*Odontaster validus*) and brittlestar (*Ophionotus victoriae*). Many species remain present in this zone but most are small, cryptic or clustered into a few ice-protected areas (Box 3.3).

Fig. 3.7. Isopods (*Antarcturus*) feeding perched on a coral. This genus broods direct developing young. © Pete Bucktrout.

Fig. 3.8. An urchin (*Ctenocidaris* sp.) brooding direct developing young. © Pete Bucktrout.

Fig. 3.9. The most noticeable Antarctic megafauna in the shallows include the common sea urchin (*Sterechinus neumayeri*), sea star (*Odontaster validus*) and brittlestar (*Ophionotus victoriae*). © Peter Enderlein.

Box 3.3. Ice-Protected Benthos

Continued

Box 3.3. Continued.

Small restricted areas in the shallows are protected from regular scour by icebergs, such as this vertical overhang in Ryder Bay, Adelaide Island, Antarctica. The dense benthos that the scuba diver (right) is observing include sponges (e.g. *Dendrilla antarctica*), anemones (e.g. *Urticiniopsis antarctica*), hydroids (e.g. *Symplectoscyphus* spp.), bryozoans (e.g. *Arachnopusia inchoata*), sea squirts (*Cnemidocarpa verrucosa*) and many other sessile animals. Most of these are suspension feeders and strain the seasonally abundant phytoplankton (such as ciliates and flagellates) from the water. There is strong competition for space in such assemblages which is very hierarchical – much like a pecking order in hens. Divers have monitored various species using tags on individuals of different sizes and ages to estimate when and how fast they grow – the answer varies with species and place but generally seems to be typically about ten times slower than equivalents elsewhere.

3.4.2 Climate change

There are two other aspects to the extreme nature of polar shallows environments, which are longer term but certainly highly influential on the structure of assemblages across time and spatial scales. The size (area) and distribution of the polar shallows has changed drastically on the time scales of glaciations and the oceanographic 'climate' is changing drastically on decadal to century scales with climate change. The area and geography of the shallows changes across the planet more than other environments because little sea level change is needed to change the area totally (little more than 20 m sea rise or fall). However, about one-third of potential polar shallows are currently covered by the grounded base of ice shelves and this is estimated to change to approximately 99% during glaciations. Thus the area of the shallows changes by a factor of three and changes horizontally as well as vertically. Disappearance of >99% of any environment and relocation of the remaining <1% for approximately 90,000 years of each 100,000 years must present considerable challenges for repeated reinvasion and establishment.

3.4.3 Water climate

The other longer term extremity aspect is one of water climate (chemistry and physics) due to rapidity of recent change, particularly around the Scotia and Bellingshausen seas. The seasonal thinning of ozone over Antarctica has, in just a decade, massively increased UV-B irradiation and wind speeds impacting the shallows. Probably more biologically significant though is more change in acidification of surface waters within a century than has occurred in the last 30 million years (Orr *et al.*, 2005; see also Glover and Neal, Chapter 24, this volume). Likewise, Meredith and King (2005) have shown a sea surface rise in temperature of about 1°C over the last half century. The crucial context of this is that marine temperatures around Antarctica have only (seasonally) varied by about 3°C over the last 4 million years. Furthermore, a boom in short term laboratory experiments on exposing Antarctic organisms from the shallows to rises in temperature has generally shown them to be highly sensitive, but care is needed in the interpretation of such data (Barnes and Peck, 2008). Thus the polar shallows represent an extreme environment for organisms along many time and spatial scales with new and accelerating 'extremes' now occurring.

3.5 Continental Shelf

Polar continental shelves are unusually wide and deep. They are the coldest and show the least seasonal temperature change of any on the planet, have the greatest seasonal variation in light penetration, primary productivity, water surface conditions (fast ice versus open water) and arguably the least direct anthropogenic impacts. Arctic and

Antarctic shelves also share the highest oxygen levels (cold fluids hold more gas), the greatest vertical mixing and have large areas where surface water sinks into the deep sea and are the youngest, still in the process of being recolonized since the last glaciation. However, the continental shelves of the two polar regions are also very different in other respects (Table 3.1). The Arctic shelf is shallow in depth and slope profile, is surrounded by continent margins and thus has many major rivers flowing into it, and much of the ice is floating. The Antarctic has the converse of these and, because of the contrasts in isolation and evolutionary history, the biotas are also very different. Both Arctic and Antarctic shelf areas represent extreme environments to organisms but they are extreme in some shared and some differing ways.

3.5.1 Disturbance

Many of the extremes described for the shallows are also true for much of the polar continental shelf but less pronounced. For example, the frequency of ice impacts is much reduced yet multibeam sonar images of continental shelves are criss-crossed with scours (Fig. 3.10). Such ice scours may be visible for thousands of years and are still evident in multibeam sonar images of the shelf around Scotland. Recovery of biological communities from such events probably takes many hundreds of years, and both the succession and timing of arrival of new colonists will be linked to the state of adjacent assemblages (which is itself mostly a function of time elapsed since scouring). Although typical shelf depths are less disturbed (by ice or other mechanical processes) than the shallows, the difference in disturbance between polar and non-polar environments is greater with depth. Until the advent of wide-scale commercial trawling, low latitude continental shelves were unlikely to be regularly disturbed (below 100m) by much on the scale of iceberg scouring (Table 3.2). There have been and remain fisheries around Antarctic continental shelves and of course there is, at some locations, fairly regular scientific trawling. To date the levels of disturbance created by such anthropogenic harvesting and research have been very minor compared with elsewhere partly because of the large (and costly) geographic distances involved in getting to theses shelves, partly due to their inaccessibility due to fast-ice or pack-ice and partly because of regulatory bodies such as CCAMLR placing restrictions on activities. So arguably the Southern Ocean continental shelf is, with respect to

Fig. 3.10. Multibeam sonar images of continental shelves are criss-crossed with scours. © Kathy Conlan.

Table 3.2. Severity, frequency and predictability of disturbance to continental shelf benthos.

	Severity	Frequency	Predictability
Storm wave/current	Severe	Subannual	Spatially predictable
Trawling >100 m	Severe	Decadal	Spatially predictable
Ice scouring >100 m	Severe	Centurial	Semi-predictable
Volcanism	Locally severe	Locally infrequent	Spatially predictable
Meteorite impacts	Locally severe	Very infrequent	Unpredictable
Mammal feeding	Moderate	Locally subannual	Predictable
Bioturbation	Minor	Frequent	Predictable

disturbance, the most naturally disturbed and the least anthropogenically disturbed.

3.5.2 Size

A key 'extreme' of the continental shelf around Antarctica is stability – temperature, salinity and other oceanographic features vary less than in any other surface environment. One of those long term conditions is high oxygen levels (cold water holds more gas). Very high oxygenation from shallow to deep waters around the polar regions has enabled some groups of animals (notably those with jointed legs, sometimes referred to as arthropods) to gain large maximum sizes (Fig. 3.11). Some of the largest sea spiders, amphipods, isopods and a few other types of animal, such as comb jellies and sponges occur on polar continental shelves. However, most representatives of these taxa are not big, and in fact some other animal groups are dominated by more smaller sizes than elsewhere in the world (for example, most bivalves are tiny, such as *Mysella charcoti*; Fig. 3.12). Being large often confers an immediate advantage (e.g. in competing for resources) so can be selected for, but getting enough oxygen to tissues becomes limiting with increased size. Thus high oxygen in polar waters provides a mechanism to support increased size where this is a selective advantage. However, the low resource levels for much of the polar year probably means that in many animal types smaller size may be selected for because there is less tissue to maintain through long periods of low food availability. Furthermore, it seems likely that selective pressures differ between glacial and interglacial phases, as for the long periods of glaciations most benthos would be far displaced from the photic zone of primary productivity but then have relative boom times between glaciations.

3.5.3 Reproduction

The many extremes involved in the (41,000 year and lately 101,000 year) alternation of glaciations has been the biggest source of environmental punctuation. As with food availability, it seems likely that this alternation of environmental conditions may have intensified selective pressures on reproductive modes, tempo and investment. Polar benthic organisms on continental shelves tend to live long lives and are record breakers in terms of reproduction; even small macrofauna may live many years, even decades, before becoming mature. For example, the brachiopod *Liothyrella uva* (Fig. 3.13) lives 10 years before first reproduction. The duration of development of eggs, larvae and/or juveniles is very slow and few (relative to low latitudes) have feeding larvae. It has been suggested that selective pressures during glaciations are for direct development or lecithotrophic larvae. This is because potential refugia where life could survive when ice sheets were maximal would be deep or underneath ice with little food. For the brief interglacials though the selective pressures would favour those with highly dispersive, feeding larvae that would enable rapid colonization of the shelves after ice retreat. Thus in animal groups such

Fig. 3.11. (a) Sea spiders such as *Decalopdia australis* reach much larger sizes than those in warm water (pictured below), © British Antarctic Survey. (b) The first work to show polar gigantism compared the sizes of amphipods from place to place, such as this *Paraceradocus miersii* from Antarctica. © David K.A. Barnes.

as the echinoids (sea urchins), very few polar representatives have planktotrophic larvae but the few that do are extremely abundant in the shallows, reflecting these contrasting pressures across evolutionary and ecological time scales.

The vast majority of polar continental shelves were covered by grounded ice sheets during glaciations such that at glacial maxima (when the ice mass was greatest) only small and patchy expanses of the seabed were ice-free. Despite polar shelves reaching 1000 m depth in places ice grounding lines from the Last Glacial Maximum are visible in multi-beam sonar images across most areas away from the shelf-break. The bulldozing of the seabed and its fauna into small fragmented refugia represents an extreme situation in itself, but the fauna in such small refuges would have been subject to very high freshening and sedimentation and very low food levels. In interglacial conditions, such as now and for the last few thousand years, conditions for benthos are very much less extreme than during the long periods of glaciations. Despite this many authors have suggested that polar benthos might have survived *in situ* and both biogeographic distribution and molecular evidence is starting to emerge which is consistent with this.

Fig. 3.12. The bivalve mollusc, *Mysella charcoti*, grows to just a few millimetres wide. © Dan Smale.

Scientific cruises during the last decade have increasingly exposed locations, such as inner shelf basins or behind impediments to ice sheets (e.g. islands), that may have been refugia for life during former glacial maxima. Of course finding areas that were not covered by grounded ice during the last, or other, glaciations, is not the same as showing that life survived at such places. One of the many stark contrasts about Antarctic shelf benthos is the apparent sensitivity to small changes in environmental conditions (and thus vulnerability to projected climate change) yet robustness to survive long term through many glaciation cycles.

3.5.4 Diversity gradients

Benthos has not merely survived glacial cycles, it is rich across taxonomic scales. For much of the last century one of the most important and obvious spatial patterns of life on Earth was considered to be the latitudinal diversity cline, such as that recorded for hermatypic corals (Fig. 3.14). This pattern was a broad decrease of species richness from low to high latitudes, or more accurately typically a decline away from an Indo-West Pacific hotspot.

Fig. 3.13. Clumps of the brachiopod *Liothyrella uva*. The species only reaches sexual maturity after 10 years. © Karen Meidlinger.

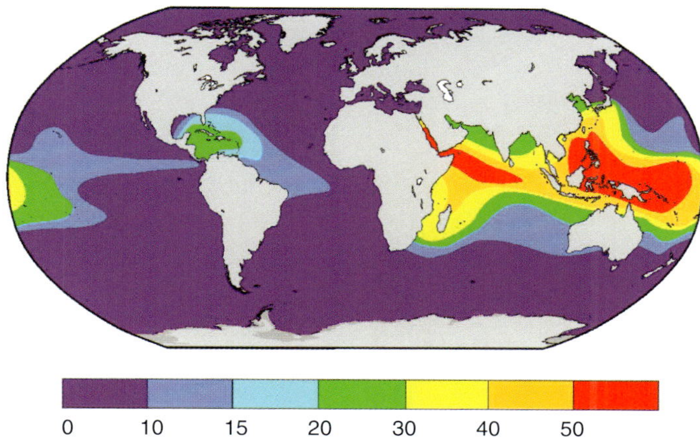

Fig. 3.14. Hermatypic coral diversity from high (red) to low (purple).

Research across taxonomic groups has supported this in a few taxa (notably in hermatypic corals, fish and crustaceans), but perhaps more typically a more complex pattern has emerged. To date decreasing diversity away from the tropics is not evident in soft sediment biotas, is much stronger in the northern than the southern hemisphere and is not evident in many taxa. Despite shelf seas being much less well studied at high latitudes, there are already more species known in many taxa (see Fig. 3.1) from Antarctica than a global average by area that these represent. Indeed brachiopods are more abundant, and sea spiders richer around Antarctica than anywhere else. Antarctic diversity has been famous for what is absent (e.g. true crabs, hermatypic corals and reptiles) or vanishingly rare (e.g. cartilaginous fish such as skates – Fig. 3.15 – and barnacles). However, perhaps the greatest extreme in diversity is more obvious – a single benthic trawl or scuba dive typically reveals representatives from half or more of the phyla (major types of organism) on Earth. This would be unusual even on a coral reef, rare on a temperate shelf and impossible in a rainforest. Therefore, at least a few representatives of most animal types are so abundant and ubiquitous in Antarctic shelf waters that a single sample taken across a range of spatial scales can encounter individuals. However, one-third of Antarctica's continental shelf (and at times most of both polar shelves) constitutes a very different environment, in which physical conditions and patterns in life differ as much from typical polar shelves, as polar shelves do from those around the rest of planet Earth.

3.6 Below Ice Shelves

The recent news that a shrimp (an amphipod) had been found in a hole that had been drilled hundreds of metres through a (floating) ice shelf that was hundreds of kilometres from open water, created considerable popular and scientific interest. However, it had been known for a number of decades that, remarkably, benthos and a number of animals had been found in such quiet, dark and isolated environments far from seasonally or permanent open water. Most recently, Census of Antarctic Marine Life (CAML) scientists drilled through more than 700 m of ice shelf (in 1200 m water depth) to find benthos 200 km from open water below the Amery Ice Shelf, Prydz Bay, East Antarctica (Census of Marine Life, 2005).

There are three major ice shelves, each in the highest latitude embayments, around Antarctica and a few smaller ones, some of

Fig. 3.15. Lower and upper surfaces of the skate *Bathyraja eatonii*. © Pete Bucktrout.

which span the entire width of the continental shelf. The relative constancy in physical conditions that characterizes polar shelves are even more constant underneath ice shelves. In part this is because there is little temporal variability in physical conditions at more than 100 m water depth anywhere around Antarctica. However, the strong seasonality of alternation of light and primary production climate, as well as sea ice cover, changes to the long term dark and very low food conditions of the least known environment on our planet's surface. Little is known about under-ice shelf currents and how much food (such as primary production) gets advected to such habitats but it can not simply be the case that life found there to date are doomed vagrants. The few expeditions which have sampled under ice shelves have found scavengers and predators but they have also shown that primary consumers, including sessile species, establish there. In some cases this has included bryozoans and sponges, which are likely to be tens to hundreds of years old. Thus these had not merely survived but grown, although to date, no analyses of growth tempo or other ecophysiological work has been possible on life in this most extreme of habitats. It is likely that some similarities will be found with cave and deep sea faunas, given similarities in constancy of conditions and permanently low food supply levels (see Hughes, Chapter 15 and Lee *et al.*, Chapter 16, this volume).

The greatest insight into this environment came with the collapse of the Larson A and B ice shelves and subsequent scientific exploration of the seabed that was 'newly exposed'. Gutt *et al.* (2011) described a fauna which seemed a mixture of an old and spare 'under ice shelf' fauna and abundant 'new pioneers'. Many individuals of a few highly dispersive and fast growing species, such as the stalked ascidian *Molgula* with planktotrophic larvae, dominated large patches. However, occasionally large Rossellid sponges were observed, which must have recruited long before ice shelf disintegration. Identification and ecological investigation of organisms collected on such voyages should drastically increase our insight into *in situ* processes and how life copes with such an environment. Collapse of an ice shelf rapidly changes one extreme environment into another very different type – analogous to a hydrothermal vent appearing or disappearing in the deep sea, but at a much larger scale. At the moment ice shelves are increasingly collapsing around the Antarctic Peninsula but extending around the Ross Sea, but how much can be ascribed to human activity (e.g. indirect through greenhouse gas generated warming) compared with normal cycles within an interglacial phase is not clear. Opening up new shelf areas can result in important new algal blooms and, through the development of faunas, sequestration of carbon to the seabed. Peck *et al.* (2010) recently described the collapse of ice shelves as a key negative feedback mechanism on climate change and releasing the second largest new carbon sink on the planet. So to date most of what we know about life under ice shelves comes from exploration immediately post-collapse and just a handful of point samples by drilling through. We know similarly little about the continental slope around Antarctica or the Arctic, despite the fact that it may have been, and may yet be, crucial to the survival of the whole polar marine fauna.

3.7 Continental Slope

3.7.1 Disturbance

The vast majority of what we know about the seabed (and life on it) is about the shelf, but half our planet's surface is abyssal plain. Between these extremely different environments is the steep continental slope that is thought to have been an important transition zone for fauna to migrate multiple times, probably from both polar shelves to the deep sea and vice versa. The steepness of slope profiles, in combination with many being along tectonic plates boundaries, means that there are likely to be (and there is evidence of) frequent cascades of rubble. Substrata are probably very unstable and earthquakes, fluid flow or shallow gas deposits can trigger massive slumps of material, especially around ultra-steep canyon systems, so fauna must be tolerant of both high disturbance and sedimentation. These rubble cascades have been referred to as 'mass wasting' and are likely to have been much more severe at high latitudes during glaciations than elsewhere. This is because grounded ice sheets probably reached most, or at times all, of the shelf break and thus bulldozed moraines over the edge to tumble down the slope. As can be seen today around the edge of ice sheets in the sea the dynamism of ice movement results in a highly mobile and turbid substratum. Therefore, at the very top of continental shelves some of the extreme nature of the environment has physical similarities with polar continental shelves. Even much of the fauna may be shared as many species in polar faunas occur across a very high depth range (eurybathy). Typically however, the conditions on continental shelves and slopes are, and would during glaciations, be very different (aside from just profile and pressure differences).

During interglacial periods, such as now, slope biota are much further removed from where food is generated than on the shelf – a particular challenge for suspension feeders (such as the bryozoan *Melicerita obliqua* on the Weddell Sea slope (Fig. 3.16; Barnes and Kuklinski, 2010), which consume plankton suspended in the water column. Also there is little seasonality in conditions as almost no light penetrates to these depths and water temperature, salinity and other variables are very constant. It is unlikely that freshening has been strong at slope depths and sinking of polar water has ensured continuous oxygenation. In contrast to the shelf during glaciations, the area of the slope available for colonization did not change as at no point was grounded ice thought to cover it.

Fig. 3.16. Gross colony of the suspension feeding bryozoan *Melicerita obliqua*. © Pete Bucktrout.

3.7.2 Carbonate compensation depth

One of the key environmental factors that typically differs between shelf and slope in polar waters is carbonate solubility. The calcium carbonate compensation (CCD) depth, that is, the point at which the water is not saturated and $CaCO_3$ dissolves, is typically near the top of the continental slope. This is shallower than elsewhere and means that it is more difficult for organisms to synthesize and maintain skeletons. High magnesium calcite skeletons dissolve more easily than those low in magnesium, and those which are mainly (or entirely) composed of the aragonite form of carbonate are especially vulnerable to dissolution. Carbonate solubility levels are naturally closest to critical around the Subarctic and Southern Ocean. The cold conditions around continental slopes, particularly at high latitudes, mean that the cost of crushing for predators (such as by crab claws) is energetically very expensive. Crushing predators are very rare on polar continental shelves but they are a bit more prevalent on continental slopes. A trawl or epibenthic sledge tow along the polar continental slope reveals that there are many species present that have carbonate skeletons, however the skeletons tend to be very thin, often so much so that they are translucent. Unlike the shelf, dead specimens of such species are very rare as when the organism dies and stops activity maintaining the skeleton it simply dissolves. Like on the abyssal plains and trenches of the deep sea, continental slope fauna lives in the extreme conditions of the environment dissolving their skeletons. This situation has differed little across the last 30 million years, however, acidity (H^+ ion concentration) of the global ocean is rapidly increasing. This will result in the CCD becoming shallower and generating this condition normally limited to the slope on polar shelves. Warming of surface waters is also likely to make crushing predation more energetically cost effective and establishment of non-indigenous species easier. So this could be viewed as life being marginal for animals such as stone crabs (Fig. 3.17) but warming may improve their prospects of colonizing continental shelves (however, any benefits by increased temperatures might be offset by detrimental acidification). How all this will influence polar continental slope biology is hard to determine.

The instability of substrata, periodic cascades of rubble and critical $CaCO_3$ solubility seem to make the environment extreme for survival of the fauna. However, there are probably few places that have had such long term constancy in conditions. The continental slope is relatively buffered against drastic changes in sea level, temperatures, food production and ocean oxygen levels. The fauna which occurs there has had tens of millions of years to evolve to make the 'best of a bad job' for this particular oceanographic setting. Entire continental shelves have been raised up to form land or sink to be deep water and the deep sea has been through many phases of anoxia, so the continental slope fauna is typically likely to be the oldest marine fauna. However, it is still unclear whether the slope

Fig. 3.17. The stone crab, *Paralomis spinosissima*, at a depth of 200 m. © David K.A. Barnes.

fauna at any given site is largely a transitional one or distinct and whether the slope is mainly a source for the deep abyssal plains fauna or a sink of it. Some studies have shown abundance decreasing across many orders of magnitude from shelf, through slope to the deep sea but others have shown that it is very patchy and that we have sampled too little to draw any general conclusions. It seems likely that the slope has a distinct fauna in some places, whereas in others it is largely a transition zone where shelf fauna grades into an abyssal one and that this too depends on the group of organisms considered (Fig. 3.18; see Kaiser *et al.*, 2010). For seamounts and young volcanic islands shelf and slope environments may be especially alike, as often they have severe topographic profiles, similar substrata, stability and ages and are isolated from colonists by deep sea. Thus the extreme nature of shelves and slopes around isolated polar islands are probably best considered together.

3.8 Isolated Islands

Arctic land masses are dominated by two continent edges, a large island (Greenland) and a number of coastal islands north of Canada and more isolated islands off Russia and Norway. The continent of Antarctica is formed by an ice sheet linking one large island (East Antarctica) with the many smaller islands of West Antarctica. The islands between Antarctica and the southern tips of South America, Africa and Australasia are amongst the most isolated on Earth. The extreme nature of the high latitude polar islands is similar in many respects to the typical shelf and slope environments already described but with some important differences resulting in quite different characters to their biotas. The richness and bias of groups present on islands typically depends on their age, size (i.e. shelf area) and distance from supply sources, such as the nearest continent margin. Older islands have had more time to be colonized and larger ones present a bigger target for colonists to find. Increasing dispersal ability is needed to reach an island with increased distance from sources of colonists. This is made slightly more complex in the Southern Ocean because of the Antarctic Circumpolar Current (ACC) fronts, the rarity of planktotrophic larvae and of the tolerance of the dispersal stage to short term temperature changes.

We consider each of these special situations in turn and thus start with the jets/fronts

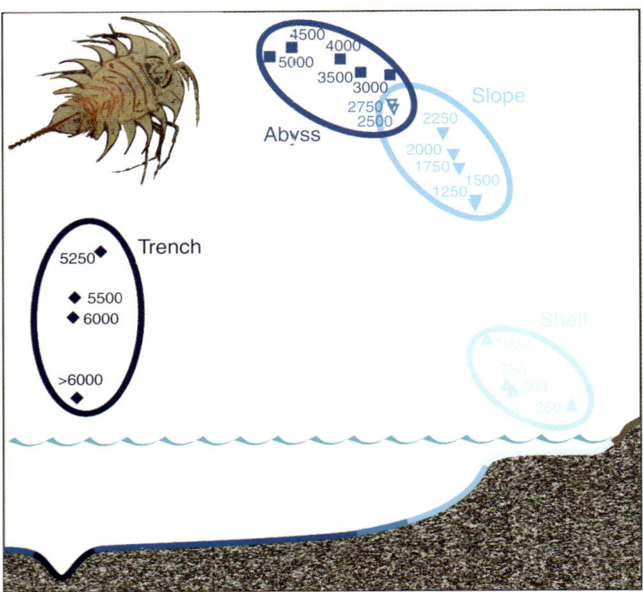

Fig. 3.18. Similarity of isopod samples with depth in metres. From data in Kaiser *et al*., 2010.

that make up the ACC. Taken together, the jets that constitute the clockwise (west to east) circulating ACC form the most powerful and fastest major current in the world and it provides a significant barrier to transport in any other direction than eastwards. The most powerful jet of this, the Polar Front, is even apparent at 4 km depth and marks a sharp discontinuity in freshness and temperature. There are many eddies which spin off both northwards and southwards and, furthermore the position of the whole ACC wobbles and migrates, but typically the distance a colonist needs to travel to reach a Southern Ocean island is greater than a simple great circle route on a map This means the realized distance from an island shelf to the nearest potential source of colonists may be longer in time than the larval duration for many species (larvae have a limited time in which they can remain in the plankton and still successfully metamorphose to young adults). In that case, reaching such an island shelf relies on adult dispersal, e.g. by rafting on kelp (e.g. by bivalve *Gaimardia trapesina*; Helmuth *et al*., 1994). The second additional factor making colonization of isolated islands harder in the Southern Ocean (and even more so in the Arctic) is the rarity of planktotrophic larvae. Gunnar Thorson's and Simon Mileikovsky's work on marine mollusc reproductive biology in the 1950s to 1970s suggested that (among other things) there was a dramatic reduction in planktonic larvae towards high latitude – this was termed 'Thorson's rule'. Much later other workers, notably Pearse *et al*. (1991), showed that this rule did not hold for echinoderms and Stanwell-Smith *et al*. (1999) who found that planktonic larvae were surprisingly quite common and abundant. Nevertheless, few polar species have feeding (planktotrophic) larvae compared with low latitudes, which is important, because these can be produced in much higher numbers and potentially travel furthest. Furthermore, many of the taxa which are unusually well represented in polar waters, such as sea spiders, isopods and amphipods have direct development and thus no larvae to disperse. Finally, the larvae that can disperse far enough to potentially reach isolated polar islands must have the acute thermal tolerance to withstand the rapid temperature changes involved. For example, the South Georgia shelf is on average about 2°C warmer than the Antarctic Peninsula shelves during the summer.

Over the last decade, a considerable body of work by the international polar biological community has examined thermal tolerances across about 20 nearshore species. The results of these studies suggest that many species cannot acclimate to a rise of even just a few degrees above their summer maximum and that a rise of 2–3°C would drastically reduce their functional competence (e.g. the limpet, *Nacella concinna*, Fig. 3.19 and the brittlestar, *O. victoriae*, Fig. 3.9; see Barnes and Peck, 2008 for summary of studies).

Colonization and establishment on isolated polar islands clearly represents a considerable challenge, and the younger, smaller and further away they are, the harder this is. However, once colonists have arrived they have to cope with different temperature regimes, often very steep profiles, periodic volcanism on some, and expansion of ice sheets over shelf areas during glaciations on others. However long the list of seemingly 'harsh' environmental conditions it is also clear that of the few polar islands well surveyed to date, some have very rich and abundant marine faunas. Fish, crabs, barnacles and others may be near absent at the South Orkney Islands but overall the benthos is as rich as has been measured in most other Atlantic islands and even comparable with Pacific archipelagos such as Galapagos. The development of rich faunas on island shelves such as those of the South Orkney's are greatly aided by the islands being old and large but crucially grounded ice did not reach the edge of the shelf during the last (and possibly other) glacial maximum. Although it is clear that life can be very rich at extremes, the main point of comparison is with shelves elsewhere which have been raked by a force more frequent and comprehensive than icebergs – commercial fishing. Strong estimates of the age spectrum, abundance and richness of life on low latitude shelves does not pre-date industrialized fishing and thus conclusions about how life copes with polar extremes relative to biotas elsewhere is seriously confounded.

3.9 Conclusions

A summary of environmental extremity in the Southern Ocean should separate across-environment points (coldest, most seasonal and most oceanographically isolated) from

Fig. 3.19. Aggregation of the limpet *Nacella concinna* spawning. © David K.A. Barnes.

those within-environment (e.g. continental shelves are the widest, deepest, most disturbed in summer and least disturbed in winter). These short term factors can only be meaningfully considered in the context of longer term changes, particularly ice sheet advance and retreat across glacial cycles and the much longer term cooling and relative constancy. More science has taken place around the polar regions in the last decade than most of the previous century and many of these extremes have begun to be quantified. These have included exactly what wind and current speeds, wave heights, sediment loads, ice scour frequencies and phytoplankton bloom densities occur around communities of organisms. Now and in the near future the Southern Ocean is and will be subject to new extremes, many of which are associated with climate change. The Bellingshausen and Scotia Seas include the fastest areas of marine warming and sea ice loss whilst the Antarctic Peninsula and outlying islands also include the most rapidly increasing air and lake temperatures and one of the greatest rates of glacier retreat and ice shelf loss. The only rival for such rates of change are parts of the Arctic where sea ice loss has been so drastic in some years that models have been unable to recreate actual data. Current CO_2 levels and rates of increase are unprecedented within the last few million years, maybe tens of millions of years, and given the very strong link between these levels and temperature (gauged from ice cores), there is a clear signal in terms of what to expect. Model projections, even conservative ones, show most regions will have increasingly rapid temperature increases, precipitation redistribution in space and time, higher dissolved CO_2 in oceans (acidification), sea level rise and higher wind speeds. The latter has largely been driven by stratospheric ozone thinning, which has caused an interaction of effects with those of greenhouse gases. Thus, the polar regions, particularly the Southern Ocean, have changed from being an extreme in terms of constancy (e.g. of stable ocean temperature) to areas of greatest tempo of change. How vulnerable the biota in the polar regions are to forthcoming changes is not clear and becoming hotly debated. We still know little about how polar life has coped with the considerable impact of glacial and interglacial onsets and all that these entail. Even over current ecological time scales we have had to revise what environmental extremes mean to life.

If many Southern Ocean shelves and possibly some continental slopes have rich and abundant biotas, does this mean that our assessment of those environments as extreme for organisms is wrong? We consider that the assessment of 'extreme' is correct because the extremity has been gauged by measurement of physical conditions in which (comparable) life occurs. However, it seems clear that, as elsewhere, life in polar marine environments thrives close to their absolute limits of tolerance of some conditions. Today some life is right on the limits of survival in the polar regions. Only a very few ectotherm-crushing predators, such as skates (Fig. 3.15), are able to make this feeding strategy work on the margins of the Southern Ocean. Likewise do anomuran crabs in water masses just above 0°C at which they effectively become anaesthetized (Fig. 3.17; see Thatje *et al.*, 2005). Remarkably, some species survive all year round in the intertidal zone despite being encased in ice for some of the year and some individuals even survive direct iceberg impacts. Polar benthos may be highly vulnerable to change in some conditions but elements have also proved tough and resilient to the many extremes in these environments. Unfortunately, given the extremes of slow development, onset of maturity and great age characterizing much polar benthos it is difficult to assess their capability to tolerate or adapt to the rapid change that is now occurring. Improving laboratory experimentation, modelling and uncovering stronger evidence of past response to interglacials such as the last (which was warmer than present) should greatly aid our understanding of how life in the two polar regions will respond to the current changes. This will be crucial as polar life response may be our best early warning system of change likely elsewhere. Furthermore, as sea ice and ice shelves retreat, and on land snow and ice cover decreases, carbon drawn down in the polar regions may be one of the biggest negative feedbacks and carbon sinks.

References

Barnes, D.K.A. and Kuklinski, P. (2010) Bryozoans of the Weddell Sea continental shelf, slope and abyss: did marine life colonize the Antarctic shelf from deep water, outlying islands or in situ refugia following glaciations? *Journal of Biogeography* 37, 1648–1656.

Barnes, D.K.A. and Peck, L.S. (2008) Vulnerability of Antarctic shelf biodiversity to predicted regional warming. *Climate Research* 37, 149–163.

Barnes, D.K.A., Kaiser, S., Griffiths, H.J. and Linse, K. (2009) Marine, intertidal, freshwater and terrestrial biodiversity of an isolated polar archipelago. *Journal of Biogeography* 36, 756–769.

Census of Marine Life (2005) Amery ice shelf. Available at: www.coml.org/discoveries/ecology/amery_ice_shelf (accessed 3 September 2010).

Clarke, A. and Johnston, N.M. (2003) Antarctic marine benthic diversity. *Oceanography and Marine Biology – An Annual Review* 41, 47–114.

Clarke, A. and Tyler, P.A. (2008) Adult Antarctic krill feeding at abyssal depths. *Current Biology* 18, 282–285.

Dayton, P.K. (1990) Polar benthos. In: Smith, W.O. (ed.) *Polar Oceanography Part B: Chemistry, Biology and Geology*. Academic Press, London, UK, pp. 631–685.

Dunton, K. (1992) Arctic biogeography: the paradox of the marine benthic fauna and flora. *Trends in Ecology and Evolution* 7, 183–189.

Gutt, J., Barratt, I., Domack, E., d'Udekem d'Acoz, C., Dimmler, W., Grémare, A., Heilmayer, O., Isla, E., Janussen, D., Jorgensen, E., Kock, K.H., Lehnert, L.S., López-Gonzáles, P., Langner, S., Linse, K., Manjón-Cabeza, E.M., Meißner, M., Montiel, A., Raes, M. and Robert, H. (2011) Biodiversity change after climate-induced ice-shelf collapse in the Antarctic. *Deep Sea Research Part II* 58, 74–83.

Helmuth, B., Veit, R.R. and Holberton, R. (1994) Long-distance dispersal of a sub-Antarctic brooding bivalve (*Gaimardia trapesina*) by kelp-rafting. *Marine Biology* 120, 421–426.

IPCC (2007) *Climate Change 2007: The Physical Science Basis*. Fourth IPCC Assessment Report Working Group 1. Cambridge University Press, Cambridge, UK.

Kaiser, S., Griffiths, H.J., Barnes, D.K.A., Brandão, S.M., Brandt, A. and O'Brien, P.E. (2010) Is there a distinct continental slope fauna in the Antarctic? *Deep Sea Research Part II* 58, 91–104.

Meredith, M.P. and King, J.C. (2005) Rapid climate change in the ocean west of the Antarctic Peninsula during the second half of the 20th century. *Geophysical Research Letters 32*, L19604.

Orr, J.C., Fabry, V.J., Aumont, O., Bopp, L., Doney, S.C., Feely, R.A., Gnanadesikan, A., Gruber, N., Ishida, A., Joos, F., Key, R.M., Lindsay, K., Maier-Reimer, E., Matear, R., Monfray, P., Mouchet, A., Najjar, R.G., Plattner, G.-K., Rodgers, K.B., Sabine, C.L., Sarmiento, J.L., Schlitzer, R., Slater, R.D., Totterdell, I.J., Weirig, M.-F., Yamanaka, Y. and Yool, A. (2005) Anthropogenic ocean acidification over the twenty-first century and its impact on calcifying organisms. *Nature* 437, 681–686.

Pearse, J.S., McClintock, J.B. and Bosch, I. (1991) Reproduction of Antarctic benthic marine invertebrates: Tempos, modes and timing. *American Zoologist* 31, 65–80.

Peck, L.S., Barnes, D.K.A., Cook, A.J., Fleming, A.H. and Clarke, A.C. (2010) Negative feedback in the cold: ice retreat produces new carbon sinks in Antarctica. *Global Change Biology* 16, 2614–2623.

Smale, D.A., Barnes, D.K.A. and Fraser, K.P.P. (2007) The influence of ice scour on benthic communities at 3 contrasting sites at Adelaide Island, Antarctica. *Austral Ecology* 32, 878–888.

Stanwell-Smith, D., Peck, L.S., Clarke, A., Murray, A.W.A. and Todd, C.D. (1999) The distribution, abundance and seasonality of pelagic marine invertebrate larvae in the maritime Antarctic. *Philosophical Transactions of the Royal Society B: Biological Sciences* 353, 1–14.

Thatje, S., Anger, K., Calcagno, J.A., Lovrich, G.A. and Pörtner, H.-O. (2005) Challenging the cold: crabs reconquer the Antarctic. *Ecology* 86, 619–625.

4 Sea Ice

David N. Thomas
School of Ocean Sciences, Bangor University, UK and Finnish Environment Institute (SYKE), Marine Research Centre, Helsinki, Finland

4.1 Introduction

Despite the fact that the Arctic and Southern Oceans are the coldest on Earth with temperatures rarely rising above freezing, life in fact thrives in these waters. These are not oceanic deserts devoid of life, but are regions of high productivity as shown by the large number of mammal and bird species that hunt and feed in the regions, many of which migrate colossal distances to exploit the extensive food stocks: polar bears, walruses in the Arctic and penguins and albatross in the Southern Ocean, as well as numerous seal and whale species in both hemispheres. These charismatic species are only there in such large numbers because there is plenty to eat, and of course ultimately this is a result of abundant seasonal growth of phytoplankton (microalgae) that sustains the large zooplankton, fish and benthic organisms on which many of these iconic organisms feed. Therefore, cold temperatures *per se* clearly do not limit biological productivity and the marine food webs of polar regions are just as complex, diverse and ultimately just as productive as many oceans and seas in most other biogeographic regions (Fig. 4.1).

As in all other aquatic systems light is the key factor controlling photosynthesis and phytoplankton growth: thus, high rates of primary production, or phytoplankton growth, occur in periods of the year when sea ice (or pack ice) is absent (Clarke *et al.*, 2008). In the polar oceans and seas this coincides with seasons when day lengths are long (up to 24 h) and the angle of the sun is high ensuring that waters are well illuminated to depth. However, typically ice-free periods, with high amounts of light, only last for a few months at best in high latitude waters, giving rise to a very intense seasonal production that contrasts with the longer seasons in which phytoplankton growth is possible at lower latitudes. The key point is that the controls on primary production are no different in the polar oceans than any other aquatic system, i.e. enough light and an adequate supply of the numerous inorganic nutrients (e.g. nitrate, phosphate, silicate and trace elements such as iron, zinc and magnesium) that are necessary for growth.

However, it is the seasons when the surface waters freeze over that makes the polar regions and some sub-polar seas (Aral, Azov, Baltic, Bohai, Caspian and Okhotsk Seas) so unique. At maximum extents about 16 million km^2 of the Arctic Ocean are frozen, and in the southern hemisphere up to 20 million km^2 of the Southern Ocean are covered by a layer of frozen seawater (Thomas and Dieckmann, 2010). Anyone

Fig. 4.1. A leopard seal hauled up on the ice. There has to be considerable primary and secondary production in the cold Antarctic waters to sustain the millions of seals, birds and whales that feed in these waters. © David N. Thomas.

who has had the privilege of journeying in these frozen realms will tell of the unworldly experience as the ship grinds and rams its way through vast expanses of ice floes, on the whole less than 1 m – but in extremes over 10 m – thick. A common, rather surprising, observation is that although the ice floes are typically covered in pristine white snow, the underlying ice is coloured with a multitude of hues from delicate translucent greens through to a rich espresso coffee brown (Fig. 4.2). These colours were noticed by the earliest of polar explorers who melted ice and looked at the samples under the microscope: Ehrenberg (1841, 1853) was the first to describe organisms within Arctic sea ice. This was quickly followed by descriptions of sea ice diatoms from Antarctic sea ice by Hooker (1847). Around the same time Sutherland (1852) reported that 'greenish slimy-looking substances' were present on the bottom of ice floes, and that under the microscope these turned out to be 'minute vegetable forms of exquisite beauty.' Even the arguably greatest polar explorer, Nansen (1897), expressed a wonder when he recorded his observations of living protists within sea ice samples collected as his ship the *Fram* drifted through the Arctic Ocean: 'and these are unicellular pieces of slime that live by the million in pools on very nearly every ice-floe all over this endless sea of ice, which we like to call a place of death! Mother earth has a strange ability to produce life everywhere. Even this ice is fertile ground to her.'

That there is life within the ice still fascinates and motivates biologists and geochemists today. Since easier access to pack ice was made possible – with dedicated research vessels and stations – since the 1970s, there has been a large international effort to study the organisms found in the ice, and how they manage to survive (Fig. 4.3).

It is important to note that all of the organisms that are found in the ice are originally recruited from the plankton (viruses, bacteria, microalgae, protozoans, small

Fig. 4.2. Overturned ice floe, revealing a dense bottom layer of sea ice diatoms that stain the ice a rich coffee brown colour. © David N. Thomas.

crustaceans and larvae) of the open water. In open water plankton live in a rather well-buffered physical and chemical environment: temperature change is dependent on the volume of water concerned, but in the oceans is rather a gradual process. Chemical change is also generally rather gradual in any large body of water. As such, although life in the open water is obviously dominated by gradients of light, temperature and dissolved chemical constituents, when exposed to any change in these there is scope for physiological, biochemical or metabolic compensation to allow the organisms to acclimate. As will become clear in the following, when encapsulated into the ice matrix the organisms are exposed to a rather unbuffered physical and chemical environment, and therein lies the potential for extreme stress.

In autumn, freezing winds sweeping off cooled land-masses rapidly cool the mixed surface layers of the ocean. When the seawater temperature falls below −1.8°C ice crystals form; seawater freezes at a lower temperature than freshwater because of the dissolved salts it contains. The first visible stage of ice formation is the accumulation of frazil ice crystals on the water's surface. These 'grease ice' layers form slicks of ice crystals a few millimetres to centimetres thick that are often blown into huge wind-rows (Fig. 4.4). It is important to note that many hundreds of square kilometres will undergo this transition from open water to grease ice in just a few hours. The ice crystals coagulate, and eventually form a closed ice cover. It is at this early stage that the ice is 'inoculated' with organisms: as they rise through the water (ice is less dense than water) the crystals effectively 'harvest' particles, including viruses, Archaea, bacteria, algae, protists, flatworms and small crustaceans; some stick to the crystals, whereas others are simply trapped in the viscous slushy ice layer. Hence a diverse group of organisms is almost instantaneously confined to a new habitat that is quite different to the one from which they were recruited. Many of the larger animals including crustaceans and fish avoid entrapment by actively swimming away.

(a)

(b)

Fig. 4.3. Only by the allocation of significant funding to support international teams of scientists with resources such as (a) ice-breaking research vessels and (b) logistics for remote ice camps, has progress in our understanding of the microbial world of sea ice been possible. Images © David N. Thomas.

Under turbulent conditions the ice crystals consolidate into ice pancakes a few centimetres in diameter (Lange *et al.*, 1989). These grow larger, becoming 20–50 cm thick super pancakes, several metres across. Wind and wave action raft the pancakes together, and often several end up lying on top of one another. They freeze together and after 1 or 2 days a closed ice cover has formed. Subsequent thickening of the ice takes place by congelation ice growth where ice crystals elongate from a skeletal ice layer on the bottom of the ice sheet. Most ice grows to, on average, 80 cm thick, although rafting and deformation processes can result in significantly thicker floes

Fig. 4.4. Windswept slicks of 'grease ice' on the surface of the Weddell Sea in autumn. This is the first stage of sea ice formation, when frazil ice crystals formed when seawater freezes at –1.8°C accumulate on the surface of the ocean. The prominent slick in the foreground is approximately 50 m wide. © David N. Thomas.

being formed. When sea ice forms under calm conditions the ice crystals, rather than forming pancakes, form uniform sheets of nilas ice. These also thicken by congelation ice growth, but also break apart to raft on top of each other (Fig. 4.5; Thomas and Dieckmann, 2010).

Ice fields, whether formed from pancake ice or nilas ice, are rarely tranquil regions and both wave and wind action result in ice floes rafting on top of each other, colliding and/or deforming to produce pressure ridges that can be tens of metres thick. Patches of open water are continually forming and closing again as large ice floes move around. These often rapidly freeze again, and so it can be seen that in any ice field there will be ice of very variable thicknesses and age (Fig. 4.6; Thomas and Dieckmann, 2010).

4.2 The Physics of Sea Ice

When ice forms from freshwater, the result is a hard brittle solid with the only inclusions being gas bubbles. In contrast, when seawater freezes the resultant ice is a semi-solid matrix, permeated by a labyrinth of brine-filled channels and pores. Dissolved substances, including salts and gases, do not enter the ice crystal structure and so as the ice forms it and other dissolved constituents of seawater are expelled and collect as a highly concentrated brine solution within the labyrinth of brine channels and pores in the ice matrix (Eicken, 1992; Weissenberger *et al.*, 1992; Golden *et al.*, 1998). The morphology of these channels and pores, the total volume of the ice occupied by them and the salinity of the brines contained within them is governed by temperature and the age of the ice. The volume of ice occupied by the brine channels is directly proportional to the temperature of the ice, as is the brine concentrations within the channels: at –6°C the brine salinity is 100, at –10°C it is 145 and at –21°C it is 216 (Fig. 4.7). The brines are not static and gravity drainage results in a continuous loss of brine due to brine expulsion and gravity drainage, resulting in a gradual reduction of bulk ice salinity. This loss of brine can be enhanced in summer sea ice when melting surface snow and ice on floe surfaces percolate down, flushing brine out.

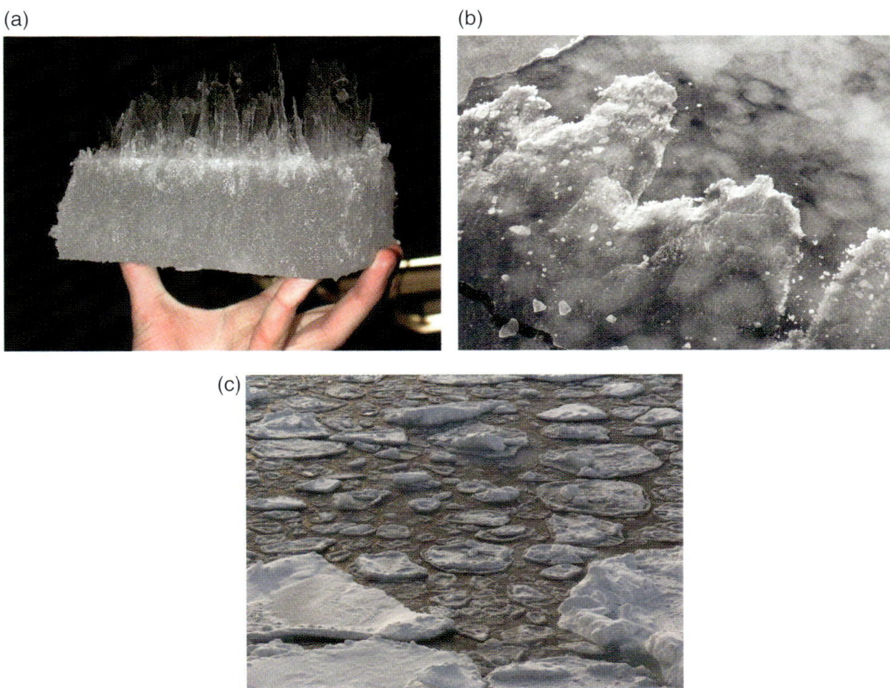

Fig. 4.5. (a) Skeletal layer on artificially produced sea ice, © Jean-Louis Tison; (b) thin sheets of nilas ice finger rafting, © David N. Thomas; and (c) pancake ice beginning to consolidate, © David N. Thomas.

The temperature at the upper surface of an ice floe is determined by the air temperature (down to −40°C) and the extent of insulating snow cover. In contrast, the temperature at the underside of an ice floe will be at or close to the freezing point of the underlying seawater. This results in gradients of temperature, brine salinity (and therefore dissolved constituents and gases) and volume of brine channels and pores throughout an ice floe (Fig. 4.8). During autumn and winter the ice is generally colder, brine salinities higher and brine volume lower in surface ice, compared with underlying ice. Naturally, as ice warms in spring and early summer these gradients break down.

4.3 Space in Ice

Clearly the key parameter as to whether an organism can live within the ice matrix is whether or not there is enough space to accommodate it, allow it to move, and whether or not the ice is porous enough to allow nutrient and gas exchange. The concept of brine channels as a habitat for organisms was greatly aided by the work of Weissenberger *et al.* (1992) who devised a resin casting technique that enabled hardened casts of internal sea ice structure to be produced which were subsequently analysed using scanning electron microscopy. Subsequently, Eicken *et al.* (2000) developed non-destructive tomographic magnetic resonance imaging (MRI) techniques to study directly the microstructure and thermal evolution of brine inclusions within sea ice. Their studies looked at ice warming from −21°C to −6°C and directly showed how pore size increased with increasing temperature, while pore number density decreased as pores merged and coalesced. Golden *et al.* (1998) and Pringle *et al.* (2009) have taken

(a)

(b)

Fig. 4.6. (a) Leads, gaps between ice floes, can be large albeit ephemeral places of rapid heat exchange between the water and air giving rise to sea smoke. (b) Huge slabs of ice can be pushed up to form pressure ridges that form effective sails to catch prevailing winds. Images © David. N. Thomas.

these studies forward using X-ray tomography of ice to investigate the thermal evolution of pore space. In an over-simplification it is worth noting the so-called 'rule of fives' proposed by Golden *et al.* (1998): bulk sea ice is essentially impermeable for brine volume fractions below 5%, above which permeability increases rapidly. Such brine volumes are typically found in ice at a critical temperature of −5°C where the bulk salinity (the resulting salinity when ice and its brine inclusions

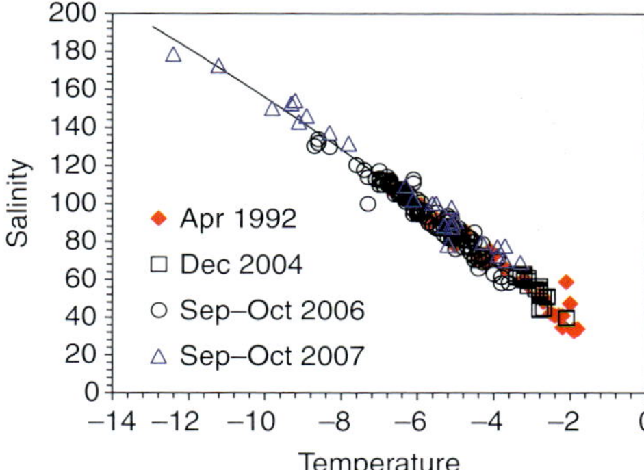

Fig. 4.7. Measured temperature and salinity of sea ice brines in first- and multi-year ice floes in the Weddell Sea, in austral autumn (April 1992; Gleitz *et al*., 1995), spring (September–October 2006, unpublished data) and summer (December 2004, Papadimitriou *et al*., 2007), and in the Indian Ocean sector of the Southern Ocean in austral spring (September–October 2007, Norman *et al*., 2011). The trend line illustrates the functional relationship of salinity with temperature in sea ice. From Thomas and Dieckmann, 2010.

melt) is 5. The practical working of this rule is well illustrated by the field observations of Tison *et al.* (2008) who show how such thresholds are critical for the seasonal development of biology in Antarctic sea ice.

Very little space is needed to accommodate most protists (Fig. 4.9), and this has been highlighted by Junge *et al.* (2001) who convincingly show viable bacteria living in sea ice at −21°C, where there is a minimum of space between ice crystals (Junge *et al.*, 2006). As important as space is for living, the porosity of the ice also limits the possibility for motile organisms to move around the ice. Bacteria, many of the pennate diatom species, flagellates, turbellarians, protozoans and copepods are motile, and have been shown to migrate through ice horizons. Naturally several of these organism classes feed on bacteria, algae and protozoans in the ice, but can only do so as long as there is space for them to move. Krembs *et al.* (2000) used glass capillaries ranging in size from 12 to 1420 µm diameter to mimic the brine channel habitat, and monitored the movement and colonization of brine channel proxies by common sea ice organisms including turbellarians, rotifers, nematodes, harpacticoid copepods, flagellates, amoebae, diatoms and bacteria. Only rotifers and turbellaria were able to traverse 'channels' significantly smaller than their body diameter. Most of the organisms simply congregated in the narrowest of tubes that they could fit into as determined by their body size. This led the researchers to conclude that pore spaces ≤200 µm in diameter are actually refugia in which bacteria, pennate diatoms, flagellates and small protozoans benefit from very much reduced grazing pressure.

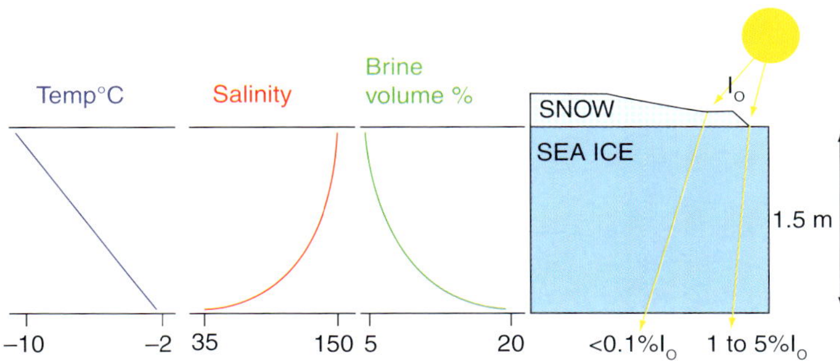

Fig. 4.8. Gradients of temperature, salinity and brine volume are established across an ice floe. The underside is always at the freezing point of seawater −1.8°C and the top of the ice close to air temperature, although this is largely dependent on snow cover. The illustration shows how snow cover can significantly reduce the amount of incident irradiance (I_o). From Thomas and Dieckmann, 2002.

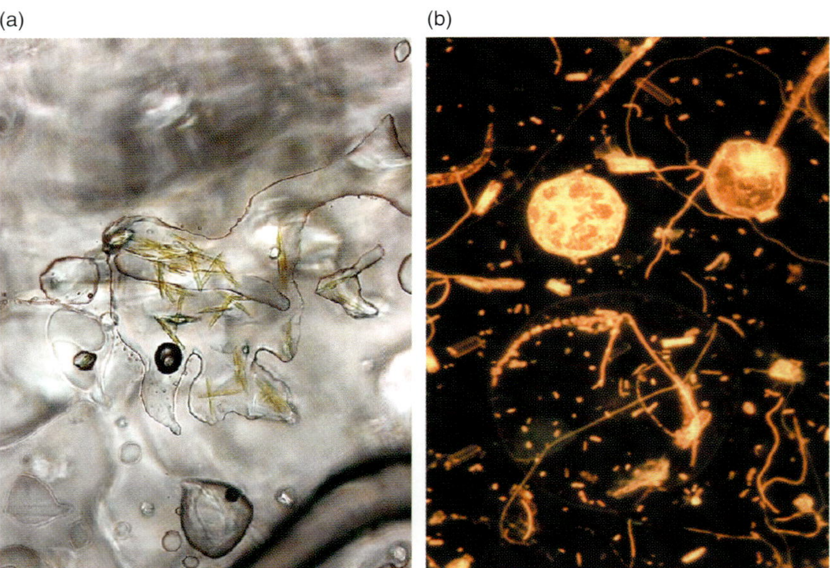

Fig. 4.9. Sea ice diatoms and bacteria. (a) Diatoms within brine inclusions of Arctic sea ice, © Christopher Krembs and Jody Deming, University of Washington. (b) Epifluorescence of labelled bacteria and algae removed from Antarctic sea ice. © Sönnke Grossmann.

Naturally it is temperature stress that most people would consider to be the major extreme stress factor when thinking about sea ice. As discussed above, across any ice floe there is a gradient of temperatures, ultimately governed by the air temperature and amount of snow insulating the ice surface from the air. With little snow cover the ice temperatures will rapidly respond to changing air temperature, and the resulting gradient through the ice floe will also change rapidly. Therefore, organisms living in the ice will need to be able to respond to temperature shifts quickly. However, it is important to note when considering temperatures that the peripheries of an ice floe in contact with seawater will always be at temperatures close to the freezing point of seawater (−1.8°C), and so under relative minor temperature stress (and little salinity and space stress either). This is one of the reasons why the most prolific biological standing stocks in sea ice systems are found on the peripheries of ice floes (Fig. 4.10). However, here they are also most vulnerable to grazing organisms.

4.4 Organisms Living in Sea Ice

As ice consolidates and gets colder, psychrophilic (cold-loving) bacteria species (minimum growth temperature ≤0°C, optimal growth temperature <15°C, and maximum growth temperature <20°C) have been shown to predominate and mesophilic species decline (Collins *et al.*, 2008). Heterotrophic bacteria are the main group of prokaryotes in sea ice, and most sea ice bacterial strains have been found to be cold-adapted and halotolerant, with both free-living and surface-associated species (Helmke and Weyland, 1995; Bowman *et al.*, 1997; Deming, 2002; Brinkmeyer *et al.*, 2003, 2004; Junge *et al.*, 2003; Abell and Bowman, 2005; Bowman, 2008). Archaea have also been found in both Arctic and Antarctic sea ice (DeLong *et al.*, 1994; Junge *et al.*, 2004; Cavicchioli, 2006; Collins *et al.*, 2010), although still in rather few studies. Cyanobacteria have been recorded in Arctic and Baltic Sea ice, but not Antarctic sea ice (Laurion *et al.*, 1995; Gradinger and Ikävalko, 1998). However, they are mostly found in

Fig. 4.10. Many bottom ice assemblages are concentrated into the bottom 5–10 cm of ice cores. © David N. Thomas.

surface melt features and melt ponds, and are most likely to be found in coastal regions influenced by freshwater runoff.

Diatoms are the most studied group of eukaryotes in sea ice. However, despite what it may seem from reading the sea ice literature, diatoms are not the only eukaryotic protists to be found in sea ice: others include prymnesiophytes, dinoflagellates, chrysophytes, cilliates, formanifera and chlorophytes amongst others (Thomas and Dieckmann, 2010). Heterotrophic nanoflagellates (HNFs) are probably the least well studied of the eukaryotes in sea ice, although potentially they are some of the most numerous (Ikävalko and Gradinger, 1997).

Bacteria from Arctic sea ice have been shown to be active at temperatures down to −20°C, and motile down to −10°C (Junge *et al.*, 2004, 2006). Sea ice diatoms have been shown to grow at low temperatures, but not down to the temperatures recorded for bacteria: Aletsee and Jahnke (1992) measured growth of the Arctic diatom *Nitzschia frigida* at −8°C, although the doubling time of 60 days was exceptionally long. In contrast, the generation times were 50–70 h at −4°C. Thomas *et al.* (1995) suggested that temperatures below −4°C may induce some sea ice diatom species to enter into a physiological resting stage (morphologically similar to the vegetative cells, but are physiologically dormant), similar to those produced by diatoms starved of nutrients or kept in the dark for long periods of time (Peters and Thomas, 1996a, b). Diatoms from freshwater and marine systems are known to produce resistant cysts, but although cyst and spore formation is well documented for sea ice dinoflagellates and chrysophytes (Garrison and Close, 1993; Stoecker *et al.*, 1997), it is not commonly reported for diatoms in sea ice (Garrison and Mathot, 1996).

Many researchers have postulated the hypothesis that sea ice algae switch from photoautotrophy to heterotrophy (based on the high levels of organic solutes in the DOM in the ice, see below) as a means for surviving long periods of winter darkness (Palmisano and Sullivan, 1985; Rivkin and Putt, 1987), although the results of experimental studies are less than convincing

(Horner and Alexander, 1972). Zaslavskaia *et al.* (2001) demonstrated the potential to transform from obligate photoautotrophy to full heterotrophy in some non-polar laboratory diatom strains. However, the fact that flagellates dominate late winter sea ice assemblages, many of which are not obligate autotrophs, and the fact that diatoms dominate the sympagic assemblages only at times, or sites, where there is light (Riedel *et al.*, 2008; Różańska *et al.*, 2009) is further evidence that switches from autotrophy to heterotrophy in diatoms may not in fact occur.

4.5 Mechanisms to Avoid Freezing

Besides avoiding freezing, the transport of essential nutrients and gases through cell membranes has to be maintained by cells undergoing a lowering of external temperature. Increase in membrane fluidity is one of the most important acclimations during freezing and is well documented, particularly for bacteria (Russell, 1997; Nichols *et al.*, 1999; Cavicchioli *et al.*, 2002). Any decrease in temperature leads to one or some combination of the following changes: an increase in the contribution of polyunsaturated fatty acids (PUFAs), a decrease in average chain length, an increase in methyl branching, and an increase in the ratio of anteiso-branching relative to iso-branching. Furthermore, PUFAs are not normally detected in temperate bacteria but are often found in significant amounts in sea ice bacteria (Russell, 1997; Nichols *et al.*, 1999). The regulation of membrane lipid composition in sea ice diatoms at low temperatures (and low irradiance and low inorganic nitrogen concentrations) has been shown to be essential for efficient electron transport in chloroplasts (Mock and Kroon, 2002a, b).

It is thought that cryoprotectants with antifreeze properties are produced by sea ice diatoms and bacteria to inhibit ice crystal formation. Moreover, there is increasing evidence that some sea ice organisms are able to reduce the formation of ice crystals (Raymond and Knight, 2003; Janech *et al.*, 2006; Raymond *et al.*, 2009; Christner, 2010). Studies have shown that some species of sea ice diatoms exude ice-binding proteins (IBPs), which cause so called 'ice-pitting', where the growth of ice crystals is reduced and the ice crystal shape is altered (Raymond *et al.*, 2009). Formerly IBPs were referred to as ice-active substances (IASs). It is postulated that the IBPs may prevent freezing injury to membranes by inhibiting the re-crystallization of ice, the process where large grains of ice grow preferentially over small grains. By having protein re-crystallization inhibitors, such as these IBPs, physical disruption of cell membranes by large crystals may be reduced. It is also speculated that when released extracellularly they will prevent the re-crystallization of the surrounding ice. Janech *et al.* (2006) described a protein sequence from a sea ice diatom with a high similarity to a protein from a snow mould that is known to reduce the freezing point of seawater. They concluded that this might represent a new class of IBP. Krell *et al.* (2008) also reported that the Antarctic ice diatom *Fragilariopsis cylindrus* contains a completely new class of IBP not known from any other algal, plant or animal species studied to date. Bayer-Giraldi *et al.* (2010) showed that there is a broad distribution of antifreeze protein (AFP) genes in a wide range of organisms (bacteria, diatoms, fungi and crustaceans), including taxa not found in cold environments. They suggest that the proteins may be multifunctional. However, in two species of ice diatom under simulated sea ice conditions, both *Fragilariopsis cylindrus* and *Fragilariopsis curta* strongly expressed selected AFP genes.

4.6 Extracellular Polymeric Substances in Sea Ice

AFPs produced by psychrophilic bacteria are possibly part of the large pool of EPS (extracellular polymeric substances) produced by many bacteria species. The Cytophaga-Flavibacterium-bacterioides (CFB) group of bacteria, often reported in winter sea ice biological assemblages (Bowman *et al.*, 1997; Brinkmeyer *et al.*, 2003, 2004; Junge *et al.*, 2004), are well known for their abundant EPS production when temperatures are lowered.

Low temperatures have been found to induce EPS production and laboratory cultures of bacteria strains isolated from Antarctic sea ice showed a 30-fold increase in EPS production at −2°C as compared to their optimal growth temperature of 20°C (Mancuso Nichols *et al.*, 2005a). The chemical structures of EPS produced by sea ice-derived bacterial isolates are also highly diverse (Mancuso Nichols *et al.*, 2005b).

EPS are also produced by diatoms and the importance of both bacterial- and algal-derived EPS for sea ice assemblages has received increasing attention. It has been well known for some time that sea ice in the Arctic, Antarctic and Baltic Sea contains high concentrations of dissolved organic matter (DOM, see citations in Norman *et al.*, 2011), exceeding surface water concentration by factors of up to 500. The nature of this organic matter has remained largely uncharacterized. Naturally, some of the DOM in the ice will be incorporated from the water from which the ice is formed (Stedmon *et al.*, 2007), although much of it is probably produced in the form of EPS produced *in situ* by algae and/or bacteria (Riedel *et al.*, 2007; Collins *et al.*, 2008; Meiners *et al.*, 2008; Underwood *et al.*, 2010). It is well known that EPS provide a protective buffer zone for cells against unfavourable shifts in aquatic habitats (Decho and Lopez, 1993), as well as aiding locomotion (Higgins *et al.*, 2003). It is thought that algae and bacteria use EPS for the same reasons in response to temperature and salinity stress in sea ice and it has been proposed that secreted EPS may even affect the physical properties and salt retention of sea ice (Krembs and Deming, 2008; Krembs *et al.*, 2011). EPS production may have a profound influence on the permeability of the ice, with implications for fluxes of brine, gases, nutrients and the accessibility of space to organisms (Wettlaufer, 1999; Krembs *et al.*, 2011). Microscopic observations have even shown that brine pockets at very low temperatures remain interconnected by fine mucus bridges, which will significantly influence the hysteresis of the pore space during temperature fluctuations (Krembs *et al.*, 2011).

Junge *et al.* (2004) demonstrated that in Arctic winter sea ice virtually all active bacteria were attached to particles, and it is now thought that many of the bacteria in sea ice may be attached to EPS particles (Meiners *et al.*, 2008). As such, different types of EPS may provide a range of ecological niches for different groups of bacteria and thereby be a vector to increase the bacterial diversity in the sea ice. It can be assumed that the larger sea ice EPS particles will contain oxygen deficient, or even anoxic, zones and so provide a habitat for anaerobic bacterial groups. This would be an explanation for the unusual anaerobic bacterial processes measured in some sea ice studies (Kaartokallio, 2001; Petri and Imhoff, 2001; Rysgaard and Glud, 2004).

4.7 Oxygen in Sea Ice

Glud *et al.* (2002), Mock *et al.* (2002) and Rysgaard *et al.* (2008) showed that oxygen is expelled along with brine during ice growth and that oxygen depletion can occur in the ice. In contrast there are many other studies that report sea ice brines to be supersaturated in oxygen. This is especially the case where there are high algal standing stocks and the oxygen produced through photosynthesis is not able to diffuse from the ice (Gleitz *et al.*, 1995; Papadimitriou *et al.*, 2007). The presence of highly localized sites supersaturated or undersaturated with oxygen creates a 'mozaic-like O_2 distribution pattern' in the sea ice (Rysgaard *et al.*, 2008). Likewise, the occurrence of microenvironments, where denitrification and its reverse microbial process, nitrification, can take place, may explain the high concentrations of nitrate and ammonium frequently measured at the same time in sea ice samples taken from close proximity (Thomas *et al.*, 1998).

4.8 Dimethylsulphonioproprionate in Sea Ice

Besides space and temperature, salinity is probably the factor that has the greatest influence on the organisms living within sea ice, especially in cold ice where the brine salinities can be more than three times that of seawater concentrations. Diatoms respond to the osmotic stress induced by increasing external salinity by increasing intracellular concentrations of ions, simple compounds such as free amino acids and sugars (Kirst and Wiencke, 1995). When external salinity is reduced, the organisms regain osmotic balance by breaking down these solutes and/or releasing ions to the external medium. High concentrations of ions such as potassium, sodium and chloride tend to be toxic to cells, and so much of the necessary osmotic adjustment is done by regulating concentrations of proline and an interesting compound, dimethylsulphoniopropionate (DMSP). Proline and DMSP are also thought to have a role as an antifreeze within some species. In addition, light and nutrient supply influence the production of DMSP, although in sea ice, salinity is the dominant factor influencing DMSP production by ice algae (Stefels *et al.*, 2007).

Sea ice diatoms, and *Phaeocystis* spp., which are common in surface ice in spring and summer, appear to be the largest producers of DMSP (Stefels *et al.*, 2007). Seawater concentrations are typically low, in the range of 0 to 50 nmol l^{-1}. Orders of magnitude higher concentrations (>2500 nmol l^{-1}) have been measured in Antarctic sea ice (Trevena and Jones, 2006; Tison *et al.*, 2010). DMSP degrades, or is broken down by various processes, to dimethylsulfide (DMS) and acrylic acid (acrylate). In remote ocean regions, DMS accounts for most of the non-sea salt sulfate in the atmosphere, and the oxidation of DMS in the atmosphere to aerosol particles and cloud condensation nuclei is part of a complex system of localized and global climate control (reviewed by Stefels *et al.*, 2007). The greatest release of DMS from sea ice regions is associated with melting ice and the corresponding reduction in ambient salinity, when cells containing high concentrations of DMSP are released into the seawater. Periods of ice melt are also times when elevated grazing activity in ice edge waters will increase the release of DMS into surface waters and therefore into the atmosphere. It has been suggested that sea ice is a major source of DMSP and DMS to polar oceans and atmosphere (Trevena and Jones, 2006; Tison *et al.*, 2010). Thus, changes in sea ice distributions, and subsequent change to populations of ice algae acclimating to sub-optimal salinity via production of DMSP, may exert important feedbacks on the climate system.

4.9 Biomarkers for Sea Ice Extent in Past Climates

Despite these claims, less than 5% of the DMS released into the water is actually released to the air (Stefels *et al.*, 2007). A proportion of MSA is oxidized to methansulphonic acid (MSA), and researchers are currently trying to use MSA in glacial ice cores as a proxy for past sea ice concentrations. The theory is a simple one: if there is lots of sea ice, there will be a lot of organisms producing DMSP. This will be broken down to DMS and subsequently MSA, which will be precipitated on the ice sheets with snow fall. This will then become incorporated into the geological record within the ice sheet as the ice thickens. Curran *et al.* (2003) were the first to measure concentrations of MSA in glacial ice cores and estimated that the previous sea ice extent in a particular sector (80°E to 140°E) off the east Antarctic coast was rather constant between 1841 and 1950 and thereafter there was a sharp decrease of about 20%. Dixon *et al.* (2005) also measured MSA concentrations in several ice sheet ice cores and estimated that the sea ice extent in the Amundson and Ross Seas were generally higher from 1800 to 1992 compared to the period 1487 to 1800. Most recently, Becagli *et al.* (2009) have shown MSA concentrations in the glacial ice of Talos Dome to be a good proxy for sea ice extent in the Ross Sea over the past 140 years.

Another example of sea ice algae producing biomarkers for elucidating past sea ice extent is the recently described C_{25} monounsaturated hydrocarbon (IP_{25}), which has been found to be produced by sea ice diatoms and to be very stable in sediments over geological time (Belt *et al.*, 2008). The marker has proven to be a robust proxy for sea ice, and the position of ice edges as far back as the past 300,000 years have been estimated from its presence in sediment cores taken from different regions of the Arctic (Massé *et al.*, 2008; Müller *et al.*, 2009; Vare *et al.*, 2010). It is a curious and captivating study to investigate the environmental cues in sea ice that induce such biomarkers to be produced, that in turn are able to elucidate so much about the paleaoclimate of the polar oceans and seas: the physiological responses of bacteria, Archaea, algae and other sea ice protists to salinity and temperature stress in the ice matrix ultimately give us the record of environmental change on colossal scales of many millions of square kilometres (Thomas and Dieckmann, 2010).

But the interest does not end with the Earth system. Some of the keenest interest in sea-ice organism biology comes from astrobiologists who are enthusiastically scrutinizing the ice-covered seas of Jupiter's moons Europa and Ganymede, and the surface of Mars for signs of life, or past life. It was not too difficult a leap of creative thought to look to sea ice organisms as potential proxies for extraterrestrial life (Marion *et al.*, 2003). Despite the tantalizing lure to compare the brown coloration observed on Europa's surface to diatom-coloured sea ice (Hoyle *et al.*, 1985; Hoover *et al.*, 1986), it has to be realized that the extraterrestrial ice systems, tens to hundreds of kilometres thick, are substantially different to the 1–10 m-thick ice we know from Earth's polar oceans. If life forms do exist, or have existed on these bodies, it seems likely that they will be quite unlike those that dominate the sea ice found on Earth today (Chyba and Phillips, 2002; Cavicchioli, 2003; Marion *et al.*, 2003).

4.10 Concluding Remarks

The amount of interest that there is in sea ice biology is highlighted in the number of comprehensive reviews that have been written about the subject over the past 20 years. Naturally these have been a source of much of this brief summary. Interested readers are encouraged to investigate the following, which are just a few of those available: Horner (1985); Garrison (1991); Eicken (1992); Palmisano and Garrison (1993); Ackley and Sullivan (1994), Garrison and Mathot (1996); Staley and Gosink (1999); Lizotte (2001); Brierley and Thomas (2002); Deming (2002); Thomas and Dieckmann (2002); Arrigo and Thomas (2004); Thomas (2004); Mock and Thomas (2005); and Deming and Eicken (2007). There are two comprehensive edited books, Thomas and Dieckmann (2010) and Eicken *et al.* (2010), which summarize state-of-the-art techniques and up-to-date understanding of sea ice research.

What is clear in reading all of this material, is that biologists have to work closely with physicists and geophysicists if they are to understand why the organisms behave and function in the way they do, and this is an exciting aspect of such research. However, just as the ice in some regions is currently undergoing unprecedented rates of change, so is the study of such systems and as a result as soon as a review is published there is a whole host of additional information in the primary literature. This is currently enhanced by the harnessing of modern molecular techniques to investigate the diversity of life in ice, and in part by investigations into how the adaptations to life in the ice may have potential for applications in biotechnological research and/or industries. Bioprospecting in polar sea ice is clearly a new force for exploring sea ice resources (Lohan and Johnston, 2005; Christner, 2010).

We are certainly part of an exciting epoch in polar research, which is just as thrilling as that enjoyed 100 years ago by the researchers mentioned at the beginning of this chapter. As then, the study of sea ice organisms has a fundamental place in the exploration of polar oceans.

References

Abell, G.C.J. and Bowman, J.P. (2005) Ecological and biogeographic relationships of class *Flavobacteria* in the Southern Ocean. *FEMS Microbiology Ecology* 51, 265–277.

Ackley, S.F. and Sullivan, C.W. (1994) Physical controls on the development and characteristics of Antarctic sea ice biological communities – a review and synthesis. *Deep-Sea Research Part I* 41, 1583–1604.

Aletsee, L. and Jahnke, J. (1992) Growth and productivity of the psychrophillic marine diatoms *Thalassiosira anatarctica* Comber and *Nitzschia frigida* Grunow in batch cultures at temperatures below the freezing point of sea water. *Polar Biology* 11, 643–647.

Arrigo, K.R. and Thomas, D.N. (2004) Large scale importance of sea ice biology in the Southern Ocean. *Antarctic Science* 16, 471–486.

Bayer-Giraldi, M., Uhlig, C., John, U., Mock, T. and Valentin, K. (2010) Antifreeze proteins in polar sea ice diatoms: diversity and gene expression in the genus *Fragilariopsis*. *Environmental Microbiology* 12, 1041–1052.

Becagli, S., Castellano, E., Cerri, O., Curran, M., Frezzotti, M., Marino, F., Morganti, A., Proposito, M., Severi, M., Traversi, R. and Udisti, R. (2009) Methanesulphonic acid (MSA) stratigraphy from a Talos Dome ice core as a tool in depicting sea ice changes and southern atmospheric circulation over the previous 140 years. *Atmospheric Environment* 43, 1051–1058.

Belt, S.T., Massé, G., Vare, L.L., Rowland, S.J., Poulin, M., Sicre, M.-A., Sampei, M. and Fortier, L. (2008) Distinctive ^{13}C isotopic signature distinguishes a novel sea ice biomarker in Arctic sediments and sediment traps. *Marine Chemistry* 112, 158–167.

Bowman, J.P. (2008) Genomic analysis of psychrophilic prokaryotes. In: Margesin, R., Schinner, F., Marx, J.-C. and Gerday, C. (eds) *Psychrophiles: from Biodiversity to Biotechnology*. Springer-Verlag, Berlin, Germany, pp. 265–284.

Bowman, J.P., McCammon, S.A., Brown, M.V., Nichols, D.S. and McMeekin, T.A. (1997) Diversity and association of psychrophilic bacteria in Antarctic sea ice. *Applied and Environmental Microbiology* 63, 3068–3078.

Brierley, A.S. and Thomas, D.N. (2002) Ecology of Southern Ocean pack ice. *Advances in Marine Biology* 43, 171–276.

Brinkmeyer, R., Knittel, K., Jugens, J., Weyland, H., Amann, R. and Helmke, E. (2003) Diversity and structure of bacterial communities in Arctic versus Antarctic pack ice. *Applied and Environmental Microbiology* 69, 6610–6619.

Brinkmeyer, R., Glockner, F.-O., Helmke, E. and Amann, R. (2004) Predominance of β-proteobacteria in summer melt pools on Arctic pack ice. *Limnology and Oceanography* 49, 1013–1021.

Cavicchioli, R. (2003) Extremophiles and the search for extraterrestrial life. *Astrobiology* 2, 281–292.

Cavicchioli, R. (2006) Cold adapted Archaea. *Nature Reviews Microbiology* 4, 331–343.

Cavicchioli, R., Siddiqui, K.S., Andrews, D. and Sowers, K.R. (2002) Low-temperature extremophiles and their applications. *Current Opinion in Biotechnology* 13, 253–261.

Christner, B.C. (2010) Bioprospecting for microbial products that affect ice crystal formation and growth. *Applied Microbiology and Biotechnology* 85, 481–489.

Chyba, F.F. and Phillips, C.B. (2002) Possible ecosystems and the search for life on Europa. *Proceedings of the National Academy of Sciences of the United States of America* 98, 801–804.

Clarke, A.C., Meredith, M.P., Wallace, M.I., Brandon, M. and Thomas, D.N. (2008) Seasonal and interannual variability in temperature, chlorophyll and macronutrients in Ryder Bay, northern Marguerite Bay, Antarctica. *Deep Sea Research Part II* 55, 1988–2006.

Collins, R.E., Carpenter, S.D. and Deming, J.W. (2008) Spatial heterogeneity and temporal dynamics of particles, bacteria, and pEPS in Arctic winter sea ice. *Journal of Marine Systems* 74, 902–917.

Collins, R.E., Rocap, G. and Deming, J.W. (2010) Persistence of bacterial and Archaeal communities in sea ice through an Arctic winter. *Environmental Microbiology* 12, 1828–1841.

Curran, M.A.J., van Ommen, T.D., Morgan, V.I., Phillips, K.L. and Palmer, A.S. (2003) Ice core evidence for Antarctic sea ice decline since the 1950s. *Science* 302, 1203–1206.

Decho, A.W. and Lopez, G.R. (1993) Exopolymer microenvironments of microbial flora – multiple and interactive effects on trophic relationships. *Limnology and Oceanography* 38, 1633–1645.

DeLong, E.F., Wu, K.Y., Prézelin, B.B. and Jovine, R.V. (1994) High abundance of Archaea in Antarctic marine picoplankton. *Nature* 371, 695–697.

Deming, J.W. (2002) Psychrophiles and polar regions. *Current Opinion in Microbiology* 3, 301–309.

Deming, J.W. and Eicken, H. (2007) Life in ice. In: Sullivan, W.T. and Baross, J.A. (eds) *Planets and Life: The Emerging Science of Astrobiology*. Cambridge University Press, Cambridge, UK, pp. 292–312.

Dixon, D., Mayewski, P.A., Kaspari, S., Kreutz, K., Hamilton, G., Maasch, K., Sneed, S.B. and Handley, M.J. (2005) A 200 year sulfate record from 16 Antarctic ice cores and associations with Southern Ocean sea-ice extent. *Annals of Glaciology* 41, 155–166.

Ehrenberg, C.G. (1841) *Nachtrag zu dem Vortrage über Verbreitung und Einfluss des mikroskipischen Lebens in Süd- und Nordamerika*. Berichte über die zur Bekanntmachung geeigneten Verhandlung der Königlich-Preussischen Akademie der Wissenschaften zu. Berlin 1841, 201–209.

Ehrenberg, C.G. (1853) *Über neue Anschauungen des kleinstein nördlichen Polarlebens*. Berichte über die zur Bekanntmachung geeigneten, Verhandlung der Königlich-Preussischen Akademie der Wissenschaften, Berlin, 1853, 522–529.

Eicken, H. (1992) The role of sea ice in structuring Antarctic ecosystems. *Polar Biology* 12, 3–13.

Eicken, H., Bock, C., Wittig, R., Miller, H. and Poertner, H.-O. (2000) Nuclear magnetic resonance imaging of sea ice pore fluids: Methods and thermal evolution of pore microstructure. *Cold Region Science and Technology* 31, 207–225.

Eicken, H., Gradinger, R., Salganek, M., Shirasawa, K., Perovich, D.K. and Leppäranta, M. (eds) (2010) *Field Techniques for Sea Ice Research*. University of Alaska Press, Fairbanks, Alaska, 588 pp.

Garrison, D.L. (1991) Antarctic sea ice biota. *American Zoologist* 31, 17–33.

Garrison, D.L. and Close, A.R. (1993) Winter ecology of the sea ice biota in Weddell Sea pack ice. *Marine Ecology Progress Series* 96, 17–31.

Garrison, D.L. and Mathot, S. (1996) Pelagic and sea ice microbial communities. In: Ross, R.M., Hofmann, E.E. and Quetin, L.B. (eds) *Foundations for Ecological Research West of the Antarctic Peninsula. American Geophysical Union, Antarctic Research Series* 70, 155–172.

Gleitz, M., Rutgers van der Loeff, M., Thomas, D.N., Dieckmann, G.S. and Millero, F.J. (1995) Comparison of summer and winter inorganic carbon, oxygen and nutrient concentrations in Antarctic sea ice brine. *Marine Chemistry* 51, 81–91.

Glud, R.N., Rysgaard, S. and Kuhl, M. (2002) A laboratory study on O_2 dynamics and photosynthesis in ice algal communities: quantification by microsensors, O_2 exchange rates, ^{14}C incubations and a PAM fluorometer. *Aquatic Microbial Ecology* 27, 301–311.

Golden, K.M., Ackley, S.F. and Lytle, V.I. (1998) The percolation phase transition in sea ice. *Science* 282, 2238–2241.

Gradinger, R. and Ikävalko, J. (1998) Organism incorporation into newly forming Arctic sea ice in the Greenland Sea. *Journal of Plankton Research* 20, 871–886.

Helmke, E. and Weyland, H. (1995) Bacteria in sea ice and underlying water of the Eastern Weddell Sea in midwinter. *Marine Ecology Progress Series* 117, 269–287.

Higgins, M.J., Sader, J.E., Mulvaney, P. and Wetherbee, R. (2003) Probing the surface of living diatoms with atomic force microscopy: the nanostructure and nanomechanical properties of the mucilage layer. *Journal of Phycology* 39, 722–734.

Hooker, J.D. (1847) *The Botany of the Antarctic voyage of H.M. Discovery ships Erebus and Terror in the years 1838-1843. Part 1. Flora Antarctica*. Reeve Brothers, London, UK.

Hoover, R.B., Hoyle, F., Wickramasinghe, N.C., Hoover, M.J. and Al-Mufti, S. (1986) Diatoms on Earth, Comets, Europa and in Interstellar Space. *Earth Moon and Planets* 35, 19–45.

Horner, R. (1985) *Sea Ice Biota*. CRC Press, Boca Raton, Florida, 224 pp.

Horner, R. and Alexander, V. (1972) Algal populations in Arctic sea ice: an investigation of heterotrophy. *Limnology and Oceanography* 17, 454–458.

Hoyle, F., Wickramasinghe, N.C. and Al-Mufti, S. (1985) The case for interstellar micro-organisms. *Astrophysics and Space Science* 110, 401–404.

Ikävalko, J. and Gradinger, R. (1997) Flagellates and heliozoans in the Greenland Sea ice studied alive using light microscopy. *Polar Biology* 17, 473–481.

Janech, M.G., Krell, A., Mock, T., Kang, J.S. and Raymond, J.A. (2006) Ice-binding proteins from sea ice diatoms (Bacillariophyceae). *Journal of Phycology* 42, 410–416.

Junge, K., Krembs, C., Deming, J., Stierle, A. and Eicken, H. (2001) A microscopic approach to investigate bacteria under in-situ conditions in sea ice samples. *Annals of Glaciology* 33, 304–310.

Junge, K., Eicken, H. and Deming, J.W. (2003) Motility of *Colwellia psychrerythraea* strain 34H at subzero temperatures. *Applied and Environmental Microbiology* 69, 4282–4284.

Junge, K., Eicken, H. and Deming, J.W. (2004) Bacterial activity at –2 to –20°C in Arctic wintertime sea ice. *Applied and Environmental Microbiology* 70, 550–557.

Junge, K., Eicken, H., Swanson, B.D. and Deming, J.W. (2006) Bacterial incorporation of leucine into protein down to –20°C with evidence for potential activity in sub-eutectic saline ice formations. *Cryobiology* 52, 417–429.

Kaartokallio, H. (2001) Evidence for active microbial nitrogen transformations in sea ice (Gulf of Bothnia, Baltic Sea) in midwinter. *Polar Biology* 24, 21–28.

Kirst, G.O. and Wiencke, C. (1995) Ecophysiology of polar algae. *Journal of Phycology* 31, 181–199.

Krell, A., Beszteri, B., Dieckmann, G., Glöckner, G., Valentin, K. and Mock, T. (2008) A new class of ice-binding proteins discovered in a salt stress induced cDNA library of the psychrophilic diatom *Fragilariopsis cylindrus* (Bacillariophyceae). *European Journal of Phycology* 43, 423–433.

Krembs, C. and Deming, J.W. (2008) The role of exopolymers in microbial adaptation to sea ice. In: Margesin, R., Schinner, F., Marx, J.-C. and Gerday, C. (eds) *Psychrophiles: from Biodiversity to Biotechnology*. Springer-Verlag, Berlin, Germany, pp. 247–264.

Krembs, C., Gradinger, R. and Spindler, M. (2000) Implications of brine channel geometry and surface area for the interaction of sympagic organisms in Arctic sea ice. *Journal of Experimental Marine Biology and Ecology* 243, 55–80.

Krembs, C., Eicken, H. and Deming, J.W. (2011) Exopolymer alteration of physical properties of sea ice and implications for ice habitability, biogeochemistry and persistence in a warming climate. *Proceedings of the National Academy of Sciences of the United States of America* 108, 3653–3658.

Lange, M.A., Ackley, S.F., Wadhams, P., Dieckmann, G.S. and Eicken, H. (1989) Development of sea ice in the Weddell Sea, Antarctica. *Annals of Glaciology* 12, 92–96.

Laurion, I., Demers, S. and Vezina, A.F. (1995) The microbial food web associated with the ice algal assemblage: biomass and bacterivory of nanoflagellate protozoans in Resolute Passage (High Canadian Arctic). *Marine Ecology Progress Series* 120, 77–89.

Lizotte, M.P. (2001) The contribution of sea ice algae to Antarctic marine primary production. *American Zoologist* 41, 57–73.

Lohan, D. and Johnston, S. (2005) *Bioprospecting in Antarctica*. United Nations University–Institute of Advanced Studies Report. Yokohama, Japan, 36 pp.

Mancuso Nichols, C., Bowman, J.P. and Guezennec, J. (2005a) Effects of incubation temperature on growth and production of exopolysaccharides by an Antarctic sea ice bacterium grown in batch culture. *Applied and Environmental Microbiology* 71, 3519–3523.

Mancuso Nichols, C., Guezennec, J. and Bowman, J.P. (2005b) Bacterial exopolysaccharides from extreme marine environments with special consideration of the Southern Ocean, sea ice, and deep-sea hydrothermal vents: a review. *Marine Biotechnology* 7, 253–271.

Marion, G.M., Fritsen, C., Eicken, H. and Payne, M.C. (2003) The search for life on Europa: limiting environmental factors, potential habitats, and Earth analogues. *Astrobiology* 3, 785–811.

Massé, G., Rowland, S.J., Sicre, M.-A., Jacob, J., Jansen, E. and Belt, S.T. (2008) Abrupt climate changes for Iceland during the last millennium: evidence from high resolution sea ice reconstructions. *Earth Planetary Science Letters* 269, 564–568.

Meiners, K., Krembs, C. and Gradinger, R. (2008) Exopolymer particles: microbial hotspots of enhanced bacterial activity in Arctic fast ice (Chukchi Sea). *Aquatic Microbial Ecology* 52, 195–207.

Mock, T. and Kroon, B.M.A. (2002a) Photosynthetic energy conversion under extreme conditions: I: Important role of lipids as structural modulators and energy sink under N-limited growth in Antarctic sea ice diatoms. *Phytochemistry* 61, 41–51.

Mock, T. and Kroon, B.M.A. (2002b) Photosynthetic energy conversion under extreme conditions: II: The significance of lipids under light limited growth in Antarctic sea ice diatoms. *Phytochemistry* 61, 53–60.

Mock, T. and Thomas, D.N. (2005) Recent advances in sea ice microbiology. *Environmental Microbiology* 7, 605–619.

Mock, T., Dieckmann, G.S., Haas, C., Krell, A., Tison, J.-L., Belem, A.L., Papadimitriou, S. and Thomas, D.N. (2002) Micro-optodes in sea ice: a new approach to investigate oxygen dynamics during sea ice formation. *Aquatic Microbial Ecology* 29, 297–306.

Müller, J., Massé, G., Stein, R. and Belt, S.T. (2009) Variability of sea ice conditions in the Fram Strait over the past 30,000 years. *Nature Geoscience* 2, 772–776.

Nansen, F. (1897) *Farthest North, Volumes 1 and 2*. Archibald Constable and Co., London, UK.

Nichols, D., Bowman, J., Sanderson, K., Mancuso Nichols, C., Lewis, T., McMeekin, T. and Nichols, P.D. (1999) Developments with Antarctic microorganisms: culture collections, bioactivity screening, taxonomy, PUFA production and cold-adapted enzymes. *Current Opinion in Biotechnology* 10, 240–246.

Norman, L., Stedmon, C., Thomas D.N., Granskog, M., Papadimitriou, S., Meiners, K.M., van der Merwe, P., Lanuzel, D. and Dieckmann, G.S. (2011) Characteristics of dissolved organic matter (DOM) and chromophoric dissolved organic matter (CDOM) in Antarctic sea ice. *Deep-Sea Research Part II* 58, 1075–1091.

Palmisano, A.C. and Garrison, D.L. (1993) Microorganisms in Antarctic sea ice. In: Friedmann, E.I. (ed.) *Antarctic Microbiology*. Wiley-Liss, New York, pp. 167–218.

Palmisano, A.C. and Sullivan, C.W. (1985) Growth, metabolism, and dark survival in sea ice microalgae. In: Horner, R.A. (ed.) *Sea Ice Biota*. CRC Press, Boca Raton, Florida, pp. 131–146.

Papadimitriou, S., Thomas, D.N., Kennedy, H., Hass, C., Kuosa, H., Krell, A. and Dieckmann, G.S. (2007) Biochemical composition of natural sea ice brines from the Weddell Sea during early austral summer. *Limnology and Oceanography* 52, 1809–1823.

Peters, E. and Thomas, D.N. (1996a) Prolonged nitrate exhaustion and diatom mortality – a comparison of polar and temperate *Thalassiosira* species. *Journal of Plankton Research* 18, 953–968.

Peters, E. and Thomas, D.N. (1996b) Prolonged darkness and diatom mortality 1: Marine Antarctic species. *Journal of Experimental Marine Biology and Ecology* 207, 25–41.

Petri, R. and Imhoff, J.F. (2001) Genetic analysis of sea ice bacterial communities of the Western Baltic Sea using an improved double gradient method. *Polar Biology* 24, 252–257.

Pringle, D.J., Miner, J.E., Eicken, H. and Golden, K.M. (2009) Pore-space percolation in sea ice single crystals. *Journal of Geophysical Research* 114, C12017, doi:10.1029/2008JC005145.

Raymond, J.A. and Knight, C.A. (2003) Ice binding, recrystallization inhibition, and cryoprotective properties of ice-active substances associated with Antarctic sea ice diatoms. *Cryobiology* 46, 174–181.

Raymond, J.A., Janech, M.G. and Fritsen, C.H. (2009) Novel ice-binding proteins from a psychrophilic Antarctic alga (Chlamydomonadaceae, Chlorophyceae). *Journal of Phycology* 45, 130–136.

Riedel, A., Michel, C., Gosselin, M. and LeBlanc, B. (2007) Enrichment of nutrients, exopolymeric substances and microorganisms in newly formed sea ice on the Mackenzie shelf. *Marine Ecology Progress Series* 342, 55–67.

Riedel, A., Michel, C. and Gosselin, M. (2008) Grazing of large-sized bacteria by sea ice heterotrophic protists on the Mackenzie Shelf during the winter–spring transition. *Aquatic Microbial Ecology* 50, 25–38.

Rivkin, R.B. and Putt, M. (1987) Heterotrophy and photoheterotrophy in Antarctic microalgae: light-dependent incorporation of amino acids and glucose. *Journal of Phycology* 23, 442–452.

Różańska, M., Poulin, M., Gosselin, M., Wiktor, J.M. and Michel, C. (2009) Influence of environmental factors on the development of bottom landfast ice diatoms, nanoflagellates and dinoflagellates in the Canadian Beaufort Sea during the winter–spring transition. *Marine Ecology Progress Series* 386, 43–59.

Russell, N.J. (1997) Psychrophilic bacteria – molecular adaptations of membrane lipids. *Comparative Biochemistry and Physiology* 118, 489–493.

Rysgaard, S. and Glud, R.N. (2004) Anaerobic N_2 production in Arctic sea ice. *Limnology and Oceanography* 49, 86–94.

Rysgaard, S., Glud, R.N., Sejr, M.K., Blicher, M.E. and Stahl, H.J. (2008) Denitrification activity and oxygen dynamics in Arctic sea ice. *Polar Biology* 31, 527–537.

Staley, J.T. and Gosink, J.J. (1999) Poles apart: biodiversity and biogeography of sea ice bacteria. *Annual Reviews in Microbiology* 53, 189–215.

Stedmon, C.A., Thomas, D.N., Granskog, M., Kaartokallio, H., Papadimitriou, S. and Kuosa, H. (2007) Characteristics of dissolved organic matter in Baltic coastal sea ice: allochthonous or autochthonous origins? *Environmental Science and Technology* 41, 7273–7279.

Stefels, J., Steinke, M., Turner, S., Malin, G. and Belviso, S. (2007) Environmental constraints on the production and removal of the chemically active gas dimethylsuphide (DMS) and implications for ecosystem modelling. *Biogeochemistry* 83, 245–275.

Stoecker, D.K., Gustafson, D.E., Merrell, J.R., Black, M.M.D. and Baier, C.T. (1997) Excystment and growth of chrysophytes and dinoflagellates at low temperatures and high salinities in Antarctic sea ice. *Journal of Phycology* 33, 585–595.

Sutherland, P.C. (1852) *Journal of a voyage in Baffin's Bay and Barrow Straits in the years 1850-51, performed by H.M. ships 'Lady Franklin' and 'Sophia,' under the command of Mr. William Penny in search of the missing crews of H.M. ships 'Erebus' and 'Terror' Volumes 1 and 2*. Longman, Brown, Green, and Longmans, London, UK.

Thomas, D.N. (2004) *Frozen Oceans – The Floating World of Pack Ice*. Natural History Museum, London, UK, 224 pp.

Thomas, D.N. and Dieckmann, G.S. (2002) Antarctic sea ice – a habitat for extremophiles. *Science* 295, 641–644.

Thomas, D.N. and Dieckmann, G.S. (eds) (2010) *Sea Ice*, 2nd edn. Wiley-Blackwell Publishing, Oxford, UK, 520 pp.

Thomas, D.N., Lara, R.J., Haas, C., Schnack-Schiel, S.B., Dieckmann, G.S., Kattner, G., Nothig, E.-M. and Mizdalski, E. (1998) Biological soup within decaying summer sea ice in the Amundsen Sea, Antarctica. In: Lizotte, M.P. and Arrigo, K.R. (eds) *Antarctic Sea Ice: Biological Processes, Interactions and Variability*. American Geophysical Union, *Antarctic Research Series* 73, 161–171.

Tison, J.-L., Worby, A., Delille, B., Brabant, F., Papadimitriou, S., Thomas, D.N., de Jong, J., Lannuzel, D. and Haas, C. (2008) Temporal evolution of decaying summer first-year sea ice in the Western Weddell Sea. *Deep-Sea Research Part II* 55, 975–987.

Tison, J.-L., Brabant, F., Dumont, I. and Stefels, J.(2010) High-resolution dimethyl sulfide and dimethylsulfoniopropionate time series profiles in decaying summer first-year sea ice at Ice Station Polarstern, western Weddell Sea, Antarctica. *Journal of Geophysical Research* 115, G04044, doi:10.1029/2010JG001427.

Trevena, A.J. and Jones, G.B. (2006) Dimethylsulphide and dimethylsulphoniopropionate in Antarctic sea ice and their release during sea ice melting. *Marine Chemistry* 98, 210–222.

Underwood, G.J.C., Fietz, S., Papadimitriou, S., Thomas, D.N. and Dieckmann, G.S. (2010) Distribution and composition of dissolved extracellular polymeric substances (EPS) in Antarctic sea ice. *Marine Ecological Progress Series* 404, 1–19.

Vare, L.L., Massé, G. and Belt, S.T. (2010) A biomarker-based reconstruction of sea ice conditions for the Barents Sea in recent centuries. *The Holocene* 20, 637–643.

Weissenberger, J., Dieckmann, G.S., Gradinger, R. and Spindler, M. (1992) Sea ice: a cast technique to examine and analyse brine pockets and channel structure. *Limnology and Oceanography* 37, 179–183.

Wettlaufer, J.S. (1999) Crystal growth, surface phase transitions and thermomolecular pressure. In: Wettlaufer, J.S., Dash, J.G. and Untersteiner, N. (eds) *Ice Physics and the Natural Environment*. NATO/ASI 56, Series I. Springer-Verlag, Berlin and New York, pp. 39–68.

Zaslavskaia, L.A., Lippmeier, J.C., Shih, C., Ehrhardt, D., Grossman, A.R. and Apt, K.E. (2001) Trophic conversion of an obligate phototrophic organism through metabolic engineering. *Science* 292, 2073–2075.

5 Polar Terrestrial Environments

Peter Convey
British Antarctic Survey, Cambridge, UK

5.1 Introduction

To our eyes, the polar regions of our planet are clearly challenging regions for life. Animals, plants and microbes that live there are exposed to a suite of environmental stresses over a range of timescales. There is chronic exposure to low temperature, high winds and solar radiation, as well as to freezing and/or desiccation stress, and to predictable or unpredictable extremes and short-term acute events. Beyond the polar circles, strong seasonality is enforced as the sun remains permanently below the horizon for days to months in winter, and the converse in summer. The absence of solar energy input in winter means that terrestrial habitats of both north and south face comparable and extremely low air temperatures, typically −40 to −60°C or lower (the lowest recorded temperature is −89.2°C, recorded at Vostok Station on the Antarctic polar plateau). However, in the summer the Antarctic experiences significantly lower temperatures than the Arctic (Convey, 1996a; Danks, 1999). This means that biological activity is constrained by the amount of energy available. As well as facing low temperatures overall, microhabitat temperature variability in polar terrestrial habitats can be similar to or even greater than that at lower latitudes (Peck *et al.*, 2006), with variations of 20–30°C not unusual on timescales of minutes to hours, and 40–60°C (even up to 100°C) over an annual cycle. Within specific microhabitats exposure to direct sunlight can lead to short-term temperature maxima of 30–40°C, even at high latitudes.

Polar terrestrial ecosystems face many physical environmental challenges in addition to temperature. Arguably the most important is the lack of liquid water (see also Hogg and Wall, Chapter 10, this volume), and water availability is recognized as being at least as important as temperature to the biota of polar ecosystems (Kennedy, 1993; Block, 1996; Hodkinson *et al.*, 1999). Across most of the continent, and for most of the time, water is biologically unavailable as ice, while typically low relative humidity leads to rapid evaporation or sublimation of any rainfall or snow. Much of the Antarctic continent is classified as a frigid desert (Sømme, 1995). The third physical variable directly impacting polar terrestrial biota is solar radiation, itself modulated at the level of the organism by other variables such as cloud, snow cover and albedo. Separately, parts of the solar radiation spectrum, particularly high energy shorter wavelength ultraviolet (UV-B) radiation, can damage biological systems. In both polar regions there have been considerable changes in exposure to UV-B radiation following anthropogenic ozone depletion (Rozema, 1999).

Environmental conditions experienced in particular in the Antarctic lie at one extreme of the range available on Earth (Peck *et al.*, 2006; Chown and Convey, 2007), a fact that has been sufficient in itself to generate a considerable body of research focusing on adaptation to environmental extremes. In recent decades it has also been recognized that parts of both polar regions are experiencing some of the most rapid rates of environmental change on the planet (particularly, but not only, in terms of temperature) (ACIA, 2004; Convey *et al.*, 2009a; Turner *et al.*, 2009). In combination with the relative structural simplicity of some of their ecosystems, the biota and ecosystems of these regions have been proposed as sensitive biological indicators of environmental change consequences (e.g. Callaghan and Jonasson, 1995; Freckman and Virginia, 1997; Walther *et al.*, 2002; Convey, 2006).

This chapter considers the terrestrial ecosystems of the Arctic and Antarctic, including terrestrial substrata temporarily or permanently free of snow or ice, their biota and the biological adaptations typified within them. Of necessity it provides a general overview, and the reader is pointed to the very considerable literature describing these ecosystems and the adaptations of their biota (e.g. Sømme, 1995; Thomas *et al.*, 2008; Denlinger and Lee, 2010). It is also recognized that these ecosystems, and the features discussed here, overlap considerably and are intimately linked with those of freshwater habitats (Laybourn-Parry and Bell, Chapter 6, this volume), polar deserts (Hogg and Wall, Chapter 10, this volume), high radiation habitats (Newsham and Davidson, Chapter 22, this volume), alpine regions (Sattler *et al.*, Chapter 8, this volume), glaciers (Bagshaw *et al.*, Chapter 9, this volume) and subglacial lakes (Pearce, Chapter 7, this volume). While being less typical, polar terrestrial ecosystems also include examples of high temperature (volcanic) habitats (Cowan *et al.*, Chapter 12, this volume) and acidic, hypersaline and anoxic environments (Chapters 18, 20, 21, this volume).

5.2 Terrestrial Ecosystems

5.2.1 Habitats

Any exposed surface offers opportunity for colonization and establishment to those organisms with suitable biochemical, physiological and life history features (Fig. 5.1). Rock surfaces on the most northerly islands in the Arctic (ca. 84 °N) and most southerly nunataks in the Antarctic (ca. 87 °S) are colonized by biota. Radiation receipt integrated over the full day in midsummer at these latitudes is greater than in the tropics, even though peak levels are lower due to the lower solar angle. Snow and ice cover provides buffering to the extreme thermal conditions – indeed a thin covering can act as a greenhouse – as well as filtering for incoming UV radiation (Cockell *et al.*, 2002). The summer warming of terrestrial substrata can lead to the melting of surrounding snow and ice, providing a resource for surrounding biota, and also penetrating the surface layers of some rock types, where endolithic (living within rocks) microbial communities can then develop (Friedmann, 1982; Hughes and Lawley, 2003). Varying widely between locations, water may be available for several months, only for a few days each year, or not at all in some years.

Many polar terrestrial habitats consist simply of rock, boulder and rubble surfaces. In the Antarctic, soils are rarely more than fragmented rock debris, with little classical horizon structure, and low nutrient and carbon content (Beyer and Bölter, 2002). Rock is broken up by glacial action, water (particularly through freeze–thaw cycles), salt weathering, insolation and wind action. More developed ‘brown’ soils are limited to extremely small areas in association with the two flowering plant species present on the Antarctic Peninsula, although soil development is more progressed on the Subantarctic islands and in much of the Arctic.

Fig. 5.1. Terrestrial habitats of the Arctic and Antarctic: (a) Isolated groups of nunataks (Behrendt Mountains) in the Antarctic. (b) Some of the largest areas of terrestrial habitat in the Antarctic are found in the 'dry valley' ecosystems of the Transantarctic Mountains (Davis Valley, Pensacola Mountains). (c) Abrasion by wind-blown debris is an additional stress facing colonizing organisms (Mars Oasis, Alexander Island). (d) Periglacial processes leading to various types of 'patterned ground' are common to both polar regions (Ny Ålesund, Svalbard). (e) Extensive tundra habitats, underlain by permafrost, are a feature of the High Arctic (Kongsfjord, Svalbard). Images © Peter Convey.

Simple unconsolidated rock debris is typically too mobile for effective colonization by plants, as biota colonizing any substratum need to be able to attach to it. In the case of epilithic (living on rock surfaces) lichens this involves penetration into the rock or boulder surface, facilitated by the secretion of a range of 'lichen compounds'. This may lead to a biological contribution to rock exfoliation and weathering, thereby also contributing to soil formation. At depth in the soil profile stabilization may be achieved by permafrost, although this does not provide a habitat for an active biological community. Permafrost is important in modifying overlying habitats, primarily through preventing drainage. In the Arctic, and maritime and coastal continental Antarctic, permafrost is usually overlain by an active layer of wet soil that freezes in winter. While clearly having potential as a biological propagule bank, the true importance of permafrost as a source of colonizing biota (or indeed that of ice itself) remains unquantified.

Abrasion by rock debris or ice crystals is a common feature of polar terrestrial habitats, and this and soil movement by cryoturbation during freeze–thaw cycles strongly limits the locations where vegetation development is possible. Other than in vegetation itself and within the simple soil matrix, terrestrial invertebrates and microbial communities are typically found in protected habitats, such as in cracks in rocks, crevices in boulder fields, and the under-surfaces of rocks. Although not unique to the polar regions, the rock itself may be colonized by endolithic microbiota, regarded as one of the apparent limits to viable life on Earth, and a model for how life may exist on other planets.

Boulder, scree and developing soil habitats may eventually become sufficiently stable to permit vegetation establishment. Stone and boulder surfaces are colonized by epilithic lichens, while crevices are occupied by green algae and mosses. Mosses also colonize soil surfaces after initial stabilization by microbial crusts, and in rare instances may go on to develop into deeper peat accumulations over centuries to millennia. Vascular plants are also an early community component in the Subantarctic and Arctic.

5.2.2 Permafrost

Permafrost consists of ice-cemented material (or solid rock) that remains continuously below 0°C over periods of at least multiple years. It is very important and extensive in the Arctic, and its southern limit is as little as 30°N in parts of Eurasia. It can also extend under the sea in the Arctic Ocean and on land can be between about 200 and 700 m in depth. The potential and actual thawing of Arctic permafrost under current rates of climate change is causing considerable concern, with wide and not fully understood impacts on areas as wide as drainage and water regimes, ecological processes, the carbon cycle (particularly stored carbon release), as well as on human populations and cultures. In Antarctica the ice-free surface area available is much smaller and less attention has been given to the distribution and influences of permafrost. However, permafrost depths of up to 1000 m have been measured in the Victoria Land Dry Valleys and it is reasonable to expect similar depths on other parts of the continent. Permafrost is also a feature of the maritime Antarctic but it is not present on the Subantarctic islands, except possibly at high mountain altitudes.

The main importance of permafrost in the context of terrestrial biology lies in its modifying of overlying habitats, in particular through preventing drainage. Permafrost is therefore typically associated with wet soils. Above the permanently frozen permafrost there is an 'active layer', rarely more than 1 m deep, which thaws in the summer months. One exception to this generalization is that the exceptionally dry soils of the Victoria Land Dry Valleys and similar habitats in Antarctica do not have an active layer. Permafrost has a strong influence on geomorphological processes and surface topography, including the overlying habitats. Perhaps one of the most spectacular illustrations of this is given by the pingos (or hydrolaccolith) that are well developed in parts of the Arctic. Pingos are mounds of earth-covered ice. The ice in their core usually originates from segregation or injection of fluid water, and can be massive. A small freshwater lake can occupy the summit of a pingo where a crater has formed from the ice melting.

5.2.3 Polar soils

Tedrow (1977) classified soils in the polar regions under four types: tundra, sub-polar desert, polar desert and cold desert. Tundra and sub-polar desert constitute large areas of the Arctic, with polar desert zone taking over at about 80 °N. In the south, tundra-like soils are only found on the Subantarctic islands, sub-polar desert is absent and polar desert is very restricted, with the remainder of the continent being cold desert, which is unique to Antarctica. Cold desert soils in the Dry Valleys of Victoria Land, Antarctica, are at least several million years old, contrasting with post-Last Glacial Maximum ages in polar desert soils of both the Arctic and Antarctic. Cold desert soils are comparable in aridity with those of hot deserts (Hogg and Wall, Chapter 10, this volume).

Sub-polar soil types are more complex, also hosting considerable biomass of plant communities including, significantly, higher plants with well-developed root systems. Here, true brown soils are found, whose developed horizons require both rooting plants and active soil processes. On the Antarctic continent, brown soils are only found with the two native vascular plants on the Antarctic Peninsula. Soil mixing organisms, including a range of invertebrates but particularly earthworms, are not found on the Antarctic continent and some of the most northern terrestrial habitats in the Arctic, but do occur in the Subantarctic.

5.3 Terrestrial Biota and Communities

5.3.1 Ice-free ecosystems

The polar regions cover large areas, and are certainly not uniform in terms of either biological or physical environmental characteristics. This has inevitably led to the development of schemes of polar biogeographical zones or regionalization. While imperfect these are still useful and, for the purposes of this chapter, the Arctic is considered in High and Low Arctic zones (Halliday, 2002), and the Antarctic in sub-, maritime and continental zones (Smith, 1984). These allow general comparison of Arctic and Antarctic plant and animal communities (Fig. 5.2; Table 5.1).

There is a general decrease in productivity and biodiversity (by whatever measure used) in terrestrial ecosystems with increasing latitude north or south. However, in detail, these trends are unclear and heavily modulated by features of history, climate, geology and topography. Furthermore, in the Antarctic it is increasingly clear that the current general regionalization does not provide an accurate biogeographical interpretation of the terrestrial ecosystems of the continent or its history (Chown and Convey, 2007; Convey and Stevens, 2007; Convey *et al.*, 2008, 2009b).

The largest ice-free ecosystems of Antarctica are found in the Dry Valley region of Victoria Land, with other significant ice-free oases in the Bunger Hills, Vestfold Hills and elsewhere in the Transantarctic Mountains (Adams *et al.*, 2006; Convey, 2007a). At the most extreme, large areas appear barren, at least of macroscopic organisms. However, it is clear that they host a range of microbes, whose diversity may have been seriously underestimated (Cowan *et al.*, 2002). Moisture content strongly influences the diversity found. In locations with available water, lichens and more limited moss communities can develop. Nematodes, tardigrades, rotifers and arthropods are similarly patchily distributed (Freckman and Virginia, 1998; Convey and McInnes, 2005; Adams *et al.*, 2006; Convey, 2007a). Given the twin stresses of chronic and extreme cold, and very limited temporal availability of liquid water, these organisms spend long periods dormant. Significant levels of species and even family level endemism are apparent even at intra-regional scales among all the terrestrial groups of invertebrate representative of the continental Antarctic (Pugh and Convey, 2008), supporting a long presence and evolutionary radiation *in situ* within the region. Similarly, recent research indicates considerable levels of microbial endemicity within continental Antarctica (De Wever *et al.*, 2009; Vyverman *et al.*, 2010), suggesting that the widely recognized 'global ubiquity hypothesis' (Finlay, 2002) may not apply well to the Antarctic continent, unlike most of the rest of the planet. Other than the larger desert ecosystems, terrestrial communities of the continental Antarctic are found on typically small 'islands' of ice-free ground, being best developed in coastal regions (where there are relatively warmer summer temperatures, water availability and sometimes enhanced nutrient input from the marine environment and marine vertebrate colonies) and the summits of inland nunataks. In terms of the detail of biodiversity knowledge, many locations remain unexamined by specialist biologists.

The maritime Antarctic and High Arctic are broadly comparable in terms of their annual temperature ranges and degree of seasonality (Convey, 1996a; Danks, 1999). Differences do remain, in particular relating to slightly higher mean monthly air temperatures in summer in parts of the High Arctic, and geographical connectivity with lower northern latitudes, both contributing to greater biomass and diversity of terrestrial biota, importantly including large grazing and predatory mammals. Seasonal snow cover is typical of terrestrial habitats in these two regions, other than exposed ridges and cliffs. This can prevent erosion and wind damage, and insulates underlying substrata maintaining temperatures often well above air temperature minima. In different circumstances, this snow cover may act as greenhouse insulation, accelerating the commencement of biological activity in the early spring. Alternatively, it can delay warming in spring. The melting of winter snow accumulation provides moisture, allowing activity of plants and animals in summer. Once again, the timing of this melt is important, with potential to reduce the growing season length if banks persist for too long, or to lead to failure of

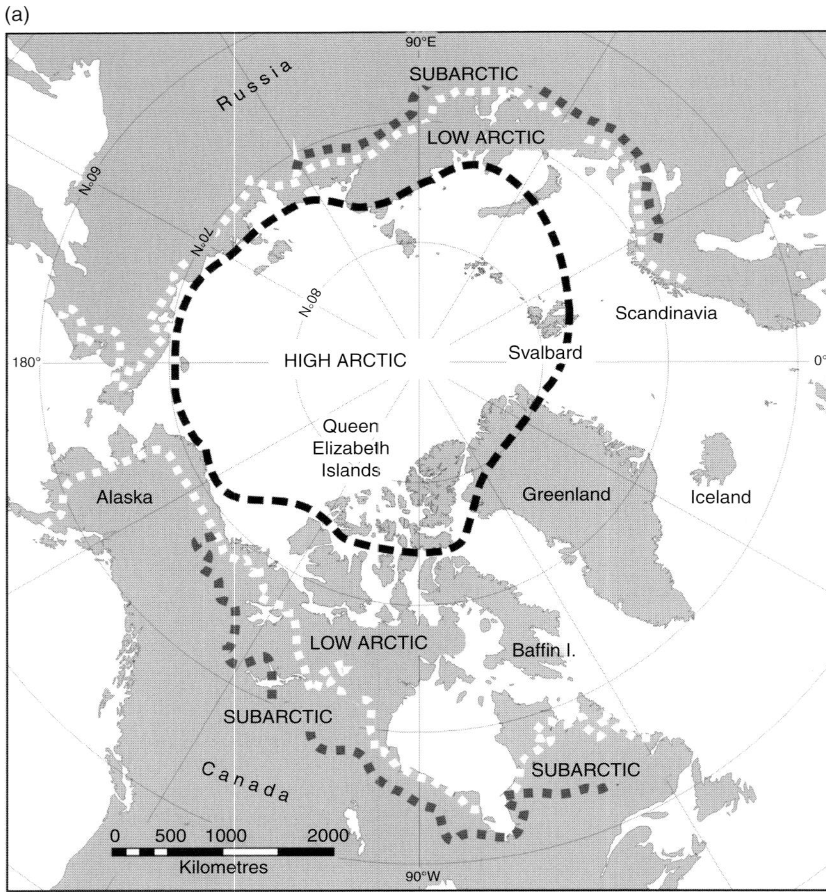

Fig. 5.2a. Regional terminology for terrestrial environments in the Arctic © Peter Convey.

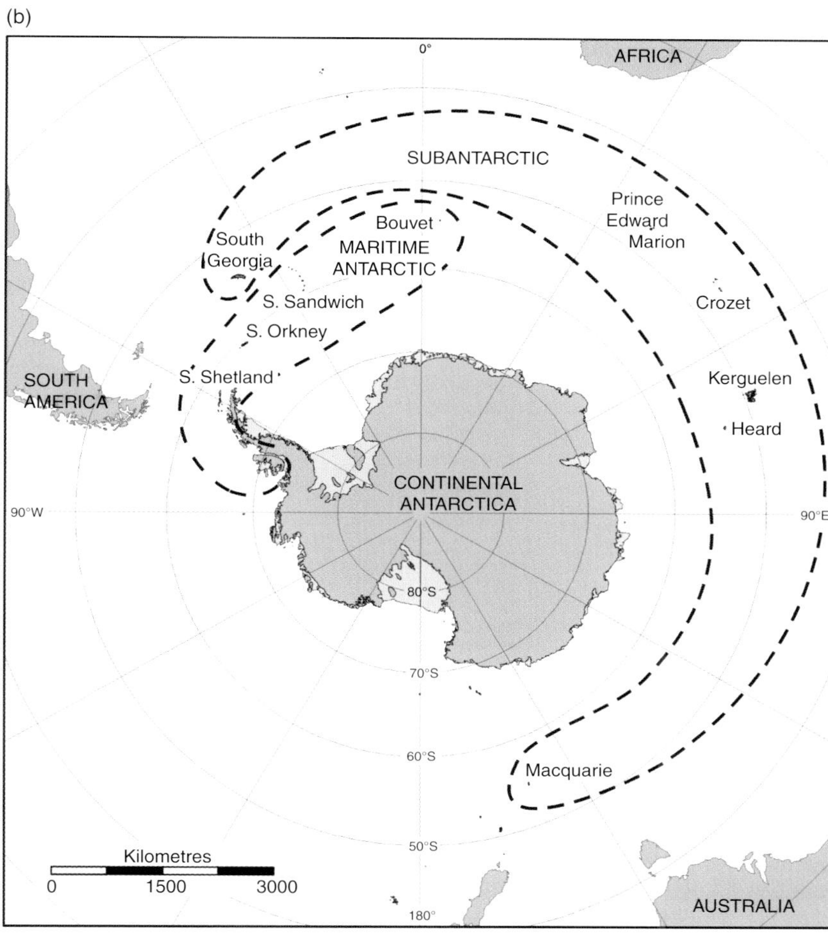

Fig. 5.2b. Regional terminology for terrestrial environments in the Antarctic. © Peter Convey.

water supply if they melt completely too early in the season.

5.3.2 Vegetation

Vegetation in these regions is generally dominated by cryptogams (plants that do not reproduce using seed: mosses, liverworts, lichens). This fellfield vegetation is particularly well developed in the maritime Antarctic (Smith, 1984). Vegetation is typically scattered, interspersed with areas of bare ground and rock, and can cover as little as 2% of the surface area. The development of this type of community on newly available ground can be very rapid – taking place in only 10–20 years on Signy Island (Smith, 1990). Most individual moss clumps are likely to be relatively short-lived, while some lichens are large enough to be 300–600 years old, and the few peat banks present in the maritime Antarctic are thought to have accumulated through continuous growth over 5000–6000 years (Fenton, 1980; Björck *et al.*, 1991).

Flowering plant diversity in the High Arctic (for instance, >100 species on Svalbard, 57 on Franz Josef Land) is far greater than that of the maritime Antarctic (two species), as well as that of the much more benign Subantarctic (Convey, 2007b). Phanerogam (flowering plant) seed banks are a feature of High Arctic soils, but are much more limited

Table 5.1. Overview of species diversity in some of the main representative terrestrial groups in the different Antarctic biogeographical zones, in comparison with three Arctic locations from which recent and partial or comprehensive compilations are available (locations/groups for which no value is given indicate that no information was located). Svalbard data from Barr (1995), Elvebakk and Hertel (1996), Frisvoll and Elvebakk (1996), Rønning (1996), Coulson and Resfeth (2004), Coulson (2007), Franz Josef Land from Barr (1995), Greenland from Jensen and Christensen (2003), Antarctic from Convey (2007a). Table updated from Thomas *et al.*, 2008.

Group	Subantarctic	Maritime Antarctic	Continental Antarctic	High Arctic (Svalbard)	High Arctic (Franz Josef Land)	Greenland
Rotifera	>59	>50	13	154		
Tardigrada	>34	26	19	83		
Nematoda	>22	28	14	111		
Platyhelminthes	4	2	0	10		
Annelida (Oligochaeta)	23	3	0	34		
Mollusca	3/4	0	0	0		2
Crustacea (non-marine)	44	10	14	33		65
Insecta	210	35	49	237		631
Collembola	>30	10	10	60		41
Araneida	20	0	0	19		60
Acarina (free-living)	140	36	29	127		127
Myriapoda	3	0	0	0		1
Mammalia	0	0	0	3	2	8
Aves	0	1	5	17	6	39
Flowering plants	0	2	60	164	57	515
Bryophytes	26	125	335	373	150	612
Lichens	150	250	250	597	>100	950

in the maritime Antarctic as flowering plants themselves are much more restricted. In the latter region there is a large bryophyte propagule bank (Smith, 1990). High Arctic vegetation can also form extensive and lush closed cover, particularly in nutrient-enriched areas such as beneath bird cliffs, a vegetation type that has no maritime Antarctic equivalent.

In the Subantarctic and on more northern cool temperate islands such as the Falkland Islands and the New Zealand shelf islands, the lack of large herbivores has led to the development of unique vegetation communities. These include tussock grasses that can be as much as 2 m high, and megaherbs such as *Stilbocarpa polaris* and *Pringlea antiscorbutica* (Meurk *et al.*, 1994; Mitchell *et al.*, 1999; Fig. 5.3). Human introduction of grazing vertebrates to most Subantarctic islands has drastically altered large areas of this native vegetation, while invertebrate introductions have had a wide range of other impacts within these islands' invertebrate-dominated ecosystems, from new detritivores releasing decay bottlenecks to predators leading to local extinction of indigenous (and endemic) insects (Frenot *et al.*, 2005).

The term 'tundra' is often used to describe terrestrial communities in both the Arctic and Antarctic. However, this is somewhat inaccurate. True tundra is present only in the Arctic, and includes habitats dominated by low vascular plants without significant tree or shrub growth (excepting dwarf prostrate forms such as Arctic willow), and is underlain by permafrost. It is present in parts of the High Arctic as well as large areas of the Low Arctic. Subantarctic habitats often appear similar in terms of the major vegetation components, but are exposed to much more benign lower latitude environmental conditions and are not underlain by permafrost (Smith, 1984). The degree of seasonality and annual pattern of variation in

Fig. 5.3. (a) In the absence of terrestrial grazing vertebrates, tussock grass is well developed on many sub- and peri-Antarctic islands. Image shows tussock grasses on Kidney Island, Falkland Islands, © Isaac Forster.

temperatures is also strikingly different between the Subantarctic and both Arctic regions (Convey, 1996a; Danks, 1999). While the highest mean monthly air temperatures in all these regions in the summer are quite similar, only the coldest Subantarctic islands have mean monthly temperatures in winter that dip below freezing, while Arctic temperatures drop well below zero, and remain there for long periods. Thus in the Subantarctic biological activity in terrestrial habitats can continue, while it ceases across the Arctic.

Primary production in Arctic tundra can vary by more than two orders of magnitude between habitats (Bliss, 1975). As also seen in the Antarctic, true herbivores are responsible for consuming relatively little of this production, with the dominant energy and nutrient flows being through decomposition. Since decomposition rates are generally low, surplus plant material accumulates as peat, much of which is currently incorporated in the permafrost. Current trends of warming across much of the Arctic, resulting in melting of permafrost, lead to concerns of increased decomposition rates and release of CO_2 to the atmosphere, in principle providing further positive feedback to global warming processes (ACIA, 2004; Callaghan *et al.*, 2004). Decomposition also releases mineral nutrients that are often scarce in polar soils. Where grazing vertebrates are present, nutrient recycling is accelerated, as they return about 70% of the plant material ingested as faeces. The absence of grazing vertebrates and invertebrates in the Antarctic imposes a severe limit on nutrient cycling, an important difference between ecosystems of the two polar regions.

5.3.3 Invertebrate fauna

The soil invertebrate fauna of these two regions is patchily distributed, and includes Protozoa, tardigrades, rotifers, nematodes, enchytraeid worms, mites and collembolans (springtails) (Block, 1984; Convey, 2007a). As with the flowering plants, the maritime Antarctic hosts a strikingly lower insect diversity of only two species of fly, while the High Arctic includes a range in particular of flies and beetles, and also other arthropods such as spiders (e.g. Jensen and Christensen, 2003; Coulson and Resfeth, 2004). The majority of soil invertebrates are thought to be generalist microbivores or detritivores, with relatively few true herbivores, possibly through energetic limitation of this feeding strategy. However, this generalization is based on few specific autecological studies (Hogg *et al.*, 2006) or, indeed, wider studies of nutrient and energy flows and ecosystem structure (excepting Davis, 1980, 1981).

5.3.4 Vertebrate fauna

A striking faunal difference between the two regions is the presence of vertebrate herbivores in the Arctic (Sage, 1986; Kovacs, 2005), such as the Arctic lemming and Arctic hare, as well as larger species such as reindeer and muskoxen. However these species are more typical of the Low Arctic and do not generally spend much time in the High Arctic (other than the permanently present Svalbard reindeer). Large migratory bird populations (geese, ptarmigan, snow bunting, wading birds, gulls) utilize the High Arctic for breeding, and in some cases forage on terrestrial vegetation and invertebrates. Vertebrate predators (polar bear, Arctic fox) are also present year-round in the High Arctic, although the former relies on the marine environment for food (Fig. 5.4). The muskox is the largest herbivore. Despite its size, its effective thermal

Fig. 5.3. Continued.
(b) Megaherbs are also well represented on some Subantarctic islands (Kerguelen), © Francoise Hennion.
(c) Specialized communities can be associated with intense nutrient enrichment. Here a black browed albatross colony on Steeple Jason Island, Falkland Islands, has a pink, guano-enriched pool in the foreground. © Isaac Forster. (d) High Arctic tundra hosts relatively rich floral diversity, © Peter Convey. (e) Dwarf willows are the world's most northern trees, © Peter Convey. (f) Nutrient enrichment under bird nesting cliffs can lead to exceptional vegetation development in the High Arctic (Stuphallet, Kongsfjord, Svalbard), © Peter Convey.

Fig. 5.4. Polar fauna. (a) Polar bears (*Ursus maritimus*) rely on the marine environment for food, © Elanor Bell. (b) Svalbard reindeer, © Peter Convey. (c) The tussock moth, *Gynaephora groenlandica*, has one of the most extended life cycles known among Arctic insects, as well as exceptional metabolic adaptations, being

insulation allows it to spend much of its time resting, permitting year-round existence in areas of low vegetation productivity, where it may have surprisingly little impact on the vegetation, removing less than 1% of the plant biomass. In contrast, at least during years of maximum population numbers, lemmings can graze up to 25% of the above-ground production, leading to widespread if temporary destruction of habitat and influencing the floristic composition of the tundra.

A feature particularly of Subantarctic and Antarctic coastal ecosystems, is concentrations of breeding or resting/moulting marine birds and mammals. The development of these vertebrate concentrations is encouraged, in a manner that does not apply so generally in the Arctic (although note the example of bird nesting cliffs given above), by the spatially very limited and island-like availability of ice-free ground. Nutrient transfer from these concentrations can fertilize and accelerate the development of vegetation and specifically associated arthropod communities. Examples of this are found even considerable distances inland in the Antarctic, e.g. at Robertskollen, a group of inland nunataks at 71°28′S 3°15′E, where local vegetation cover is about seven times greater than on nunataks with birds than those without (Ryan and Watkins, 1989).

5.3.5 Exceptional ecosystems

Both the Arctic and Antarctic also contain examples of regionally unusual terrestrial ecosystems, most striking of which are those associated with geothermal habitats. Such habitats are given more general consideration in Cowan *et al.* (Chapter 12, this volume), but their Antarctic context is considered briefly here. Geothermally active locations are very limited in ice-free parts of the Arctic and Antarctic regions considered here. Active sites (including fumaroles, warmed ground, heated pools or springs) are know from the shores of Bockfjord (Svalbard) and Disko Island (west Greenland) in the High and Low Arctic, on Mounts Erebus, Melbourne and Rittman in Victoria Land in the continental Antarctic, and on the South Shetland and South Sandwich islands and Bouvetøya in the maritime Antarctic. Three Subantarctic islands are also active, although no studies of associated biological communities have been published: Heard and Macdonald Islands being currently continuously active, and Marion Island, which has been active throughout at least the Pleistocene, although no current areas of activity are known. Conditions at these locations are typified by warmth and a constant supply of liquid water in streams or through condensation, both of which promote biological activity, intense chemical weathering producing clay minerals and direct chemical challenges such as low pH, and high mineral and sulfur levels. In Greenland, warm springs host some species otherwise unrecorded in the region, and regional species that are at their northern limits (Heide-Jørgensen and Kristensen, 1999; Jensen and Christensen, 2003), as is also seen on the South Sandwich and South Shetland Islands in the maritime Antarctic (Convey *et al.*, 2000a, b; Smith, 2005) and around exceptional fumaroles at high altitude on the continental Antarctic volcanoes (Broady, 1993; Bargagli *et al.*, 2004; Fig. 5.5).

Fig. 5.4. Continued.
one of the largest insect larvae to undergo anhydriobiosis as a survival strategy in the Arctic winter, © R.E. Lee. (d) *Eretmoptera murphyi* adult, a chironomid midge endemic to Subantarctic South Georgia, © British Antarctic Survey. (e) Mites are the dominant terrestrial arthropod in many Antarctic habitats. The dark oribatid mite, *Alaskozetes antarcticus*, typically forms dense aggregations on rocks and vegetation near the soil surface and is one of the largest herbivores present. The orange predatory mesostigmatid mite, *Gamasellus racovitzai*, is one of the few predators in these ecosystems, © Ian Collinge, British Antarctic Survey slide collection. (f) A group of at least several hundred springtails (*Cryptopygus antarcticus*) temporarily trapped on a water surface of a waterlogged moss, © Peter Convey. (g) Soil meiofauna (here illustrated a nematode worm) are found in most polar soils, © British Antarctic Survey.

Fig. 5.5. Exceptional geothermal terrestrial habitats in Antarctica: (a) Mount Erebus is one of Antarctica's

5.4 Physiology and Ecology

5.4.1 Life history strategies

Life history strategies of polar terrestrial invertebrates and plants are generally described as 'adversity' or 'stress' selected (Convey, 1996b; Peck *et al.*, 2006; see also Panikov (1995) for similar concepts in tundra soil microbial communities). These typically involve significant physiological/biochemical investment in stress tolerance strategies, much reduced competitive and dispersal abilities, low reproductive investment, and extended life cycles. Commonly there is considerable flexibility in important aspects of life history and physiology (Crawford, 1995; Convey, 1996b; Hodkinson and Wookey, 1999; Hodkinson, 2005), which is advantageous in environments whose conditions lie close to limits for biological activity, and experience typically rapid or unpredictable variation in these variables.

The general life history models on which these descriptions are based have an evolutionary basis, and it is assumed that variation in the relevant life history features has an underlying heritable component. While no explicit studies of polar biota have attempted to address this, there is some strong circumstantial evidence, such as the typically large investment in biochemical cold and desiccation tolerance strategies that must be an enhancement of ancestral capabilities. However, there are also areas of life history where an evolutionary explanation may be inappropriate for features seen in polar biota, even where these are consistent with adversity or stress selection. For instance, the life cycle extension often reported in polar biota may simply be a result of the low levels of energy available, and extended period of winter inactivity, together providing sufficient constraint to development (Norton, 1994).

It is important to recognize that the suite of environmental stressors, and their interactions, faced by polar biota is wide and their effects are complex. Somewhat counter-intuitively, the effectiveness of responses to stresses experienced during the summer and in preparation for winter or other periods of specific extremes are likely to be more important to organism fitness than are events during winter itself. This is based on the fact that, other than in the Subantarctic, winter temperatures in the polar zones under consideration are low

Fig. 5.5. Continued.
highest mountains, and overlooks the American McMurdo Station and New Zealand Scott Base on Ross Island in Victoria Land. It is one of very few active volcanoes on the Antarctic continent, and fumaroles close to its summit are home to unique microbial and lower plant communities. (b) Deception Island (South Shetland Islands) is a flooded caldera that last erupted in the late 1960s, destroying two of the then active research stations on the island. Today the island is still active, with areas of fumaroles and heated ground both on land and the shallow floor of the caldera. This photograph illustrates the warmed summit ridge of Cerro Caliente, overlooking the current Argentine and Spanish research stations on the island. (c) There are extensive areas of geothermally active ground, fumaroles, etc. on the several of the maritime Antarctic South Sandwich Islands, some of which are richly vegetated as here within the crater wall on Bellingshausen Island. The surface temperature of bare patches of ground may be 45–100°C, with slightly cooler ground then being colonized by concentric rings of mosses and liverworts depending on their thermal tolerance, many of which are unique to these locations. (d) Mixed community of mosses and lichens developed close to the mouth of an active fumarole on Bellinghausen Island. (e) Mosses, particularly of the genus *Campylopus* (which can withstand temperatures as high as 45°C at the soil surface as little as 2–3 cm below the growing shoot tips), are amongst the first colonists of heated ground near the summit of Lucifer Hill, Candlemas Island, South Sandwich Islands. (f) A fumarole in Clinker Gulch, Candlemas Island, provides the catalyst for the development of a rich surrounding liverwort and moss community on an otherwise barren rubble slope. (g) Macroscopic fungi are generally rare in Antarctic terrestrial ecosystems, but may develop abundantly in the vicinity of warmed ground (Lucifer Hill, Candlemas Island). (h) Volcanic eruptions are a perennial threat for organisms on islands such as the South Sandwich Islands, meaning that any specific community or colonized location is effectively ephemeral. Even active vertebrates are not immune, as illustrated by this penguin buried in volcanic ash and debris on Zavodovski Island. Images © Peter Convey.

enough to preclude active biological processes, while it is also self-evident that the biota present have the appropriate abilities to survive through these periods. Thus, natural selection will act on the many processes involved in growth, maintenance, reproduction, metabolism, etc. that take place actively during the summer months.

5.4.2 Life under extreme stress

One often-reported feature of polar invertebrates, microbes and plants is that they are less sensitive to low temperature than are similar species from lower latitudes – being psychrotolerant or psychophilic (e.g. Longton, 1988; Vincent, 1988; Peck *et al.*, 2006). In simple terms, the annually integrated production from any given biological process in a polar organism (e.g. fixation of carbon in photosynthesis, biomass increase) is a function of the length of the growing season. Thus, short-term growth rate of polar plants and invertebrates can be at least comparable to those of similar temperate species living in environments 15–20°C warmer, even though they typically achieve much lower annual growth rates through spending much more of the year inactive.

Studies of adaptation to environmental extremes have been carried out at various scales of organization, from the genomic, biochemical and metabolic to the behavioural and life history, in polar terrestrial organisms, contributing to stress tolerance, maintenance of metabolic activity and survival in the face of low temperatures and other environmental challenges (see reviews of and introductions to various elements of this vast research area in Block, 1990; Convey, 1996b; Schmidt-Nielsen, 1997; Hennion *et al.*, 2006; Peck *et al.*, 2006; Thomas *et al.*, 2008; Denlinger and Lee, 2009). Among the terrestrial vertebrates of the Arctic, the Arctic fox possess the most effective insulating fur of any mammal, despite its relatively small size. They and the polar bear (that relies on a marine food source – seals) are basically the only mammals to overwinter in the High Arctic, with the bears having a very effective hibernation strategy timing with the females giving birth. The majority of grazing mammals, and the large summer breeding bird population, leave the High Arctic and migrate south for the winter months.

Among the invertebrates, plants (and microbes) that contribute the vast majority of species biodiversity north and south, migration is not an option, and resistance adaptations to cold, desiccation and radiation stress are well developed. Two general strategies are apparent in the survival of cold stress: freeze avoidance, whereby the organism accumulates chemical antifreezes that maintain its body contents in the liquid state (typically allowing survival to −25 to −35°C in polar invertebrates), and freeze tolerance, where ice nucleating agents are used to seed ice formation in carefully controlled (extracellular) parts of the body, drawing water from the cells and thereby lowering their own freezing point. Some Antarctic nematodes are currently the only invertebrates thought to be able to survive intracellular freezing, although several are known to survive to at least −60°C when frozen in water, and thus survival of intracellular freezing may be more widespread than thought.

As noted, in many respects desiccation is a more important threat to organism survival than cold in polar terrestrial habitats. Desiccation tolerance or resistance is accordingly well developed across invertebrate, plant and microbial groups. Amongst the invertebrates, anhydrobiosis (the ability to survive the loss of virtually all body water content) is a feature of groups such as nematodes and tardigrades – whilst in this state some members of these groups can survive a range of biologically unrealistic stresses such as immersion in liquid nitrogen (even in liquid helium in the case of one tardigrade), and (again for tardigrades) exposure to the levels of vacuum and radiation exposure experienced in space. Many lower plants (mosses), lichens, algae, fungi and other microbes can also tolerate equivalent levels of dehydration and demonstrate poikilohydry.

5.4.3 Life cycles

The short active season presents a particular challenge for completion of the life cycle, especially for obligate annual life cycles. Biennials are better represented, while most plants are perennials and most invertebrates also have multi-year development. Roughly half of Arctic flowering plants, and virtually all Subantarctic and Antarctic ones, are wind-pollinated. Bryophytes also rely entirely on aerial or water transfer mechanisms for fertilization in those species requiring it. The current lack of pollinating insects and insect-pollinated plants in the Antarctic highlights a vulnerability of these ecosystems in current scenarios of global warming combined with sometimes intensive human contact (Frenot *et al.*, 2005). At present, a large proportion of established non-indigenous plants on the Subantarctic islands are of persistent rather than invasive status, in part probably because they require pollination by insects for successful seed set. Should anthropogenic assistance result in the introduction of suitable pollinators to these locations, as may have happened recently on South Georgia (Convey *et al.*, 2010), there is likely to be a synergy between non-indigenous plant and invertebrate species, and a step-change in the rate of spread of the plants, with associated impacts on the native ecosystem structure and function.

5.5 Colonization of the Polar Regions by Terrestrial Biota

It is not unreasonable to assume that many of the terrestrial biota encountered in the polar regions are relatively recent colonists of most of the specific locations included in their contemporary distributions. As recently as the Last Glacial Maximum (LGM) 20,000 years ago, as well as during previous Pliocene and Miocene glaciations, much currently exposed land was covered by more extensive and thicker ice sheets. In the northern hemisphere, it is well documented that these extended southwards covering much of the northern states of the USA, most land north of the Alps in Europe and far into the taiga in Asia, with glacial centres also radiating from other major mountain ranges in Europe and Asia. Today Antarctica is almost 99.7% ice covered but, even here, analogous ice expansion and thickening occurred, extending offshore to the continental shelf drop-off. On subsequent ice retreat, the large geographical differences between the Antarctic and Arctic presented very different challenges to biota recolonizing these regions. In the Arctic, land continuity through the large northern continents easily permitted small scale and stepwise movements both southwards as ice sheets expanded and northwards as habitat became available again. In contrast, the large expanse of the Southern Ocean, with its isolating ocean currents and atmospheric circulation (Barnes *et al.*, 2006) provides few if any such stepping stones to the other southern continents or refuges en route, apparently leading to a greater likelihood of extinction during glacial expansion phases, and considerably greater challenges to recolonization or new colonization during retreat phases.

Understandably this history has led to a widely held view that the terrestrial biota of the Antarctic continent in particular must simply have been exterminated and that the current biota is therefore largely (with the exception of certain taxa associated with higher nunataks; e.g. Marshall and Pugh, 1996) the result of post-Pleistocene recolonization. However, recent molecular phylogeographic studies in combination with classical biogeographic analyses have led to a rethinking of this view, and it is now clear that many if not most elements of the contemporary terrestrial fauna of Antarctica have had a long-term presence on the continent over multi-million-year timescales at least (reviewed by Convey and Stevens, 2007; Convey *et al.*, 2008; Pugh and Convey, 2008; Vyverman *et al.*, 2010). With few exceptions (e.g. Maslen and Convey, 2006) the precise locations of glacial refugia cannot be tightly hypothesized, but it is clear that incorporating their presence must challenge more accurate glaciological reconstructions and models of the continent's history. Similarly, even with the north–south connectivity typical of Arctic latitudes, some biogeographical features are difficult to explain without recognition of a refugial element (Halliday, 2002; Jensen and Christensen, 2003).

Notwithstanding the presence of refugia, it remains clear that the majority of currently exposed land especially in coastal regions of Antarctica would have been ice-covered at the LGM, and that new land continues to be exposed to the current day, in the Antarctic Peninsula and Subantarctic regions being accelerated by recent trends of anthropogenic regional warming (Cook *et al.*, 2005, 2010). That such areas of land are rapidly colonized by terrestrial biota representative of the wider region in which they are located indicates that effective dispersal and colonization are possible. The colonization process has been thought of in three elements in the polar regions (Hughes *et al.*, 2006): (i) transfer, requiring survival of the various stresses faced by propagules, and in effect providing a biodiversity filter accepting only those biota with the necessary features; (ii) establishment, in which propagules are stochastically deposited at a suitable location and time of year, again with appropriate capacities for the environment at this sink location; and (iii) the ability to reach a stage of population development that permits establishment in the long term. Four general transfer mechanisms are often hypothesized to facilitate the arrival of colonists in the Arctic and Antarctic, although there are very few explicit demonstrations of specific colonization events: transport in air or water currents (Vincent, 1988; Marshall, 1996; Coulson *et al.*, 2002), zoochory (i.e. in association with other biota) (Schlichting *et al.*, 1978; Pugh, 1997), in water associated with natural or anthropogenic debris (Barnes and Fraser, 2003) and in association with human activities (Frenot *et al.*, 2005).

Many of the terrestrial biota characteristic of the Antarctic (lichens, mosses, tardigrades, nematodes, various microbiota) have small propagules or resistant life stages, which can survive the stresses faced in the air column. Unusual atmospheric events such as storms carry greater concentrations of biological particles, and more rapidly across greater distances, than normal (Marshall, 1996). Meteorological 'back-track' analyses can trace the origin and transfer timescale of these air masses and their associated biota (Marshall, 1996; Greenslade *et al.*, 1999; Convey, 2005). There is also strong circumstantial evidence consistent with colonization following wind dispersal (Muñoz *et al.*, 2004; Chown and Convey, 2007). As discussed above, the presence of exotic biota associated with small and extremely isolated sites of geothermal activity also demonstrates the occurrence of long-distance transfer of their propagules.

5.6 Arctic and Antarctic Comparisons

The most striking differences in terrestrial biology between the planet's two polar regions are driven by the fundamental differences in geography of the two hemispheres – in simple terms the Arctic is a relatively small polar ocean surrounded by the northern elements of the large continental landmasses of Europe, Asia and North America, while the Antarctic is a large and high polar continent separated from all other southern continents by a vast expanse of deep, cold ocean. This leads to differences in climate between the two regions such that similar biological processes and communities are found on land at latitudes roughly 10–15° higher in the Arctic than in the Antarctic, and thus that there is no Arctic equivalent to much of the continental Antarctic.

In the Antarctic and the more extreme elements of the Arctic, terrestrial communities appear similar, being dominated by lower plants and photosynthetic microbial groups, and lower groups of micro-arthropods (springtails, mites) and microfauna (nematodes, tardigrades, rotifers). These share well-developed abilities to tolerate the various extreme environmental stresses presented, as well as possessing considerable life history flexibility (Convey, 1996b). Among the bryophytes, lichens and microbes of both poles there are a number of cosmopolitan or bipolar species (Øvstedal and Smith, 2001; Pearce *et al.*, 2007; Ochyra *et al.*, 2008) and, although in most cases yet to be examined in molecular studies, this supports long distance connectivity between these two remote regions. Endolithic communities can grow under the most severe terrestrial conditions that either pole can inflict, giving one extreme of the spectrum of biological habitats on the planet. However, striking differences are apparent between Arctic and Antarctic terrestrial communities. In terms of overall diversity, there are ~900 species of flowering plants in the Arctic and only two on the Antarctic continent, and in contrast with the Arctic's 48 species of land mammals, Antarctica has none. It should be remembered that biodiversity data remain incomplete and in many cases unreliable for many of the smaller and non-charismatic groups of invertebrates, lower plants and microbes, and that recent advances in molecular taxonomic techniques are only now starting to be used in these groups and regions. It is clear that the physical isolation of Antarctica is an important factor underlying these differences.

A further contrast between the two regions lies in the relatively low number of non-indigenous species recorded from Arctic locations such as Svalbard in comparison with the Subantarctic (compare Rønning (1996) and Coulson (2007) with Frenot *et al.* (2005)) – many Antarctic non-indigenous species are in fact cosmopolitan or boreal weeds, possibly indicating that they or their equivalents have already reached northern polar latitudes by natural means, in contrast with the south where human assistance has facilitated overcoming the isolation barrier.

Very strong seasonality is a feature of high latitudes in both hemispheres. Biota are adapted to take advantage of short favourable periods for growth, while in the case of invertebrates possessing increased life history flexibility permitting repeated overwintering in different life stages. The Antarctic generally faces a continuously greater risk of exposure to desiccating or freezing conditions than does the Arctic, underlying high investment in protective strategies even in the summer months, and the loss of certain life history characteristics such as true diapause that are widespread in Arctic invertebrates (Convey, 2010a). Within this overall description, the Subantarctic is distinct, with much reduced thermal seasonality, longer and chronically cool and wet growing seasons with often continuous biological activity.

The Arctic terrestrial environment provides an important feeding and breeding resource for terrestrial vertebrates originating from lower latitudes, in a manner that has no southern equivalent. The productive wetlands of the Arctic support large populations of migrant animals – for instance, while only 11 bird species remain in the Arctic over winter, around 90 species arrive to take advantage of summer food supplies. A large biomass of migratory mammals (primarily reindeer or caribou) similarly move into mostly Low Arctic latitudes in summer. Most vertebrate movements from Antarctic locations are fundamentally different, involving marine species that simply spend the non-breeding season at sea in the Southern Ocean.

Finally, a fundamental difference between Arctic and Antarctic is that, since retreat of ice following the LGM, the former has been occupied by indigenous populations of humans. From their earliest presence these have had an important impact on the biota and ecosystems surrounding them. Now these peoples and ecosystems face many direct challenges, from climate and other environmental change, through economic development and exploitation, to societal change, and have an uncertain future.

References

ACIA (2004) *Impacts of a Warming Arctic.* The Arctic Climate Impact Assessment, Cambridge University Press, Cambridge, UK, 144 pp.

Adams, B., Bardgett, R.D., Ayres, E., Wall, D.H., Aislabie, J., Bamforth, S., Bargagli, R., Cary, C., Cavacini, P., Connell, L., Convey, P., Fell, J., Frati, F., Hogg, I., Newsham, N., O'Donnell, A., Russell, N., Seppelt, R. and Stevens, M.I. (2006) Diversity and distribution of Victoria Land biota. *Soil Biology and Biochemistry* 38, 3003–3018.

Bargagli, R., Skotnicki, M.L., Marri, L., Pepi, M., Mackenzie, A. and Agnorelli, C. (2004) New record of moss and thermophilic bacteria species and physico-chemical properties of geothermal soils on the northwest slope of Mt Melbourne (Antarctica). *Polar Biology* 27, 423–431.

Barnes, D.K.A. and Fraser, K.P.P. (2003) Rafting by five phyla on man-made flotsam in the Southern Ocean. *Marine Ecology Progress Series* 262, 289–291.

Barnes, D.K.A., Hodgson, D.A., Convey, P., Allen, C. and Clarke, A. (2006) Incursion and excursion of Antarctic biota: past, present and future. *Global Ecology and Biogeography* 15, 121–142.

Barr, S. (ed.) (1995) *Franz Josef Land*. Norsk Polarinstitutt, Tromsø, Norway, 173 pp.

Beyer, L. and Bölter, M. (eds) (2002) *Geoecology of Antarctic Ice-free Coastal Landscapes*. Ecological Studies Vol. 154, Springer, Berlin, Germany, 451 pp.

Björck, S., Malmer, N., Hjort, C., Sandgren, P., Ingolfsson, O., Wallen, B., Smith, R.L. and Jonsson, B. (1991) Stratigraphic and paleoclimatic studies of a 5500-year-old moss bank on Elephant Island, Antarctica. *Arctic and Alpine Research* 23, 361–374.

Bliss, L.C. (1975) Devon Island, Canada. In: Rosswall, T. and Heal, O.W. (eds) *Structure and Function of Tundra Ecosystems*. Swedish Natural Science Research Council, Stockholm, Sweden, pp. 17–60.

Block, W. (1984) Terrestrial microbiology, invertebrates and ecosystems. In: Laws, R.M. (ed.) *Antarctic Ecology*. Academic Press, London, UK, pp. 163–236.

Block, W. (1990) Cold tolerance of insects and other arthropods. *Philosophical Transactions of the Royal Society of London, Series B* 326, 613–633.

Block, W. (1996) Cold or drought – the lesser of two evils for terrestrial arthropods? *European Journal of Entomology* 93, 325–339.

Broady, P.A. (1993) Soils heated by volcanism. In: Friedmann, E.I. (ed.) *Antarctic Microbiology*. Wiley-Liss, New York, pp. 413–432.

Callaghan, T.V. and Jonasson, S. (1995) Arctic terrestrial ecosystems and environmental change. *Philosophical Transactions of the Royal Society of London, Series A: Mathematical, Physical and Engineering Sciences* 352, 259–276.

Callaghan, T.V., Björn, L.O., Chernov, Y., Chapin, T., Christensen, T., Huntley, B., Ims, R.A., Johansson, M., Jolly, D., Jonasson, S., Matveyeva, N., Panikov, N., Oechel, W., Shaver, G., Schaphoff, S., Sitch, S. and Zöckler, C. (2004) Climate change and UV-B impacts on Arctic tundra and polar desert ecosystems. Key findings and extended summaries. *Ambio* 33, 386–392.

Chown, S.L. and Convey, P. (2007) Spatial and temporal variability across life's hierarchies in the terrestrial Antarctic. *Philosophical Transactions of the Royal Society of London, Series B: Biological Sciences* 362, 2307–2331.

Cockell, C.S., Rettberg, P., Horneck, G., Wynn-Williams, D.D., Scherer, K. and Gugg-Helminger, A. (2002) Influence of ice and snow covers on the UV exposure of terrestrial microbial communities: dosimetric studies. *Journal of Photochemistry and Photobiology B: Biology* 68, 23–32.

Convey, P. (1996a) Overwintering strategies of terrestrial invertebrates from Antarctica – the significance of flexibility in extremely seasonal environments. *European Journal of Entomology* 93, 489–505.

Convey, P. (1996b) The influence of environmental characteristics on life history attributes of Antarctic terrestrial biota. *Biological Reviews* 71, 191–225.

Convey, P. (2005) Recent lepidopteran records from sub-Antarctic South Georgia. *Polar Biology* 28, 108–110.

Convey, P. (2006) Antarctic climate change and its influences on terrestrial ecosystems. In: Bergstrom, D.M., Convey, P. and Huiskes, A.H.L. (eds) *Trends in Antarctic Terrestrial and Limnetic Ecosystems: Antarctica as a Global Indicator*. Springer, Dordrecht, The Netherlands, pp. 253–272.

Convey, P. (2007a) Antarctic ecosystems. In: Levin, S.A. (ed.) *Encyclopedia of Biodiversity*, 2nd (online) edn. Elsevier, San Diego, doi:10.1016/B0-12-226865-2/00014-6.

Convey, P. (2007b) Influences on and origins of terrestrial biodiversity of the sub-Antarctic islands. *Papers and Proceedings of the Royal Society of Tasmania* 141, 83–93.

Convey, P. (2010a) Life history adaptations to polar and alpine environments. In: Denlinger, D.L. and Lee, R.E., Jr (eds) *Low Temperature Biology of Insects*. Cambridge University Press, Cambridge, pp. 297–321.

Convey, P. (2010b) Terrestrial biodiversity in Antarctica – recent advances and future challenges. *Polar Science* 4, 135–147.

Convey, P. and McInnes, S.J. (2005) Exceptional, tardigrade dominated, ecosystems from Ellsworth Land, Antarctica. *Ecology* 86, 519–527.

Convey, P. and Stevens, M.I. (2007) Antarctic biodiversity. *Science* 317, 1877–1878.

Convey, P., Greenslade, P. and Pugh, P.J.A. (2000a) Terrestrial fauna of the South Sandwich Islands. *Journal of Natural History* 34, 597–609.

Convey, P., Smith, R.I.L., Hodgson, D.A. and Peat, H.J. (2000b) The flora of the South Sandwich Islands, with particular reference to the influence of geothermal heating. *Journal of Biogeography* 27, 1279–1295.

Convey, P., Gibson, J., Hillenbrand, C.-D., Hodgson, D.A., Pugh, P.J.A., Smellie, J.L. and Stevens, M.I. (2008) Antarctic terrestrial life – challenging the history of the frozen continent? *Biological Reviews* 83, 103–117.

Convey, P., Bindschadler, R.A., di Prisco, G., Fahrbach, E., Gutt, J., Hodgson, D.A., Mayewski, P., Summerhayes, C.P. and Turner, J. (2009a) Antarctic climate change and the environment. *Antarctic Science* 21, 541–563.

Convey, P., Stevens, M.I., Hodgson, D.A., Smellie, J.L., Hillenbrand, C.-D., Barnes, D.K.A., Clarke, A., Pugh, P.J.A., Linse, K. and Cary, S.C. (2009b) Exploring biological constraints on the glacial history of Antarctica. *Quaternary Science Reviews* 28, 3035–3048.

Convey, P., Key, R.S. and Key, R.J.D. (2010) The establishment of a new ecological guild of pollinating insects on sub-Antarctic South Georgia. *Antarctic Science* doi: 10.1017/S095410201000057X.

Cook, A.J., Fox, A.J., Vaughan, D.G. and Ferrigno, J.G. (2005) Retreating glacier fronts on the Antarctic Peninsula over the past half-century. *Science* 308, 541–544.

Cook, A.J., Poncet, S., Cooper, A.P.R., Herbert, D.J. and Christie, D. (2010) Glacier retreat on South Georgia and implications for the spread of rats. *Antarctic Science* doi:10.1017/S0954102010000064.

Coulson, S.J. (2007) The terrestrial and freshwater invertebrate fauna of the High Arctic archipelago of Svalbard. *Zootaxa* 1448, 41–58.

Coulson, S.J. and Resfeth, D. (2004) The terrestrial and freshwater fauna of Svalbard (and Jan Mayen). In: Prestrud, P., Strøm, H. and Goldman, H.V. (eds) *A Catalogue of the Terrestrial and Marine Animals of Svalbard*. Norwegian Polar Institute, Trømso, Norway, pp. 57–122.

Coulson, S.J., Hodkinson, I.D., Webb, N.R. and Harrison, J.A. (2002) Survival of terrestrial soil-dwelling arthropods on and in seawater: implications for trans-oceanic dispersal. *Functional Ecology* 16, 353–356.

Cowan, D.A., Russell, N.J., Mamais, A. and Sheppard, D.M. (2002) Antarctic Dry Valley mineral soils contain unexpectedly high levels of microbial biomass. *Extremophiles* 6, 431–436.

Crawford, R.M.M. (1995) Plant survival in the High Arctic. *Biologist* 42, 101–105.

Danks, H.V. (1999) Life cycles in polar arthropods – flexible or programmed? *European Journal of Entomology* 96, 83–102.

Davis, R.C. (1980) Peat respiration and decomposition in Antarctic terrestrial moss communities. *Biological Journal of the Linnean Society* 14, 39–49.

Davis, R.C. (1981) Structure and function of two Antarctic terrestrial moss communities. *Ecological Monographs* 51, 125–143.

De Wever, A., Leliaert, F., Verleyen, E., Vanormelingen, P., Van der Gucht, K., Hodgson, D.A., Sabbe, K. and Vyverman, W. (2009) Hidden levels of phylodiversity in Antarctic green algae: further evidence for the existence of glacial refugia. *Proceedings of the Royal Society* 276, 3591–3599.

Denlinger, D.L. and Lee, R.E. (eds) (2009) *Low Temperature Biology of Insects*. Cambridge University Press, Cambridge, 404 pp.

Elvebakk, A. and Hertel, H. (1996) Part 6. Lichens. In: Elvebakk, A. and Prestrud, P. (eds) *A Catalogue of Svalbard Plants, Fungi and Cyanobacteria*. Norwegian Polar Institute Skrifter 198. Norwegian Polar Institute, Tromsø, pp. 271–359.

Fenton, J.H.C. (1980) The rate of peat accumulation in Antarctic moss banks. *Journal of Ecology* 68, 211–228.

Finlay, B.J. (2002) Global dispersal of free-living microbial eukaryote species. *Science* 296, 1061–1063.

Freckman, D.W. and Virginia, R.A. (1997) Low-diversity Antarctic soil nematode communities: distribution and response to disturbance. *Ecology* 78, 363–369.

Freckman, D.W. and Virginia, R.A. (1998) Soil biodiversity and community structure in the McMurdo Dry Valleys, Antarctica. *Antarctic Research Series* 72, 323–336.

Frenot, Y., Chown, S.L., Whinam, J., Selkirk, P., Convey, P., Skotnicki, M. and Bergstrom, D. (2005) Biological invasions in the Antarctic: extent, impacts and implications. *Biological Reviews* 80, 45–72.

Friedmann, E.I. (1982) Endolithic microorganisms in the Antarctic cold desert. *Science* 215, 1045–1053.

Frisvoll, A.A. and Elvebakk, A. (1996) Part 2. Bryophytes. In: Elvebakk, A. and Prestrud, P. (eds) *A Catalogue of Svalbard Plants, Fungi and Cyanobacteria*. Norwegian Polar Institute Skrifter 198. Norwegian Polar Institute, Tromsø, Norway, pp. 57–172.

Greenslade, P., Farrow, R.A. and Smith, J.M.B. (1999) Long distance migration of insects to a subantarctic island. *Journal of Biogeography* 26, 1161–1167.

Halliday, G. (2002) The British flora in the Arctic. *Watsonia* 24, 133–144.

Heide-Jørgensen, H.S. and Kristensen, R.M. (1999) Puilassoq, the warmest homothermal spring of Disko Island. *Berichte zur Polarforschung* 330, 32–43.

Hennion, F., Huiskes, A., Robinson, S. and Convey, P. (2006) Physiological traits of organisms in a changing environment. In: Bergstrom, D.M., Convey, P. and Huiskes, A.H.L. (eds) *Trends in Antarctic Terrestrial and Limnetic Ecosystems: Antarctica as a Global Indicator*. Springer, Dordrecht, The Netherlands, pp. 129–159.

Hodkinson, I.D. (2005) Adaptations of invertebrates to terrestrial Arctic environments. *Det Konelige Norske Videnskabers Selskab, Skrifter*, 45 pp.

Hodkinson, I.D. and Wookey, P.A. (1999) Functional ecology of soil organisms in tundra ecosystems: towards the future. *Applied Soil Ecology* 11, 111–126.

Hodkinson, I.D., Webb, N.R., Bale, J.S. and Block, W. (1999) Hydrology, water availability and tundra ecosystem function in a changing climate: the need for a closer integration of ideas? *Global Change Biology* 5, 359–369.

Hogg, I.D., Cary, S.C., Convey, P., Newsham, K., O'Donnell, T., Adams, B.J., Aislabie, J., Frati, F.F., Stevens, M.I. and Wall, D.H. (2006) Biotic interactions in Antarctic terrestrial ecosystems: are they a factor? *Soil Biology and Biochemistry* 38, 3035–3040.

Hughes, K.A. and Lawley, B. (2003) A novel Antarctic microbial endolithic community within gypsum crusts. *Environmental Microbiology* 5, 555–565.

Hughes, K., Ott, S., Bölter, M. and Convey, P. (2006) Colonisation processes. In: Bergstrom, D.M., Convey, P. and Huiskes, A.H.L. (eds) *Trends in Antarctic Terrestrial and Limnetic Ecosystems: Antarctica as a Global Indicator*. Springer, Dordrecht, the Netherlands, pp. 35–54.

Jensen, D.B. and Christensen, K.D. (eds) (2003) The biodiversity of Greenland – a country study. *Technical Report 55*. Pinngortitaleriffik, Grønlands Naturinstitut, Nuuk, Greenland, 165 pp.

Kennedy, A.D. (1993) Water as a limiting factor in the Antarctic terrestrial environment: a biogeographical synthesis. *Arctic and Alpine Research* 25, 308–315.

Kovacs, K.M. (ed.) (2005) *Birds and Mammals of Svalbard*. Norsk Polarinstitutt, Tromsø, Norway, 203 pp.

Longton, R.E. (1988) *The Biology of Polar Bryophytes and Lichens*. Cambridge University Press, Cambridge, UK, 399 pp.

Marshall, D.J. and Pugh, P.J.A. (1996) Origin of the inland Acari of continental Antarctica, with particular reference to Dronning Maud Land. *Zoological Journal of the Linnean Society* 118, 101–118.

Marshall, W.A. (1996) Biological particles over Antarctica. *Nature* 383, 680.

Maslen, N.R. and Convey, P. (2006) Nematode diversity and distribution in the southern maritime Antarctic – clues to history? *Soil Biology and Biochemistry* 38, 3141–3151.

Meurk, C.D., Foggo, M.N. and Wilson, J.B. (1994) The vegetation of subantarctic Campbell Island. *New Zealand Journal of Ecology* 18, 123–168.

Mitchell, A.D., Meurk, C.D. and Wagstaff, S.J. (1999) Evolution of *Stilbocarpa*, a megaherb from New Zealand's sub-antarctic islands. *New Zealand Journal of Botany* 37, 205–211.

Muñoz, J., Felicísimo, A.M., Cabezas, F., Burgaz, A.R. and Martínez, I. (2004) Wind as a long-distance dispersal vehicle in the Southern Hemisphere. *Science* 304, 1144–1147.

Norton, R.A. (1994) Evolutionary aspects of oribatid mite life histories and consequences for the origin of the Astigmata. In: Houck, M. (ed.) *Ecological and Evolutionary Analyses of Life-History Patterns*. Chapman & Hall, New York, pp. 99–135.

Ochyra, R., Smith, R.I.L. and Bednarek-Ochyra, H. (2008) *The Illustrated Moss Flora of Antarctica*. Cambridge University Press, Cambridge, UK, 704 pp.

Øvstedal, D.O. and Smith, R.I.L. (2001) *Lichens of Antarctica and South Georgia. A Guide to Their Identification and Ecology*. Cambridge University Press, Cambridge, UK, 424 pp.

Panikov, N.S. (1995) *Microbial Growth Kinetics*. Chapman and Hall, London, UK, 392 pp.

Pearce, D.A., Cockell, C.S., Lindström, E.S. and Tranvik, L.J. (2007) First evidence for a bipolar distribution of dominant freshwater lake bacterioplankton. *Antarctic Science* 19, 245–252.

Peck, L.S., Convey, P. and Barnes, D.K.A. (2006) Environmental constraints on life histories in Antarctic ecosystems: tempos, timings and predictability. *Biological Reviews* 81, 75–109.

Pugh, P.J.A. (1997) Acarine colonization of Antarctica and the islands of the Southern Ocean: the role of zoohoria. *Polar Record* 33, 113–122.

Pugh, P.J.A. and Convey, P. (2008) 'Surviving out in the cold': Antarctic endemic invertebrates and their refugia. *Journal of Biogeography* 35, 2176–2186.

Rønning, O.I. (1996) *The Flora of Svalbard*. Norsk Polarinstitut, Oslo, Norway, 184 pp.

Rozema, J. (ed.) (1999) *Stratospheric Ozone Depletion, the Effects of Enhanced UV-B Radiation on Terrestrial Ecosystems*. Backhuys, Leiden, Germany, 355 pp.

Ryan, P.G. and Watkins, B.P. (1989) The influence of physical factors and ornithogenic products on plant and arthropod abundance at an inland nunatak group in Antarctica. *Polar Biology* 10, 151–160.

Sage, B. (1986) *The Arctic and its Wildlife*. Croom Helm, Beckenham, London, UK, 187 pp.

Schlichting, H.E., Speziale, B.J. and Zink, R.M. (1978) Dispersal of algae and protozoa by Antarctic flying birds. *Antarctic Journal of the United States of America* 13, 147–149.

Schmidt-Nielsen, K. (1997) *Animal Physiology: Adaptation and Environment*, 5th edn. Cambridge University Press, Cambridge, UK, 617 pp.

Smith, R.I.L. (1984) Terrestrial plant biology of the sub-Antarctic and Antarctic. In: Laws, R.M. (ed.) *Antarctic Ecology*. Academic Press, London, pp. 61–162.

Smith, R.I.L. (1990) Signy Island as a paradigm of biological and environmental change in Antarctic terrestrial ecosystems. In: Kerry, K.R. and Hempel, G. (eds) *Antarctic Ecosystems, Ecological Change and Conservation*. Springer-Verlag, Berlin, pp. 32–50.

Smith, R.I.L. (2005) The thermophilic bryoflora of Deception Island: unique plant communities as a criterion for designating an Antarctic Specially Protected Area. *Antarctic Science* 17, 17–27.

Sømme, L. (1995) *Invertebrates in Hot and Cold Arid Environments*. Springer-Verlag, Berlin, Germany, 275 pp.

Tedrow, J.C.F. (1977) *Soils of the Polar Landscapes*. Rutgers University Press, New Brunswick, 638 pp.

Thomas, D.N., Fogg, G., Convey, P., Fritsen, C., Gilli, J.-M., Gradinger, R., Laybourne-Parry, J., Reid, K. and Walton, D.W.H. (2008) *The Biology of Polar Habitats*. Oxford University Press, Oxford, UK, 394 pp.

Turner, J., Bindschadler, R., Convey, P., di Prisco, G., Fahrbach, E., Gutt, J., Hodgson, D., Mayewski, P. and Summerhayes, C. (eds) (2009) *Antarctic Climate Change and the Environment*. Scientific Committee on Antarctic Research, Cambridge, UK, 526 pp.

Vincent, W.F. (1988) *Microbial Ecosystems of Antarctica*. Cambridge University Press, Cambridge, UK, 320 pp.

Vyverman, W., Verleyen, E., Wilmotte, A., Hodgson, D.A., Willems, A., Peeters, K., de Vijver, B.A., De Wever, A., Leliaert, F. and Sabbe, K. (2010) Evidence for wide-spread endemism among Antarctic micro-organisms. *Polar Science* doi:10.1016/j.polar.2010.03.006.

Walther, G.-R., Post, E., Convey, P., Menel, A., Parmesan, C., Beebee, T.J.C., Fromentin, J.-M., Hoegh-Guldberg, O. and Bairlein, F. (2002) Ecological responses to recent climate change. *Nature* 416, 389–395.

6 High Altitude and Latitude Lakes

Johanna Laybourn-Parry[1] and Elanor M. Bell[2]
[1]*Bristol Glaciology Centre, University of Bristol, UK;*
[2]*Scottish Institute for Marine Science, Dunbeg, UK*

6.1 Introduction

Alpine lakes lie above the treeline and have catchments with sparse vegetation. Thus the degree of allochthonous input is relatively small. Subalpine lakes, which can be at high altitude, lie in more vegetated catchments and have greater potential for allochthonous carbon and nutrient inputs. Lakes are fed by snow and glacial melt. Spring snow melt is a major hydrological driving force in high altitude lakes, and plays an important role in controlling phytoplankton dynamics and interspecific resource-based competition (McKnight *et al.*, 1990). Depending on altitude, montane lakes are ice covered for a significant portion of the year (5–9 months), but unlike polar lakes they receive higher annual levels of photosynthetically active radiation (PAR), which increases with decreasing latitude. Moreover, they are subject to significant UV radiation penetration. UV increases with elevation up to 11% per 1000 m for 320 nm wavelengths and 24% per 1000 m for 3000 nm (Rose *et al.*, 2009). Inevitably the impact of UV radiation on biological and biogeochemical processes has attracted considerable attention in alpine lakes.

The Arctic has a vast array of lake types. Permafrost thaw lakes (thermokarst lakes and ponds) are the most abundant lake type and often form a mosaic of water bodies. A good example is the Mackenzie River delta that has around 45,000 such lakes and ponds (Vincent *et al.*, 2008). One of the major differences between the Arctic and Antarctic are allochthonous (external) inputs. Unless Antarctic systems are adjacent to sea wallows or penguin rookeries they receive no nutrient inputs from their catchments as the Antarctic is virtually devoid of vegetation. In contrast, many Arctic lakes are in vegetated catchments and receive significant allochthonous inputs of carbon and nutrients. The Antarctic has few streams or rivers, the lakes are fed directly by glacial or snow melt, while in contrast the Arctic has some of the largest river catchments in the world (e.g. Pechora, Ob, Yenisei, Lena, Kolyma and Mackenzie) that are fed by snow melt, as well as many wetlands and lower order flowing water systems (Vincent *et al.*, 2008).

The vast majority of the Antarctic continent is covered by an ice cap. Despite this, the ice-free areas around the coast and the inland McMurdo Dry Valleys contain an amazing diversity of lakes that range from freshwater to hypersaline (seven times the conductivity of seawater; Fig. 6.1). The most southerly lakes, such as those of the Dry Valleys and lakes abutting glaciers, have perennial ice-covers (Fig. 6.2), while many of the coastal lakes lose part or all of their ice covers for a number of weeks each summer.

Fig. 6.1. Deep Lake, Vestfold Hills, Antarctica. The lake is seven times saltier than seawater and as a result has never been known to completely freeze. © Elanor M. Bell.

Fig. 6.2. The west lobe of Lake Bonney, McMurdo Dry Valleys, Antarctica, has a perennial ice cover 3–4 m thick. Lake Bonney field camp is visible on the far shore of the lake. © Elanor M. Bell.

Antarctica also has many tidal epishelf lakes, freshwater systems that overlie colder denser seawater between ice shelves and the continent, as well as subglacial lakes lying many kilometres beneath the ice cap (see Pearce, Chapter 7, this volume). Both saline and epishelf lakes are less common in the Arctic. Antarctic lakes experience continuous low temperatures and low annual levels of PAR, and rank among the most extreme lacustrine ecosystems on the planet.

All of these extreme lacustrine ecosystems are poorly researched compared with temperate and low altitude lakes, for obvious logistic reasons, but all offer us a fascinating insight into aquatic life at extremes.

6.2 Trophic Structure

Antarctic lakes are characterized by simple truncated food webs dominated by microbial plankton with few or no metazoans. In contrast, Arctic lakes can be regarded as an extension of temperate limnology (Fig. 6.3). In the low Arctic the biological, physical and chemical processes and the species makeup is very similar to that seen in oligotrophic temperate lakes. As one moves north the systems become more oligotrophic and entire trophic levels drop out until the lakes resemble those of the Antarctic (Hobbie and Laybourn-Parry, 2008). Maritime Antarctic lakes such as those of Signy Island possess a distinct zooplankton and illustrate the same pattern. Similarly, alpine and subalpine lakes have trophic structures similar to temperate lakes, but as oligotrophic/ultra-oligotrophic systems they have low productivity. Alpine lakes typically lack fish but some have been subject to fish introductions to provide recreational fisheries. Such introductions have allowed the impact of fish predation on community structure to be assessed in North American

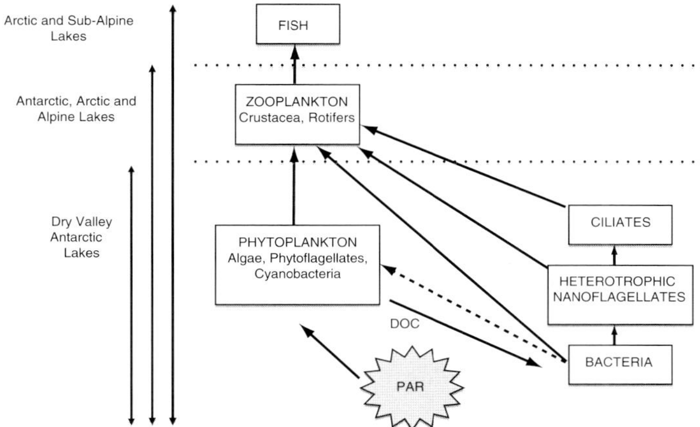

Fig. 6.3. Trophic structure of the plankton in polar and alpine lakes. Arrows on left indicate typical number of trophic levels. Dotted arrow indicates mixotrophy by phytoflagellates on bacteria. DOC = dissolved organic carbon exuded by the phytoplankton. © Johanna Laybourn-Parry.

lakes (Knapp *et al.*, 2001). Predictably the larger crustacean zooplankton (>1 mm) declined, for example *Hesperodiaptomus shoshone* and *Daphnia middendorffiana*, while smaller species (<1 mm) were unaffected. *Daphnia middendorffiana* distribution is also significantly controlled by predation pressure from fish in Arctic lakes, although its growth and survival indicate that it would survive well in fishless lakes (Yurista and O'Brien, 2001).

The simplicity of polar lake food webs, in particular, offer scientists a fantastic opportunity to study community and food-web dynamics *in situ*. For this reason, many alpine and polar lakes have been used intensively as living laboratories and some form the focus of Long-Term Ecological Research programmes (LTERs), e.g. Lake Bonney, Antarctica (Fig. 6.2) and Toolik Lake, Arctic (Section 6.3.2).

6.3 Seasonal Patterns of Primary Productivity

6.3.1 Alpine lake primary production

Alpine and subalpine lakes have clear seasonal patterns of productivity worldwide. All except the shallow lakes are dimictic (having two phases of thermal stratification each year). Lakes in the Rocky Mountain National Park had a bloom of the pennate diatom *Asterionella formosa* at spring snow melt, and an autumn bloom of the cyanobacterium *Oscillatoria limnetica*, with an intervening phase of chlorophytic algal abundance. The highest productivity occurred in spring (McKnight *et al.*, 1990) (Table 6.1). A study of nine remote alpine lakes in Europe revealed two phases of phytoplankton productivity in lakes with longer summer stratification. Once the ice forms in autumn it is clear and allows good PAR transmission to the water column resulting in episodic high productivity under developing ice-cover (Catalan *et al.*, 2002). However, once snow falls productivity is curtailed as snow cover strongly attenuates light. Alpine lakes can have snow covers of a metre or more. Thus, in spring it is not until snow melt occurs that primary production can commence. The phytoplankton communities of alpine lakes are characterized by diatoms, chrysophytes, dinoflagellates and cryptophytes. Diatoms are dominant in many lakes (McKnight *et al.*, 1990; Sommaruga, 2001; Saros *et al.*, 2005). In all of these systems chlorophyll *a* levels are low (Table 6.1).

Table 6.1. Chlorophyll *a* concentrations and rates of photosynthesis in Arctic, Antarctic and high altitude lakes. Note, alpine lakes are those located above the treeline. ND = No data.

Lake	Chlorophyll *a* μg l^{-1}	Primary production μg C l^{-1} day^{-1}	Source
ALPINE/SUBALPINE			
The Loch (3048 m) N. America	1.2–2.3	22.8–1464 June–Sept	McKnight *et al.*, 1990
7 lakes in N. America (2460–3162 m)	8.0–16.0	ND	Saros *et al.*, 2005
57 lakes in Austrian Alps (1970–2770 m)	0.2–10.6	ND	Sommaruga *et al.*, 1999
ARCTIC			
Char Lake, High Arctic, Canada	0.46–0.78 summer	27.6–115.2 summer	Markager *et al.*, 1999
Meretta Lake, High Arctic, Canada	0.96–1.11 summer	3.36–15.6 summer	Markager *et al.*, 1999
Lake Sophia, High Arctic, Canada	0.31–1.24 summer	12.0–79.2 summer	Markager *et al.*, 1999
Barrow Ponds, Alaska	~2.2 summer	200 summer	Alexander *et al.*, 1980
Toolik Lake, Alaska	1.8 summer	Approx max 40	O'Brien *et al.*, 1992
Lake at Ossian, Sarsfjella, Svalbard	0.2–0.3 summer	3.1–7.9 summer	Laybourn-Parry and Marshall, 2003
ANTARCTIC			
Crooked Lake, Vestfold Hills	0.5–0.05 over year	0–38.4 over year	Henshaw and Laybourn-Parry, 2002
Ace Lake, Vestfold Hills	0.7–5.74 over year	0–748 over year	Laybourn-Parry *et al.*, 2005
Highway Lake, Vestfold Hills	4.27–12.66 over year	13.44–303.8 over year	Laybourn-Parry *et al.*, 2005
Beaver Lake, MacRobertson Land	0.2–2.7 summer	2.1–6.7 summer	Laybourn-Parry *et al.*, 2006
Lake Bonney, Dry Valleys	0.03–3.8 summer	0–3.2 summer	Lizotte *et al.*, 1996
Lake Fryxell, Dry Valleys	<1.0–8.0 summer	up to 30 summer	Spaulding *et al.*, 1994
Lake Vanda, Dry Valleys	0.05–0.93 summer	0.63–2.1 summer	Vincent, 1981

One feature of alpine lakes that appears global is the development of deep chlorophyll *a* maxima (DCM), subsurface concentrations of chlorophyll-containing organisms. Much attention has been devoted to the factor(s) responsible for DCM formation. Avoidance of damaging UV radiation has been identified as the major factor in the formation of the DCM (Sommaruga and Psenner, 1997; Sommaruga and Augustin, 2006; Rose *et al.*, 2009). However, in some lakes other factors may operate. For example, in a number of lakes in the Beartooth Mountains (Montana/Wyoming) the DCM was strongly correlated with a suite of variables related to nitrogen, while there was no clear correlation with UV radiation levels (Saros *et al.*, 2005). DCM in temperate lakes are often associated with the thermocline, and in meromictic lakes, including those in Antarctica, with the chemocline, where nutrients diffusing upwards from the lower more nutrient-rich waters support primary production in an environment with reduced PAR. In the Beartooth Mountain lakes, the chrysophyte (golden alga) *Dinobryon* dominated the phytoplankton assemblage in June and July (Saros *et al.*, 2005). This species is highly motile and capable of mixotrophy (combining heterotrophic and autotrophic nutritional strategies).

6.3.2 Arctic lake primary production

Like alpine lakes Arctic lakes exhibit seasonal patterns of productivity and are dominated by chrysophytes. In many Arctic locations the seasonal ice is covered by snow that precludes primary production until melt, even when 24 h daylight prevails. In the high Arctic (Svalbard 78°N) snow and ice melt did not occur until late July. In the two study lakes chlorophyll *a* concentrations never exceeded 2 μg l^{-1} and were usually below 1 μg l^{-1} (Laybourn-Parry and Marshall, 2003). In both lakes highest primary production occurred in mid to late July with a further smaller peak in late August (Table 6.1). Toolik Lake (68°N) is the site of a US LTER and is probably one of the most intensively studied Arctic lakes. Data collected over many years indicates that after ice break-out primary production is low and variable, and starts under ice. This is the phase of highest phytoplankton production, followed by low levels for the open water period due to low levels of inorganic nitrogen (O'Brien *et al.*, 1997; Table 6.1). Tundra ponds of the Barrow region in Alaska (71°N) are shallow thermokarst ponds. Their study was part of the International Biological Programme (IBP) conducted in the 1970s. Phytoplankton biomass as chlorophyll *a* is relatively low in these water bodies (Table 6.1). Primary production exhibited a seasonal pattern with a peak in mid to late June/early July followed by a reduction to a low level in late July/ early August, with a further peak in late August that exceeded the earlier peak (Hobbie *et al.*, 1980). In contrast, high Arctic polar desert lakes (73–75°N) are much more extreme and are characterized by low chlorophyll *a* concentrations (typically <1.0 μg l^{-1}) and low levels of primary production (Markager *et al.*, 1999; Table 6.1).

6.3.3 Antarctic lake primary production

Clear seasonal patterns in primary production are apparent in maritime Antarctic systems such as those of Signy Island (60°S). In Heywood Lake there was a clear spring peak of chlorophyll *a* and primary production in October and November that occurred under ice that lacked snow cover. A further smaller peak occurred towards the onset of winter in February and March (Hawes, 1983). In much more oligotrophic Sombre Lake, the peaks were less pronounced, but there was an increase in summer (Hawes, 1983). Both lakes were mixed in summer. The lakes had ice covers with snow on top until the end of September, and this undoubtedly inhibited any earlier increase in photosynthesis.

Conditions on the Antarctic continent are much more severe. There are limited annual data sets for Antarctic lakes. The most complete come from the Vestfold Hills in eastern Antarctica (68°S). This coastal oasis has a wide spectrum of lakes ranging from ultra-oligotrophic freshwater lakes to hypersaline lakes. The phytoplankton is dominated by phytoflagellates, but in the brackish and saline lakes cryptophyte algae are a conspicuous component, as are dinoflagellates. Diatoms are rare. Large freshwater lakes (Crooked Lake and Lake Druzhby) exhibit very low chlorophyll *a* concentrations (<1.0 μg l^{-1}) and photosynthesis increases in spring and summer (Table 6.1). However, these lakes are covered by highly transparent ice that lacks snow accumulations because it is rapidly blown away by the katabatic winds that originate on the ice cap (Fig. 6.4). Once the light returns in July measurable carbon fixation occurs (Bayliss *et al.*, 1997; Henshaw and Laybourn-Parry, 2002). These lakes are amictic and do not develop thermal stratification. Their waters are below 4°C.

The marine-derived saline lakes of the Vestfold Hills are more productive than the freshwater lakes and consequently display higher levels of primary production. Brackish Highway Lake (4–5%) and meromictic Ace Lake (mixolimnion 18‰) had measurable photosynthesis during winter, but both showed increased rates during spring and summer with highest productivity in late summer during January (Laybourn-Parry *et al.*, 2005; Table 6.1). The peak in production appears to vary from year to year.

Fig. 6.4. The transparent ice cover on Crooked Lake in the Vestfold Hills, eastern Antarctica. © Elanor M. Bell.

In the summer of 1999/2000 maximum production occurred in December in both Highway Lake and Ace Lake (Laybourn-Parry *et al.*, 2002). In this unusually cold summer the ice cover persisted. Antarctic lakes are characterized by considerable inter-annual variations. For example, Pendant Lake had unusually high chlorophyll *a* concentrations under ice cover during the summer of 1999/2000, up to 29 µg l^{-1} resulting from a bloom of the green alga *Stichococcus bacillaris* (Laybourn-Parry *et al.*, 2002), while in most years summer chlorophyll *a* levels are around 7.5 µg l^{-1} (J. Laybourn-Parry, 2011, unpublished results).

The Dry Valley lakes lie inland and are much further south (76°S to 78°S) than those of the Vestfold Hills. They possess perennial ice-covers up to 5 m thick. The ice contains considerable wind-blown debris derived from the surrounding exposed rocks and soils of the Dry Valleys. Consequently light is severely attenuated by the ice-covers, so that only around 0.5–3.3% of incident PAR is available to support primary production immediately under the ice of lakes in the Taylor Valley, while in Lake Vanda in the Wright Valley up to 20% of PAR penetrates the ice (Howard-Williams *et al.*, 1998). Owing to the extremity of the environment and logistic constraints, there are no annual data sets for these lakes. Studies are confined mainly to the months of November to January. However, in 1991 an investigation of the spring–summer transition was undertaken commencing on 9 September, and in 2008 the research season was extended into April. Meromictic Lake Bonney (Taylor Valley) had measurable photosynthesis on 9 September 1991, indicating that primary production had commenced in late winter (Lizotte *et al.*, 1996; Table 6.1). This is consistent with what occurs in the lakes of the Vestfold Hills. As winter encroached in 2008 chlorophyll *a* continued to increase after photosynthesis had ceased. Cryptophytes, one of the major algal groups in Lake Bonney, are able to switch to using mixotrophy as a nutritional strategy and this may account for the persistence of chlorophyll *a* despite the measured decrease in primary production (Priscu, 2008). Rates of primary production in the Dry Valley lakes are generally lower than those recorded in meromictic coastal lakes like those of the Vestfold Hills (Table 6.1). However, phytoplankton biodiversity of Dry Valley lakes and the brackish and saline lakes of the Vestfold Hills are very similar.

Epishelf lakes are not well researched and, due to their distance from Antarctic stations, data are confined to the summer. Beaver Lake is the largest epishelf lake in Antarctica, Lying adjacent to the Amery Ice Shelf. Beaver Lake has perennial ice cover up to 5 m thick. The ice is highly transparent allowing 8.2–14.8% of surface PAR to penetrate the water column immediately under the ice. The ultra-oligtrophic waters allow PAR to penetrate to 100 m on occasions. Chlorophyll *a* values were very low (<0.55 µg l^{-1}) with correspondingly low rates of photosynthesis (Table 6.1). It is likely that low temperatures (<1.5°C) and nutrient limitation severely limit carbon fixation in what can be described as an end member system (Laybourn-Parry *et al.*, 2006).

There are a number of factors that control primary production. The major factor limiting photosynthesis in Antarctica is water temperature and in Antarctic lakes temperatures are usually much lower than in Arctic and high altitude lakes. Other factors include low transmission of PAR through persistent, often debris-containing ice covers, low annual PAR levels and periodic nutrient limitation. The lack of any significant zooplankton in continental Antarctic lakes removes predation pressure on the phytoplankton, whereas in many Arctic lakes and high altitude lakes this may be a factor, together with the impact of thermal stratification on nutrient availability and the attenuation of light through snow-covered ice.

6.4 Secondary Production: Bacteria and Viruses

In general, bacterial production is limited by the availability of labile dissolved organic carbon (DOC), inorganic nutrients (nitrogen and phosphorus) and temperature. There are few studies of Arctic bacterioplankton. In Toolik Lake bacterial production varied considerably down the water column during summer with a major peak occurring in late May to early June associated with a major influx of DOC from the catchment. The community was composed of persistent populations that occurred throughout the year and transient populations that appeared and disappeared (Crump *et al.*, 2003). Bacterial cell concentrations in the Arctic appear of the same order of magnitude, within the constraints of limited data, with those reported from the Antarctic and alpine lakes (Table 6.2). High Arctic lakes in Svalbard showed a progressive increase in summer with a peak in August. Bacterial abundances in both study lakes were similar with maxima of $28.8 \times 10^{8}\ l^{-1}$ in Tvillingvatnet, and $23.6 \times 10^{8}\ l^{-1}$ in a lake at Ossian Sars (Laybourn-Parry and Marshall, 2003; Table 6.2). There are considerably more data for Antarctic lakes as shown in Table 6.2. The range in cell densities is wide, reflecting the productivity of the lakes. The meromictic lakes of the Vestfold Hills and Dry Valleys support higher biomass than extreme freshwater systems like ultra-oligotrophic Beaver Lake, a large epishelf lake. Generally the bacterial cell concentrations are similar to those reported for oligotrophic temperate lakes.

Bacterial production data for high altitude and Arctic lakes are limited. Much more information is available for Antarctic systems, including data sets that cover an entire annual cycle (Table 6.2). In ultra-oligotrophic systems like Beaver Lake bacterial production represents less than 10% of primary production, suggesting that exudation of photosynthate from the phytoplankton is low. Certainly the DOC pool in this lake was low, always less than $0.5\ mg\ l^{-1}$. On occasion DOC levels may have been limiting to bacterial growth (Laybourn-Parry *et al.*, 2006). In the epishelf lake, Beaver Lake, the DOC pool is probably almost entirely derived form autochthonous (internally produced) primary production. In contrast a carbon budget for the Dry Valley lakes showed a discrepancy between bacterial respiratory demand and the available carbon pool. Based on a range of measured inputs bacterial respiration demand exceeded the available pool by three to eight times (Takacs *et al.*, 2001). The conclusion was that the major source of carbon for bacterial production comes from draw-down of bulk DOC or the decomposition of particulate material. Like Lakes Bonney and Lake Fryxell in the Dry Valleys, Ace Lake in the Vestfold Hills is a brackish meromictic lake. During 2003 bacterial production exceeded primary production during January by 34%, but at all other times bacterial production was lower than primary production. In December it was 26.8% of primary production. In neighbouring Pendant Lake, an unstratified lake with a salinity of 16‰, bacterial and primary production were equal in January, but like Ace lake bacterial production was usually lower than primary production, for example in December it was 16% (Laybourn-Parry *et al.*, 2007). However, one of the possible sources not accounted for in the study of Dry Valley lakes was the

Table 6.2. Bacterial concentrations and production in polar and high altitude lakes.

Lake	Bacterial numbers $\times 10^8$ l^{-1}	Production µg C l^{-1} day^{-1}	Source
ALPINE/SUBALPINE			
Gossenköllesee (2417 m) Austria	1.8–11.8 summer	–	Hofer and Sommaruga, 2001
Gossenköllesee (2417 m) Austria	1.1–5.5 over year		Pernthaler *et al.*, 1998
ARCTIC			
Tvillingvatnet, Svalbard	9.1–28.8 summer	–	Laybourn-Parry and Marshall, 2003
Toolik Lake, Alaska	10–22 summer	36 pmol l^{-1} h^{-1} max* summer	O'Brien *et al.*, 1992
ANTARCTIC			
East Lobe Lake Bonney, Dry Valleys	0.3–81.8 summer	max 4.9 summer	Takacs and Priscu, 1998
Lake Fryxell, Dry Valleys	5.3–436 summer	max 9.0 summer	Takacs and Priscu, 1998
Lake Druzhby, Vestfold Hills	1.1–3.3 over year	0–8.5 over year	Laybourn-Parry *et al.*, 2004
Crooked Lake, Vestfold Hills	1.6–4.7 over year	0–11.4 over year	Laybourn-Parry *et al.*, 2004
Ace Lake mixolimnion, Vestfold Hills	5–15 over year	4.8–74.4 over year	Madan *et al.*, 2005 Laybourn-Parry *et al.*, 2007
Beaver Lake, MacRobertson Land	0.9–1.4 summer	0.05–0.29 summer	Laybourn-Parry *et al.*, 2006

*Tritiated thymidine

potential release and recycling of DOC from the lysis of bacteria by viruses (bacteriophage). In an extensive review of prokaryote viruses Weinbauer (2004) indicates that the products of viral lysis might be a significant source of DOC. Current indications are that 23% of bacterial demand can be met by viral lysis in Lake Bonney, while in Crooked Lake and Lake Druzhby 0.8% to 69% of the DOC pool, respectively, is supplied by viral lysis of bacterioplankton over the annual cycle (Säwström *et al.*, 2008).

Despite the conventional thinking that bacterial production is curtailed by low temperatures, bacterial growth continues over winter in Antarctic lakes. Data are now emerging to show that bacterial production is also measurable in very extreme environments like cryoconites on glaciers where temperatures are very close to 0°C (Anesio *et al.*, 2007; Bagshaw *et al.*, Chapter 9, this volume). During winter DOC recycled by viral lysis may be an important source of substrate to the bacterioplankton at a time when primary production is low or absent. The limited data suggest that viruses are common in Antarctic lakes. In Vestfold Hills brackish lakes virus to bacterial ratios ranged from 30.6 to 80.0 in Ace Lake, 18.63 to 126.7 in Highway Lake and 30.0 to 96.7 in Pendant Lake (Madan *et al.*, 2005). Most of the viruses appeared to be bacteriophages. Although the lytic cycle predominated in winter with a maximum 32% in Pendant Lake and 71% in Ace Lake, virus numbers in the water remained high. Bacterioplankton in freshwater lakes appear to have high rates of viral infection, in Lake Druzhby up to 22.7% and in Crooked Lake up to 34.2% (Säwström *et al.*, 2007). In lower latitude lakes levels of infection are significantly lower ranging from 0.6% to 4.2%. Low temperatures and low rates of bacterial production mean that infected cells have a limited capacity to produce virus particles, so that burst sizes are low with a mean around 4 compared with from 9 to 71 in lower latitude systems (Säwström *et al.*, 2007). Estimates suggest that in the winter months 60% of the carbon supplied to the DOC pool originated from viral lysis in these freshwater lakes. In the more southerly Dry Valley lakes, virus numbers in the water were lower than in the Vestfold Hills brackish lakes: 2.26×10^5 ml^{-1} to 5.56×10^7 ml^{-1} compared with 0.89×10^7 ml^{-1} to 12.02×10^7 ml^{-1}, respectively (Lisle and Priscu, 2004; Madan *et al.*, 2005; Fig. 6.5). The indications are that viruses may play a more prominent role in biogeochemical cycling in microbially dominated Antarctic lakes than in lower latitude lacustrine ecosystems. This is an area that demands further investigation. Viruses also play an important role in transferring genetic material between bacteria, but this aspect of their biology has not been considered in an Antarctic context to date.

Alpine lakes appear to have lower virus numbers based on a study of Gossenköllesee (2417 m). Here concentrations ranged between 0 to 4.6×10^6 ml^{-1}. Maximum abundance occurred under ice in May with a second peak occurring in October/November (Hofer and Sommaruga, 2001). The lower numbers may, however, be the result of a TEM technique. The Antarctic data given above were derived from SYBR gold or SYBR green staining and epifluorescence microscopy. The TEM technique does allow examination of the morphology of the viruses and in Gossenköllesee there was a wide range of morphologies including filamentous forms that accounted for up to 100% of the virioplankton. These have not been described from other lakes. Levels of infection ranged between 0.9 to 2.3% and burst sizes ranged between means of from 7 to 17. Information on viruses in Arctic lakes is equally sparse. A pro-glacial lake in Svalbard had virus concentrations of 0.65×10^6 ml^{-1} and a virus to bacteria ratio of 2.2 (Anesio *et al.*, 2007). These values differ considerably from those reported from Antarctic freshwater lakes, but as yet there are insufficient data to draw any firm conclusions.

Studies on the biodiversity of polar and high altitude bacterial communities are relatively recent, but are beginning to provide an interesting picture of the adaptation and

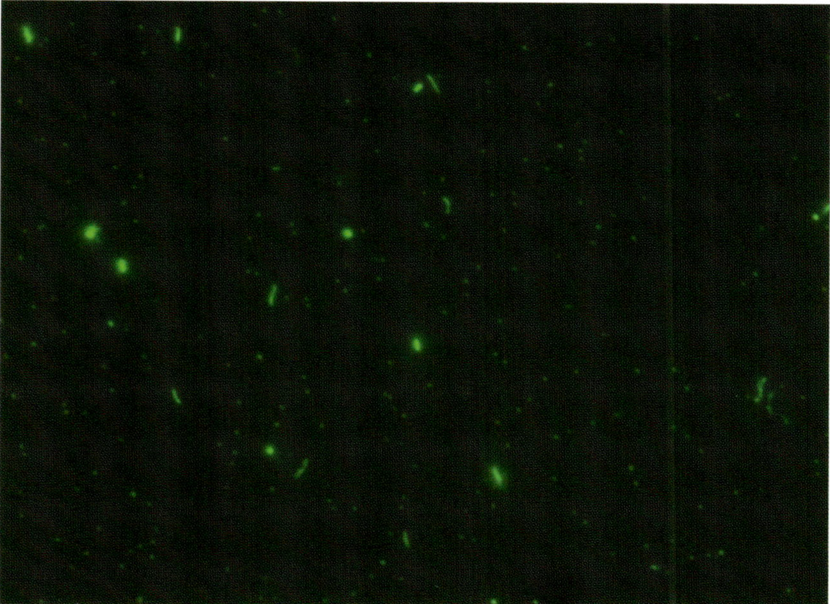

Fig. 6.5. Epifluorescent micrograph of virus-like particles and bacteria collected from Lake Bonney, Dry Valleys, Antarctica, and stained with SYBR Gold nucleic acid gel stain. © Elanor M. Bell.

evolution of extreme populations. Among the Proteobacteria, the Betaproteobacteria are particularly well represented in polar lakes. Studies in Antarctica suggest some degree of endemicity; for example, a large number of isolates from Lake Fryxell and Lake Bonney were not closely related to previously described species (Stingl *et al.*, 2008). The Dry Valley lakes have a long history and have undergone periods of dry down so this is not surprising. In contrast the relatively young (10,000 years), marine-derived saline lakes of the Vestfold Hills have many species that are closely related to marine groups (Franzmann and Dobson, 1993).

6.5 Secondary Producers: Protozoa and Invertebrates

The bacterial community is predated by the protozooplankton, particularly the heterotrophic nanoflagellates (HNAN) (Fig. 6.3). Our knowledge of the biodiversity of the HNAN is currently limited to classic taxonomic identifications. In Antarctic freshwater lakes *Paraphysomonas* sp., *Heteromita* sp., and *Monosiga consociata* occur, while in the marine derived lakes of the Vestfold Hills marine choanoflagellate species such as *Diaphanoeca grandis* may dominate (Tong *et al.*, 1997; Hobbie and Laybourn-Parry, 2008; Fig. 6.6). Similar HNAN are found in Arctic lakes. Many of the photosynthetic nanoflagellates that occur in polar and alpine lakes are mixotrophic, for example *Ochromonas*, *Pyramimonas* and cryptophytes. *Dinobryon* is a common mixotrophic genus in Arctic lakes and occurs in alpine lakes. These phytoflagellates exploit bacteria as an energy source to varying degrees and may also take up DOC. At times mixotrophic nanoflagellates can outgraze the HNAN, as was the case in Lakes Hoare and Fryxell in the Dry Valleys (Roberts and Laybourn-Parry, 1999).

Ciliated protozoan communities are much more diverse in the older lakes of the Dry Valleys. Many are mixotrophic by virtue of the sequestration of plastids from their phytoflagellate prey or by the possession of symbiotic zoochlorellae. The community includes predatory suctorian ciliates that prey on other ciliates (James *et al.*, 1998). The freshwater lakes of the Vestfold Hills have a very low species diversity including mixotrophic oligotrichs, scuticociliates and *Askenasia*. Similarly the brackish lakes are also characterized by low ciliate species diversity and are dominated by the ubiquitous marine ciliate *Mesodinium rubrum* (Perriss *et al.*, 1995; Bell and Laybourn-Parry, 1999; Fig. 6.7). *Mesodinium*

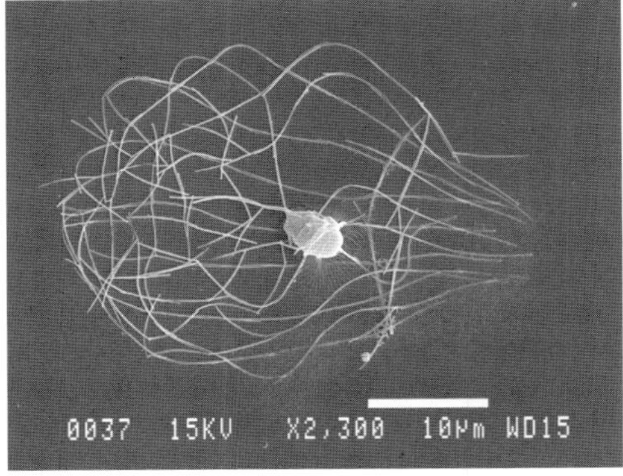

Fig. 6.6. *Diaphanoeca grandis*, a choanoflagellate that dominates the heterotrophic nanoflagellate (HNAN) community of Highway Lake, Vestfold Hills, Antarctica. © Gerry Nash, Australian Antarctic Division.

Fig. 6.7. The mixotrophic, ciliated protozoan *Mesodinium rubrum*, collected from Ace Lake, Vestfold Hills, Antarctica. © Elanor M. Bell.

occurs in lakes with salinities that range from 4‰ to 60‰. It is a mixotroph that either sequesters the plastids of its cryptophyte prey or has an endosymbiotic cryptophycean. In Arctic lakes ciliate communities are dominated by oligotrichs (some mixotrophic) *Urotricha* and *Askenasia* (Hobbie and Laybourn-Parry, 2008).

The impact of HNAN and mixotrophic phytoflagellate grazing on bacterioplankton production varies considerably across the Arctic and Antarctic (Table 6.3). In extreme Antarctic systems the HNAN and mixotrophic photosynthetic nanoflagellates (PNAN) remove a relatively small percentage of bacterial biomass or production in contrast to over 200% in fertilized corrals in Toolik Lake, Arctic. Generally, mixotrophic PNAN have little impact on the bacterioplankton except in Lake Fryxell and Lake Hoare, Dry Valleys, Antarctica, where they outnumber the HNAN and exert a greater grazing impact, removing up to 13% of bacterial biomass in Lake Fryxell but only up to 3% in Lake Hoare (Table 6.3).

Zooplankton are sparse in Antarctic lakes. The Dry Valley lakes lack crustaceans but do support rotifers, mostly members of the genus *Philodina* (Laybourn-Parry *et al.*, 1997). The less extreme coastal lakes of the Vestfold Hills have both rotifers and Crustacea. The freshwater lakes have *Daphniopsis studeri* as the sole planktonic species (Fig. 6.8). Populations are sparse, less than 1 individual l^{-1}. This species has invaded the slightly brackish lakes, such as Highway Lake, where it has probably replaced the marine calanoid copepod *Paralabidocera antarctica* found in the brackish to saline lakes such as Ace Lake. In this lake *P. antarctica* can reach concentrations in excess of 250 l^{-1} in summer (Bell and Laybourn-Parry, 1999). In Crooked Lake, *D. studeri* feeds on bacteria and the phytoplankton and can on occasion have a significant grazing impact on bacterioplankton during winter, when it can remove up to 34% of bacterial production (Säwström *et al.*, 2009). In this lake viral lysis of bacteria appears to impose a greater impact on bacterial mortality than grazing by zooplankton and nanoflagellates.

Epishelf, Beaver Lake, Antarctica, supports a very sparse population of a dwarf subspecies of the copepod *Boeckella poppei* and the rotifer *Notholca*. Egg sacs seen on a few females contained only a few eggs, suggesting that fecundity is low. Undoubtedly development in the extremely cold waters of this large, ultra-oligotrophic lake is slow. *Boeckella poppei* occurs in western Antarctica and South America but is not widespread in eastern Antarctica and seems to be confined to lakes in MacRobertson Land (Laybourn-Parry *et al.*, 2006).

Arctic lakes and alpine lakes support a much more diverse zooplankton of crustaceans and rotifers, which reflects higher phytoplankton biomass (Table 6.1). Alpine and subalpine lakes in the Sierra Nevada of the USA support the copepods *Hesperodiaptomus shoshone, Leptodiaptomus signicauda* and the cladocerans *Daphnia middendorffiana, Daphnia rosea* and *Chydorus sphaericus*. In addition, five genera of rotifers occur (Knapp *et al.*, 2001). Lake Tovel (1178 m) in the Adamello Brenta National Park (Italy) has a zooplankton community of the cladocerans *Bosmina longirostris* and *Daphnia longispina*, the former dominating the assemblage, and the cyclopoid copepod *Cyclops strenuus* (Obertegger *et al.*, 2007). Water residence time was an important factor in structuring zooplankton succession in Lake Tovel. Crustacean biomass was directly controlled by water residence time, while in contrast rotifer biomass was controlled by exploitative competition with crustaceans for phytoplankton. Alpine lakes are oligotrophic and consequently abundances of zooplankton are low, for example in Lake Tovel the most common species *B. longirostris* ranged between 1 and 37 individuals l^{-1}, while *C. strenuus* ranged between 1 and 7 individuals l^{-1}.

Toolik Lake has a diverse zooplankton community. In 1975 it contained two species of *Daphnia* (*D. middendorffiana* and *D. longiremis*), *Holopedium gibberum, Bosmina longirotris, Heterocope septentrionalis, Diaptomus pribiliofensis* and *Cyclops scutifer*. Two of the larger species, *D. middendorffiana* and *H. gibberum*, had

Table 6.3. Protozoan grazing rates and the grazing impact on the bacterioplankton of Arctic and Antarctic lakes.

Lake*	Temp °C	Clearance rate nl cell^{-1} h^{-1}	Grazing rate Bact. cell^{-1} h^{-1}	Bacterial biomass grazed day^{-1}	Source
ARCTIC					
Lake at Ossian Sars, Svalbard					
HNAN	4	2.1–7.2	0.79–16.9	up to 20%	Laybourn-Parry and Marshall, 2003
PNAN	4	0.78–4.2	0.28–1.12	<1%	
Toolik Lake fertilized corral					
HNAN and PNAN	11–13	2–20	3–160	50–200%	Hobbie and Helfrich, 1988
Barrow Pond HNAN				20% of production	Hobbie *et al.*, 1980
ANTARCTIC					
Crooked Lake, Vestfold Hills HNAN	2–4	–	0.2	9.7% of production	Laybourn-Parry *et al.*, 1995
Beaver Lake, MacRobertson Land HNAN	1	1.11	0.113–0.15	up to 100% of production	Laybourn-Parry *et al.*, 2006
Highway Lake, Vestfold Hills					
HNAN	2	0.02–1.8	0.02–2.4	0.5–5.8%	Laybourn-Parry *et al.*, 2005
PNAN	2	0.04–1.05	0.02–1.56	1% max	
Ace Lake, Vestfold Hills					
HNAN	2	0.04–0.37	–1.45	1%	Laybourn-Parry *et al.*, 2005
PNAN	2	0.002–0.21	0.003–1.04	<1%	
Lake Fryxell, Dry Valleys					
HNAN	0.41	–	4.0–5.4	1–4%	Roberts and Laybourn-Parry, 1999
PNAN	0.41		1.6–3.6	5–13%	
Lake Hoare, Dry Valleys					
HNAN	2	–	6.2–6.7	1–2%	Roberts and Laybourn-Parry, 1999
PNAN	2		0.2–1.0	<1–3%	

*HNAN, heterotrophic nanoflagellates; PNAN, photosynthetic nanoflagellates engaging in mixotrophy

virtually disappeared by the mid-1990s. Their disappearance is difficult to explain, but there is evidence that changes in the dominant fish species, possibly resulting from bird predation, may have imposed greater predation pressure on the larger species. Despite the high diversity abundances are low, for example *Daphnia* and *Bosmina* usually occur at around 1 individual l^{-1} and the copepods *Diaptomus* and *Cyclops* have maxima around 10 individuals l^{-1} (O'Brien *et al.*, 1997). Four species of rotifer dominate the community with five other species occurring occasionally.

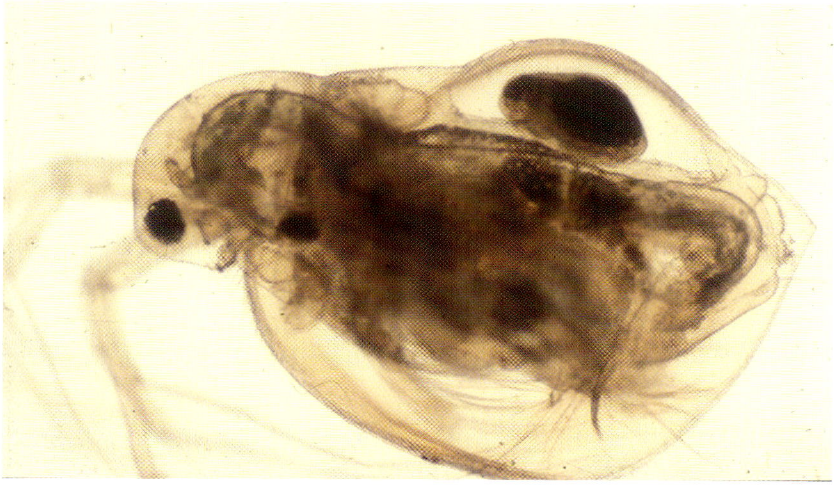

Fig. 6.8. *Daphniopsis studeri* (Crustacea; Cladocera) from Crooked Lake, Vestfold Hills, Antarctica. © Johanna Laybourn-Parry.

Their concentrations ranged from 30 to 300 l^{-1} in surface waters (Rublee, 1992). Toolik Lake is surrounded by vegetation, but other Arctic lakes lie in polar deserts, such as Lake Hazen on Ellesmere Island. It contains only one fish species, the Arctic char (Johnson, 1983), while in Toolik Lake, sculpin, grayling and trout are found. Zooplankton diversity is lower with *D. middendorffiana* and *C. scutifer* and a few rotifers (McLaren, 1964). While Toolik Lake and Lake Hazen are at similar latitudes, the latter is in a polar desert lacking the allochthonous inputs that Toolik Lake receives.

Many Arctic lakes possess well developed benthic communities (Thomas *et al.*, 2008 and references therein). Toolik Lake benthos includes chironomid larvae, caddis larvae, molluscs and microcrustacea. Char Lake (74°N) was the subject of an International Biological Programme investigation in the 1970s. It lies in a polar desert and has extremely low plankton production. The majority (80%) of the primary production is achieved by the benthic vegetation. The plankton supports only one microcrustacean, the copepod *Limnocalanus macrurus*, and there is only one fish species, the Arctic char, that sustains very slow growth rates of approximately 0.03 C m^{-2} $year^{-1}$.

6.6 Survival Strategies and Adaptation in Extreme Lakes

The Antarctic continent is subject to a more extreme climate than the Arctic and lake water temperatures are usually below 4°C. Many Arctic lakes and alpine lakes experience temperatures in excess of 4°C during the summer. Toolik Lake and the Barrow Ponds can rise to 15°C and in the high Arctic lakes can reach 7°C. The austral summer is short, but the lack of extensive snow on the lake ice-covers allows primary production as soon as light returns after the winter darkness. As discussed above, annual investigations of Vestfold Hills lakes in Antarctica have shown that the plankton of both freshwater and saline lakes continues to function over winter enabling populations to 'hit the deck running' when summer arrives. Evidence from studies that have followed the transition from winter to spring and summer to winter in the Dry Valleys suggests that these lakes also function in winter. Further evidence comes from Lugol's iodine fixed samples collected over winter by an automated sampling device in Lake Fryxell, Dry Valleys. Vegetative cells were the most common form of all species in samples. Two cryptophyte species were more abundant in winter than in summer and the chlorophyte *Stichococcus* was only observed in winter (McKnight *et al.*, 2000).

Many of the species of phytoplankton in Antarctic lakes are mixotrophic, i.e. able to use both autotrophic and heterotrophic modes of nutrition. For example, cryptophyte algae and the prasinophyte alga *Pyramimonas gelidicola* (Bell and Laybourn-Parry, 2003; Laybourn-Parry *et al.*, 2005; Figs 6.9 and 6.10) dominate in some Vestfold Hills lakes and remain active over winter by grazing on bacteria. This ability to feed on bacteria or, in some cases, take up DOC (a strategy termed osmotrophy) enables these phytoflagellates to sustain biomass when PAR is limited or absent. Not only does mixotrophy supply a carbon source, it is also a means of accessing phosphorous and nitrogen from ingested bacteria when ambient concentrations are low.

Mixotrophic cryptophytes also dominate the austral summer phytoplankton of Lakes Fryxell and Hoare. In summer they do not become entirely photosynthetic, at midsummer they continued to ingest bacteria at low rates (Roberts and Laybourn-Parry, 1999). The contribution to the cryptophyte carbon budgets derived from grazing decreased from November to December. In Lake Fryxell for example, it ranged from 8% to 31% in November, varying with depth in the water column, to 2% to 24% in December (Marshall and Laybourn-Parry, 2002). Based on evidence from McKnight *et al.* (2000) it is likely that the cryptophyte populations that occurred in winter were heavily dependent on grazing to sustain their numbers. In the brackish lakes of the Vestfold Hills both cryptophytes and *Pyramimonas* were mixotrophic throughout winter; in Highway Lake (Laybourn-Parry *et al.*, 2005) and Ace Lake (E.M. Bell, 2004, unpublished results) they also exploited DOC ranging from 4 kDa to 500 kDa.

Photosynthetic dinoflagellates in Vestfold Hills lakes encyst in winter, but there are distinct seasonal differences between lakes (Bell and Laybourn-Parry, 1999; Rengefors *et al.*, 2008). In Ace Lake and Pendant Lake numbers remained high in winter, particularly *Gyrodinium*, while in Highway Lake vegetative cells disappeared. Those species that remain active in winter are either relying on endogenous energy reserves or are able to adopt a mixotrophic strategy. The dominant ciliate of the saline lakes, *Mesodinium rubrum*, has a two-pronged survival strategy. Part of the population encysts, while a proportion remains active surviving on endogenous starch

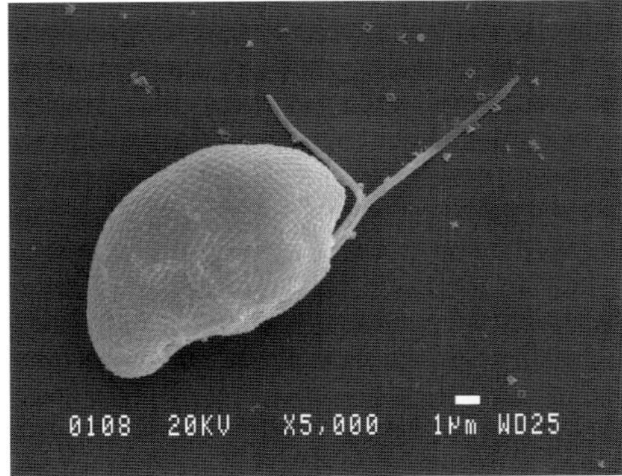

Fig. 6.9. A scanning electron micrograph of the mixotrophic cryptophyte *Geminigera cryophila*, from Ace Lake, Vestfold Hills, Antarctica. © Elanor M. Bell.

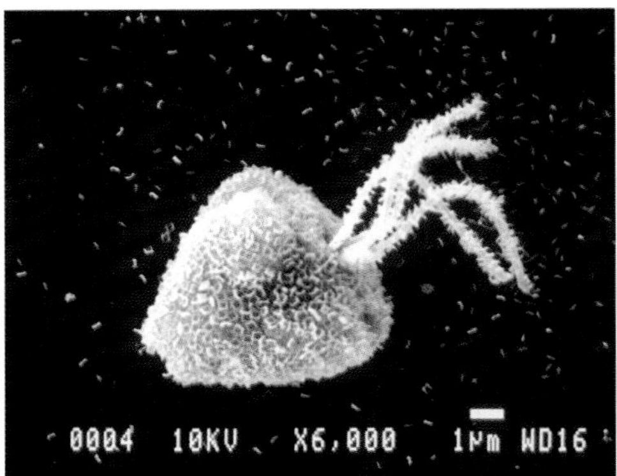

Fig. 6.10. A scanning electron micrograph of the mixotrophic prasinophyte *Pyramimonas gelidicola*, from Ace Lake, Vestfold Hills, Antarctica. © Elanor M. Bell.

reserves (Bell and Laybourn-Parry, 1999). The encysted individuals excyst in summer when favourable conditions return, augmenting the actively growing population.

Winter data for Arctic lakes are unavailable. Where lake ice is covered by snow phytoplankton are likely to enter resting stages as they do in temperate lakes. Similarly, during phases of low food resources many zooplankton species may enter diapause or overwinter as resting eggs. In contrast, in Antarctic lakes the limited crustacean zooplankton remains active throughout winter.

Bacteria that inhabit extremely cold environments have evolved a range of adaptations such as low temperature enzymes and ice active substances. These novel biochemicals have attracted the interest of biotechnologists as they have potential industrial and medical applications. Examples include a cold active alkaline phosphatase and a low temperature lipase (Yang *et al.*, 2004; Dhaked *et al.*, 2005; see also Arora and Bell, Chapter 25, this volume). Antifreeze proteins (AFP) have been identified in a range of planktonic bacteria from saline lakes in the Vestfold Hills (Gilbert *et al.*, 2004) including a hyperactive Ca^{2+} dependent AFP from the psychrophilic bacterium *Marimomonas primoryensis*, isolated from Ace Lake (Gilbert *et al.*, 2005).

6.7 Future Directions

Polar and alpine lakes are often logistically difficult to study. The growing technology in remote sensing offers considerable opportunity to acquire large continuous data sets (see a review by Laybourn-Parry and Vincent, 2008). Surface imagery involving satellites can be of particular value in assessing long term changes in the extent and timing of ice cover. Catchment characteristics such as vegetation abundance can be assessed. This is an important determinant of coloured dissolved organic matter (CDOM) entering lakes from their catchments. CDOM can have a major impact on PAR and UV radiation transmission in a water column. Unmanned aerial vehicles (UAVs) have been used to map the abundance of cryoconite holes, which are effectively mini-lakes on an Arctic glacier surface (see also Sattler *et al.*, Chapter 8, this volume). The earliest UAVs were developed in World War I as unmanned aerial combat vehicles and with the maturing and miniaturization of applicable technologies in the 1980s and 1990s began to be used for strategic military reconnaissance. Now UAVs are used in a wide range of civil applications. While such applications are invaluable, there is still the need for an element of ground-truthing. A fast-moving area of development in remote sensing at present is Synthetic Aperture Radar (SAR). The Canadian Radarsat -1 pace-borne SAR system was one of the first to be applied to the cryosphere. It has the advantage that imagery is unaffected by cloud cover and is optimized for snow, ice and water. For example, SAR can detect bubble ice that implies spatial variations in biological activity and convective motions.

Wireless networks represent another form of remote sensing. They are systems that collect and transmit data from remote environments. For example, a platform that was deployed on the surface of Crooked Lake (Antarctica) measured water temperature, PAR and UV-B radiation at a range of depths in the underlying water column, as well as ice thickness. Surface sensors collected meteorological data, and surface PAR and UV-B levels. The platform was solar and wind powered. The data were uploaded periodically via Iridium satellite phone by researchers at Davis Station, Vestfold Hills, Antarctica (Palethorpe *et al.*, 2004). This platform logged data continuously every 5 min, providing a detailed picture of total daily PAR and UV-B transmission to the water column.

While molecular techniques are now being more widely applied to assess biodiversity in extreme lakes, particularly for prokaryotes, the next challenge is to determine which organisms contribute to crucially important processes such as biogeochemical cycling. Functional genomics offers a means of determining the mechanisms that enable organisms to function in the natural environment. This approach has identified novel biochemicals such as degradation enzymes (Healy *et al.*, 1995) and cold-adapted proteins (Saunders *et al.*, 2003). Proteomics offers another important tool in understanding adaptation to life in extreme conditions. For example, Goodchild *et al.* (2004) identified proteins specific to growth at 4°C versus growth at the optimum growth temperature of 24°C of the archaeal species *Methanococcoides burtonii*, isolated from Ace Lake. Crucial features of cold adaptation involved transcription, protein folding and metabolism including the specific activity of RNA polymerase subunit E, a response regulator and preptidyl-prolyl *cis/trans* isomerase. The technology is being refined at an increasing pace and will provide exciting insights into the evolution and functioning of life in extreme environments and potentially life on other extraterrestrial bodies.

References

Alexander, V., Stanley, D.W., Daley, R.J. and McRoy, C.P. (1980) Primary producers. In: Hobbie, J.E. (ed.) *Limnology of Tundra Ponds, Barrow, Alaska*. Dowden, Hutchinson & Ross, Stroudsburg, Pennsylvania, pp. 179–250.

Anesio, A.M., Mindl, B., Laybourn-Parry, J., Hodson, A.J. and Sattler, B. (2007) Viral dynamics in cryoconite holes on a high Arctic glacier (Svalbard). *Journal of Geophysical Research* 112, G04S31, doi:10.1029/2006JG000350.

Bayliss, P., Ellis-Evans, J.C. and Laybourn-Parry, J. (1997) Temporal patterns of primary production in a large ultra-oligtrophic Antarctic freshwater lake. *Polar Biology* 18, 363–370.

Bell, E.M. and Laybourn-Parry, J. (1999) Annual plankton dynamics in an Antarctic saline lake. *Freshwater Biology* 41, 507–519.

Bell, E.M. and Laybourn-Parry, J. (2003) Mixotrophy in the Antarctic phytoflagellate *Pyramimonas gelidicola* McFadden (Chlorophyta: Prasinophyceae). *Journal of Phycology* 39, 644–649.

Catalan, J., Ventura, M., Brancelj, A., Granados, I., Thies, H., Nickus, U., Korhola, A., Lotter, A.F., Barbieri, A., Stuchlik, E., Lien, L., Bitusik, P., Buchaca, T., Camarero, L., Goudsmit, G.H., Kopacek, J., Lemcke, G., Livingstone, D.M., Mueller, B., Rautio, M., Sisko, M., Sorvari, S., Sporka, F., Strunecky, O. and Toro, M. (2002) Seasonal ecosystem variability in remote mountain lakes: implications for detecting climatic signals in sediment records. *Journal of Paleolimnology* 28, 25–46.

Crump, B.C., Kling, G.W., Bahr, M. and Hobbie, J.E. (2003) Bacterioplankton community shifts in an Arctic lake correlate with seasonal changes in organic matter source. *Applied and Environmental Microbiology* 69, 2253–2268.

Dhaked, R.K., Alam, S.I., Dixit, A. and Singh, L. (2005) Purification and characterization of thermo-labile alkaline phosphatase from an Antarctic psychrotolerant *Bacillus* sp. P9. *Enzyme Microbiological Technology* 36, 855–861.

Franzmann, P.D. and Dobson, S.J. (1993) The phylogeny of bacteria from a modern Antarctic refuge. *Antarctic Science* 5, 267–270.

Gilbert, J.A., Hill, P.J., Dodd, C.E.R. and Laybourn-Parry, J. (2004) Demonstration of antifreeze protein activity in Antarctic lake bacteria. *Microbiology* 150, 171–180.

Gilbert, J.A., Davies, P. and Laybourn-Parry, J. (2005) A hyperactive calcium dependent antifreeze protein in an Antarctic bacterium. *FEMS Microbiology Letters* 245, 67–72.

Goodchild, A., Ertan, H., Raftery, M., Guilhaus, M., Curmi, P.M.G. and Cavicchioli, R. (2004) A proteomic determination of cold adaptation in the Antarctic archaeon *Methanococcoides burtonii*. *Molecular Microbiology* 53, 309–321.

Hawes, I. (1983) Nutrients and their effects on phytoplankton populations in lakes on Signy Island, Antarctica. *Polar Biology* 2, 115–126.

Healy, F.G., Ray, R.M., Aldrich, H.C., Wilkes, A.C. and Shanmuggan, K.T. (1995) Direct isolation of functional genes encoding cellulases from microbial consortia in a thermophilic, anaerobic digester maintained on lignocellulose. *Applied Microbiology and Biotechnology* 43, 667–674.

Henshaw, T. and Laybourn-Parry, J. (2002) The annual patterns of photosynthesis in two large, freshwater, ultra-oligotrophic Antarctic lakes. *Polar Biology* 25, 744–752.

Hobbie, J.E. and Helfrich, J.V.K. (1988) The effect of grazing by microprotozoans on production of bacteria. *Archiv für Hydrobiologie* 31, 21–288.

Hobbie, J.E. and Laybourn-Parry, J. (2008) Heterotrophic microbial processes in polar lakes. In: Vincent, W.F. and Laybourn-Parry, J. (eds) *Polar Lakes and Rivers: Limnology of Arctic and Antarctic Aquatic Ecosystems*. Oxford University Press, Oxford, UK, pp. 197–210.

Hobbie, J.E., Stanley, D.W., Daley, R.J. and McRoy, C.P. (1980) Primary producers. In: Hobbie J.E. (ed.) *Limnology of Tundra Ponds Barrow, Alaska*. Dowden, Hutchinson and Ross, Stroudsberg, USA, pp. 179–250.

Hofer, J.S. and Sommaruga, R. (2001) Seasonal dynamics of viruses in an alpine lake: importance of filamentous forms. *Aquatic Microbial Ecology* 26, 1–11.

Howard-Williams, C., Schwarz, A.-M., Hawes, I. and Priscu, J.C. (1998) Optical properties of the McMurdo Dry Valley lakes, Antarctica. In: Priscu, J.C. (ed.) *Ecosystem Dynamics in a Polar Desert: the McMurdo Dry Valleys, Antarctica*. American Geophysical Union, Antarctic Research Series Vol. 72, Washington, DC, pp. 189–204.

James, M.R., Hall, J.A. and Laybourn-Parry, J. (1998) Protozooplankton and microplankton ecologyin lakes of the Dry valleys, Southern Victoria Land. In: Priscu, J.C. (ed.) *Ecosystem Dynamics in a Polar Desert: the McMurdo Dry Valleys, Antarctica*. American Geophysical Union, Antarctic Research Series Vol. 72, Washington, DC, pp. 255–267.

Johnson, L. (1983) Homeostatic mechanisms in a single species arctic fish populations. *Canadian Journal for Fisheries and Aquatic Science* 40, 987-1-24.

Knapp, R.A., Matthews, K.R. and Sarnelle, O. (2001) Resistance and resilience of alpine lake fauna to fish introductions. *Ecological Monographs* 71, 401–421.

Laybourn-Parry, J. and Marshall, W.A. (2003) Photosynthesis, mixotrophy and microbial plankton dynamics in two high Arctic lakes during summer. *Polar Biology* 26, 517–524.

Laybourn-Parry, J. and Vincent, W.F. (2008) Future directions in polar limnology. In: Vincent, W.F. and Laybourn-Parry, J. (eds) *Polar Lakes and Rivers: Limnology of Arctic and Antarctic Aquatic Ecosystems*. Oxford University Press, Oxford, UK, pp. 307–316.

Laybourn-Parry, J., Bayliss, P. and Ellis-Evans, J.C. (1995) The dynamics of heterotrophic nanoflagellates and bacterioplankton in a large ultra-oligotrophic Antarctic lake. *Journal of Plankton Research* 17, 1835–1850.

Laybourn-Parry, J., James, M.R., McKnight, D.M., Priscu, J.C., Spaulding, S. and Sheil, R. (1997) The microbial plankton of Lake Fryxell, Southern Victoria Land, Antarctica during the summers of 1992 and 1994. *Polar Biology* 17, 54–61.

Laybourn-Parry, J., Henshaw, T. and Quayle, W.C. (2002) The evolution and biology of Antarctic saline lakes in relation to salinity and trophy. *Polar Biology* 25, 542–552.

Laybourn-Parry, J., Henshaw, T., Jones, D.J. and Quayle, W. (2004) Bacterioplankton production in freshwater Antarctic lakes. *Freshwater Biology* 49, 735–744.

Laybourn-Parry, J., Marshall, W.A. and Marchant, H.J. (2005) Flagellate nutritional versatility as a key to survival in two contrasting Antarctic saline lakes. *Freshwater Biology* 50, 830–838.

Laybourn-Parry, J., Madan, N.J., Marshall, W.A., Marchant, H.J. and Wright, S.W. (2006) Carbon dynamics in an ultra-oligotrophic epishelf lake (Beaver Lake, Antarctica) in summer. *Freshwater Biology* 51, 1116–1130.

Laybourn-Parry, J., Marshall, W.A. and Madan, N.J. (2007) Viral dynamics and patterns of lysogeny in saline Antarctic lakes. *Polar Biology* 30, 351–358.

Lisle, J.T. and Priscu, J.C. (2004) The occurrence of lysogenic bacteria and microbial aggregates in the lakes of the McMurdo Dry Valleys. *Microbial Ecology* 47, 427–439.

Lizotte, M.P., Sharp, T.R. and Priscu, J.C. (1996) Phytoplankton dynamics in the stratified water column of Lake Bonney, Antarctica I. Biomass and productivity during the winter–spring transition. *Polar Biology* 16, 155–162.

Madan, N.J., Marshall, W.A. and Laybourn-Parry, J. (2005) Viral and microbial loop dynamics over an annual cycle in three contrasting Antarctic lakes. *Freshwater Biology* 50, 1291–1300.

Markager, S., Vincent, W.F. and Tang, E.P.Y. (1999) Carbon fixation in high Arctic lakes: implications of low temperature for photosynthesis. *Limnology and Oceanography* 44, 597–607.

Marshall, W.A. and Laybourn-Parry, J. (2002) The balance between photosynthesis and grazing in Antarctic mixotrophic cryptophytes during summer. *Freshwater Biology* 47, 2060–2070.

McKnight, D.M., Smith, R.L. and Bradbury, J.P. (1990) Phytoplankton dynamics in three Rocky Mountain lakes, Colorado, USA. *Arctic, Antarctic and Alpine Research* 22, 264–274.

McKnight, D.M., Howes, B.L., Taylor, C.D. and Goehringer, D.D. (2000) Phytoplankton dynamics in a stably stratified Antarctic lake during winter darkness. *Journal of Phycology* 36, 852–861.

McLaren, I.A. (1964) Zooplankton of Lake Hazen, Ellesmere Island, and a nearby pond with special reference to the copepod *Cyclops scutifer* Sars. *Canadian Journal of Zoology* 42, 613–629.

Obertegger, U., Flaim, G., Branioni, M.G., Sommaruga, R., Corradini, F. and Borsato, A. (2007) Water residence time as a driving force of zooplankton structure and succession. *Aquatic Science* 69, 575–583.

O'Brien, J.W., Hershey, A.E., Hobbie, J.E., Hullar, M.A., Kipphut, G.W., Miller, M.C., Moller, B. and Vestal, J.R. (1992) Control mechanisms of arctic lake ecosystems: a limnocorral experiment. *Hydrobiologia* 240, 143–188.

O'Brien, J.W., Bahr, M., Hershey, A.E., Hobbie, J.E., Kipphut, G.W., Kling, G.W., Kling, H., McDonald, M., Miller, M.C., Rublee, P.A. and Vestal, J.R. (1997) The limnology of Toolik Lake. In: Milner, A.M. and Oswood, M.W. (eds) *Freshwaters of Alaska: Ecological Synthesis*. Springer, New York, pp. 61–106.

Palethorpe, B., Hayes-Gill, B., Crowe, J., Sumner, M., Crout, N., Foster, M., Reid, T., Benford, S., Greenhalgh, C. and Laybourn-Parry, J. (2004) Real time physical data acquisition through a remote sensing platform on a polar lake. *Limnology and Oceanography Method* 2, 191–201.

Pernthaler, J., Glöckner, F.-O., Uterholzner, S., Alfreider, A., Psenner, R. and Amann, R. (1998) Seasonal community and population dynamics of pelagic bacteria and Archaea in a high mountain lake. *Applied and Environmental Microbiology* 64, 4299–4306.

Perriss, S.J., Laybourn-Parry, J. and Marchant, H.J. (1995) The widespread occurrence of the autotrophic ciliate *Mesodinium rubrum* (Ciliophora: Haptorida) in the freshwater and brackish lakes of the Vestfold Hills, Eastern Antarctica. *Polar Biology* 15, 423–428.

Priscu, J.C. (2008) Plankton dynamics in the McMurdo Dry Valley lakes during the transition to polar night – a project contributing to EBA. *SCAR Evolution and Biodiversity in Antarctica, Newsletter Issue 2*, 3 pp.

Rengefors, K., Laybourn-Parry, J., Logares, R., Marshall, W.A. and Hansen, G. (2008) Marine derived dinoflagellates in Antarctic saline lakes: community composition and annual dynamics. *Journal of Phycology* 44, 592–604.

Roberts, E.C. and Laybourn-Parry, J. (1999) Mixotrophic cryptophytes and their predators in the Dry Valley lakes of Antarctica. *Freshwater Biology* 41, 737–746.

Rose, K.C., Williamson, C.F., Saros, J.E., Sommaruga, R. and Fischer, J.M. (2009) Differences in UV transparency and thermal structure between alpine and subalpine lakes: implications for organisms. *Photochemical and Photobiological Sciences* 8, 1244–1256.

Rublee, P.A. (1992) Community structure and bottom-up regulation of heterotrophic microplankton in Arctic LTER lakes. *Hydrobiologia* 240, 133–141.

Saros, J.E., Interlandi, S.J., Doyle, S., Michel, T.J. and Williamson, C.E. (2005) Are the deep chlorophyll maxima in alpine lakes primarily induced by nutrient availability, not UV avoidance? *Arctic, Antarctic and Alpine Research* 37, 557–563.

Saunders, N.F.W., Thomas, T., Curmi, P.M.G., Mattick, J.S., Kuczek, E., Slade, R., Davis, J., Franzmann, P.D., Boone, D., Rusterholtz, K., Feldman, R., Gates, C., Bench, S., Sowers, K., Kadner, K., Aerts, A., Dehal, P., Detter, C., Glavina, T., Lucas, S., Richardson, P., Larimer, F., Hauser, L., Land, M. and Cavicchioli, R. (2003) Mechanisms of thermal adaptation revealed from genomes of the Antarctic Archaea *Methanogenium frigidum* and *Methanococcoides burtonii*. *Genome Research* 13, 1580–1588.

Säwström, C., Anesio, A.M., Granéli, W. and Laybourn-Parry, J. (2007) Seasonal viral loop dynamics in two ultra-oligotrophic Antarctic freshwater lakes. *Microbial Ecology* 53, 1–11.

Säwström, C., Lisle, J., Anesio, A.M., Priscu, J.C. and Laybourn-Parry, J. (2008) Bacterioplankton in polar inland waters. *Extremophiles* 12, 167–175.

Säwström, C., Karlsson, J., Laybourn-Parry, J. and Granéli, W. (2009) Zooplankton feeding on algae and bacteria under ice in Lake Druzhby, East Antarctica. *Polar Biology* 32, 1195–1202.

Sommaruga, R. (2001) The role of solar UV radiation in the ecology of alpine lakes. *Journal of Photochemistry and Photobiology B: Biology* 47, 262–272.

Sommaruga, R. and Augustin, G. (2006) Seasonality in UV transparency of an alpine lake is associated to changes in phytoplankton biomass. *Aquatic Science* 68, 129–141.

Sommaruga, R. and Psenner, R. (1997) Ultraviolet radiation in a high mountain lake of the Austrian Alps and underwater measurements. *Journal of Photochemistry and Photobiology B: Biology* 65, 957–963.

Sommaruga, R., Psenner, R., Schafferer, E., Koinig, K.A. and Sommaruga-Wögrath, S. (1999) Dissolved organic carbon concentration and phytoplankton biomass in high-mountain lakes of the Austrian Alps: potential effect of climatic warming on UV underwater attenuation. *Arctic, Antarctic and Alpine Research* 31, 247–253.

Spaulding, S.A., McKnight, D.M., Smith, R.L. and Dufford, R. (1994) Phytoplankton population dynamics in perennially ice-covered Lake Fryxell, Antarctica. *Journal of Plankton Research* 16, 527–541.

Stingl, U., Cho, J.-C., Foo, W., Vergin, K.L., Lanoil, B. and Giovannoni, S.J. (2008) Dilution-to-extinction culturing of psychrotolerant planktonic bacteria from permanently ice-covered lakes in the McMurdo Dry Valleys, Antarctica. *Microbial Ecology* 55, 395–405.

Takacs, C.D. and Priscu, J.C. (1998) Bacterioplankton dynamics in the McMurdo Dry Valley lakes, Antarctica, production and biomass loss over four seasons. *Microbial Ecology* 36, 239–250.

Takacs, C.D., Priscu, J.C. and McKnight, D.M. (2001) Bacterial dissolved organic carbon demand in McMurdo Dry Valley lakes, Antarctica. *Limnology and Oceanography* 46, 1189–1194.

Thomas, D.N., Fogg, G.E., Convey, P., Fritsen, C.H., Gili, J.-M., Gradinger, R., Laybourn-Parry, J., Reid, K. and Walton, D.W.H. (2008) *The Biology of the Polar Regions*. Oxford University Press, Oxford, UK, 416 pp.

Tong, S., Vørs, N. and Patterson, D.J. (1997) Heterotrophic flagellates, centroheloid Heliozoa and filose amoebae from marine and freshwater sites in Antarctica. *Polar Biology* 18, 91–106.

Vincent, W.F. (1981) Production strategies in Antarctic polar waters: phytoplankton eco-physiology in a permanently ice-covered lake. *Ecology* 62, 1215–1224.

Vincent, W.F., Hobbie, J.E. and Laybourn-Parry, J. (2008) Introduction to the limnology of high latitude lake and river ecosystems. In: Vincent W.F. and Laybourn-Parry, J. (eds) *Polar Lakes and Rivers: Limnology of Arctic and Antarctic Aquatic Ecosystems*. Oxford University Press, Oxford, UK, pp. 1–23.

Weinbauer, M.G. (2004) Ecology of prokaryote viruses. *FEMS Microbiology Reviews* 28, 127–181.

Yang, X.X., Lin, X.Z., Bian, J., Sun, X.Q. and Hunag, X.H. (2004) Identification of five strains of Antarctic bacteria producing low temperature lipase. *Acta Oceanologica Sinica* 23, 717–723.

Yurista, P.M. and O'Brien, W.J. (2001) Growth, survivorship and reproduction of *Daphnia middendorffiana* in several Arctic lakes and ponds. *Journal of Plankton Research* 23, 733–744.

7 Subglacial Lakes

David A. Pearce
British Antarctic Survey, Natural Environment Research Council, Cambridge, UK

7.1 Introduction

Antarctic subglacial lakes are one of the few remaining unexplored environments on Earth, and from a microbiological perspective, perhaps one of the most interesting. First identified by Robin *et al.* (1970) after airborne radio-echo sounding (RES) investigations of the ice-sheet interior, the scale and potential variety of Antarctic subglacial lake ecosystems is enormous. The total volume of water held beneath the Antarctic ice sheet alone is believed to be between 4000 and 12,000 km^3 (Siegert, 2000) and the known characteristics and depths of these subglacial lakes suggest that they represent significant aquatic environments. However, we still know very little about them as, to date, they have yet to be accessed directly. They have the potential to be one of the most extreme environments on Earth, and are certainly one of the least accessible, with combined stresses of high pressure, low but probably stable temperature, permanent darkness, predominantly low-nutrient availability (potentially one of the most oligotrophic environments on Earth, particularly in the absence of geothermal activity) and highly variable oxygen concentrations (derived from the ice that provided the original melt water). Indeed, the predominant mode of nutrition is likely to be chemoautotrophic, i.e. any organisms that exist will obtain nourishment through the oxidation of chemical compounds (Siegert *et al.*, 2003). Existing chemical analysis of both ice and sediment samples suggests a very dilute environment for microbial life, so where present, there will probably be very low standing stocks of viable organisms that could be relatively inactive.

Many questions have already been asked of these extreme environments, such as can they even sustain life, and if present, what type of organism inhabits or even thrives under these conditions? What specific niche might the organisms occupy and how do they overcome intense and often adverse selection pressures? What can any organisms found tell us about the distribution and evolution of microbial life elsewhere? In addition, from a geological and glaciological perspective, is any potential historical climate change record locked within the sediments? How do Antarctic subglacial lakes interact with and influence the rest of the overlaying ice and indeed the rest of the biosphere, i.e. do they influence global biogeochemical cycles? Are these ecosystems stable and how do they function? Through the existing analyses of analogues and proxies, we are now beginning to discover what life might be like in these remarkable ecosystems.

7.2 Geographic Range, History and Distribution

Although the existence of subglacial lakes has been predicted for more than 50 years, it is only recently that we have started to gain a better understanding of their characteristics and impact. Only 70 Antarctic subglacial lakes were originally identified in 2003 (Dowdeswell and Siegert, 2003). This rapidly increased to 145 by 2005 (Siegert *et al.*, 2005), then 190 by 2008 (Forieri *et al.*, 2008), and the latest estimates suggest that there are at least 387 (Siegert *et al.*, 2011). In addition, new lakes are being discovered or described all of the time, for example, those recently located in the area between the Belgica Highlands and the Concordia Trench, Antarctica (Forieri *et al.*, 2008).

Subglacial lakes are found where rock comes into contact with ice, predominantly beneath Arctic and alpine glaciers and under the Antarctic ice sheet, the key difference being one of scale. The largest concentration and highest diversity of subglacial lake types are the widely distributed lakes beneath both the East and West Antarctic ice sheets. Here, the bedrock topography of the ice-sheet interior is essentially characterized by large subglacial basins separated by mountain ranges. The lakes are found in areas of relatively low bed relief, in and on the margins of subglacial basins (Dowdeswell and Siegert, 2003).

The largest subglacial lake exists beneath Vostok Station in Antarctica, and is 14,000 km^2 in area, approximately the size of Lake Ontario, USA. The sheer scale of Lake Vostok is highlighted by the second largest subglacial lake, Lake Concordia, located in East Antarctica near Dome Concordia beneath approximately 4000 m ice and a mere 800 km^2 in area. Lake Concordia is believed to have a volume of 31 km^3, a maximum water column thickness of 126 m and a water residence time of 16,800 ± 7600 years, which is significantly shorter than Lake Vostok (Tikku *et al.*, 2004; Thoma *et al.*, 2009). However, the combined area of additional subglacial lakes beneath Dome Concordia is 15,000 km^2 (Siegert, 2000).

As might be expected when viewing the basal topography of the Antarctic ice sheet, evidence exists for significant interconnectivity between the different components of Antarctic subglacial lake ecosystems, including direct exchanges between the base of the ice sheet and the lake water (Siegert, 2000), the identification of rapid discharge events (Wingham *et al.*, 2006), specific connections (Ericker and Scambos, 2009) and the description of warm-based fast-flowing ice streams (Dowdeswell and Siegert, 2003). Subglacial lake water can move between lakes and rapidly drain, causing catastrophic floods. Indeed, it has been predicted that a widespread, dynamic subglacial water system in the Antarctic may exert an important control on ice flow and mass balance (Fricker *et al.*, 2007). On a continental scale, the locations of Antarctic subglacial lakes match the modelled distribution of pressure melting at the ice-sheet bed (Dowdeswell and Siegert, 2003). Indeed, water plays a crucial role in ice-sheet stability and the onset of ice streams (Bell *et al.*, 2007). In Iceland, active volcanoes and hydrothermal systems underlie the ice caps. Glacier–volcano interactions produce melt water that either drains toward the glacier margin or accumulates in subglacial lakes (Björnsson, 2003), sometimes allowing discharge of material into the atmosphere, as in the 2010 eruptions of Eyjafjallajökull, Iceland.

7.3 The Extreme Characteristics of Subglacial Lakes

Subglacial ecosystems are subject to some of the most extreme conditions on Earth. Temperatures in subglacial ecosystems are generally below 0°C, however, microbes have shown activity in the Antarctic at −17°C (Carpenter *et al.*, 2000) and if geothermal activity is found to be present, could in places be much higher than this. Despite the relatively low temperatures, as long as liquid water is present, life is theoretically possible. Even the coldest environments on Earth have enough liquid water to sustain life (Laybourn-Parry, 2009). This includes inside the ice itself, where a network of micron-diameter veins and a nanometre thick film of unfrozen water can remain (Price, 2006). For subglacial lake ecosystems, some have a significant measured water depth of the order of hundreds of metres, and a melting ice–water interface is indicated by the presence of accretion ice. Heat can also be added to the system through friction and geothermal energy. Calculations indicate that the geothermal heat flux varies spatially over the Antarctic Plate between 37 and 65 mW m^{-2} (Siegert, 2000). It is likely that some form of physiological plasticity would be advantageous here. Skidmore *et al.* (2005) suggested that microbes inhabiting this type of environment must be capable of regular freezing and thawing and possibly seasonal cycling between highly oxygenated and anoxic conditions. The pressure inside a subglacial lake depends on a number of factors, especially the depth of ice on top of the lake. Some have only a thin overlying ice column, however, most are subject to high pressure, and pressures of ~350 atmospheres (Siegert *et al.*, 2001) or ≤400 bars (1 bar = 100 kPa; Price, 2000) have been quoted. However, microbial life has been shown to function at gigapascal pressures (Sharma *et al.*, 2002) and bacteria recovered from the deep ocean at around 4000 m have been shown to retain both structural integrity and metabolic activity (Pearce, 2009) (Fig. 7.1).

The absence of sunlight is a key difference from the surface polar lakes. The effect of sunlight could, however, be felt if photosynthetically fixed carbon enters the lake as a result of migration through the ice, and there are plenty of microorganisms in the air over Antarctica, including phototrophic bacteria (Hughes *et al.*, 2004; Pearce *et al.*, 2010). Oxygen, however, could range from absent to super-saturated. Importantly, dissolved oxygen is available at least at the lake surface, from equilibration with air hydrates released from melting basal glacier ice (Siegert *et al.*, 2001). Chemical analysis of different subglacial systems (Table 7.1) has shown a high degree of variability, even strongly acidic or saline solutions are possible, depending on the inflow and geology (De Angelis *et al.*, 2004). The presence of thermophilic bacteria

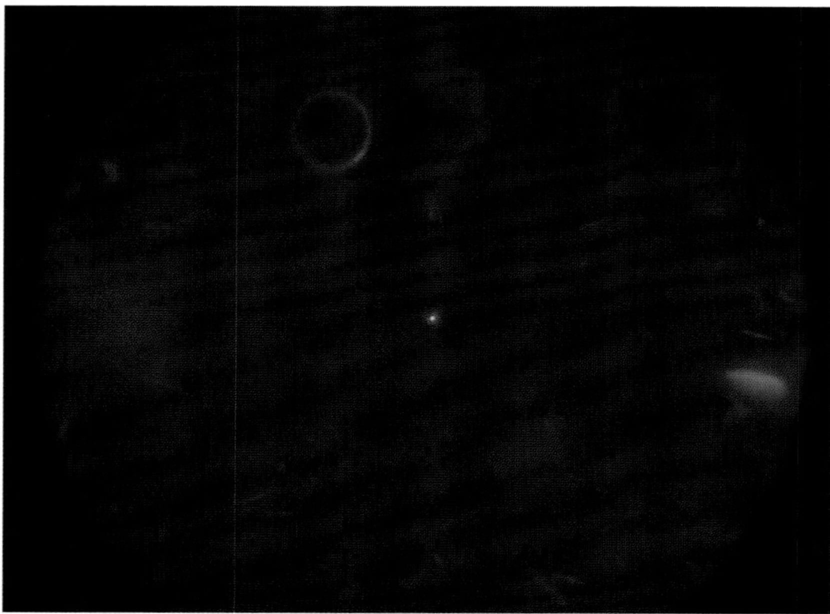

Fig. 7.1. A viable organism recovered from the seabed at 4000 m, as shown by CTC staining.

Table 7.1. Suggested likely chemistry for Lake Ellsworth, Antarctica. (Lake Ellsworth Consortium, 2007).

Units: μeq l^{-1}	H^+	pH	Ca^{2+}	Mg^{2+}	Na^+	K^+	NH_4^+	Cl^-	SO_4^{2-}	NO_3^-	HCO_3^-
Average Byrd Ice Core	1.8	5.7	~1.0	0.4	1.5	0.05	0.13	2.0	1.0	0.7	~1.2
Inferred Lake Ellsworth	<140	>3.9	>80	30	120	1	<10	160	80	<50	>100
Lake Ellsworth (including contribution from glacial flour)	~10	~6	~420	~90	120	3	1	160	240	1	240

suggests that a geothermal system exists beneath the cold water body of Lake Vostok, and that subglacial lake ecosystems could indeed sustain thermophilic chemolithoautotrophic microbial communities (Bulat *et al.*, 2004). Foght *et al.* (2004) found substantial aerobic mineralization of ^{14}C-acetate at 8°C, indicating that sufficient nutrients and viable psychrotolerant microbes were present to support metabolism in sediment samples from glaciers in New Zealand. Elsewhere, an analysis of bulk melt water chemistry from Arctic glaciers (Skidmore *et al.*, 2005) has shown that sulfide oxidation and carbonate dissolution accounted for the majority of solute flux from the Bench Glacier, Alaska, whereas gypsum/anhydrite and carbonate dissolution accounted for the majority of the flux from the John Evans Glacier, Ellesmere Island, Canada.

The physical characteristics of the growth substrate are also important. Where present, sediment often has a higher biomass, this also applies to the pore spaces between ice crystals (Thomas and Dieckmann, 2002). Although for Lake Ellsworth it is estimated that there is >2 m of sediment below the lake floor, sedimentation rates are low.

7.4 The Predicted Diversity and Community Composition of Life in Subglacial Lakes

In reality, we have no idea which extremophilic microbes may inhabit subglacial lake ecosystems. We can only guess based upon what has been found in other cryospheric ecosystems. We do not even know whether they are likely to be psychrophilic (cold-loving) or mesophilic (organisms that grow best in moderate temperatures) but tolerating temperatures below their optimum for growth and reproduction. What we can do, is describe what population densities might be theoretically possible, although there are a very wide range of estimates published in the literature (Table 7.2). We can also speculate what types of organisms might be found, given what we know of the nature of the environment (Table 7.3). For example, subglacial lakes in Iceland are dominated by a few groups of putative chemotrophic bacteria whose closest cultivated relatives use sulfide, sulfur or hydrogen as electron donors, and oxygen, sulfate or carbon dioxide as electron acceptors (Gaidos *et al.*, 2009). We can also guess at the likely activity where we know about the likely chemical and physical composition of the subglacial lake waters, for example, where known geothermal activity exists.

There is also no knowledge of the stability of these environments microbiologically; this stability may well depend on flow through and turnover rates (the lake system acting as a giant chemostat). Growth rates can be estimated from leucine and thymidine incorporation studies, but as yet, none have been done *in situ*, although estimates can be made from recovered accretion ice. There is an almost complete lack of knowledge about different trophic levels or the trophic web; although it is likely that

Table 7.2. Estimates of cell density.

≥1 cell cm^{-3}	In acid veins 4 × 10^5-year-old colder glacial ice	Price, 2000
<1 cfu ml^{-1}	Polar ice	Abyzov *et al.*, 1982
1–5 cfu ml^{-1}	Glacial ice from Canadian High Arctic	Dancer *et al.*, 1997
10–50 cells ml^{-1}	Meltwater	Bulat *et al.*, 2002
100 cells ml^{-1}	Glacial ice from Vostok	Priscu *et al.*, 2007
2.8 × 10^3 and 3.6 × 10^4 cell ml^{-1}	Melted ice	Priscu *et al.*, 1999
6 × 10^7 cells ml^{-1}	Greenland ice core	Sheridan *et al.*, 2003
2–7 × 10^6 cells g^{-1} dry weight sediment[a]	Fox Glacier and Franz Josef Glacier in the Southern Alps of New Zealand	Foght *et al.*, 2004
6–9 × 10^5 colony-forming units g^{-1} dry weight		
2–7× higher bacterial density	In Lake Vostok accretion ice than the overlying glacial ice	Christner *et al.*, 2006
2.8 × 10^3 to 3.6 × 10^4 cells ml^{-1}[b]	Accretion ice Lake Vostok	Priscu *et al.*, 1999
~10^7 cells g^{-1}	Kamb Ice Stream, West Antarctic Ice Sheet – sediment environment beneath the Antarctic Ice Sheet	Lanoil *et al.*, 2009
1.3 × 10^2 up to 9.6 × 10^2 cl ml^{-1}	Upper, younger horizons Lake Vostok	
5 × 10^5 cells ml^{-1}[c]	Anoxic bottom waters of a volcanic lake beneath the Vatnajökull ice cap Iceland	Gaidos *et al.*, 2009
10^3 to 10^4 cells cm^{-3}[d]	Lake Vostok accretion ice	Abyzov *et al.*, 2004
3 × 10^3 to 4 × 10^4 bacterial cells ml^{-1}[e]	Melted accretion ice	Priscu *et al.*, 1999
2 × 10^2 and 3 × 10^2 bacterial cells ml^{-1}	Lake Vostok accretion ice	Karl *et al.*, 1999
4 × 10^7 g^{-1}	Tephra sediments	Gaidos *et al.*, 2004
2 × 10^4 ml^{-1}	Lake water column, Iceland	Gaidos *et al.*, 2004
3.32 × 10^5 cells ml^{-1}	BAS ice core DAPI count Antarctic glacial ice core	D.A. Pearce, unpublished results
1–16 CFU ml^{-1}	BAS ice core cultures Antarctic sub glacial lake	D.A. Pearce, unpublished results

[a]Viable counts in the glacier ice typically were 3–4 orders of magnitude smaller than in sediment
[b]No biological incorporation of selected organic substrates or bicarbonate was detected
[c]Essentially all bacteria, with no detectable Archaea
[d]Consumption of a ^{14}C-labeled protein hydrolysate (from 0.0044 mg l^{-1} h^{-1} at 1665 m to 0.0002 mg l^{-1} h^{-1} at 2750 m)
[e]Which were not culturable and did not incorporate radiocarbon during 52 h in air at 1 bar

full ecological interactions, predation, competition, symbiosis, etc. occur, we simply do not know at present. What little we do know and what we can predict are summarized below.

7.4.1 Analogues for life in subglacial lakes

Although direct study of life in Antarctic subglacial lakes has yet to be achieved, many analogues exist for the potential life that might be found there, for example, in accretion ice above subglacial lakes, glacial ice cores, glacial foreland, subglacial sediments, Antarctic marine and lake waters, supraglacial lakes, Siberian and Antarctic permafrost and in other uncultured organisms derived from cold or frozen environments. Antarctica also has a remarkable diversity of other lake types, ranging from freshwater, brackish, saline, hypersaline (up to 10 times the conductivity of seawater) and epishelf lakes (Laybourn-Parry, 2002; Laybourn-Parry and Bell, Chapter 6, this volume).

Table 7.3. Types of metabolism.

Lake Vostok accretion ice	*Brachybacteria, Methylobacteria, Paenibacillus* and *Sphinogmonas*	Christner *et al.*, 2001
Southern hemisphere glaciers	*Polaromonas vacuolata* and *Rhodoferax antarcticus*	Foght *et al.*, 2004
(Fox and Franz Josef Glaciers)	Nitrate-reducing and ferric iron-reducing bacteria	Foght *et al.*, 2004
(Fox Glacier sediment)	Nitrogen-fixing bacteria	Foght *et al.*, 2004
Beneath the Antarctic ice sheet	Low diversity (only five phylotypes) Fe and S oxidizing metabolisms	Lanoil *et al.*, 2009
Greenland ice core (GISP2)	Methanogens and Fe reducing anaerobes	Price, 2006
	β-, γ- and δ-Proteobacteria	Christner *et al.*, 2006
Glaciers	β-Proteobacteria	Skidmore *et al.*, 2005
Blood Falls, Antarctica	*Thiomicrospira arctica*, β-, γ- and δ-Proteobacteria, Bacteroidetes	Mikucki and Priscu, 2007
Glacial ice, Antarctica	Sulfate reducers	Sjöling and Cowan, 2003
Lake Vostok accretion ice	α- and β-Proteobacteria, low- and high-G+C Gram-positive, CFB α- and β-Proteobacteria and the Actinomycetes	Christner *et al.*, 2008
Lake Vostok accreted ice	Pro- and eukaryotic microorganisms	Poglazova *et al.*, 2001
Lake Vostok accreted ice	*Hydrogenophilus thermoluteolus*	Bulat *et al.*, 2004

Antarctic subglacial lakes may share some of the characteristics with other Antarctic or indeed Arctic lakes, for example, Antarctic lakes are unproductive, with continuous low temperatures and the lack of any significant input of inorganic nutrients and organic carbon (Laybourn-Parry, 2002). A wide range of aerobic and anaerobic microbial metabolisms are also likely to be present, and these could be reflected in other polar lake analogues. However, the key difference will be in the absence of photosynthesis as a key mechanism of carbon fixation in subglacial lake ecosystems, due to the lack of light. Encouragingly, viable microbes have already been detected beneath several geographically distant glaciers, even those underlain by different bedrock (Skidmore *et al.*, 2005). Furthermore, life present in the snow itself may provide an analogue: Carpenter *et al.*, (2000) found 200–5000 cells ml^{-1} in South Pole snow and between 1.47 (±0.41) × 10^3 and 1.68 (±0.31) × 10^3 cells ml^{-1} have been counted in snow melted from the site of Lake Ellsworth (D.A. Pearce, 2011, unpublished results).

7.4.2 Challenges for the study of life in subglacial lakes

A key challenge in accessing the microbial communities in subglacial lakes following first access, is that the microorganisms present in the snow and ice around these ecosystems are already adapted to survive in these conditions. It is therefore essential to know which microorganisms have come from the lake itself, and which might have come from the surrounding environment and been introduced during the sampling process. To do this, it is important to either be sure that lake access is completely sterile (although this is almost impossible to achieve) or to be sure that anything likely to be introduced is known. For example, DNA isolated from South Pole snow yielded ribosomal DNA sequences similar to those of several psychrophilic bacteria and a bacterium which aligned closely with members of the genus *Deinococcus*, an ionizing-radiation- and desiccation-resistant genus (Carpenter *et al.*, 2000). This particular study also obtained evidence of low rates of bacterial DNA and protein synthesis, which suggested that the organisms were metabolizing at ambient subzero temperatures (−12 to −17°C).

7.4.3 Likely organism groups

The potential exists for all forms of microbial life to exist in subglacial lake ecosystems, with the possible exception of photoautotrophs (photosynthesizers), which could potentially still be present, transmitted through the ice, but inactive. However, there are unlikely to be any Metazoa due to the pressures of transmission through the ice, and any community present is almost certainly dominated by the microbial loop. Microorganism groups are likely to include: (i) viruses – it is now well established that viruses are ubiquitous in aquatic ecosystems worldwide (Wilson *et al.*, 2000), and bacteriophage are common in polar inland waters (Säwström *et al.*, 2007); (ii) fungi – Basidiomycetous yeasts have been isolated from glacial and subglacial waters of northwest Patagonia (Brizzio *et al.*, 2007) and *Penicillium* in Arctic subglacial ice (Sonjak *et al.*, 2006); (iii) Archaea – Archaea, eubacteria and viruses have all been isolated from the basal ice of the Greenland Ice Core (GISP2 core; Tung *et al.*, 2005) and Vostok accretion ice (Christner *et al.*, 2001; D'Elia *et al.*, 2008) has been found to contain similar organisms (e.g. those related to *Methylobacterium* and *Sphingomonas*); (iv) eukaryotes – Willerslev *et al.* (1999) used PCR amplification of fragments of the 18S rRNA gene to identify a diversity of fungi, plants, algae and protists extracted from 2000- and 4000-year-old ice-core samples from North Greenland. Indeed, subglacial lakes are also thought to be ancient systems that may contain exotic biota (Karl *et al.*, 1999; Priscu *et al.*, 1999; Bulat *et al.*, 2004). Antarctic subglacial lakes may, therefore, provide a good source of novel organisms and model virus–host associations (which could, though very unlikely in theory, act as potential reservoirs of disease), particularly the potential 'seed bank' trapped in glacial ice and within subglacial lakes (Pearce and Wilson, 2003). Subglacial Lake Vostok may be a unique reservoir of genetic material and it may contain as yet unknown organisms with distinct adaptations, particularly as the vast majority of sequences derived from even soil-borne Antarctic bacteria are from as yet uncultivated species. However, all this is still speculative as they have yet to be explored directly (Bell *et al.*, 2002).

7.5 Existing Studies

Much of what we know or suspect about life in subglacial environments has been derived from studies in analogues or in accessory environments, for example, glaciers, accretion ice, melt water, sediment, inflows and outflows, life in the ice itself, in the snow and in sub-lithic studies. Bidle *et al.* (2007) reported the potential metabolic activity of microbes in the oldest known ice on Earth, derived from the Mullins and upper Beacon Valleys, Antarctica (the minimum ages of which are 100,000 years and ~8 million years, respectively). Radio-labelled substrates were incorporated into macromolecules, and microbes grew in nutrient-enriched melt waters, but metabolic activity and cell viability were critically compromised by age. Their 16S rDNA-based community reconstruction suggested relatively low bacterial sequence diversity, and metagenomic analyses of community DNA revealed many diverse orthologs (genes in different species that have evolved from a common ancestral form via speciation) to extant metabolic genes. Analyses of ice samples demonstrated an exponential decline in the average community DNA size with a half-life of ~1.1 million years, thereby constraining the geological preservation of microbes in icy environments and the possible exchange of genetic material to the oceans. In more recent samples, Sjöling and Cowan (2003) found a high 16S rDNA bacterial diversity in glacial melt water and lake sediment from Bratina Island, Antarctica. In addition, their results revealed a vertical stratification of bacterial species, probably reflecting a vertical gradient of temperature, oxygen content, and redox potential of the sediment. Each of these characteristics might well be important in subglacial lake environments. Gaidos *et al.* (2004) drilled into and sampled a lake deep beneath a glacier in Iceland. The lake, buried under 300 m of ice and maintained by the heat of an active volcano, was 100 m deep at the sampling location. The lake community was distinct from the assemblages of organisms in both the borehole water (before penetration) and from the overlying ice and snow. Encouragingly, sequencing of selected DGGE bands obtained from the samples gave sequences that were highly similar to known psychrophilic organisms or cloned DNA from other cold environments. Significant uptake of ^{14}C-labelled bicarbonate occurred in the dark, low-temperature incubations of lake water samples, indicating the presence of autotrophs, and acetylene reduction assays under similar incubation conditions showed no significant nitrogen fixation potential by lake water samples.

At the retreating margins of the ice sheet, there are a number of locations where former subglacial lakes are emerging from under the ice but remain perennially ice covered. A preliminary search for microbial life in one of these lakes, Lake Hodgson (72°00.549′S, 68°27.708′W) has been successful (Fig. 7.2). Lake Hodgson is a perennially ice-covered freshwater lake, with dimensions of approximately 2 × 1.5 km, and a 93.4 m deep water column under 3.6–4.0 m of perennial lake ice (Hodgson *et al.*, 2009a). The waters, isolated from the atmosphere, have a chemical composition consistent with subglacial melting of catchment ice. The lake is ultra-oligotrophic with nutrient concentrations within the ranges of those found in the accreted lake ice of subglacial Lake Vostok. Total organic carbon and dissolved organic carbon are present, but at lower concentrations than typically recorded in continental rain. Critically, no organisms and no pigments associated with photosynthetic or bacterial activity were detected using light microscopy and high performance liquid chromatography. Subsequent analysis has shown DAPI and FISH stained cells from surface sediment samples (D.A. Pearce, 2011, unpublished results). Increases in SO_4 and cation

Fig. 7.2. Lake Hodgson, Antarctica. © D. Hodgson.

concentrations at depth and declines in O_2 provide some evidence for sulfide oxidation and very minor bacterial demand upon O_2 that result in small, perhaps undetectable changes in the carbon biogeochemistry (Hodgson *et al.*, 2009b).

Lanoil *et al.* (2009) reported 16S rRNA gene and isolate diversity in sediments collected from beneath the Kamb Ice Stream, West Antarctic Ice Sheet and stored for 15 months at 4°C. This is the first report of microbes in samples from the sediment environment beneath the Antarctic Ice Sheet. Other potential analogues include: cryoconites, glacial ice (Bagshaw *et al.*, Chapter 9, this volume), ice fields (valley or alpine glaciers; Sattler *et al.*, Chapter 8, this volume) and ice caps (if a volcano or mountain range is completely glaciated), basal sediments beneath glaciers, Blood Falls, a subglacial outflow from the Taylor Glacier, Antarctica (Mikucki and Priscu, 2007; McGenity and Oren, Chapter 20, this volume) and material excavated from mines (Pokorný *et al.*, 2005). However, there are limitations associated with each of the current studies and these include scant preliminary knowledge, lack of a background database (present in other microbes in Antarctic studies), slow growth rates, reduced viability (important in revival/culture), the potential for contamination and, ultimately, the cost.

7.6 Methods Employed to Study Subglacial Lakes

7.6.1 Physical

Antarctic subglacial lakes have been identified through a combination of airborne radio-echo sounding, seismic reflection, ice thickness and surface elevation data and GPS measurement of ice flow. The depth and shape of subglacial Lake Vostok's water cavity was determined from aerogravity data. Recently, satellite altimetry has been used to measure anomalous near-flat regions on the ice-sheet surface. It is also possible to map such regions from Space. Satellite laser altimeter elevation profiles from 2003 to 2006 collected over the lower parts of Whillans and Mercer ice streams, West Antarctica, revealed 14 regions of temporally varying elevation, which were interpreted as the surface expression of subglacial water movement. Vertical motion and the spatial extent of two of the largest regions were confirmed by satellite image differencing.

7.6.2 Biological

The equipment used to detect life within subglacial ecosystems is critical. A variety of techniques is currently being used or adapted, which include: adenosine triphosphate (ATP) determination (which is present in all living cells), the measurement of endogenous fluorescence, direct observation (using light and scanning electron microscopy), fluorescence microscopy tuned to detect NADH in live organisms (Price, 2000), raman spectroscopy, an antibody microarray device currently under development or a life marker chip. Karl *et al.* (1999) adopted a polyphasic approach to analyse a sample of the accreted Lake Vostok ice for: (i) microbial cell enumeration by epifluorescence microscopy, scanning electron microscopy (SEM) and dual laser flow cytometry; (ii) microbial biomass with ATP and lipopolysaccharide (LPS); (iii) microbial cell viability and potential metabolic activity by analysis of rates of ^{14}C-CO_2 production and ^{14}C-incorporation into macromolecules after timed incubations with exogenous ^{14}C-labelled organic substrates; and (iv) the presence of potential carbon and nitrogen growth substrates. Such an approach using a variety of general techniques can generally be classified by approach; for example: (i) microscopy; fluorescent and electron microscopy (often used with specific gene probes); (ii) biochemistry or biogeochemical cycling assays, for example, infrared Raman used to detect specific biomolecules; (iii) molecular biology; targeting genomic DNA (using gene probes coupled with fluorescent *in situ* hybridization) or by direct extraction from material obtained and used to construct a metagenomic library to screen for novel physiologies. Alternatively, the approach could be to target the identification of specific organisms or groups. Lavire *et al.* (2006) established the presence of the thermophilic bacterium *Hydrogenophilus thermoluteolus* DNA in accretion ice in subglacial Lake Vostok, Antarctica, using *rrs, cbb* and *hox* genes. The results pointed to the presence of thermophilic chemoautotrophic microorganisms in Lake Vostok accretion ice. They presumably originated from deep faults in the bedrock cavity containing the lake in which episodes of seismotectonic activity would release debris along with microbial cells.

7.6.3 Chemical

Subglacial lake chemistry can be determined through a combination of *in situ* devices and remote assays. Depth, pressure, conductivity, temperature, pH levels, biomolecules (using life marker chips), anions and nitrogen isotopes can all be determined *in situ* once the lakes' water columns are accessed. It is also possible to visualize the environment (using cameras and light sources), measure dissolved gases (using chromatography), and examine the morphology of the lake floor and sediment structures (using sonar). It is likely the chemical composition could resemble the chemistry recorded in the Byrd Ice Core (Table 7.1) or in Lake Vostok (Table 7.2). Hydrological controls on microbial communities in subglacial environments are reviewed by Tranter *et al.* (2005).

7.6.4 Planned access

There are currently three immediate candidates for subglacial lake access:

1. The largest Antarctic subglacial lake, Lake Vostok, has been most extensively surveyed using both satellite remote sensing and radio-echo sounding data (Kapitsa *et al.*, 1996). Microbes, some of which may be viable, have been found in ice cores drilled at Vostok Station at depths down to ~3600 m (Price, 2000). Drilling operations at Vostok were suspended at a depth of 3720.47 m within 5G-2 borehole, approximately 30 m from the ice–water interface. Full access is anticipated in the 2011/12 season (S.A. Bulat, Petersburg, 2011, personal communication).

2. The Whillans Ice Stream Subglacial Access Research Drilling (WISSARD) project is focused on the hydrological continuum beneath the Whillans Ice Stream in glaciological, microbiological, geochemical and oceanographic contexts. Research will be conducted through three complementary projects: GBASE: Geomicrobiology of Antarctic Subglacial Environments; LISSARD: Lake and Ice Stream Subglacial Access Research Drilling; and RAGES: Robotic Access to Grounding-zones for Exploration and Science. It is expected that access will also be achieved in 2011–13.

3. The third is Lake Ellsworth, located at 78°58′34″S, 90°31′04″W, near the Ellsworth Mountains in West Antarctica, 20 km from the ice divide. The lake is located within a distinct topographic hollow which is 1.5 km deeper than the surrounding bed (Siegert *et al.*, 2004; Lake Ellsworth Consortium, 2007, 2011). It lies beneath 3.4 km of ice, is 12 km long, 3 km wide and up to 160 m deep. Access to Lake Ellsworth is planned for the 2012/13 field season. Taken together, these three studies will represent the centre of the continent, near the ice divide (Lake Ellsworth), the main Antarctic plateau (Lake Vostok) and the extensive delta regions near the ice margins (WISSARD).

Key considerations prior to any lake access include an environmental impact assessment, the development of controlled access protocols and appropriate sample recovery techniques, further technological development, a determination of assay detection limits, full environmental characterization and the verification of clean access technologies. For Lake Ellsworth, the minimum requirements for in-field analysis have been established; these include a sterile containment unit, on-site power and preliminary in-field analysis at an Antarctic station (Rothera) to measure activity including radiotracers and culture work. Ellsworth ice will be sampled with depth to determine the ice input contribution, and any possible contribution from the operators. Equipment and the aerial environment will also be sampled for potential contaminants.

7.6.5 Cleanliness

Antarctic subglacial lakes are pristine and represent a new extreme environment in which to search for microorganisms that have been isolated from the rest of the biosphere for a long time. With this in mind, it is critical that non-indigenous microorganisms are not introduced during the drilling or sample recovery process. The cleaning, sterilization and validation technologies require particular attention. A review of the state-of-the-art in current standards for cleanliness (for example in the pharmaceutical, space and high-tech manufacturing industries), methods of calculating permissible microbial load, microbial loading in West Antarctic ice sheets and implication for subglacial lakes, microbial control measures used in analogous projects (e.g. planetary protection), underwater engineering for microbial control (design and materials), methods of sterilization for engineering structures and materials, methods of cleaning for engineering structures and materials and of maintaining microbial control during field experiments is reviewed in Mowlem *et al.* (2011). There are three distinct microbial control areas: (i) during manufacture and transport; (ii) the drill site environment and activity; and (iii) the ice through which the probe penetrates. Key remaining unknowns are first, what the effect of microorganism concentrations below the detection levels of current technology might be and second, the ecological consequences of introducing even a single microorganism.

7.6.6 Detection limits

Advances in molecular technology have vastly improved life detection limits, such that microscopy and molecular techniques such as polymerase chain reaction (PCR) are now capable of detecting individual cells per millilitre, or the DNA itself at 0.1–0.2 ml^{-1}, although realistically, the actual PCR detection limit is likely to be at least an order of magnitude higher than this. This is well within the predicted minimum estimated for Lake Vostok of 1 cell ml^{-1} (Bulat *et al.*, 2002). To date, 16S rDNA-based community reconstruction has shown up to 93 sequences from Lake Vostok accretion ice (although this figure is known to include potential contaminants). Adopting a culture-based approach from Antarctic ice cores, 0, 2 and 10 colony-forming units (cfu) ml^{-1} have been isolated from Dyer Plateau, Siple Station and Taylor Dome respectively (Christner *et al.*, 2000), and 1–16 cfu ml^{-1} from a Dronning Maud Land ice core (D.A. Pearce, 2011, unpublished results). Radiolabelled substrates can yield uptake rates at the level of several hundred cells (Karl *et al.*, 1999). The cell concentration of 2×10^2 ml^{-1} by Sheridan *et al.* (2003) implies that a 20 ml sample from Lake Ellsworth would contain 4×10^3 cells. However, no one approach is likely to provide a complete unbiased picture of the microorganisms residing in a sample or their relative numbers, and the design of specific, clean sampling strategies is extremely important (Pearce, 2009).

7.7 Subglacial Lake Extremophiles and Biotechnology

It has been estimated that Lake Vostok has been isolated from the atmosphere for between 420,000 (Christner *et al.*, 2001) and >15 million years (Christner *et al.*, 2006). However, the time estimated for ice to traverse into Lake Vostok is 16,000–20,000 years (Bell *et al.*, 2002), and this ice could contain viable microorganisms. Bidle *et al.* (2007) have shown the potential viability of microorganisms preserved in ice up to ~300,000 years. Calculations by J.M. Tiedje suggest that one would expect on average a change in only one base pair per gene in the

1 million years since Lake Vostok microbes have been isolated from their relatives and species level differentiation may take at least 10–100 million years.

As a result of this isolation and novel selection pressures in Antarctic lake systems, there is a potential for the discovery of novel extremophilic species, biochemical pathways and physiological adaptations (Pearce and Laybourn-Parry, in press). Any extremophiles found in the subglacial lakes are a potential source for novel enzymes and interest is likely to focus on low-temperature enzyme activity, cold shock induction and ice-active substances (Ferrer *et al.*, 2007). Novel physiological adaptations could also suggest evolutionary separation – biofilm formation and synergy may be two physiological strategies for nutrient acquisition in the subglacial systems. We already know that low concentrations of nutrients in other Antarctic freshwater habitats have led to nitrogen fixation levels of 1 g m^{-2} $year^{-1}$ in cyanobacterial mats, so nitrogen availability is likely to be a key nutritional factor controlling microbial production (Olson *et al.*, 1998). It has also been suggested that low temperatures might induce the viable but non-culturable (VBNC) state in Antarctic lake microorganisms and the VBNC state of some bacteria, collected from Antarctic lakes, has been reported (Chattopadhyay, 2000). A bacterial strain recovered from glacial ice at a depth of 3519 m, just above the accreted ice from subglacial Lake Vostok, was found to produce a 54 kDa ice-binding protein (GenBank EU694412) that is similar to ice-binding proteins previously found in sea ice diatoms, a snow mould and a sea ice bacterium. The protein has the ability to inhibit the recrystallization of ice, a phenotype that has clear advantages for survival in ice (Raymond *et al.*, 2009) and has potential food industry and pharmaceutical applications. As part of a project aimed at the selection of cold-adapted yeasts expressing biotechnologically interesting features, the extracellular enzymatic activity (EEA) of 91 basidiomycetous yeasts isolated from glacial and subglacial waters of north-west Patagonia (Argentina) has also been investigated (Brizzio *et al.*, 2007). The cold-adapted yeasts are currently considered a potential source of industrially relevant cold-active enzymes.

7.8 Conclusions

Antarctic subglacial lakes represent one of the last frontiers for exploration on Earth and hold the exciting prospect of novel organisms, particularly the potential 'seed bank' trapped within overlying glacial ice. There could be differences between systems found at the ice divide, deep within the continental interior and associated with coastal features. As potentially isolated systems, the microorganisms they might contain will tell us a great deal about microbial biogeography, the microbial ubiquity hypothesis and the potential for everything to be everywhere. Microorganisms isolated from subglacial lake ecosystems may allow us to determine the response threshold for individual stresses in Antarctic environments, the extent of psychrophily, and gain new insights into life at low temperatures. They may even provide clues to the origin of life itself. A detailed analysis of any community present, both in the water column and the underlying sediments could provide a link to past microbial communities, and metagenomic analysis may allow a study of ecosystem function.

Further questions we might address are: whether individual species are important (i.e. do the same biogeochemical activities happen regardless of which species are present) and the link between the active microorganisms and the gene pool in the 'seed bank' of microbial propagules. We could also monitor seasonal fluctuations in population structure and target increases and collapses in species richness/diversity to identify potential causes and possible links to biogeochemical cycling, as well as investigating the role of viruses in ecosystem function. Furthermore, it is hoped that Antarctic subglacial lakes will provide us with an opportunity to test technologies that may be applied to the future exploration for life on other planets with a cryosphere, for example, on Mars, Saturn's moon Titan, and Jupiter's moons Europa and Callisto (Abyzov *et al.*, 2001; Price *et al.*, 2002).

In summary, it has been hypothesized that Antarctic subglacial lakes are extreme, yet viable habitats, which may host unique microbial assemblages that have been potentially isolated for millions of years (Siegert *et al.*, 2001). Price (2000) stated that we do not yet know of any species capable of tolerating all of the following conditions: no sunlight, no oxygen, high pressure (400 bar), temperature below 0°C, and a highly acidic or saline medium. It is therefore possible that microbes living in liquid veins in ancient ice would be different from any yet known. However, the microbes detected so far, in the two Antarctic subglacial systems sampled to date (accreted ice from subglacial Lake Vostok in East Antarctica and saturated till from beneath ice streams draining the West Antarctic Ice Sheet) (e.g. Karl *et al.*, 1999; Priscu *et al.*, 1999; Christner *et al.*, 2006; Bulat *et al.*, 2009, Lanoil *et al.*, 2009), have DNA profiles similar to those of contemporary surface microbes. Microbiologists expect, however, that Lake Vostok, and other subglacial lakes will harbour unique species, particularly within the deeper waters and associated sediments (Siegert *et al.*, 2003).

Useful Websites

http://english.ruvr.ru/2011/01/25/41230102.html
http://www.sale.scar.org/key-programs/subglacial-hydrology/
http://wissard.org/
http://salegos-scar.montana.edu/
http://www.geos.ed.ac.uk/research/ellsworth/
http://www.ellsworth.org.uk

References

Abyzov, S.S., Lipenkov, V.Y., Bobin, N.E. and Kudryashov, B.B. (1982) Microflora of central Antarctic glacier and methods for sterile ice-core sampling for microbiological analyses. *Biological Bulletin of the Academy of Sciences USSR* 9, 304–349.

Abyzov, S.S., Mitskevich, I.N., Poglazova, M.N., Barkov, I.N., Lipenkov, V.Y., Bobin, N.E., Koudryashov, B.B., Pashkevich, V.M. and Ivanov, M.V. (2001) Microflora in the basal strata at Antarctic ice core above the Vostok lake. *Advances in Space Research* 28, 701–706.

Abyzov, S.S., Hoover, R.B., Imura, S., Mitskevicha, I.N., Naganuma, T., Poglazova, M.N. and Ivanov, M.V. (2004) Use of different methods for discovery of ice-entrapped microorganisms in ancient layers of the Antarctic glacier. *Advances in Space Research* 33, 1222–1230.

Bell, R.E., Studinger, M., Tikku, A.A., Clarke, G.K.C., Gutner, M.M. and Meertens, C. (2002) Origin and fate of Lake Vostok water frozen to the base of the East Antarctic ice sheet. *Nature* 416, 307–310.

Bell, R.E., Studinger, M., Shuman, C.A., Fahnestock, M.A. and Joughin, I. (2007) Large subglacial lakes in East Antarctica at the onset of fast-flowing ice streams. *Nature* 445, 904–907.

Bidle, K.D., Lee, S.H., Marchant, D.R. and Falkowski, P.G. (2007) Fossil genes and microbes in the oldest ice on Earth. *Proceedings of the National Academy of Sciences of the United States of America* 104, 13455–13460.

Björnsson, H. (2003) Subglacial lakes and jökulhlaups in Iceland. *Global and Planetary Change* 35, 255–271.

Brizzio, S., Turchetti, B., García, V. de, Libkind, D., Buzzini, P. and Broock, M.V. (2007) Extracellular enzymatic activities of basidiomycetous yeasts isolated from glacial and subglacial waters of northwest Patagonia (Argentina). *Canadian Journal of Microbiology* 53, 519–525.

Bulat, S.A., Alekhina, I.A., Krylenkov, V.A. and Lukin, V.V. (2002) Molecular biological studies of microbiota in subglacial Lake Vostok (the Antarctic). *Advances in Current Biology* 122, 211–221.

Bulat, S.A., Alekhina, I.A., Blot, M., Petit, J.-R., de Angelis, M., Wagenbach, D., Lipenkov, V.Y., Vasilyeva, L.P., Wloch, D.M., Raynaud, D.R. and Lukin, V.V. (2004) DNA signature of thermophilic bacteria from the aged accretion ice of Lake Vostok, Antarctica: implications for searching for life in extreme icy environments. *International Journal of Astrobiology* 3, 1–12.

Bulat, S.A., Alekhina, I.A., Lipenkov, V. Ya., Lukin, V.V., Marie, D. and Petit, J.R. (2009) Cell concentrations of microorganisms in glacial and lake ice of the Vostok ice core, East Antarctica. *Microbiology* 78, 808–810.

Carpenter, E.J., Lin, S. and Capone, D.G. (2000) Bacterial activity in South Pole snow. *Applied and Environmental Microbiology* 66, 4514–4517.

Chattopadhyay, M.K. (2000) Cold adaptation of Antarctic microorganisms – possible involvement of viable but non culturable state. *Polar Biology* 23, 223–224.

Christner, B.C., Mosley-Thompson, E., Thompson, L.G., Zagorodnov, V., Sandman, K. and Reeve, J.N. (2000) Recovery and identification of viable bacteria immured in glacial ice. *Icarus* 144, 479–485.

Christner, B.C., Mosley-Thompson, E., Thompson, L.G. and Reeve, J.N. (2001) Isolation of bacteria and 16S rDNAs from Lake Vostok accretion ice. *Environmental Microbiology* 3, 570–577.

Christner, B.C., Royston-Bishop, G., Foreman, C.M., Arnold, B.R., Tranter, M., Welch, K.A., Berry Lyons, W., Tsapin, A.I., Studinger, M. and Priscu, J.C. (2006) Limnological conditions in subglacial Lake Vostok, Antarctica. *Limnology and Oceanography* 51, 2485–2501.

Christner, B.C., Mosley-Thompson, E., Thompson, L.G. and Reeve, J.N. (2008) Isolation of bacteria and 16S rDNAs from Lake Vostok accretion ice. *Environmental Microbiology* 3, 570–577.

Dancer, S.J., Shears, P. and Platt, D.J. (1997) Isolation and characterization of coliforms from glacial ice and water in Canada's High Arctic. *Journal of Applied Microbiology* 82, 597–609.

De Angelis, M., Petit, J.-R., Savarino, J., Souchez, R. and Thiemens, M.H. (2004) Contributions of an ancient evaporitic-type reservoir to subglacial Lake Vostok chemistry. *Earth and Planetary Science Letters* 222, 751–765.

D'Elia, T., Veerapaneni, R. and Rogers, S.O. (2008) Isolation of microbes from Lake Vostok accretion ice. *Applied and Environmental Microbiology* 74, 4962–4965.

Dowdeswell, J.A. and Siegert, M.J. (2003) The physiography of modern Antarctic subglacial lakes. *Global and Planetary Change* 35, 221–236.

Ferrer, M., Golyshina, O., Beloqui, A. and Golyshin, P.N. (2007) Mining enzymes from extreme environments. *Current Opinion in Microbiology* 10, 207–214.

Foght, J., Aislabie, J., Turner, S., Brown, C.E., Ryburn, J., Saul, D.J. and Lawson, W. (2004) Culturable bacteria in subglacial sediments and ice from two southern hemisphere glaciers. *Microbial Ecology* 47, 329–340.

Forieri, A., Tabacco, I.E., Cafarella, L., Urbini, S., Bianchi, C. and Zirizzotti, A. (2008) Evidence for possible new subglacial lakes along a radar transect crossing the Belgica Highlands and the Concordia Trench. *Terra Antarctica Reports* 14, 209–212.

Fricker, H.A. and Scambos, T. (2009) Connected subglacial lake activity on lower Mercer and Whillans Ice Streams, West Antarctica, 2003–2008. *Journal of Glaciology* 55, 303–315.

Fricker, H.A., Scambos, T., Bindschadler, R. and Padman, L. (2007) An active subglacial water system in West Antarctica mapped from space. *Science* 315, 1544–1548.

Gaidos, E., Lanoil, B., Thorsteinsson, T., Graham, A., Skidmore, M., Han, S.-K., Rust, T. and Popp, B. (2004) A viable microbial community in a subglacial volcanic crater lake, Iceland. *Astrobiology* 4, 327–344.

Gaidos, E., Marteinsson, V., Thorsteinsson, T., Jóhannesson, T., Rúnarsson, A.R., Stefansson, A., Glazer, B., Lanoil, B., Skidmore, M., Han, S., Miller, M., Rusch, A. and Foo, W. (2009) An oligarchic microbial assemblage in the anoxic bottom waters of a volcanic subglacial lake. *ISME Journal* 3, 486–497.

Hodgson, D.A., Roberts, S.J., Bentley, M.J., Smith, J.A., Johnson, J.S., Verleyen, E., Vyvermanc, W., Hodson, A.J., Lenge, M.J., Cziferszkya, A., Fox, A.J. and Sanderson, D.C.W. (2009a) Exploring former subglacial Hodgson Lake, Antarctica. Paper I: site description, geomorphology and limnology. *Quaternary Science Reviews* 28, 2295–2309.

Hodgson, D.A., Roberts, S.J., Bentley, M.J., Carmichael, E.L., Smith, J.A., Verleyen, E., Vyverman, W., Geissler, P., Leng, M.J. and Sanderson, D.C.W. (2009b) Exploring former subglacial Hodgson Lake, Antarctica Paper II. *Quaternary Science Reviews* 28, 2310–2325.

Hughes, K.A., McCartney, H.A., Lachlan-Cope, T.A. and Pearce, D.A. (2004) A preliminary study of airborne biodiversity over peninsular Antarctica. *Cellular and Molecular Biology* 50, 537–542.

Kapitsa, A.P., Ridley, J.K., Robin, G. de Q., Siegert, M.J. and Zotikov, I.A. (1996) A large deep freshwater lake beneath the ice of central East Antarctica. *Nature* 381, 684–686.

Karl, D.M., Bird, D.F., Björkman, K., Houlihan, T., Shackelford, R. and Tupas, L. (1999) Microorganisms in the accreted ice of Lake Vostok, Antarctica. *Science* 286, 2144–2147.

Lake Ellsworth Consortium (2007) Exploration of Ellsworth Subglacial Lake: a concept paper on the development, organisation and execution of an experiment to explore, measure and sample the environment of a West Antarctic subglacial lake. *Reviews in Environmental Science and Biotechnology* 6, 1569–1705.

Lake Ellsworth Consortium (Bentley, M.J., Blake, D., Capper, L., Clarke, R., Cockell, C.S., Corr, H.F.J., Harris, W., Hill, C., Hindmarsh, R.C.A., Hodgson, D.A., King, E.C., Lamb, H., Maher, B., Makinson, K., Mowlem, M., Parnell, J., Pearce, D.A., Priscu, J., Rivera, A., Ross, N., Siegert, M.J., Smith, A.M., Tait, A., Tranter, M., Wadham, J.L., Whalley, W.B. and Woodward, J.) (2011) Ellsworth Subglacial Lake, West Antarctica: a review of its history and recent field campaigns. In: Siegert, M.J., Kennicutt, M.C. and Bindschadler, R. (eds) *Antarctic Subglacial Aquatic Environment*. AGU Geophysical Monograph Series, Volume 192, 254 pp.

Lanoil, B., Skidmore, M., Priscu, J.C., Han, S., Foo, W., Vogel, S.W., Tulaczyk, S. and Engelhardt, H. (2009) Bacteria beneath the West Antarctic Ice Sheet. *Environmental Microbiology* 11, 609–615.

Laybourn-Parry, J. (2002) Survival mechanisms in Antarctic lakes. *Philosophical Transactions of the Royal Society London, Series B: Biological Sciences* 357, 863–869.

Laybourn-Parry, J. (2009) No place too cold. *Science* 324, 1521–1522.

Lavire, C., Normand, P., Alekhina, I., Bulat, S., Prieur, D., Birrien, J.-L., Fournier, P., Hänni, C. and Petit, J.R. (2006) Presence of *Hydrogenophilus thermoluteolus* DNA in accretion ice in the subglacial Lake Vostok, Antarctica, assessed using rrs, cbb and hox. *Environmental Microbiology* 8, 2106–2114.

Mikucki, J.A. and Priscu, J.C. (2007) Bacterial diversity associated with Blood Falls, a subglacial outflow from the Taylor Glacier, Antarctica. *Applied and Environmental Microbiology* 73, 4029–4039.

Mowlem, M.C., Tsaloglou, M.-N., Waugh, E.M., Floquet, C.F.A., Saw, K., Fowler, L., Brown, R., Pearce, D., Wyatt, J.B., Beaton, A.D., Brito, M.P., Hodgson, D.A., Griffiths, G. and the Lake Ellsworth Consortium (Bentley, M., Blake, D., Campbell, H., Capper, L., Clarke, R., Cockell, C., Corr, H., Harris, W., Hill, C., Hindmarsh, R., King, E., Lamb, H., Maher, B., Makinson, K., Parnell, J., Priscu, J., Rivera, A., Ross, N., Siegert, M.J., Smith, A., Tait, A., Tranter, M., Wadham, J., Whalley, B. and Woodward, J.) (2011) Probe technology for the direct measurement and sampling of subglacial Lake Ellsworth. *Environments and the Search for Life Beyond the Earth* (in press).

Olson, J.B., Steppe, T.F., Litaker, R.W. and Paerl, H.W. (1998) N2 fixing microbial consortia associated with the ice cover of Lake Bonney, Antarctica. *Microbial Ecology* 36, 231–238.

Pearce, D.A. (2009) Antarctic subglacial lakes – a new frontier in microbial ecology. *ISME Journal* 3, 877–880.

Pearce, D.A. and Laybourn-Parry, J. (2011) Antarctic lakes as models for the study of microbial biodiversity, biogeography and evolution. In: Rogers *et al.* (eds) *Antarctic Ecosystems: An Extreme Environment in a Changing World* (in press).

Pearce, D.A. and Wilson, W.H. (2003) Viruses in Antarctic ecosystems. *Antarctic Science* 15, 319–331.

Pearce, D.A., Hughes, K.A., Harangozo, S.A., Lachlan-Cope, T.A. and Jones, A.E. (2010) Biodiversity of air-borne microorganisms at Halley station, Antarctica. *Extremophiles* 14, 145–159.

Poglazova, M.N., Mitskevich, I.N., Abyzov, S.S. and Ivanov, M.V. (2001) Microbiological characterization of the accreted ice of subglacial Lake Vostok, Antarctica. *Microbiology* 70, 723–730.

Pokorný, R., Olejníková, P., Balog, M., Zifčák, P., Hölker, U., Janssen, M., Bend, J., Höfer, M., Holienčin, R., Hudecová, D. and Varečka, L. (2005) Characterization of microorganisms isolated from lignite excavated from the Záhorie coal mine (southwestern Slovakia). *Research in Microbiology* 156, 932–943.

Price, P.B. (2000) A habitat for psychrophiles in deep Antarctic ice. *Proceedings of the National Academy of Sciences of the United States of America* 97, 1247–1251.

Price, P.B. (2006) Microbial life in glacial ice and implications for a cold origin of life. *FEMS Microbiology Ecology* 59, 217–231.

Price, P.B., Nagornov, O.V., Bay, R., Chirkin, D., He, Y., Miocinovic, P., Richards, A., Woschnagg, K., Koci, B. and Zagorodnov, V. (2002) Temperature profile for glacial ice at the South Pole: implications for life in a nearby subglacial lake. *Proceedings of the National Academy of Sciences of the United States of America* 99, 7844–7847.

Priscu, J.C., Adams, E.E., Lyons, W.B., Voytek, M.A., Mogk, D.W., Brown, R.L., McKay, C.P., Takacs, C.D., Welch, K.A., Wolf, C.F., Kirshtein, J.D. and Avci, R. (1999) Geomicrobiology of subglacial ice above Lake Vostok, Antarctica. *Science* 286, 2141–2144.

Priscu, J.C., Christner, B.C., Foreman, C.F. and Royston-Bishop, G. (2007) Biological material in ice cores. In: Elias, S.A. (ed.) *Encyclopaedia of Quaternary Science*, Vol. 2. Elsevier, UK, pp. 1156–1166.

Raymond, J.A., Christner, B.C. and Schuster, S.C. (2009) A bacterial ice-binding protein from the Vostok ice core. *Extremophiles* 12, 713–717.

Robin, G.D.Q., Swithinbank, C.W.M. and Smith, B.M.E. (1970) Radio echo exploration of the Antarctic ice sheet. International Symposium on Antarctic Glaciological Exploration (ISAGE), Hanover, New Hampshire, United States, 3–7 September 1968.

Säwström, C., Anesio, M.A., Granéli, W. and Laybourn-Parry, J. (2007) Seasonal viral loop dynamics in two large ultraoligotrophic Antarctic freshwater lakes. *Microbial Ecology* 53, 1–11.

Sharma, A., Scott, J.H., Cody, G.D., Fogel, M.L., Hazen, R.M., Hemley, R.J. and Huntress, W.T. (2002) Microbial activity at gigapascal pressures. *Science* 295, 1514–1516.

Sheridan, P.P., Loveland-Curtze, J., Miteva, V.I. and Brenchley, J.E. (2003) *Rhodoglobus vestalii* gen. nov. sp. nov., a novel psychrophilic organism isolated from an Antarctic Dry Valley Lake. *International Journal of Systematic and Evolutionary Microbiology* 53, 985–994.

Siegert, M.J. (2000) Antarctic subglacial lakes. *Earth-Science Reviews* 50, 29–50.

Siegert, M.J., Ellis-Evans, J.C., Tranter, M., Mayer, C., Petit, J.-R., Salamatin, A. and Priscu, J.C. (2001) Physical, chemical and biological processes in Lake Vostok and other Antarctic subglacial lakes. *Nature* 414, 603–609.

Siegert, M.J., Tranter, M., Ellis-Evans, J.C., Priscu, J.C. and Berry Lyons, W. (2003) The hydrochemistry of Lake Vostok and the potential for life in Antarctic subglacial lakes. *Hydrological Processes* 17, 795–814.

Siegert, M.J., Hindmarsh, R.C.A., Corr, H.F.J., Smith, A.M., Woodward, J., King, E.C., Payne, A.J. and Joughin, I. (2004) Subglacial Lake Ellsworth: a candidate for *in situ* exploration in West Antarctica. *Geophysical Research Letters* 31, L23403, doi:10.1029/2004GL021477.

Siegert, M.J., Carter, S., Tabacco, I., Popov, S. and Blankenship, D.D. (2005) A revised inventory of Antarctica subglacial lakes. *Antarctic Science* 17, 453–460.

Siegert, M.J., Kennicutt II, M.C. and Bindschadler, R.A. (2011) Antarctic subglacial aquatic environments. *Geophysical Monograph Series*, Vol. 192, 254 pp.

Sjöling, S. and Cowan, D.A. (2003) High 16S rDNA bacterial diversity in glacial meltwater lake sediment, Bratina Island, Antarctica. *Extremophiles* 7, 275–282.

Skidmore, M., Anderson, S.P., Sharp, M., Foght, J. and Lanoil, B.D. (2005) Comparison of microbial community compositions of two subglacial environments reveals a possible role for microbes in chemical weathering processes. *Applied and Environmental Microbiology* 71, 6986–6997.

Sonjak, S., Frisvad, J.C. and Gunde-Cimerman, N. (2006) *Penicillium mycobiota* in Arctic subglacial ice. *Microbial Ecology* 52, 207–216.

Thoma, M., Grosfeld, K., Filinac, I. and Mayer, C. (2009) Modelling flow and accreted ice in subglacial Lake Concordia, Antarctica. *Earth and Planetary Science Letters* 286, 278–284.

Thomas, D.N. and Dieckmann, G.S. (2002) Antarctic sea ice – a habitat for extremophiles. *Science* 295, 641–644.

Tikku, A.A., Bell, R.E., Studinger, M. and Clarke, G.K.C. (2004) Ice flow field over Lake Vostok, East Antarctica, inferred by structure tracking. *Earth and Planetary Science Letters* 227, 249–261.

Tranter, M., Skidmore, M. and Wadham, J. (2005) Hydrological controls on microbial communities in subglacial environments. *Hydrological Processes* 19, 995–998.

Tung, H.C., Bramall, N.E. and Price, P.B. (2005) Microbial origin of excess methane in glacial ice and implications for life on Mars. *Proceedings of the National Academy of Sciences of the United States of America* 102, 18292–18296.

Willerslev, E., Hansen, A.J., Christensen, B., Steffensen, J.P. and Arctander, P. (1999) Diversity of Holocene life-forms in fossil glacier ice. *Proceedings of the National Academy of Sciences of the United States of America* 96, 8017–8021.

Wilson, W.H., Lane, D., Pearce, D.A. and Ellis-Evans, J.C. (2000) Transmission electron microscope analysis of virus-like particles in the freshwater lakes of Signy Island, Antarctica. *Polar Biology* 23, 657–660.

Wingham, D.J., Siegert, M.J., Shepherd, A. and Muir, A.S. (2006) Rapid discharge connects Antarctic subglacial lakes. *Nature* 440, 1033–1036.

Wright, A. and Siegert, M.J. (2011) Studies of other subglacial lakes in IPY. In: *Understanding Earth's Polar Challenges: International Polar Year 2007–2008*. IPY Joint Committee, 277 pp.

8 Cold Alpine Regions

Birgit Sattler,[1] Barbara Post,[1] Daniel Remias,[2] Cornelius Lütz,[3] Herbert Lettner[4] and Roland Psenner[1]

[1]*University of Innsbruck, Institute of Ecology, Austria;*
[2]*University of Innsbruck, Institute of Pharmacy, Austria;*
[3]*University of Innsbruck, Institute of Botany, Austria;*
[4]*Division of Physics and Biophysics, Department of Materials Engineering and Physics, University of Salzburg, Austria*

8.1 Introduction

When describing life in extreme environments, temperature is one of the major factors for consideration. The general perception of life in the cold is mostly associated with polar regions rather than high altitude environments. However, based on the United Nations Environment Programme World Conservation Monitoring Centre's Classes 1–7 of mountain regions, the global mountain area covers almost 40 million km^2 (approximately 27% of the Earth's surface). The proportion of land area that is mountainous is approximately 24% compared to ca. 14% considered to be polar, excluding Antarctica. Thus alpine regions are worthy of close consideration.

There are numerous parallels between high-latitude versus high-altitude regions. The parameters that change with latitude in polar regions are altered by altitude in alpine areas. In general, air pressure, temperature and humidity decrease with increasing altitude whereas solar radiation (especially UV) and wind speed increase. At 1500 m, the partial pressure of oxygen is about 84 % of the value at sea level, falling to 75 % at 2500 m and 63 % at 3500 m. On average, air temperature decreases about 5.5°C for every 1000 m increase in altitude (but varies diurnally, seasonally, latitudinally and from region to region). Moreover, air holds less water vapour as temperatures fall with increasing altitude. All these factors substantially reduce the duration of the productive biological period.

Alpine regions exert a major influence on weather, climates and hydrological resources, as well as providing recreational areas for human societies. They are also the most sensitive environments to changing processes such as climate, making them prime areas of environmental concern. Unfortunately, over the past 50 years a severe change in alpine regions has been observed due to an accumulation of global changes. Examples include industrialization, tourism and pollution.

Alpine regions host a variety of ecosystems that sustain life, including nearly all compartments of the cryosphere containing freshwater (Fig. 8.1), such as glaciers (including glacial streams, beds and glacier forefields), high mountain lakes, snow areas and permafrost. The overarching system which influences these cryospheric compartments is the atmosphere in which supercooled cloud droplets inoculate remote areas with airborne microorganisms (Sattler *et al.*, 2001; Pearce *et al.*, 2009).

Temperature is the main driver governing the availability of liquid water – the prerequisite for life. Ice and snow covers on

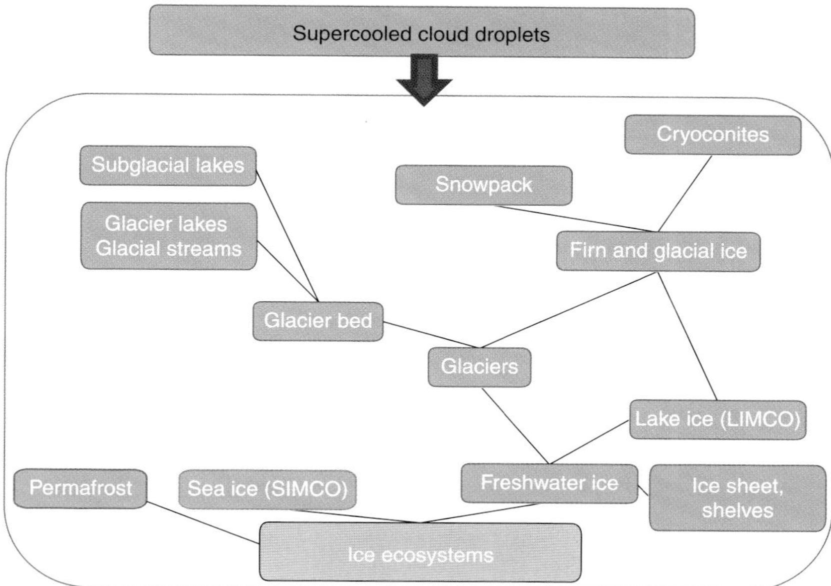

Fig. 8.1. The various components of the cryosphere with the atmosphere containing supercooled cloud droplets as the overarching system. © Roland Psenner and Birgit Sattler.

solid ground can – if they are not too thick and solar radiation is strong enough – initiate the formation of liquid water layers. Solar radiation is necessary to produce heat to melt the ice, and to provide radiation for photosynthesis. Hence, active metabolism in solid ice depends either on direct sunlight and/or exclusion of ions during the freezing process to create veins inside a matrix filled with brine. The formation of brine channels is also the prerequisite for the numerously reported metabolic rates within sea ice where organisms can thrive during the entire ice-covered season (see Thomas, Chapter 4, this volume). In the course of discussions about possible life forms in Lake Vostok (a subglacial lake under 4 km of ice in the Eastern Antarctic, referred to in Pearce, Chapter 7, this volume), Buford Price (2000) established solid ice, formed by the transition from snow to glacial ice, as another potential microbial habitat (Price, 2000): 'I propose a habitat consisting of interconnected liquid veins along three-grain boundaries in ice in which psychrophilic (cold loving) bacteria can move and obtain energy and carbon from ions in solution'.

Compared to the size of the well described brine channels in sea ice (Thomas, Chapter 4, this volume), veins in alpine ice are relatively small with a diameter of only a few micrometres. Hence, microbial abundances and metabolic rates will not reach values as high as those observed in the ice cover of the polar ocean. Polar lake ice has boundary layers between ice crystals as well, however the metabolic 'hot spots' are located in sediment clusters which have sunk into the ice due to the absorption of solar energy (Priscu *et al.*, 1998, 2002). The sinking distance is a function of the solar heat and the radius of the sediment particle.

8.2 Alpine Cryospheric Compartments and the Diversity of Life Within Them

8.2.1 Ice covers on high mountain lakes

The ice cover on high-latitude and high-altitude freshwater bodies plays a major role in determining the environmental properties of the pelagic world underneath by reducing or completely inhibiting heat and gas exchange, and wind-generated mixing of the water column (Catalan, 1992). Consequently, the liquid water remains quiescent and is dominated by molecular processes (i.e. turbulent transport processes become insignificant). For these reasons, ice does not only influence the transfer of energy and matter between the atmosphere and the water, but within the pelagic system itself. Light transmission is severely limited by snow layers and can be reduced by up to 99.9% depending on the thickness and structure of the snow cover (Felip *et al.*, 1995, 2002; Psenner and Sattler, 1998; Psenner *et al.*, 1999, 2003). In alpine regions with regular snowfall, winter covers have a completely different structure to those in polar regions with virtually no precipitation, exemplified by the Dry Valleys in Antarctica (see Hogg and Wall, Chapter 10, this volume). Due to repeated events of snowfall in high alpine regions the weight of the snow on top of lake ice is exerting pressure which causes a flooding of the ice sheet with lake water. This in turn evokes the formation of alternating layers of slush and white ice that are so characteristic of alpine lakes. Winter lake covers in areas without snowfall, as is the case for the McMurdo Dry Valleys, are generally composed of uniform ice with inclusions of water lenses, but no horizontal layering (Psenner and Sattler, 1998). Moreover, freshwater lakes in polar regions are mostly characterized by perennial ice covers (see Laybourn-Parry and Bell, Chapter 6, this volume) whereas high mountain lakes are generally seasonal.

Formation and characterization of alpine lake ice

The entrapment of atmospheric gases in ice-covered lakes can result in severe oxygen deficiency and strong over-saturation of CO_2 and other gases in the water beneath. Nutrient input from the atmosphere is also inhibited by the ice cover, and sedimentation fluxes can become very low. Thus, the formation of an ice cover is a major factor limiting the productivity in the pelagic zone of alpine lakes, and can turn them into heterotrophic systems for long periods of time. In addition to this, ice covers add a new structural element where active and diverse microbial communities can develop. Dynamic changes in the physical structure of seasonal ice covers are driven by snowfall, melting, freezing and flooding – similar to those processes creating active microbial habitats on the surface of sea ice (Thomas, Chapter 4, this volume). Ice sheets on alpine lakes lasting for half a year or longer are generally characterized by slush layers located between solid ice, created through dynamic flood–freeze cycles. Slush may best be compared to sandy sediments, consisting of ice crystals rather than sediment grains. In general the seasonal ice cover starts to develop on a cold autumn night, with the formation of black ice which is optically clear and hard and virtually devoid of air bubbles. Snowfall on the black ice pushes it downward and lake water infiltrates the base of the snow cover (which is an excellent thermal isolator) through cracks. Freezing of the infiltrated water can create a solid layer consisting of snow crystals and frozen lake water (called white ice) on the surface of the black ice. Incomplete freezing of the infiltrated snow and subsequent flooding and freezing, depending on the hydrostatic adjustment, creates a sandwich-like structure consisting of alternating layers of snow, white ice, and slush on top of black ice (Fig. 8.2).

During spring the seasonal ice cover becomes thinner and structurally weaker. This, and the formation and widening of cracks, enhance gas exchange with the atmosphere and increase sedimentation and light transmission. The increase in light penetration can enhance photosynthesis of phytoplankton and, consequently, bacterial activity in the water column. However, the influence of the ice cover on pelagic life may not stop upon complete melting, because the microbial communities that develop in the ice can provide an inoculum to the water column. This can influence the development and composition of microbial communities in the open lake (Alfreider *et al.*, 1996). It should be noted that this effect

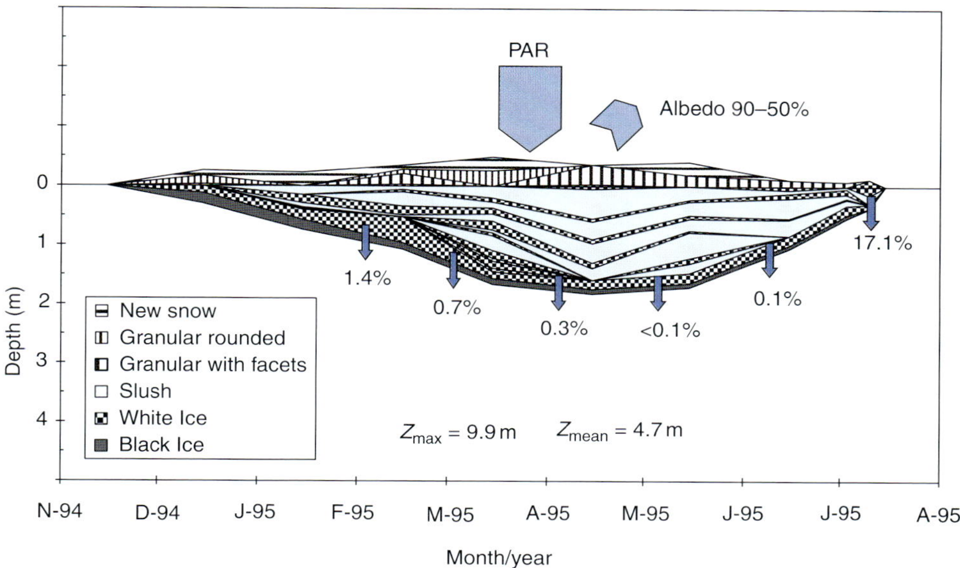

Fig. 8.2. Winter cover of an alpine high mountain lake showing the characteristic layers of black ice, white ice and slush. © Birgit Sattler and Anton Wille.

may be restricted to superficial water layers and short periods, since the disappearance of the ice cover is accompanied by rapid flushing of the lake with meltwater.

Associated with the physical changes in the ice cover are chemical changes within the different layers. This can lead to either concentration or dilution of dissolved and particulate matter by partial melting, wash-out, or entrainment of material from the atmosphere, the catchment and the lake water. As a result, slush layers enriched in nutrients (phosphorus, ammonium and nitrate) are common, and at the end of winter, when light attenuation by snow decreases with thawing, light and nutrient levels can be sufficient for net photosynthesis. In addition, seepage water from melting snowfields in the drainage area and atmospheric deposition can deliver organic matter and nutrients to the cover.

The infiltration of lake water into the ice cover transports organisms, inorganic nutrients and other lake water constituents into higher strata of the winter cover where they often encounter more light than below the ice. Microorganisms found in slush layers, however, originate also from atmospheric and terrestrial sources. Depending on the presence of soils and the degree of vegetation in the catchment area during the ablation (the removal of material from the ice surface by vaporization or other erosive processes) period, the diversity of terrestrial communities which are introduced into the winter cover increases.

Microbial communities in alpine lake ice

Food webs inside the layers of alpine lake ice are truncated and mainly dominated by microbes consisting of bacteria, viruses, ciliates and algae. On very rare occasions one can find Metazoa such as rotifers, which are trapped during the freezing process. Apart from numbers and dominances they resemble the food webs of the pelagic zone of the underlying lake.

Prokaryotes in slush layers tend to dominate in terms of abundance and biomass. Bacterial communities are morphologically diverse in slush layers and lake water (Alfreider *et al.*, 1996; Psenner *et al.*, 1999, 2003). Water-column bacteria are generally small free-living cells, mostly short rods and cocci with volumes of ca. 0.05–0.1 μm³. However, in slush layers, along with these small forms, long thin filaments appear, sometimes longer than 100 μm, which are the dominant organisms regarding bacterial biomass. Long branched filaments and stalked bacteria are not uncommon, and in a few cases *Ancalomicrobium*-like morphologies are found. To obtain information concerning bacterial community structure, *in situ* hybridization with fluorescent labelled rRNA-targeted oligonucleotide probes was carried out by using probes specific for all members of the domain bacteria, for the alpha, beta and gamma subclasses of the class Proteobacteria, and for the *Cytophaga-Flavobacterium* group (Alfreider *et al.*, 1996). The results showed a distinct bacterial community composition in the different snow, slush and pelagic layers.

Similarly discrete communities of eukaryotic Protozoa are also found in the different ice layers. Felip *et al.* (2002) showed that the species compositions in three distinct habitats (slush layers, surface pools and lake water) in Estany Redó, the Pyrenees, Andorra, were clearly different from each other and this difference persisted over different years. The only autotrophic species occurring only in the winter cover are those found in surface pools (the green algae *Chlamydomonas nivalis* and an unidentified *Chlamydomonas* sp., and an unidentified yellow-brown alga, *Chromulina* sp.). Non-flagellated organisms seldom appear in the slush layers. Other autotrophs found in slush are the Cryptophyceae, *Rhodomonas minuta*, the dinoflagellate, *Gymnodinium uberrimum* and, centric diatom, *Cyclotella* sp., which appear also in the pelagic zone of the lake. Ciliated Protozoa are common in high alpine lake water and typically include species such as *Askenasia chlorelligera*, *Urotricha* sp. and *Balanion planctonicum*, whereas slush layers are characterized by nonplanktonic genera such as *Urosoma*, *Lacrymaria* and *Dileptus* and some unidentified species. In addition, rotifers such as *Synchaeta* sp. appear, but these are also observed in the pelagic zone of the lake.

At the end of the winter, the upper slush layers, and especially the newly formed pools at the surface, are rich in allochthonous (externally derived) material, such as pollen, conidia (non-motile fungal spores), leaves, stem debris and amorphous organic material. Bacterial secondary production and respiration in slush layers generally increases with growth and development of the winter cover, and ranges from values characteristic for ultraoligotrophic to eutrophic lakes.

It seems paradoxical that metabolic efficiencies of bacterial secondary production reach higher values at freezing point than in the warmer pelagic zone. Although organisms living at low temperatures have higher nutrient requirements to sustain metabolic activities, growth efficiencies may be higher (e.g. Pomeroy and Wiebe, 2001) for marine bacteria. Marine bacteria living at 0°C can be substrate limited and their growth can be stimulated by the addition of nutrients and organic compounds. Therefore, high nutrient concentrations may be necessary in cold-water and icy environments in order to compensate for the low temperature constraints on metabolic processes and growth. Experiments to determine temperature optima of water column and ice bacteria indicate that bacteria living in slush layers are adapted to temperatures from 0 to 5°C and their production decreased dramatically with rising temperatures. By contrast lake-water bacteria show the opposite trend – their preferred temperature range is around 15°C. Hence, there seems to be not only a taxonomic but also a physiological difference between the bacterial assemblages in the winter ice cover and those in the underlying lake water.

High nutrient concentrations in slush layers could stimulate bacterial growth. Slush layers are inhabited by algae that support bacteria through extracellular release of carbon during photosynthetic activity. From the formation of the ice cover to its decay there is a shift from a system dominated by heterotrophs to an assemblage dominated by autotrophic organisms. This progression has consequences for the water column after the ice breaks out delivering a high biomass inoculum into the pelagic zone (Psenner *et al.*, 2003).

Another effect favouring high microbial biomasses inside slush is the very low density of protozoan and metazoan grazers. Microbial interactions in the ice cover are dominated mainly by bacteria and algae, thus being more similar to Antarctic lake ice than to the alpine lake-water itself. Although Protozoa such as heterotrophic nanoflagellates and ciliates are observed, their abundance is generally so low that they exert little grazing pressure. One possible reason for the low number of grazers might be the sediment-like structure of slush layers, which reduces mobility and prevents effective grazing. It is not known if the low temperature also limits protistan activity. As mentioned before, beside the occurrence of metazoans (mostly rotifers), only ciliate species typically found in sediments occur in the slush, which has led to the assumption that the slush habitat resembles conditions in benthic strata or sandy sediments rather than in the pelagic zone.

Winter plankton dynamics in high mountain lakes is clearly affected by the occurrence of an ice and snow cover, which changes the physico-chemical properties substantially. While plankton has a dominant influence on microbial communities in the ice cover during its formation and growth phases, the catchment plays a major role in characterizing this unique environment with its assemblages during the ablation and snow melt period. The slush assemblages exert a seeding effect on lake communities shortly before the final disappearance of the last ice which does not necessarily affect the whole water column. However, there is a clear continuum between the lake and its catchment area during the ablation period where the pelagic zone acts like a trap for all incoming material resulting from a sudden release during ice melt, whereas the ice cover itself has been serving as a trap during winter time.

8.2.2 Snow

Snow covers are key components in global biogeochemical cycles and they play essential roles in the seasonal dynamics of terrestrial ecosystems. They influence glaciers, terrestrial ecosystems, and also lakes and rivers in a significant manner. Snow stores and releases latent heat of fusion, is place for sublimation and crystal bonding forces, provides shielding from radiation and acts as an insulator for underlying terrestrial and aquatic ecosystems. Moreover, it is a huge water reservoir, shallow but of enormous dimensions (i.e. millions of square kilometres) and thus of major importance for a large number of people. Finally, it is also a transport medium both laterally and vertically. Snow has unique physical properties, and ice crystals, liquid water and water vapour coexist in a number of combinations. In addition to its passive physical properties, the snow cover interacts with the atmosphere (gaseous exchange, dust and wet deposition), with the soil or the underlying water body and, last but not least, with the organisms thriving in the melting snowpack.

Microbial communities in snow

According to Bauer (1819), the first accounts of microbial communities that cause red snow are found in the writings of Aristotle. De Saussure (1786) was convinced that the coloration of snow was due to dust depositions. Others thought it was due to mineral deposits or the oxidation products leached from rocks. This assumption has since been revised and we now know that the coloration is of plant origin. Initially, it was considered that the colour came from the spores of moss or pollen until the reddish fungi *Uredo nivalis* was identified (Bauer, 1819). Wille (1903) was the first to describe the 'red snow' as the ubiquitous microalga *Chlamydomonas nivalis* (Fig. 8.3). After this discovery, snow microbiology received no further attention until the 20th century when investigations centred almost entirely on the algae living on the surface of snow layers (Kol, 1968). Studies of the fungi and bacteria living on snow appeared much later (Hoham *et al.*, 1993; Sattler *et al.*, 2002; Miteva, 2008).

Fig. 8.3. Red snow caused by a microalga (most likely *Chlamydomonas nivalis*) on the surface of the snow cover in high altitude regions with a close up in (Inset a). After snow melt the ice alga *Mesotaenium bergrenii* (Zygnemaphyceae, Chlorophyta) appears at glacial surfaces, often hidden in greyish slush (Inset b). © Birgit Sattler and Daniel Remias.

During late spring, the superficial snow in alpine regions becomes highly pigmented 'red' or 'watermelon' snow. The aforementioned snow algae, like *C. nivalis*, are responsible for this. *Chlamydomonas nivalis* is a species of flagellated green alga that contains the secondary red carotinoid pigment, astaxanthin, in addition to chlorophyll. *Chlamydomonas nivalis* is cryophilic (cold-loving) and thrives in freezing water. The algae are extremely efficient primary producers and well adapted to resist the harming UV radiation (multiplied by reflection of snow crystals) and repeated freeze–thaw cycles that accompany life in snow. Astaxanthin protects the chloroplast from intense visible and UV radiation, as well as absorbing heat, which in turn provides the alga with liquid water as the snow melts around it. Individual cells measure 20–30 μm and massive blooms of 1 million cells ml^{-1} of snow water equivalent can extend to a depth of 25 cm in the snow cover. These blooms are triggered by the first appearance of liquid water in the snowpack, which stimulates the *C. nivalis* cells to swim upwards to the snow surface. However, in order to avoid excessively high doses of radiation, the algae do not move straight to the surface. After their reproduction they lose their flagella and encyst forming aplanospores (thick-walled resting cells/cysts), which are immobile but still metabolize (Remias *et al.*, 2010). These cysts remain dormant in the snow over winter ready for the subsequent spring and appearance of liquid water, which stimulates the cycle once more. *Chlamydomonas nivalis* is an important food source for many other snow-dwelling species, such as ciliates, rotifers, nematodes, ice worms (Section 8.2.3) and springtails.

Prokaryotic microorganisms living in snow benefit from the extracellular release of dissolved organic carbon (DOC) from snow algae. Hence, bacterial secondary production is enhanced by the occurrence of snow algae at the surface and can reach values comparable to eutrophic lakes (Psenner *et al.*, 1999, 2003). In general, wet snow holds more liquid water and thus favours microbial activity. However, snow melt running off a snowpack is generally very dilute and contains very low nutrient and microorganism concentrations.

Snow covers are in continuous transformation, from first deposition until melting (in temperate zones) or conversion into firn (partially compacted névé, a type of snow left over from previous seasons and recrystallized into a firm structure) and glacial ice (at high elevation and in polar regions). Wind redistribution is not only responsible for the patchy appearance of snow covers in open areas but also for sublimation and compaction. Radiation, sensible and latent heat fluxes, melting–freezing dynamics, gas exchange, wet and dry deposition and migration of ions and solutes convert snowpacks into ever changing habitats whereby not only the structure of the snow crystals is affected but also the water content and the concentration of dissolved substances in the thin water film surrounding the ice crystals transforming them into round grains. If grains are close together, the small ones evaporate and distil on to large ones. Any temperature gradient, for instance from the warm base to the cold surface, causes diffusion of water molecules from warm to cold grains. Melting, freezing and recrystallization bring about a fractionation of nutrients, because only a small fraction of the dissolved chemical species will be retained in the ice. The distribution coefficient is in the order of 10^{-4} for most ions (K, $NaNO_3$), but significantly larger (6×10^{-3}) for ammonium, which can lead to a 27-fold increase from the original concentration (Gross, 2003), meaning that ammonium will be incorporated much better into the ice during freezing process than Ka and $NaNO_3$. When the first melting occurs (which can happen at air temperatures below 0°C, when solar radiation is intensive), a highly concentrated solution that surrounds the ice crystals enters deeper snow layers where it may refreeze again. A series of such melt–freeze cycles will increase the concentration of solutes until a highly concentrated liquid finally reaches soils, streams or lakes. If this happens on glacial areas it has consequences for proglacial ecosystems by influencing, for instance, their nutrient concentration, turbidity and microbial composition. Once the accumulation area of a glacier retreats to higher altitudes due to summer melting, new environments start to form or to become active again. These include glacier forefields, which are an important habitat for pioneer plants and animals which slowly resettle recently ablated areas.

8.2.3 Alpine glaciers

Generally, glaciers worldwide show a retreating trend, however, polar glaciers are less subject to ablation than those in alpine regions. According to Zemp *et al.* (2006) alpine glaciers lost 35% of their total area between 1850 and the 1970s, and almost 50% by 2000. Total alpine glacier volume was estimated to be 200 km^3 in ca. 1850 but is now only approximately one-third of this value. Furthermore, it is predicted that an increase of 3°C in summer air temperature would reduce current alpine glacier cover by some 80%, i.e. to 10% of the glacier extent of 1850 (Zemp *et al.*, 2006). In the event of a 5°C temperature increase, alpine glaciers would be threatened by extinction and the European Alps would become almost completely ice-free. However, as paradoxical as it sounds, increasing temperatures provide more liquid water, which in turn affects the microbial activity of supraglacial communities.

A quote from Ludwig Lang (1927) demonstrates the general awareness of glaciers as a 'body of dead nature with very few morphological changes' (Lang, 1927). This perception survived for more than 30 years until there was finally evidence that glaciers provide a variety of habitats such as supraglacial, englacial and subglacial environments that are able to support living communities such as microbes and metazoans (see also Bagshaw *et al.*, Chapter 9, this volume). To be entirely correct, a 'glacial ecosystem' also includes glacial streams, moraines, ponds and the proglacial areas (Hodson *et al.*, 2007a).

These compartments are all interconnected with each other but are characterized by very distinct features and hence exhibit different microbial activities. Photosynthesis in these environments presents a fascinating opportunity for investigating feedbacks between physical and biological systems, not least because the absorption of solar radiation by low-albedo organic matter also causes melting to occur, providing liquid water. Thus, following melt, dark organic matter in sun cups upon snow, cryoconite holes, streams and moraines are flushed by meltwater and begin to sequester nutrients from adjacent snow, ice, debris, and even directly from the atmosphere (e.g. Tranter *et al.*, 2004). Presently, only the snowpack and the ice surface has received attention in this context and there are very few accounts of supraglacial stream ecology, although Fortner *et al.* (2005) have shown that supraglacial stream hydrochemistry in the McMurdo Dry Valleys, Antarctica, bears the distinct traits of nitrogen uptake via microbial activity. However, for proglacial areas supraglacial streams are of immense importance for the input of nutrients (Mindl *et al.*, 2007).

Ice worms (*Mesenchytraceus solifugus*, phylum Annelida) are so far a poorly investigated but charismatic member of the glacial food chain. They have been found on glaciers in alpine regions of the USA where several thousands can cover 1 m^2 of glacial surface. The Latin name describes their behaviour of avoiding sunlight and its associated increases in temperature and irradiance; after dawn the ice worms move downwards, literally into the ice. They are very sensitive to temperature and even slight increases can cause them to liquefy (Hartzell *et al.*, 2005). The mechanisms that enable the few-centimetre-long annelids to travel inside the ice are so far poorly understood. However, it is thought that ice worms can move along veins and fissures in the ice or can even excrete antifreeze-like chemicals to lower the freezing point of the ice, creating localized liquid water and, thus, conduits to move through.

One of the major ecosystem components of glaciers are the cryoconite holes which occur on glaciers worldwide (see Bagshaw *et al.*, Chapter 9, this volume). Cryoconites are small, water-filled depressions (typically <1 m in diameter and usually <0.3 m deep) that form on the surface of glaciers when solar-heated inorganic and organic debris melts into the ice. They can cover the ablation zone of a glacier from 0.1 to 10.0%, depending on the catchment area for organic and inorganic matter that can be deposited on the ice. These individual ecosystems, which resemble mini lakes with a sediment layer and a water column, have distinct boundaries, energy flow and nutrient cycling. Nutrients from allothonous (external) or autochthonous (internal) sources are made available by the presence of liquid water either from the sediments or inflows. Throughout the winter, or even during a day–night cycle, the supernatant water in a hole is frozen subjecting cryoconite communities to extremely harsh conditions.

Life in alpine cryoconite holes

Cryoconite holes harbour a diverse community of microorganisms and Metazoa. Supraglacial communities in particular, i.e. cryoconite holes, contribute substantially to the carbon budget of cold ecosystems (Anesio *et al.*, 2009). These special environments are settled by mainly microbial communities (viruses, bacteria, microalgae and Protozoa). Cyanobacteria often dominate the community and show high primary production rates (Anesio *et al.*, 2009). It is also known that Metazoa such as tardigrades, rotifers and nematodes inhabit cryoconite holes. However, very little information is available about the role of metazoans (De Smet and van Rompu, 1994) since most of the studies to date have focused on microbial processes (e.g. Säwström *et al.*, 2002; Porazinska *et al.*, 2004; Anesio *et al.*, 2007; Foreman *et al.*, 2007; Hodson *et al.*, 2007b; Edwards *et al.*, 2011). From a geographical point of view there is also more literature available from polar areas than alpine regions. During the short summer period when the water in cryoconite holes is mostly liquid and active metabolism is possible, the abundance of Metazoa is very high and can reach several thousands per cryoconite hole (B. Post, 2011, unpublished results). To date, the biological activity within cryoconite holes has been attributed to microbial communities and Metazoa have been underestimated. However, due to the high numbers of metazoans in alpine regions, it is likely that higher amounts of CO_2 than previously estimated from polar glaciers are produced, which would require carbon budgets within cryoconite holes and the composition of the food web of alpine supraglacial environments to be reconsidered.

During ice ages, cryoconite holes were most likely refugia for microorganisms and Metazoa. Charlesworth (1957) speculated that some microorganisms, algae, Protozoa, rotifers and tardigrades probably persisted during the Pleistocene in cryoconite holes and on nunataks.

So far 12 tardigrade species have been identified living in cryoconite holes worldwide (excluding questionable identifications), but only one species has been described in the cryoconite holes of the Austrian Alps, *Hypsibius klebelsbergi* Mihelčič, 1959 (family Hypsibiidae; Dastych *et al.* 2003; Greven *et al.*, 2005; Fig. 8.4). These animals range in size from 162 to 532 μm long. Virtually nothing is known about the ingestion of them or their role in the food web (Steinböck, 1936). *Hypsibius klebelsbergi*

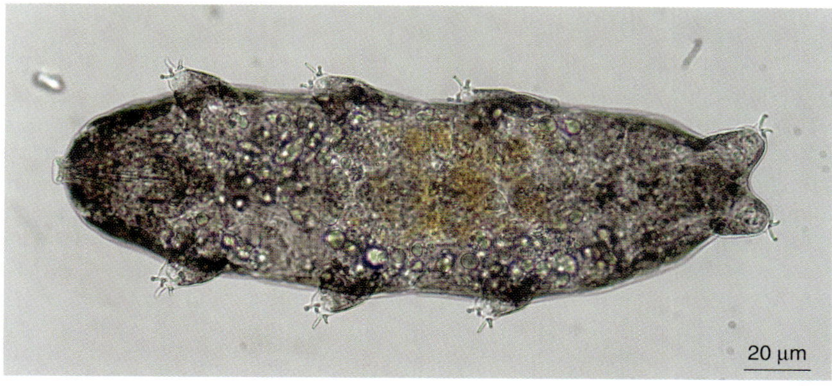

Fig. 8.4. A tardigrade (*Hypsibius klebelsbergi*) from a cryoconite hole of Stubacher Sonnblickkees, Austrian Alps. © Barbara Post.

are generally present in large numbers and have an unusual dark pigmentation, distinct to cryoconite tardigrades in the Austrian Alps (Somme, 1996). The as yet unidentified pigment only occurs in the epidermis and is produced by tiny, blackish-brown granules that fill up epidermal cells (Kraus, 1977). It is possible that the pigment confers protection against UV radiation. *Hypsibius klebelsbergi* and *H. janetscheki* (a Himalayan species that is also highly pigmented) have only been reported from the surface of the glaciers on which they live. The data available clearly suggest that the two taxa are true cryobionts.

Our understanding of the physiology of *H. klebelsbergi* is limited, but it is hypothesized that in its active form it has a narrow temperature optimum range and is adapted to low temperatures (0.1°C). This low temperature tolerance may be one of the factors that limit the occurrence of this species in other habitats. Tardigrades (with the exception of marine Arthrotardigrada), have the ability to form a tun which is a state of cryptobiosis and is caused by unfavourable conditions such as dehydration and anhydrobiosis. Their body water drops to between 1% and 3% of that in their active state (Horikawa *et al.*, 2008) causing their body to shrink. In this ametabolic dry state these animals show no visible sign of life, but become active again if rehydrated. While in anhydrobiosis, tardigrades can be tolerant of a wide variety of extreme environmental conditions (Kinchin, 1994; Jönsson and Bertolani, 2001); they are known to be able to withstand freezing at −22°C (Grøngaard, 1995).

The springtail, *Desoria saltans* (Collembola), has attracted a lot of interest as a permanent inhabitant of snow and glacial surfaces. The black-coloured springtail (Fig. 8.5) is 1.5–2.5 mm long and feeds on pollen, cryoconite material and plant remains. Collembola are known to be freezing susceptible and depend on supercooling (avoidance of ice formation) to survive low temperatures; during metabolic activity at low temperature, low molecular weight antifreeze proteins produce a hysteresis enabling them to survive in subzero conditions (Kopeszki, 1988).

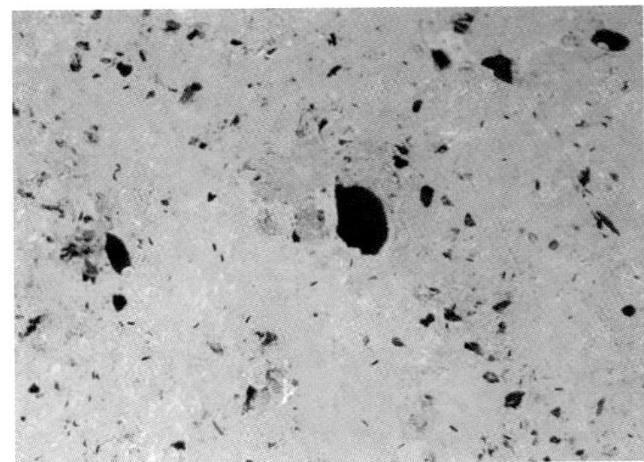

Fig. 8.5. Snow surfaces (here at Gaisbergferner, Austrian Alps) are often covered with seemingly black debris, which are Collembola (*Desoria saltans*) up to 2.5 mm in length. © Barbara Post.

Rotifers are also known to play a role in the food web of alpine cryoconite holes. Steinböck (1958) described the species *Philodina roseola* Ehrenberg (1832) (Bdelloidea) from cryoconite holes on a glacier in the Stubaier Alps, Austria (Fig. 8.6). Commonly known as wheel animals, rotifers are characterized by their ability to reproduce exclusively by parthenogenesis (without males being present). Rotifers respond to environmental stresses such as temperature or desiccation by entering a state of dormancy (anhydrobiosis), which protects them against rapid dehydration (Ricci, 1998). They are capable of remaining in this dormant state until optimal environmental conditions reoccur. They can also produce offspring that are in an unhatched, dormant state and the young will only hatch when environmental conditions are optimal. These forms of dormancy are also known as cryptobiosis or quiescence (Crowe, 1972).

Depending on the altitude, season and day/night cycles, cryoconite holes may be closed by an ice lid during winter and even on cold nights during other seasons. This condition resembles the winter cover of a polar lake with all associated consequences (Laybourn-Parry and Bell, Chapter 6, this volume). The composition of cyanobacterial, algal and phytoflagellate communities within open and closed cryoconite holes has been shown to differ: a gradient analysis between community characteristics and environmental

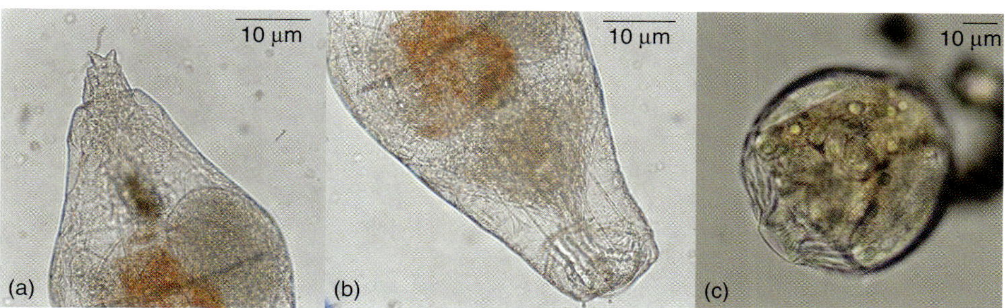

Fig. 8.6. Bdelloidea are scarcely found in alpine cryoconite holes (here from Stubacher Sonnblickkees): (a) shows a well developed corona; (b) depicts the rectum; and (c) a Bdelloidea in anhydrobiosis. © Barbara Post.

variables (largely the physical and chemical attributes of the holes and the climate of the sites) was undertaken for the Canada Glacier (McMurdo Dry Valleys, Antarctica) and the White Glacier (Axel Heiberg Island, Canada). The results of this bipolar study showed that variability in the community structure was dependent upon the environmental conditions at Canada Glacier, while on White Glacier, continuous flushing by meltwater was thought to repeatedly reset the community structure present within the holes (Mueller and Pollard, 2004). Thus, it is assumed that inter-hole mixing leads to a more homogeneous habitat in the Arctic, while isolation leads to more optimized communities in areas with different environmental pressures in the colder Antarctic system. The constant flushing of a cryoconite hole leads to the formation of spherical granules at the bottom of the holes, which collect organic and inorganic material simply by being rolled and trapping further organic material (Takeuchi *et al.*, 2001).

Anesio *et al.* (2009) demonstrated that the microbial community in a variety of cryoconites on an Arctic glacier are dominated by cyanobacteria and, typically, autotrophic systems. Alpine glaciers seem to be more influenced by solar radiation during seasons and therefore this autotrophic dominance is not apparent. Fitz (2005) measured primary and secondary production in cryoconites throughout the snow-free season and demonstrated a clear shift from a hetero- to an autotrophic state during the summer, followed by a return to heterotrophy during the autumn. Snowfall impeded winter measurements.

Independent of cryoconite hole ecosystems, alpine glacier surfaces are frequently populated by the ice alga *Mesotaenium berggrenii* (Remias *et al.*, 2009). Hence large areas of wet surrounding ice contain considerable amounts of typical ice algae, which, in contrast to snow algae, normally do not grow on snow. It is difficult to recognize them in the field because of their less intense coloration compared to true snow algae. They contain a brownish-reddish cell sap which protects them from the high irradiation in their environment (Fig. 8.3b, inset). The physiology of their extreme climate adaptation is not well understood but is under investigation (Remias *et al.*, 2009).

The implications of glacial melt on alpine glaciers

Glacial retreat as a result of global temperature rises has extremely unpleasant implications. In the summer of 2003, Alpine glaciers in the European Alps looked very 'dirty'; due to the unusually hot summer, large amounts of snow and ice, approximating to 10% of glacier mass, melted exposing dark sediments on the glacier surface. They appeared to consist mainly of locally eroded rock, soil and plant material, but also fine particles from long-range transport, impressively demonstrated by a deposition episode of brownish Saharan dust on to Alpine glaciers in November 2002. The dark colour of the cryoconites leads to decreased albedo on the glacial surface and enhances melting, therefore they form part of a positive feedback process (Frey *et al.*, 2001; Tieber *et al.*, 2009).

Radionuclides attached to aerosols were deposited in the Alps after the atmospheric nuclear bomb explosions (mainly 1950s to 1960s) and in the days after the Chernobyl accident (27 April 1986). The resulting contamination of snow and ice layers has been located by various researchers in Austria (Ambach *et al.*, 1969, 1989), Switzerland (Häberli *et al.*, 1988), France, Greenland and Spitsbergen (Pourchet *et al.*, 1986, 1995). However, over longer periods the often very complex patterns of glacier ice movement will generally distort such layers. On temperate glaciers (e.g. most Austrian glaciers) soluble radionuclides are expected to be washed out with meltwater soon after fallout, but particulate fallout may stay where it has been deposited, with the exception of redistribution in the course of the cryoconite formation process. Investigating the spatial pattern of cryoconite occurrence and composition may therefore help to understand small scale redistribution processes and airborne assessments are likely the most efficient.

Radionuclides as representatives of different chemical elements behave differently in a chemically reactive environment. Therefore, the fractions of various radionuclides in an environmental medium are not identical to those in the fallout from where they originate. Transport and transfer processes may differentiate between elements. Different ions, metals, polychlorinated substances and other substances have different solubilities in water. For cryoconites, very little is understood about such processes so far, although high radionuclide concentrations make them interesting, but potentially hazardous, candidates for scientific investigations (Tieber *et al.*, 2009). The highest activity concentrations of artificial radionuclides outside nuclear test sites have been found in Alpine cryoconites. Apparently, radioactive substances attached to aerosols and airborne dust, once deposited on the glacier surface, are not further depleted or removed by physical, chemical and biological processes. This property makes cryoconites unique habitats in nature.

Dilution, however, can occur due to the addition of new matter to an existing cryoconite, since input of dust and eroded material continues constantly, even if it is isolated by snow or ice. Likewise, melting of an ice barrier which has separated two recently isolated cryoconite pockets, leads to merging and subsequent mixing of the two. Other transport processes that are part of glacier dynamics can, of course, also have an effect. By investigating cryoconite samples from different locations on the surface of the Dachstein glacier in the Austrian Alps, mixing of global and Chernobyl fallouts could be traced using the isotopic ratios of caesium and plutonium as a consequence of the cryoconite formation dynamics. Samples of similar mixing ratio (old = global fallout versus new = Chernobyl fallout) appear to be spatially aggregated. A reliable assumption of the spatial pattern of cryoconites is also helpful for estimating the 'age' of cryoconite material (as quantified by the mixing ratio). As a comparison between glaciers in the less burdened High Arctic and the Austrian Central Alps has shown (Antlinger, 2007), cryoconite material with a higher radioactive load revealed less bacterial production than that of other glaciers. However, one has to take into account that the chemical composition of cryoconite sediment, the proportion of organic to inorganic material, and sediment size, also influence the productivity rate of a cryoconite hole. Hence, the long term consequences of the high radionuclide accumulation rates on glacial food webs are not entirely understood and require further investigation.

8.3 Extremotolerant Organisms in Biotechnology and Astrobiology

Cold-adapted organisms from Alpine and polar environments are not only crucial players in the global carbon cycle but also in biotechnological applications. Due to the production of anti-nucleating proteins which prevent ice crystals forming within a cell, psychrophilic organisms can withstand repeated freeze–thaw cycles without damage. These protein complexes with antifreeze properties are of immense interest to researchers in the fields of food technology, pharmacology, biomedical research, cosmetology and even biofuel development (Huston, 2008; see also Arora and Bell, Chapter 25, this volume).

Within the cryosphere there is a huge variety of organisms with adaptations to the living conditions in the cold, which are often accompanied with dryness, extreme pressure regimes, heavy winds, long periods of darkness, etc. So far the emphasis has been on unicellular organisms. Within multicellular organisms the herbivore tardigrade or 'water bear', *Ramazottius varieornatus*, seems to be suitable as a model animal for astrobiological studies to check for extremotolerant characteristics (Horikawa *et al.*, 2008). When in a dehydrated state, the tardigrades show higher resistance to stress factors. Due to their ability to form a tun they can withstand extreme conditions, for example, desiccation, radioactive radiation (up to 4000 Gy of heavy ions such as ^{4}He), coldness (−196°C), heat (90°C) or organic solvents (acetonitrile). Thus, tardigrades offer us valuable insights into the strategies that allow survival in extreme environments on Earth and potentially on other planets and extraterrestrial bodies (see also Gómez *et al.*, Chapter 26, this volume).

8.4 Conclusions

Despite their remoteness and extreme living conditions (comparable to polar regions), the large variety of cryospheric compartments in alpine environments are highly active ecosystems. With continued global temperature increases some of these environments will recede to higher altitudes, reduce in size or will slowly vanish. Some alpine ecosystems are crucial for environments at lower altitudes due to meltwater connections, such as glacier forefields, which can be colonized by former cryoseston. However, elevated temperatures can to a certain degree foster the growth of microbes and Metazoa as a result of the higher availability of liquid water. Our increasing awareness of extreme life forms and their niches in high altitude regions can serve as indicators of environmental stress and help us to predict the consequences of global environmental change.

Acknowledgements

The authors are grateful to the Alpine Research Station Obergurgl (AFO), the bm.wf (Sparkling Science Program) and the TAWANI Foundation.

References

Alfreider, A., Pernthaler, J., Amann, R., Sattler, B., Wille, A. and Psenner, R. (1996) Community analysis on the bacterial assemblages in the winter cover and pelagic layers of a high mountain lake by in situ hybdridization. *Applied and Environmental Microbiology* 62, 2138–2144.

Ambach, W., Eisner, H., Prantl, A. and Slupetzky, H. (1969) Studies on vertical total-beta-activity profile of fission products in the accumulation area of the Stubacher Sonnblickkees (Hohe Tauern, Salzburg, Austria). *Pure Applied Geophysics* 74, 83–91.

Ambach, W., Rehwald, W., Blumthaler, M., Eisner, H. and Brunner, P. (1989) Chernobyl fallout on alpine glaciers. *Health Physics* 56, 27–31.

Anesio, A.M., Mindl, B., Laybourn-Parry, J., Hodson, A.J. and Sattler, B. (2007) Viral dynamics in cryoconite holes on a high Arctic glacier (Svalbard). *Journal of Geophysical Research Biogeosciences* 112, G04S31, doi:10.1029/2006JG000350.

Anesio, A.M., Sattler, B., Hodson, A.J., Fritz, A. and Psenner, R. (2009) High microbial activities on glaciers: importance to the global cycle. *Global Change Biology* 15, 955–960.

Antlinger, I. (2007) Bakterielle Aktivität und Diversität in Kryokoniten alpiner und arktischer Gletscher. MSc thesis. The University of Innsbruck, Innsbruck, Austria.

Bauer, F. (1819) Microscopical observations on the red snow. *Quarterly Journal of Literature, Science and Arts* 7, 222–229.

Catalan, J. (1992) Evolution of dissolved and particulate matter during the ice-covered period in a deep, high mountain lake. *Canadian Journal of Fisheries and Aquatic Sciences* 49, 945–955.

Charlesworth, J.K. (1957) *The Quaternary Era*, Vol. II. Edward Arnold, London, 1105 pp.

Crowe, J.H. (1972) Evaporative water loss by tardigrades under controlled relative humidity. *Biological Bulletin* 142, 407–416.

Dastych, H., Kraus, H. and Thaler, K. (2003) Redescription and notes on the biology of the glacier tardigrade *Hypsibius klebelsbergi* Mihelčič, 1959 (Tardigrada), based on material from the Ötztal Alps, Austria. *Mitteilungen Hamburgisches Zoologisches Museum und Institut* 100, 73–100.

De Saussure, H.B. (1786) Voyage dans les Alpes, Neuchâtel.

De Smet, W.H. and van Rompu, E.A. (1994) Rotifera and Tardigrada from some cryoconite holes on a Spitsbergen (Svalbard) glacier. *Belgian Journal of Zoology* 124, 27–37.

Edwards, A., Rassner, S., Anesio, A.M., Sattler, B., Perkins, W.T., Hubbard, B.P., Young, M. and Griffith, G.W. (2011) Possible interactions between bacterial diversity, microbial activity and supraglacial hydrology of cryoconite holes in Svalbard. *International Society for Microbial Ecology Journal* 5, 150–160.

Felip, M., Sattler, B., Psenner, R. and Catalan, J. (1995) Highly active microbial communities in the ice and snow cover of high mountain lakes. *Applied and Environmental Microbiology* 61, 2394–2401.

Felip, M., Wille, A., Sattler, B. and Psenner, R. (2002) Microbial communities in the winter cover and the water column of an alpine lake: system connectivity and uncoupling. *Aquatic Microbial Ecology* 29, 123–134.

Fitz, G. (2005) Kryokonite alpiner Gletscher. MSc thesis. The University of Innsbruck, Innsbruck, Austria.

Foreman, C.F., Sattler, B., Mickuchi, J.A., Porazinska, D.L. and Priscu, J.C. (2007) Metabolic activity and diversity of cryoconites in the Taylor Valley, Antarctica. *Journal of Geophysical Research* 112, G04S32, doi:10.1029/2006JG000358.

Fortner, S.K., Tranter, M., Fountain, A., Lyons, W.B. and Welch, K.A. (2005) The geochemistry of supraglacial streams of Canada Glacier, Taylor Valley (Antarctica) and their evolution into proglacial waters. *Aquatic Geochemistry* 11, 391–412.

Frey K., Eicken, H., Perovich, D.K., Grenfell, T.C., Light, B., Shapiro, L.H. and Stierle, A.P. (2001) Heat budget and decay of clean and sediment-laden sea ice off the northern coast of Alaska, Port and ocean engineering in the Arctic conference (POAC'1) proceedings, 3, Ottawa, Canada, pp. 1405–1412.

Greven, H., Dastych, H. and Kraus, H. (2005) Notes on the integument of the glacier-dwelling tardigrade *Hypsibius klebelsbergi* Mihelcic, 1959 (Tardigrada). *Mitteilungen aus dem Hamburgischen Zoologischen Museum und Institut*, 102, 11–20.

Gross, G.W. (2003) Nitrates in ice: uptake; dielectric response by the layered capacitor method. *Canadian Journal of Physics* 81, 439–450.

Grøngaard, A. (1995) The invertebrate fauna on and in the inland ice of Greenland. Prize Essay. University of Copenhagen, Copenhagen, Denmark, 43 pp.

Häberli, W., Gäggeler, H., Baltensperger, U., Jost, D. and Schotterer, U. (1988) The signal from the Chernobyl accident in high-altitude firn areas of the Swiss Alps. *Annals of Glaciology* 10, 48–51.

Hartzell, P.L., Nghiem, J.V., Richio, K.J. and Shain, D.H. (2005) Distribution and phylogeny of glacier ice worms (*Mesenchytraeus solifugus* and *Mesenchytraeus solifugus rainierensis*). *Canadian Journal of Zoology* 83(9), 1206–1213.

Hodson, A., Anesio, A.M., Tranter, M., Fountain, A., Osborn, M., Priscu, J.C., Laybourn-Parry, J. and Sattler, B. (2007a) Glacial ecosystems. *Ecological Monographs* 78, 41–67.

Hodson, A., Anesio, A.M., Ng, F., Watson, R., Quirk, J., Irvine-Fynn, T., Dye, A., Clark, C., McCloy, P., Kohler, J. and Sattler, B. (2007b) A glacier respires: quantifying the distribution and respiration CO_2 flux of cryoconite across an entire Arctic supraglacial ecosystem. *Journal of Geophysical Research* 112, G04S36, doi:10.1029/2007JG000452.

Hoham, R.W., Laursen, A.W., Clive, S.O. and Duval, B. (1993) Snow algae and other microbes in several alpine areas in New England. In: Ferrick, M. (ed.) *Proceedings of the 50th Annual Eastern Snow Conference*. Quebec City, Canada, pp. 165–173.

Horikawa, D.D., Kunieda, T., Abe, W., Watanabe, M., Nakahara, Y., Yukuhiro, F., Sakashita, T., Hamada, N., Wada, S., Funayama, T., Katagiri, C., Kobayashi, Y., Higashi, S. and Okuda, T. (2008) Establishment of a rearing system of the extremotolerant tardigrade *Ramazzottius varieornatus*: a new model animal for astrobiology. *Astrobiology* 8, 549–556.

Huston, A. (2008) Biotechnological aspects of cold-adapted enzymes. In: Margesin, R., Schinner, F., Marx, J.C. and Gerday, C. (eds) *Psychrophiles: from Biodiversity to Biotechnology*. Springer Verlag, Berlin-Heidelberg, Germany, pp. 347–363.

Jönsson, K.I. and Bertolani, R. (2001) Facts and fiction about long-term survival in tardigrades. *Journal of Zoology* 255, 121–123.

Kinchin, I.M. (1994) *The Biology of Tardigrades*. Portland Press Ltd., London, 240 pp.

Kol, E. (1968) Kryobiologie. Biologie und Limnologie des Schnees und Eises. I. Kryovegetation. In: Elster, H.J. and Ohle, W. (eds) *Die Binnengewasser*. E. Schweizerbart'sche Verlagsbuchhandlung, Stuttgart, Germany, 216 pp.

Kopeszki, H. (1988) Zur Biologie zweier hochalpiner Collembolen – *Isotomurus palticeps* (UZEL, 1891) und *Isotoma saltans* (NICOLET, 1841). *Zoologischer Jahresbericht der Systematik* 115, 405–439.

Kraus, H. (1977) *Hypsibius (Hypsibius) klebelsbergi* Mihelčič, 1959 (Tardigrada) aus dem Kryokonit des Rotmoosfernes. PhD thesis. The University of Innsbruck, Innsbruck, Austria.

Lang, L. (1927) *Gletschereis*. Kosmos Verlag, Stuttgart, Germany, 76 pp.

Mindl, B., Anesio, A.M., Meirer, K., Hodson, A.J., Laybourn-Parry, J., Sommaruga, R. and Sattler, B. (2007) Factors influencing bacterial dynamics along a transect from supraglacial runoff to proglacial lakes of a high Arctic glacier. *FEMS Microbiology Ecology* 59, 307–317.

Miteva, V. (2008) Bacteria in snow and ice. In: Margesin, R. and Schinner, F. (eds) *Psychrophiles: from Biodiversity to Biotechnology*. Springer Verlag, Heidelberg, Germany, pp. 31-50.

Mueller, D.R. and Pollard, W.H. (2004) Gradient analysis of cryoconite ecosystems from two polar glaciers. *Polar Biology* 27, 66–74.

Pearce, D.A., Bridge, P.D., Hughes, K., Sattler, B., Psenner, R. and Russell, N.J. (2009) Microorganisms in the atmosphere over Antarctica. *FEMS Microbiology Ecology* 69, 1–15.

Pomeroy, L.R. and Wiebe, W.J. (2001) Temperature and substrates as interactive limiting factors for marine heterotrophic bacteria. *Aquatic Microbial Ecology* 23, 187–204.

Porazinska, D.L., Fountain, A.G., Nylen, T.H., Tranter, M., Virginia, R.A. and Wall, D.H. (2004) The biodiversity and biogeochemistry of cryoconite holes from McMurdo Dry Valley glaciers, Antarctica. *Arctic Antarctic and Alpine Research* 36, 84–91.

Pourchet, M., Pinglot, J.F. and Gascard, J.C. (1986) Accumulation in Svalbard glaciers deduced from ice cores with nuclear tests and Chernobyl reference layers. *Nature* 323, 676.

Pourchet, M., Lefaconnier, B., Pinglot, J.F. and Hagen, O. (1995) Mean net accumulation of ten glacier basins in Svalbard estimated from detection of radioactive layers in shallow ice cores. *Zeitschrift für Gletscherkunde und Glazialgeologie* 3, 73–84.

Price, P.B. (2000) A habitat for psychrophiles in deep Antarctic ice. *Proceedings of the National Academy of Sciences of the United States of America* 97, 1247–1251.

Priscu, J.C., Fritsen, C.H., Adams, E.E., Giovannoni, S.J., Paerl, H.W., McKay, C.P., Doran, P.T., Gordon, D.A., Lanoil, B.D. and Pinckney, J.L. (1998) Perennial Antarctic lake ice: an oasis for life in a polar desert. *Science* 280, 2095–2098.

Priscu, J.P., Adams, E.E., Pearl, H.W., Fritsen, C.H., Dore, J.E., Lisle, J.T., Wolf, C.F. and Mikucki, J.A. (2002) Perennial Antarctic Lake Ice: a refuge for cyanobacteria in an extreme environment. In: Rogers, S. and Castello, J. (eds) *Life in Ancient Ice*. Princeton Press, Princeton, pp. 22–49.

Psenner, R. and Sattler, B. (1998) Life at the freezing point. *Science* 280, 2073–2074.

Psenner, R., Sattler, B., Wille, A., Fritsen, C., Priscu, J., Felip, M. and Catalan, J. (1999) Lake Ice Microbial Communities (LIMCOs) in Alpine and Antarctic Lakes. In: Margesin, R. and Schinner, F. (eds) *Cold Adapted Organisms – Ecology, Physiology, Enzymology and Molecular Biology*. Springer-Verlag, Heidelberg, Germany, pp. 17–31.

Psenner, R., Wille, A., Priscu, J.C., Felip, M., Wagenbach, D. and Sattler, B. (2003) Low temperature environments and biodiversity. In: Extremophiles: Ice Ecosystems and Biodiversity. In: *Knowledge for Sustainable Development. An Insight into the Encyclopaedia of Life Support Systems*, Vol. III. UNESCO Publishing-Eolss Publishers, Oxford, UK, pp. 573–598.

Remias, D., Holzinger, A. and Lütz, C. (2009) Physiology, ultrastructure and habitat of the ice alga *Mesotaenium berggrenii* (Zygnemaphyceae, Chlorophyta) from glaciers in the European Alps. *Phycologia* 48, 302–312.

Remias, D., Karsten, U., Lütz, C. and Leya, T. (2010) Physiological and morphological processes in the Alpine snow alga *Chloromonas nivalis* (Chlorophyceae) during cyst formation. *Protoplasma* 243, 73–86.

Ricci, C. (1998) Anhydrobiotic capabilities of bdelloid rotifers. *Hydrobiologia* 387, 321–326.

Sattler, B., Puxbaum, H. and Psenner, R. (2001) Bacterial growth in supercooled cloud droplets. *Geophysical Research Letters* 28, 239–242.

Sattler, B., Wille, A., Waldhuber, S., Sipiera, P. and Psenner, R. (2002) Various ice ecosystems in alpine and polar regions – an overview. In: Lacoste, H. (ed.) *Proceedings of the First European Workshop on Exo-Astrobiology*. ESA Publications Division, Noordwijk, the Netherlands, pp. 16–19.

Säwström, C., Mumford, P., Marshall, W., Hodson, A. and Laybourn-Parry, J. (2002) The microbial communities and primary productivity of cryoconite holes in an Arctic glacier (Svalbard 79 degrees N). *Polar Biology* 25, 591–596.

Somme, L. (1996) Anhydrobiosis and cold tolerance in tardigrades. *European Journal of Entemology* 93, 349–357.

Steinböck, O. (1936) Cryoconite holes and their biological significance. *Zeitschrift für Gletscherkunde* 24, 1–21.

Steinböck, O. (1958) De Natura Tirolensi: Fragmenta limnologica alpine. *Schlernschriften* 188, 113–144.

Takeuchi, N., Kohshima, S. and Seko, K. (2001) Structure, formation, darkening process of albedo reducing material (cryoconite) on a Himalayan glacier: a granular algal mat growing on the glacier. *Arctic, Antarctic and Alpine Research* 33, 115–122.

Tieber, A., Lettner, H., Hubmer, P., Sattler, B. and Hofmann, W. (2009) Accumulation of anthropogenic radionuclides in cryoconites on Alpine glaciers. *Journal of Environmental Radioactivity* 100, 590–598.

Tranter, M., Fountain, A., Fritsen, C., Lyons, B., Statham, P. and Welch, K. (2004) Extreme hydrochemical conditions in natural microcosms entombed within Antarctic Ice. *Hydrological Processes* 18, 379–387.

Wille, N. (1903) Algologische Notizen IX-XIV. *Nyt Mag Naturvidenskab* 41, 88–189.

Zemp, M., Haeberli, W., Hoelzle, M. and Paul, F. (2006) Alpine glaciers to disappear within decades? *Geophysical Research Letters* 33, L13504, doi:10.1029/2006GL026319.

9 Glacier Surface Habitats

Elizabeth A. Bagshaw, Marek Stibal, Alexandre M. Anesio, Chris Bellas, Martyn Tranter, Jon Telling and Jemma L. Wadham
Bristol Glaciology Centre, School of Geographical Sciences, University of Bristol, UK

9.1 Introduction to Glacier Surface Habitats

Glaciers are masses of ice which deform under their own weight and flow under the influence of gravity. They exist across the world, but the largest ice masses are found in the Earth's polar regions, which are home to the Antarctic and Greenland ice sheets. These environments are extremely cold, subject to extremes of solar radiation (24 h daylight in the summer and 24 h darkness in the winter) and frequently experience high wind speeds. The interior regions experience little precipitation and as such are classified as deserts. This suite of extreme conditions makes an unlikely habitat for life, and indeed glacial regions were considered devoid of life until the last decade. However, it is now known that a wide range of microbial life exists on, within and beneath glaciers. This chapter will concentrate on life on glacier surfaces, in the supraglacial environment. The subglacial (beneath) and englacial (within) environments are also home to microbial populations, but they will not be discussed in detail here.

Glacier surfaces are host to a variety of microorganisms, which inhabit a diverse range of habitats including cryoconite holes, cryolakes, supraglacial mantles, kames (irregular hills or mounds comprised of sediment) and other surface sedimentary deposits. These vary in scale and significance for the functioning of the glacial ecosystem as a whole, and are believed to display tele-connections to neighbouring ecosystems via the export of water and nutrient. Surface ecosystems are of interest for three primary reasons.

1. Glacier surface ecosystems are believed to be important in sustaining the productivity of downstream ecosystems via the export of labile dissolved organic carbon and nutrients (Anesio *et al.*, 2009; Hood *et al.*, 2009). Biotic communities in many glacial and proglacial environments are limited by harsh physical conditions, limited access to organic matter and low nutrient availability (Barrett *et al.*, 2007; Zeglin *et al.*, 2009). Any process which serves to maximize organic matter or nutrient storage in the presence of liquid water is thus important on an ecosystem-wide scale. We show that processes which occur in cryoconite holes and other glacier surface habitats can have significant impacts on the structure of the supraglacial ecosystem, and potentially on the surrounding environment.
2. Glaciers may serve as 'early warning systems' for climate-induced changes in polar regions. Polar ecosystems are sensitive

to climate change because small temperature perturbations have large impacts on hydrological and biological processes via changes in rates of ice and snow melt (Kennedy, 1993; Fountain *et al.*, 1999; Doran *et al.*, 2002; Foreman *et al.*, 2004; Esposito *et al.*, 2006; Wall, 2007). Glacier surface habitats are considered effective model ecosystems for assessing the impacts of climate warming on cold, nutrient-limited environments. Temperatures in these systems are close to the 0°C isotherm and the low, spatially-limited surface melting (Fountain *et al.*, 1999; Hoffman *et al.*, 2008) is representative of large sectors of the Earth's cryosphere (e.g. ice sheet interiors), which could potentially 'switch on' and produce meltwater during climate warming.

3. Glaciers represent an analogue for potential ice-bound, extraterrestrial life habitats. They provide shelter from extreme environmental conditions but retain access to liquid water, carbon and other nutrients, thus enabling biological activity even when they are sealed off from the surrounding atmosphere. Analogous conditions may exist on icy moons such as Jupiter's moons, Europa and Ganymede, and Saturn's moon, Rhea (Priscu and Christner, 2002; Teolis *et al.*, 2010), and potentially on the Martian surface during its geological history (Tranter *et al.*, 2010).

This chapter will examine the types of ecosystems that exist on glacier surfaces worldwide, and the key differences between these habitat types in different climatic zones. The presence of liquid water on the glacier surface throughout the ablation season (the time period when snow, ice or water is removed from the surface of the glacier by evaporation, erosion, sublimation or calving) supports diverse microbial consortia. We will discuss the dominant species found in glacial surface ecosystems, and explain the biogeochemical processes which prevail on glacier surfaces as a result of this microbial activity. Finally, we will discuss the potential impacts of the microbial processes in glacier surface ecological niches on the surrounding environment and on ecosystems that are fed by glacier melt.

9.2 Structures and Types of Supraglacial Environments

Glacier surface ecosystems comprise debris, microorganisms and glacial meltwater. The debris is generally sourced from the surrounding environment (Porazinska *et al.*, 2004; Nkem *et al.*, 2006; Bøggild *et al.*, 2010) and is typically deposited on the glacier by wind. The same is true for the microorganisms that inhabit glacier surface habitats. The presence of dark debris on the light, high albedo glacier surface causes local regions of enhanced melting and the formation of water-filled depressions. It is these depressions (e.g. cryoconite holes, cryolakes) which together with more dispersed debris accumulations, constitute supraglacial life habitats. They cover varying amounts of the glacier surface, typically between 3 and 15% (Fountain *et al.*, 2004; Hodson *et al.*, 2008).

9.2.1 Cryoconite holes

The most widely studied of the supraglacial habitats is cryoconite holes. These ice-bound mesocosms are small, cylindrical depressions, which form as a result of enhanced surface melting around supraglacial debris. The debris melts down into the ice, forming a water-filled hole with a near uniform layer of debris at its base (Fig. 9.3).

Cryoconite holes have been documented in the ablation zone of many glaciers across the globe (Fig. 9.2), since they were first discovered in 1870 (Nordenskjöld, 1875). Those in different climatic regions have contrasting characteristics according to the ice they form in, the solar radiation they receive, the contents of the cryoconite material and local climate conditions. The morphology of cryoconite holes differs greatly according to the melt regime of the parent glacier. The magnitude of precipitation and meltwater production controls the persistence of discrete cryoconite holes (Takeuchi *et al.*, 2000). In more temperate regions, cryoconite debris is only visible at the glacier surface when the snowpack thaws in early summer (Margesin *et al.*, 2002). By contrast, the sediment in cryoconite holes in the McMurdo Dry Valleys (MCMDV), Antarctica, remains sealed below the ice surface throughout the year (Fountain *et al.*, 2004; Tranter *et al.*, 2004; Bagshaw *et al.*, 2007). In Arctic regions, the degree of exposure of debris to the atmosphere depends upon the hydrological regime, and cryoconite holes may additionally form following the deposition of debris by fluvial, mass movement and/or melt-out processes (Hodson *et al.*, 2008). Cryoconite holes are thus divided into two main types: open and closed. In open holes, the water within the hole remains open to the atmosphere and frequently exchanges water and solute with other supraglacial water bodies, such as streams or lakes. Open holes can be subdivided into discrete and submerged types, according to the holes' morphology (Fig. 9.1), where submerged holes may exist within a supraglacial stream or lake. In closed holes, the water body remains sealed off from the atmosphere by an ice lid throughout the ablation season.

Open holes

Open holes exist on ice masses where the surface air temperature is high enough to enable widespread melting throughout the summer thaw period. The water in an open hole has free access to the atmosphere. The sediment is frequently flushed out of the holes and redistributed elsewhere on the glacier surface, commonly forming other cryoconite holes. The speed and volume of redistribution varies according to the melt rate of the glacier surface, thus these holes are typically found on temperate and polythermal glaciers where there is widespread summer melting in the glacier ablation zone (Hodson *et al.*, 2008).

The constant hydrological activity of open cryoconite holes has implications for

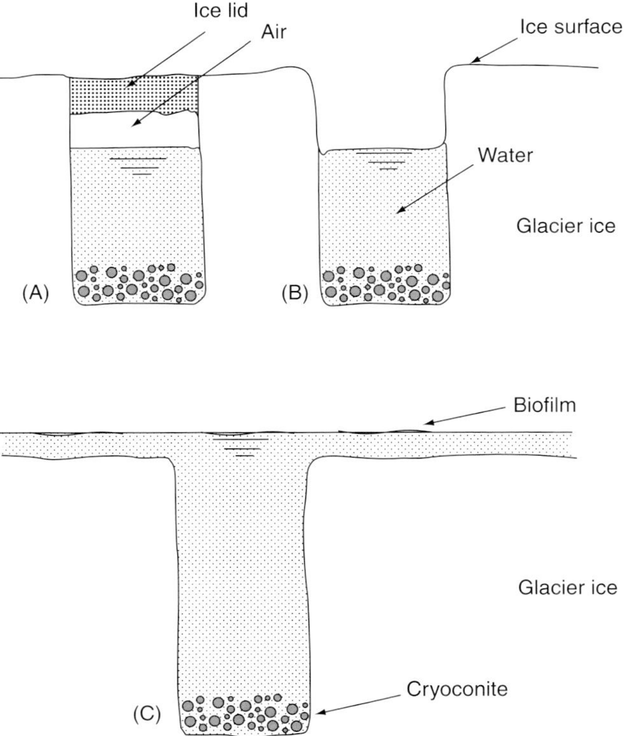

Fig. 9.1. Structure of (A) closed, (B) open and (C) submerged cryoconite holes (Hodson *et al.*, 2008).

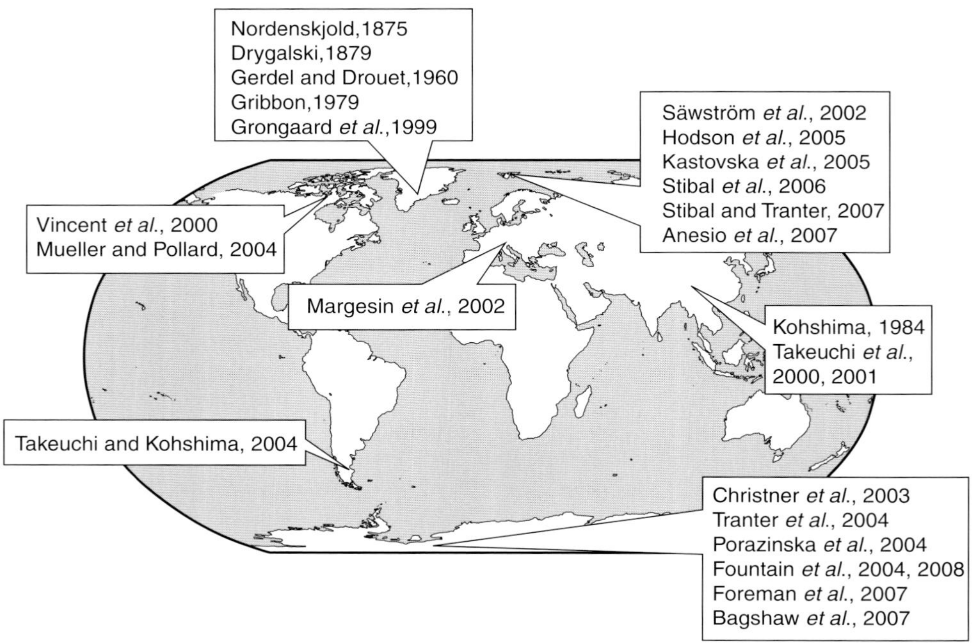

Fig. 9.2. Location of a selection of key studies on cryoconite holes across the globe.

the organisms which inhabit them. They must cope not only with low temperatures, oligotrophic nutrient conditions and the varying availability of liquid water, but also with regular redistribution of the microbial community. The inherently unstable nature of melting glacier surfaces means that cryoconite hole coverage can change rapidly over short timespans in the summer months. Hodson *et al.* (2010) observed a reduction in cryoconite cover over 1 m^2 of Longyearbreen Glacier, Svalbard, over the course of 20 days at the height of the ablation season, with coverage decreasing from ~12% of the monitored area to <7%. The redistributed cryoconite is flushed into supraglacial streams, from where its progress has been likened to the movement of tumbleweed (Hodson *et al.*, 2010).

The constant motion of the debris in and between open cryoconite holes causes the development of aggregate particles, which are bound together by biogenic films (Takeuchi *et al.*, 2001; Hodson *et al.*, 2010). These granules, typically <1 cm in diameter (Irvine-Fynn *et al.*, 2010), are quasi-spherical, mineral rich, and frequently contain microorganisms. Cyanobacteria tend to concentrate on the circumference of the granule, where they can photosynthesize on and just below the surface of the aggregate. The carbon that they fix cements the community together by the production of an extracellular polymeric substance (EPS) by both cyanobacteria and other non-filamentous micro-algae (Hodson *et al.*, 2010).

Closed holes

Closed holes are covered by an ice lid, which refreezes over the cryoconite debris and restricts access to the atmosphere. Closed holes are more stable than their open counterparts, existing in much colder ice and on glacier surfaces with very low ablation rates, which promotes hydrological disconnection of the hole from the surrounding environment. They are most common in very cold, dry ice, and the most studied examples are found in the MCMDV, Antarctica. Here, mean ice temperatures are −18°C, melt rates are 1–3 cm water equivalent (w.e.) per year and accumulation is less than 10 cm w.e. year^{-1} (Fountain *et al.*, 1998, 1999). This constitutes one of the coldest and driest environments on Earth, where microorganisms must adapt to extreme condition in order to survive.

The ice lid prevents the exchange of water, gases and solute between the cryoconite hole and the atmosphere and surrounding glacier surface drainage system. The debris layer continues to absorb solar radiation, and a pocket of meltwater develops below the ice lid (Fig. 9.3). Liquid water is thus maintained even when air temperatures are well below zero. Much of the supraglacial drainage structure on these cold, polar glaciers is located just below the surface ice, in the top 0.5 m of the glacier. This near-surface drainage system consists of a network of veins, cracks and pipes that commonly form along old crevasse traces, which convey meltwater and which intersect cryoconite holes, supraglacial streams and cryolakes across the glacier surface. The glacier surface drainage system frequently conveys water at pressures that exceed atmospheric values, which may be responsible for driving connections between previously isolated cryoconite holes (Fountain *et al.*, 2004). We therefore see evidence that closed cryoconite holes may become connected to the surrounding drainage system while remaining isolated from the atmosphere. For clarification, these systems are referred to as 'hydrologically connected, closed cryoconite holes'.

Occasionally the closed holes lose their ice lids, for example, during a period of warmer than average air temperatures (Foreman *et al.*, 2004; Wall, 2007; Barrett *et al.*, 2008). Under these circumstances, the majority of cryoconite holes across polar glaciers become hydrologically connected, and their ice lids frequently melt out. The contents of the cryoconite holes, both water and sediment, are redistributed across the glacier, via a similar mechanism to that which operates on open cryoconite holes in more temperate regions (Hodson *et al.*, 2008). Following the last warm period in Taylor Valley (MCMDV) in 2001/02, large

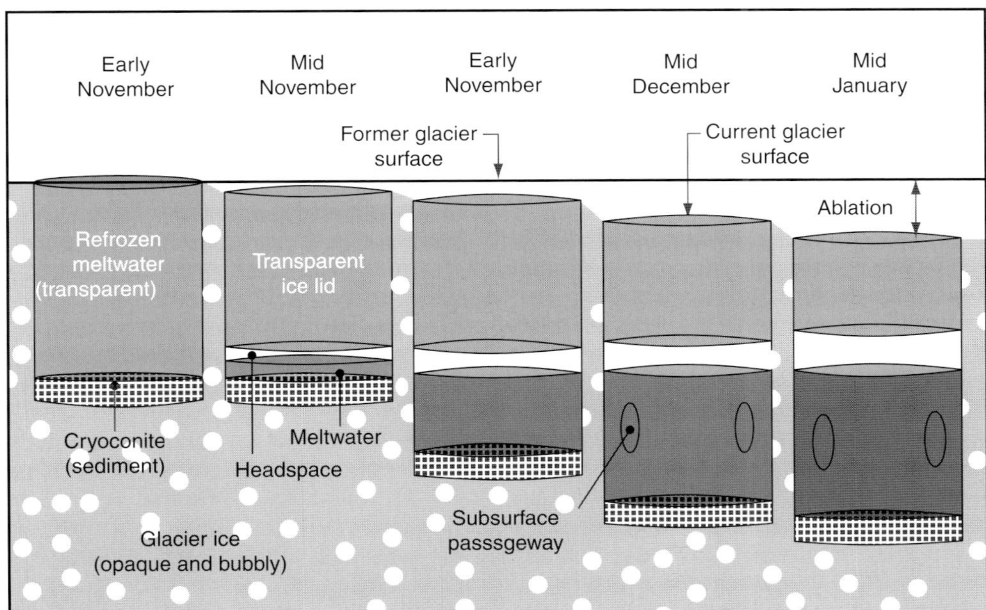

Fig. 9.3. Schematic of the annual cycle of a closed cryoconite hole, showing subsurface melting and potential for hydrological connection (Fountain *et al.*, 2008).

sediment conglomerations developed on the surface of the Canada and Taylor Glaciers as the cryoconite debris was deposited in surface depressions where meltwater collected. The presence of this new mass of sediment trapped further windblown and water-borne debris in channels (Fig. 9.4a) and supraglacial lake basins (Fig. 9.4b), and created new habitats for life on the glacier surface. As the basins grow, they scavenge the surrounding cryoconite holes and accumulate more debris, causing them to melt down even further and create steep ice cliffs. The high incident angle of radiation to these cliffs enhances melting further (Fig. 9.4c), forming deeply incised cryolakes.

9.2.2 Cryolakes

Cryolakes are surface melt ponds, which form on cold, polar glaciers as a result of sediment accumulation and associated melting. Water cannot drain from the lakes as a result of restricted outflow channels, thus water accumulates on the surface. They range in diameter from 10–50 m and function as large-scale, hydrologically connected, closed cryoconite holes (Tranter *et al.*, 2010). The debris layer at the base of these lakes ranges from 0.5 to 5 cm deep, and is the locus of enhanced biogeochemical reactions via microbial activity. Meltwater ponds in the lakes and does not completely freeze, because solar radiation continues to heat debris in the cryolake beds. Hence, biogeochemical reactions can continue to take place within the lakes, as in unconnected cryoconite holes. The relatively large water volumes that collect in these features can have implications for surface meltwater drainage from the glacier. Evidence for the episodic flooding and drainage of cryolakes has been observed over relatively short timescales in MCMDV (Bagshaw *et al.*, 2010). Organisms that inhabit these surface ecosystems therefore have to cope with low temperatures and fluctuating availability of liquid water and nutrients.

Similar structures form seasonally near the margins of the Greenland ice sheet, with surface areas ranging from a few metres to 10 km^2 (Echelmeyer *et al.*, 1991). These supraglacial lakes form in the early melt season, usually in the same location, and drain rapidly via moulins (or glacier mills: tubular chutes, holes or crevasses) in the mid to late ablation season (Catania and Neumann, 2010). The location of the lake is generally associated with a bedrock depression (Catania and Neumann, 2010) and the water frequently drains to the bed of the ice sheet. The lakes rarely contain significant volumes of sediment, thus they constitute a very different, and less ecologically significant, environment to Antarctic cryolake features.

9.2.3 Other sediment forms

Cryolake-type features are less common on more temperate glaciers where meltwater

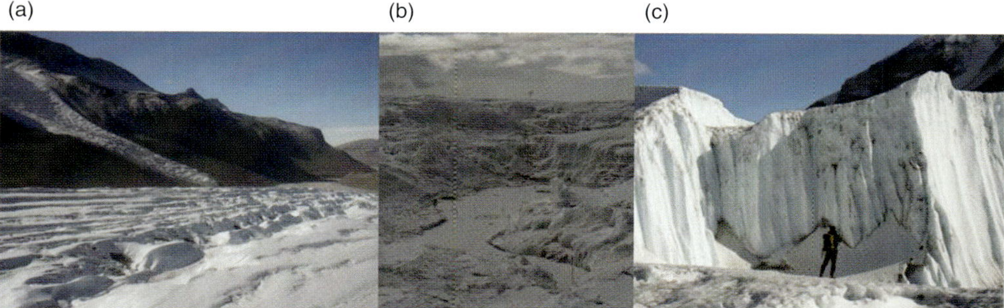

Fig. 9.4. The presence of debris on cold ice surfaces forms deeply incised (a) meltwater channels and (b) cryolakes, and forms (c) ice cliffs, which are zones of enhanced melting. Images © Elizabeth A. Bagshaw.

is either routed off the glacier through a relatively efficient surface drainage system (temperate and polythermal glaciers), or can access the bed via crevasses, moulins and cracks (temperate glaciers and temperate zones of polythermal glaciers). Large accumulations of sediment do still occur, particularly as a result of the constant redistribution and reworking of cryoconite debris. The sediment instead forms large, dry deposits such as supraglacial kames (Fig. 9.5a) (Stibal *et al.*, 2006) or cryoconite surface mantles (Fig. 9.5b) (Hodson *et al.*, 2008). These can become sufficiently thick as to allow the development of warm, anaerobic conditions within the sediment. In these areas, very different biogeochemical and mechanical processes occur. The sediment layer is sufficiently thick to suppress rather than enhance the ablation of the underlying ice, acting as an insulating blanket of debris. The sediment surface becomes cracked (Fig. 9.5b), allowing oxygen to ingress into the sediment. This creates a diversity of redox conditions within the sediment, permitting a range of biogeochemical processes to occur (Hodson *et al.*, 2008).

In summary, the structure of supraglacial habitats is influenced by wind, water and solar radiation, thus habitats on glaciers in different climatic zones may take on very different forms. The mixing of debris, meltwater and microorganisms is, however, common to all glacier surfaces, and it is these processes that govern the development and activity of these small, yet significant ecosystems.

9.3 Extreme Conditions in Supraglacial Environments

Biogeochemical conditions in glacier surface ecosystems control the microbial processes that can occur, and these microbial processes in turn influence the biogeochemistry. In this section, we discuss the biogeochemical processes that occur and how they affect the life which exists in these unique habitats.

9.3.1 Biogeochemical processes in cryoconite holes

Oxygen

Photosynthesis and respiration are the two main microbial processes that prevail on glacier surfaces. Net photosynthesis (where fixation of organic carbon by autotrophic organisms exceeds consumption of organic carbon by heterotrophic organisms) occurs in regions of relatively high irradiation, normally close to the glacier ice surface and particularly where debris layers are thin.

Fig. 9.5. Concentrated and dispersed debris deposits on the surface of glaciers also constitute microbial habitats, which form structures including (a) kames, © Marek Stibal and (b) supraglacial mantles, © Andy Hodson.

It leads to an accumulation of O_2 in solution, which in turn may result in degassing of O_2 to the atmosphere if the hole is open, in addition to diffusion of O_2 into underlying sediment.

Net respiration (where consumption of organic carbon by heterotrophic organisms exceeds carbon fixation by photosynthesis) dominates over photosynthesis in the deeper sediment where irradiation is limited or absent and results in the consumption of oxygen, driving the aquatic environment towards anoxia if the system is closed or access to atmospheric O_2 is restricted. This may be due to kinetic reasons, for example if diffusion of O_2 into the sediment is slower than the rate of oxygen consumption, or because the ice lid of the cryoconite hole is restricting access to the atmosphere.

Organic matter

During the degradation of organic matter some of the organic matter is converted to dissolved organic carbon (DOC) compounds, as shown in Eqn 9.1:

$$2CH_2O\ (s) + O_2\ (aq) \leftrightarrow xCO_2\ (aq) + yH_2O\ (l) + C_{(2-x)}H_{(2-y)}O_{(2-2x-y)}\ (aq) \quad (9.1)$$

where $C_{(2-x)}H_{(2-y)}O_{(2-2x-y)}$ is a simplistic representation of DOC.

Nutrient cycling

Nutrient cycling within surface glacial ecosystems may have significant impacts on the speciation of dissolved organic compounds in glacier melt. Nitrogen (N) and phosphorus (P) are essential macronutrients required for microbial activity. Dissolved inorganic (DI) N and P species are rapidly consumed, and the ratio of C:N:P in the ensuing organic matter often approaches a ratio of ~106:16:1, depending on the type of organic matter and other environmental stresses (Redfield, 1958; Barrett *et al.*, 2007). Of the DIN species, NH_4^+ in particular is rapidly utilized. For example, ^{15}N-isotope dilution experiments (J.C. Priscu, 2011, unpublished results) reveal that NH_4^+ uptake and regeneration in Antarctic cryoconite debris ranged from 3.8 to 18.3% day^{-1} and 1.5 to 9.7% day^{-1}, respectively. The uptake:regeneration molar ratios always exceeded 1 (with a range of 1.6–3.2), suggesting that some of the NH_4^+ is converted into different DI species, such as NO_3^-, or DON. DIP is similarly rapidly utilized. Successive cycles of microbial photosynthesis and respiration convert DIC, DIN and DIP into DOC, DON and DOP (Fig. 9.6). Such cycling produces a molar DON:DOC ratio of ~0.09 in the waters of closed cryoconite holes in the Dry Valleys of Antarctica (Tranter *et al.*, 2004), similar to that found in particulate organic matter in a wide variety of natural waters (Guildford and Hecky, 2000). At least 80%, and often >95–100%, of the total dissolved N and P is present as DON and DOP, respectively (Tranter *et al.*, 2004). Glacial DOC, DON and DOP are reported to be labile (Hodson *et al.*, 2005; Tranter *et al.*, 2005; Hodson, 2006).

pH

Extreme pH conditions can arise in the closed cryoconite holes of polar glaciers, found most commonly on Antarctic glaciers. Photosynthesis drives up pH, since concentrations of free CO_2 in solution decline, and the equilibria of other DIC species, such as HCO_3^- and CO_3^{2-}, shift to compensate, so producing OH^- (Eqn 9.2).

$$CO_3^{2-} + H_2O \leftrightarrow HCO_3^- + OH^- \leftrightarrow CO_2 + 2OH^- \quad (9.2)$$

The pH of the cryoconite holes waters may rise to exceptionally high values, approaching pH 11, and the pCO_2 may fall from atmospheric values of $10^{-3.6}$ to $10^{-7.6}$ atm at high pH, drawing into solution most of the CO_2 in the air space beneath the ice lid (Tranter *et al.*, 2004; Bagshaw *et al.*, 2011a, b). Alkaline pHs of >10 are some of the highest values recorded in pristine, natural aquatic ecosystems at or near the earth surface (Tranter *et al.*, 2004), and are commonly only recorded in arid soda lakes (Drever, 1997; see Bell, Chapter 19, this volume), or in domestic cleaning products. These extreme conditions may also impose geochemical limitations upon further

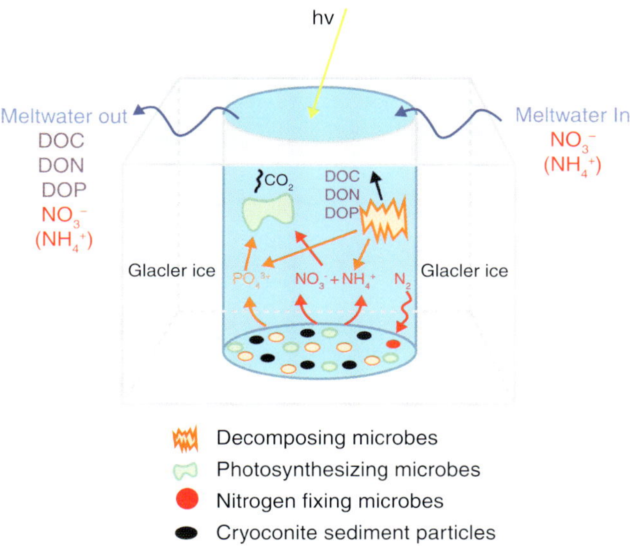

Fig. 9.6. Carbon and nutrient cycling processes that occur in cryoconite holes as a result of photosynthesis, respiration and the degradation of organic matter by the microorganisms that inhabit them.

photosynthesis, thus slowing rates of organic matter accumulation within the holes. This could potentially explain why Antarctic cryoconite debris has lower organic carbon composition (0.5%, Barrett *et al.*, 2007) and lower gross photosynthesis rates (average of 2.4 μg C g^{-1} day^{-1}; Hodson *et al.*, 2010) than Arctic cryoconite debris.

9.3.2 Measuring biological activity in cryoconite holes

Despite the extreme conditions, there is some compelling evidence that high rates of microbial activity occur on glacier surfaces, and that they are able to accumulate organic matter and release microbially labile (readily available) substrates to adjacent habitats. Rates of gross photosynthesis and community respiration in cryoconite holes have recently been measured for a number of glaciers in the Arctic (Säwström *et al.*, 2002; Stibal *et al.*, 2008; Anesio *et al.*, 2009; Hodson *et al.*, 2010; Telling *et al.*, 2010b) and the Antarctic (Hodson *et al.*, 2010; Bagshaw *et al.*, 2011a, b). The balance between gross photosynthesis (i.e. the total conversion of CO_2 to organic carbon) and respiration (i.e. the total oxidation of organic carbon back to CO_2) can give an indication of the potential for glacier surfaces to accumulate and export organic matter to adjacent habitats (e.g. glacial forefields and lakes). It can also explain the origin of large, organic-rich accumulations of debris on the glacier surface (Fig. 9.5). If community respiration dominates over gross photosynthesis (i.e. the system is considered net heterotrophic), this means that glacial surfaces must receive allochthonous (external) contributions of organic matter in order to sustain heterotrophic processes, just as can be observed for the majority of freshwaters around the globe (Cole *et al.*, 2007). There are three hypotheses: (i) that organic C on the glacier surface accumulates as a result of autotrophic production exceeding heterotrophic activity; (ii) that any excess in C is from allochthonous sources and that autotrophy is insignificant; or (iii) that both processes are occurring, and that accumulation and consumption of C are approximately in balance.

The organic carbon content of supraglacial debris far exceeds the typical organic carbon concentration of debris from basal ice and ice-marginal moraines (0.5%, Kastovska *et al.*, 2005). There are two possible explanations for this: (i) there is *in situ* production of carbon via photosynthesis; or (ii) wind transport deposits allochthonous material on to the glacier surface which contains organic carbon (OC). There are only few measurements of OC in airborne particles, but those few indicate that airborne particles have low OC content (e.g. 0.5% OC, Takeuchi *et al.*, 2001; 0.2% OC, Sabacka *et al.*, 2012). However, it is hypothesized that the lighter organic fraction of soils is preferentially transported by winds and it is thus this more organically enriched material that is deposited on glacier surfaces (Stibal *et al.*, 2012). Average values reported across different studies reveal that OC at the glacier surface can be highly autochthonous (produced *in situ*; Anesio *et al.*, 2009), highly allochthonous (from external sources; Stibal *et al.*, 2008) or reflect a balance between autochthonous and allochothonous processes (Hodson *et al.*, 2007, 2010; Telling *et al.*, 2010b). The reason for this variation is still poorly understood, but differences in methodological settings during incubation experiments, and *in situ* variations in light penetration within the debris of cryoconite holes are believed to play a major role in determining the rates of gross photosynthesis. For instance, Telling *et al.* (2010a) showed that the thickness of debris within cryoconite material on Svalbard glaciers exerts a key control on gross photosynthesis rates. In their study, net metabolism in cryoconite material switched from net autotrophic to net heterotrophic once the debris reached a thickness of ca. 2.5 mm or more. They argue that rates of gross photosynthesis in cryoconite holes can be inhibited by the shading effect of thick accumulations of cryoconite in holes, since this promotes heterotrophy in the deeper parts of the sediment. The balance between production of carbon through autotrophy and consumption of carbon via heterotrophic activity in the cryoconite hole community is thus tipped towards net carbon consumption as the sediment layer thickens.

In locations where C-fixation exceeds C-oxidation there will be net production of OC (Hodson *et al.*, 2010; Telling *et al.*, 2010a). If more OC is produced by the cryoconite communities than is consumed, then there will be an excess of OC (Anesio *et al.*, 2009). This OC may be redistributed, either to other areas of the glacier surface, or to ice-marginal areas surrounding the glacier which receive the runoff (Bagshaw *et al.*, 2007). The carbon is likely labile, and can be easily consumed by heterotrophic organisms living downstream of the melt (Section 9.5). For example, recent evidence shows that dissolved OC in glacial runoff has a strong microbial component, constituting microbially produced proteinaceous DOC species (Lafreniere and Sharp, 2004; Hood *et al.*, 2009).

9.4 The Diversity of Supraglacial Communities

Despite the extreme conditions in supraglacial habitats, life has been shown to thrive in them. The wind-blown debris which drives the development of many supraglacial ecosystems is also a source of biological material to glacier surfaces (Stibal and Elster, 2005). The most likely sources of airborne cells, spores and other propagules which inoculate glacier surface habitats are nearby terrestrial and freshwater environments, such as soils, streams, lakes and wetlands (Wharton *et al.*, 1981; Christner *et al.*, 2003; Porazinska *et al.*, 2004; Stibal *et al.*, 2006). There is now a wealth of studies from glaciers and ice sheets worldwide showing that prokaryotic microbes dominate the supraglacial ecosystem, although eukaryotic microorganisms and metazoans are also found and may be important for some biogeochemical transformations on glacier surfaces (Wharton *et al.*, 1981; Yoshimura *et al.*, 1997; Mueller *et al.*, 2001; Anesio *et al.*, 2007; Edwards *et al.*, 2011; Uetake *et al.*, 2010).

9.4.1 Cryoconite hole communities

Life on glaciers is mostly concentrated in cryoconite holes, due to their persistence within a dynamic and changing environment, and the presence of liquid water and debris which provides the essential nutrients for life. Cryoconite hole communities are structured into truncated food webs (Fig. 9.7), comprising primary producers, decomposers and grazers (Mueller *et al.*, 2001; Säwström *et al.*, 2002; Porazinska *et al.*, 2004). Viruses also have an important role in the microbial loop (Anesio *et al.*, 2007; Säwström *et al.*, 2007).

Primary producers

The base level of the cryoconite hole trophic community structure consists of photoautotrophic (photosynthetic) microorganisms, such as cyanobacteria and microscopic algae. Photoautotrophs in cryoconite holes are believed to be psychrotrophic (tolerant of low temperatures), rather than truly psychrophilic (cold-loving) extremophiles. Laboratory experiments with photoautotrophs in Svalbard cryoconite (Stibal and Tranter, 2007) have determined that their physiological optima are at higher temperatures than those that occur in cryoconite holes. This, in concert with the notion that most Arctic and Antarctic freshwater cyanobacteria are psychrotrophic (Tang *et al.*, 1997), suggests that the photoautotroph inoculum in cryoconite holes is likely to originate from nearby higher-temperature aquatic habitats such as wetlands and streams (Porazinska *et al.*, 2004; Stibal *et al.*, 2006). Based upon their relatively simple identification by microscopy, the diversity of photoautotrophs in cryoconite holes has been well described (Gerdel and Drouet, 1960; Wharton *et al.*, 1981; Mueller *et al.*, 2001; Säwström *et al.*, 2002; Porazinska *et al.*, 2004; Stibal *et al.*, 2006), and supported by later molecular analyses (Christner *et al.*, 2003; Edwards *et al.*, 2011). The cryoconite photoautotroph assemblages are almost always dominated by filamentous cyanobacteria. Species of the

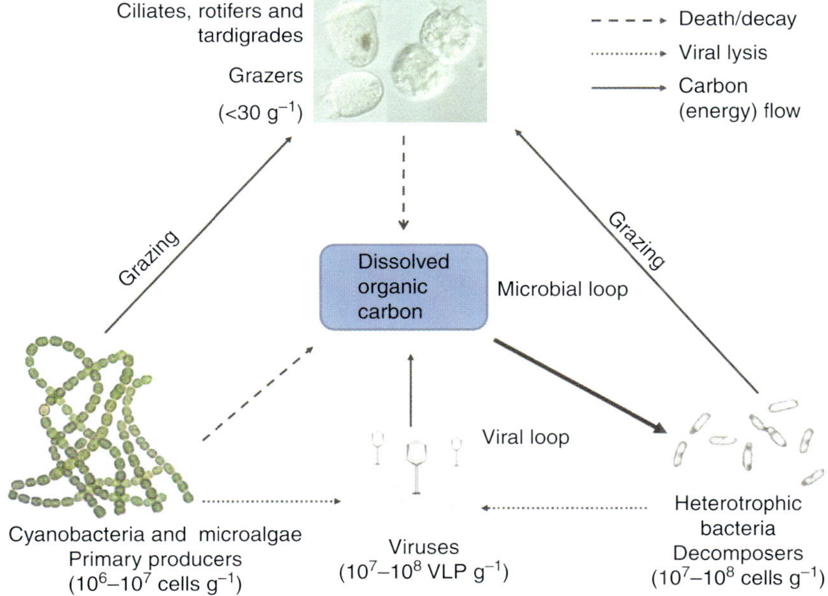

Fig. 9.7. The dominant members of cryoconite ecosystems. The food web is truncated, being comprised of primary producers, decomposers and very few grazers. Values shown represent measured abundances of organisms from cryoconite sediment on Svalbard glaciers. The viral loop depicts the recycling of particulate organic carbon (POC) into dissolved organic carbon (DOC) by the virally mediated destruction of bacterial cells. DOC and nutrients are then made available to the uninfected microbial community. Protist © Yuuji Tsukii.

Oscillatorialean genera *Leptolyngbya*, *Phormidium* and the heterocystous *Nostoc* are usually dominant. Other cyanobacterial genera found in cryoconite holes include unicellular *Aphanothece*, *Chroococcus* and *Chamaesiphon*, and filamentous *Oscillatoria*, *Microcoleus*, *Schizothrix*, *Anabaena*, *Calothrix*, *Crinalium* and *Plectonema* (Gerdel and Drouet, 1960; Wharton *et al.*, 1981; Mueller *et al.*, 2001; Säwström *et al.*, 2002; Mueller and Pollard, 2004; Porazinska *et al.*, 2004; Stibal *et al.*, 2006). Eukaryotic microalgae from the groups Chlorophyceae, Zygnematophyceae and Bacillariophyceae (diatoms) are also frequently found in cryoconite holes (Takeuchi *et al.*, 2001; Säwström *et al.*, 2002; Mueller and Pollard, 2004; Stibal *et al.*, 2006; Van de Vijver *et al.*, 2010; Yallop and Anesio, 2010). However, it is likely that some of them are washed into cryoconite holes from their original habitats such as melting snow (e.g. *Chlamydomonas* and *Chloromonas* species) or glacier ice (glacier ice desmids *Ancylonema*, *Cylindrocystis* and *Mesotaenium*). An exceptional case of moss (*Racomitrium* sp.) growth based on supraglacial debris was described from an Iceland glacier (Porter *et al.*, 2008).

Typical abundances of photoautotrophic microbes within cryoconite debris range from between 10^6 and 10^7 cells g^{-1} of sediment (wet weight), or between 2 and 10 $\times 10^2$ cells ml^{-1}. Their contribution to the total biomass of the cryoconite microbial community is usually 1–5% by mass (Säwström *et al.*, 2002; Stibal *et al.*, 2006, 2008). Filamentous cyanobacteria have been found to account for 90–99% of photoautotrophic cell numbers in cryoconite holes on Svalbard, although their contribution to the total biomass of the phototroph community may be lower due to the usually larger volume of eukaryotic algae (Stibal *et al.*, 2006; Stibal and Tranter, 2007).

Decomposers

Heterotrophic bacteria and, to a lesser extent, fungi, represent the decomposing component of cryoconite hole communities (Wharton *et al.*, 1985; Margesin *et al.*, 2002; Christner *et al.*, 2003; Foreman *et al.*, 2007; Turchetti *et al.*, 2008; Edwards *et al.*, 2011). Due to their more difficult identification – requiring molecular approaches – the diversity of heterotrophic bacteria in cryoconite holes is less well known than that of photoautotrophic microbes. Members of most bacterial lineages – Acidobacteria, Actinobacteria, Bacteroidetes, Chloroflexi, Cyanobacteria, Firmicutes, Gemmatimonadetes, OP10, Planctomycetes, Proteobacteria and Verrucomicrobia – have been detected in Arctic, Antarctic and alpine cryoconite debris (Christner *et al.*, 2003; Foreman *et al.*, 2007; Pradhan *et al.*, 2010; Edwards *et al.*, 2011). Obligately psychrophilic bacteria were identified and isolated from cryoconite in Antarctica (Christner *et al.*, 2003). This suggests that the heterotrophs inhabiting cryoconite holes may be better adapted to low temperature than the photoautotrophs, particularly since heterotrophic microbes from cryoconite holes can utilize a wide range of substrates. These include various simple compounds, such as amino acids, carboxylic acids and carbohydrates, as well as more complex compounds like cellulose and lignin (Margesin *et al.*, 2002; Foreman *et al.*, 2007). Utilization of aromatic hydrocarbons, diesel oil or keratin by cryoconite heterotrophs has also been reported. However, most bacterial sequences identified in cryoconite holes on several Svalbard glaciers were related to temperate organisms (Edwards *et al.*, 2011). Hence, it is possible that the identification of psychrophiles among heterotrophs and the lack thereof among photoautotrophs may simply reflect the considerably higher diversity of the former in the cryoconite communities.

Proteobacteria, especially Sphingomonadaceae (α-Proteobacteria), were the dominant bacterial group in cryoconite holes on several Svalbard glaciers, followed by cyanobacteria, Bacteroidetes and Actinobacteria (Edwards *et al.*, 2011). β-Proteobacteria and Bacteroidetes were also the main components of the bacterial communities on several glaciers in the Antarctic Dry Valleys (Foreman *et al.*, 2007). Actinobacteria and Firmicutes were dominant in glacier debris on a Himalayan glacier (Pradhan *et al.*, 2010). The majority of strains isolated from cryoconite collected on an Austrian glacier belonged in the genera *Pseudomonas* (γ-Proteobacteria) and *Sphingomonas* (Margesin *et al.*, 2002). Novel psychrophilic species of *Sphingomonas* and *Pedobacter* (Bacteroidetes), and a novel genus *Glaciimonas* (Oxalobacteraceae, β-Proteobacteria) has been described from this glacier (Margesin *et al.*, 2003; Zhang *et al.*, 2010a, b). Typical bacterial abundances reported from cryoconite holes are between 10^7 and 10^8 cells g^{-1} of cryoconite or between 10^4 and 10^5 cells ml^{-1} (Fig. 9.7) (Säwström *et al.*, 2002, 2007; Stibal *et al.*, 2006, 2008; Foreman *et al.*, 2007). Margesin *et al.* (2002) obtained 2–100 × 10^4 colony forming units (cfu) of heterotrophic bacteria per gram of debris when culturing cryoconite samples from an Austrian glacier at 2°C.

Relatively less attention has been paid to fungi in supraglacial environments, and very little is known about the abundance, diversity and activity of fungi in cryoconite holes. Gerdel and Drouet (1960) found species of the genera *Penicillium, Aspergillus, Cephalothecium, Circinella* and *Alternaria* (Ascomycota) in cryoconite holes in Greenland, and a fungal sequence related to *Choiromyces* (Ascomycota) was obtained in a molecular study of an Antarctic cryoconite hole (Christner *et al.*, 2003). Margesin *et al.* (2002) isolated 2–160 × 10^3 cfu g^{-1} of yeast from cryoconite samples from an Austrian glacier incubated at 2°C, and novel psychrophilic species of the genus *Rhodotorula* (Basidiomycota) were described from these samples (Margesin *et al.*, 2007). Turchetti *et al.* (2008) isolated 10–20 cfu g^{-1} of yeast from supraglacial sediment from two Italian glaciers, and identified species of *Rhodotorula, Cryptococcus* and *Mrakia* (Basidiomycota) and *Aureobasidium* and *Naganishia* (Ascomycota).

Grazers

Protozoa (ciliates) and Metazoa (rotifers, tardigrades, and even an insect) have been found in cryoconite holes on Arctic, Antarctic and alpine glaciers (Gerdel and Drouet, 1960; Koshima, 1984; Desmet and Vanrompus, 1994; Pugh and McInnes, 1998; Grongaard *et al.*, 1999; Säwström *et al.*, 2002; Porazinska *et al.*, 2004). They feed on microbial cells and organic carbon produced by the primary producers or deposited by the wind and so form the top level of the trophic food web within cryoconite holes (Porazinska *et al.*, 2004; Fig. 9.7). The abundance of ciliates was ~10 cells ml^{-1} in cryoconite holes on a Svalbard glacier, with *Monodinium* (Haptorida) and *Halteria* and *Strombidium* (Oligotrichia) species dominating the community (Säwström *et al.*, 2002). The numbers of metazoans are also very low (<30 individuals g^{-1} of cryoconite), as is their diversity. Only a few species of rotifers in a few genera have been identified in cryoconite holes in Antarctica (*Philodina, Cephalodella*; Porazinska *et al.*, 2004), in Greenland (*Philodina, Philodinavus*; Gerdel and Drouet, 1960; Grongaard *et al.*, 1999) and on Svalbard (*Macrotrachela, Philodina, Polyarthra*; Desmet and Vanrompus, 1994; Säwström *et al.*, 2002), and ~10 species of tardigrades have been reported from cryoconite holes worldwide, all belonging to the family Hypsibiidae (Pugh and McInnes, 1998). A new species of non-flying midge (*Diamesa*, Chironomidae) was found on a Himalayan glacier, living in cryoconite holes and supraglacial channels and feeding on bacteria and algae (Koshima, 1984).

Viruses

Prokaryotic viruses (bacteriophages) are abundant in cryoconite holes, infecting local bacteria, and are believed to play an important role in the recycling of organic carbon within the holes (Anesio *et al.*, 2007; Säwström *et al.*, 2007). The abundances of virus-like particles (VLP) in the sediment of cryoconite holes on two Svalbard glaciers were between 10^7 and 10^8 VLP ml^{-1} (Fig. 9.7), and the virus:bacterium ratios showed great variation, with ratios ranging from 0.3 to 1.5 in the sediment and 7.3 to 31 in the water (Anesio *et al.*, 2007). The viral community can be sustained in cryoconite holes despite the low host density due to high infection rates, as supported by the high fraction of visibly infected bacterial cells (5–20%) and the low burst size (2–6) measured on Svalbard glaciers (Säwström *et al.*, 2007).

9.4.2 Glacier ice and snow communities

The glacier surface physical environment outside cryoconite holes is relatively hostile to life, due to the shortage of liquid water and sediment-bound nutrients. However, microbial communities largely consisting of bacteria and microscopic algae have been found in melting snow and glacier ice in the absence of cryoconite debris on glaciers worldwide (Yoshimura *et al.*, 1997; Takeuchi and Kohshima, 2004; Amato *et al.*, 2007; Liu *et al.*, 2009; Segawa *et al.*, 2010; Uetake *et al.*, 2010).

Amato *et al.* (2007) reported bacterial abundances of 1–4 × 10^4 cells ml^{-1} in the snow cover and 2 × 10^5 cells ml^{-1} in the surface layer of the glacier ice on a Svalbard glacier, and isolated ten strains belonging to α-, β- and γ-Proteobacteria, Firmicutes and Actinobacteria. Liu *et al.* (2009) found bacterial abundances ranging from ~0.1 to 70 × 10^4 cells ml^{-1} in the snow on several Tibetan Plateau glaciers, with α-, β- and γ-Proteobacteria, Bacteroidetes and Actinobacteria dominating the communities and *Sphingomonas* (α-Proteobacteria) and *Polaromonas* (β-Proteobacteria) present at all glaciers investigated. Bacteroidetes and α- and β-Proteobacteria were found to be dominant in the bacterial community (1–5 × 10^6 cells ml^{-1}) in the snow and ice on an Alaskan glacier (Segawa *et al.*, 2010).

Communities of algae have also been reported from a number of high-latitude and -altitude glaciers (Yoshimura *et al.*, 1997; Takeuchi and Kohshima, 2004; Uetake *et al.*, 2010). They typically consist of snow algae (*Chloromonas, Chlamydomonas*) and saccoderm desmids (*Ancylonema, Cylindrocystis, Mesotaenium*), and their abundances are usually between 10^4 and 10^5 cells ml^{-1}. Whereas active snow algae are mostly confined to melting snow and only appear as resting stages in ice and cryoconite holes, the desmids seem to thrive on bare ice (Yoshimura *et al.*, 1997; see also Sattler *et al.*, Chapter 8, this volume).

9.5 Impacts on Surrounding Environments

Glacier surface ecosystems certainly have some impact on microorganisms in glacier-marginal locations, by providing microbially labile organic carbon and macronutrients. The importance of surface ecosystem production to surrounding biological communities is, as yet, largely unquantified. However, it is speculated that the newly produced organic material is readily available for microbes and is an important resource for downstream environments in nutrient-limited ecosystems (Hodson *et al.*, 2004; Bagshaw *et al.*, 2007; Anesio *et al.*, 2009). As meltwater is flushed off the surface of glaciers, so too are the contents of supraglacial ecosystems, and thus via this export mechanism, biological processes which occur on glacier surfaces have the potential to influence their surrounding environment.

Fountain *et al.* (2004, 2008) showed that in the McMurdo Dry Valleys of Antarctica, cryoconite holes contribute 13–15% of the meltwater that constitutes runoff from Canada Glacier. Long-term monitoring of the electrical conductivity (EC, used as a proxy for total dissolved solids) of a network of cryoconite holes and of the runoff from the glacier showed that as the EC of hydrologically connected cryoconite holes decreased over the ablation season, so too did the EC of the runoff (Fig. 9.8). This supports the hypothesis that solute within cryoconite holes, including dissolved C, N and P species, can be flushed off the glacier. Figure 9.9 shows DOC concentrations in cryoconite holes, glacier ice, runoff and meltwater-fed streams (Canada Stream and Andersen Creek) at Canada Glacier. DOC concentrations in runoff and the streams frequently exceed that in melted glacier ice, particularly in the early part of the ablation season. There is minimal subglacial melting in this region because the glaciers remain

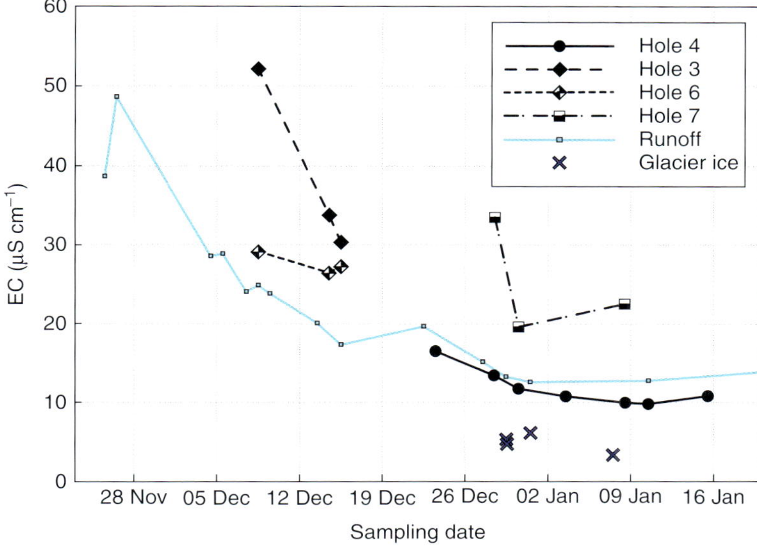

Fig. 9.8. Electrical conductivity (EC) of hydrologically connected (Hole 4), closed cryoconite holes and runoff from Canada Glacier, McMurdo Dry Valleys, Antarctica (Fountain *et al.*, 2008). EC decreases in both holes and runoff over the ablation season, showing progressive flushing of solutes in the supraglacial drainage system into runoff.

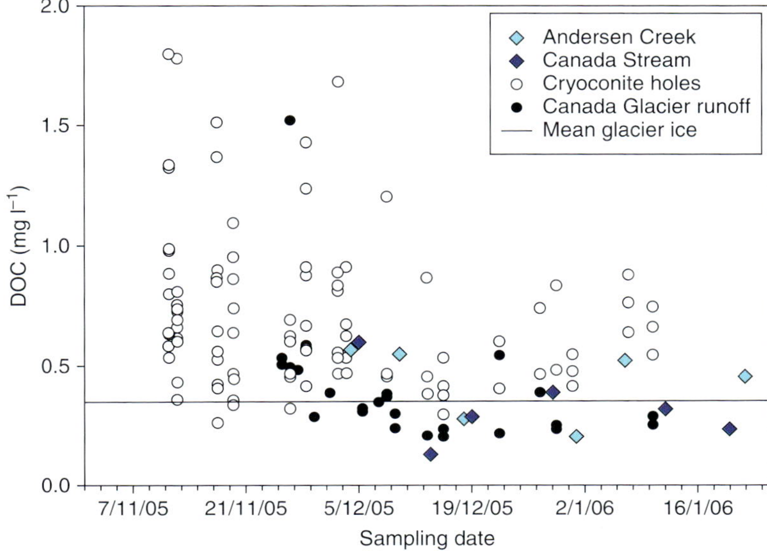

Fig. 9.9. Dissolved organic carbon (DOC) concentrations in cryoconite holes, runoff and two glacier-fed streams in Taylor Valley, McMurdo Dry Valleys, Antarctica. The black line indicates the mean DOC concentration of glacier ice; most cryoconite holes and approximately half of the runoff exceed this concentration. See Bagshaw *et al.* (2007) for methods of sample collection and analysis.

frozen to their beds year-round. All meltwater thus passes through the supraglacial drainage system. This suggests that glacier surface ecosystems do indeed contribute organic matter to the runoff from glaciers, in agreement with several other studies (Foreman *et al.*, 2004; Lafreniere and Sharp, 2004; Stibal and Tranter, 2007). The DOC is likely to be labile, since a proportion will be freshly cycled via the processes described above (Anesio *et al.*, 2009; Hood *et al.*, 2009).

In certain circumstances, glacier surface ecosystems can export sufficient N to impact on surrounding ecosystems. N and P, sourced from dust and organic matter on the glacier surface, are utilized by microbes for growth in glacier surface ecosystems. Decomposition of the subsequent necromass results in the release of DON and DOP to solution. The protein- and acetate-rich DOC that glacial microbes produce (Hood *et al.*, 2009) is more available than the relatively refractive, aromatic- and lignin-rich DOC that results from the degradation of terrestrial organic matter (Holmes, 2008), and is readily utilizable by heterotrophic bacteria in surrounding environments (Berggren *et al.*, 2009; Fellman *et al.*, 2010). An example of this can be found in the oligotrophic MCMDV, where soil carbon and nutrients are frequently unavailable to organisms over a viable timescale since they are not in immediately bioavailable forms (Barrett *et al.*, 2007). By contrast, the cryoconite holes on cold, polar glaciers contain 'new' C, N and P in readily available forms (Tranter *et al.*, 2004). When they connect to the drainage system, the solute contained within the holes is flushed to the ephemeral streams and lakes that are supported by annual glacier melt (Fountain *et al.*, 2004), where it is consumed by heterotrophic organisms or used to support further primary production (Barrett *et al.*, 2007).

The magnitude of this input has not been quantified, but the importance of the supraglacial ecosystem as a resource can be estimated by studying an extreme melt event caused by abnormally warm air temperatures in Taylor Valley (MCMDV) in the austral summer of 2001/02 (Foreman *et al.*, 2004). Ephemeral stream discharge was at a record high, lake levels rose (Doran *et al.*, 2008), soil moisture increased (Barrett *et al.*, 2008), supraglacial streams were widespread (Fortner *et al.*, 2005), and the majority of cryoconite holes lost their ice lids and connected to the supraglacial drainage system (Fountain *et al.*, 2004). In the year following the flood, primary production (PP) in the top 5 m of Lake Fryxell, one of the perennially ice-covered lakes fed by glacier melt, increased by 12.8 μg C l^{-1} day^{-1} from the long term average of 2.4 μg C l^{-1} day^{-1} (Foreman *et al.*, 2004). Foreman *et al.* (2004) hypothesize that the excess carbon and nutrient required to support this increase was sourced from the cryoconite holes on the glaciers which feed Lake Fryxell. Simple calculations, based on the mean P and N content of the holes and the percentage of the glacier surfaces that they cover, supports this hypothesis. Assuming that the excess autotrophic production of carbon in the lake is produced in accordance with the Redfield ratio (106:16 C:N; Redfield, 1958), the extra C which is produced in 2.8×10^{10} l of water requires some 55 kg N day^{-1}, or 1640 kg N over the 30 day ablation season. By calculating the mean N concentration in the sediment and water in the cryoconite holes on Canada and Commonwealth Glaciers, it is estimated that they contain of the order of 100 and 1600 kg N, respectively (E.A. Bagshaw and M. Tranter, 2011, unpublished results). For comparison, the total nitrogen recommended for a maize-based crop is of the order of 110 kg ha^{-1}, and the total effective amount is nearer 67 kg ha^{-1}. The N stored in the cryoconite holes on these two glaciers could therefore fertilize approximately 24 ha of maize (Gallagher, 2010). Hence, the input of N to Lake Fryxell from cryoconite hole flushing, when coupled with inputs from ephemeral streams and meltwater seeps, and with subsequent N recycling in the lake water column, is of the correct order of magnitude to support the observed increase in primary production.

9.6 Implications for Surrounding Regions and Life on Other Planets

Glacier surface ecosystem processes are considered as analogues for life on other icy terrestrial planets in our solar system, in common with other extreme habitats for life (see Gómez *et al.*, Chapter 26, this volume). These processes are also important for ice-marginal communities in nutrient-poor environments, and may also impact other communities that rely on ice-melt, including coastal waters in the Arctic (Hood *et al.*, 2009). This becomes pertinent in a warming climate, where meltwater fluxes increase and higher rates of flushing of surface ecosystems are predicted to occur. The impact of these changes on fluxes of DOC, N and P from glacier surfaces to downstream ecosystems has yet to be determined, but could have significance for regional carbon and nutrient cycling. Hence, processes that occur in glacier surface habitats are likely to become more significant in completing our understanding of biogeochemical fluxes and cycling that occurs in the polar regions.

References

Amato, P., Hennebelle, R., Magand, O., Sancelme, M., Delort, A.M., Barbante, C., Boutron, C. and Ferrari, C. (2007) Bacterial characterization of the snow cover at Spitzberg, Svalbard. *Fems Microbiology Ecology* 59, 255–264.

Anesio, A.M., Mindl, B., Laybourn-Parry, J., Hodson, A. and Sattler, B. (2007) Viral dynamics in cryoconite holes on a high Arctic glacier (Svalbard). *Journal of Geophysical Research* 112, G04S31.

Anesio, A.M., Hodson, A.J., Fritz, A., Psenner, R. and Sattler, B. (2009) High microbial activity on glaciers: importance to the global carbon cycle. *Global Change Biology* 15, 955–960.

Bagshaw, E.A., Tranter, M., Fountain, A.G., Welch, K.A., Basagic, H. and Lyons, W.B. (2007) Biogeochemical evolution of cryoconite holes on Canada Glacier, Taylor Valley, Antarctica. *Journal of Geophysical Research* 112, G04S35, doi:10.1029/2007JG000442.

Bagshaw, E.A., Tranter, M., Wadham, J.L., Fountain, A.G. and Basagic, H. (2010) Dynamic behaviour of supraglacial lakes on cold polar glaciers: Canada Glacier, McMurdo Dry Valleys, Antarctica. *Journal of Glaciology* 56, 366–368.

Bagshaw, E.A., Tranter, M., Wadham, J., Fountain, A.G. and Mowlem, M. (2011a) High resolution monitoring reveals dissolved oxygen dynamics in an Antarctic cryoconite hole. *Hydrological Processes*, DOI: 10.1002/hyp.8049.

Bagshaw, E.A., Wadham, J.L., Mowlem, M., Eveness, J., Fountain, A.G., Tranter, M. and Telling, J. (2011b) Determination of dissolved oxygen in the cryosphere: a comprehensive laboratory and field evaluation of fibre optic sensors. *Environmental Science and Technology* 45, 700–705.

Barrett, J.E., Virginia, R.A., Lyons, W.B., McKnight, D., Priscu, J., Doran, P.T., Fountain, A., Wall, D.H. and Moorhead, D.L. (2007) Biogeochemical stochiometry of Antarctic Dry Valley ecosystems. *Journal of Geophysical Research* 112, G01010, doi:10.1029/2005JG000141.

Barrett, J.E., Virginia, R.A., Wall, D.H., Doran, P.T., Fountain, A.G., Welch, K.A. and Lyons, W.B. (2008) Persistent effects of a discrete warming event on a polar desert ecosystem. *Global Change Biology* 14, 2249–2261.

Berggren, M., Laudon, H., Haei, M., Strom, L. and Jansson, M. (2009) Efficient aquatic bacterial metabolism of dissolved low-molecular-weight compounds from terrestrial sources. *ISME Journal* 4, 408–416.

Bøggild, C.E., Brandt, R.E., Brown, K.J. and Warren, S.G. (2010) The ablation zone in northeast Greenland: ice types, albedos and impurities. *Journal of Glaciology* 56, 101–113.

Catania, G.A. and Neumann, T.A. (2010) Persistent englacial drainage features in the Greenland Ice Sheet. *Geophysical Research Letters* 37, doi: L0250110.1029/2009gl041108.

Christner, B.C., Kvitko, B.H. and Reeve, J.N. (2003) Molecular identification of Bacteria and Eukarya inhabiting an Antarctic cryoconite hole. *Extremophiles* 7, 177–183.

Cole, J.J., Prairie, Y.T., Caraco, N.F., McDowell, W.H., Tranvik, L.J., Striegl, R.G., Duarte, C.M., Kortelainen, P., Downing, J.A., Middelburg, J.J. and Melack, J. (2007) Plumbing the global carbon cycle: Integrating inland waters into the terrestrial carbon budget. *Ecosystems* 10(1): 171–184 doi: 10.1007/s10021-006-9013-8.

Desmet, W.H. and Vanrompus, E.A. (1994) Rotifera and tardigrada from some cryoconite holes on a Spitsbergern (Svalbard) Glacier. *Belgian Journal of Zoology* 124, 27–37.

Doran, P.T., McKay, C.P., Clow, G.D., Dana, G.L., Fountain, A.G., Nylen, T. and Lyons, W.B. (2002) Valley floor climate observations from the McMurdo dry valleys, Antarctica, 1986–2000. *Journal of Geophysical Research-Atmospheres* 107, doi:10.1029/2001JD002045.

Doran, P.T., McKay, C.P., Fountain, A.G., Nylen, T., McKnight, D.M., Jaros, C. and Barrett, J.E. (2008) Hydrologic response to extreme warm and cold summers in the McMurdo Dry Valleys, East Antarctica. *Antarctic Science* 20, 499–509.

Drever, J.I. (1997) *The Geochemistry of Natural Waters: Surface and Groundwater Environments*. Prentice Hall, Englewood Cliffs, New Jersey, 436 pp.

Drygalski, E.V. (1897) Die Kryoconitlocher. *Gronland-Expedition der Gesellschaft fur Erdkunde zu Berlin 1891–1893*. Kuhl, W.H., Berlin, Germany, pp. 93–103.

Echelmeyer, K., Clarke, T.S. and Harrison, W.D. (1991) Surficial glaciology of Jakobshavn Isbrae, West Greenland. 1. Surface-morphology. *Journal of Glaciology* 37, 368–382.

Edwards, A., Anesio, A.M., Rassner, S.M., Sattler. B., Hubbard, B.P., Perkins, W.T., Young, M. and Griffith, G.W. (2011) Possible interactions between bacterial diversity, microbial activity and supraglacial hydrology of cryoconite holes in Svalbard. *ISME* 5, 15–160.

Esposito, R.M.M., Horn, S.L., McKnight, D.M., Cox, M.J., Grant, M.C., Spaulding, S.A., Doran, P.T. and Cozzetto, K.D. (2006) Antarctic climate cooling and response of diatoms in glacial meltwater streams. *Geophysical Research Letters* 33, L07406, doi:10.1029/2006GL025903.

Fellman, J.B., Spencer, R.G.M., Hernes, P.J., Edwards, R.T., D'Amore, D.V. and Hood, E. (2010) The impact of glacier runoff on the biodegradability and biochemical composition of terrigenous dissolved organic matter in near-shore marine ecosystems. *Marine Chemistry* 121, 112–122.

Foreman, C.M., Wolf, C.F. and Priscu, J.C. (2004) Impact of episodic warming events on the physical, chemical and biological relationships of lakes in the McMurdo Dry Valleys, Antarctica. *Aquatic Geochemistry* 10, 239–268.

Foreman, C.M., Sattler, B., Mickuki, J., Porazinska, D.L. and Priscu, J.C. (2007) Metabolic activity of cryoconites in the Taylor Valley, Antarctica. *Journal of Geophysical Research: Biogeosciences* 112, G04S32, doi:10.1029/2006JG000358.

Fortner, S.K., Tranter, M., Fountain, A., Lyons, W.B. and Welch, K.A. (2005) The geochemistry of supraglacial streams of Canada Glacier, Taylor Valley (Antarctica), and their evolution into proglacial waters. *Aquatic Geochemistry* 11, 391–412.

Fountain, A.G., Dana, G.L., Lewis, K.J., Vaughn, B.H. and McKnight, D. (1998) Glaciers of the McMurdo Dry Valleys, Southern Victoria Land, Antarctica. In: Priscu, J.C. (ed.) *Ecosystem Dynamics in a Polar Desert: The McMurdo Dry Valleys, Antarctica*. American Geophysical Union, Washington, DC, pp. 65–75.

Fountain, A.G., Lyons, W.B., Burkins, M.B., Dana, G.L., Doran, P.T., Lewis, K.J., McKnight, D.M., Moorhead, D.L., Parsons, A.N., Priscu, J.C., Wall, D.H., Wharton, R.A. and Virginia, R.A. (1999) Physical controls on the Taylor Valley ecosystem, Antarctica. *Bioscience* 49, 961–971.

Fountain, A.G., Tranter, M., Nylen, T.H., Lewis, K.J. and Mueller, D.R. (2004) Evolution of cryoconite holes and their contribution to meltwater runoff from glaciers in the McMurdo Dry Valleys, Antarctica. *Journal of Glaciology* 50, 35–45.

Fountain, A.G., Nylen, T., Tranter, M. and Bagshaw, E.A. (2008) Temporal variation of cryoconite holes on Canada Glacier, McMurdo Dry Valleys, Antarctica. *Journal of Geophysical Research* 112, G01S92, doi:10.1029/2007JG000430.

Gallagher, M. (2010) *Environmental Aspects of Bioenergy Production*. AGU Fall Meeting, San Francisco.

Gerdel, R.W. and Drouet, F. (1960) The cryoconite of the Thule Area, Greenland. *Transactions of the American Microscopical Society* 79, 256–272.

Gribbon, P.W. (1979) Cryoconite holes on Sermikaysak, West Greenland. *Journal of Glaciology* 22, 177–181.

Grongaard, A., Pugh, P.J.A. and McInnes, S.J. (1999) Tardigrades, and other cryoconite biota, on the Greenland Ice Sheet. *Zoologischer Anzeiger* 238, 211–214.

Guildford, S.J. and Hecky, R.E. (2000) Total nitrogen, total phosphorus, and nutrient limitation in lakes and oceans: is there a common relationship? *Limnology and Oceanography* 45, 1213–1223.

Hodson, A., Murnford, P. and Lister, D. (2004) Suspended sediment and phosphorus in proglacial rivers: bioavailability and potential impacts upon the P status of ice-marginal receiving waters. *Hydrological Processes* 18, 2409–2422.

Hodson, A., Anesio, A.M., Ng, F., Watson, R., Quirk, J., Irvine-Fynn, T., Dye, A., Clark, C., McCloy, P., Kohler, J. and Sattler, B. (2007) A glacier respires: Quantifying the distribution and respiration CO_2 flux of cryoconite across an entire Arctic supraglacial ecosystem. *Journal of Geophysical Research-Biogeosciences* 112, G04S36, doi:10.1029/2007jg000452.

Hodson, A., Anesio, A.M., Tranter, M., Fountain, A., Osborn, M., Priscu, J., Laybourn-Parry, J. and Sattler, B. (2008) Glacial ecosystems. *Ecological Monographs* 78, 41–67.

Hodson, A., Cameron, K., Boggild, C., Irvine-Fynn, T., Langford, H., Pearce, D. and Banwart, S. (2010) The structure, biological activity and biogeochemistry of cryoconite aggregates upon an Arctic valley glacier: Longyearbreen, Svalbard. *Journal of Glaciology* 56, 349–362.

Hodson, A.J. (2006) Biogeochemistry of snowmelt in an Antarctic glacial ecosystem. *Water Resources Research* 42, doi: 10.1029/2005WR004311.

Hodson, A.J., Mumford, P.N., Kohler, J. and Wynn, P.M. (2005) The High Arctic glacial ecosystem: new insights from nutrient budgets. *Biogeochemistry* 72, 233–256.

Hoffman, M.J., Fountain, A.G. and Liston, G.E. (2008) Surface energy balance and melt thresholds over 11 years at Taylor Glacier, Antarctica. *Journal of Geophysical Research-Earth Surface* 113, 12 doi: 10.1029/2008jf001029.

Holmes, R.M. (2008) Lability of DOC transported by Alaskan Rivers to the Arctic Ocean. *Geophyical Research Letters* 35, L03402, doi:10.1029/2007GL032837.

Hood, E., Fellman, J., Spencer, R.G.M., Hernes, P.J., Edwards, R., D'Amore, D. and Scott, D. (2009) Glaciers as a source of ancient and labile organic matter to the marine environment. *Nature* 462, 1044–1047.

Irvine-Fynn, T.D.L., Bridge, J.W. and Hodson, A.J. (2010) Rapid quantification of cryoconite: granule geometry and *in situ* supraglacial extents, using examples from Svalbard and Greenland. *Journal of Glaciology* 56, 297–308.

Kastovska, K., Elster, J., Stibal, M. and Santruckova, H. (2005) Microbial assemblages in soil microbial succession after glacial retreat in Svalbard (high Arctic) *Microbial Ecology* 50(3), 396–407.

Kennedy, A.D. (1993) Water as a limiting factor in the Antarctic Terrestrial Environment – a biogeographical synthesis. *Arctic and Alpine Research* 25, 308–315.

Koshima, S. (1984) A novel cold-tolerant insect found in a Himalayan glacier. *Nature* 310, 225–227.

Lafreniere, M.J. and Sharp, M.J. (2004) The concentration and fluorescence of dissolved organic carbon (DOC) in glacial and nonglacial catchments: interpreting hydrological flow routing and DOC sources. *Arctic Antarctic and Alpine Research* 36, 156–165.

Liu, Y., Yao, T., Jiao, N., Kang, S., Xu, B., Zeng, Y., Huang, S. and Liu, X. (2009) Bacterial diversity in the snow over Tibetan Plateau Glaciers. *Extremophiles* 13, 411–423.

Margesin, R., Zacke, G. and Schinner, F. (2002) Characterization of heterotrophic microorganisms in alpine glacier cryoconite. *Arctic Antarctic and Alpine Research* 34, 88–93.

Margesin, R., Sproer, C., Schumann, P. and Schinner, F. (2003) *Pedobacter cryoconitis* sp nov., a facultative psychrophile from alpine glacier cryoconite. *International Journal of Systematic and Evolutionary Microbiology* 53, 1291–1296.

Margesin, R., Fonteyne, P.A., Schinner, F. and Sampaio, J.P. (2007) *Rhodotorula psychrophila* sp nov., *Rhodotorula psychrophenolica* sp. nov. and *Rhodotorula glacialis* sp nov., novel psychrophilic basidiomycetous yeast species isolated from alpine environments. *International Journal of Systematic and Evolutionary Microbiology* 57, 2179–2184.

Mueller, D.R. and Pollard, W.H. (2004) Gradient analysis of cryoconite ecosystems from two polar glaciers. *Polar Biology* 27, 66–74.

Mueller, D.R., Vincent, W.F., Pollard, W.H. and Fritsen, C.H. (2001) Glacial cryoconite ecosystems: a bipolar comparison of algal communities and habitats. *Nova Hedwigia* 123, 173–197.

Nkem, J.N., Wall, D.H., Virginia, R.A., Barrett, J.E., Broos, E.J., Porazinska, D.L. and Adams, B.J. (2006) Wind dispersal of soil invertebrates in the McMurdo Dry valleys, Antarctica. *Polar Biology* 29, 346–352.

Nordenskjöld (1875) *Geological Magazine* Decade 2, 157–167.

Porazinska, D.L., Fountain, A.G., Nylen, T.H., Tranter, M., Virginia, R.A. and Wall, D.H. (2004) The biodiversity and biogeochemistry of cryoconite holes from McMurdo Dry Valley glaciers, Antarctica. *Arctic Antarctic and Alpine Research* 36, 84–91.

Porter, P.R., Evans, A.J., Hodson, A.J., Lowe, A.T. and Crabtree, M.D. (2008) Sediment-moss interactions on a temperate glacier: Falljokull, Iceland. *Annals of Glaciology* 48, 25–31.

Pradhan, S., Srinivas, T.N.R., Pindi, P.K., Kishore, K.H., Begum, Z., Singh, P.K., Singh, A.K., Pratibha, M.S., Yasala, A.K., Reddy, G.S.N. and Shivaji, S. (2010) Bacterial biodiversity from Roopkund Glacier, Himalayan mountain ranges, India. *Extremophiles* 14, 377–395.

Priscu, J.C. and Christner, B.C. (2002) Earth's icy biosphere. In: Bull, A. (ed.) *Microbial Biodiversity and Microprospecting*. American Society of Microbiology Press, Washington, DC, pp. 130–145.

Pugh, P.J.A. and McInnes, S.J. (1998) The origin of Arctic terrestrial and freshwater tardigrades. *Polar Biology* 19, 177–182.

Redfield, A.C. (1958) The biological control of chemical factors in the environment. *American Scientist* 46, 205–221.

Sabacka, M., Priscu, J., Basagic, H., Fountain, A., Wall, D.H., Virginia, R.A. and Greenwood, M. (2012) Aeolian flux of biotic and abiotic material in Taylor Valley, Antarctica. *Geomorphology*, in press.

Säwström, C., Mumford, P., Marshall, W., Hodson. A. and Laybourn-Parry, J. (2002) The microbial communities and primary productivity of cryoconite holes in an Arctic glacier (Svalbard 79 degrees N). *Polar Biology* 25, 591–596.

Säwström, C., Graneli, W., Laybourn-Parry, J. and Anesio, A.M. (2007) High viral infection rates in Antarctic and Arctic bacterioplankton. *Environmental Microbiology* 9, 250–255.

Segawa, T., Takeuchi, N., Ushida, K., Kanda, H. and Kohshima, S. (2010) Altitudinal changes in a bacterial community on Gulkana Glacier in Alaska. *Microbes and Environments* 25, 171–182.

Stibal, M. and Elster, J. (2005) Growth and morphology variation as a response to changing environmental factors in two Arctic species of *Raphidonema* (Trebouxiophyceae) from snow and soil. *Polar Biology* 28, 558–567.

Stibal, M. and Tranter, M. (2007) Laboratory investigation of inorganic carbon uptake by cryoconite debris from Werenskioldbreen, Svalbard. *Journal of Geophysical Research-Biogeosciences* 112, doi: G04s33 10.1029/2007jg000429.

Stibal, M., Sabacka, M. and Kastovska, K. (2006) Microbial communities on glacier surfaces in Svalbard: impact of physical and chemical properties on abundance and structure of cyanobacteria and algae. *Microbial Ecology* 52, 644–654.

Stibal, M., Tranter, M., Benning, L.G. and Rehak, J. (2008) Microbial primary production on an Arctic glacier is insignificant in comparison with allochthonous organic carbon input. *Environmental Microbiology* 10, 2172–2178.

Stibal, M., Lawson, E.C., Lis, G.P., Mak, K.M., Wadham, J.L. and Anesio, A.M. (2010) Organic matter content and quality in supraglacial debris across the ablation zone of the Greenland ice sheet. *Annals of Glaciology* 51, 1–8.

Takeuchi, N. and Kohshima, S. (2004) A snow algal community on Tyndall Glacier in the Southern Patagonia Icefield, Chile. *Arctic, Antarctic, and Alpine Research* 36, 92–99.

Takeuchi, N., Koshima, S., Yoshimura, Y., Seko, K. and Fujita, K. (2000) Characteristics of cryoconite holes on a Himalayan glacier, Yala Glacier Central Nepal. *Bulletin of Glaciological Research* 17, 51–59.

Takeuchi, N., Kohshima, S. and Seko, K. (2001) Structure, formation, and darkening process of albedo-reducing material (cryoconite) on a Himalayan glacier: a granular algal mat growing on the glacier. *Arctic Antarctic and Alpine Research* 33, 115–122.

Tang, E.P.Y., Tremblay, R. and Vincent, W.F. (1997) Cyanobacterial dominance of polar freshwater ecosystems: are high-latitude mat-formers adapted to low temperature? *Journal of Phycology* 33, 171–181.

Telling, J., Anesio, A.M., Hawkings, J., Tranter, M., Irvine-Fynn, T., Hodson, A., Butler, C., Yallop, M.L. and Wadham, J. (2010a) Influence of sediment thickness and shading on rates of net community production in cryoconite holes, Svalbard. *Journal of Geophysical Research: Biogeosciences*, in review.

Telling, J., Anesio, A.M., Hawkings, J., Tranter, M., Wadham, J., Hodson, A., Irvine-Fynn, T. and Yallop, M.L. (2010b) Measuring rates of gross photosynthesis and net community production in cryoconite holes: a comparison of field methods. *Annals of Glaciology* 51, 56, 135–144.

Teolis, B.D., Jones, G.H., Miles, P.F., Tokar, R.L., Magee, B.A., Waite, J.H., Roussos, E., Young, D.T., Crary, F.J., Coates, A.J., Johnson, R.E., Tseng, W.L. and Baragiola, R.A. (2010) Cassini finds an oxygen-carbon dioxide atmosphere at Saturn's icy moon Rhea. *Science* 330, 1813–1815.

Tranter, M., Fountain, A.G., Fritsen, C.H., Lyons, W.B., Priscu, J.C., Statham, P.J. and Welch, K.A. (2004) Extreme hydrochemical conditions in natural microcosms entombed within Antarctic ice. *Hydrological Processes* 18, 379–387.

Tranter, M., Fountain, A.G., Lyons, W.B., Nylen. T.H. and Welch, K.A. (2005) The chemical composition of runoff from Canada Glacier, Antarctica: implications for glacier hydrology during a cool summer. *Annals of Glaciology* 40, 15–19.

Tranter, M., Bagshaw, E.A., Fountain, A.G. and Foreman, C.M. (2010) The biogeochemistry and hydrology of McMurdo Dry Valley glaciers: is there life on martian ice now? In: Doran, P.T., Lyons, W.B. and McKnight, D.M. (eds) *Life in Antarctic Deserts and other Cold, Dry Environments: Astrobiological Analogues*. Cambridge University Press, Cambridge, UK, 320 pp.

Turchetti, B., Buzzini, P., Goretti, M., Branda, E., Diolaiuti, G., D'Agata, C., Smiraglia, C. and Vaughan-Martini, A. (2008) Psychrophilic yeasts in glacial environments of Alpine glaciers. *FEMS Microbial Ecology* 63, 73–83.

Uetake, J., Naganuma, T., Hebsgaard, M.B., Kanda, H. and Koshima, S. (2010) Communities of algae and cyanobacteria on glaciers in west Greenland. *Polar Science* 4, 71–80.

Van de Vijver, B., Mataloni, G., Stanish, L. and Spaulding, S.A. (2010) New and interesting species of the genus *Muelleria* (Bacillariophyta) from the Antarctic region and South Africa. *Phycologia* 49, 22–41.

Vincent, W.F., Gibson, J., Pienitz, R., Villeneuve, V., Broady, P., Hamilton, P. and Howard-Williams, C. (2000) Ice shelf microbial ecosystems in the High-Arctic and implications for life on Snowball Earth. *Naturwissenschaften* 87, 137–141.

Wall, D.H. (2007) Global change tipping points: above- and below-ground biotic interactions in a low diversity ecosystem. *Philosophical Transactions of the Royal Society B: Biological Sciences* 362, 2291–2306.

Wharton, R.A., Vinyard, W.C., Parker, B.C., Simmons, G.M. and Seaburg, K.G. (1981) Algae in cryoconite holes on Canada Glacier in Southern Victorialand, Antarctica. *Phycologia* 20, 208–211.

Wharton, R.A., McKay, C.P., Simmons, G.M. and Parker, B.C. (1985) Cryoconite holes on glaciers. *Bioscience* 35, 499–503.

Yallop, M.L. and Anesio, A.M. (2010) Benthic diatom flora in supra-glacial habitats: a generic level comparison. *Annals of Glaciology* 51, 15–22.

Yoshimura, Y., Kohshima, S. and Ohtani, S. (1997) A community of snow algae on Himalayan glacier: change of algal biomass and community structure with altitude. *Arctic and Alpine Research* 29, 126–137.

Zeglin, L.H., Sinsabaugh, R.L., Barrett, J.E., Gooseff, M.N. and Takacs-Vesbach, C.D. (2009) Landscape distribution of microbial activity in the McMurdo Dry Valleys: linked biotic processes, hydrology, and geochemistry in a Cold Desert Ecosystem. *Ecosystems* 12, 562–573.

Zhang, D.C., Busse, H.J., Liu, H.C., Zhou, Y.G., Schinner, F. and Margesin, R. (2010a) *Sphingomonas glacialis* sp. nov., a psychrophilic bacterium isolated from alpine glacier cryoconite. *International Journal of Systematic and Evolutionary Microbiology*, doi: doi:ijs.0.023135-0.

Zhang, D.C., Redzic, M., Schinner, F. and Margesin, R. (2010b) *Glaciimonas immobilis* gen. nov., sp. nov., a novel member of the family Oxalobacteraceae isolated from alpine glacier cryoconite. *International Journal of Systematic and Evolutionary Microbiology*, doi: doi:ijs.0.028001-0.

10 Polar Deserts

Ian D. Hogg[1] and Diana H. Wall[2]
[1]*Department of Biological Sciences, University of Waikato, New Zealand;*
[2]*Department of Biology, and Natural Resource Ecology Laboratory, Colorado State University, USA*

10.1 Introduction: a Cold, Dry Environment

As extreme environments go, polar deserts may be among the most extreme and include some of the coldest, driest habitats on the planet. They are commonly defined as having less than 250 mm of annual precipitation and with maximum temperatures of below 10°C (Fountain *et al.*, 1999; Doran *et al.*, 2002a). Using these criteria, habitats include those found in both the high Arctic as well as on continental Antarctica. However, few extreme environments match the Dry Valleys of the Ross Sea region in Antarctica (Fig. 10.1). There, cold, dry air from the polar plateau is gravity-fed – the katabatic winds – on to an area of roughly 4800 km^2 of mountains and exposed valley floors along the southern Victoria Land coast of the Ross Sea region (Campbell and Claridge, 2006).

Captain Scott, on discovering the Dry Valleys in 1903, commented on their extreme nature by observing that 'We have seen no living thing, not even a moss or a lichen'. However, despite these initial observations and the harsh, dry conditions, a range of microbes, non-vascular plants and animals manage to eek out an existence here. These include mosses, lichens, microscopic invertebrates such as nematode worms, tardigrades and microarthropods (all <1 mm). Indeed, the largest year-round animals are the 1.3 mm springtails (Fig. 10.2), which survive on the remedial autotrophic (algal) and heterotrophic (bacterial, fungal) organisms available. Even for these residents, activity is limited to a few months a year and usually only during the austral 'summer' period (i.e. November to March).

We use the Dry Valleys as an example of an 'extreme' polar desert in order to describe life in these habitats. To avoid overlap with other chapters, we restrict our focus to the terrestrial realm of polar desert life. A more general overview of polar terrestrial environments can be found in Convey (Chapter 5, this volume), and Laybourn-Parry and Bell (Chapter 6, this volume) provide a focus on polar freshwaters. Here, we continue with a more detailed look at the physical environment of the Dry Valleys, before turning our attention to the biotic components, their survival mechanisms, trophic interactions as well as food webs and conclude with a section on the consequences of environmental changes for polar desert habitats.

10.2 The Physical Realm

The Dry Valleys of Southern Victoria Land shown in Fig. 10.1, are part of the

Fig. 10.1. Satellite image of the Dry Valleys and the, predominantly ice covered, surrounding region. Cartography by Brad Herried.

Transantarctic mountains and occur over a latitude of approximately 77–78°30′S. They comprise the largest ice-free area on the Antarctic continent (Ugolini and Bockheim, 2008). Within the main area (about 77°S), three large valley systems are found (Taylor, Wright and Victoria Valleys). There are also numerous smaller valleys such as the McKelvey, Balham and Barwick Valleys in the vicinity of these larger systems. In addition, another series of small valleys are immediately to the south of the main area (approx. 78°30′), collectively known as the 'Southern Dry Valleys' these include Garwood, Marshall, Miers and Hidden Valleys (see Fig. 10.1). All areas have been the subject of geological and biological studies beginning with the Scott expeditions of the early 1900s. Mountains comprising the Dry Valleys have elevations ranging from sea level to 800 m and are surrounded by mountains reaching nearly 2000 m. The sub-zero temperatures (average air temperature ~−17°C), low snowfall (ranging from ~3 mm water equivalent year^{-1} inland to ~50 mm year^{-1} water equivalent near the coast), high ablation rates and high winds contribute to an extreme ecosystem of low biological diversity and activity. The Dry Valleys have remained relatively stable for the past few million years compared to recent anthropogenic modifications of landscapes in other temperate deserts (Brown *et al.*, 1991; Fountain *et al.*, 1999; Virginia and Wall, 1999), providing an opportunity for understanding the physical controls on biological activity and diversity in an extreme ecosystem.

Fig. 10.2. The springtail *Gomphiocephalus hodgsoni* (Collembola) is the largest year-round inhabitant of the Dry Valleys (actual size 1.3 mm). Note the considerably reduced furcula (spring) on the underside of the abdomen. © Barry O'Brien.

The Dry Valley landscapes are composed of barren soils, glaciers, perennially ice-covered lakes and ephemeral meltstreams (Fig. 10.3) (Fountain *et al.*, 1999). Arid soils in these valleys are considered the oldest, driest and coldest on earth (Campbell *et al.*, 1998), and comprise the largest land feature of the Dry Valleys, occupying about 95% of glacier-ice-free surfaces below 1000 m (Burkins *et al.*, 2000). The dominant soil features are soil polygons (Fig. 10.4) that are formed by the freezing and thawing of soils and rock (Campbell and Claridge, 1987). Soils are derived from bedrock and tills composed of granites, sandstones, dolerites and meta-sedimentary rocks that range from Holocene to Miocene in age (Denton *et al.*, 1989; Bockheim, 1997; Hall and Denton, 2000). These desert soils have high salinity that generally relates to soil surface age (Campbell and Claridge, 1987; Bockheim, 2002), a neutral to high pH, in general a coarse texture (95–99% sand), extremely low organic matter content (0.01–0.03% organic carbon by weight; Campbell and Claridge, 1987; Burkins *et al.*,

Fig. 10.3. Typical Dry Valley landscape: the lower Garwood Valley looking up-valley towards the Garwood Glacier. The Joyce Glacier is visible in the background. © Philip Ross.

Fig. 10.4. Patterned ground (soil polygons) visible along the eastern face of the Canada Glacier in Taylor Valley. © Leo Sancho.

2000) and permafrost at 10–30 cm depth (Pastor and Bockheim, 1980; Bockheim, 1997; Ugolini and Bockheim, 2008).

10.3 The Diversity of Life in the Dry Valleys Polar Desert

The extreme environment limits life forms within the Dry Valleys. Seasonal visitors to the Dry Valleys include tourists, researchers, occasional Adélie penguins (*Pygoscelis adeliae*) and skuas (*Catharacta maccormicki*) flying overhead. The mummified remains of crabeater (*Lobodon carcinophagus*), Weddell (*Leptonychotes weddellii*) and leopard seals (*Hydrurga leptonyx*), as well as the odd penguin (Fig. 10.5), suggest that not all of these individuals are able to leave the Dry Valleys. Resident Dry Valley life is most commonly found where there is at least some access to available water. Examples include, near the feet of glaciers, snow patches, meltwater streams or areas of permafrost thaw, or for aquatic soil animals (nematodes, rotifers and tardigrades) where soil moisture or soil relative humidity is high enough to maintain activity. No vascular plants occur in the Dry Valleys, but algae, bryophytes (mosses) and colourful lichens (Fig. 10.6) are found where there is adequate moisture, such as at higher altitudes receiving snow, or near glacial meltstreams or flush zones.

All Antarctic terrestrial diversity (e.g. number of species or taxa) is lower than that found in temperate and tropical ecosystems. However, the terrestrial diversity of the Dry Valleys is considered low, even relative to local coastal habitats, such as near the Mackay Glacier where, for example, lichen and moss diversity is noticeably higher (Seppelt *et al.*, 2010). Overall, soil biodiversity is also low relative to non-Dry Valley, Antarctic sites both further north (Northern Victoria Land, 74°S; Barrett *et al.*, 2006) and further south (Queen Maud Mountains, 83°S; Green *et al.*, 2011). Most invertebrate species appear to be endemic, although some taxa (e.g. algae)

Fig. 10.5. Hidden Valley, one of the 'Southern Dry Valleys', looking towards the Adams Glacier (visible in the background). A dead penguin is visible in the foreground. Penguins and seals occasionally wander into the Dry Valleys, but either due to starvation or disorientation seldom make it back out again. This provides one of the few sources of external nutrients to these ecosystems. © Ian Hogg.

Fig. 10.6. Lichens (*Acarospora*) growing on the surface of a rock (pencil provided for scale). © Ian Hogg.

are a mixture of endemic and cosmopolitan species (Esposito *et al.*, 2006).

The distribution of taxa across the Dry Valleys is irregular and patchy as some soils are not suitable habitats for organisms (Virginia and Wall, 1999). Soils at small spatial scales (<1 m^2) have no visible algae or mosses, nor invertebrates due to high soil salinity, low carbon, and the high heterogeneity in other chemical and physical properties (Virginia and Wall, 1999; Courtright *et al.*, 2001; Elberling *et al.*, 2006). Biotic communities of the Dry Valleys generally have high numbers of a single invertebrate species in dry barren soils (e.g. 10–50 individuals g^{-1} dry weight soil), but more species and increased complexity of communities (more than two taxa) near meltstreams (Barrett *et al.*, 2004). For example, soils of ephemeral meltstreams are generally richer in carbon and nitrogen compared to dry soils, and support greater abundances and higher taxonomic diversities (e.g. two to three species each of nematodes and arthropods as well as tardigrades, rotifers and several species of mosses and algae).

To understand the functioning of the Dry Valley ecosystem, it is helpful to know something of the key life forms and their survival mechanisms. Below, we describe these biotic components of the Dry Valleys along with an indication of their diversity and biogeography as well as notes on their natural history.

10.3.1 Plant and fungal life

Mosses

Mosses or bryophytes are among the more noticeable of life forms in the Dry Valleys where they occasionally form lush cushions in areas of high soil moisture, such as in glacial flush zones (Fig. 10.7). On the northern margin of the Dry Valleys (e.g. Granite Harbour near the Mackay Glacier), there are at least ten species of moss (Seppelt *et al.*, 2010). However, within

Fig. 10.7. Flush area in front of the Canada Glacier in Taylor Valley. Extensive moss beds are found in the darker areas visible in the middle of the photograph. © Ian Hogg.

the Dry Valleys proper (e.g. Canada Glacier in Taylor Valley), there are only three (Seppelt *et al.*, 2010). This can be partly attributed to the harsher conditions within the Dry Valleys, but also the greater access to nutrients (e.g. skua guano) and moisture at coastal sites (e.g. Granite Harbour).

Mosses can provide substrate for lichens, as well as maintaining a moist habitat for a range of invertebrates (e.g. rotifers, springtails, nematodes; Simmons *et al.*, 2009a). Dead and decomposing mosses also provide valuable nutrients to developing 'soil' communities.

Lichens

Lichens are symbiotic organisms consisting of an algal and fungal component that are widely distributed in temperate desert and other ecosystems, but their range in the Dry Valleys is much more limited. In temperate regions, the relationship between the symbionts is usually one-to-one (i.e. a specific algal species matched with a specific fungal species). However, in Antarctica where access to the required or even preferred symbiont may be limited, it has recently been discovered that there may be some flexibility in this relationship (Wirtz *et al.*, 2003).

Similar to the mosses, there are far fewer species found within the Dry Valleys than at nearby coastal sites (e.g. Granite Harbour) with only two species easily visible on the Dry Valley floors and restricted to small <1 m² patches of moist rock within glacial flushes (Green *et al.*, 1992). In total, there are roughly 15 species of lichen found in the Dry Valleys with most species more prevalent on valley rock sides and at higher altitude (T.G.A. Green, University of Waikato, 2010, personal communication). However, although lichen occurrences within the valleys are small and very sporadic (e.g. Fig. 10.6), they are much more common and visible on the mountains between the valleys (Fig. 10.8). This is due in part to the increased stability, larger substrate size and corresponding habitat heterogeneity, as well as the increased moisture levels at higher altitudes (e.g. cloud effect, snow retention).

Fig. 10.8. *Buellia frigida*, a lichen species commonly found growing on rocks in higher elevation areas throughout the Dry Valleys. © Rod Seppelt.

Algae

Approximately 80 species of algae are found within the Dry Valleys, the majority being diatoms found in the lake and stream ecosystems (Broady, 2005; Esposito *et al.*, 2006, 2008). However, a few species inhabit soils and these are often observed as hypoliths (growing under rocks) and endoliths (growing within rocks). Taxa are predominantly 'blue-green algae' (cyanobacteria) and usually represented by a single species, *Oscillatoriales* sp. (Cowan *et al.*, 2010). Large mats of cyanobacteria are also frequently observed around the margins of lakes, small ponds and streams, where they can become exposed (and part of the terrestrial realm) when water levels decline. The depth of water covering the algal mats in stream channels may only be a few centimetres, which is sufficient for growth of the desiccation-tolerant filamentous cyanobacteria and associated diatoms which comprise the mats (Alger *et al.*, 1997). Even in cold summers, soil temperatures will rise above freezing for at least a few hours per day over several weeks (Barrett *et al.*, 2008), although overall biological activity is constrained by low water availability rather than temperature (Ball *et al.*, 2009; Treonis *et al.*, 2000).

Fungi

The globally ubiquitous fungi are a critical component of the Dry Valley ecosystems. They feed off the limited nutrients present in these habitats and are important in the development of 'living' soils. They can frequently be observed as hypoliths occurring under and around the margins of rocks – particularly translucent quartz rocks, which moderate the local thermal microenvironment and reduce ambient light levels (Cowan *et al.*, 2010). Cowan *et al.* (2010) have questioned whether the fungal presence in these habitats is parallel with, or successional to, the algal hypoliths (i.e. taking advantage of their initial productivity). Nevertheless, they serve to stabilize the soil surface and subsurface and also provide a valuable food source to many of the animal components described below.

10.3.2 Animals

Protozoa

Although no longer included as part of the animal kingdom, protozoans are microscopic single-celled organisms found in a range of aquatic and soil habitats. Very little study has been undertaken on the protozoans of the Dry Valleys, although representatives include flagellates, small amoebae and ciliates (Bamforth *et al.*, 2005). However, they are thought to be widespread and Bamforth *et al.* (2005) found individuals in 92% of soil samples collected from the Dry Valleys. Protozoans are also likely to contribute to the diets of multicellular animals such as rotifers, tardigrades and nematodes (Bardgett, 2005).

Tardigrades

Tardigrades or 'water bears' are common inhabitants of soil and aquatic habitats throughout Antarctica. Within the Dry Valleys, roughly four species have been identified (Adams *et al.*, 2006). However, there has been much confusion and 'synonymization' of previous species designations. Ongoing morphological and molecular studies will help to resolve this issue further as molecular studies have shown there are at least seven species of tardigrades in the Taylor Valley (B.J. Adams, Brigham Young University, 2010, personal communication).

Rotifers

Rotifers are minute animals (<0.01 mm), most commonly known as a component of the plankton in freshwater lakes but also moist soils. Even in the Dry Valley region, at least 15 species are associated with lakes and ponds (Adams *et al.*, 2006). At least three species are associated with moist soils and with mosses and lichens where they can access the higher levels of relative humidity which they require to remain in an active state (Schwarz *et al.*, 1992). In the absence of moisture, rotifers can produce drought-resistant eggs, allowing them to survive prolonged dry periods and also to exploit episodic melting of snow and glaciers. Little is known of their diets in the Dry Valleys. However, elsewhere, they are known to consume bacteria, algae and protozoans, all of which are present in the Dry Valleys.

Nematodes

Nematodes are clearly among the most widespread of multicellular animals on the planet. As a consequence, they can exist almost anywhere and different species exhibit various lifestyles from fully free-living to obligate parasitic. The Dry Valley soils are home to at least four genera of free-living nematodes: *Scottnema, Plectus, Geomonhystera*, all consumers of bacteria and *Eudorylaimus*, an algal consumer (Wall, 2005; Adams *et al.*, 2006). Most are represented by only a single species. However, ongoing molecular analyses may revise these totals slightly, particularly for *Plectus* (B.J. Adams *et al.*, 2011, unpublished results). By far, the most commonly encountered species of nematode in the Dry Valleys is *Scottnema lindsayae* (Fig. 10.9), which occurs in a range of different soil types and chemistries (Wall Freckman and Virginia, 1998).

Collembola

Collembola or springtails (e.g. Fig. 10.2) are small, primitive, insect-like animals, usually <2 mm in body length, although some, such as the 'giant' springtails of New Zealand are >5 mm. Springtails were once classed as insects but are now considered sufficiently different to have been placed in their own class (Ellipura). Springtails are primarily decomposers found in a variety of soil and aquatic detrital habitats. Globally, they are the most widely distributed hexapods with representatives found from the tropics to within a few hundred kilometres of both poles. They are also the most diverse hexapods in both polar regions. As their common name suggests, many species are known for their propulsive, spring-like 'tails' (furcula), which are used for dispersal and predator avoidance. However, such devices are usually confined to surface-dwelling species, and those species found lower in the soil profile have reduced or absent furcula as well as shortened legs and antennae. Few of the Antarctic species have fully developed furcula and the single species common in the Dry Valleys (*Gomphiocephalus hodgsoni*) has a considerably reduced furcula (see Fig. 10.2). The diet of *G. hodgsoni* consists predominantly of fungal, bacterial and algal material (Davidson and Broady, 1996).

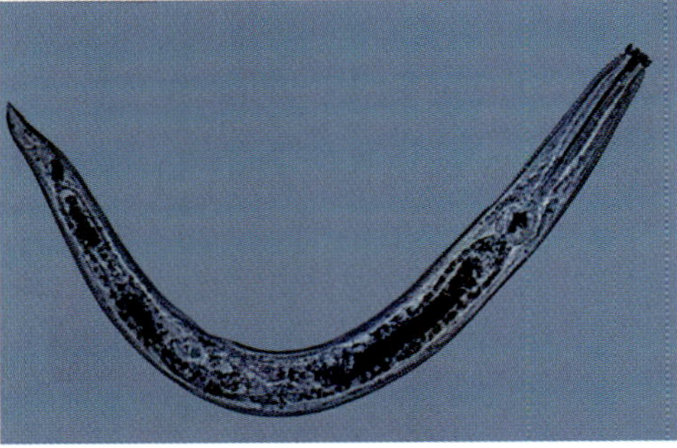

Fig. 10.9. *Scottnema lindsayae*, a widespread Antarctic nematode and the most common soil invertebrate found in the Dry Valleys. © Manuel Mundo.

Mites

Mites (Acari) are one of the more recognizable and common inhabitants of the Dry Valleys and, alongside the springtails, are the only other free-living terrestrial arthropods. Where present, they are usually found on the underside of small, dark rocks or occasionally scurrying about on the surface. They move relatively quickly for their small size and this usually alerts the observer to their presence. They may have greater dispersal abilities relative to springtails, and as a consequence, tend to be found in a wider range of locations (Stevens and Hogg, 2002, 2006). They also seem to be able to tolerate slightly wetter habitats than those of the springtails and are often found within glacial flush zones (Sinclair and Stevens, 2006). Within the Dry Valleys, two species are encountered, *Stereotydeus mollis* and *Nanorchestes antarcticus* (Strandtmann, 1967). They are both small (<1 mm), and are presumed to have very similar diets to *G. hodgsoni* (above).

10.3.3 Microbes

Bacteria are perhaps the most widespread and numerous of life forms on the planet. They perform a number of roles within ecosystems including nutrient cycling (e.g. decomposition) as well as occasionally serving as vectors of disease. Despite this, comparatively little work has been undertaken on the microbiology of the Dry Valleys terrestrial environments. Initial work had suggested that the microbial diversity of Dry Valley soils was limited and that most of the taxa were psychrotrophic (cold-adapted) varieties unique to Antarctica (Smith *et al.*, 2006). However, more recent work using advanced molecular techniques has suggested the opposite. Indeed, the diversity of microbes in Dry Valley soils may actually rival that of more temperate habitats (Cary *et al.*, 2010). The other surprising finding has been that instead of having a 'typical' microbial flora throughout the Antarctic, there are actually regional and local differences (Barrett *et al.*, 2006). In addition to their important role in the breakdown of available nutrients, microbes also provide a valuable component of the diets for several of the invertebrates (e.g. nematodes) in the Dry Valleys.

10.4 Mechanisms for Survival in Polar Deserts

Some soil animals of the Dry Valleys appear to be specifically adapted to exist in the cold dry environment. For example, the most common nematode of the Dry Valleys, *Scottnema lindsayae*, survives and reproduces better at 10°C than it does at 15°C (Overhoff *et al.*, 1993). However, in order to endure the cold, dry conditions, most animals must possess at least some survival mechanisms to respond to the extreme and potentially rapid (minute-by-minute scale) fluctuations in moisture and temperature.

10.4.1 Desiccation and anhydrobiosis

Nematodes, rotifers and tardigrades are often active in water films around soil particles. When adverse environmental conditions occur, such as low soil relative humidity or rapid sublimation rates, the animals can desiccate and enter what is termed anhydrobiosis (an ametabolic state entered during extreme desiccation; Wharton and Barclay, 1993). They achieve this by altering their morphology to reduce surface area and become ametabolic (Wharton, 2003). However, within minutes to hours of a return to favourable environmental conditions, anhydrobiotic invertebrates can activate (Treonis and Wall, 2005). Algae, mosses and lichens also tolerate desiccation and can survive prolonged periods without access to water, yet within minutes can return to an active metabolic state when favourable conditions occur (Pannewitz *et al.*, 2003; Schlensog *et al.*, 2004; McKnight *et al.*, 2007). In many cases, these ametabolic states can last for months or years until favourable environmental conditions return (Browne *et al.*, 2002; Treonis and Wall, 2005). For example, a species of nematode, *Panagrolaimus*, collected from Armenia has been reported to have survived more than 8 years in an anhydrobiotic state (Aroian *et al.*, 1993).

10.4.2 Freeze tolerance, freeze avoidance and 'supercooling'

Prior to entering an anhydrobiotic state, species such as nematodes can also change their biochemistry to produce antifreeze compounds (e.g. trehalose, inositol; Crowe and Madin, 1974). All of these responses appear to be under genetic control (Adhikari *et al.*, 2009), and can allow animals to tolerate temperatures much lower than would be possible in the 'normal' active state. Remarkably, for one species of Antarctic nematode (*Panagrolaimus davidi*), this can allow them to survive temperatures as low as −80°C (Wharton and Brown, 1991). The triggering of cold-tolerant genes and corresponding responses can also allow nematodes to cope with more minor fluctuations in temperature and moisture (Adhikari *et al.*, 2010).

In contrast, arthropod species, such as springtails (Collembola) and mites, lack the ability to enter an anhydrobiotic state and for them, freezing is always lethal (Sømme, 1981). Instead, they rely on freeze-avoidance techniques such as the minimizing of ice-nucleating compounds (e.g. food) within their bodies at times when temperatures are likely to be decreasing (e.g. onset of winter; Sinclair and Sjursen, 2001). Springtails are also capable of lowering their freezing points through the production of antifreeze proteins such as glycol. For the common Dry Valley springtail (*G. hodgsoni*), such adaptations allow individual animals to 'supercool' to as low as −35°C before freezing occurs (Sinclair and Sjursen, 2001).

10.5 Trophic Structure and Food Webs

Trophic structure in Dry Valley ecosystems is very simple. Indeed, they are often viewed as one of the few examples of an ecosystem that may be driven entirely by abiotic factors (Hogg *et al.*, 2006). However, this may also be the result of research focused at inappropriate spatial or trophic levels, and the generality of this view has been questioned (e.g. Caruso *et al.*, 2007). Regardless, they are certainly among the simplest of ecosystems on the planet.

One of the major factors limiting life in the Dry Valleys is access to nutrients. Along the coastlines and on offshore islands (e.g. Beaufort, Ross Islands), skua colonies and/or penguin rookeries as well as other marine sources provide an ample (and often overwhelming) supply of nutrients. However, within the Dry Valleys proper, which can be a considerable distance from the sea, these sources decrease with distance from the coast (Burkins *et al.*, 2000) and include both autochthonous (produced within) or allochthonous (external) sources (Barrett *et al.*, 2010). Hopkins *et al.* (2006) provided a theoretical analysis of the sources of carbon available in Dry Valley ecosystems and a simplified representation of these sources is provided in Fig. 10.10. Allochthonous sources are primarily marine in origin and include nutrients that have been blown in (e.g. marine sediments), walked in (e.g. dead and dying seals and penguins; see Fig. 10.5), or are legacies of past events (e.g. seawater

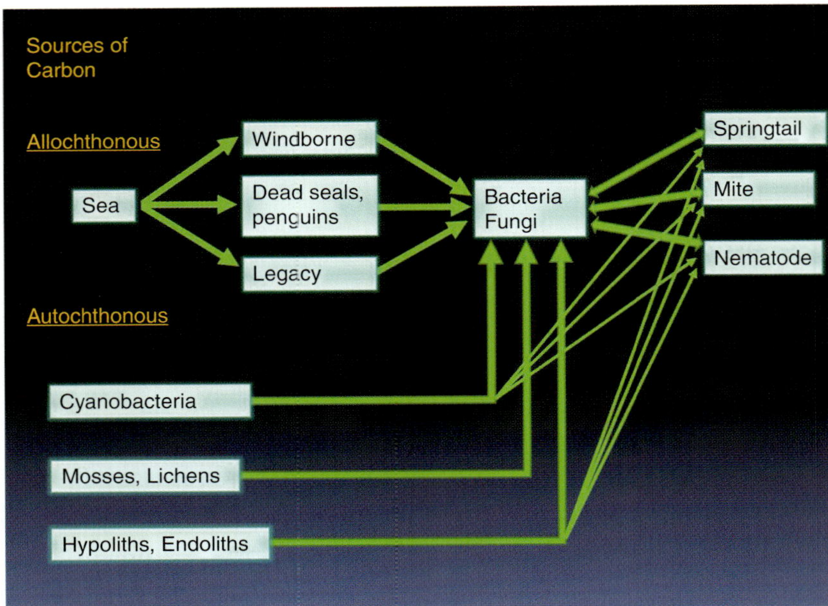

Fig. 10.10. Sources of carbon in the Dry Valleys. © Ian Hogg and Diana Wall.

intrusions). Autochthonous sources include the primary producers, primarily cyanobacteria and other algal taxa found in lakes, ponds and soils. Species found in lakes often produce large 'mats' that can be exposed when water levels drop. These then dry out and can be blown about the Valleys spreading the nutrients over a much wider area. There are also legacy effects in areas where lakes and ponds once existed. These old lake beds are repositories of nutrients produced when the lake was active. Controls over soil carbon balance are also influenced by events occurring over glacial time scales (e.g. lake inundation), which is in turn influenced by temperature (Burkins *et al.*, 2000).

Most taxa are thought to be predominantly involved in the decomposition/detritus pathway as there are no vascular plants for herbivores such nematodes, mites and springtails, and there are no known predators (Wall, 2005). Nevertheless, recent observations have suggested that a previously unidentified species of tardigrade may be a facultative predator on other tardigrade species (U. Nielsen, Colorado State University, 2010, personal communication). Further work, which is ongoing, is required to confirm this. However, in the interim, Dry Valley food webs can be thought to have primarily three trophic levels: producers; consumers; and decomposers. A simplified view of a theoretical food web is provided in Fig. 10.11. Cyanobacteria (blue-green algae), as well as eukaryotic algae (e.g. diatoms), are a major food source for virtually all of the consumers (Wall, 2005). There is no evidence that lichens or mosses are directly consumed, and instead are more often used as suitable habitats where access to other food sources (e.g. algae, fungi) is facilitated (Simmons *et al.*, 2009b). However, once dead, all components of the food web (including mosses and lichens) would be processed by the decomposers. In turn, both bacterial and fungal

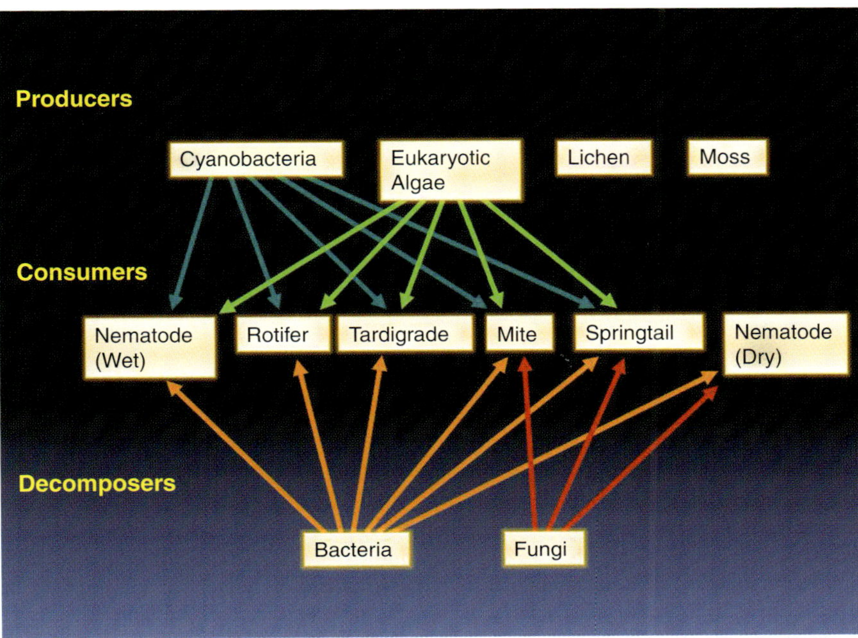

Fig. 10.11. A simplified version of the Dry Valley food web showing primary producer, consumer and decomposer trophic levels. All components feed into the decomposers; however, these pathways have not been added for simplicity. © Ian Hogg and Diana Wall.

components provide a major food source to the consumers – bacterial components being more important for the smaller consumers, such as protozoa and nematodes.

10.6 Environmental Changes and Biotic Response

10.6.1 Climate changes

Global atmospheric changes, such as global warming, are providing some of the most serious challenges facing the Earth's biota. Mean global air temperatures are expected to increase in the range of 1.4–6.4°C within the next 50–100 years (Solomon *et al.*, 2007). In both polar regions, temperature increases may be double the projected global means, due to the melting of snow/ice cover and the resulting decrease in albedo (reflecting of the sun's rays by snow). Furthermore, the rates at which increases occur are very likely to exceed any rates previously experienced by the biota. In addition to predicted increases in the mean temperatures and the rates of increase, the increased incidence of extreme-weather events such as droughts, floods and storm events is likely to provide additional challenges (see also Psenner and Sattler, Chapter 23, and Glover and Neal, Chapter 24, this volume).

In contrast to the Antarctic Peninsula where warming has occurred more rapidly in the past 50 years than any location on the planet (~3°C increase from 1950 to 2000, Doran *et al.*, 2002b; Montes-Hugo *et al.*, 2009; Turner *et al.*, 2009), the Dry Valley region has cooled within the past 20 years (Doran *et al.*, 2002b; Thompson and Solomon, 2002). This cooling climate has more recently been punctuated by annual warming events (Doran *et al.*, 2008). In contrast, and on a broader scale, detailed climate data collected by NASA over the last 50 years and covering the vicinity of the Dry Valleys show a noticeable warming trend of around 0.05°C year^{-1} for the area of West Antarctica (Fig. 10.12). Indeed, for areas of the Western Antarctic Ice Sheet (WAIS), these increases have been closer to a staggering 0.1°C year^{-1}.

10.6.2 Biotic responses to change

One of the surprising aspects of the Dry Valley desert system is the mosaic of diversity that occurs within and among valleys. Throughout the Antarctic, 'higher' level taxa such as springtails and mites often show remarkable genetic differences even over distances as small as a few kilometres (e.g. Fanciulli *et al.*, 2001; McGaughran *et al.*, 2008; Hawes *et al.*, 2010). This suggests that instead of being dictated solely by climatic variables, the distribution of organisms is determined by historical events and chance survival of organisms within particular habitats.

These past influences can even be observed on relatively small spatial scales. For example, in Taylor Valley, two different maternal lineages (mitochondrial DNA haplotypes) of the springtail, *G. hodgsoni*, were found to trace an ancient, glacial lake shoreline in Taylor Valley from when temperatures were warmer (Nolan *et al.*, 2006). The current explanation is that ancestral individuals would have been separated during glacial advances and then genetically diverged in isolation. One haplotype represented the resident haplotype of Taylor Valley while the other represented a neighbouring area. Towards the end of the last glacial cycle (roughly 20,000 years ago), warmer temperatures and melting of the glacial lake may have facilitated dispersal of individuals via 'rafting' (*sensu* Hawes *et al.*, 2008) on meltwater 'moats' around the edge of the lake. Here, the two lineages were reunited and have co-existed to the present. What remains unknown is whether this co-existence is benign, or whether there is any local, competitive advantage (or disadvantage). Specifically, although the 'neighbouring' haplotype traces the shoreline, it does not occur at other, up-valley sites. An alternative explanation is, simply, that their absence 'up valley' reflects the limited dispersal abilities of springtails (e.g. Stevens and Hogg, 2003).

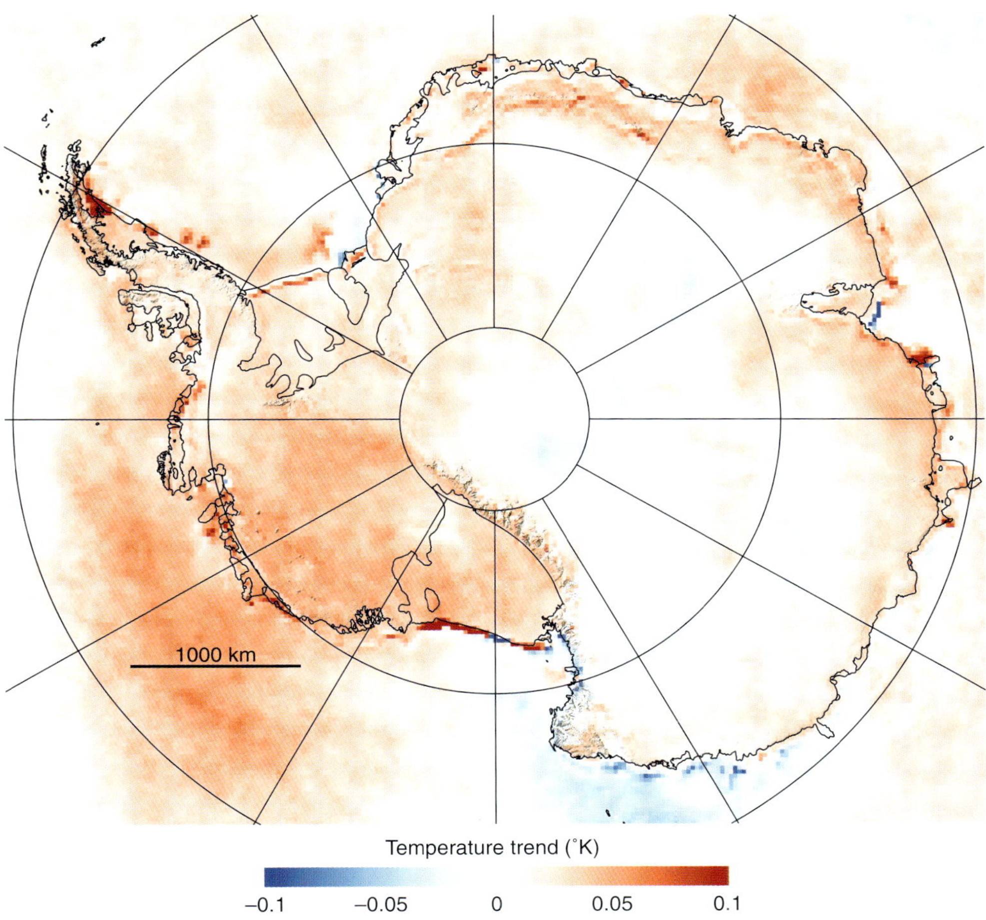

Fig. 10.12. A heat map of Antarctica showing surface temperature changes based on temperature records collected over the previous 50 years (1957–2007). Of particular note are the 'red' areas over West Antarctica. © NASA 2007, used with permission.

Although these local genetic differences are perhaps not surprising for arthropods with restricted dispersal abilities, the same may hold true for other taxa such as lichens (Green *et al.*, 2011) and even microbes (e.g. Pointing *et al.*, 2009). Lichens that can disperse themselves by small spores can be easily be blown about by the wind. So, one would expect these organisms to disperse very easily. In fact, the opposite seems to be true. Genetic analyses of a common lichen species in the Dry Valleys have shown that isolated populations are in fact genetically distinct from each other (Ruprecht *et al.*, 2010).

Microbial communities of the Dry Valleys also show a remarkably high level of diversity, sometimes rivaling that of more temperate zones (Cary *et al.*, 2010). There is also tremendous habitat complexity reflecting a legacy of geological history and physical conditions that create soil heterogeneity. Consequently, even for the microbes, different valleys are likely to house different communities that have developed in isolation over hundreds of thousands of years (Pointing *et al.*, 2009).

This suggests that dispersal or colonization is not as easy or common as previously thought, nor is it solely responsible for the distribution of microbes. Instead, the Dry Valley landscape has been tiled with a mosaic of genetically distinctive populations reflecting past and present conditions at these sites. Some of the likely consequences of global climate changes will be shifts in wind patterns, retreating glaciers, increased stream flow from melting glaciers and the melting of icesheets exposing 'new' ground. All of these events have the potential to increase the dispersal capacities of organisms. On one hand, the inherent genetic diversity of Antarctic organisms could be advantageous as this variability may provide some potential to respond to environmental changes. Most Dry Valley life should presumably have little trouble exploiting new, suitable habitat as they are by their very nature, early successional taxa. However, not all soil habitats are, or will be, capable of supporting life (Courtright *et al.*, 2001; Wall, 2005). Alternatively, and of more certainty, is that the mixing up of local populations has the potential to destroy possibly millions of years of history and local adaptation.

The decreases in air temperatures recorded in the Dry Valleys in the late 1990s resulted in many interconnected effects on the ecosystem including decreased stream flow, increased ice thickness on lakes, and a decline in nematode populations (Doran *et al.*, 2002b; Barrett *et al.*, 2008), as well as changing the composition of diatom species in streams (Esposito *et al.*, 2006). Manipulative experiments and natural climate variation showed that warming alters soil chemistry and hydrology increasing soil moisture and modifying soil habitats resulting in a decline in the dominant nematode species (Simmons *et al.*, 2009b). Warming events also affected biota and chemistry of streams and lakes (Foreman *et al.*, 2004; Lyons *et al.*, 2005). Collectively, these results indicate that the Dry Valley ecosystems and their biota are sensitive to even small changes in seasonal air temperatures.

10.7 The Human Connection

Aside from the detrimental impacts imposed on Dry Valley ecosystems by climate change, humans can also have more direct influences resulting from their activities and presence in Antarctica. These include physical disturbances (e.g. hiking, construction activities), chemical contamination (e.g. fuel spills) and the introduction of alien species. However, humans also have the opportunity and responsibility to ensure that these unique extreme habitats are protected for future generations. Given the potentially negative consequences of environmental changes, formulating appropriate strategies to mitigate these changes is one of the many challenges facing those responsible for protecting Antarctica's polar desert habitats.

All activities in Antarctica are dictated by the Antarctic Treaty and it is the responsibility of all nations working in Antarctica to ensure that obligations under the treaty system are met. The Scientific Committee on Antarctic Research (SCAR) provides scientific advice on these issues. In the case of the Dry Valleys this has been addressed through the formation of an 'Antarctic Specially Managed Area' (ASMA). Ongoing research supported by the various national programmes and funding agencies is a key component of this decision-making process.

Perhaps one of the more serious threats to Antarctic biodiversity is the introduction and establishment of alien species. Many of these organisms are brought in on clothing, footwear, fresh foods and transportation vectors (aircraft, ships). Current and projected temperature increases as well as an increase in the potential numbers of visitors to the Dry Valleys (e.g. tourists, researchers) all increase the probability of alien species being introduced. A concerted effort through several of the national programmes (e.g. United States Antarctic Program; Antarctica New Zealand) is now underway to ensure that all visitors adhere to the strictest of standards to minimize these risks.

Of further concern is the longer distance movement and introduction of species already within Antarctica, or indeed from outside of Antarctica, to the Dry Valleys. The competitive response of Dry Valley organisms when confronted with such new arrivals is entirely uncertain. Experience from the Subantarctic

South Georgia Island is not reassuring. For example, Convey *et al.* (1999) suggested that introduced species of springtail have indeed displaced the local resident species.

Even walking to and from study sites can have a negative effect. Ayres *et al.* (2008) found that minimal trampling disturbance resulted in measureable differences in the ratios of living to dead nematodes in underlying soils. In cases of well-used walking trails, mortality rates of nematodes were as high as 76%. As the numbers of visitors to the Antarctic Dry Valleys increase (e.g. through research activities, tourism), these effects will be exacerbated. Guidelines outlining proper procedures when working and visiting the Dry Valleys are now required reading for all visitors to the Dry Valleys. Strategies include avoiding particularly sensitive areas (e.g. moss beds), and using previously established trails.

Other, often localized, effects include chemical contamination, particularly through the use of hydrocarbon-based products such as gasoline and oil-based lubricants (Aislabie *et al.*, 2004). Most contaminated sites are the result of spills during construction or transportation events. However, some are the result of earlier scientific activities such as the Dry Valley Drilling Project where diesel fuel was used as a drilling fluid at Lake Vida in the Victoria Valley (Aislabie *et al.*, 2004). Such activities are now strictly monitored and all chemicals used in the Dry Valleys must pass a Preliminary Environmental Evaluation (PEE) before the project is even allowed to proceed. However, any number of guidelines and restrictions are still not always enough to protect fragile extreme ecosystems. In these cases, education is often the best strategy. The greater the awareness of the value of extreme environments as a functional and integral component of global ecosystems, the more likely they will be used in an appropriate manner.

In summary, we have used the Dry Valleys of Antarctica as an example of an extreme polar desert. Accumulated knowledge continues to lead to a greater understanding of the organisms, their interactions and their survival capacity. There is a wealth of evidence building that these ecosystems are also experiencing rapid rates of environmental changes that could significantly affect the responses of biota and functioning of the ecosystems. As Antarctica has many unexplored terrestrial areas and the fewest species of any continent, there is a need to move forward on research priorities such as determining the composition and functioning of present ecosystems prior to changes resulting from global warming and invasive species. An enhanced understanding of the response of these relatively simple polar desert ecosystems to global environmental changes will provide critical insights for predicting the consequences for more complex ecosystems elsewhere – the so-called global barometer. These polar deserts may also provide valuable analogues for past and present life on other planets such as Mars (e.g. Doran *et al.*, 2010). Ongoing international cooperation, as well as the integration of research and monitoring, will be necessary to reap these benefits as well as to help protect these fragile ecosystems.

References

Adams, B.J., Bardgett, R.D., Ayres, E., Wall, D.H., Aislabie, J., Bamforth, S., Bargagli, R., Cary, C., Cavacini, P., Connell, L., Convey, P., Fell, J.W., Frati, F., Hogg, I.D., Newsham, K.K., O'Donnell, A., Russell, N., Seppelt, R.D. and Stevens, M.I. (2006) Diversity and distribution of Victoria land biota. *Soil Biology and Biochemistry* 38, 3003–3018.

Adhikari, B.N., Wall, D.H. and Adams, B.J. (2009) Desiccation survival in an Antarctic nematode: molecular analysis using expressed sequenced tags. *BMC Genomics* 10, 69.

Adhikari, B.N., Wall, D.H. and Adams, B.J. (2010) Effect of slow desiccation and freezing on gene transcription and stress survival of an Antarctic nematode. *Journal of Experimental Biology* 213, 1803–1812.

Aislabie, J.M., Balks, M.R., Foght, J.M. and Waterhouse, E.J. (2004) Hydrocarbon spills on Antarctic soils: effects and management. *Environmental Science and Technology* 38, 1265–1274.

Alger, A.S., McKnight, D.M., Spaulding, S.A., Tata, C.M., Shupe, G.H., Welch, K.A., Edwards, R. and House, H.R. (1997) Ecological processes in a cold desert ecosystem: the abundance and species of algal mats in glacial meltwater streams in Taylor Valley, Antarctica. *INSTAAR Occasional Paper* 51, 108 pp.

Aroian, R.V., Carta, L., Kaloshian, I. and Sternberg, P.W. (1993) A free-living *Panagrolaimus* sp. from Armenia can last over eight years in the anhydrobiobic state. *Journal of Nematology* 25, 500–502.

Ayres, E., Nkem, J.N., Wall, D.H., Adams, B.J., Barrett, J.E., Broos, E.J., Parsons, A.N., Powers, L.E., Simmons, L.E. and Virginia, R.A. (2008) Effects of human trampling on populations of soil fauna in the McMurdo Dry Valleys, Antarctica. *Conservation Biology* 22, 1544–1551.

Ball, B.A., Virginia, R.A., Barrett, J.E., Parsons, A.N. and Wall, D.H. (2009) Interactions between physical and biotic factors influence CO_2 flux in Antarctic dry valley soils. *Soil Biology and Biochemistry* 41, 1510–1517.

Bamforth, S.S., Wall, D.H. and Virginia, R.A. (2005) Distribution and diversity of soil protozoa in the McMurdo Dry Valleys of Antarctica. *Polar Biology* 28, 756–762.

Bardgett, R.D. (2005) *The Biology of Soil: A Community and Ecosystem Approach*. Oxford University Press, Oxford, UK, 254 pp.

Barrett, J.E., Wall, D.H., Virginia, R.A., Parsons, A.N., Powers, L.E. and Burkins, M.B. (2004) Variation in biogeochemistry and soil biodiversity across spatial scales in a polar desert ecosystem. *Ecology* 85, 3105–3118.

Barrett, J.E., Virginia, R.A., Wall, D.H., Cary, S.C., Adams, B.J., Hacker, A.L. and Aislabie, J.M. (2006) Co-variation in soil biodiversity and biogeochemistry in northern and southern Victoria Land, Antarctica. *Antarctic Science* 18, 535–548.

Barrett, J.E., Virginia, R.A., Wall, D.H., Doran, P.T., Fountain, A.G., Welch, K.A. and Lyons, W.B. (2008) Persistent effects of a discrete warming event on a polar desert ecosystem. *Global Change Biology* 14, 2249–2261.

Barrett, J.E., Poage, M.A., Gooseff, M.N. and Takacs-Vesbach, C. (2010) The legacy of aqueous environments on soils of the McMurdo Dry Valleys: contexts for future exploration of martian soils. In: Doran, P.T., Lyons, W.B. and McKnight, D.M. (eds) *Life in Antarctic Deserts and Other Cold Dry Environments*. Cambridge University Press, Cambridge, UK, pp. 139–159.

Bockheim, J.G. (1997) Properties and classification of cold desert soils from Antarctica. *Soil Science Society of America Journal* 61, 224–231.

Bockheim, J.G. (2002) Landform and soil development in the McMurdo Dry Valleys, Antarctica: a regional synthesis. *Arctic, Antarctic and Alpine Research* 34, 308–317.

Broady, P.A. (2005) The distribution of terrestrial and hydro-terrestrial algal association at three contrasting locations in southern Victoria Land, Antarctica. *Algalogical Studies* 118, 95–112.

Brown, E.T., Edmond, J.M., Raisbeck, G.M., Yiou, F., Kurz, M. and Brook, E.J. (1991) Examination of surface exposure ages of Antarctic moraines using in-situ produced ^{10}Be and ^{26}Al. *Geochimica et Cosmochimica Acta* 55, 2269–2283.

Browne, J., Tunnacliffe, A. and Burnell, A. (2002) Plant desiccation gene found in nematode. *Nature* 416, 38.

Burkins, M.B., Virginia, R.A., Chamberlain, C.P. and Wall, D.H. (2000) Origin and distribution of soil organic matter in Taylor Valley, Antarctica. *Ecology* 81, 2377–2391.

Campbell, I.B. and Claridge, G.G.C. (1987) *Antarctica: Soils, Weathering Processes and Environment*. Elsevier, Amsterdam, 368 pp.

Campbell, I.B. and Claridge, G.G.C. (2006) Permafrost properties, patterns and processes in the Transantarctic Mountains region. *Permafrost and Periglacial Processes* 17, 215–232.

Campbell, I.B., Claridge, G.G.C., Campbell, D.I. and Balks, M.R. (1998) The soil environment of the McMurdo Dry Valleys, Antarctica. In: Priscu, J.C. (ed.) *Ecosystem Dynamics in a Polar Desert. The McMurdo Dry Valleys, Antarctica*. American Geophysical Union, Washington, DC, pp. 297–322.

Caruso, T., Borghini, F., Bucci, C., Colacevich, A. and Bargagli, R. (2007) Modelling local-scale determinants and the probability of microarthropod species occurrence in Antarctic soils. *Soil Biology and Biochemistry* 39, 2949–2956.

Cary, S.C., McDonald, I.R., Barrett, J.E. and Cowan, D.A. (2010) On the rocks: the microbiology of Antarctic Dry Valley soils. *Nature Reviews Microbiology* 8, 129–138.

Convey, P., Greenslade, P., Arnold, R.J. and Block, W. (1999) Collembola of sub-Antarctic South Georgia. *Polar Biology* 22, 1–6.

Courtright, E.M., Wall, D.H. and Virginia, R.A. (2001) Determining habitat suitability for soil invertebrates in an extreme environment, the McMurdo Dry Valleys, Antarctica. *Antarctic Science* 13, 9–17.

Cowan, D.A., Khan, N., Pointing, S.B. and Cary, S.C. (2010) Diverse hypolithic refuge communities in the McMurdo Dry Valleys. *Antarctic Science* 22, 714–720.

Crowe, J.H. and Madin, K.A. (1974) Anhydrobiosis in tardigrades and nematodes. *Transactions of the American Microscopical Society* 93, 513–524.

Davidson, M.M. and Broady, P.A. (1996) Analysis of gut contents of *Gomphiocephalus hodgsoni* Carpenter (Collembola: Hypogastruridae) at Cape Geology, Antarctica. *Polar Biology* 16, 463–467.

Denton, G.H., Bockheim, J.G., Wilson, S.C. and Stuiver, M. (1989) Late Wisconsin and early Holocene glacial history, inner Ross Embayment, Antarctica. *Quaternary Research* 31, 151–182.

Doran, P.T., McKay, C.P., Clow, G.D., Dana, G.L., Fountain, A.G., Nylen, T. and Lyons, W.B. (2002a) Valley floor climate observations from the McMurdo dry valleys, Antarctica, 1986–2000. *Journal of Geophysical Research-Atmospheres* 107, 4772.

Doran, P.T., Priscu, J.C., Lyons, W.B., Walsh, J.E., Fountain, A.G., McKnight, D.M., Moorhead, D.L., Virginia, R.A., Wall, D.H., Clow, G.D., Fritsen, C.H., McKay, C.P. and Parsons, A.N. (2002b) Antarctic climate cooling and terrestrial ecosystem response. *Nature* 415, 517–520.

Doran, P.T., McKay, C.P., Fountain, A.G., Nylen, T., McKnight, D.M., Jaros, C. and Barrett, J.E. (2008) Hydrologic response to extreme warm and cold summers in the McMurdo Dry Valleys, East Antarctica. *Antarctic Science* 20, 499–509.

Doran, P.T., Lyons, W.B. and McKnight, D.M. (eds) (2010) *Life in Antarctic Deserts and other Cold Dry Environments: Astrobiological Analogs*. Cambridge University Press, Cambridge, UK, 320 pp.

Elberling, B., Gregorich, E.G., Hopkins, D.W., Sparrow, A.D., Novis, P. and Greenfield, L.G. (2006) Distribution and dynamics of soil organic matter in an Antarctic dry valley. *Soil Biology and Biochemistry* 38, 3095–3106.

Esposito, R.M.M., Horn, S.L., McKnight, D.M., Cox, M.J., Grant, M.C., Spaulding, S.A., Doran, P.T. and Cozzetto, K.D. (2006) Antarctic climate cooling and response of diatoms in glacial meltwater streams. *Geophysical Research Letters* 33, L07406, doi:10.1029/2006GL025903.

Esposito, R.M.M., Spaulding, S.A., McKnight, D.M., de Vijver, B.V., Kopalova, K., Lubinski, D., Hall, B. and Whittaker, T. (2008) Inland diatoms from the McMurdo Dry Valleys and James Ross Island, Antarctica. *Botany-Botanique* 86, 1378–1392.

Fanciulli, P.P., Summa, D., Dallai, R. and Frati, F. (2001) High levels of genetic variability and population differentiation in *Gressittacantha terranova* (Collembola, Hexapoda) from Victoria Land, Antarctica. *Antarctic Science* 13, 246–254.

Foreman, C.M., Wolf, C.F. and Priscu, J.C. (2004) Impact of episodic warming events on the physical, chemical and biological relationships of lakes in the McMurdo Dry Valleys, Antarctica. *Aquatic Geochemistry* 10, 239–268.

Fountain, A.G., Lyons, W.B., Burkins, M.B., Dana, G.L., Doran, P.T., Lewis, K.J., McKnight, D.M., Moorhead, D.L., Parsons, A.N., Priscu, J.C., Wall, D.H., Wharton, R.A. and Virginia, R.A. (1999) Physical controls on the Taylor Valley ecosystem, Antarctica. *Bioscience* 49, 961–971.

Green, T.G.A., Seppelt, R.D. and Schwartz, A.-M.J. (1992) Epilithic lichens on the floor of the Taylor Valley, Ross Dependency, Antarctica. *Lichenologist* 24, 57–61.

Green, T.G.A., Sancho, L.G., Türk, R., Seppelt, R.D. and Hogg, I.D. (2011) High diversity of lichens at 84°S, Queen Maud Mountains, suggests preglacial survival of species in the Ross Sea region, Antarctica. *Polar Biology*, doi:10.1007/s00300-011-0982-5.

Hall, B.L. and Denton, G.H. (2000) Radiocarbon chronology of the Ross Sea drift, eastern Taylor Valley, Antarctica. *Geografiska Annaler Series A* 82, 305–336.

Hawes, T.C., Worland, M.R., Bale, J.S. and Convey, P. (2008) Rafting in Antarctic Collembola. *Journal of Zoology* 274, 44–50.

Hawes, T.C., Torricelli, G. and Stevens, M.I. (2010) Haplotype diversity in the Antarctic springtail *Gressittacantha terranova* at fine spatial scales – a Holocene twist to a Pliocene tale. *Antarctic Science* 21, 766–773.

Hogg, I.D., Cary, S.C., Convey, P., Newsham, K., O'Donnell, A., Adams, B.J., Aislabie, J., Frati, F., Stevens, M.I. and Wall, D.H. (2006) Biotic interactions in Antarctic terrestrial ecosystems: are they a factor? *Soil Biology and Biochemistry* 38, 3035–3040.

Hopkins, D.W., Sparrow, A.D., Novis, P.M., Gregorich, E.G., Elberling, B. and Greenfield, L.G. (2006) Controls on the distribution of productivity and organic resources in Antarctic Dry Valley soils. *Proceedings of the Royal Society B: Biological Sciences* 273, 2687–2695.

Lyons, W.B., Welch, K.A., Carey, A.E., Doran, P.T., Wall, D.H., Virginia, R.A., Fountain, A.G., Csatho, B.M. and Tremper, C.M. (2005) Groundwater seeps in Taylor Valley Antarctica: an example of a subsurface melt event. *Annals of Glaciology* 40, 200–206.

McGaughran, A., Hogg, I.D. and Stevens, M.I. (2008) Phylogeographic patterns for springtails and mites in southern Victoria Land, Antarctica suggest a Pleistocene and Holocene legacy of glacial refugia and range expansion. *Molecular Phylogenetics and Evolution* 46, 606–618.

McKnight, D.M., Tate, C.M., Andrews, E.D., Niyogi, D.K., Cozzetto, K., Welch, K., Lyons, W.B. and Capone, D.G. (2007) Reactivation of a cryptobiotic stream ecosystem in the McMurdo Dry Valleys, Antarctica: A long-term geomorphological experiment. *Geomorphology* 89, 186–204.

Montes-Hugo, M., Doney, S.C., Ducklow, H.W., Fraser, W., Martinson, D., Stammerjohn, S.E. and SchoWeld, O. (2009) Recent changes in phytoplankton communities associated with rapid regional climate change along the western Antarctic Peninsula. *Science* 323, 1470–1473.

Nolan, L., Hogg, I.D., Stevens, M.I. and Haase, M. (2006) Fine scale distribution of mtDNA haplotypes for the springtail *Gomphiocephalus hodgsoni* (Collembola) corresponds to an ancient shoreline in Taylor Valley, continental Antarctica. *Polar Biology* 29, 813–819.

Overhoff, A., Freckman, D.W. and Virginia, R.A. (1993) Life cycle of the microbivorous Antarctic Dry Valley nematode *Scottnema lindsayae* (Timm 1971). *Polar Biology* 13, 151–156.

Pannewitz, S., Green, T.G.A., Scheidegger, C., Schlensog, M. and Schroeter, B. (2003) Activity pattern of the moss *Hennediella heimii* (Hedw.) Zand. in the Dry Valleys, Southern Victoria Land, Antarctica during the mid-austral summer. *Polar Biology* 26, 545–551.

Pastor, J. and Bockheim, J.G. (1980) Soil development on moraines of Taylor Glacier, lower Taylor Valley, Antarctica. *Soil Sciences Society of America Journal* 44, 341–348.

Pointing, S.B., Chan, Y., Lacap, D.C., Lau, M.C.Y., Jurgens, J.A. and Farrell, R.L. (2009) Highly specialized microbial diversity in hyper-arid polar desert. *Proceedings of the National Academy of Sciences of the United States of America* 106, 19964–19969.

Ruprecht, U., Lumbsch, H.T., Brunauer, G., Green, T.G.A. and Turk, R. (2010) Diversity of *Lecidea* (Lecideaceae, Ascomycota) species revealed by molecular data and morphological characters. *Antarctic Science* 22, 727–741.

Schlensog, M., Pannewitz, S., Green, T.G.A. and Schroeter, B. (2004) Metabolic recovery of continental Antarctic cryptogams after winter. *Polar Biology* 27, 399–408.

Schwarz, A.M.J., Green, T.G.A. and Seppelt, R.D. (1992) Terrestrial vegetation at Canada Glacier, Southern Victoria Land, Antarctica. *Polar Biology* 12, 397–404.

Seppelt, R.D., Türk, R., Green, T.G.A., Moser, G., Pannewitz, S., Sancho, L.G. and Schroeter, B. (2010) Lichen and moss communities of Botany Bay, Granite Harbour, Ross Sea, Antarctica. *Antarctic Science* 22, 691–702.

Simmons, B.L., Wall, D.H., Adams, B.J., Ayres, E., Barrett, J.E. and Virginia, R.A. (2009a) Mesofauna communities in above- and below-ground habitats in mosses and algal mats in Taylor Valley, Antarctica. *Polar Biology* 32, 1549–1558.

Simmons, B.L., Wall, D.H., Adams, B.J., Ayres, E., Barrett, J.E. and Virginia, R.A. (2009b) Long-term experimental warming reduces soil nematode populations in the McMurdo Dry Valleys, Antarctica. *Soil Biology and Biochemistry* 41, 2052–2060.

Sinclair, B.J. and Sjursen, H. (2001) Cold tolerance of the Antarctic springtail *Gomphiocephalus hodgsoni* (Collembola, Hypogastruridae). *Antarctic Science* 13, 271–279.

Sinclair, B.J. and Stevens, M.I. (2006) Terrestrial microarthropods of Victoria Land and Queen Maud Mountains, Antarctica: implications of climate change. *Soil Biology and Biochemistry* 38, 3158–3170.

Smith, J.J., Tow, L.A., Stafford, W., Cary, S.C. and Cowan, D.A. (2006) Bacterial diversity in three different Antarctic cold desert mineral soils. *Microbial Ecology* 51, 413–421.

Solomon, S., Qin, D., Manning, M., Chen, Z., Marquis, M., Averyt, K.B., Tignor, M. and Miller, H.L. (2007) *Climate Change 2007: The Physical Science Basis. Contribution of Working Group I to the Fourth Assessment Report of the Intergovernmental Panel on Climate Change.* Cambridge University Press, Cambridge, UK and New York, 996 pp.

Sømme, L. (1981) Cold tolerance of Alpine, Arctic and Antarctic Collembola and mites. *Cryobiology* 18, 212–220.

Stevens, M.I. and Hogg, I.D. (2002) Expanded distributional records of Collembola and Acari in southern Victoria Land, Antarctica. *Pedobiologia* 36, 485–496.

Stevens, M.I. and Hogg, I.D. (2003) Long-term isolation and recent range expansion revealed for the endemic springtail *Gomphiocephalus hodgsoni* from southern Victoria Land, Antarctica. *Molecular Ecology* 12, 2357–2369.

Stevens, M.I. and Hogg, I.D. (2006) Contrasting levels of mitochondrial DNA variability between mites (Penthalodidae) and springtails (Hypogastruridae) from the Trans-Antarctic Mountains suggest long-term effects of glaciation and life history on substitution rates, and speciation processes. *Soil Biology and Biochemistry* 38, 3171–3180.

Strandtmann, R.W. (1967) Terrestrial Prostigmata (Trombidiform mites). *Antarctic Research Series* 10, 51–95.

Thompson, D.W. and Solomon, S. (2002) Interpretation of recent Southern Hemisphere climate change. *Science* 296, 895–899.

Treonis, A.M. and Wall, D.H. (2005) Soil nematodes and desiccation survival in the extreme arid environment of the Antarctic Dry Valleys. *Integrative and Comparative Biology* 45, 741–750.

Treonis, A.M., Wall, D.H. and Virginia, R.A. (2000) The use of anhydrobiosis by soil nematodes in the Antarctic Dry Valleys. *Functional Ecology* 14, 460–467.

Turner, J., Bindschadler, R., Convey, P., di Prisco, G., Fahrbach, E., Gutt, J., Hodgson, D., Mayewski, P. and Summerhayes, C. (2009) *Antarctic Climate Change and the Environment*. SCAR and Scott Polar Research Institute, Cambridge, UK, 526 pp.

Ugolini, F.C. and Bockheim, J.G. (2008) Antarctic soils and soil formation in a changing environment: a review. *Geoderma* 144, 1–8.

Virginia, R.A. and Wall, D.H. (1999) How soils structure communities in the Antarctic Dry Valleys. *BioScience* 49, 973–983.

Wall, D.H. (2005) Biodiversity and ecosystem functioning in terrestrial habitats of Antarctica. *Antarctic Science* 17, 523–531.

Wall Freckman, D. and Virginia, R.A. (1998) Soil biodiversity and community structure in the McMurdo Dry Valleys, Antarctica. In: Priscu, J.C. (ed.) *Ecosystem Dynamics in a Polar Desert. The McMurdo Dry Valleys, Antarctica*. American Geophysical Union, Washington, DC, pp. 323–336.

Wharton, D.A. (2003) The environmental physiology of Antarctic terrestrial nematodes: a review. *Journal of Comparative Physiology B: Biochemical, Systemic, and Environmental Physiology* 173, 621–628.

Wharton, D.A. and Barclay, S. (1993) Anhydrobiosis in the free-living Antarctic nematode *Panagrolaimus davidi* (Nematoda, Rhabditida). *Fundamentals of Applied Nematology* 16, 17–22.

Wharton, D.A. and Brown, I.M. (1991) Cold tolerance mechanisms of the Antarctic nematode *Panagrolaimus davidi*. *Journal of Experimental Biology* 155, 629–641.

Wirtz, N., Lumbsch, H.T., Green, T.G.A., Türk, R., Pintado, A., Sancho, L. and Schroeter, B. (2003) Lichen fungi have low cyanobiont selectivity in maritime Antarctica. *New Phytologist* 160, 177–183.

11 Hot Desert Environments

Boris Rewald, Amir Eppel, Oren Shelef, Amber Hill, Asfaw Degu, Avital Friedjung and Shimon Rachmilevitch
French Associates Institute for Agriculture and Biotechnology of Drylands, Blaustein Institutes for Desert Research, Ben-Gurion University of the Negev, Israel

11.1 The Hot Desert Environment

11.1.1 Formation and geographical distribution

Hot deserts cover between 14 and 20% of the Earth's surface, approximately 19–25 million km^2 (see Middleton and Thomas, 1997; Peel *et al.*, 2007 for distribution of deserts). Most hot deserts, such as the Sahara of North Africa and the deserts of the south-western USA, Mexico and Australia, occur along the tropics in both the northern and southern hemispheres (between approximately 10° and 30–40° latitude). They are created as a result of global Hadley air circulation (Warner, 2004). The sun's radiation causes hot air to rise and the accumulation of moisture around the equator. As the air moves away it cools, starts to descend and at this point all of the moisture is lost as rainfall in the tropics. As the air subsides and becomes compressed it also becomes warmer and, consequently, the relative humidity in desert air decreases, even though the absolute amount of water vapour held may be substantial, as evidenced by dew-fall in the cold hours before dawn (Parsons and Abrahams, 2009). Because most of the water in the atmosphere is derived from evaporation from the seas, there is often an aridity gradient on large continents, i.e. the land closer to the sea often receives a larger share of this water. Regions lying deep within a continent may become drylands simply because the air currents that reach them have already traversed vast land distances and lost most of their moisture (Warner, 2004). Continentality is a major factor, especially for the Taklimakan Desert in central East Asia and the Australian deserts. Further factors accounting for the distribution of deserts are a low inland relief and high coastal ranges (i.e. creating a 'rain shadow') and the presence of cool ocean water close to the shore and/or aloft in a tropical jet-stream, which lead to the formation of coastal deserts (Goudie and Wilkinson, 1977). A good example is the Atacama Desert, Chile, where the air circulation from the sea is drying out the land. Notably, coastal deserts can be only found on the western fringes of continents and are not as hot as subtropical deserts (Warner, 2004). The current boundaries and extents of hot deserts are likely to continue to expand over the next century due to a combination of global warming and poor land management (e.g. overgrazing, indiscriminate clearing of the vegetation cover and salinization; Laity, 2008).

11.1.2 Types of hot deserts

A defining characteristic of a hot desert is aridity. According to the Koppen-Geiger climate classification, deserts are regions with an annual precipitation of less than 250 mm (Peel *et al.*, 2007). However, annual precipitation can be misleading because water loss is just as important a component of the water budget. Thus, the United Nations Environmental Programme's definition of desert is an annual moisture deficit under normal climatic conditions, where the potential evapotranspiration (PET) is over five times higher than actual precipitation (Middleton and Thomas, 1997). The high PET prevails because, owing to the lack of cloud cover, incoming solar energy approaches a maximum in arid regions.

Deserts can be divided into two types according to their level of aridity: hyperarid deserts have an aridity index (P/PET) of <0.05 and arid deserts have P/PET between 0.05 and 0.2. As such, deserts are distinguished from semi-arid drylands (P/PET 0.2–0.5) and dry sub-humid drylands (P/PET 0.5–0.65). The diurnal temperature variation in deserts is very pronounced, with high day time and low night time temperatures (Woodward, 2003). Due to the high surface temperature and temperature differences most deserts are also high wind energy environments (Parsons and Abrahams, 2009).

Because of these extreme conditions, hot deserts are stereotypically portrayed as vast seas of sand dunes with cacti and circling vultures (Fig. 11.1). In fact there are many different kinds of hot deserts with varying landforms, altitudes and life-forms (Parsons and Abrahams, 2009). For example, only approximately 15–20% of deserts are covered by sand dunes, instead rocky plateau or mountain desert landscapes prevail (Ward, 2009). Soil properties influence the degree of aridity and thus plant productivity. There are arid climates with >500 mm of annual rain that falls in intense events on hard soil or rock, and the water runs off horizontally or evaporates quickly (Warner, 2004). At the other extreme there are soils in so-called edaphic deserts which are extremely porous and have such a low water-holding capacity that the annual precipitation drains through them so rapidly that it is virtually unavailable for vegetation. Related to water, plant productivity in deserts

Fig. 11.1. The heterogenic desert landscape contains a diversity of landforms, soils and various rock covers. Kasui Dune, Negev Hills, Israel. © Oren Shelef.

can also be nutrient limited. For example, nitrogen is a key limiting nutrient in most deserts (West and Skujins, 1978), while phosphorous is the most limiting nutrient in Australian deserts (Ward, 2009). Soil nutrients and organic matter tend to be concentrated in the upper 2–5 cm of the soil and the greatest concentrations are measured underneath the canopies of desert shrubs or trees in so called 'islands of fertility' (Ward, 2009). Due to high evaporation rates many desert soils are also affected by accumulated salt ions. A detailed description of saline habitats appears in McGenity and Oren, Chapter 20, this volume.

Fig. 11.2. *Sternbergia clusiana* bloom in the autumn in the Negev Highlands, Israel. Desert plants have evolved adaptations to cope with the lack of water, enabling their survival and allowing them to complete their life cycles in dry conditions. *Sternbergia clusiana* is a geophyte that relies on a bulb in order to bloom before the rainy season. © Oren Shelef.

11.2 Abiotic Stress Factors in Hot Deserts and Mechanisms of Stress Resistance

When one thinks of hot deserts one immediately envisages extremely hot, sunny, dry environments. Indeed, hot deserts present the organisms that live in them with a number of abiotic extremes: high levels of radiation, high temperatures, aridity, and often high salinity, strong winds and periodic dust storms. However, deserts are not sterile wastelands, instead they host a wide diversity of organisms specially adapted to face the extreme challenges that desert dwelling presents them with (Fig. 11.2). One might predict natural selection to drive desert organisms along convergent paths to a common suite of adaptive traits, however, organisms inhabiting deserts may actually show a wide range of growth forms and adaptations.

A biological response to stress, termed stress resistance, usually involves both avoidance and tolerance mechanisms (Levitt, 1972). Stress avoidance is partially or completely excluding the stress by means of a physical barrier or timing of behaviour and/or appearance, while stress tolerance is withstanding the stress using either repair or prevention techniques.

According to their high levels of stress resistance, many desert organisms can be classified as xerophiles (i.e. organisms adapted for life and growth in very dry surroundings) and/or thermophiles (i.e. species adapted to withstand hot air and soil temperatures). Xerohalophiles are able to survive in the saline micro-environments that often occur in hot deserts.

In this chapter we will: (i) introduce the major abiotic constraints that species face in hot deserts and discuss how organisms, primarily from the plant kingdom, have adapted to circumvent these constraints; and (ii) shed light on some of the biotic interactions that take place in desert environments.

11.2.1 Radiation and temperature

While aspects of topography can significantly affect local temperature in deserts, radiation is the main factor. Radiation in deserts can be as high as 840 GJ km^{-2} year^{-1} (Laity, 2008), more than twice as much as in temperate ecosystems. This is for two main reasons: (i) deserts are located in latitudes that are relatively close to the equator; and (ii) low levels of cloud cover allow large amounts of radiation to reach the Earth's surface. The broad band extending across the Sahara and the Arabian deserts is the largest area to receive radiation of this magnitude. Deserts are renowned for their large diurnal temperature fluctuations; low vegetation cover within deserts means that the sun heats up the area quickly, but at night heat radiates rapidly away. While the temperature of winter nights in hot deserts may fall below 0°C, daytime air temperatures may remain as high as 40°C. The highest air temperature ever recorded was 57°C in the Libyan part of the Sahara. However, the temperature on the soil surface can be considerably higher than the air temperature; as high as 75–80°C (Ward, 2009).

Both high radiation and high temperature significantly affect the physiology of organisms, especially if water is scarce. In plants and some cyanobacteria, the visible part of the radiation spectrum (photosynthetically active radiation – PAR – wavelength 400–700 nm) is captured by pigments (chlorophylls and carotenoids) and they convert its energy, via photochemical processes, into reductive chemical products. The products are mainly used to produce sugar in a process termed the Calvin Cycle. In hot deserts, the high level of irradiance is usually combined with high temperature and shortage of water, which restricts the activity of the Calvin Cycle. Therefore, high irradiance might create an overflow of excess reductive power in the chemical form of reactive oxygen species (ROS). These molecules can react and cause damage to DNA, proteins and lipid membranes, which are crucial to plant survival (Akashi *et al.*, 2008). In higher plants, one approach to dealing with ROS excess is to reduce light absorbance by the leaf. Light absorbance is strongly affected by leaf orientation and reflectance, therefore, although sun-tracking leaves are advantageous in maximizing photon absorbance, leaves that are not arranged perpendicular to the sun's rays may be more advantageous to desert plants by reducing irradiance and heat accumulation (Gibson, 1996). Plants with fixed vertical and very steeply angled leaves, and with azimuth east–west orientation of leaves and branches, have been observed in deserts (Nobel *et al.*, 1993). These leaves receive most light in early morning and late afternoon while avoiding intercepting excess levels of radiation during midday.

Leaf hairs and deposition of wax on the leaf surface can also serve as barriers that reflect excess light, resulting in the evolution of whitish, silvery and greyish leaf types (Holmes and Keiller, 2002). It has been shown that leaf hairs on the desert plant *Encelia farinose* decreased the visible light absorption by as much as 56% compared to *E. californica*, an inhabitant of coastal ecosystems that does not possess leaf hairs (Ehleringer *et al.*, 1976). Unlike radiation in the visible part of the spectrum, ultraviolet (UV) radiation (wavelength 10–400 nm) causes damage to all living creatures directly via harmful reactions with DNA, proteins and lipids, or indirectly by producing ROS (see Newsham and Davidson, Chapter 22, this volume). In order to cope with UV radiation, desert plants and cyanobacteria produce phenolic compounds, such as flavonoids, which trap the UV radiation and prevent damage to other molecules. In higher plants, these compounds are mainly distributed on the outer layers of the leaves (epidermis) thus protecting the inner layers from damage (Caldwell *et al.*, 1998). The aforementioned plant adaptations to avoid excessive amounts of visible light are also helpful in dealing with high UV radiation (Holmes and Keiller, 2002).

Plants' tolerance to high radiation involves the activity of the photosynthetic process itself. One such adaptation is photorespiration; in this process the reductive power generated by the light reaction is used to reduce oxygen ions to produce carbon dioxide (CO_2), thus removing the excess energy absorbed by the plants (Kozaki and Tabeka, 1996). The use of photorespiration in photo-protection has been demonstrated in desert plants; the perennial semi-shrub *Reaumuria soongorica*, increases its rate of photorespiration under drought conditions to deal with excess radiation (Bai *et al.*, 2008). Photorespiration is considered an 'energetically wasteful' process, which means it consumes oxygen and emits CO_2. Consequently, many plant species in hot, dry and high light environments have an alternative photosynthetic mechanism, known as the C_4 photosynthetic pathway. It is designed to concentrate high levels of CO_2 in the proximity of the Calvin Cycle process in the chloroplast where carbon fixation occurs. This high concentration of CO_2 in chloroplasts means that C_4 plants can utilize excess light for photosynthesis without needing photorespiration (Sage, 2004).

A different strategy to deal with excess light is non-photochemical quenching (NPQ). NPQ is a universal mechanism that diverts the absorbed excess light into heat and not to the photochemical processes (Müller *et al.*, 2001). While it has been shown that this process occurs in desert plants grown under high irradiance in the winter, the importance of this mechanism during the hot summers is less clear due to the unfortunate side effect of leaf heating that it causes (Barker *et al.*, 2002).

Heat resistance by photosynthetic organs requires a balance between heat influx due to radiation (solar and environmental-infrared) and heat efflux (Gibson, 1996). High temperature is, thermodynamically speaking, a high level of molecule movement in a medium, resulting in increased entropy. When a leaf-cell's temperature exceeds approximately 46°C, which is a common midday summer temperature in hot deserts, several physiological processes may begin to break down (Berry and Björkman, 1980). For example, increased entropy causes disruption of protein structures and functions. Membrane stability is also endangered, disrupting water and solute concentrations, damaging cellular homeostasis and leading to cell death (Larcher, 2001). To avoid tissue damage, heat can be dissipated via increased conduction, convection, and transpirational cooling in plants (e.g. Levitt, 1980) or evaporational cooling (sweating) in animals. Unfortunately, the high temperature in deserts correlates with reduced water availability, thus transpiration/sweating for cooling purposes can further increase dehydration.

Many desert animals are relatively pale in colour, possibly preventing their bodies from absorbing too much solar radiation during the day (e.g. Hetem *et al.*, 2009). In addition, the fur of mammals often forms an efficient barrier against heat gain from the environment (Schmidt-Nielsen *et al.*, 1956). However, for desert mammals body overheating is generally a result of heavy exercise (by metabolic heat production in muscles) rather than passive heat gain during resting, therefore, desert animals often reduce their level of exertion on hot days (Vaughan *et al.*, 2010). Furthermore, in contrast to sessile plants, desert mammals, reptiles (which are exothermic and thus greatly affected by the heat sources around them) and others, can avoid the hottest midday sun by retreating to burrows, the shade of plants, or even by becoming fully nocturnal (i.e. 'behavioural thermoregulation'; e.g. Vaughan *et al.*, 2010; Wilms *et al.*, 2011).

Because the brain tissue of mammals is especially heat sensitive, some species have specialized cooling devices. For example, in some antelopes like the oryx (Fig. 11.3) the blood going to the brain is cooler than most of the rest of the body because of a counter-current heat exchange between veins and arteries and evaporative cooling of veins in the nasal cavity; a clever system called 'carotid rete' (Vaughan *et al.*, 2010). In general, desert animals can be classified into heat 'avoiders', 'evaporators' and 'endurers' (i.e. the body temperature of 'endurers' can increase by several degrees without tissue damage occurring; e.g. camels; Fig. 11.4). However, there is a lot of overlap between heat 'avoiders' and 'evaporators' (Ward, 2009).

In plants and animals alike, conduction and convection are strongly affected by the boundary layer thickness – the boundary layer being the layer of unmixed air above the organ/body surface – and its thickness is related to organ/body shape and size, and wind speed. Desert environments often host smaller animals with larger extremities

Fig. 11.3. Herd of oryx antelopes (*Oryx gazella*) in the Central Namib Desert, south-west Africa. © Elsita Maria Kiekebusch.

Fig. 11.4. Herd of Mongolian camels (*Camelus bactrianus*) in the Gobi Desert, Mongolia. © Uri Shapira.

compared to colder climates (the so-called 'Bergman rule'), increasing the cooling rate due to reduced thermal inertia and increased surface area:body mass ratios (Willmer *et al.*, 2004). Additionally, bare skin areas, e.g. found on some vulture heads, play an important role in thermoregulation by reducing the boundary layer compared to feathered skin areas (Ward, 2009).

Perennial desert plants are either drought-deciduous or aphyllous, i.e. with green photosynthetic stems, or possess small and narrow, even microphyllous and heteroplastic leaves. This allows leaf temperature to be maintained very close to ambient temperature without high transpirational costs (Smith, 1978). However, this is not true for many succulent plants; their large size largely prevents convective heat exchange and often means that the temperature of sunlit surfaces increases 10–15°C above ambient air temperature. As a result some desert cacti must be able to tolerate tissue temperatures approaching 60°C, a tolerance level unsurpassed in hydrated vascular plants and the primary physiological cost borne by succulents (Smith *et al.*, 1997). On the other hand, desert species with broad leaves, e.g. fan palms, *Washingtonia* spp., cool their leaves via high transpiration rates, explaining why these species are commonly found in areas that have accessible groundwater, e.g. oases.

Heat tolerance is achieved in plants by various cellular and physiological mechanisms. One common response to increased temperature is the expression of heat shock proteins (HSP). These proteins allow processes such as protein folding to occur at higher than optimal temperatures and also enhance membrane stability (Larcher, 2001). In the heat tolerant, broad leaf succulent, *Agave tequilana*, extreme heat resistance correlated with both high levels of transpiration and the expression of HSPs (Lujan *et al.*, 2009).

At the physiological level, photosynthesis is more vulnerable to heat stress than respiration, which tends to increase with temperature. At high temperatures, photorespiration is enhanced at the expense of photosynthesis, depleting a plant's energy budget. C_4 plant photosynthesis is much less affected by temperature than C_3 plants using the Calvin Cycle; thus C_4 plants are more common in deserts than in temperate zone ecosystems (Sage, 2004). In plants that possess this type of metabolic pathway there is almost no photorespiration, and they are therefore able to perform photosynthesis at higher temperatures than C_3 plants, which do perform photorespiration (Ehleringer and Björkman, 1977). So how do C_3 plants adjust their photosynthetic activity to cope with the hot and dry desert environment? One mechanism is displayed by the evergreen desert shrub *Retama raetam*: in this plant the lower, shaded and cooler canopy continues photosynthesis throughout the hot dry summer period, while the upper, sun exposed and warmer canopy remains photosynthetically dormant. The upper canopy maintains a certain amount of mRNA coding for photosynthetic proteins. When water becomes available, these mRNA are translated and allow for fast recovery of photosynthetic activity in the upper canopy (Mittler *et al.*, 2001).

While a wide variety of soil mesofauna can survive exposure to very high temperatures, including nematodes, tardigrades and rotifers, the only truly thermophilic organisms in deserts are soil microbes (Jeffery *et al.*, 2010).

11.2.2 Water scarcity

Deserts are 'water-controlled ecosystems' (Noy-Meir, 1973); available water for desert organism survival relies mainly on the balance between water contribution and water loss. Surface flow, precipitation, groundwater and atmospheric vapour are the natural sources of water for organisms in arid regions (Flanagan and Ehleringer, 1991). However, perennial or ephemeral streams are limited in availability in most (hyper-)arid environments and groundwater is often saline (Akram and Chandio, 1998). Dew accumulation may contribute up to 30% of the annual precipitation in some deserts (Zangvil, 1996) while in hyperarid ecosystems the amount of condensation is scarce but might exceed precipitation. While the Chilean Atacama Desert receives <10 mm of annual precipitation mainly in the form of condensed fog, the primary source of freshwater in most deserts is rainfall falling in a relatively predictable season but as an unpredictable amount (Noy-Meir, 1973). The seasonality and the time distribution of precipitation affects the 'bioactivity' of water; for example, rain that falls in the cooler winter months is less likely to be evaporated and is therefore more available to organisms. The northern part of the Sahara and Arabian Deserts, including the Negev Desert, receive winter rains (50–100 mm year^{-1}) while the southern part of the Sahara Desert receives summer rain (sometimes <5 mm year^{-1}; Gutterman, 2002a, b).

Deserts that receive monsoonal thunderstorms, and in particular those regions within or adjacent to mountain ranges, are subjected to sudden flash floods. Flash floods have huge discharges and velocities >2 m s^{-1} (Fisher and Minckley, 1978), therefore, riparian species in dry riverbeds must withstand the severe physical agitation of these infrequent but very forceful natural disasters, often resulting in uprooting or partially uncovered root systems (Fig. 11.5). Different water harvesting systems, including the collection and storage of runoff during these floods in open ponds, are being used in arid ecosystems worldwide to grow grass and trees (for fodder, fruits and firewood), and for recreation (Lavee *et al.*, 1997; Fig. 11.6).

The coincidence of low annual rainfall with extremely high evapotranspiration rates due to the high summer temperatures leads to the rapid depletion of water reserves (Ceballos *et al.*, 2004). In order to cope with this a plant may employ both avoidance and tolerance mechanisms. In general, certain plant growth forms can confer a selective advantage in desert environments by facilitating tradeoffs between maximal net carbon gain and tolerance of environmental stress. For example, long-lived perennial shrubs and trees have relatively low net carbon gains but are buffered against environmental stress, whereas annuals possess high rates of net carbon gain but restrict their active growth to relatively short periods when soil moisture is high (Gibson, 1996). A simple classification concept was developed for the Negev Desert; plants were described as: (i) able to recover from dehydration (desiccation-tolerant); (ii) inactive during the dry season (arido-passive); or (iii) plants that are active during the dry season (arido-active; Evenari *et al.*, 1971).

Desiccation-tolerant desert plants have the supreme ability to withstand 'drying without dying'. The most extreme examples are the so-called resurrection plants. Resurrection plants are mostly poikilohydrous, which means that their water content adjusts with the relative humidity in the environment. They are able to stay in a dehydrated state until water becomes available, allowing them to rehydrate and to resume full physiological activities. They can revive from an air-dried state as low as 0% (v/v) relative humidity (Gaff, 1987). Resurrection plants take immediate advantage of rainfall after dry periods: they resurrect and restore photosynthetic activities often within 24 h of rehydration, and can subsequently grow and reproduce before other species are able to (Scott, 2000). Two types of resurrection plants can be distinguished, (i) poikilochlorophyllous plants, these plants lose chlorophyll and the thylakoid membranes are at least partially degraded during water loss; and (ii) homoiochlorophyllous plants, such as *Craterostigma plantagineum*, which retain chlorophyll and retain the intact photosynthetic structures although structural changes occur (Bartels, 2005). While most resurrection plants are herbaceous, the largest known resurrection plant is the woody shrub *Myrothamnus flabellifolia* (Sherwin *et al.*, 1998).

Arido-passive species are classic examples of drought avoiding ('drought escaping') plants. They have insufficient mechanisms to combat water stress and must complete their life cycle only when ample free soil water is available. Ephemeral or annual plants are found even in the extreme deserts; their seeds may remain viable in the dry soil for many years (Gutterman, 2002a, b). Variable, delayed germination of seeds is an adaptation

Fig. 11.5. Desert trees count on developed root systems that allow them to utilize water from deep soil layers or, like this *Pistacia atlantica* (a), exploit protected water trapped in crevices. (b) Developed roots of *Tamarix aphylla* excavated by strong seasonal streams, Negev Highlands, Israel. Images © Oren Shelef.

that allows desert ephemerals to avoid conditions when soils are too dry or temperatures are too high for survival until the reproduction stage is reached. For example, cool season ephemerals in North American deserts typically germinate in the autumn and persist in the winter as leaf rosettes; in early spring, these rosettes quickly produce flowers and seeds (Whitford, 2002). The annual grass *Schismus arabicus* (Fig. 11.7) demonstrates annual periodicity as an adaptation to hot and dry deserts. Results by Gutterman *et al.* (2010) indicate that survival of *Schismus* seedlings depends on the

stems, twigs and petioles can cause cavitation in xylem vessels, i.e. the occurrence of air bubbles ('emboli') that prevent further water conductance. Many species adapted to more xeric conditions reduce their vulnerability to cavitation, e.g. via reduced pore sizes in pit membrane or different hydraulic strategies (Rosenthal *et al.*, 2010). Desert shrubs that are summer active but leafless in the winter (e.g. *Flourensia* spp., *Krameria* spp.) are especially prone to hydraulic xylem failure due to the high evaporative demand but may have advantages in leaf cooling if water is available.

Other examples of the trade-off between adaptational mechanisms come from deciduous and evergreen shrubs. A habitat in which both shrub types grow is the Mojave Desert, California, USA, which is dominated by the evergreen *Larrea tridentate* on mounds and the deciduous *Ambrosia dumosa* in soil depressions and temporal waterways. Deciduous shrubs have twice the photosynthetic capacity of evergreens under optimal conditions, but can only 'operate' during the most favourable parts of the year. It was calculated that the primary cost of the deciduous 'strategy' is the time lag to produce new leaves after the onset of the wet season, thus evergreen shrubs are at an advantage relative to deciduous shrubs and annuals when the rainy season or water availability is generally short and unpredictable (Comstock and Ehleringer, 1986).

Phreatophytes are desert species that avoid water stress by developing very deep root systems able to tap into reliable moisture sources such as groundwater. These species, trees or shrubs, show little tendency for water conservation and are therefore termed pseudo-xerophytes, clearly distinguishing them from the xerophytes described above (Fig. 11.5). Arborescent (tree-like) phreatophytes, such as *Alhagi sparsifolia*, *Populus euphratica* and diverse *Tamarix* spp., can develop taproots that can sometimes extend to a depth of >50 m (Phillips, 1963; Thomas *et al.*, 2008). Extensive clonal growth below ground has also been observed and is often the basis for the occupation of space and of vegetative regeneration once the groundwater is too deep to be reached by seedlings (Bruelheide *et al.*, 2004).

Herds of grazing and browsing animals often migrate to find water and food in response to seasonal and year-to-year variations in rainfall and food availability (Miller and Spoolman, 2008; Fig. 11.3). Animals that cannot migrate often obtain nearly all their water from plants (Brown and Ernest, 2002) or store water in their bodies, which is gradually depleted until the body is once again rehydrated. Furthermore, some insects can absorb moisture from the air.

An excellent example of drought-evading animals are Australian desert frogs, such as *Cycloana platycephala*, which spend the dry parts of the year (>6 months) in aestivation (a form of dormancy) inside a burrow and additionally build cocoons to reduce water loss (Tracy *et al.*, 2007). Another mechanism to reduce water loss is increased body temperature; for example, the body temperature of camels at rest can increase from about 34°C to >40°C in the summer. The variations in temperature are of great significance in water conservation in two ways: (i) the increase in body temperature means that heat is stored in the body instead of being dissipated via the evaporation of water. Subsequently, at night the excess heat can be given off without water expenditure; and (ii) heat gain from the hot environment is reduced because the temperature gradient is reduced (Schmidt-Nielsen *et al.*, 1956). Insects, land snails and many reptiles and birds excrete uric acid as a major nitrogenous waste, while mammals excrete water-soluble urea. Because uric acid is excreted as a paste, it is associated with very little water loss (Ward, 2009). In contrast, xeric-adapted mammals have often very long loops of Henle in their kidneys to concentrate ions and maximize water conservation.

In the most extreme hot deserts, e.g. parts of the Atacama Desert, higher plants or invertebrates are virtually absent. However, cyanobacteria were found to grow beneath translucent quartz rocks (rainfall/fog >1 mm; 'hypolithic') or even beneath the surface of rocks ('endolithic'). The translucent rock crystals allow light to penetrate and support photosynthesis and the soils beneath the quartz contain more moisture than the exposed surface soils, providing a habitat for, e.g. cyanobacteria (Azúa-Bustos *et al.*, 2011). Since organic materials are scarce inside rock some are also adapted to utilize inorganic materials to produce energy (chemolithotrophy), e.g. bacteria found in a rock in the Atacama Desert can oxidize inorganic arsenite (Campos *et al.*, 2009).

11.2.3 Salinity

Saline soils cover 6% of the continental surface (Larcher, 2001). The excessive accumulation of salts in arid soils occurs mainly due to the evaporation of water derived from precipitation, irrigation or from solute movement in shallow groundwater. Arid regions are subject to high evaporation and low precipitation and are, therefore, especially prone to high soil salinity. Saline soil environments are associated with osmotic stress, toxic ions and ion imbalances (Lambers *et al.*, 2008). Because water uptake is driven by a passive transport which relies upon a gradient of water pressure (Taiz and Zeiger, 1998), water moves into plant tissues only when the water potential of fine roots is more negative than the soil solution. In saline soils water is osmotically held and becomes less accessible to the plant creating a so-called 'physiological drought' if plants are not adapted. An excess of ions, specifically Na^+ and Cl^-, also has negative effects on enzymes and membranes which impedes various systems in the plant, including energy balance, nitrogen assimilation and protein metabolism (Larcher, 2001).

There is an immense diversity of organisms that can be called halophiles (McGenity and Oren, Chapter 20, this volume). While all plants are endowed with certain mechanisms to tolerate salinity to some extent, halophytes (about 1% of the world's flora) are plants that are able to complete their life cycle in elevated salinities of 200 mM NaCl or more (Flowers and Colmer, 2008). In contrast, glycophytes (plants not adapted to salinity) can only withstand salinities up to 10 mM NaCl (Nilsen and Orcutt, 2000). Obligatory halophytes require saline conditions for growth, whereas facultative halophytes can also live under freshwater conditions (Waisel, 1972). Examples of xerohalophytes include members of the Chenopodiaceae, e.g. *Atriplex halimus* and *Salsola* spp., and Tamaricaceae, such as *Tamarix* spp. and *Reaumuria* spp. (Feinbrun-Dothan and Danin, 1998). An example of a halotropic xerohalophyte is *Bassia indica*, which is a facultative xerohalophyte. In the Negev Desert, Israel, *Bassia indica* roots exhibit a unique salt tolerance mechanism; the plants develop thick horizontal roots, originating from the main taproot, and these grow toward saline regions. This phenomenon was termed halotropism, growth cued by and towards salt (Shelef *et al.*, 2010).

The major salinity tolerance mechanisms that have evolved in desert plants are: spatial avoidance, changes in root morphogenesis, retention of toxic ions in vacuoles or granular compartments, cellular ion regulation and metabolite synthesis for osmotic adjustment. More physiological salt tolerance strategies are reviewed by Parida and Das (2005) including: (i) limitation of salt uptake by roots and transport into leaves; (ii) alterations in photosynthetic pathways; (iii) changes in membrane structure; (iv) induction of antioxidative enzymes; and (v) induction of hormones. In addition, salt may be diluted by reverse flux from shoots to roots via the phloem (Cheeseman, 1988) and plants may possess special mechanisms for seed preservation, protection and germination. In this sense, salt dwelling is sometimes beneficial since the hygroscopic traits of salt mean that saline soils have higher water contents than soils without salts. This link is prominent in water-limited ecosystems where plants are often exposed to drought conditions combined with salinity.

For desert animals, excess salts can be particularly troublesome because they can exacerbate dehydration. However, some desert-adapted animals, like the spiny mouse (*Acomys cahirinus*), tolerate a significantly higher plasma osmolarity after a salt diet than similar lab mice

(Dickinson *et al.*, 2007). The efficiency of the spiny mouse kidney in filtering and excreting a higher concentration of salt resulted in their reduced water consumption.

11.2.4 Strong wind and dust storms

Most deserts are associated with weak horizontal pressure gradients and are therefore subject to weak, large-scale winds. However, weather disturbances intruding from the tropics or mid-latitudes occasionally generate high wind speeds, and the high convectional overturning of air boundary layers governs wind speeds from aloft down to the desert surface. For example, the winds in the Somali-Chalbi Desert are among the strongest and most sustained wind systems of the world; wind speeds of over 14 m s^{-1} (>50 km h^{-1}) prevail on more than 50 days of the year (Warner, 2004). High wind conditions initiate soil particle movement, peaking in dust storms, which are an important part of the desert climate (Fig. 11.8). Because sand grains form relatively small soil aggregates, sandy soils are especially susceptible to erosion by strong wind (Skidmore, 1989). For example, threshold wind speeds inducing dust storms were calculated as 6.7 m s^{-1} in the sandy Taklimakan Desert, China, contrasting with 13.8 m s^{-1} in the stony Gobi Desert, Mongolia (Kurosaki and Mikami, 2007). Unfortunately, since the 1950s, wind-blown dust storms in the Sahara desert have increased ten-fold due to overgrazing and drought caused by climate change and population growth (Miller and Spoolman, 2008).

Wind and/or wind-blown sand have been hypothesized to be important abiotic components shaping plant phenotypes, community composition and subsequently reducing yield in sand dune and other wind-prone ecosystems, such as semi-arid crop lands (Grace, 1988). As well as increasing evapotranspiration rates by reducing the boundary layers on leaf and soil surfaces (Caldwell, 1970), wind exerts direct mechanical pressures on plants, including increased motion as a result of both the turbulence and the drag force of the wind, uprooting of plants when the wind's force exceeds the stem or root/soil strength, and physical leaf damage arising from leaf tearing, stripping and abrasion.

A plant's resistance to breakage or overturning in windy climates depends largely

Fig. 11.8. Strong wind may elevate heavy dust to heights of several dozen metres causing dust storms. This photo was taken during a dust storm in Mali, Africa. © Ziv Sherzer.

on structural modifications for mechanical strength. Plant growth responses to wind movement, termed 'thigmomorphogenesis', include changes in branch and foliar development, stem shape and biomass allocation (Jaffe, 1973). For example, a *Tamarix* species growing in the Taklimakan Desert was found to increase the rigidity and length of roots and increase the thickness of stems on the leeward side. These measures probably prevent plants from overturning in the windy habitat (Liu *et al.*, 2008). Aboveground adaptations that reduce wind susceptibility further include a reduced plant height, increased numbers of lignified cells, and decreased leaf numbers or leaf areas (Grace, 1988; Eugster, 2008).

Another factor is sand movement, which is a characteristic of dunes. Several studies concluded that the difference in the tolerance to sand movement determines the distribution of plant species in areas where accumulation or erosion of soil frequently occurs (Avis and Lubke, 1985). Many psammophilic ('sand-loving') species are tolerant of sand movement and their growth and vigour can even be stimulated by it, contrasting with species preferentially growing on rocks or clay soils. Sand movement tolerance mechanisms include high shoot growth rates (Gilbert and Ripley, 2010).

Leaf stripping and obvious tears, abrasions, lesions or breakages of laminae seem to be less of a threat in desert species with microphyllous leaves which tend to have few or no petioles, additionally reducing torsional shear stresses on laminae (Niklas, 1992). However, wind rubbing of adjacent branches and leaves against each other and soil particles carried by wind ('sandblasting') can cause surface damage to plant tissues. For example, Thompson (1974) grew the grass *Festuca arundinacea* under relatively low wind speeds (~3 m s^{-1}) but reported microscopic damage to the epidermis. The damage included rupture of epidermal cells, cracking of the cuticle, and smoothing and redistribution of the wax deposits which cover the surface. In a parallel experiment it was demonstrated that leaves which had been damaged in this way lost much of their capacity to control water loss (Grace, 1974). Macroscopic damage features can be more evident when the wind carries particles of sand (Armbrust, 1982). In addition to having direct effects on plant growth and survival, sand abrasion significantly increases plant susceptibility to bacterial infections (Pohronezny *et al.*, 1992). The extent to which a given species is affected by abrasion depends *inter alia* on the structure of the cuticle (Yura and Ogura, 2006). While desert plants are known to have thicker cuticles to reduce transpiration, long and bending leaves growing approximately horizontally are thought to reduce the direct collision of wind-borne sand particles (Yura and Ogura, 2006).

Animals will probably try to reduce the impact of dust storms with behavioural changes, e.g. hiding, although physiological adaptations exist (e.g. a camel can close its nostrils to keep sand out of its nose). Dust storms were also found to lead to oxidative damage, mediated by pro-oxidant/antioxidant imbalance or excess free radicals, of different degrees in lungs, hearts and livers of rats (Meng and Zhang, 2006). However, further studies are required to clarify the toxicological role of sand storms on mammals.

11.3 Biotic Interactions in Deserts

Abiotic factors have long been considered more important than biotic variables in structuring desert plant communities (Grime, 1977) because constituent populations often exhibit episodic germination, recruitment and mortality as a direct result of fluctuations in environmental variables (Fowler, 1986). Some have even argued that environmental fluctuations prevent desert plant populations from ever reaching equilibrium with resource availability, thereby minimizing the importance of competition (Fowler, 1986 and references therein). However, research in recent decades has found evidence that biotic interactions are potentially of major functional importance in desert communities.

In deserts, distributions suggestive of positive interactions (i.e. facilitation) between species have generally been attributed to the amelioration of harsh environmental factors by one species, thereby benefiting the germination or survival of another; whereas distributions suggestive of negative species interactions (i.e. competition) have been attributed to the intensification of environmental stresses by neighbouring plants by resource depletion (Fonteyn and Mahall, 1981). The presence or absence of neighbouring plants can be a structuring force in desert plant communities (Allen *et al.*, 2008). Earlier research concluded that in the desert: (i) plant cover decreases with a concomitant increase in lateral root growth, until almost all of the available space is fully occupied; and (ii) regular spacing in areas of low rainfall is caused and maintained by root competition for available water (e.g. Walter, 1962). However, spatial arrangement and occurrence of plant species is recognized to be additionally influenced by factors such as habitat heterogeneity, disturbance by animals (e.g. herbivores) and direct plant–plant competition via allelochemicals (Friedman and Waller, 1987).

Plant cover in water-limited ecosystems is usually less than 60%, and most of it is a two-phase mosaic complex (Aguiar and Sala, 1999). The mosaic is often composed of woody vegetation patches and inter-shrub patches. Woody vegetation patches are characterized by perennial shrubs or trees and the open patches are characterized by crusted soil with microphytic communities (Boeken and Shachak, 1994). Shrubs in drylands modulate the flow of resources by modifying the impacts of wind and runoff water, increasing water infiltration and deposition of sediments (Eldridge *et al.*, 2000), trapping organic matter and nutrients (Thompson *et al.*, 2006), and ameliorating microclimatic conditions, which in turn increase animal and microbial activity (Zaady *et al.*, 1996). 'Soil mounds' with loose soil are formed under the shrubs and alter soil properties in the shrub patch. The soil beneath shrubs in drylands tends to be enriched with water, carbon, nitrogen and phosphate (Thompson *et al.*, 2006). As a consequence, shrub patches in drylands are considered 'islands of fertility' creating improved microhabitat conditions for seedling, sub-shrubs, and annuals and other organisms (Boeken, 2008). However, xerohalophytes may increase the soil salinity under the canopy due to litter decomposition (*Zygophyllum* spp.; Sarig and Steinberger, 1994) or excrete excess salt from special salt glands (*Tamarix* spp.; Thomson and Liu, 1967), thus altering soil properties and creating a microhabitat that accommodates specialized species assemblages (Sarig and Steinberger, 1994).

Interactions are not limited to plants; desert plants and animals interact in ways that have strongly influenced their respective evolutionary trajectories (Ward, 2009). It has been argued (Oksanen *et al.*, 1981; Ayal *et al.*, 2005) that herbivores are not a key factor in deserts, due to low levels of primary production. However, herbivory does play an important role in many aspects of the design and function of desert ecosystems. Herbivores remove valuable tissues from plants whose productivity is already limited by the harsh environment, with potential consequences for the plants' reproductive capacity. An estimated 2–10% of leaf and stem biomass is consumed by herbivores in deserts (Mares, 1999). In desert plants two defence mechanisms are common: (i) production of chemical compounds; and (ii) mechanical protection by a thick outer skin, spines or thorns. Herbaceous desert plants commonly contain alkaloids while higher concentrations of terpenes, amines or resins can be found in woody desert plants. The possession of spines or thorns may also act to deter vertebrate herbivores, or at least slow their feeding rate (Young, 1987). Desert plants may also minimize grazing damage via the timing of their growth (mostly ephemeral), or hiding. The curious 'living stones', genus *Lithops*, in the Namib Desert disguise themselves as pebbles to avoid being eaten by herbivores (Fig. 11.9). The most common herbivores are mammals and insects. The desert locust, *Schistocerca gregaria*, was documented as early as in the Bible, and forms large swarms that can consume more than 1000 t day^{-1} of plant matter (Mares, 1999).

However, desert animals do not only have negative impacts on vegetation they can

Fig. 11.9. The curious 'living stones', genus *Lithops*, are succulents that look similar to pebbles, a strategy that helps them to avoid herbivory, Central Namib Desert, south-west Africa. The translucent cover of the leaves allows photosynthesis to take place while allowing excellent camouflage. © Elsita Maria Kiekebusch.

be integral parts of the ecosystem, e.g. by facilitating pollination, seed dispersal or water infiltration rates. Often mutualistic interactions between species can be recognized, e.g. plants provide shade and food whilst the animals disperse their seeds. A classic example of this is the highly complex three-way interaction between bruchid beetles (Bruchidae), *Acacia* spp. and large mammalian herbivores, such as the Dorcas gazelles (*Gazella dorcas*) and ibex (*Capra ibex*). These mammals are both predators and dispersers of *Acacia* seeds; while some seeds are destroyed, others are defecated unharmed. Ingestion by large herbivores facilitates germination by scarification of the seedcoat. While infestation by bruchid beetles reduces *Acacia* germination, herbivores may reduce bruchid infestation: (i) due to their stomach acids; (ii) crushing by the herbivore's teeth; or (iii) by removing seeds prior to (re-)infestation (Or and Ward, 2003).

Another example is the fat sand rat (*Psammomys obesus*; Fig. 11.10), which has developed a digestive mechanism that allows them to feed solely on desert plants of the family Chenopodiaceae, which contain large amounts of salts in the leaves (Palgi *et al.*, 2005).

Apart from their trophic effect, shrubs modulate environmental parameters for animals: woody vegetation in drylands provides shade, a cool refuge during the extreme heat of the day and warmer temperatures during the cool nights. In this manner they change movement behaviour (Shelef and Groner, 2011) and distribution of ground-dwelling insects (Mazia *et al.*, 2006), which in turn affects the behaviour and spatial distribution of their predators. Shrubs may also provide defence from bird predation for small animals (Groner and Ayal, 2001), although, conversely, shrubs may attract predators that seek their prey mainly in shrub patches. However, for the predator, the risk of injury as a result of contact with a potentially well defended, e.g. spiny, shrub is an important trade-off factor

Fig. 11.10. Fat sand rat (*Psammomys obesus*). © Oren Shelef.

(Berger-Tal *et al.*, 2009) tipping the advantage in favour of their small, sheltering prey.

11.4 Concluding Remarks

Hot deserts impose constant and changing stresses on plants, both throughout a daily cycle, from day to night, and on an annual cycle, between seasons. Stress in hot deserts includes mainly abiotic stresses such as high radiation, extreme temperatures, water scarcity, salinity, wind and dust storms. Nevertheless, biotic stress and the biotic interactions discussed also play an important role in hot desert ecology. Plant plasticity can overcome the disadvantages they encounter, especially being sessile organisms, and enables them to adapt and acclimate to the harsh and extreme environment. Similarly, animals possess a wide range of clever mechanisms that allow them to survive and often thrive in hot deserts.

References

Aguiar, M.R. and Sala, O.E. (1999) Patch structure, dynamics and implications for the functioning of arid ecosystems. *Trends in Ecology and Evolution* 14, 273–277.

Akashi, K., Yoshimura, K., Nanasato, Y., Takahara, K., Munekage, Y. and Yokota, A. (2008) Wild plant resources for studying molecular mechanisms of drought/strong light stress tolerance. *Plant Biotechnology Journal* 25, 257–263.

Akram, M. and Chandio, B.A. (1998) Conjunctive use of rainwater and saline groundwater for desertification control in Pakistan through agroforestry and range management. *Journal of Arid Land Studies* 7, 161–164.

Allen, A.P., Pockman, W.T., Restrepo, C. and Milne, B.T. (2008) Allometry, growth and population regulation of the desert shrub *Larrea tridentata*. *Functional Ecology* 22, 197–204.

Anderson, E.F. (2001) *The Cactus Family*. Timber Press, Cambridge, UK, 776 pp.

Arizaga, S. and Ezcurra, E. (2002) Propagation mechanisms in *Agave macroacantha* (Agavaceae), a tropical arid-land succulent rosette. *American Journal of Botany* 89, 632–641.

Armbrust, D.V. (1982) Physiological-responses to wind and sandblast damage by grain *Sorgum* plants. *Agronomy Journal* 74, 133–135.

Avis, A.M. and Lubke, R.A. (1985) The effect of wind-borne sand and salt spray on the growth of *Scirpus nodosus* in a mobile dune system. *South African Journal of Botany* 51, 100–110.

Ayal, Y., Polis, G.A., Lubin, Y., Goldberg, D.E., Shachak, M., Gosz, J.R., Pickett, S.T.A. and Perevolotsky, A. (2005) How can high animal diversity be supported in low-productivity deserts? The role of macrodetritivory and habitat physiognomy. In: Shachak, M., Gosz, G.R. and Pickett, S.T.A. (eds) *Biodiversity in Drylands: Toward a unified framework*. Oxford University Press, New York, pp. 15–29.

Azúa-Bustos, A., González-Silva, C., Mancilla, R., Salas, L., Gómez-Silva, B., McKay, C. and Vicuña, R. (2011) Hypolithic cyanobacteria supported mainly by fog in the coastal range of the Atacama Desert. *Microbial Ecology* 61, 568–581.

Bai, J., Xu, D.H., Kang, H.M., Chen, K. and Wang, G. (2008) Photoprotective function of photorespiration in *Reaumuria soongorica* during different levels of drought stress in natural high irradiance. *Photosynthetica* 46, 232–237.

Barker, D.H., Adams, W.W., Demmig-Adams, B.. Logan, B.A., Verhoeven, A.S. and Smith, S.D. (2002) Nocturnally retained zeaxanthin does not remain engaged in a state primed for energy dissipation during the summer in two Yucca species growing in the Mojave Desert. *Plant Cell and Environment* 25, 95–103.

Bartels, D. (2005) Desiccation tolerance studied in the resurrection plant *Craterostigma plantagineum*. *Integrative and Comparative Biology* 45, 696–701.

Berger-Tal, O., Mukherjee, S., Kotler, B.P. and Brown, J.S. (2009) Look before you leap: is risk of injury a foraging cost? *Behavioural Ecology and Sociobiology* 63, 1821–1827.

Berry, J.A. and Björkman, O. (1980) Photosynthetic response and adaptation to temperature in higher plants. *Annual Review of Plant Physiology* 31, 491–534.

Boeken, B. (2008) The role of seedlings in the dynamics of dryland ecosystems – their response to and involvement in dryland heterogeneity, degradation and restoration. In: Leck, M., Parker, V.T. and Simpson, R. (eds) *Seedling Ecology and Evolution*. Cambridge University Press, Cambridge, UK, pp. 307–330.

Boeken, B. and Shachak, M. (1994) Desert plant-communities in human-made patches – implications for management. *Ecological Applications* 4, 702–716.

Brown, J.H. and Ernest, S.K.M. (2002) Rain and rodents: complex dynamics of desert consumers. *Bioscience* 52, 979–987.

Bruelheide, H., Manegold, M. and Jandt, U. (2004) The genetical structure of *Populus euphratica* and *Alhagi sparsifolia* stands in the Taklimakan desert. In: Runge, M. and Zhang, X. (eds) *Ecophysiology and Habitat Requirements of Perennial Plant Species in the Taklimakan Desert*. Shaker, Aachen, Germany, pp. 153–160.

Caldwell, M.M. (1970) Plant gas exchange at high wind speeds. *Plant Physiology* 46, 535–537.

Caldwell, M.M., Björn, L.O., Bornman, J.F., Flint, S.D., Kulandaivelu, G., Teramura, A.H. and Tevini, M. (1998) Effects of increased solar ultraviolet radiation on terrestrial ecosystems. *Journal of Photochemistry and Photobiology B: Biology* 46, 40–52.

Campos, V.L., Escalante, G., Jañez, J., Zaror, C.A. and Mondaca, A.M. (2009) Isolation of arsenite-oxidizing bacteria from a natural biofilm associated tc volcanic rocks of Atacama Desert, Chile. *Journal of Basic Microbiology* 49, 93–97.

Ceballos, A., Martínez-Fernández, J. and Luengo-Ugidos, M.A. (2004) Analysis of rainfall trends and dry periods on a pluviometric gradient representative of Mediterranean climate in the Duero Basin, Spain. *Journal of Arid Environments* 64, 215–233.

Cheeseman, J.M. (1988) Mechanisms of salinity tolerance in plants. *Plant Physiology* 87, 547–550.

Comstock, J. and Ehleringer, J. (1986) Canopy dynamics and carbon gain in response to soil water availability in *Encelia frutescens* Gray, a drought-deciduous shrub. *Oecologia* 68, 271–278.

Dickinson, H., Moritz, K., Wintour, E.M., Walker, D.W. and Kett, M.M. (2007) A comparative study of renal function in the desert-adapted spiny mouse and the laboratory-adapted C57BL/6 mouse: response to dietary salt load. *American Journal of Physiclogy - Renal Physiology* 293, F1093–F1098.

Ehleringer, J. and Björkman, O. (1977) Quantum yields for CO_2 uptake in C_3 and C_4 plants. *Plant Physiology* 59, 86–90.

Ehleringer, J.R., Björkman, O. and Mooney, H.A. (1976) Leaf pubescence: effects on absorptance and photosynthesis in a desert shrub. *Science* 192, 376–377.

Eldridge, D.J., Zaady, E. and Shachak, M. (2000) Infiltration through three contrasting biological soil crusts in patterned landscapes in the Negev, Israel. *Catena* 40, 323–336.

Eugster, W. (2008) Wind effects. In: Jørgensen, S.E. and Fath, B. (eds) *Encyclopedia of Ecology*. Academic Press, Oxford, pp. 3794–3803.

Eugster, W., Jørgensen, S.E. and Fath, B. (2008) Wind effects. In: Jørgensen, S.E. and Fath, B. (eds) *Encyclopedia of Ecology*. Academic Press, Oxford, UK, pp. 3794–3803.

Evenari, M., Shanan, L. and Tadmor, N. (1971) *The Negev: the challenge of a desert*. Harvard University Press, Cambridge, USA, 357 pp.

Evenari, M., Lange, O.L., Schulze, E.D., Buschbom, U. and Kappen, L. (1975) Adaptive mechanisms in desert plants. In: Vemberg, F.J. (ed.) *Physiological Adaptation to the Environment*. Intext Press, Platteville, USA, pp. 111–129.

Feinbrun-Dothan, N. and Danin, A. (1998) *Analytical Flora of Eretz-Israel*. Carta, Jerusalem, Israel.

Fisher, S.G. and Minckley, W.L. (1978) Chemical characteristics of desert stream in flash flood. *Journal of Arid Environments* 1, 25–33.

Flanagan, L.B. and Ehleringer, J.R. (1991) Stable isotope compositions of stems and leaf water applications to the study of plant water use. *Functional Ecology* 5, 270–277.

Flowers, T.J. and Colmer, T.D. (2008) Salinity tolerance in halophytes. *New Phytologist* 179, 945–963.

Fonteyn, P.J. and Mahall, B.E. (1981) An experimental-analysis of structure in a desert plant community. *Journal of Ecology* 69, 883–896.

Fowler, N. (1986) The role of competition in plant communities in arid and semiarid regions. *Annual Review of Ecology and Systematics* 17, 89–110.

Friedman, J. and Waller, G.R. (1987) Allelopathy in desert ecosystems. In: Waller, G.R. (ed.) *Allelochemicals: role in agriculture and forestry*. American Chemical Society, Washington, DC, pp. 53–68.

Gaff, D.F. (1987) Desiccation tolerant plants in South America. *Oecologia* 74, 133–136.

Gibson, A.C. (1996) *Structure-Function Relations of Warm Desert Plants*. Springer, Heidelberg, Germany, 216 pp.

Gibson, A.C. and Nobel, P.S. (1990) *The Cactus Primer*. Harvard University Press, Boston, Massachusetts, 296 pp.

Gibson, C. (1998) Photosynthetic organs of desert plants. *Bioscience* 48, 911–920.

Gilbert, M.E. and Ripley, B.S. (2010) Resolving the differences in plant burial responses. *Austral Ecology* 35, 53–59.

Goudie, A. and Wilkinson, J. (1977) *The Warm Desert Environment*. Cambridge University Press, London, 95 pp.

Grace, J. (1974) The effect of wind on grasses. I. Cuticular and stomatal transpiration. *Journal of Experimental Botany* 25, 542–551.

Grace, J. (1988) Plant-response to wind. *Agriculture Ecosystems and Environment* 22, 71–88.

Grime, J.P. (1977) Evidence for the existence of three primary strategies in plants and its relevance to ecological and evolutionary theory. *The American Naturalist* 111, 1169–1194.

Groner, E. and Ayal, Y. (2001) The interaction between bird predation and plant cover in determining habitat occupancy of darkling beetles. *Oikos* 93, 22–31.

Guralnick, L.J. and Ting, I.P. (1987) Physiological changes in *Portulacaria afra* (L.) Jacq. during a summer drought and rewatering. *Plant Physiology* 85, 481–486.

Gutterman, Y. (2002a) *Survival Strategies of Annual Desert Plants*. Springer, Berlin, Germany, 368 pp.

Gutterman, Y. (2002b) Survival adaptations and strategies of annuals occurring in the Judean and Negev Deserts of Israel. *Israel Journal of Plant Sciences* 50, 165–175.

Gutterman, Y., Gendler, T. and Rachmilevitch, S. (2010) Survival of *Schismus arabicus* seedlings exposed to desiccation depends on annual periodicity. *Planta* 231, 1475–1482.

Hetem, R.S., de Witt, B.A., Fick, L.G., Fuller, A., Kerley, G.I.H., Meyer, L.C.R., Mitchell, D. and Maloney, S.K. (2009) Body temperature, thermoregulatory behaviour and pelt characteristics of three colour morphs of springbok (*Antidorcas marsupialis*). *Comparative Biochemistry and Physiology, Part A: Molecular and Integrative Physiology* 152, 379–388.

Holmes, M.G. and Keiller, D.R. (2002) Effects of pubescence and waxes on the reflectance of leaves in the ultraviolet and photosynthetic wavebands: a comparison of a range of species. *Plant Cell and Environment* 25, 85–93.

Jaffe, M.F. (1973) Thigmomorphogenesis: the response of plant growth and development to mechanical stimulation. *Planta* 114, 143–157.

Jeffery, S., Gardi, C., Jones, A., Montanarella, L., Marmo, L., Miko, L., Ritz, K., Peres, G., Römbke, J. and van der Putten, W.H. (2010) *European Atlas of Soil Biodiversity*. European Commission, Publications Office of the European Union, Luxembourg, 128 pp.

Kozaki, A. and Tabeka, G. (1996) Photorespiration protects C_3 plants from photooxidation. *Nature* 384, 557–560.

Kurosaki, Y. and Mikami, M. (2007) Threshold wind speed for dust emission in east Asia and its seasonal variations. *Journal of Geophysical Research* 112, D17202.

Laity, J. (2008) *Deserts and Desert Environments*. Wiley-Blackwell, Oxford, UK, 360 pp.

Lambers, H., Pons, T.L. and Chapin III, F.S.C. (2008) Mineral nutrition – nutrient acquisition from 'toxic' or 'extreme' soils. In: Lambers, H., Pons, T.L. and Chapin III, F.S. (eds) *Plant Physiological Ecology*, 2nd edn. Springer, New York, pp. 284–301.

Larcher, W. (2001) *Ökophysiologie der Pflanzen*. Ulmer, Stuttgart, Germany, 408 pp.

Lavee, H., Poesen, J. and Yair, A. (1997) Evidence of high efficiency water harvesting by ancient farmers in the Negev desert, Israel. *Journal of Arid Environments* 35, 341–348.

Levitt, J. (1972) *Responses of Plants to Environmental Stresses*. Academic Press, New York.

Levitt, J. (1980) *Responses of Plants to Environmental Stresses*, Vol. I. *Chilling, freezing and high temperature stresses*. Academic Press, New York, 497 pp.

Liu, G., Zhang, X., Li, X., Wei, J. and Shan, L. (2008) Adaptive growth of *Tamarix taklamakanensis* root systems in response to wind action. *Chinese Science Bulletin* 53, 164–168.

Lujan, R., Lledias, F., Martinez, L.M., Barreto, R., Cassab, G.I. and Nieto-Sotelo, J. (2009) Small heat-shock proteins and leaf cooling capacity account for the unusual heat tolerance of the central spike leaves in *Agave tequilana* var. Weber. *Plant Cell and Environment* 32, 1791–1803.

Mares, M.A. (1999) *Encyclopedia of Deserts*. University of Oklahoma Press, Oklahoma, 654 pp.

Mazia, C.N., Chaneton, E.J. and Kitzberger, T. (2006) Small-scale habitat use and assemblage structure of ground-dwelling beetles in a Patagonian shrub steppe. *Journal of Arid Environments* 67, 177–194.

Meng, Z. and Zhang, Q. (2006) Oxidative damage of dust storm fine particles instillation on lungs, hearts and livers of rats. *Environmental Toxicology and Pharmacology* 22, 277–282.

Middleton, N. and Thomas, D. (1997) *World Atlas of Desertification*. Arnold, London, 192 pp.

Miller, G.T. and Spoolman, S.E. (2008) *Essentials of Ecology*. Brooks/Cole, Pacific Grove, USA, 544 pp.

Mittler, R., Merquiol, E., Hallak-Herr, E., Rachmilevitch, S., Kaplan, A. and Cohen, M. (2001) Living under a 'dormant' canopy: a molecular acclimation mechanism of the desert plant *Retama raetam*. *Plant Journal* 25, 407–416.

Müller, P., Li, X.P. and Niyogi, K.K. (2001) Non-photochemical quenching. A response to excess light energy. *Plant Physiology* 125, 1558.

Niklas, K.J. (1992) Petiole mechanics, light interception by lamina, and 'Economy in design'. *Oecologia* 90, 518–526.

Nilsen, E.T. and Orcutt, D.M. (2000) *The Physiology of Plants Under Stress. Abiotic factors*. John Wiley, New Jersey, 704 pp.

Nobel, P.S., Forseth, I.N., Long, S.P., Hall, D.O., Scurlock, J.M.O., Bolhár-Nordenkampf, H.R., Leegood, R.C. and Long, S.P. (1993) Canopy structure and light interception. In: Hall, D.O., Scurlock, J.M.O., Long, S.P., Leegood, R.C. and Bolhár-Nordenkampf, H.R. (eds) *Photosynthesis and Production in a Changing Environment. A field and laboratory manual*. Chapman and Hall, London, pp. 79–90.

Noy-Meir, I. (1973) Desert ecosystems: environment and producers. *Annual Review of Ecological Systems* 4, 25–51.

Oksanen, L., Fretwell, S.D., Arruda, J. and Niemela, P. (1981) Exploitation ecosystems in gradients of primary productivity. *American Naturalist* 118, 240–261.

Or, K. and Ward, D. (2003) Three-way interactions between *Acacia*, large mammalian herbivores and bruchid beetles – a review. *African Journal of Ecology* 41, 257–265.

Palgi, N., Vatnick, I. and Pinshow, B. (2005) Oxalate, calcium and ash intake and excretion balances in fat sand rats (*Psammomys obesus*) feeding on two different diets. *Comparative Biochemistry and Physiology A: Molecular and Integrative Physiology* 141, 48–53.

Parida, A.K. and Das, A.B. (2005) Salt tolerance and salinity effects on plants: a review. *Ecotoxicology and Environmental Safety* 60, 324–349.

Parsons, A.J. and Abrahams, A.D. (2009) *Geomorphology of Desert Environments*. Springer, New York, 834 pp.

Peel, M.C., Finlayson, B.L. and McMahon, T.A. (2007) Updated world map of the Koppen-Geiger climate classification. *Hydrology and Earth System Sciences* 11, 1633–1644.

Phillips, S.J. and Comus, P.W. (2000) *A Natural History of the Sonoran Desert*. Arizona-Sonora Desert Museum Press, Tucson, Arizona, 628 pp.

Phillips, W.S. (1963) Depth of roots in soil. *Ecology* 44, 424.

Pohronezny, K., Hewitt, M., Infante, J. and Datnoff, L. (1992) Wind and wind-generated sand injury as factors in infection of pepper by *Xanthomonas campestris* pv. *vesicatoria*. *Plant Disease* 76, 1036–1039.

Rosenthal, D.M., Stiller, V., Sperry, J.S. and Donovan, L.A. (2010) Contrasting drought tolerance strategies in two desert annuals of hybrid origin. *Journal of Experimental Botany* 61, 2769–2778.

Sage, R.F. (2004) The evolution of C-4 photosynthesis. *New Phytologist* 161, 341–370.

Sarig, S. and Steinberger, Y. (1994) Microbial biomass response to seasonal fluctuation in soil-salinity under the canopy of desert halophytes. *Soil Biology and Biochemistry* 26, 1405–1408.

Schmidt-Nielsen, K., Schmidt-Nielsen, B., Jarnum, S.A. and Houpt, T.R. (1956) Body temperature of the camel and its relation to water economy. *American Journal of Physiology* 188, 103–112.

Scott, P. (2000) Resurrection plants and the secrets of eternal leaf. *Annals of Botany* 85, 159–166.

Shelef, O. and Groner, E. (2011) Linking landscape and species: effect of shrubs on patch preference and movement patterns of beetles in arid and semiarid ecosystems. *Journal of Arid Environments,* doi:10.1016/j.jaridenv.2011.04.016.

Shelef, O., Lazarovitch, N., Rewald, B., Golan-Goldhirsh, A. and Rachmilevitch, S. (2010) Root halotropism: salinity effects on *Bassia indica* root. *Plant Biosystems* 144, 471–478.

Sherwin, H.W., Pammenter, N.W., February, E., van der Willigen, C. and Farrant, J.M. (1998) Xylem hydraulic characteristics, water relations and wood anatomy of the resurrection plant *Myrothamnus flabellifolius* Welw. *Annals of Botany* 81, 567–575.

Skidmore, E.L. (1989) Wind erosion in deserts: surface susceptibility and climatic erosivity. *Journal of Agriculture Rijks University*, Ghent, pp. 327–336.

Smith, S.D., Monson, R.K. and Anderson, J.E. (1997) *Physiological Ecology of North American Desert Plants.* Springer, New York, 286 pp.

Smith, W.K. (1978) Temperatures of desert plants – another perspective on adaptability of leaf size. *Science* 201, 614–616.

Sperry, J.S. and Hacke, U.G. (2002) Desert shrub water relations with respect to soil characteristics and plant functional type. *Functional Ecology* 16, 367–378.

Taiz, L. and Zeiger, E. (1998) Water balance of the plant. In: Taiz, L. and Zeiger, E. (eds) *Plant Physiology,* 2nd edn. Sinauer Associates, Sunderland, Massachusetts, pp. 81–101.

Thomas, F.M., Foetzki, A., Gries, D., Bruelheide, H., Li, X.Y., Zeng, F.J. and Zhang, X.M. (2008) Regulation of the water status in three co-occurring phreatophytes at the southern fringe of the Taklamakan Desert. *Journal of Plant Ecology* 1, 227–235.

Thompson, J.R. (1974) The effect of wind on grasses. II. Mechanical damage in *Festuca arundinacea* Schreb. *Journal of Experimental Botany* 25, 965–972.

Thompson, T.L., Zaady, E., Huancheng, P., Wilson, T.B. and Martens, D.A. (2006) Soil C and N pools in patchy shrublands of the Negev and Chihuahuan Deserts. *Soil Biology and Biochemistry* 38, 1943–1955.

Thomson, W.W. and Liu, L.L. (1967) Ultrastructural features of salt gland of *Tamarix aphylla* L. *Planta* 73, 201–220.

Tracy, C.R., Reynolds, S.J., McArthur, L., Tracy, C.R. and Christian, K.A. (2007) Ecology of aestivation in a cocoon-forming frog, *Cyclorana australis* (Hylidae). *Copeia* 2007, 901–912.

Vaughan, T.A., Ryan, J.M. and Czaplewski, N.J. (2010) Aspects of physiology. In: Vaughan, T.A., Ryan, J.M. and Czaplewski, N.J. (eds) *Mammalogy*. Jones and Bartlett Learning, Sudbury, USA, pp. 421–466.

Veste, M. (2008) Temporal and spatial variability of plant water status and leaf gas exchange. In: Breckle, S.W., Yair, A. and Veste, M. (eds) *Arid Dune Ecosystems*. Springer, Berlin, pp. 367–375.

Waisel, Y. (1972) *Biology of Halophytes*. Academic Press, New York, 395 pp.

Walter, H. (1962) *Einführung in die Phytologie. III. Grundlagen der Pflanzenverbreitung. 1.Teil: Standortslehre.* Thieme, Stuttgart, Germany.

Ward, D. (2009) *The Biology of Deserts*. Oxford University Press, New York, 304 pp.

Warner, T.T. (2004) *Desert Meteorology*. Cambridge University Press, Cambridge, UK, 612 pp.

West, N.E. and Skujins, J.J. (1978) *Nitrogen in Desert Ecosystems*. Dowden, Hutchinson & Ross, Stroudsburg, USA.

Whitford, W.G. (2002) *Ecology of Desert Systems*. Academic Press, London, 343 pp.

Willmer, P., Stone, G. and Johnston, I.A. (2004) *Environmental Physiology of Animals*, 2nd edn. Wiley-Blackwell, New York.

Wilms, T.M., Wagner, P., Shobrak, M., Rödder, D. and Böhme, W. (2011) Living on the edge? On the thermobiology and activity pattern of the large herbivorous desert lizard *Uromastyx aegyptia microlepis* Blanford, 1875 at Mahazat as-Sayd Protected Area, Saudi Arabia. *Journal of Arid Environments* 75, 636–647.

Woodward, S.L. (2003) *Biomes of Earth: terrestrial, aquatic, and human-dominated.* Greenwood Press, Westport, Connecticut, 456 pp.

Young, T.P. (1987) Increased thorn length in *Acacia depranolobium* – an induced response to browsing. *Oecologia* 71, 436–438.

Yura, H. and Ogura, A. (2006) Sandblasting as a possible factor controlling the distribution of plants on a coastal dune system. *Plant Ecology* 185, 199–208.

Zaady, E., Groffman, P.M. and Shachak, M. (1996) Litter as a regulator of N and C dynamics in macrophytic patches in Negev desert soils. *Soil Biology and Biochemistry* 28, 39–46.

Zangvil, A. (1996) Six years of dew observations in the Negev Desert, Israel. *Journal of Arid Environments* 32, 361–371.

12 Terrestrial Hydrothermal Environments

Don Cowan, Marla Tuffin, Inonge Mulako and James Cass
Institute for Microbial Biotechnology and Metagenomics, University of the Western Cape, Cape Town, South Africa

12.1 Introduction to Terrestrial Hydrothermal Environments

Hydrothermal sites are surprisingly common and widely distributed across the terrestrial surfaces of planet Earth. For one of the most comprehensive distribution surveys of global terrestrial hydrothermal resources, see Waring (1965). Such sites are variously known by different names (Table 12.1), most of which broadly reflect the physical appearance of the hydrothermal locality.

Hydrothermal sites have multiple geological origins, but may be broadly divided into those which are associated with volcanic areas, and those that are not.

Higher temperature hydrothermal environments are usually associated with volcanic features (calderas, fault lines, tectonic plate boundaries, back-arc basins), which allow magma to penetrate upward to a depth where it can interact directly with deep groundwater. At such sites, where the depth of the magma chamber may be as shallow as 2–5 km below ground level, the thermal gradient may be as high as 150–200°C km^{-1} (Sabadell and Axtmann, 1975). The groundwater is super-heated (for example, to approx. 350°C at 300 m depth) and forced back to the surface at high temperature and under considerable pressure. In some instances this pressure is retained to the surface, resulting in the intermittent ejection of superheated water (100–105°C) in the form of geysers.

In some cases hot pools support a high load of viscous mud and/or clay in the form of a suspended slurry. These 'mud pools' are typically anaerobic, often near boiling point, and many show the effects of outgassing (Fig. 12.1a).

The heated water streams may contain substantial concentrations of carbon dioxide (CO_2), nitrogen (N_2), hydrogen (H_2) and hydrogen sulfide (H_2S), and the buffering of the weak acids CO_2 ($pK_a = 6.3$) and H_2S ($pK_a = 7.3$) gives a near neutral pH. The high temperature of the source water accelerates erosion and hot spring waters can carry high concentrations of silicates. The latter readily precipitate on cooling giving the thermal fields the silicate 'terrace' structures (Fig. 12.1b) characteristic of many high temperature hydrothermal fields. High temperature subterranean erosion processes can also result in high levels of soluble metals and metalloids (Table 12.2), and both the bulk water and the resulting precipitates (Fig. 12.1c) can far exceed human health limits. The implications of these concentrations for the physiology, survival and resistance strategies of microbial populations are issues of considerable relevance.

Table 12.1. Terminology for terrestrial hydrothermal sites.

Name	Origin	Example location	Typical characteristics
Geothermal area	A generic term for a region of geological instability (caldera, fault line etc.) containing hydrothermal features	Yellowstone National Park, USA	Multiple hydrothermal sites
Hot pool	'A pool of water which is appreciably hotter than the surroundings'	Numerous (e.g. Yellowstone National Park, USA; Whakarewarewa, NZ)	Aquatic, wide range of temperatures and pH values
Hot spring	'A spring which brings water to the surface which is hotter than the surroundings'	Numerous (e.g. Yellowstone National Park, USA; Whakarewarewa, NZ)	Aquatic; wide range of temperatures and pH values; sometimes with substantial outflow
Solfatara/ solfatare	Italian: 'sulfur earth'	Solfatara, Pozzuoli, Italy	Steam-heated earth, low water flow, S-rich, acidic
Souffrière	French: 'sulfur outlet'	Montserrat, other Caribbean islands	Steam-heated earth, low water flow, S-rich, acidic
Geyser	Icelandic *geysa*: 'to gush'	Geysir, Iceland	Intermittent discharge of water and steam, often to considerable heights
Fumarole	Latin: 'smoke'	Most volcanoes	Steam (with CO_2, H_2S, SO_2)-heated soil, little or no liquid flow; often acidic
Mud pool, mud pot, paint pot	–	Numerous (e.g. Yellowstone National Park, USA; Wai-o-tapu, NZ)	Viscous mud/clay slurry, often near boiling point
Mud 'volcano'	–	Numerous (e.g. Yellowstone National Park, USA; Wai-o-tapu, NZ)	As above, but with significant gas evolution, ejecting mud vertically to distances of up to 1–2 m

12.2 Physical and Chemical Characteristics of Terrestrial Hydrothermal Environments

Low temperature hydrothermal sites, which are typical of geologically old and or stable zones (such as the UK and much of South Africa), exist where surface water penetrates to sufficient depth to be heated as a result of the normal crustal thermal gradient (30–40°C km^{-1} depth; Sabadell and Axtmann, 1975), and is forced to the surface by hydrothermal expansion. Such hot (or warm) pools and springs are frequently near neutrality and in the temperature range of 40°C to 60°C. The famous Bath City spa, UK, is one such example.

In rare instances, large subterranean water bodies may be heated, as in the case of the Australian Great Artesian Basin. The temperature of this water body, which is estimated to extend to a depth of up to 3 km and over a surface area of 1.7 million km^2, ranges from 30°C to 100°C.

The pH values of hydrothermal features range from less than 0 to above 10. However, thermal waters may be conveniently divided into three dominant classes: acidic (pH 1–5), neutral (pH 6–7.5) and alkaline (>pH 7.5) (Pentacost, 1996). The pH distribution in the acid–neutral range is largely the product of two buffering systems: carbonate/bicarbonate (pK_a, 6.3) and sulfuric acid/bisulfate (pK_2, 1.92). Neutral waters tend to be high in the alkaline earth metals (e.g. calcium) and may deposit large amounts of travertine ($CaCO_3$), as seen in Mammoth Hot Springs, Yellowstone National Park, USA (Fig. 12.1d). Alkaline hot springs are stabilized by silicate ($pK_n = 9.7$) and carbonate

Fig. 12.1. Thermal sites across the pH range. (a) Bubbling mud pool at Wai-o-tapu thermal field, Rotorua, New Zealand, © Don A. Cowan. (b) Silica-rich hot pool, Tokaanu, New Zealand, © Don A. Cowan. (c) Arsenic-rich deposits in Champagne Pool, Wai-o-tapu, New Zealand, © Don A. Cowan. (d) Travertine deposits at Mammoth Hot Springs, Yellowstone National Park, USA, © David Monniaux. (e) Sinter deposits at Whakarewarewa thermal field, Rotorua, New Zealand, © Don A. Cowan. (f) Steam-heated acid soils and pools at Wai-o-tapu thermal field, New Zealand, © Don A. Cowan. (g) Crystalline sulfur deposits around a steam vent at Wai-o-tapu thermal field, New Zealand, © Don A. Cowan.

(f)

(g)

Fig. 12.1. Continued.

Table 12.2. Metals and metalloid concentrations in New Zealand hot pool water precipitates. Data from Weissberg, 1969.

	Elemental concentration (ppm)	
Element	Champagne Pool, Wai-o-tapu	Ohaki Pool, Broadlands
Au	80	85
Ag	175	500
As	2%	400
Sb	2%	~10%
Hg	170	2000
Tl	320	630
Pb	15	25
Zn	50	50

($pK_2 = 10.25$; Kristjansson and Stetter, 1991) and often deposit extensive regions of sinter (Fig. 12.1e).

Lower pH hydrothermal sites are the product of more complex, near surface processes and typically exhibit low liquid flow rates. Acidification is the result of chemical and biological oxidation of H_2S to sulfuric acid ($pK_a = 1.92$) in shallow groundwater and biological oxidation of S by thermophilic aerobic chemoautotrophs such as *Sulfolobus* (Mosser *et al.*, 1973). The acidity of the hydrothermal water stream reflects the degree of acid groundwater input to neutral or alkaline hydrothermal water streams. These complex mixing processes lead to surface pools and springs across the full pH spectrum but with a bimodal distribution reflecting the two principal buffering systems (Brock, 1971). The complexities of flow paths and mixing processes also account for the presence of pools of widely differing pH values sometimes only metres apart.

Where hydrothermal fields are dominated by steam emission rather than water flow (as in Solfatara, Naples), such sites are characterized by hot, acidic, sulfur-rich soils (Fig. 12.1f) or even crystallized sulfur deposits (Fig. 12.1g). Such soils frequently exhibit a layered structure (Stetter, 1998) – the upper layer is largely aerobic and often brick-red in colour due to the presence of ferric iron (Fe^{3+}). The lower horizon is typically blue-black due to the presence of ferrous iron (Fe^{2+}), nearer neutrality and strongly reduced.

The chemistry of different thermal water sources is highly variable, being dependent on local geology, water flow rate and temperature, together with factors such as mixing, degassing, oxidation and precipitation. These processes are overlaid by biological redox and uptake processes. In addition, water chemistries vary on both short and long temporal scales, as the result of seasonal precipitation patterns and long term changes in tectonic activity (Nordstrom *et al.*, 2005). Despite the high level of variability, virtually all thermal water streams are characterized by low nutrient levels (particularly available nitrogen and DOC) and can be considered to be essentially oligotrophic environments for heterotrophic microorganisms.

12.3 The Diversity of Terrestrial Hydrothermal Environments

Despite their widespread occurrence across the terrestrial world, not all hydrothermal sites have received equivalent attention from microbiologists. Studies of microbial ecology and diversity have tended to reflect both the ease of physical access to sites and the proximity of major research groups. Accordingly, the hydrothermal features of Yellowstone National Park, being both readily accessible and the focus of a number of major US research teams (initiated by Thomas Brock in the late 1960s), have attracted more attention than those of any other thermal area. Table 12.3 lists the most extensively studied terrestrial hydrothermal fields and some of the key researchers who have made substantial contributions to understanding their microbial ecology.

Organisms belonging to the all three domains (Eukaryal, Bacterial and Archaeal) have been identified in terrestrial hydrothermal environments. The diversity of these organisms is primarily determined by temperature (Kristjansson and Hreggvidsson, 1995). Organisms that grow optimally above 50–60°C are termed thermophiles (Brock, 1978; Kristjansson and Hreggvidsson, 1995), which can be further divided into three broad categories; moderate thermophiles, extreme thermophiles and hyperthermophiles. Moderate thermophiles have an optimum growth temperature of 50–60°C (Lebedinsky *et al.*, 2007). Extreme and hyperthermophilic organisms grow optimally at 70°C and 80°C, respectively (Reysenbach and Shock, 2002; Lebedinsky *et al.*, 2007). Although temperature is the defining parameter of this extreme environment, other variables (pH, organic matter and chemical composition) vary significantly in terrestrial hydrothermal environments. For instance, pH values can vary from pH 1 to pH 9 (Blank *et al.*, 2002; Purcell *et al.*, 2006; Mathur *et al.*, 2007; Costa *et al.*, 2009; Stout *et al.*, 2009). These factors have a significant impact on the diversity of microorganisms, and microorganisms inhabiting terrestrial hydrothermal environments, therefore, survive in the presence of several extreme conditions. These microorganisms are known as polyextremophiles.

12.3.1 Microbiology of terrestrial thermal habitats: neutral and alkaline habitats

Bacterial diversity in neutral and alkaline environments

Extensive investigations of neutral and alkaline environments have shown that bacterial diversity varies with both temperature and pH (Kristjansson and Hreggvidsson, 1995). While a diverse range of bacterial species inhabits moderately thermophilic environments, photoautotrophs can come to dominate the microbial community and form macroscopic mat structures. These photosynthetic mats are present on the surface of most neutral and alkaline (>pH 6.0) hot springs at temperatures of 40–71°C with cyanobacteria established as the dominant primary producers (Papke *et al.*, 2003; Van der Meer *et al.*, 2005; Allewalt *et al.*, 2006; Ward *et al.*, 2006; Portillo *et al.*, 2009). At temperatures

Table 12.3. Key sites and researchers in the evolution of the microbiology of terrestrial hydrothermal systems.

Hydrothermal field	Key researchers	Current affiliations	Representative references
Yellowstone National Park, USA	Thomas Brock	University of Wisconsin, WI, USA	Brock, 1978
	Susan Barns	Los Alamos National Laboratory, NM, USA	Barns *et al.*, 1994
	Norman Pace	University of Colorado, CO, USA	
	Anna-Louise Reysenbach	Portland State University, OR, USA	Reysenbach *et al.*, 1994
Pisciarella solfatara, Naples, Italy	Mario De Rosa	ICMIB, CNR, Naples, IT	De Rosa *et al.*, 1975
	Agata Gambacorta	ICMIB, CNR, Naples, IT	
	Mosè Rossi	IPBE, CNR, Naples, IT	
Rotorua and other NZ thermal sites	Hugh Morgan, Roy Daniel	University of Waikato, Hamilton, NZ	Niederberger *et al.*, 2007
	Bharat Patel	Griffith University, Brisbane, AU	
Great Artesian Basin, Australia	Bharat Patel	Griffith University, Brisbane, AU	Ogg and Patel, 2010
Krisuvik, Hveragerdi-Hengill, Biskupstungur-Geysir, Krafla and other Icelandic thermal sites	Jacob Kristjansson	Matis-Prokaria, Reykjavik, Iceland	Kristjansson, 1991
	Karl Stetter,	University of Regensberg, DE and University of California, CA, USA	Huber and Stetter, 2001
	Robert Huber	University of Regensberg, DE	
Uzon Valley and other sites, Kamchatka, Russia	Elizaveta Bonch-Osmolovskaya	Winogradsky Institute of Microbiology, Moscow, Russia	Lebedinski *et al.*, 2007
	Anna-Louise Reysenbach	Portland State University, OR, USA	
	Juergen Wiegel	University of Georgia, GA, USA	Wiegel and Adams, 1998

above 55°C, the photosynthetic mats are predominantly inhabited by unicellular cyanobacteria such as *Synechococcus* sp. (Papke *et al.*, 2003; Ward, 2006). The distribution of *Synechococcus* is influenced by geography, with species from the A/B lineage only found in hot springs in North America, while C lineage species are more widely distributed in North America, New Zealand and Japan. Their distribution is additionally defined by temperature, where A/B and C lineages are found at temperatures of 55–71°C and 55–63°C, respectively. *Synechococcus* are absent in thermophilic mats in Alaska, the Azores and Iceland (Papke *et al.*, 2003), and other cyanobacteria dominate in these.

Photosynthetic microbial mats can also become predominately inhabited by bacteria from the genera *Chloroflexus* and *Roseiflexus*, both members of the order Chloroflexales. When associated with cyanobacteria, *Chloroflexus* and *Roseiflexus* species grow optimally under photoheterotrophic conditions (Hanada, 2003; Broomer *et al.*, 2009; Portillo *et al.*, 2009). In addition, a diverse group of bacterial species associated with photosynthetic mats have been shown to contain both aerobic and anaerobic heterotrophic organisms, and include members of the Planctomycetes, Candidate Division OP10, Bacteroides, Proteobacteria, Fermicutes and Thermus-Deinococcus phyla (Ward, 1998; Portillo *et al.*, 2009). Microbial mats in moderately thermophilic (50–70°C) hot springs which lack cyanobacteria persist and are formed predominantly by *Chloroflexus* (Giovannoni *et al.*, 1987; Skirnisdottir *et al.*, 2000).

In hyperthermophilic (72–98°C) neutral and alkaline hot springs photosynthesis is absent allowing chemolithoautotrophic bacteria to predominate (Blank *et al.*, 2002; Spear *et al.*, 2005). Chemolithoautotrophic organisms belonging to the phylum Aquificales are ubiquitous to all continents (Reysenbach *et al.*, 2000, 2009; Blank *et al.*, 2002; Eder and Huber, 2002). These micorganisms can oxidize hydrogen or molecular sulfur with oxygen as an electron acceptor and fix carbon via the reductive tricarboxylic acid (rTCA) pathway (Reysenbach *et al.*, 2009). Furthermore, these chemolithoautotrophic organisms are believed to be primary producers at temperatures above 70°C. As with cyanobacteria, species of the Aquificales phyla vary with geographic location. Organisms from the genus *Thermocrinus*, particularly *Thermocrinis ruber*-like phylotypes, are only found in hot springs in northern America (Blank *et al.*, 2002; Purcell *et al.*, 2006) while *Hydrogenobacter* inhabit hot springs in Thailand, America, Japan and Russia (Purcell *et al.*, 2006).

Within the temperature range of 70–89°C the diversity of thermophilic bacteria extends to its greater reaches. Members of the Fermicutes (*Bacillus* sp.) and Proteobacteria phyla are dominant at pH values ranging from 6.5 to 9.0 (Blank *et al.*, 2002; Purcell *et al.*, 2006; Costa *et al.*, 2009), but members of the Chloroflexi, Bacteriodetes (*Rhodothermus marinus*), Thermus-Deinococcus (*Thermus thermophilus, Deinococcus aquaticus*) and Thermotogae (*Thermotoga hypogea, Thermotoga petrophila*) have also been identified (Blank *et al.*, 2002; Purcell *et al.*, 2006; Costa *et al.*, 2009). Once temperatures exceed this boundary, bacterial diversity decreases with increasing temperatures. Above 90°C Aquificales and Thermotogales are the only members of the Bacteria domain identified (Blank *et al.*, 2002).

Archaeal diversity in neutral and alkaline habitats

The majority of cultured and uncultured Archaea identified in neutral and alkaline thermal environments belong to the Crenarchaeota phyla (Kvist *et al.*, 2007; Perevalova *et al.*, 2008). Metagenomic analysis has highlighted a diverse group of unknown species within the Crenarchaeota phyla that are present under moderate thermophilic conditions (55–72°C). Species belonging to the Crenarchaeota also dominate at temperatures above 70°C with uncultured species belonging to the group pSL12, which have been identified in various geographic terrestrial hydrothermal environments (Perevalova *et al.*, 2008; Song *et al.*, 2010). Species belonging to the order Thermoproteales (*Thermofilum pendens*) and Desulfurococcales (*Thermosphaera aggregans, Stetteria hydrogenophila*) have been commonly identified in neutral hot pools between 73 and 77°C (Costa *et al.*, 2009). In addition, an autotrophic ammonia-oxidizer, '*Nitrosocaldus yellowstonii*', has been isolated from neutral thermal environments (De la Torre *et al.*, 2008; Costa *et al.*, 2009; Erguder *et al.*, 2009). Although isolated at temperatures between 73 and 77°C, '*Nitrosocaldus yellowstonii*' is only capable of oxidizing ammonia aerobically below 74°C (De la Torre *et al.*, 2008). These organisms are thought to play an important role in nitrogen cycling in extremophilic habitats and their distribution in terrestrial hot springs has greatly extended the upper temperature limit of nitrification.

At temperatures above 80°C, anaerobes within the Crenarchaeota predominate. Hyperthermophilic Thermoproteales species distantly related to *Thermoproteus neutrophilus, Vulcanisaeta distributa* and *Thermofilum pendens* have been identified (Kvist *et al.*, 2007; Costa *et al.*, 2009). These strict anaerobes reduce elemental sulfur using hydrogen as an electron donor to generate energy (Stetter *et al.*, 1990). In addition, autotrophs related to *Aeropyrum pernix,*

Desulfurococcus mobilis, *Ignisphaera aggregans* (order Desulfurococcales) have been identified, which oxidize hydrogen and use nitrate and nitrite as electron acceptors (Barns *et al.*, 1994; Reysenbach *et al.*, 2000; Kvist *et al.*, 2007; Costa *et al.*, 2009). Other uncultured species belonging to the group pJP41 are dominant in both neutral and alkaline hot springs above 80°C (Perevalova *et al.*, 2008; Song *et al.*, 2010), but have also been reported in more moderate thermophilic environments in the range of 60 and 70°C (Barns *et al.*, 1994; Perevalova *et al.*, 2008).

Few members of the phylum Euryarchaeota have been identified in environments above 70°C. Species closely related to *Archaeoglobus fulgidus* were identified in the US Great Basin in low abundance at temperatures of 77°C (Costa *et al.*, 2009), and an uncultured phylotype distantly related to the Halobacteriales has been identified in Tunisian geothermal springs with temperature and pH values of 71°C and 6.5, respectively (Sayeh *et al.*, 2010).

The Korarchaeota are restricted to hyperthermophilic sites and are possibly the least widespread of the archaeal phyla. Korarchaeal phylotypes have been reported in two hydrothermal sites in Yellowstone National Park and the Great Basin (Reysenbach *et al.*, 2000; Costa *et al.*, 2009), but appear to be absent from many high temperature aquatic habitats. Analysis of the genome of the sole Korarchaeal isolate suggests a very deep branching ancestry (Elkins *et al.*, 2008), consistent with the classification of the Korarchaea as an independent phylum.

Eukaryote diversity in neutral and alkaline environments

The diversity and abundance of eukaryotes dramatically decreases at temperatures above 50°C compared to prokaryotes. Under such extreme temperature conditions only unicellular eukaryotic organisms have been reported. Nevertheless, 102 thermophilic fungal strains have been identified in Tengchong Rehai National Park, China (conditions at the sampling sites varied between 47 and 71°C, and pH from 7.5 to 9.0) (Pan *et al.*, 2010). *Thermomyces lanuginosus* and *Scytalidium thermophilum* dominated the eukaryotic community, representing 34.8% and 28.3% of isolates, respectively. Previously, *T. lanuginosus* was typically associated with composting heaps within the temperature range of 50–60°C (Singh *et al.*, 2003). Thermophilic protists have also been described. In Agnano Terme, Italy, and Yellowstone National Park, a thermophilic amoeba, *Echinamoeba thermarum*, was identified at temperatures between 50°C and 60°C (Baumgartner *et al.*, 2003). Other moderately thermophilic amoebae include *Marinamoeba thermophilia* (De Jonckheere *et al.*, 2009) and *Oramoeba fumarolia* (De Jonckheere *et al.*, 2010) with optimal growth temperatures of 50°C and 54°C, respectively. However, despite eukaryotic organisms inhabiting hydrothermal environments under moderately thermophilic conditions, none have been observed at temperatures of above 60°C.

12.3.2 Microbiology of terrestrial thermal habitats: acidic habitats

Bacterial diversity in thermacidophilic habitats

The variations in local conditions within a single terrestrial geothermal active hot spring are limited compared to that of deep-sea hydrothermal vents (Prokofeva *et al.*, 2005; Lebedinsky *et al.*, 2007; see also Lutz, Chapter 13, this volume). Such relatively stable conditions may allow for the establishment of static microbial communities which show little variation with time (Wilson *et al.*, 2008). While temperature may be the defining abiotic factor used to differentiate the particular habitats and organisms present, the local water/sediment chemistry is also a major factor influencing microbial community structure (Mathur *et al.*, 2007; Inskeep *et al.*, 2010).

Due to the relatively constant conditions of terrestrial thermoacidophilic environments, distinct phyla may become dominant members of the microbial community (Prokofeva *et al.*, 2005). In two soil types from the Yellowstone National Park with pH values in the range 2.8–3.8, over 75% of clones belonged to heterotrophic acidophilic bacteria, including *Acidiosphaera* and *Acidiphilium* species, members of the α-Proteobacteria (Hamamura *et al.*, 2005). In samples from a water column of geothermal spring water from the same region but with a temperature range of 60–70°C and pH values of 1–1.5, *Hydrogenobaculum* spp. was the predominant bacteria (Mathur *et al.*, 2007). *Hydrogenobaculum*-like organisms have also been shown to be dominant in geothermal pools (70–80°C, pH 5.5) in St Lucia, Lesser Antilles, where they accounted for between 90 and 100% of the clones identified in a 16S rRNA sequence survey (Stout *et al.*, 2009). Many other geographically distant sites (with possibly different chemistries) show the presence of a dominant organism, which include *Thermoanaerobacter* spp. in Kamchatka, Russia, and *Moorella* spp. in Montserrat (Burton and Norris, 2000; Prokofeva *et al.*, 2005; Mathur *et al.*, 2007). Despite the dominance at the individual sites, a highly diverse structure at the genus and species level is retained in the microbial communities (Hamamura *et al.*, 2005; Lebedinsky *et al.*, 2007; Mathur *et al.*, 2007; Wilson *et al.*, 2008; Stout *et al.*, 2009).

The phylogenetic diversity of thermoacidophilic bacteria is matched by the diverse metabolic pathways utilized by microorganisms in these extreme environments (Stetter *et al.*, 1990; Kaksonen *et al.*, 2006; Boyd *et al.*, 2009). For example, a wide range of thermoacidophilic sulfate reducers have been described, including the Gram-positive *Desulfotomaculum* and *Thermodesulfobium* species (Campbell and Postgate, 1965; Mori *et al.*, 2003) and the Gram-negative bacteria *Thermodesulfator* spp. (Mori *et al.*, 2003). Phylogenetic analysis of sulfate reducers in a geothermally active underground mine in Japan identified ten thermophilic sulfate reducers showing 16S rRNA gene sequence similarities in the range of 89–99%, potentially representing eight different bacterial genera (Kaksonen *et al.*, 2006). A novel genus and species, *Desulfovirgula thermocuniculi*, was also isolated from the same site (Kaksonen *et al.*, 2007). Thermophilic sulfate-reducers have attracted considerable biotechnological interest due to their potential use in biodesulfurization and biohydrometallurgical processes (Kaksonen *et al.*, 2006, 2007).

Iron oxidizers and reducers are also important components of thermophilic microbial communities (Stetter *et al.*, 1990). Iron-oxidizing *Sulfobacillus* spp. from the island of Montserrat, with maximum growth temperatures of 65°C, have been described (Atkinson *et al.*, 2000) and mixotrophic *Sulfobacillus disulfidooxidans* has been isolated at higher temperatures (75°C) (Mathur *et al.*, 2007). Many novel iron oxidizers/reducers observed in the temperature range of 60–75°C have been described as Gram-positive acidophiles but remain to be fully characterized (Mathur *et al.*, 2007). As the upper temperature limit for photosynthetic activity in acidic geothermal environments is only 56°C, these primary producers are essential members of the microbial community (Toplin *et al.*, 2008). However, CO_2 fixation has also been shown to occur at higher temperatures and lower pH values, outside the growth range of iron-oxidizing autotrophs (Boyd *et al.*, 2009). Surprisingly, CO_2 uptake was an order of magnitude higher in a pH 3.1 microbial mat community dominated by *Hydrogenobaculum* spp. compared to neutral and alkaline microbial mats dominated by the cyanobacterium *Synechococcus* (Boyd *et al.*, 2009). It was also shown that CO_2 fixation by *Hydrogenobaculum* spp. was a light-independent chemolithoautotrophic process.

While iron- and sulfur-oxidizers/reducers are important members of the microbial community, other diverse metabolically important processes such as methanotrophy and methanogenesis can also be observed in geothermal areas (Kristjansson and Hreggvidsson, 1995). A number of thermophilic and thermotolerant acidophilic methanotrophs have been isolated and described (Trotsenko *et al.*, 2009). *Methylokorus infernorum*, an obligate thermoacidophilic methanotroph with a growth optimum at pH 2–2.5 and 60°C, was isolated from New Zealand (Dunfield *et al.*, 2007), while *Methyloacida kamchatkensis*, with an optimal growth at 55°C and pH 2–5, was isolated from the Uzon Caldera, Russia (Islam *et al.*, 2008). Both species belong to the phylum Verrucomicrobia, a diverse group of taxa consisting of many uncultured and unknown genotypes (Trotsenko *et al.*, 2009).

Archaeal diversity in thermoacidophilic habitats

The earliest archaeal phylum indentified as moderately thermophilic was the Crenarchaeota (Lebedinsky *et al.*, 2007). Currently, cultivated thermophilic Crenarchaeota are assigned to a single class: Thermoprotei. This class consists of five orders: the Thermoproteales and Desulfurococcales, containing anaerobic hyperthermophiles (both lithoautotrophic and organotrophic), and the Sulfolobales, mainly containing aerobic thermophiles which oxidize sulfur as an energy substrate. The recently proposed novel order of Fervidicoccales (Perevalova *et al.*, 2010) and Acidilobales (Prokofeva *et al.*, 2009) also include thermophilic and hyperthermophilic Archaea. The other principal archaeal phylum, the Euryarchaeota, contains eight classes of which three exclusively contain hyperthermophiles: the Thermococci, Archaeoglobi and Methanopyri (Lebedinsky *et al.*, 2007).

The ability to grow at low pH values is common among thermophilic Archaea. Among them, organisms with a respiratory metabolism predominate (Johnson, 1998) and include autotrophs of the genera *Sulfolobus, Sulfurococcus, Metallosphaera, Acidianus, Sulfurisphaera, Picrophilus, Thermoplasma* and the microaerophiles *Thermocladium* and *Caldivirga* (Kristjansson and Hreggvidsson, 1995; Johnson, 1998). Such organisms predominate on or near the surface of both thermophilic soil and aquatic systems, where O_2 concentrations are higher. Elemental sulfur is an important component of metabolic activity acting as an electron acceptor in respiration or facilitated fermentation in the majority of Thermoproteales and Desulfurcoccales (Miroshnichenko *et al.*, 1994). Oxidation of H_2S sequentially to sulfur and sulfuric acid by thermacidophilic Archaea is essential to the overall microbial community (Kristjansson and Hreggvidsson, 1995; Lebedinsky *et al.*, 2007). As well as providing organic matter for aerobic and anaerobic thermoacidophilic heterotrophs, the sulfur-oxidizers actively lower the pH of the environment.

Iron-oxidation and reduction are also common metabolic processes in thermophilic Archaea, including *Sulfolobus* and *Metallosphaera* species (Johnson, 1998; Kozubal *et al.*, 2008). Other redox reaction couples used by thermoacidophilic Archaea have been described but, generally, metabolic diversity remains low compared to lower temperature and neutrophilic environments (Kristjansson and Hreggvidsson, 1995; Johnson, 1998).

The phylogenetic diversity of Archaea in terrestrial geothermal environments is highly varied (Prokofeva *et al.*, 2005; Lebedinsky *et al.*, 2007; Aditiawati *et al.*, 2009). For example, several archaeal genera have been identified in thermoacidophilic pools in Kamchatka, Russia, whereas only a single genus was identified in thermal pools in St Lucia (Prokofeva *et al.*, 2005). Boiling Springs Lake in Lassen Volcanic National Park, USA, hosted an even greater diversity of Archaea (Wilson *et al.*, 2008), while other sites were dominated by a single genus or even a single archaeal species (Burton and Norris, 2000; Prokofeva *et al.*, 2005). The pre-dominance of a single phylotype has also been noted in man-made thermoacidophilic habitats (Mikkelsen *et al.*, 2006, 2009), where *Sulfolobus shibatae* and *Sulfurisphaera* accounted for 69 and 74% of total clones recovered from two separate South African bioleaching reactors (Mikkelsen *et al.*, 2006). As with bacteria, archaeal diversity in thermoacidophilic environments remains generally low compared to that found in less 'extreme' habitats (Kristjansson and Hreggvidsson, 1995).

Eukaryote diversity in thermoacidophilic habitats

As with neutral and alkaline hydrothermal habitats, eukaryotic life in thermophilic acidophilic environments is largely limited to unicellular forms (Brown and Wolfe, 2006). Nevertheless, eukaryotic organisms are important components of the microbial community and are associated with microbial mats and biofilms (Brown and Wolfe, 2006; Aguilera *et al.*, 2010). Members of algal, protozoan and fungal taxa have been identified in thermoacidophilic environments (Owen *et al.*, 2008; Baumgartner *et al.*, 2009; Aguilera *et al.*, 2010).

Algal diatoms, *Pinnularia* sp., have been recorded at two geothermal hot springs with temperatures of 60°C and 85°C and pH values of 2.4 and 3.7, respectively (Owen *et al.*, 2008). However, growth (and probably survival) at 85°C is very unlikely for this organism, although the formation of microniches that provide conditions more suitable for growth at such high temperatures may be an essential element (Owen *et al.*, 2008). Algae have also been isolated in Lassen Volcanic National Park, USA (Brown and Wolfe, 2006). A single pool with a pH of 1.8 and temperature of 68°C was shown to contain a diverse range of algae including both red (*Cyanidioschyzon merolae*) and green algae (species of the family Chylamydomonadaceae). A number of fungi (two potential species) and Protozoa, including species of the Vampyrellidae, were also recovered from the same pool (Brown and Wolfe, 2006). Similarly, phylotypes potentially representing members of the red (Rhodophyta) and green (Chlorophyta) algae were identified in biofilms at a geothermal hot spring in Seltun, Iceland (Aguilera *et al.*, 2010). A number of organisms closely related to uncultured eukaryotes were also present in the acidic Rio Tinto, Spain (Aguilera *et al.*, 2010). The phototrophic Cyanidiales have been shown to be globally distributed in terrestrial geothermal environments at temperatures of 56°C and below and at pHs as low as 0.5 (Toplin *et al.*, 2008). Three algal genera (*Cyanidium, Galdieria* and *Cyanidioschyzon*) were observed to be dominant in distant locations including Yellowstone National Park, Owakudani (Japan) and New Zealand (Toplin *et al.*, 2008).

Thermophilic Protozoa have been observed in both man-made and natural geothermal environments (Rohr *et al.*, 1998; Sheehan *et al.*, 2003; Brown and Wolfe, 2006). A recently described thermoacidophilic amoeboflagellate protozoan, *Tetramitus thermacidophilus*, present in Kamchatka and Pisciarelli Solfatara is able to grow over a pH range of 1.2–5 and temperatures of 28–54°C (Baumgartner *et al.*, 2009). Parasitic Protozoa such as *Naegleria fowleri* and *Acanthamoeba* spp. have also been detected in the geothermal waters of Yellowstone and Grand Teton National Parks in the USA (Sheehan *et al.*, 2003). In these extreme environments Protozoa are routinely associated with microbial mats, presumably using the mats as a source of prey (Sheehan *et al.*, 2003; Brown and Wolfe, 2006; Baumgartner *et al.*, 2009).

Thermotolerant and thermophilic acidophilic fungi have been isolated from both Sinokawara Hot Spring, Japan, and Yellowstone National Park, USA (Cullings and Makhija, 2001; Yamazaki *et al.*, 2010). A novel thermotolerant fungal species (*Teratosphaeria acidotherma*) isolated from the Sinokawara hot spring was recovered from water temperatures which fluctuated in the range of 30–70°C. Optimal pH conditions for growth were in the pH range of 2–4 (Yamazaki *et al.*, 2010). Although the authors refer to the novel species as 'thermophilic', no hyphal growth was observed at 50°C and therefore the organism might be more accurately termed 'thermotolerant'. Sixteen fungal species have been recovered from geothermally heated soils although, of the 16 isolates recovered, only two species, *Decatylaria constrictum* var. *gallopava* and *Acremonium alabamense*, showed growth at 50°C (Redman *et al.*, 1999).

12.4 Mechanisms for Survival in Terrestrial Hydrothermal Environments

Subjection of mesophilic microorganisms to high temperature leads to the denaturation and degradation of cellular macromolecules, whereas the integrity of membranes, genetic material and biochemical function of proteins and enzymes is maintained in thermophiles (Grosjean and Oshima, 2007). Thermophiles have developed adaptive mechanisms for survival and growth under harsh conditions, often experiencing several factors simultaneously (e.g. pH and temperature). For extensive reviews of molecular and physiological adaptations of thermophiles the readers are directed to Robb and Newby (2007), Driessen and Albers (2007) and Canganella and Wiegel (2011). The adaptations can be broadly summarized in three categories below.

12.4.1 Genetic stability at high temperatures

Comparative whole genome sequence analyses suggest that thermophile genomes have characteristic codon usage and high percentage G + C content (Bhaya *et al.*, 2007; Barabote *et al.*, 2009; Miralles, 2010). However, this is not ubiquitous to all thermophiles. Interestingly, there is a high representation of genes coding for DNA repair mechanisms within the genomes of thermophiles (Makarova *et al.*, 2002). Furthermore, all hyperthermophilic and a few thermophilic bacteria and Archaea encode for a reverse gyrase (in some cases multiple copies), proposed to be essential for genetic material stability (Brochier-Armanet and Forterre, 2006; Li *et al.*, 2011). The reverse gyrase prevents chemical and heat denaturation of genomic DNA at high temperatures, thereby reducing the propensity of spontaneous mutations being incorporated during replication. Other genetic adaptive mechanisms include the accumulation of protective compatible solutes such as polyamines. Interestingly, Archaea and bacterial hyperthermophiles (such as *Aeropyrum pernix, Aquifix pyrophilum* and *Methonococcus jannaschii*) contain high concentrations of longer polyamines such as Tris(3-aminopropyl) amine, Caldohexamine and tetrakis-(3-aminopropyl) ammonium (Oshima, 2007). These longer polyamines are found exclusively in hyperthermophiles and contribute to thermostability of DNA, RNA and translation at temperatures above 80°C (Terui *et al.*, 2005).

12.4.2 Protein stability

A high proportion of protein sequences in thermophiles contain a higher frequency of charged residues such as Asp, Glu, Lys and Arg, as well as the hydrophobic amino acids Val and Tyr, to provide protein stabilization through ion bonds (Suhre and Claverie, 2003). In addition, proteins from thermophilic organisms have shorter loops, N and C terminal chains and an increase in disulfide and salt bridges resulting in more compact and rigid proteins (Sterner and Liebl, 2001; Vieille and Zeikus, 2001; Chakravarty and Varadarajan, 2002; Mallick *et al.*, 2002; Hickey and Singer, 2004).

12.4.3 Membranes and intracellular pH homeostasis

Stability of bacterial cell membranes under high temperature is maintained by increased branching and length of fatty acid chains, such as phytanyl and biphytamyl chains. In contrast, archaeal cell membrane stability is maintained through the incorporation of tetraether lipids which contain caldarchaeol, isocaldarchaeol, calditoglycerocaldarcaeol and crenarchaeol as core structures (Macalady *et al.*, 2004; Driessen and Albers, 2007; Ulrih *et al.*, 2009). Archaeal membrane composition, however, appears additionally to be determined by the pH of the environment. While a high percentage (95–100%) of tetraether lipids are found in thermophilic Archaea in acidic thermophilic environments (eg. *Ferroplasma acidarmanus*), Archaea growing optimally at neutral to near-neutral pH (5.5–8) (e.g. *Archaeoglobus fungidus*) contain tetraether lipids in low concentrations, and those in alkaline environments have membranes composed primarily of diether core lipids.

Various mechanisms to maintain intracellular pH homeostasis are essential to thermophiles due to the pH range they may experience. Adaptations include an increase in impermeability of cell membranes to protons (Baker-Austin and Dopson, 2007) and through the use of passive and active ion transporters and antiporters present on the cell membranes to control ions such as $K^+/Na^+/H^+$ (Canganella and Wiegel, 2011). The molecular and physiological adaptations for survival in polyextremophilic environments (thermoalkaliphiles and thermoacidophiles) are not very well understood.

12.5 Evolutionary Implications

What evolutionary processes have occurred for these thermophilic phenotypes to be developed? Did the last common ancestor (LCA) originate with these properties, or did thermophily develop as a result of later adaptation? The debate regarding the nature of the LCA has been controversial between scientists for a long time. Despite the advancements made toward ordering microbial taxonomy, a final answer to this question is still not possible, as there is evidence to support each of the fundamentally different views regarding the origin of thermophily in prokaryotes.

From a geological perspective, records indicate that about 3 billion years ago Earth's hydrosphere was hot and subject to repeated local sterilization from impact events (Kasting, 1993; Klenk *et al.*, 2004). The survival of hyperthermophiles, therefore, would be a likely scenario. However, due to the rapid decomposition of most biochemical molecules at high temperatures as well as the instability of RNA under thermophilic conditions, a hyperthermophilic origin is strongly refuted, and a more likely temperature of about 50°C for the origin of life has been postulated (Miller and Lazcano, 1995; Moulton *et al.*, 2000). In support of this, analysis of the G+C content of rRNA sequences of various species representing the major lineages of life resulted in an inferred G+C content of the last universal ancestor of all cellular life that would be incompatible with survival at high temperature (Galtier *et al.*, 1999). Chemical evolution at hyperthermophilic temperatures has, however, been addressed. Saline environments, thought to have been possible in early Earth, appear to stabilize DNA/RNA at hyperthermophilic temperatures (Marguet and Forterre, 1998; Tehi *et al.*, 2002). In addition, the abiotic synthesis of amino acids under hydrothermal conditions has been experimentally observed and shown to be thermodynamically favourable (Hennet *et al.*, 1992; Marshall, 1994; Shock *et al.*, 1998). Using a thermophily index based on the propensity of particular amino acids to be used more frequently in hyperthermophile proteins, it was concluded that the late stage of genetic code structuring (the stereochemical interactions between anticodons/codons and amino acids) must have taken place in a hyperthermophile (Di Giulio, 2000, 2003; Schwartzman and Lineweaver, 2004). In addition, the absence of any living eukaryote with a T_{max} above 60°C, despite the 2 billion years of opportunity to adapt to hyperthermophily, is supportive of a hyperthermophilic LCA and the asymmetric evolution of hyperthermophiles to mesophiles (Schwartzman and Lineweaver, 2004).

Molecular phylogenetic analyses of the rRNA locate hyperthermophiles, in particular the hyperthermophilic Archaea, at the deepest positions within the prokaryotic domains, implying a hyperthermophilic LCA (Achenbach-Richter *et al.*, 1987; Caetano-Anolles, 2002; Matte-Tailliez *et al.*, 2002). Global and concatenated (linked) phylogenies often confirm the deep phylogenetic location of the hyperthermophilic lineages within the kingdom Bacteria (Brown, 2001; Wolf *et al.*, 2001; Klenk *et al.*, 2004). However, the phylogenetic system proposed by Cavalier-Smith implies an alternative scenario where thermophily is a late evolutionary invention, with Archaea emerging as an adaptation to thermal (and acidic) environments (Cavalier-Smith, 2002). Although this theory contradicts a very large catalogue of existing phylogenetic analyses and genome comparisons (Lebedinsky *et al.*, 2007), the concept of adaptation to thermophily is not unsupported.

The detection of a high percentage of archaeal genes in the genomes of hyperthermophilic bacteria suggests that these genes were vertically inherited from the last hyperthermophilic ancestor of Archaea and Bacteria, and have been lost in mesophilic bacteria (Aravind *et al.*, 1998; Kyrpides and Olsen, 1999; Nelson *et al.*, 1999). However, many of these archaeal genes are clustered, supporting the hypothesis that they were horizontally transferred as an adaptation to life at high temperatures (Aravind *et al.*, 1999). If this was the case, then the question of whether the adaptation was transferred from Archaea to hyperthermophilic bacteria or vice versa, is important (Forterre *et al.*, 2000). The answer to this may lie in reverse gyrase (*rgy*), the only gene that is absolutely characteristic of thermophiles (Archaea and Bacteria) (Forterre, 1996, 2002; Forterre *et al.*, 2000). All bacterial *rgy* genes are located on archaeal-like gene clusters (Forterre *et al.*, 2000) and analysis of the open reading frames both upstream and downstream of *rgy* genes suggests that these gene clusters were not formed under selection pressure, suggesting gene transfer as a likely mechanism for their acquisition. Furthermore, the occurrence of *rgy* in both the Crenarchaeota and Euryarchaeota, as well as the phylogeny of all known reverse gyrases, suggests that the transfer occurred from Archaea to Bacteria (likely via one or two lateral gene transfer events, followed by a secondary distribution within bacteria) and not the other way around (Brochier-Armanet and Forterre, 2006). This would imply that all Archaea share a hyperthermophilic ancestor and adapted to life at high temperatures before Bacteria. This being the case, the LCA of all prokaryotes would not have been a hyperthermophile.

Further support for the acquisition of hyperthermophily as a secondary trait has been derived from analysis of the genome sequence of *Thermus thermophilus*, a bacterium that thrives under thermal stress (Omelchenko *et al.*, 2005). It is proposed that *T. thermophilus* evolved from a mesophilic bacterium, and acquired a distinct set of new genes, through horizontal gene transfer, to confer the thermophilic phenotype (Omelchenko *et al.*, 2005). Using a flexible approach to phyletic pattern analysis, 290 clusters of orthologous groups of proteins (COGs) have been identified as putative thermophilic determinants (Makarova *et al.*, 2003). *Thermus thermophilus* acquired 23 of these gene families, which included proteins belonging to the predicted mobile DNA repair system characteristic of thermophiles (Jansen *et al.*, 2002; Makarova *et al.*, 2002), as well as several archaeal genes that have never been identified in any mesophilic bacterial genome (Omelchenko *et al.*, 2005).

Several recent studies, using comparative genomics methods, have shown that there is a relationship between amino acid composition and optimal growth temperature. A thermophilic amino acid signature, associated with thermophily, has been identified from these comparisons (Suhre and Claverie, 2003; Pe'er *et al.*, 2004; Takami *et al.*, 2004; Tekaia and Yeramian, 2006; Zeldovich *et al.*, 2007), and enables (hyper) thermophiles to be distinguished from other species (Puigbo *et al.*, 2008). When these analyses are superimposed with the position of the (hyper)thermophiles in the tree of life, several evolutionary scenarios become apparent. Some mesophilic species which cluster with (hyper)thermophiles are found to have the thermophilic amino acid signature. This suggests a scenario where transversion to mesophily has occurred in these species, probably as a recent event. Conversely, some thermophiles which are taxonomically related to mesophiles do not have the thermophilic signature. These could represent examples of recent transversions to thermophily. A third scenario is proposed based on the identification of thermophiles which have the thermophilic signature, but are taxonomically related to mesophiles. These could represent examples of an ancient acquisition of thermophily (Omelchenko *et al.*, 2005; Puigbo *et al.*, 2008). The gain and loss of thermophily most probably occurs at differential rates, which in turn reflect the differences in selection intensities (Puigbo *et al.*, 2008).

It is clear that answering the fundamental question of whether life originated at elevated temperatures is controversial and the arguments presented are not universally accepted. What we can conclude is that insight and perspective from a range of disciplines is required, and that despite the availability of a wealth of new genetic information, this question is yet to be resolved. However, collectively the data show that over relatively short evolutionary periods, prokaryotes have the capacity to not only adapt to thermophily but also to lose such adaptations.

12.6 Biotechnology: Commercial Utilization of Thermophiles

The molecular, genomic and cellular adaptations of thermophiles are of great industrial interest, particularly thermostable enzymes (Haki and Rakshit, 2003; see also Arora and Bell, Chapter 25, this volume). While the global market for enzymes was estimated to be worth US$2.3 billion in 2007 (Thakore, 2008), penetration of thermophilic enzymes has been historically low, with the notable exception of *Taq* polymerase and a handful of *Bacillus* spp. enzymes (Sharp *et al.*, 1992). Other successful examples include amylases and xylanses for sweeteners and paper bleaching, and proteases used in food processing, baking, brewing and detergents (Haki and Rakshit, 2003; Canganella and Wiegel, 2011).

The use of thermophilic organisms as whole cell biocatalysts is also an attractive option for the industrial sector, particularly as a platform for second generation bioethanol production (Taylor *et al.*, 2009). Engineered pryogeny of *Geobacillus thermoglucosidasius, Thermoanaerobacterium saccharolyticum* and *Thermoanerobacter mathranii* are currently being exploited by several new biofuel companies. Recent developments in the UK have led to a US$500 million contract between TMO Renewables and US utilities to build an industrial scale commercial bio-ethanol production facility utilizing whole cell thermophiles to process municipal waste. Thus commercialization of thermophilic ethanologens in the near future remains a distinct possibility (Taylor *et al.*, 2009).

Thermophiles have been successfully applied in the biomining industry, where low grade ores are processed to yield valuable metals. Sulfate-oxidizing Archaea such as *Acidianus* spp., *Sulfolobus* spp., *Sulfurusphaera* spp. and *Metallosphaera* spp., dominate the commercial thermophilic desulfurication processes where operational temperatures approach 80°C (Mikkelsen *et al.*, 2006, 2009). The potential of this process was demonstrated in a 300 m^3 industrial test reactor in South Africa where 20 kilotons of cathode copper per annum was produced (Batty and Younger, 2007). Application of the desulfurization process by thermophiles may also prove commercially useful in coal decontamination and general bioremediation of contaminated materials (Boogerd *et al.*, 1990).

12.7 Possible Analogues for Life on Early Earth and in Exobiology

Terrestrial hydrothermal environments are widely considered to be a suitable analogue to conditions that may have given rise to early life on planet Earth (Shock *et al.*, 1998; Stetter, 1998; Wiegel and Adams, 1998; Villar and Edwards, 2006; see also Gómez *et al.*, Chapter 26, this volume). Hyperthermophilic organisms are contained in the deepest, least evolved branches of the universal phylogenetic tree, and have metabolic capabilities to utilize substrates which may have dominated primordial terra. This suggests a potential hyperthermophilic last common ancestor (Cady and Noffke, 2009; Canganella and Wiegel, 2011), although opposing theories do persist (Section 12.5; Glansdorf *et al.*, 2008). In addition, most extremophilic environments on earth, such as terrestrial hydrothermal habitats, are likely to be analogues for conditions on extraterrestrial worlds (Sharp *et al.*, 1992). Hydrothermal processes are thought to explain a number of features and observations on Mars, such as the deuteron:hydrogen (D/H) ratios of water extracted from Martian meteorites and iron oxide-rich spectral units on the floors of some rifted basins (Farmer, 1996). Therefore, the conditions that support thermophiles here on Earth may have been present on Mars and may exist elsewhere in the universe (Wiegel and Adams, 1998). Consequently, in addition to an interest in thermophiles from a microbiological and industrial aspect, the fields of astrobiology and geobiology have also taken an interest in determining signs of fossilized life on Earth to search for life on other planets (Cady and Noffke, 2009; Canganella and Wiegel, 2011).

12.8 Conclusion

While Eukaryotes, Archaea and Bacteria are all represented within terrestrial hydrothermal environments, their diversity and ability to withstand the conditions characteristic of these environments varies greatly. Eukaryotes are restricted to temperatures of less than 65°C (Toplin *et al.*, 2008) while both Bacteria and Archaea have optimal growth temperatures greatly exceeding 65°C (Lebedinsky *et al.*, 2007; Mathur *et al.*, 2007). The relative distribution of members of the three domains is broadly summarized in the temperature–pH plot shown in Fig. 12.2. All domains of life show a decreasing diversity as temperatures and pH values increase. While many authors have identified dominant microorganisms in particular thermophilic habitats (Löhr *et al.*, 2006; Wilson *et al.*, 2008; Stout *et al.*, 2009), such data should always be treated with some caution. Sampling, culturing and molecular techniques are all known to introduce biases which may skew the apparent abundance of a particular group of microorganisms (Stout *et al.*, 2009). Nevertheless, research on thermophilic

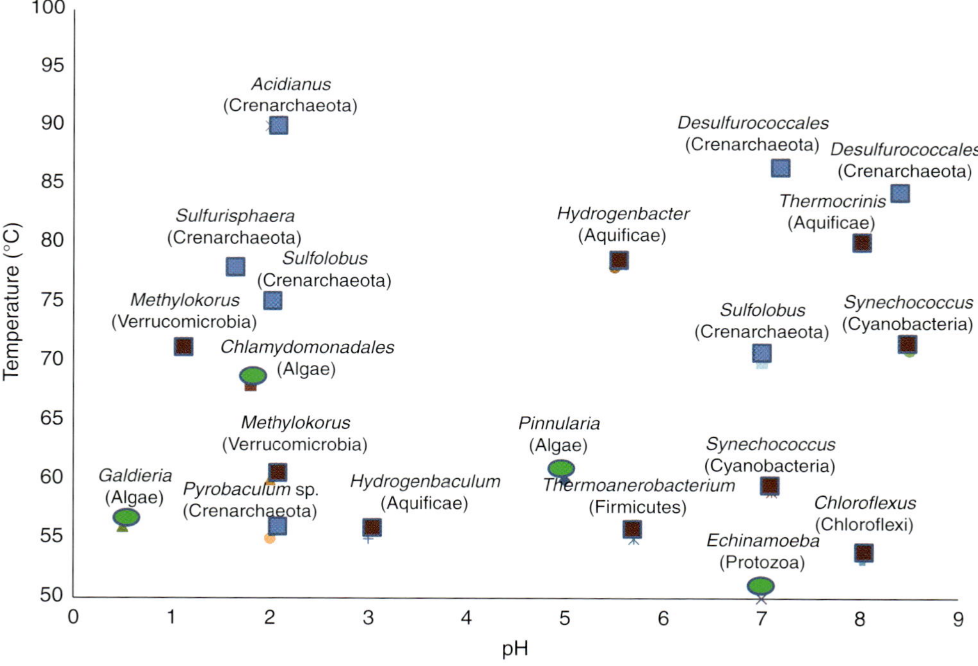

Fig. 12.2. Summary of the relative distribution of members of the three domains, Eukaryotes (green), Archaea (blue) and Bacteria (purple), with respect to the temperature and pH conditions that they can withstand. © Don A. Cowan.

microorganisms is increasing rapidly due to their biological 'eccentricities'. The possible exploitation of their thermophilic properties and thermostable proteins for biotechnological application has been realized and thermophiles have become increasingly important targets for bio-prospecting. In addition, the developments that occur in the field of thermophilic microbial research will continue to offer insight into the evolution of all living organisms, and possibly to the origin of life itself.

References

Achenbach-Richter, L., Gupta, R., Stetter, K.O. and Woese, C.R. (1987) Were the original eubacteria thermophiles? *Systematic and Applied Microbiology* 9, 34–39.

Aditiawati, P., Yohandini, H., Madayanti, F. and Akhmaloka (2009) Microbial diversity of Acidic Hot Spring (Kawah Hujan B) in Geothermal Field of Kamojang area, West Java-Indonesia. *Open Microbiology Journal* 3, 58–66.

Aguilera, A., Souza-Egipsy, V., González-Toril, E., Rendueles, O. and Amils, R. (2010) Eukaryotic microbial diversity of phototrophic microbial mats in two Icelandic geothermal hot springs. *International Microbiology* 13, 21–32.

Allewalt, J.P., Bateson, M.M., Reysenbach, N.P., Slack, K. and Ward, D.M. (2006) Effect of temperature and light on growth of and photosynthesis by *Synechococcus* isolates typical of those predominating in the Octopus Spring microbial mat community of Yellowstone National Park. *Applied and Environmental Microbiology* 72, 544–550.

Aravind, L., Tatusov, R.L., Wolf, Y.I., Walker, D.R. and Koonin, E.V. (1998) Evidence for massive gene exchange between Archaeal and bacterial hyperthermophiles. *Trends in Genetics* 14, 442–444.

Aravind, L., Tatusov, R.L., Wolf, Y.I., Walker, D.R. and Koonin, E.V. (1999) Reply. *Trends in Genetics* 15, 299–300.

Atkinson, T., Cairns, S., Cowan, D.A., Danson, M.J., Hough, D.W., Johnson, D.B., Norris, P.R., Raven, N., Robinson, C., Robson, R. and Sharp, R.J. (2000) A microbiological survey of Montserrat Island hydrothermal biotopes. *Extremophiles* 4, 305–313.

Baker-Austin, C. and Dopson, M. (2007) Life in acid: pH homeostasis in acidophiles. *Trends in Microbiology* 15, 165–171.

Barabote, R.D., Xie, G., Leu, D.H., Normand, P., Necsulea, A., Daubin, V., Medigue, C., Adney, W.S., Xu, X.C., Lapidus, A., Parales, R.E., Detter, C., Pujic, P., Bruce, D., Lavire, C., Challacombe, J.F., Brettin, T.S and Berry, A.M. (2009) Complete genome of the cellutytic thermophile *Acidothermus cellulolyticus* !!B provides insights into its ecophysiological and evolutionary adaptations. *Genome Research* 19, 1033–1043.

Barns, S.M., Fundyga, R.E., Jeffries, M.W. and Pace, N.R. (1994) Remarkable Archaeal diversity detected in a Yellowstone National Park hot spring environment. *Proceedings of the National Academy of Sciences of the United States of America* 91, 1609–1613.

Batty, L.C. and Younger, P.L. (2007) The effect of pH on plant litter decomposition and metal cycling in wetland mesocosms supplied with mine drainage. *Chemosphere* 66, 158–164.

Baumgartner, M., Yapi, A., Gröbner-Ferreira, R. and Stetter, K.O. (2003) Cultivation and properties of *Eschinamoaeba thermarum* n. sp., an extremely thermophilic amoeba thriving in hot springs. *Extremophiles* 7, 267–274.

Baumgartner, M., Eberhardt, S., De Jonckheere, J.F. and Stetter, K.O. (2009) *Tetramitus thermacidophilus* nov. sp., an amoeboflagellate from acidic hot springs. *Journal of Eukaryotic Microbiology* 56, 201–206.

Bhaya, D., Grossman, A.R., Steunou, A.S., Khuri, N., Cohan, F.M., Hamamura, N., Melendrez, M.C., Bateson, M.M., Ward, D.M. and Heidelberg, J.F. (2007) Population level functional diversity in a microbial community revealed by comparative genomic and metagenomic analyses. *ISME Journal* 1, 703–718.

Blank, C.E., Cady, S.L. and Pace, N.R. (2002) Microbial composition of near-neutral silica-depositing thermal springs throughout Yellowstone National Park. *Applied and Environmental Microbiology* 68, 5123–5135.

Boogerd, F.C., Bos, P., Kuenen, J.G., Heijnen, J.J. and van der Lans, R.G. (1990) Oxygen and carbon dioxide mass transfer and the aerobic, autotrophic cultivation of moderate and extreme thermophiles: a case study related to the microbial desulfurization of coal. *Biotechnology Bioengineering* 35, 1111–1119.

Boyd, E.S., Leavitt, W.D. and Geesey, G.G. (2009) CO_2 uptake and fixation by a thermoacidophilic microbial community attached to precipitated sulfur in a geothermal spring. *Applied and Environmental Microbiology* 75, 4289–4296.

Brochier-Armanet, C. and Forterre, P. (2006) Widespread distribution of Archaeal reverse gyrase in thermophilic bacteria suggests a complex history of vertical inheritance and lateral gene transfers. *Archaea* 2, 83–93.

Brock, T.D. (1971) Bimodal distribution of pH values of thermal springs across the world. *Geological Society of America Bulletin* 82, 1393–1396.

Brock, T.D. (1978) *Thermophilic Microorganisms and Life at High Temperatures*. Springer-Verlag, New York, 378 pp.

Broomer, S.M., Noll, K.L., Geesey, G.G. and Datton, B.E. (2009) Formation of multilayered photosynthetic biofilms in an alkaline thermal spring in Yellowstone National Park, Wyoming. *Applied and Environmental Microbiology* 75, 2464–2475.

Brown, J.R. (2001) Genomic and phylogenetic perspectives on the evolution of prokaryotes. *Systematic Biology* 50, 497–512.

Brown, P.B. and Wolfe, G.V. (2006) Protist genetic diversity in the acidic hydrothermal environments of Lassen Volcanic National Park, USA. *Journal of Eukaryotic Microbiology* 53, 420–431.

Burton, N.P. and Norris, P.R. (2000) Microbiology of acidic, geothermal springs of Montserrat: environmental rDNA analysis. *Extremophiles* 4, 315–320.

Cady, S.L. and Noffke, N. (2009) Geobiology: evidence for life on Earth and the search for life on other planets. *GSA Today* 19, 4–10.

Caetano-Anolles, G. (2002) Evolved RNA secondary structure and the rooting of the universal tree of life. *Journal of Molecular Evolution* 54, 333–345.

Campbell, L.L. and Postgate, J.R. (1965) Classification of the spore-forming sulfate-reducing bacteria. *Bacteriological Reviews* 29, 359–363.

Canganella, F. and Wiegel, J. (2011) Extremophiles: from abyssal to terrestrial ecosystems and possibly beyond. *Naturwissenschaften* 98, 253–279.

Cavalier-Smith, T. (2002) The neomuran origin of archaebacteria, the negibacterial root of the universal tree and bacterial megaclassification. *International Journal of Systematic and Evolutionary Microbiology* 52, 7–76.

Chakravarty, S. and Varadarajan, R. (2002) Elucidation of factors responsible for enhanced thermal stability of proteins: a structural genomics based study. *Biochemistry* 41, 8152–8161.

Costa, K.C., Navarro, J.B., Shock, E.L., Zhang, C.L., Soukup, D. and Heldlund, B.P. (2009) Microbiology and geochemistry of great boiling and mud hot springs in the United States Great Basin. *Extremophiles* 13, 447–459.

Cullings, K. and Makhija, S. (2001) Ectomycorrhizal fungal associates of *Pinus contorta* in soils associated with a hot spring in Norris Geyser Basin, Yellowstone National Park, Wyoming. *Applied and Environmental Microbiology* 67, 5538–5543.

De Jonckheere, J.F., Baumgartner, M., Opperdoes, F.R. and Stetter, K.O. (2009) *Marinamoeba thermophilia*, a new marine heteroblasosean amoeba growing at 50 degrees C. *European Journal of Protistology* 45, 231–236.

De Jonckheere, J.F., Baumgartner, M., Eberhardt, S., Opperdoes, F.R. and Stetter, K.O. (2010) *Oramoeba fumarolia* gen. nov., sp. nov., a new marine heterolobosean amoeboflagellate growing at 54°C. *European Journal of Protistology* 47, 16–23.

De la Torre, J.R., Walker, C.B., Walker, C.B., Ingalls, A.E., Konneke, M. and Stahl, D.A. (2008) Cultivation of a thermophilic ammonia oxidizing archaeon synthesizing crenarchaeol. *Environmental Microbiology* 10, 810–818.

de Rosa, M., Gambacorta, A. and Bu'lock, J.D. (1975) Extremely thermophilic acidophilic bacteria convergent with *Sulfolobus acidocaldarius*. *Journal of General Microbiology* 86, 156–164.

Di Giulio, M. (2000) The late stage of genetic code structuring took place at a high temperature. *Gene* 261, 189–195.

Di Giulio, M. (2003) The ancestor of the Bacteria domain was a hyperthermophile. *Journal of Theoretical Biology* 221, 425–436.

Driessen, A.J.M. and Albers, S.-V. (2007) Membrane adaptations of (hyper)thermophilesto high temperatures. In: Gerday, C. and Glandsdorff, N. (eds) *Physiology and Biochemistry of Extremophiles*. ASM Press, Washington, DC, pp. 104–116.

Dunfield, P.F., Yuryev, A., Senin, P., Smirnova, A.V., Stott, M.B., Hou, S., Ly, B., Saw, J.H., Zhou, Z., Ren, Y., Wang, J., Mountain, B.W., Crowe, M.A., Weatherby, T.M., Bodelier, P.L.E., Liesack, W., Feng, L., Wang, L. and Alam, M. (2007) Methane oxidation by an extremely acidophilic bacterium of the phylum Verrucomicrobia. *Nature* 450, 879–882.

Eder, W. and Huber, R. (2002) New isolates and physiological properties of the Aquificales and description of *Thermocrinis albus* sp. nov. *Extremophiles* 6, 309–318.

Elkins, J.G., Podar, M., Graham, D.E., Makarova, K.S., Wolf, Y., Randau, L., Hedlund, B.P., Brochier-Armanet, C., Kunin, V., Anderson, I., Lapidus, A., Goltsman, E., Barry, K., Koonin, E.V., Hugenholtz, P., Kyrpides, N., Wanner, G., Richardson, P., Keller, M. and Stetter, K.O. (2008) A korArchaeal genome reveals insights into the evolution of the Archaea. *Proceedings of the National Academy of Sciences of the United States of America* 105, 8805–8806.

Erguder, T.H., Boon, N., Wittebolle, L., Marzorati, M. and Verstraete, W. (2009) Environmental factors shaping the ecological niches of ammonia-oxidizing Archaea. *FEMS Microbial Reviews* 33, 855–869.

Farmer, J.D. (1996) Hydrothermal systems on Mars: an assessment of present evidence. In: *Evolution of Hydrothermal Ecosystems on Earth (and Mars?)*. Wiley and Sons, Chichester, West Sussex, UK, 346 pp.

Forterre, P. (1996) A hot topic: the origin of hyperthermophiles. *Cell* 85, 789–792.

Forterre, P. (2002) A hot story from comparative genomics: reverse gyrase is the only hyperthermophile-specific protein. *Trends in Genetics* 18, 236–237.

Forterre, P., Bouthier de la Tour, C., Philippe, H. and Duguet, M. (2000) Reverse gyrase from hyperthermophiles: probable transfer of a thermoadaptation trait from Archaea to Bacteria. *Trends in Genetics* 16, 152–154.

Galtier, N., Tourasse, N. and Gouy, M. (1999) A nonhyperthermophilic common ancestor to extant life forms. *Science* 283, 220–221.

Giovannoni, S.J., Revsbech, N.P., Ward, D.M. and Castenholz, R.W. (1987) Obligately phototrophic *Chloroflexus*: primary production in anaerobic hot spring microbial mats. *Archives of Microbiology* 147, 80–87.

Glansdorff, N., Xu, Y. and Labedan, B. (2008) The Last Universal Common Ancestor: emergence, constitution and genetic legacy of an elusive forerunner. *Biology Direct* 3, 29.

Grosjean, H. and Oshima, T. (2007) Fuctional genomics in thermophilic microorganisms. In: Gerday, C. and Glandsdorff, N. (eds) *Physiology and Biochemistry of Extremophiles*. ASM Press, Washington, DC, pp. 39–53.

Haki, G.D. and Rakshit, S.K. (2003) Developments in industrially important thermostable enzymes: a review. *Bioresource Technolgy* 89, 17–34.

Hamamura, N., Olson, S.H., Ward, D.M. and Inskeep, W.P. (2005) Diversity and functional analysis of bacterial communities associated with natural hydrocarbon seeps in acidic soils at Rainbow Springs, Yellowstone National Park. *Applied and Environmental Microbiology* 71, 5943–5950.

Hanada, S. (2003) Filamentous anoxygenic phototrophs in hot springs. *Microbes and Environments* 12, 51–61.

Hennet, R.J.-C., Holm, N.G. and Engel, M.H. (1992) Abiotic synthesis of amino acids under hydrothermal conditions and the origin of life: a perpetual phenomenon? *Die Naturwissenschaften* 79, 361–365.

Hickey, D.A. and Singer, G.A. (2004) Genomic and proteomic adaptations to growth at high temperature. *Genome Biology* 5, 117.

Huber, R. and Stetter, K.O. (2001) Discovery of hyperthermophilic microorganisms. *Methods in Enzymology* 330, 11–24.

Inskeep, W.P., Rusch, D.B., Jay, Z.J., Herrgard, M.J., Kozubal, M.A., Richardson, T.H., Macur, R.E., Hamamura, N., Jennings, R.D., Fouke, B.W., Reysenbach, A.L., Roberto, F., Young, M., Schwartz, A., Boyd, E.S., Badger, J.H., Mathur, E.J., Ortmann, A.C., Bateson, M., Geesey, G. and Frazier, M. (2010) Metagenomes from high-temperature chemotrophic systems reveal geochemical controls on microbial community structure and function. *PLoS ONE* 5, e9773.

Islam, T., Jensen, S., Reigstad, L.J., Larsen, O. and Birkeland, N.K. (2008) Methane oxidation at 55°C and pH 2 by a thermoacidophilic bacterium belonging to the Verrucomicrobia phylum. *Proceedings of the National Academy of Sciences of the United States of America* 105, 300–304.

Jansen, R., Embden, J.D., Gaastra, W. and Schouls, L.M. (2002) Identification of genes that are associated with DNA repeats in prokaryotes. *Molecular Microbiology* 43, 1565–1575.

Johnson, D.B. (1998) Biodiversity and ecology of acidophilic microorganisms. *FEMS Microbiology Ecology* 27, 307–317.

Kaksonen, A.H., Plumb, J.J., Robertson, W.J., Spring, S., Schumann, P., Franzmann, P.D. and Puhakka, J.A. (2006) Novel thermophilic sulfate-reducing bacteria from a geothermally active underground mine in Japan. *Applied and Environmental Microbiology* 72, 3759–3762.

Kaksonen, A.H., Spring, S., Schumann, P., Kroppenstedt, R.M. and Puhakka, J.A. (2007) *Desulfovirgula thermocuniculi* gen. nov., sp. nov., a thermophilic sulfate-reducer isolated from a geothermal underground mine in Japan. *International Journal of Systematic and Evolutionary Microbiology* 57, 98–102.

Kasting, J.F. (1993) Earth's early atmosphere. *Science* 259, 920–926.

Klenk, H.-P., Spitzer, M., Ochsenreiter, T. and Fuellen, G. (2004) Phylogenomics of hyperthermophilic Archaea and Bacteria. *Biochemical Society Transactions* 32, 175–178.

Kozubal, M., Macur, R.E., Korf, S., Taylor, W.P., Ackerman, G.G., Nagy, A. and Inskeep, W.P. (2008) Isolation and distribution of a novel iron-oxidizing crenarchaeon from acidic geothermal springs in Yellowstone National Park. *Applied and Environmental Microbiology* 74, 942–949.

Kristjansson, J.K. (1991) *Thermophilic Bacteria*. CRC Press, Boca Raton, Florida, 228 pp.

Kristjansson, J.K. and Hreggvidsson, G.O. (1995) Ecology and habitats of extremophiles. *World Journal of Microbiology and Biotechnology* 11, 17–25.

Kristjansson, K.J. and Stetter, K.O. (1991) Thermophilic bacteria. In: Kristjansson, K.J. (ed.) *Thermophilic Bacteria*. CRC Press, Boca Raton, Florida, pp. 2–13.

Kvist, T., Ahring, B.K. and Westermann, P. (2007) Archaeal diversity in Icelandic hot springs. *FEMS Microbiology Ecology* 59, 71–80.

Kyrpides, N.C. and Olsen, G.J. (1999) Archaeal and bacterial hyperthermophiles: horizontal gene exchange or common ancestry? *Trends in Genetics* 15, 298–299.

Lebedinsky, A.V., Chernyh, N.A. and Bonch-Osmolovskaya, E.A. (2007) Phylogenetic systematics of microorganisms inhabiting thermal environments. *Biochemistry (Moscow)* 72, 1299–1312.

Li, J., Liu, J., Zhou, J. and Xiang, H. (2011) Functional evaluation of four putative DNA-binding regions in *Thermoanaerobic tengcongensis* reverse gyrase. *Extremophiles* 12, 281–291.

Löhr, A.J., Laverman, A.M., Braster, M., Straalen, N.M. and Röling, W.F.M. (2006) Microbial communities in the world's largest acidic volcanic lake, Kawah Ijen in Indonesia, and in the Banyupahit River originating from it. *Microbial Ecology* 52, 609–618.

Macalady, J.L., Vestling, M.M., Baumler, D., Boekelheide, N., Kaspar, C.W. and Banfield, J.F. (2004) Tetraether-linked membrane monolayers in *Ferroplasma* spp: a key to survival in acid. *Extremophiles* 8, 411–419.

Makarova, K.S., Aravind, L., Grishin, N.V., Rogozin, I.B. and Koonin, E.V. (2002) A DNA repair system specific for thermophilic Archaea and bacteria predicted by genomic context analysis. *Nucleic Acids Research* 30, 482–496.

Makarova, K.S., Wolf, Y.I. and Koonin, E.V. (2003) Potential genomic determinants of hyperthermophily. *Trends in Genetics* 19, 172–176.

Mallick, P., Boutz, D.R., Eisenberg, D. and Yeates, T.O. (2002) Genomic evidence that the intracellular proteins of Archaeal microbes contain disulfide bonds. *Proceedings of the National Academy of Sciences of the United States of America* 99, 9679–9684.

Marguet, E. and Forterre, P. (1998) Protection of DNA by salts against thermodegradation at temperatures typical for hyperthermophiles. *Extremophiles* 2, 115–122.

Marshall, W.L. (1994) Hydrothermal synthesis of amino acids. *Geochimica Cosmochimica Acta* 58, 2099–2106.

Mathur, J., Bizzoco, R.W., Ellis, D.G., Lipson, D.A., Poole, A.W., Levine, R. and Kelley, S.T. (2007) Effects of abiotic factors on the phylogenetic diversity of bacterial communities in acidic thermal springs. *Applied and Environmental Microbiology* 73, 2612–2623.

Matte-Tailliez, O., Brochier, C., Forterre, P. and Philippe, H. (2002) Archaeal phylogeny based on ribosomal proteins. *Molecular Biology and Evolution* 19, 631–639.

Mikkelsen, D., Kappler, U., McEwan, A.G. and Sly, L.I. (2006) Archaeal diversity in two thermophilic chalcopyrite bioleaching reactors. *Environmental Microbiology* 8, 2050–2056.

Mikkelsen, D., Kappler, U., McEwan, A.G. and Sly, L.I. (2009) Probing the Archaeal diversity of a mixed thermophilic bioleaching culture by TGGE and FISH. *Systematic and Applied Microbiology* 32, 501–513.

Miller, S. and Lazcano, A. (1995) The origin of life – did it occur at high temperatures? *Journal of Molecular Evolution* 41, 689–692.

Miralles, F. (2010) Compositional properties and thermal adaptation of SRP-RNA in Bacteria and Archaea. *Journal of Molecular Evolution* 70, 181–189.

Miroshnichenko, M.L., Gongadze, G.A., Lysenko, A.M. and Bonch-Osmolovskaya, E.A. (1994) *Desulfurella multipotens* sp. nov., a new sulfur-respiring thermophilic eubacterium from Raoul Island (Kermadec archipelago, New Zealand). *Archives of Microbiology* 161, 88–93.

Mori, K., Kim, H., Kakegawa, T. and Hanada, S. (2003) A novel lineage of sulfate-reducing microorganisms: Thermodesulfobiaceae fam. nov., *Thermodesulfobium narugense*, gen. nov., sp. nov., a new thermophilic isolate from a hot spring. *Extremophiles* 7, 283–290.

Mosser, J.L., Mosser, A.G. and Brock, T.D. (1973) Bacterial origin of sulfuric acid. *Science* 79, 1323–1324.

Moulton, V., Gardner, P.P., Pointon, R.F., Creamer, L.K., Jameson, G.B. and Penny, D. (2000) RNA folding argues against a hot-start origin of life. *Journal of Molecular Evolution* 51, 416–421.

Nelson, K.E., Clayton, R.A., Gill, S.R., Gwinn, M.L., Dodson, R.J., Haft, D.H., Hickey, E.K., Peterson, J.D., Nelson, W.C., Ketchum, K.A., McDonald, L., Utterback, T.R., Malek, J.A., Linher, K.D., Garrett, M.M., Stewart, A.M., Cotton, M.D., Pratt, M.S., Phillips, C.A., Richardson, D., Heidelberg, J., Sutton, G.G., Fleischmann, R.D., Eisen, J.A., White, O., Salzberg, S.L., Smith, H.O., Venter, J.C. and Fraser, C.M. (1999) Evidence for lateral gene transfer between Archaea and Bacteria from genome sequence of *Thermotoga maritima*. *Nature* 399, 323–329.

Niederberger, T.D., Ronimus, R.S. and Morgan, H.W. (2007) The microbial ecology of a high-temperature near-neutral spring situated in Rotorua, New Zealand. *Microbiological Research* 163, 594–603.

Nordstrom, D.K., Ball, J.W. and McCleskey, R.B. (2005) Groundwater to Surface Water: Chemistry of Thermal Outflows in Yellowstone National Park. *United States Geological Survey Publication*, pp. 73–94.

Ogg, C.D. and Patel, B.K. (2010) *Fervidicella metallireducens* gen. nov., sp. nov., a thermophilic, anaerobic bacterium from geothermal waters. *International Journal of Systematic and Evolutionary Microbiology* 60, 394–400.

Omelchenko, M.V., Wolf, Y.I., Gaidamakova, E.K., Matrosova, V.Y., Vasilenko, A., Zhai, M., Daly, M.J., Koonin, E.V. and Makarova, K.S. (2005) Comparative genomics of *Thermus thermophilus* and *Deinococcus radiodurans*: divergent routes of adaptation to thermophily and radiation resistance. *BMC Evolutionary Biology* 5, 57.

Oshima, T. (2007) Unique polyamines produced by an extreme thermophile, *Thermus thermophilus*. *Amino Acids* 33, 367–372.

Owen, R., Renaut, R. and Jones, B. (2008) Geothermal diatoms: a comparative study of floras in hot spring systems of Iceland, New Zealand and Kenya. *Hydrobiologia* 610, 175–192.

Pan, W.Z., Huang, X.W., Wei, K.B., Zhang, C.M., Yang, D.M., Ding, J.M. and Zhang, K.Q. (2010) Diversity of thermophilic fungi in Tengchong Rehai National Park revealed by ITS nucleotide sequence analysis. *Korean Journal of Microbiology* 48, 146–152.

Papke, R.T., Ramsing, N.B., Bateson, M.M. and Ward, D.M. (2003) Geography isolation in hot spring cyanobacteria. *Environmental Microbiology* 5, 650–659.

Pe'er, I., Felder, C.E., Man, O., Silman, I., Sussman, J.L. and Beckmann, J.S. (2004) Proteomic signatures: amino acid and oligopeptide compositions differentiate among phyla. *Proteins* 54, 20–40.

Pentacost, A. (1996) High temperature ecosystems and their chemical interactions with the environment. In: *Evolution of Hydrothermal Ecosystems on Earth (and Mars?)*. CIBA Foundation Symposium 202. Wiley and Sons, New York, pp. 108–126.

Perevalova, A.A., Kolganova, T.V., Birkeland, N.-K., Schleper, C., Bronch-Osmolovskaya, E.A. and Lebedinsky, A.V. (2008) Distribution of Crenarchaeota representatives in terrestrial hot springs of Russia and Iceland. *Applied and Environmental Microbiology* 74, 7620–7628.

Perevalova, A.A., Bidzhieva, S.K., Kublanov, I.V., Hinrichs, K.U., Liu, X.L., Mardanov, A.V., Lebedinsky, A.V. and Bonch-Osmolovskaya, E.A. (2010) *Fervidicoccus fontis* gen. nov., sp. nov., a novel anaerobic thermophilic crenarchaeote from hot springs in Kamchatka, and proposal of Fervidicoccaceae fam. nov. and Fervidicoccales ord. nov. *International Journal of Systematic and Evolutionary Microbiology* 60, 2082–2088.

Portillo, M.C., Sririn, V., Kanoksilapatham, W. and Gonzalez, J.M. (2009) Pigment profiles and bacterial communities from Thailand thermal mats. *Antonie van Leeuwenhoek* 96, 559–567.

Prokofeva, M., Kublanov, I., Nercessian, O., Tourova, T., Kolganova, T., Lebedinsky, A., Bonch-Osmolovskaya, E., Spring, S. and Jeanthon, C. (2005) Cultivated anaerobic acidophilic/acidotolerant thermophiles from terrestrial and deep-sea hydrothermal habitats. *Extremophiles* 9, 437–448.

Prokofeva, M.I., Kostrikina, N.A., Kolganova, T.V., Tourova, T.P., Lysenko, A.M., Lebedinsky, A.V. and Bonch-Osmolovskaya, E.A. (2009) Isolation of the anaerobic thermoacidophilic crenarchaeote *Acidilobus saccharovorans* sp. nov. and proposal of Acidilobales ord. nov., including Acidilobaceae fam. nov. and Caldisphaeraceae fam. nov. *International Journal of Systematic and Evolutionary Microbiology* 59, 3116–3122.

Puigbò, P., Pasamontes, A. and Garcia-Vallve, S. (2008) Gaining and losing the thermophilic adaptation in prokaryotes. *Trends in Genetics* 24, 10–14.

Purcell, D., Sompong, U., Yim, L.C. and Barraclough, T.G. (2006) The effects of temperature, pH and sulphide on the community structure of hyperthermophilic streamers in hot springs of northern Thailand. *FEMS Microbiology Ecology* 60, 456–466.

Redman, R.S., Litvintseva, A., Sheehan, K.B., Henson, J.M. and Rodriguez, R.J. (1999) Fungi from geothermal soils in Yellowstone National Park. *Applied and Environmental Microbiology* 65, 5193–5197.

Reysenbach, A.-L. and Shock, E. (2002) Merging genomes with geochemistry in hydrothermal environments. *Science* 296, 1077–1082.

Reysenbach, A.-L., Wickham, G.S. and Pace, N.R. (1994) Phylogenetic analysis of the hyperthermophilic pink filament community in Octopus Spring, Yellowstone National Park. *Applied and Environmental Microbiology* 60, 2113–2119.

Reysenbach, A.-L., Ehringer, M. and Hershberger, K. (2000) Microbial diversity of 83°C in Calcite Springs, Yellowstone National Park: another environment where the Aquificales and 'Korarchaeota' coexist. *Extremophiles* 4, 61–67.

Reysenbach, A.-L., Hamamura, N., Podar, M., Griffith, E., Ferreira, S., Hochstein, R., Heidelberg, J., Mead, J.J.D., Pohorille, A., Sarmiento, M., Schweighofer, K., Seshadri, R. and Voytek, M.A. (2009) Complete draft genome sequences of six members of the Aquificales. *Journal of Bacteriology* 191, 1992–1993.

Robb, F.T. and Newby, D.T. (2007) Fuctional genomics in thermophilic microorganisms. In: Gerday, C. and Glandsdorff, N. (eds) *Physiology and Biochemistry of Extremophiles.* ASM Press, Washington, DC, pp. 30–38.

Rohr, U., Weber, S., Michel, R., Selenka, F. and Wilhelm, M. (1998) Comparison of free-living amoebae in hot water systems of hospitals with isolates from moist sanitary areas by identifying genera and determining temperature tolerance. *Applied and Environmental Microbiology* 64, 1822–1824.

Sabadell, J.E. and Axtmann, R.C. (1975) Heavy metal contamination from geothermal sources. *Environmental Health Perspectives* 12, 1–7.

Sayeh, R., Birrien, J.L., Alain, K., Barbier, G., Hamdi, M. and Prieur, D. (2010) Microbial diversity in Tunisian geothermal springs as detected by molecular and culture based approaches. *Extremophiles* 12, 501–514.

Schwartzman, D.W. and Lineweaver, C.H. (2004) The hyperthermophilic origin of life revisited. *Biochemical Society Transactions* 32, 168–171.

Sharp, R.J., Riley, P.W. and White, D. (1992) Hetrotrophic thermophilic *Bacilli*. In: Kristjansson, J.K. (ed.) *Thermophilic Bacteria*. CRC Press, Florida, pp. 19–51.

Sheehan, K.B., Ferris, M.J. and Henson, J.M. (2003) Detection of *Naegleria* sp. in a thermal, acidic stream in Yellowstone National Park. *Journal of Eukaryotic Microbiology* 50, 263–265.

Shock, E.L., McCollom, T. and Schulte, M.D. (1998) The emergence of metabolism from within hydrothermal systems. In: Wiegel, J. and Adams, M.W.W. (eds) *Thermophiles: the Keys to Molecular Evolution and the Origin of Life?* Taylor and Francis, Philadelphia, pp. 59–76.

Singh, S., Madlala, A.M. and Prior, B.A. (2003) *Thermomyces lanuginosus*: properties of strains and their hemicellulases. *FEMS Microbiology Reviews* 27, 2–16.

Skirnisdottir, S., Hreggvidsson, G.O., Hjorleifsdottir, S., Marteinsson, V.T., Petursdottir, S.K., Holst, O. and Kristjansson, J.K. (2000) Influence of sulphide and temperature on species composition and community structure of hot spring microbial mats. *Applied and Environmental Microbiology* 66, 2835–2841.

Song, Z.-Q., Chen, J.Q., Jiang, H.-C., Zhou, E.-M., Tang, S.-K., Zhi, X.-Y., Zhang, L.-X., Zhang, C.-L.L. and Li, W.-J. (2010) Diversity of Crenarchaeota in terrestrial hot springs in Tengchong, China. *Extremophiles* 14, 287–296.

Spear, J.R., Walker, J.J., McCollom, T.M. and Pace, N.R. (2005) Hydrogen and bioenergetic in Yellowstone geothermal ecosystem. *Proceedings of the National Academy of Sciences of the United States of America* 102, 2555–2560.

Sterner, R. and Liebl, W. (2001) Thermophilic adaptation of proteins. *Critical Reviews in Biochemistry and Molecular Biology* 36, 39–106.

Stetter, K.O. (1998) Hyperthermophiles; isolation, classification and properties. In: Horikoshi, K. and Grant, W.D. (eds) *Extremophiles: Microbial Life in Extreme Environments*. Wiley, New York, pp. 1–24.

Stetter, K.O., Fiala, G., Huber, G., Huber, R. and Segerer, A. (1990) Hyperthermophilic microorganisms. *FEMS Microbiology Letters* 75, 117–124.

Stout, L.M., Blake, R.E., Greenwood, J.P., Martini, A.M. and Rose, E.C. (2009) Microbial diversity of boron-rich volcanic hot springs of St. Lucia, Lesser Antilles. *FEMS Microbiology Ecology* 70, 402–412.

Suhre, K. and Claverie, J.M. (2003) Genomic correlates of hyperthermostability, an update. *Journal Biological Chemistry* 278, 17198–17202.

Takami, H., Takaki, Y., Chee, G.J., Nishi, S., Shimamura, S., Suzuki, H., Matsui, S. and Uchiyama, I. (2004) Thermoadaptation trait revealed by the genome sequence of thermophilic *Geobacillus kaustophilus*. *Nucleic Acids Research* 32, 6292–6303.

Taylor, M.P., Eley, K.L., Martin, S., Tuffin, M.I., Burton, S.G. and Cowan, D.A. (2009) Thermophilic ethanologenesis: future prospects for second-generation bioethanol production. *Trends in Biotechnology* 27, 398–405.

Tehi, M., Franzetti, B., Maurel, M.-C., Vergne, J., Hountondji, C. and Zaccai, G. (2002) The search for traces of life: the protective effect of salt on biological macromolecules. *Extremophiles* 6, 427–430.

Tekaia, F. and Yeramian, E. (2006) Evolution of proteomes: fundamental signatures and global trends in amino acid compositions. *BMC Genomics* 7, 307.

Terui, Y., Ohnuma, M., Hiraga, K., Kawashima, E. and Oshima, T. (2005) Stabilization of nucleic acids by unusual polyamines produced by an extreme thermophile, *Thermus thermophilus*. *Biochemistry Journal* 388, 427–433.

Thakore, Y. (2008) Enzymes for industrial applications BIO030E. Available at: www.bccresearch.com/report/BIO030E.html (accessed 2 April 2011).

Toplin, J.A., Norris, T.B., Lehr, C.R., McDermott, T.R. and Castenholz, R.W. (2008) Biogeographic and phylogenetic diversity of thermoacidophilic Cyanidiales in Yellowstone National Park, Japan and New Zealand. *Applied and Environmental Microbiology* 74, 2822–2833.

Trotsenko, Y., Medvedkova, K., Khmelenina, V. and Eshinimayev, B. (2009) Thermophilic and thermotolerant aerobic methanotrophs. *Microbiology* 78, 387–401.

Ulrih, N.P., Gmajner, D. and Raspor, P. (2009) Structural and physiochemical properties of polar lipids from thermophilic Archaea. *Applied Microbial Biotechnology* 84, 249–260.

Van der Meer, M.T.J., Schouten, S., Bateson, M.M., Nubel, U., Wieland, A., Kuhl, M., de Leeuw, J.W., Sinninghe-Damste, J.S. and Ward, D.M. (2005) Diel variations in carbon metabolism by green nonsulfur-like bacteria in alkaline siliceous hot spring microbial mats from Yellowstone National Park. *Applied and Environmental Microbiology* 71, 3978–3986.

Vieille, C. and Zeikus, G.J. (2001) Hyperthermophilic enzymes: sources, uses, and molecular mechanisms for thermostability. *Microbiology and Molecular Biology Reviews* 65, 1–43.

Villar, J.S. and Edwards, H. (2006) Raman spectroscopy in astrobiology. *Analytical and Bioanalytical Chemistry* 384 (1), 100–113.

Ward, D.M. (1998) A natural species concept for prokaryotes. *Current Opinion in Microbiology* 1, 271–277.

Ward, D.M. (2006) Microbial diversity in natural environments: focusing on fundamental questions. *Antonie van Leeuwenhoek* 90, 309–324.

Ward, D.M., Bateson, M.M., Ferris, M.J., Kühl, M., Wieland, A., Koeppel, A. and Cohan, F.M. (2006) Cyanobacterial ecosystem in the microbial mat community of Mushroom Spring (Yellowstone National Park, Wyoming) as species-like units linking microbial community composition, structure and function. *Philosophical Transactions of the Royal Society B: Biological Sciences* 361, 1997–2008.

Waring, G.A. (1965) *Thermal springs of the United States and other countries of the world – a summary. Geological Survey Professional Paper* 492. United States Government Printing Office, 383 pp.

Weissberg, B.G. (1969) Gold-silver ore-grade precipitates from New Zealand thermal waters. *Economic Geology* 64, 95.

Wiegel, J. and Adams, M.W.W. (1998) *Thermophiles: the keys to molecular evolution and the origin of life.* Taylor and Francis, London, 346 pp.

Wilson, M., Siering, P., White, C., Hauser, M. and Bartles, A. (2008) Novel Archaea and bacteria dominate stable microbial communities in North America's largest hot spring. *Microbial Ecology* 56, 292–305.

Wolf, Y.I., Rogozin, I.B., Grishin, N.V., Tatusov, R.L. and Koonin, E.V. (2001) Genome trees constructed using five different approaches suggest new major bacterial clades. *BMC Evolutionary Biology* 1, 8–29.

Yamazaki, A., Toyama, K. and Nakagiri, A. (2010) A new acidophilic fungus *Teratosphaeria acidotherma* (Capnodiales, Ascomycota) from a hot spring. *Mycoscience* 51, 443–455.

Zeldovich, K.B., Berezovsky, I.N. and Shakhnovich, E.I. (2007) Protein and DNA sequence determinants of thermophilic adaptation. *PLoS Computational Biology* 3, e5.

13 Deep-Sea Hydrothermal Vents

Richard A. Lutz
Institute of Marine and Coastal Sciences, Rutgers University, New Brunswick, USA

13.1 Introduction

With temperatures ranging from 2°C to over 400°C, pressures in excess of several thousand pounds per square inch (psi), and high concentrations of toxic hydrogen sulfide, deep-sea hydrothermal vents are arguably the most extreme environments on our planet. The present chapter summarizes a spectrum of exciting discoveries that have been made since the first expedition to these sites in 1977 and the implications that these discoveries have in providing insights into the origin of life on our own planet, as well as on extraterrestrial bodies. The chapter is also designed to impart to the reader an appreciation for the phenomenal rates at which both biological and geological processes are occurring in these unusual deep-sea ecosystems, despite their extremity.

The Earth's surface is divided into a series of huge geological plates that move around relative to one another. At the boundaries of these plates, three possible things happen: (i) the plates move apart, in which case you have a 'spreading centre'; (ii) one plate moves underneath another plate, in which case you have a 'subduction zone'; or (iii) one plate slides past another plate. Most vents on which this chapter focuses are located along spreading centres.

13.2 Diversity of Life at Deep-sea Hydrothermal Vent Sites

It was in 1977 that a group of geologists and geochemists set sail with the deep diving submersible *DSV Alvin* (Fig. 13.1) to a spreading centre located about 320 km off the coast of the Galapagos Islands without a single biologist on board the ship. Yet they were to make one of the most significant biological oceanographic discoveries of the century, if not of all time. They descended 2500 m to the crest of an undersea volcanic ridge known as the Galapagos Rift and discovered an ecosystem based not on photosynthesis, as every other ecosystem that we know on Earth, but rather one based on chemosynthesis, where microbes, serving as the lower link of the food chain, were deriving their energy from geothermal heat and chemicals deep within the Earth's interior. Every organism that they saw during that first dive to the Galapagos Rift (Fig. 13.2) and every organism depicted in this chapter was unknown to science in 1977.

In order to place in perspective the significance of this exciting discovery, it is important to understand our perception of life in the deep sea prior to this time. Essentially all deep-sea environments visited prior to 1977 were characterized by extremely low biomass (Fig. 13.3) and

Fig. 13.1. *DSV Alvin* and its mother ship, *R/V Atlantis*. © Emory Kristof.

Fig. 13.2. Photo taken in 1977 during the expedition to hydrothermal vents along the Galapagos Rift. Organisms within the image include: vestimentiferan tubeworms with dark red plumes, a bythitid fish (*Bythites hollisi*) in the centre of the photo, brachyuran crabs (*Bythograea thermydron*) (bottom centre and top right), serpulid polychaetes (lower left), and numerous species of archaeogastropod limpets attached to basaltic surfaces throughout the image. © John M. Edmond.

considered too extreme for much activity to proceed (see also Hughes, Chapter 15, this volume). However, one of the first insights into the rates at which biological processes were occurring in these environments was gleaned from a fortuitous experiment involving the same submersible (*Alvin*) that first visited the vents.

Fig. 13.3. A typical soft-sediment, deep-sea environment at a depth of 2500 m in the eastern Pacific. © Richard A. Lutz.

During the commencement of a dive in October 1968, while *Alvin* was being lowered down on its elevator between the pontoons of its mother ship, the *R/V Lulu*, the steel cables holding the elevator snapped and *Alvin* went down below the surface and started filling with water – the pilot and the two scientists scrambled out of the submersible just before it completely filled with water and sank to a depth of 1500 m. *Alvin* remained on the bottom with its hatch open for 10 months, after which time it was finally retrieved. One of the things that went down with *Alvin* was the food that would have been the lunch for the three occupants of the submersible. When *Alvin* was finally brought back to the surface and the recovery team opened the lunch box, they found a bologna sandwich and an apple that appeared remarkably intact, with little or no degradation – in fact, scientists even took a bite out of the bologna sandwich. This led to the hypothesis that microbiological processes were occurring at extremely low rates in deep sea (Jannasch *et al.*, 1971). Shortly after this, the shell of a small clam, *Tindaria callistiformis*, only 8 mm long and collected from a depth of approximately 4000 m in the Atlantic, was analysed radiometrically and estimated to be 100 years old (Turekian *et al.*, 1975). Thus, as late as 1975, there were two independent sources of evidence that biological processes in the deep sea were proceeding very slowly. As will become apparent later in this chapter, this is in marked contrast to the rates at which biological processes are proceeding at deep-sea hydrothermal vents.

That first dive to the Galapagos Rift vent field in 1977 revealed a spectrum of highly unusual megafauna. There were extensive beds of large white clams (*Calyptogena magnifica*), with the shells of some individuals reaching lengths in excess of 30 cm. They had dark red haemoglobin (Fig. 13.4) – actually intracellular haemoglobin very similar to that of human blood – an unusual, but not unique, feature within certain families of bivalve molluscs. The clams had a large, rugose foot that they extended into cracks and crevices where the water temperatures reached approximately 20°C (ambient temperatures 2–4°C; Fig. 13.5). These warm waters were loaded with hydrogen sulfide, present at concentrations up to about 200 μM. Subsequent studies would reveal that this hydrogen sulfide provides nutrition for chemosynthetic

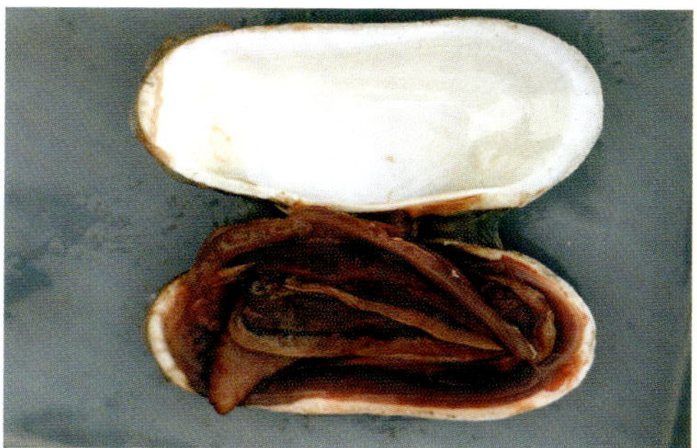

Fig. 13.4. The disarticulated valves and associated soft tissues of a 26 cm-long vesicomyid clam (*Calyptogena magnifica*) from a hydrothermal vent located at a depth of 2500 m along the Galapagos Rift. The dark red coloration of the soft tissues is due to the presence of intracellular haemoglobin. © Richard A. Lutz.

Fig. 13.5. A vesicomyid clam (*Calyptogena magnifica*) with an extended, large rugose foot as seen at a hydrothermal vent at a depth of 2500 m along the Galapagos Rift. © James J. Childress.

bacteria that serve as the lower link of the food chain.

There were large mussels (*Bathymodiolus thermophilus*), over 10 cm in length (Fig. 13.6), that were also nestled within the cracks and crevices of the basaltic lava where venting fluids reached temperatures up to 18°C. They were often seen clustered in large beds that were frequently surrounded by numerous galatheid crabs (*Munidopsis subsquamosa*) (Fig. 13.7).

Perhaps the most unusual organism encountered during that first dive was the vestimentiferan tubeworm *Riftia pachyptila*. Subsequent dives to a nearby vent field

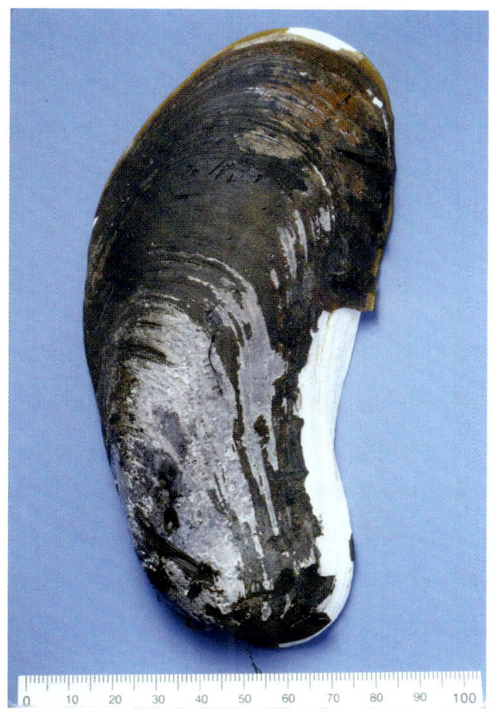

Fig. 13.6. The vent mussel *Bathymodiolus thermophilus* from a Galapagos Rift vent field (depth = 2500 m). © Richard A. Lutz.

along the Galapagos Rift known as 'Rose Garden' revealed large colonies of this species, with individuals reaching lengths of up to 2 m (Fig. 13.8). These tubeworms have no mouth, no stomach, no anus (i.e. no digestive system) – they are a completely closed sack with no opening to the external environment. The bright red colour of the plume at the tip is due to the presence of extracellular haemoglobin. Hydrogen sulfide is transported across the cell membrane associated with the filaments of this plume and is transported via the extracellular haemoglobin to a specialized 'tissue' called the trophosome, which is comprised entirely of symbiotic chemosynthetic bacteria. In essence, the tubeworms have an internal 'garden' of bacteria that is nourished by the hydrogen sulfide and provides the 'food' for the tubeworm – an extraordinary adaptation. These unusual organisms were subsequently placed in a new phylum,

Fig. 13.7. A large bed of mussels (*Bathymodiolus thermophilus*) (upper right), together with numerous galatheid crabs (*Munidopsis subsquamosa*) (lower left). © Richard A. Lutz.

Fig. 13.8. A colony of vestimentiferan tubeworms (*Riftia pachyptila*) at the 'Rose Garden' hydrothermal vent along the Galapagos Rift (depth = 2500 m). © J. Frederick Grassle.

Vestimentifera (Jones, 1985), although more recently they found their way into the phylum Annelida (Southward *et al.*, 2005).

Porcelain-white brachyuran crabs were common throughout the Galapagos Rift vent fields. They are scavengers, feeding on everything from microbes to molluscs to the plumes of tubeworms. The crabs (*Bythograea thermydron*) were subsequently placed in a new genus by Williams (1980) (Fig. 13.2).

Serpulid polychaetes, *Laminatubus alvini*, were abundant annelids in peripheral areas of various vents along the Galapagos Rift. Their tubes reach lengths of between 5 and 10 cm and are comprised of calcite. A 'feather-like' plume at the end is apparent on living specimens, hence, the name 'feather-dusters', which is often applied to them (Fig. 13.9).

Fig. 13.9. Serpulid polychaetes (*Laminatubus alvini*) in a peripheral area of a hydrothermal vent field along the Galapagos Rift (depth = 2500 m). © J. Frederick Grassle.

Enteroptneusts or 'spaghetti worms' (*Saxipendium coronatum*) were also abundant in the peripheral areas of the Galapagos Rift hydrothermal fields (Fig. 13.10). These organisms have not been commonly encountered at other vent sites throughout the world's oceans, although they were observed as new colonizers at 9°50'N along the East Pacific Rise following a massive volcanic eruption in the region (see description later in this chapter of biological succession following this ridge crest eruptive event; Lutz *et al.*, 2001).

Colonial siphonophores (called 'dandelions' when first encountered by the geologists and geochemists in 1977) were also abundant in the peripheral areas of the vent site (Fig. 13.11). The colony consists of several hundred polyps (all with specific functions) – it was subsequently designated a new genus of coelenterate (*Thermopalia taraxaca*).

Fig. 13.10. Enteroptneusts or 'spaghetti worms' (*Saxipendium coronatum*) in a peripheral area of a hydrothermal vent field along the Galapagos Rift (depth = 2500 m). © James J. Childress.

Fig. 13.11. Colonial siphonophores (*Thermopalia taraxaca*) in a peripheral area of a hydrothermal vent field along the Galapagos Rift (depth = 2500 m). © J. Frederick Grassle.

All of the megafaunal organisms shown in the images presented thus far were encountered at the Galapagos Rift vent fields. At other vent fields visited over the past three decades in the eastern Pacific, many of these same species were encountered, as were numerous other new species. In 1979, a group of geologists (once again without a single biologist as part of the expedition) discovered a site at 21 °N along the East Pacific Rise with a remarkably similar fauna to that encountered along the Galapagos Rift, despite the distance of several thousand kilometres between the two sites. There were large fields of vesicomyid clams (*C. magnifica*) (Fig. 13.12) and tubeworms (*R. pachyptila*). There were also spectacular polymetallic sulfide chimneys, called 'black smokers', spewing forth mineral-laden waters at temperatures of 350°C, the boiling point of water at a depth of 2500 m (Fig. 13.13). The term 'polymetallic chimney' comes from the fact that these structures are comprised of some of the world's richest mineral deposits, with high concentrations of iron, copper, zinc and a wide spectrum of trace metals. The black 'smoke' is actually not smoke at all, but is precipitating minerals, in particular iron sulfide (pyrite). Many other similar chimneys have been found elsewhere along spreading centres throughout the world's oceans. Some of these can get to be quite large; one known as 'Godzilla' reached a height of approximately 50 m before collapsing, presumably during an undersea earthquake. A large chimney (the height of a five-storey building) known as 'Rebecca's Roost' (named after the daughter of the author of this chapter), located in Guaymas Basin in the Gulf of California, is depicted in Fig. 13.14. A polychaete annelid (*Alvinella pompejana*), named after the submersible *Alvin*, is also a common inhabitant of the sides of many of these chimneys found in the eastern Pacific and is claimed to be one of the most eurythermal organisms (capable of tolerating a wide range of temperatures) on the planet, withstanding temperatures ranging from approximately 2°C to over 50°C (Fig. 13.15).

At a site located at 13 °N along the East Pacific Rise, that was first visited by French scientists diving in the submersible *Cyana* in the early 1980s, a spectacular grove of

Fig. 13.12. A large field of vesicomyid clams (*C. magnifica*) at a vent field known as 'Clam Acres' at 21 °N along the East Pacific Rise (depth = 2500 m). © Richard A. Lutz.

Fig. 13.13. A polymetallic sulfide chimney (a 'black smoker') spewing forth mineral-laden waters at a temperature of 350°C at a hydrothermal vent field at 21 °N along the East Pacific Rise (depth = 2500 m). © Richard A. Lutz.

tubeworms (*R. pachyptila*) (Fig. 13.16) was encountered in an area that, just a few years earlier, was filled with only dead or dying vent organisms. The site was named 'Genesis', reflecting the spectacular 'rebirth' of an entire hydrothermal vent field.

13.3 How Fast Do Biological and Geological Processes Occur at Deep-sea Hydrothermal Vents?

One of the fundamental questions concerning deep-sea hydrothermal vents is 'what

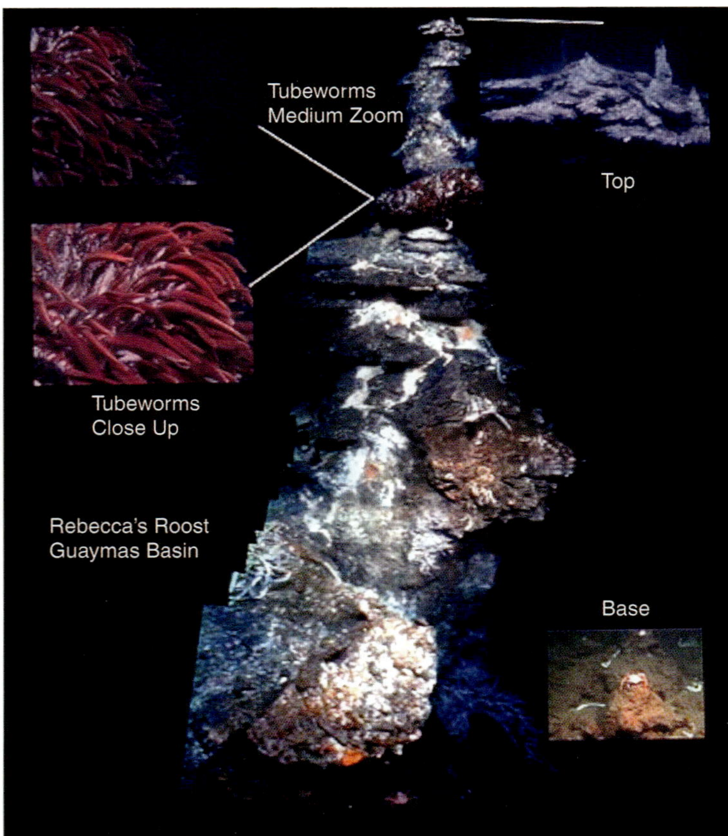

Fig. 13.14. A large polymetallic sulfide chimney known as 'Rebecca's Roost' in Guaymas Basin in the Gulf of California, Mexico. Mosaic constructed from images collected in April 1998 and compiled by Hanumant Singh and Dana Yoerger of the Woods Hole Oceanographic Institution.

Fig. 13.15. A polychaete annelid (*Alvinella pompejana*) on the side of a black smoker at a hydrothermal vent field at 9°50′N along the East Pacific Rise (depth = 2500 m). © Richard A. Lutz.

Fig. 13.16. A massive colony of vestimentiferans (*Riftia pachyptila*) at a hydrothermal vent field known as 'Genesis' located at 13 °N along the East Pacific Rise. © Richard A. Lutz.

are the rates of biological and geological processes at the vents?' For example, how rapidly do the massive polymetallic sulfide chimneys, such as the one depicted in Fig. 13.14, form? Similarly, how old are the organisms in communities such as the one depicted in Fig. 13.16 and how fast do the organisms grow? Earlier in this chapter, reference was made to a bologna sandwich that remained edible after sitting on the bottom of the deep-sea floor for a period of nearly a year, suggesting that microbiological processes occur extremely slowly in the deep sea. It was also pointed out that an 8 mm long clam from a depth of approximately 4000 m in the Atlantic was estimated to be about 100 years old. These two independent sources of evidence, coupled with an intuitive sense that it should take a while to form a five-storey high polymetallic sulfide chimmney suggest that the rates of biological processes in a 'typical' deep-sea environment are extremely slow. However, over the past few decades a lot more has been learned about the rates at which biological and geological processes are proceeding at deep-sea hydrothermal vents.

In November/December 1989, Rachel Haymon of the University of California at Santa Barbara led an expedition during which the near-bottom ARGO imaging system was used to survey the East Pacific Rise from 9 ° to 10 °N, a distance of about 60 nautical miles (Haymon *et al.*, 1991). On the crest of this ridge axis is a feature known as the Axial Summit Caldera (ASC), subsequently named the Axial Summit Trough (AST), which is like a narrow highway. Its width ranges from a few metres to several hundred metres and the height of the walls bordering the highway range from less than 1 m to approximately 10 m high. This 'highway' represents the middle of a tectonically- and volcanically-active spreading centre. During the course of the cruise, Dr Haymon and her colleagues surveyed approximately 80% of the ridge crest between 9 ° and 10 °N and documented the numerous geological and biological features associated with hydrothermal activity in this region. Figure 13.17 is typical of the images they took during the survey, with this particular one (taken at a height of 9.7 m above the seafloor) depicting a massive biological community.

In April 1991, Dr Haymon led another expedition to the region, this time armed with the human-occupied submersible *DSV*

Alvin, in order to visit and sample the geological and biological features that had been documented during the November/December 1989 expedition. The image depicted in Fig. 13.18 was taken in exactly the same region as the image in Fig. 13.17 and depicts a massive colony of tubeworms (*R. pachyptila*) atop what appeared to be a lava pillar – the structure was named 'Tubeworm Pillar' and would be the focus

Fig. 13.17. A massive biological community photographed using the near-bottom ARGO imaging system in November/December 1989 at a depth of 2500 m at 49.6 °N along the East Pacific Rise. © Rachel M. Haymon.

Fig. 13.18. An extensive colony of tubeworms (*R. pachyptila*) on top of a lava structure at a depth of 2500 m at 49.6 °N along the East Pacific Rise photographed using an external camera system mounted on *DSV Alvin* in April, 1991. The photograph was taken in the exact same region as that depicted in Fig. 13.17. The structure was named 'Tubeworm Pillar'. © Rachel M. Haymon.

of numerous studies over the course of subsequent years (see Shank *et al.*, 1998).

During the April 1991 expedition, Haymon and her colleagues visited all of the biological communities that had been seen during the November/December 1989 expedition. During the first few *Alvin* dives, they encountered both biological communities and geological structures that were anticipated based on the images taken in 1989. There were massive groves of tubeworms, large beds of mussels, large fields of vesicomyid clams (*Calyptogena magnifica*), zoarcid (*Thermarces andersoni*) (Fig. 13.19) and bythitid (*Bythites hollisi*) (Fig. 15.20) fish (Rosenblatt and Cohen, 1986; Cohen *et al.*, 1990), alvinellid polychaetes (*A. pompejana*), numerous species of limpets and coiled archaeogastropods, and large numbers of brachyuran (*Bythograea thermydron*) (Fig. 13.19) and galatheid (*M. subsquamosa*) crabs (Fig. 13.21) – assemblages of hydrothermal vent fauna very similar to those that had been seen at numerous other sites in the

Fig. 13.19. The zoarcid fish *Thermarces andersoni* amidst the tubeworms *Riftia pachyptila* and *Tevnia jerichonana* at a hydrothermal vent located at 9°50′N along the East Pacific Rise (depth 2500 m). Several vent crabs (*Bythograea thermydron*) are also within the field of view. © Richard A. Lutz.

Fig. 13.20. The bythitid fish *Bythites hollisi* with a black smoker in the background in a hydrothermal vent field at 9°50′N along the East Pacific Rise (depth 2500 m). © Richard A. Lutz.

eastern Pacific over the years. One thing that they encountered that had not been seen at previous hydrothermal vent sites was a large field of stalked barnacles (*Neolepas zevinae*) (Figs 13.22 and 13.23); only a few isolated specimens of this species had been previously encountered at vent fields along the Galapagos Rift and at 21°N along the East Pacific Rise. There were also numerous smokers with fluids exiting at 350°C.

Fig. 13.21. Galatheid crabs (*Munidopsis subsquamosa*) lining a fissure near 9°50′N along the East Pacific Rise (depth 2500 m). Photo taken with a camera mounted on the sponson of *DSV Alvin*; courtesy of Woods Hole Oceanographic Institution.

Fig. 13.22. A large field of stalked barnacles (*Neolepas zevinae*) at a hydrothermal vent at 9°52′N along the East Pacific Rise (depth 2500 m). © Richard A. Lutz.

All of the biological assemblages and geological features described in the previous paragraph were things that were expected to be encountered during the April 1991 expedition based on what had been seen in the images taken during the November/ December 1989 expedition. What the scientists did not expect to encounter was the scene depicted in Fig. 13.24. In 1989, a thriving biological community inhabited this

Fig. 13.23. Stalked barnacles (*Neolepas zevinae*) collected from a hydrothermal vent at 9°52′N along the East Pacific Rise (depth 2500 m). © Richard A. Lutz.

Fig. 13.24. A site known as 'Tubeworm Barbeque' at 9°51′N along the East Pacific Rise as seen in April, 1991 (depth, 2500 m). Dead tubeworms and mussels litter the seafloor and there is a fine, grey ash coating shiny, fresh lava. In November 1989 this site had been colonized by a thriving biological community. Photo taken with a camera mounted on the sponson of *DSV Alvin*; courtesy of Woods Hole Oceanographic Institution.

region. It was now the scene of death and destruction. Dead tubeworms and mussels littered the seafloor and there was a fine, grey ash coating shiny, fresh lava. Some of the tubeworms were actually encased in lava in much the same manner as trees are often encased in fresh lava in places like Hawaii (Fig. 13.25). A number of dead tubeworms were retrieved from the site. In Fig. 13.26, the base of one tubeworm is depicted with what at first appeared to be a piece of lava, as the tubeworms attach directly to the basalt.

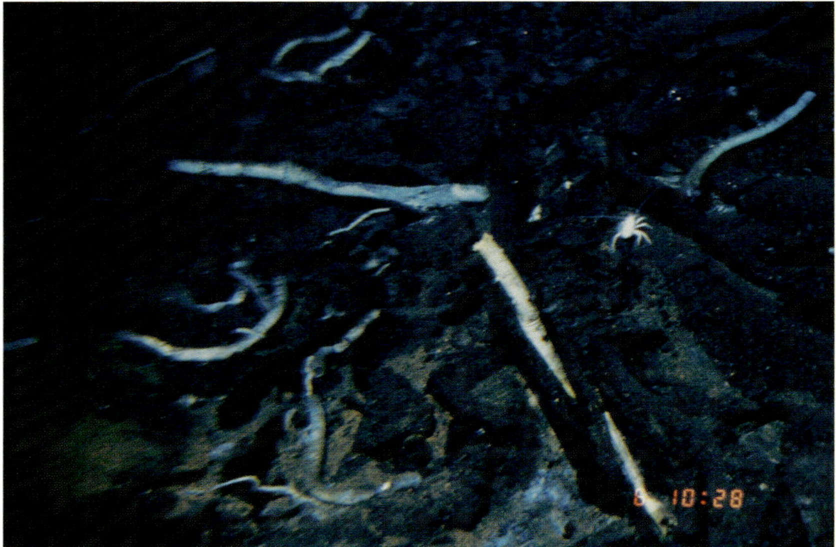

Fig. 13.25. Dead tubeworms encased in fresh lava at a site known as 'Tubeworm Barbeque' at 9°51′N along the East Pacific Rise as seen in April 1991 (depth, 2500 m). © Rachel M. Haymon.

Fig. 13.26. The base of a tubeworm retrieved in April 1991 from a site known as 'Tubeworm Barbeque' at 9°51′N along the East Pacific Rise (depth, 2500 m). The blackened material is seared trophosome, the 'tissue' containing the tubeworm's symbiotic bacteria. © Richard A. Lutz.

However, closer examination under the dissecting microscope revealed that the blackened material was not lava at all, but that it was seared trophosome, the 'tissue' containing the tubeworm's symbiotic bacteria. It seemed that the tubeworm had literally been 'cooked'. Limpets that were peeled off the outside of the tube had been similarly seared and cooked (Fig. 13.27). The scientists appropriately named the site 'Tubeworm Barbeque'. When a site slightly south of the 'Tubeworm Barbeque' area was visited, it became clear what had happened: a massive, new lava flow had completely blanketed the area, killing all of the organisms, with the exception of those that were high enough above the flow to escape destruction (Fig. 13.28). In this same general region were a number of high-temperature vents nicknamed 'snow-blowers', where a massive amount of material of microbial origin was exiting from the earth's interior (Fig. 13.29), a sight never seen before at deep-sea vents previously visited. The entire floor of the axial summit caldera in this region was covered with a microbial mat consisting of bacteria, anywhere from 5 to 10 cm thick (Fig. 13.30).

Fig. 13.27. Limpets removed from the external surface of a tube of a dead tubeworm retrieved in April 1991 from a site known as 'Tubeworm Barbeque' at 9°51′N along the East Pacific Rise. (depth, 2500 m). Note the 'seared and cooked' nature of the soft tissues of the limpets. © Richard A. Lutz.

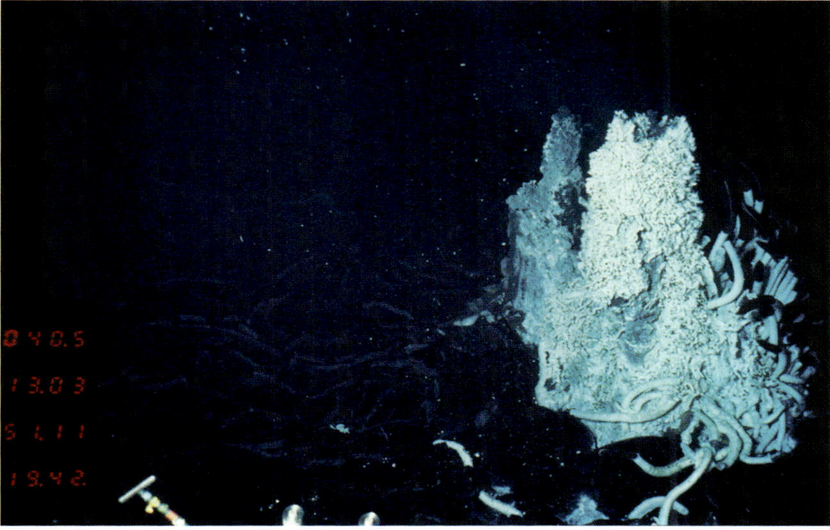

Fig. 13.28. The upper portion of a sulfide edifice covered with alvinellid polychaetes and tubeworms at 9°49.7′N along the East Pacific Rise as seen in April 1991 (depth, 2500 m). The bottom portion of the edifice, as well as any surrounding biological assemblages, has been buried by a fresh lava flow. Photo taken with a camera mounted on the sponson of *DSV Alvin*; courtesy of Woods Hole Oceanographic Institution.

Among the most startling discoveries during the course of the April 1991 expedition were areas where 'black smoke' was exiting from bare basalt at a temperature of 403°C, over 50°C above the theoretical maximum temperature possible at this depth. The scientists were in fact witnessing a never-before-seen phenomenon in the deep sea known as 'vapour phase separation' (Fig. 13.31). No sulfide chimneys were associated with this

Fig. 13.29. A 'snow-blower' vent at 9°50′N along the East Pacific Rise as seen in April 1991 (depth, 2500 m). Massive amounts of material of microbial origin are seen exiting from a vent opening in the ocean floor. Photo taken with a camera mounted on the sponson of *DSV Alvin*; courtesy of Woods Hole Oceanographic Institution.

Fig. 13.30. The floor of the axial summit caldera covered with a microbial mat 5–10 cm thick at 9°50′N along the East Pacific Rise as seen in April 1991 (depth, 2500 m). Photo taken with a camera mounted on the sponson of *DSV Alvin*; courtesy of Woods Hole Oceanographic Institution.

hot exiting fluid and it was at this point that the scientists on board realized that they were witnessing the birth of a hydrothermal vent and had an opportunity to follow biological and geological changes from the initiation of such a hydrothermal system.

In this region, there was also an area into which the *Alvin*'s pilots were afraid to descend; a cloudy cauldron lined with microbial mats that they nicknamed 'Hole to Hell' (Fig. 13.32). Here they suspected that hot magma lay at the bottom of the 'hole'.

Fig. 13.31. Hot water exiting from bare basalt at 9°50′N along the East Pacific Rise as seen in April 1991 (depth, 2500 m). The measured temperature of the exiting water was 403°C, over 50°C above the theoretical maximum temperature possible at this depth.

Fig. 13.32. A region known as 'Hole to Hell' at 9°50′N along the East Pacific Rise as seen in April 1991 (depth, 2500 m). Photo taken with a camera mounted on the sponson of *DSV Alvin*; courtesy of Woods Hole Oceanographic Institution.

Fig. 13.33. A photo taken in March 1992 in the same region ('Hole to Hell' at 9° 50′N along the East Pacific Rise) as the photo depicted in Fig. 13.32. Photo taken with a camera mounted on the sponson of *DSV Alvin*; courtesy of Woods Hole Oceanographic Institution.

The scene depicted in Fig. 13.32 provides a convenient springboard to propel the reader forward in time to the next expedition to the area that took place 11 months later in March 1992. Figure 13.33, taken during this new expedition, depicts the same region. There was a marked reduction in the amount of microbial material in the area, the water was less cloudy, and a variety of typical vent organisms had colonized the area. There were large numbers of amphipods, copepods and brachyuran crabs and, perhaps most impressively, large numbers of an early colonizing tubeworm known as *Tevnia jerichonana* (Fig. 13.34), a different genus of tubeworm to the giant tubeworm (*Rifia pachyptila*) that was common throughout the region in November/December 1989. Where the scientists encountered 403°C water in April 1991 (Fig. 13.31) exiting from bare basalt, there were now sulfide chimneys about 1–2 m high (Fig. 13.35).

Fig. 13.34. Specimens of the tubeworm *Tevnia jerichonana* collected in March 1992 from the region depicted in Fig. 13.33 ('Hole to Hell' at 9°50′N along the East Pacific Rise). © Richard A. Lutz.

Numerous markers were deployed at various sites in the region. Figure 13.36 depicts a hydrothermally active region as it appeared in March 1992 with a number of such markers (notably markers 9, 10, 16 and 17) in the field of view. Within this region there were numerous *T. jerichonana* tubeworms but a notable absence of *R. pachyptila* tubeworms. In marked contrast, 21 months later a tremendous grove of *R. pachyptila* tubeworms with individual tubes up

Fig. 13.35. A photo taken in March 1992 in the same region (at 9°50′N along the East Pacific Rise) as depicted in Fig. 13.31. © Richard A. Lutz.

Fig. 13.36. A hydrothermally active region ('Hole to Hell' at 9°50′N along the East Pacific Rise; depth, 2500 m) as seen in March 1992. Note numerous markers (9, 10, 16 and 17) deployed with an eye towards documenting future changes occurring at this site. © Richard A. Lutz.

to 1.5 m long was apparent near marker 16 (Fig. 13.37 is taken at the lower left-hand corner of the previous image). This translates to a minimum average growth rate of some of the tubes of 2–3 mm day^{-1} over the 21-month period. This is not only fast growth for an organism in the deep sea, but it is the fastest growing marine invertebrate known on the planet (Lutz *et al.*, 1994), quite a contrast to the low rates of biological processes surmised from the slow microbial activity degradation of that bologna sandwich!

Equally, if not more, impressive were the rates of geological processes that occurred over the course of less than 3 years. Figure 13.38 depicts a microbial-covered area as it appeared in April 1991, with water exiting directly from a crack in the basalt, with no associated sulfide deposits. Figure 13.39 is the same region in December 1993 depicting the base of a polymetallic sulfide chimney that is over 7 m high, with numerous tubes of the polychaete *A. pompejana* coating the chimney's exterior. In the course

Fig. 13.37. An image taken in December 1993 in exactly the same area as the photo depicted in Fig. 13.36. Marker #16, which is also seen in Fig. 13.36, is apparent in the lower left-hand corner of this image, surrounded by a tremendous grove of *Rifta pachyptila* tubeworms with individual tubes up to 1.5 m long. © Richard A. Lutz.

Fig. 13.38. A microbial-covered area at 9°50′N along the East Pacific Rise (depth, 2500 m) as it appeared in April 1991, with water exiting directly from a crack in the basalt, with no associated sulfide deposits. Photo taken with a camera mounted on the sponson of *DSV Alvin*; courtesy of Woods Hole Oceanographic Institution.

of collecting water samples from the top of the chimney during the December 1993 expedition, the chimney was accidentally knocked over. When scientists returned to the site 3 months later in March 1994, a chimney >5 m high had re-grown (Fig. 13.40).

During the decade following the December 1993 expedition to the region,

Fig. 13.39. A photo taken in December 1993 in the same area as the photo depicted in Fig. 13.38. The photo shows the base of a polymetallic sulfide chimney that is over 7 m high, with numerous tubes of the polychaete *Alvinella pompejana* coating the chimney's exterior. Photo taken with a camera mounted on the sponson of *DSV Alvin*; courtesy of Woods Hole Oceanographic Institution.

Fig. 13.40. A photo taken in March 1994 in the same area as depicted in Figs 13.38 and 13.39. The chimney depicted in Fig. 13.39 was accidentally knocked over in December 1993 during routine sampling procedures; the chimney seen here with a height over 5 m had re-grown at the site over a period of less than 4 months. Photo taken with a camera mounted on the sponson of *DSV Alvin*; courtesy of Woods Hole Oceanographic Institution.

Fig. 13.41. Succession in biological community structure at a site (Marker #119) located at 9°49.9′N along the East Pacific Rise (depth, 2500 m) from 1991 to 2004. Particularly noteworthy is the lack of mussels

Fig. 13.42. Shrimp (*Rimicaris exoculata*) at the TAG hydrothermal vent field along the Mid-Atlantic Ridge. © Richard A. Lutz.

numerous subsequent expeditions documented many other spectacular changes in biological community structure (Lutz *et al.*, 2001). Figure 13.41 depicts the biological successional changes at one such site (where marker #119 was located) that occurred over the 13-year period from April 1991 to April 2004.

Thus far in this chapter, the focus has been on deep-sea hydrothermal systems in the eastern Pacific. As one travels throughout the world's oceans to other sites along the Mid-Ocean Ridge Spreading Centre and Back-Arc Spreading Centres, vastly different types of biological communities are encountered. For example, many of the biological communities along the Mid-Atlantic Ridge are dominated by shrimp (Fig. 13.42). While mussels within the genus *Bathymodiolus* and vesicomyid clams are also occasional inhabitants of these sites, there is a noticeable lack of vestimentiferan tubeworms. Many of the same families and genera of organisms are found at hydrothermal sites in both the Atlantic and Pacific, but no species are shared at hydrothermal sites in these two ocean basins. Similarly, while several of the same genera and families of organisms encountered in the eastern Pacific and Atlantic Ocean basins area are found in the Back-Arc Spreading Centres in the western Pacific, all of the species collected at these sites are distinct from those found at the eastern Pacific and Atlantic vent fields visited to date. It is not yet fully understood why.

13.4 Organism Strategies and Adaptations at Deep-sea Hydrothermal Vents

The unusual organisms inhabiting deep-sea hydrothermal vents have developed a variety of strategies and adaptations that enable them to colonize, survive at, and disperse to and from deep-sea hydrothermal

Fig. 13.41. Continued.
(*Bathymodiolus thermophilus*) in 1995 and the predominance of this species in both 2000 and 2004 concomitant with the decline in the abundance of tubeworms (both *Tevnia jerichonana* and *Riftia pachyptila*). © Richard A. Lutz.

vents. Some of the more notable of such strategies and adaptations are summarized below.

13.4.1 Larval dispersal

Virtually all of the invertebrates encountered in deep-sea vent environments have larval stages (either planktotrophic or non-planktotrophic) that appear to be capable of remaining within the water column for extended periods of time (on the order of weeks or months) (Lutz *et al.*, 1984, 1986; Lutz, 1988). This enables the larvae to disperse tremendous distances (hundreds of kilometres), facilitating colonization of geographically-isolated vents.

13.4.2 Larval settlement

How vent organisms locate and colonize hydrothermal sites remains a question fundamental to our understanding of vent ecology. Key to addressing this question is an understanding of the variety of cues that trigger larval settlement. Of the spectrum of such cues that have been identified over the years to stimulate larval settlement (especially the settlement of marine invertebrate larvae), three would seem to play a significant role in organism recruitment in vent environments: temperature, sulfide and bacterial films (see Crisp (1974) for a review of larval invertebrate settlement cues). Acute temperature increase has long been recognized as a stimulus to settlement in certain species of molluscs (Lutz *et al.*, 1970) and larval forms passively carried by currents in the vicinity of vent environments would certainly be exposed to such increased temperatures. In a detailed analysis of the prodissoconch morphology of a mytilid (*Bathymodiolus thermophilus*) from a hydrothermal vent environment along the Galapagos Rift, Lutz *et al.* (1980) suggest that high microbial densities, elevated temperatures and hydrogen sulfide concentrations may play significant roles in stimulating larval settlement in deep-sea vent systems.

13.4.3 Sulfide stabilization

High concentrations of sulfide are generally extremely toxic to aerobic marine invertebrates. Thus, the discovery that the tubeworm *Riftia pachyptila* had blood sulfide concentrations as high as several mM raised many questions concerning how this organism was capable of surviving given such enormously high concentrations of this toxic substance (Childress *et al.*, 1984). Fisher and Childress (1984) found that haemoglobin within the blood is capable of stabilizing sulfide and oxygen in concentrations up to several mM, thus preventing poisoning of the cytochrome-c-oxidase enzyme of this unusual vent invertebrate and facilitating survival in this extreme chemical environment (Powell and Somero, 1983, 1986).

13.4.4 Thermal tolerance

Many microbes isolated from deep-sea hydrothermal vents have been found to live at extremely high temperatures. One such microbe, known as 'Strain 121' (isolated from a deep-sea hydrothermal vent along the Juan de Fuca Ridge, off the coast of Washington State, USA), has been reported to survive at temperatures of up to 121°C, extending the upper temperature limit at which life is known to exist on Earth (Kashefi and Lovley, 2003). There have also been numerous reports that the annelid *A. pompejana*, found at many vents along the East Pacific Rise, can withstand extreme temperatures for various lengths of time. A temperature probe that came in contact with one *A. pompejana* individual recorded a temperature of 105°C while other temperature recordings made inside the tubes of several specimens of this species ranged from 40 to 80°C (Chevaldonne *et al.*, 1992). Temperature measurements reported by Cary *et al.* (1998) suggested that this species may be capable of surviving at temperatures of up to 60°C for several hours. The results of these and other studies suggest that *A. pompejana* (which can also survive at temperatures below 2°C for extended periods of time) may be the most eurythermal metazoan on the planet (Luther *et al*, 2001; Shillito *et al.*, 2001).

13.4.5 Symbionts

As mentioned in the text above, the tubeworm *Riftia pachyptila* (Figs 13.8 and 13.19) is one of the most spectacular and unusual organisms encountered at vent sites along both the Galapagos Rift and the East Pacific Rise. The organism derives its nutrition from chemosynthetic sulfur-oxidizing bacteria (bacteria able to produce biomass from single carbon molecules) within a specialized 'tissue' known as the trophosome. The mechanism associated with the transfer of energy from these symbiotic microbes to the host is not well understood, but the energy transfer system appears to be remarkably efficient and such efficiency probably accounts for the remarkably rapid rates of growth of *R. pachyptila*. It is known that both the vesicomyid clam *Calyptogena magnifica* (Figs 13.5 and 13.12) and the mussel *Bathymodiolus thermophilus* (Figs 13.6 and 13.7), found along the Galapagos Rift and East Pacific Rise, also have chemosynthetic bacterial symbionts (associated with their gill tissues) and both of these species also exhibit unusually rapid rates of growth (Rhoads *et al.*, 1981; Lutz *et al.*, 1985, 1988).

13.4.6 Food supply

Most or all of the other organisms at deep-sea hydrothermal vents are directly or indirectly dependent for nutrition upon chemosynthetic microbes at the base of the food chain. Limpets and coiled archaeogastropods graze on chemosynthetic microbes on a wide variety of surfaces ranging from basalt to the shells of other invertebrates to tubeworm tubes in the vent region. Crabs and fish are scavengers, feeding either directly on microbes or other invertebrates (parts or whole specimens) that are ultimately dependent upon microbes at the base of the food chain. Many of the annelids present at the vents filter or graze on microbes in the water column or on a wide variety of substrates.

13.5 Implications for Origin of Life and Extraterrestrial Life Forms

Scientists generally agree that life on Earth originated approximately 4 billion years ago. Key to understanding how life originated is an understanding of the conditions that existed on Earth prior to this time. Of central importance in this context is the predominance of anaerobic conditions prior to 4 billion years ago during which time the prebiotic synthesis of biomolecules likely occurred, as did chemical synthesis. It is believed by many that life originated under extreme conditions; for example, lightning discharges into shallow seas or extreme temperatures in marine environments have both been evoked as potentially playing significant roles in the origin of life (see also Lomax, Chapter 2, this volume).

Of all extreme marine environments, deep-sea hydrothermal vents have received perhaps the most attention in discussions concerning where life may have originated. In these environments, high temperature fluids that emanate from polymetallic sulfide edifices, such as that depicted in Fig. 13.13, are devoid of oxygen, creating anaerobic conditions within the walls of the chimneys. When pieces of chimneys have been sampled and brought back to the surface for analysis, numerous species

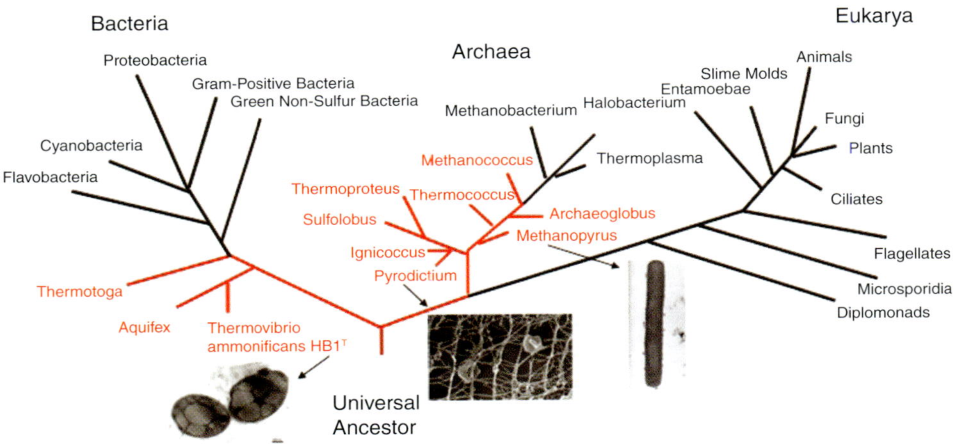

Fig. 13.43. A 'phylogenetic tree of life'. Comparative phylogenetic analyses of genes and genomes from representatives of the three domains of life suggest that extant thermophilic bacteria and Archaea (red branches) are very ancient and are the closest relatives to the Universal Ancestor. Diagram courtesy of Costantino Vetriani.

of thermophilic microorganisms (both Bacteria and Archaea) have been isolated and subsequently cultured. For example, an anaerobic thermophilic Archaeon (*Methanococcus*) and an anaerobic thermophilic bacteria (*Caminibacter*) were both isolated from chimneys along the East Pacific Rise. Another thermophilic Archaeon (*Methanopyus*) that was isolated from a deep-sea vent in Guaymas Basin in the Gulf of California has an optimal temperature of 100°C and can survive at temperatures as high as 110°C. It is strictly anaerobic and reduces carbon dioxide to methane. It also has very unusual lipids, which contain unsaturated side chains, which may represent an ancestral form of lipids. So what are the implications that these discoveries have in providing insights into the origin of life? Figure 13.43 depicts a 'phylogenetic tree of life', with the microorganisms mentioned above, as well as other anaerobic thermophilic microorganisms (both Bacteria and Archaea) living today and isolated from deep-sea hydrothermal vents, occupying the most primitive nodes on this tree. This provides compelling evidence that such organisms may be the closest relatives to the Universal Ancestor.

On a final note, it is worth reflecting on the implications that the above discoveries at deep-sea hydrothermal vents on our own planet have for the possibility of life on extraterrestrial bodies. For scientists working at deep-sea hydrothermal vents over the past few decades, perhaps the most intriguing extraterrestrial body in our own solar system is Europa, one of Jupiter's 63 confirmed moons (Greenberg, 2005). The surface of Europa is completely covered with a layer of ice estimated to be between 3 and 30 km thick, beneath which is an ocean that may be about 100 km deep (Schenk *et al.*, 2004). At the bottom of that ocean are volcanoes. An ocean with volcanoes on its floor has all of the ingredients for hydrothermal circulation and hydrothermal circulation generates all of the ingredients for life as we know it at deep-sea hydrothermal vents on our own planet. Earth's deep-sea scientific community would not be surprised to find life at vents under the ocean on Europa, it would actually be surprised *not* to find life there!

References

Cary, C.S., Shank, T. and Stein, J. (1998) Worms bask in extreme temperatures. *Nature* 391, 545–546.

Chevaldonne, P., Desbruyeres, D. and Childress, J.J. (1992) Some like it hot... and some like it even hotter. *Nature* 359, 593–594.

Childress, J.J., Arp, A.J. and Fisher, C.R. (1984) Metabolic and blood characteristics of the hydrothermal vent tubeworm *Riftia pachyptila*. *Marine Biology* 83, 109–124.

Cohen, D.M., Rosenblatt, R.H. and Moser, H.G. (1990) Biology and description of a bythitid fish from deep-sea thermal vents in the tropical eastern Pacific. *Deep-Sea Research* 37, 267–283.

Crisp, D.J. (1974) Factors influencing the settlement of marine invertebrate larvae. In: Grant, P.T. and Mackie, A.M. (eds) *Chemoreception in Marine Organisms*. Academic Press, London, pp. 177–265.

Fisher, C.R. and Childress, J.J. (1984) Substrate oxidation by trophosome tissue from *Riftia pachyptila* Jones (Phylum Pogonophora). *Marine Biology Letters* 5, 171–184.

Greenberg, R. (2005) *Europa: The Ocean Moon: Search for an Alien Biosphere*. Springer Praxis Books, Berlin, Heidelberg, Germany, 380 pp.

Haymon, R.M., Fornari, D.J., Edwards, M.H., Carbotte, S., Wright, D. and Macdonald, K.C. (1991) Hydrothermal vent distribution along the East Pacific Rise crest (9°09′ - 54′N) and its relationship to magmatic and tectonic processes on fast-spreading mid-ocean ridges. *Earth and Planetary Science Letters* 104, 513–534.

Jannasch, H.W., Elmhjellen, K., Wirsen, C.O. and Farmanfamaian, A. (1971) Microbial degradation of organic matter in the deep sea. *Science* 171, 672–675.

Jones, M.L. (1985) On the vestimentiferan, new phylum: six new species, and other taxa, from hydrothermal vents and elsewhere. *Bulletin of the Biological Society of Washington* 6, 117–158.

Kashefi, K. and Lovley, D.R. (2003) Extending the upper temperature limit for life. *Science* 301, 934.

Luther, G., Rozan, T.F., Tallefert, M., Nuzzio, D.B., Di Meo, C., Shank, T.M., Lutz, R.A. and Cary, S.C. (2001) Chemical speciation drives hydrothermal vent ecology. *Nature* 410, 813–815.

Lutz, R.A. (1988) Dispersal of organisms at deep-sea hydrothermal vents: a review. *Oceanologica Acta* 8, 23–29.

Lutz, R.A., Hidu, H. and Drobeck, K.G. (1970) Acute temperature increase as a stimulus to setting in the American oyster, *Crassostrea virginica* (Grelin). *Proceedings of the National Shellfisheries Association* 60, 68–71.

Lutz, R.A., Jablonski, D., Rhoads, D.C. and Turner, R.D. (1980) Larval dispersal of a deep-sea hydrothermal vent bivalve from the Galapagos Rift. *Marine Biology* 57, 127–133.

Lutz, R.A., Turner, R.D. and Jablonski, D. (1984) Larval development and dispersal at deep-sea hydrothermal vents. *Science* 226, 1451–1454.

Lutz, R.A., Fritz, L.W. and Rhoads, D.C. (1985) Molluscan growth at deep-sea hydrothermal vents. *Bulletin of the Biological Society of Washington* 6, 199–210.

Lutz, R.A., Bouchet, P., Jablonski, D., Turner, R.D. and Waren, A. (1986) Larval ecology of mollusks at deep-sea hydrothermal vents. *American Malacological Bulletin* 4, 49–54.

Lutz, R.A., Fritz, L.W. and Cerrato, R.M. (1988) A comparison of bivalve (*Calyptogena magnifica*) growth at two deep-sea hydrothermal vents in the eastern Pacific. *Deep-Sea Research* 35, 1793–1810.

Lutz, R.A., Shank, T.M., Fornari, D.J., Haymon, R.M., Lilley, M.D., Von Damm, K.L. and Desbruyeres, D. (1994) Rapid growth at deep-sea vents. *Nature* 371, 663–664.

Lutz, R.A., Shank, T.M. and Evans, R. (2001) Life after death in the deep sea. *American Scientist* 89, 422–431.

Powell, M.A. and Somero, G.N. (1983) Blood components prevent sulfide poisoning of respiration of the hydrothermal vent tube worm *Riftia pachyptila*. *Science* 219, 297–299.

Powell, M.A. and Somero, G.N. (1986) Adaptations to sulfide by hydrothermal vent animals: sites and mechanisms of detoxification and metabolism. *Biological Bulletin* 171, 274–290.

Rhoads, D.C., Lutz, R.A., Revelas, E.C. and Cerrato, R.M. (1981) Growth of bivalves at deep-sea hydrothermal vents along the Galapagos Rift. *Science* 214, 911–913.

Rosenblatt, R.H. and Cohen, D.M. (1986) Fishes living in deep-sea thermal vents in the tropical eastern Pacific, with descriptions of a new genus and two new species of eelpouts (Zoarcidae). *Transactions of the San Diego Society of Natural History* 21, 71–79.

Schenk, P.M., Chapman, C.R., Zahnle, K. and Moore, J.M. (2004) Ages and interiors: the cratering record of the Galilean satellites. In: Bagenal, F., Dowling, T.E. and McKinnon, W.B. (eds) *Jupiter: The Planet, Satellites and Magnetosphere*. Cambridge University Press, Cambridge, UK, 732 pp.

Shank, T.M., Fornari, D.J., Von Damm, K.L., Lilley, M.D., Haymon, R.M. and Lutz, R.A. (1998) Temporal and spatial patterns of biological community development at nascent deep-sea hydrothermal vents along the East Pacific Rise. *Deep-Sea Research* 45, 465–515.

Shillito, B., Jollivet, D., Sarradin, P.-M., Rodier, P., Lallier, F., Desbruyeres, D. and Gaill, F. (2001) Temperature resistane of *Hesiolyra bergi,* a polychaetous annelid living on deep-sea vent smoker walls. *Marine Biology Progress Series* 216, 141–149.

Southward, E.C., Schulze, A. and Gardiner, S.L. (2005) Pogonophora (Annelida): form and function. *Hydrobiologia* 535/536, 227–251.

Turekian, K.K., Cochran, K., Kharkar, D.P., Cerrato, R.M., Vaisnys, J.R., Sanders, H.L., Grassle, J.F. and Allen, J.A. (1975) Slow growth rate of a deep-sea clam determined by 228Ra chronology. *Proceedings of the National Academy of Sciences* 72, 2829–2832.

Williams, A. (1980) A new crab family from the vicinity of submarine thermal vents on the Galapagos Rift (Crustacea: Decapoda: Brachyura). *Proceedings of the Biological Society of Washington* 93, 443–472.

14 High Hydrostatic Pressure Environments

Florence Pradillon
Département Ressources Physiques et Écosystèmes fond de mer, Etude des Écosystèmes Profonds, Laboratoire Environnements Profonds, Ifremer Centre de Brest, Plouzané, France

14.1 Introduction

All biological systems on Earth are exposed to various levels of pressure. On land, we are all exposed to pressure of 1.03 kg cm^{-2} of our body surface (= 1 atmosphere (atm) = 0.101 MegaPascal or MPa; see Table 14.1), but we usually do not pay attention to the pressure load exerted on us because it is remarkably constant. However, cells in our hip joint cartilages experience pressure increases of 100 to 200 times that of atmospheric pressure (~10–20 MPa) each time we take a step forward (Hall *et al.*, 1993). What we consider pressure extremes tend to be measured at high altitude or in the ocean depths. At an altitude of about 5800 m, pressure is half that encountered at sea level. Conversely, in aquatic environments, the weight of the water column results in a linear increase of hydrostatic pressure by 1 atm for every 10 m increase in depth. In the deepest part of the oceans, hydrostatic pressure may reach more than 1000 times that of atmospheric pressure (over 110 MPa at the deepest known point of the ocean, The Challenger Deep in the Mariana Trench, at a depth of 10,911 m). Recently, live microorganisms were even discovered underground in continental or subseafloor systems at depth where pressure is likely to be over 150 MPa (Roussel *et al.*, 2008; Oger and Jebbar, 2010).

Since the realization that even the greatest depths provide a habitat for many piezophilic (pressure-loving) life forms (Whitman *et al.*, 1998; Todo *et al.*, 2005; Lutz, Chapter 13 and Hughes, Chapter 15, this volume), there has been much interest in the physiological and ecological importance of pressure as a limiting factor in the environment (Somero, 1990). Besides the pressure conditions at these extremes, very modest pressure variations, in the order of tenths of an atmosphere, may also have significant influences on an organism's life history strategies. Many shallow water marine organisms respond to small pressure variations due to waves, tides, winds or atmospheric pressure variations (Forward and Bourla, 2008; Fraser, 2010). If we want to understand how life forms respond and adapt to extreme pressure conditions, it is also very useful to understand pressure response mechanisms in less challenging conditions.

But what should we define as an 'extreme' pressure condition? From an evolutionary perspective, one might consider the earliest environment for life as 'normal'. Recent scenarios developed for Earth's early history predict that life might have emerged under high pressure at the bottom of the oceans, because conditions were too unstable at the surface (Daniel *et al.*, 2006).

Table 14.1. Conversion of the different pressure units used in the literature. In aquatic environments, pressure values are usually expressed as relative pressure, i.e. the difference between atmospheric pressure at sea level and the pressure at a given depth. For example, at 1000 m depth in the ocean, pressure is 10.06 MPa above the atmospheric pressure, or 10.16 MPa in absolute. Since atmospheric pressure is negligible most of the time in comparison to hydrostatic pressure, relative pressure is commonly used.

	atmosphere	kg cm^{-2}	bar	Pa	MPa	psi	mmHg	cmH_2O
1 atm (atmosphere)	1	1.03323	1.01325	101 325	0.10132	14.69595	760	1007
1 kg cm^{-2}	0.96784	1	0.98066	98 066.5	0.09807	14.22334	736	975
1 bar	0.98692	1.01972	1	100 000	0.1	14.50377	750	993
1 Pascal (Pa)	9.87×10^{-6}	1×10^{-5}	1×10^{-5}	1	1×10^{-6}	0.00015	0.0075	0.00994
1 MPa	9.86923	10.19716	10	10^6	1	145.038	7501	9940
1 psi (pounds $inch^{-2}$)	0.06805	0.07031	0.06895	6 894.76	0.00689	1	51.715	68,526
1 mmHg (=1 Torr)	0.00132	0.00136	0.00133	133.322	0.00013	0.01934	1	1.32506
1 cmH_2O (seawater)	0.00099	0.001026	0.001007	100.616	0.00010	0.01459	0.75468	1

Atmospheric pressure could therefore be viewed as the extreme. A more physical definition of 'extreme' is when conditions make it difficult for organisms to function (Rothschild and Mancinelli, 2001). Accordingly, high pressure in the deep ocean is an extreme condition for surface dwellers, whereas it is atmospheric pressure that appears extreme for piezophilic organisms. As surface dwellers, we usually consider high pressure environments as 'extreme', so how do we define the limits of this high pressure biosphere?

Oceans are the most widespread 'high pressure' environments on the Earth's surface, with an average depth of 3800 m and an average pressure of 38 MPa. In subsurface systems (marine sediments, continental and oceanic crusts), pressure increases by about 2 to 3 atm every 10 m (Oger and Jebbar, 2010; see Bell and Heuer, Chapter 17, this volume). Whereas temperature is low and constant in the deep ocean (2°C with some exceptions such as the deep Mediterranean Sea and hydrothermal vents), temperature increases by 25–30°C for every kilometre with depth underground. Based on our current knowledge of the highest temperature and pressure records for life, respectively 122°C for the hyperthermophilic, methanogenic archaeon *Methanopyrus kandlerii* (Takai *et al.*, 2008) and 120 MPa for the obligate piezophilic, hyperthermophilic archaeon isolated from a deep-sea hydrothermal vent, *Pyrococcus yayanosii* (Zeng *et al.*, 2009), the high pressure biosphere may extend as deep as 5 km below ground.

The maximum pressures encountered by organisms living on Earth (<200 MPa) are not sufficient to break biomolecules. However, much lower pressures affect the conformation and function of many of them, leading to loss of function and/or death. High hydrostatic pressure, in the range of tens of MPa, is generally assumed to be non-lethal but exerts adverse effects on the growth of organisms that are adapted to atmospheric pressure. Neuronal transmission, motility and cell division are impaired at pressures above 10–20 MPa (Somero, 1992; Grossman *et al.*, 2010). On the other hand, organisms living deeper than 2000 m in the oceans do not survive collection from their deep-sea environment and transport to the surface if they are not maintained close to their native pressure during collection or returned to it shortly after collection (Yayanos, 1978, 1981; Treude *et al.*, 2002). Further, there is evidence to suggest that, even at depths of 1000 m, substantial adaptation to high pressure regimes characterizes many resident deep-sea invertebrates (Gibbs and Somero, 1989).

Hence, 1000 m depth (or 10 MPa) is considered as the upper limit of the high pressure biosphere by several authors. The term 'piezosphere' has been coined to refer to the volume of the biosphere having pressure greater than 10 MPa (Macdonald, 1997; Fang *et al.*, 2010). This represents 79% of the volume of the marine biosphere and 62% of the volume of the Earth's biosphere (not including the subsurface biosphere; Bell and Heuer, Chapter 17, this volume). Thus, both in terms of biosphere volume and cell number, the majority of life forms on Earth are flourishing under high pressure (Whitman *et al.*, 1998).

Examination of both atmospheric pressure adapted and high pressure adapted organisms provides useful information to help us understanding pressure effects on living organisms. Mesophiles' (atmospheric pressure loving organisms) exposure to pressure aids in the characterization of fundamental cellular processes while piezophiles may provide the essence of the adaptation of life to high pressure environment. However, due to technological limitations, investigations on the effects of high pressure were initially only conducted on atmospheric pressure adapted organisms submitted to pressure. These early studies focused on the gel-sol properties of organism cytoskeletons (the protein 'scaffolding' or 'skeleton' contained within the cytoplasm of cells) and revealed profound effects of high pressure at the cellular level (Marsland and Landau, 1954; Zimmerman and Marsland, 1964; Zobell and Cobet, 1964; Salmon, 1975). Later, when dedicated technology became available to retrieve deep-sea organisms from their habitat, it appeared that decompression causes similar cellular defects (Yayanos and Dietz, 1983).

Today, we have a fairly good overview of the pressure effects at organism, tissue, cell and molecule levels. However, the interplay between these effects at different levels and the mechanisms and pathways involved in the pressure response are not yet fully identified. Transcriptome level studies recently provided a large amount of data, and they usually show that, for sublethal pressure, part of the pressure response is in fact a general stress response (Fernandes *et al.*, 2004).

Difficulties in interpretation arise from several aspects. First, other factors such as temperature, pH, osmotic pressure, to name but a few, may modify pressure response, acting synergically or counteracting it. Second, pressure sensitivity depends on the life stage of an organism. Bacterial cultures are much more sensitive to pressure in the growth phase than in the stationary phase (Pagán and Mackey, 2000). Another more dramatic example is given by anydrobiotic invertebrates, such as Tardigrades, that resist pressure that would kill them in their ordinary hydrated state (Seki and Toyoshima, 1998; Horikawa *et al.*, 2009). Third, there are methodological issues arising from the fact that different studies have used protocols in which pressure magnitudes, regimes and durations were different, making it difficult to provide meaningful comparisons. According to the pressure range applied, the pathways involved in the pressure response may be radically different. Small pressure variations in the range of 0.1 MPa can be detected and are important for many coastal species. Larger pressure variations in the range of 10–50 MPa are sub-lethal for most mesophiles, but might trigger acclimation mechanisms in many cases. Pressure variations of 50–200 MPa, although remaining mostly in the range encountered by life on Earth, are lethal for most mesophiles and are the ranges exploited by piezophiles. Pressures above 200 MPa may be considered as irrelevant since life on Earth was probably never exposed to such pressure; nevertheless, they are used in the biotechnology field to kill pathogens, or at the molecular level to unravel the structure of biomolecules.

This chapter will first explain the basic mechanisms underlying pressure effects on biological systems and then review these effects at different organization levels, from molecules to organisms. Due to the interplay of pressure with other factors, and due to the fact that pressure impact arises ubiquitously everywhere in an organism and at all levels, it is still very difficult to get an accurate picture of the mechanisms and pathways involved in pressure sensing and response. After a brief description of some of the technological advances that have progressively allowed pressure research to go ahead, the chapter will go on to discuss how organisms living at depth are adapted to high pressure, and how pressure variations (both high pressure variations and slight pressure changes) are felt by organisms and influence their life history strategies.

14.2 Pressure Effects on Biological Systems

14.2.1 Principle of pressure effects on biological systems

In aquatic environments, hydrostatic pressure results from the weight of the water column on an immersed object. Similarly, in subsurface habitats, pressure is generated by the weight of the overlaying sediment layers and/or oceanic or continental crusts. Hydrostatic pressure acts all around an organism and is in balance with no net force unlike a differential pressure. Pressure affects chemical equilibria and reaction rates. The behaviour of all systems under high pressure is governed by 'Le Châtelier's principle', which predicts that the application of pressure shifts equilibrium towards the state that occupies a smaller volume, and accelerates processes for which the transition state has a smaller volume than the ground state (Smeller, 2002). Thus, processes or reactions that proceed with a volume reduction ($\Delta V < 0$) are favoured by pressure, whereas processes or reactions accompanied by a volume increase are inhibited by pressure, and those with no significant volume change are relatively insensitive to pressure.

Pressure effects on living systems are largely influenced by water, which is the solvent of most biomolecules. Water has low compressibility and many of the effects of pressure on molecular systems are due to solvation (dissolution) with water. Indeed, pressure favours the solvation of charged and apolar groups, and the entry of water into void cavities within the hydrophobic core of proteins, thus causing unfolding. In fact, when water is withdrawn, molecular systems become much more resistant to pressure (Oliviera *et al.*, 1994).

Living organisms on Earth might encounter pressure within the range ~0.1–200 MPa. Such pressures only change intermolecular distances and affect conformations, but do not change covalent bond distances or bond angles. The covalent structure of small molecules and the primary structure of macromolecules are not perturbed by pressure up to 2 GPa. Indeed, pressure below 200 MPa predominantly affects weak bonds, and acts on the conformation and supramolecular structures of biological systems, and thus on their functionality in the cells.

The way pressure affects biological molecules depends on the nature and number of weak interactions involved in maintaining their structure and the compressibility of the molecule, which depends on the number of void cavities they have. Weak interactions affected by pressure are ionic interactions, hydrophobic interactions, hydrogen bonds, Van der Waals forces and staking interactions. Pressure usually favours the disruption of ionic interactions (short range interactions that occur between negatively and positively charged residues, also called salt bridges). The dissociation of chemical species into their ionic forms favours the contraction of water molecules around the charged groups. This is called electrostriction and results in a global volume reduction. Since both the tertiary and quaternary structure of proteins are generally stabilized by ionic interactions, pressure usually destabilizes multimers and monomers.

High pressure also usually disrupts hydrophobic interactions and promotes the solvation of hydrophobic groups with water. Pressure forces the penetration of water molecules into the hydrophobic cavities within the protein (compression effect), which usually leads to the unfolding of the protein. Hydrogen bonds are slightly stabilized, or unaffected, by pressure because of the weak volume reduction that accompanies their formation. The DNA double helix is stabilized under pressure because its maintenance involves many hydrogen bonds. Van der Waals forces (weak interactions produced by all atoms and molecules as a result of interactions related to induced polarization effects at close distance) are likely to be favoured by pressure since they tend to maximize the packing density of proteins, thus reducing volume. However, they may in some cases contribute to the destabilization of proteins under pressure. Stalking interactions, which result from Van der Waals forces, are also stabilized by pressure.

Although the basic principle of pressure effects is very simple, it remains difficult to explain their impact on complex molecular assemblages and reaction networks. Pressure instantly and ubiquitously penetrates cells, tissues and whole organisms, thereby imposing its effect on the entirety of the intracellular components and their molecular interactions. Many individual reactions with different sensitivity to pressure arise simultaneously, either synergistically or counteracting each other, making it very difficult to explain pressure effects at higher organization levels.

14.2.2 Pressure effects on molecular assemblages in cells

Pressure effects on proteins

Many biological functions involve protein–protein interactions: multimeric enzymes, ribosomes, cytoskeleton, signal transduction pathways, antigen–antibody interactions. Pressures of approximately 500 MPa are required to cause denaturation, but lower pressures of 100–200 MPa can promote dissociation of oligomeric proteins that leads to modifications in enzymatic reactions as well as protein interactions and functionality. Dissociation at such pressure may lead to reversible effects. This is the case for cytoskeleton, which usually shows dissociation well below 100 MPa (Begg *et al.*, 1983; Bourns *et al.*, 1988; Arai *et al.*, 2008).

Protein assemblage sensitivity to pressure depends not only on the nature of the interactions involved in the maintenance of their structure, i.e. their conformational stability to compensate for the loss of weak interactions, but also on the size of the cavities within which water molecules can penetrate. Hydrostatic pressure typically causes dissociation of multimeric proteins because the processes are usually accompanied by negative volume changes due to the solvation of the charged groups that form salt bridges, or by the exposure of non-polar groups to the solvent (Boonyaratanakornkit *et al.*, 2002). Hydration can account for major pressure effects on proteins: pressure causes infiltration of water molecules into the hydrophobic core of the protein. Moreover, protein structures that contain a large fraction of void volume will be highly sensitive to high hydrostatic pressure. For example, mature amyloid fibrils are extremely stable and cannot be dissociated by high hydrostatic pressure of more than 1 GPa, due to the probable involvement of hydrogen bonds and a very densely packed structure, whereas proto-fibrils (precursors involving mainly hydrophobic and electrostatic interactions and exhibiting a number of voids) are disaggregated by hydrostatic pressure in the 200–300 MPa range (Meersman and Dobson, 2006).

In enzyme–substrate interactions, substrate binding to the catalytic site of the enzyme mainly involves ionic interactions, and pressure may weaken substrate binding. Protein synthesis is highly susceptible to increasing pressure because the dissociation of ribosome subunits is accompanied by a large negative volume change (Gross *et al.*, 1993). The destabilization of ribosomes and inhibition of DNA synthesis are factors limiting cell growth at high pressure (Erijman and Clegg, 1995, 1998; Kawano *et al.*, 2004). DNA synthesis is one of the most piezo-sensitive cellular processes, particularly the initiation of DNA replication rather than polymerization itself. In contrast, RNA synthesis is relatively pressure resistant.

Pressure effects on biomembranes

Biological membranes are one of the most pressure sensitive cellular components.

The basic structural element of biological membranes consists of a lamellar phospholipid bilayer matrix (Winter and Dzwolak, 2005). Upon compression, the lipids adapt to volume restriction by changing their conformation and packing. The compression of the phospholipid bilayer is anisotropic. Under high pressure conditions, the acyl chains straighten, which results in a lateral shrinking and an increase in thickness. This phenomenon is also accompanied by a phase transition from the liquid-crystalline to the gel phase. The consequence is a decrease in cell membrane fluidity. This lack of fluidity affects trans-membrane ions and small molecule fluxes, as well as the conformation and function of many membrane proteins.

An increase of 100 MPa in hydrostatic pressure has the same effect on biomembranes as a decrease of 13–21°C in temperature, depending on the composition of the lipid-based system (Cossins and Macdonald, 1989; Somero, 1992). Several studies have demonstrated that organisms can modulate the physical state of their membranes in response to changes in the external environment (pressure and/or temperature) by regulating fractions of the lipid components in the cell membrane varying in chain length, chain unsaturation or headgroup structure ('homeoviscous adaption'; DeLong and Yayanos, 1985). In fact, membranes are significantly more fluid in barophilic and/or psychrophilic (cold temperature-loving) species. This is principally a consequence of an increase in the unsaturated/saturated lipid ratio, but also of an adjustment of the membrane concentration of sterols (e.g. cholesterol), if present, or proteins. Lipid biomembranes are often considered the main target in the pressure inactivation of microorganisms because of their sensitivity to pressure.

Pressure effects on nucleic acids

The covalent structure of nucleic acids is stable up to at least 1.2 GPa (Macgregor, 2002; Girard *et al.*, 2007). In addition, high pressure tends to stabilize DNA hydrogen bonds and stacking interactions, and thus increases the duplex-single strand transition temperature. Therefore, DNA itself is usually not affected by pressures encountered by life on Earth. However, *in vivo* exposure to high pressure often indirectly affects DNA integrity by triggering strand breaks (Aertsen and Michiels, 2005), different methylation patterns (Long *et al.*, 2006) or mobilization of genetic elements (Aertsen and Michiels, 2005; Lin *et al.*, 2006). In rice, a 20 min exposure of germinating seeds to a high pressure of 80–100 MPa induces mobilization of transposable elements and insertion at novel sites within the somatic tissues of the plant (Lin *et al.*, 2006).

Protein–DNA interactions are disturbed by hydrostatic pressure, which can either destabilize the interaction due to pressure-induced structural modification of the protein leading to a loss of affinity for its binding site on DNA (Lynch and Sligar, 2002), or stabilize the interaction which may interfere with transcription regulation processes, thus altering gene expression.

Pressure effects on chemical equilibrium and enzymatic reactions

Hydrostatic pressure affects any reaction that occurs with a volume change and shifts equilibrium towards the state that occupies the smaller volume. The difference between the final and initial volume in the entire system at equilibrium, or between the initial and transition volume, governs the magnitude and direction (acceleration or inhibition) of the effect of pressure on the reaction (Northrop, 2002). When a reaction is accompanied by a volume increase, it is inhibited by increasing pressure. When a reaction is accompanied by a volume decrease, it is facilitated by increasing pressure. Pressure accelerates, inhibits or does not affect reactions depending on the direction and the magnitude of the volume change, whereas temperature simply accelerates reactions. For example, pressure decreases the self-assembly rate of globular actin into filamentous actin in shallow-living fishes, because the polymerization process occurs with a global volume increase (Swezey and Somero, 1985).

The ionization constants of charged molecules are also affected by pressure, and pressure favours ionization of weak acids because this leads to a volume reduction by condensing water molecules. In the yeast *Saccharomyces cerevisiae*, cytoplasms and vacuoles become more acidic in proportion to increasing hydrostatic pressure as a consequence of weak acid dissociation (Abe *et al.*, 1999). Increases in pH may in turn perturb many cellular functions.

14.2.3 Thresholds of pressure effects and interactions with other physico-chemical factors

Figure 14.1 gives examples of the biological processes affected by hydrostatic pressure at the cellular level. However, all these processes are not equally perturbed by pressure. Membrane fluidity is one of the most sensitive cellular components to pressure, and together with that, all processes mediated via the membrane are affected. Pressure in the range of 10 MPa or even less is sufficient to perturb membrane-based functions, such as synaptic transmission, neural and muscular functioning, leading to locomotion alteration, loss of equilibrium and paralysis (Grossman *et al.*, 2010; see Table 14.2).

Protein/protein assemblages are sensitive to pressure of about 40 MPa, leading to cytoskeleton anomalies and with that, perturbations in all processes involving the cytoskeleton, such as cell division (Molina-Höppner *et al.*, 2003). Protein/nucleic acid interactions are also quite sensitive to pressure;

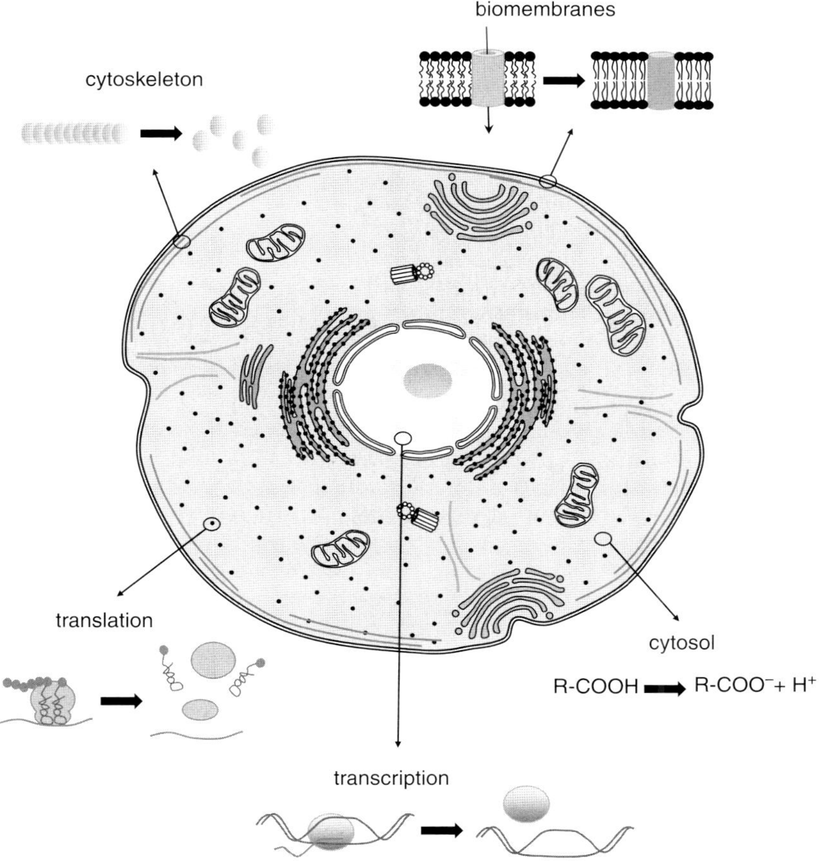

Fig. 14.1. Effects of high hydrostatic pressure on cells and cellular processes in 1-atm-adapted organisms.

Table 14.2. Examples of biological processes affected by hydrostatic pressure in 1 atm-adapted organisms.

Pressure (MPa)	Effect	Organism	Reference
0.1–10	Reduced mobility	Hermit crab	Thatje *et al.*, 2010
	Inhibition of synaptic transmission	Rat, shallow water fishes	Somero, 1992 and references therein
	Neural and muscular function inhibition	Shallow water organisms	Somero, 1992 and references therein
	Heartbeat rate increase	Sublittoral spider crab	Robinson *et al.*, 2009
	Transmembrane sodium efflux	Human erythrocytes	Goldinger *et al.*, 1980
	Sodium uptake	Freshwater crayfish	Roer and Shelton, 1982
	Dehydrogenases enzyme cofactor binding	Shallow water fish	Somero, 1990
	Abnormal cell division	Shallow water sea urchin embryos	Tyler and Dixon, 2000
	Loss of viability (long term pressure exposure)	Shallow water eukaryotes	Somero, 1992
10–50	Loss of mobility	Copepod (*Bosmina longirostris*)	Bao *et al.*, 2010
	Arrest of cell cycle	Bacteria (*Escherichia coli*)	Yayanos and Pollard, 1969; Bartlett, 2002
	Arrest of cell cycle	Yeast (*Saccharomyces cerevisiae*)	Abe and Horikoshi, 1995, 2000
	Depolymerization of cytoskeleton *in vivo*	Rabbit, human	Begg *et al.*, 1983; Bourns *et al.*, 1988
	Depolymerization of cytoskeleton *in vivo*	Bacteria (*E. coli*)	Ishii *et al.*, 2004
	Inhibition of amino acids uptake	Yeast (*S. cerevisiae*)	Abe and Horikoshi, 2000
	Replication	Bacteria (*E. coli*)	Yayanos and Pollard, 1969; Abe, 2007
	Transcription initiation	Bacteria (*E. coli*)	Erijman and Clegg, 1995
	Translation	Human chondrocytes	Elo *et al.*, 2000
	Uncharged ribosomes dissociation (no mRNA, no tRNA)	Bacteria (*E. coli*)	Gross *et al.*, 1993
	Enzyme-DNA complex dissociation	*Bam HI in vitro*	Lynch and Sligar, 2002
	Changes in protein expression profiles	Yeast (*S. cerevisiae*)	Iwahashi *et al.*, 2003, 2005; Fernandes *et al.*, 2004
	Changes in protein expression profiles	Human chondrocytes	Kaarniranta *et al.*, 1998; Elo *et al.*, 2000
50–100	Reduced viability (short term pressure exposure)	Mice blastocysts, pig oocytes	Pribenszky *et al.*, 2005, 2008
	Transcription elongation	Bacteria (*E. coli*)	Erijman and Clegg, 1998; Bartlett, 2002; Kawano *et al.*, 2004
	Translation	Bacteria (*E. coli*)	Yayanos and Pollard, 1969; Abe, 2007
	Charged ribosomes dissociation	Bacteria (*E. coli*)	Gross *et al.*, 1993
	Cytoplasmic and vacuolar acidification	Yeast (*S. cerevisiae*)	Abe and Horikoshi, 1995, 2000
	Inhibition of ethanol fermentation	Yeast (*S. cerevisiae*)	Abe and Horikoshi, 1998
	Depolymerization of cytoskeleton *in vivo*	Yeast (*Schizosaccharomyces pombe*)	Arai *et al.*, 2008
	Depolymerization of cytoskeleton *in vitro*	Chicken	Begg *et al.*, 1983
	Changes in protein expression profiles	Bacteria (*E. coli*)	Welch *et al.*, 1993

	Acquired piezotolerance	Yeast (*S. cerevisiae*)	Iwahashi *et al.*, 1997, 2000, 2001; Palhano *et al.*, 2004; Domitrovic *et al.*, 2006
	Transposable elements mobilization	Germinating seeds of rice	Lin *et al.*, 2006
100–500	Reduced viability (short term pressure exposure)	Yeast (*S. cerevisiae*)	Sato *et al.*, 1996, 1999; Bravim *et al.*, 2010
	Reduced viability (short term pressure exposure)	Bacteria (*E. coli*)	Pagán and Mackey, 2000
	Reduced viability (short term pressure exposure)	Mammalian cells	Frey *et al.*, 2004
	Reduced viability (exposure time <1 min)	Nematode eggs (*Ascaris*)	Rosypal *et al.*, 2007
	Apoptosis	Human lymphoblasts	Takano *et al.*, 1997
	Stress inducible gene expression	Yeast (*S. cerevisiae*)	Iwahashi *et al.*, 2003; Fernandes *et al.*, 2004
	Nuclear membrane perturbation	Yeast (*S. cerevisiae*)	Kobori *et al.*, 1995
	Leakage of internal structures	Yeast (*S. cerevisiae*)	Shimada *et al.*, 1993
	Protein structure (monomers, tertiary structure)	Bacteria (*E. coli*)	Gross and Jaenicke, 1994
	Dissociation of amyloid fibrils	Purified fibrils *in vitro*	Foguel *et al.*, 2003
>500	Reduced viability of anhydrobiotic invertebrates	Tardigrades, nematodes, insects, crustaceans	Seki and Toyoshima, 1998; Horikawa *et al.*, 2009
	Disruption of DNA covalent structure		Macgregor, 2002; Girard *et al.*, 2007
	Protein secondary and primary structure		Boonyaratanakornkit *et al.*, 2002

in particular, ribosomes dissociate under pressure (Gross *et al.*, 1993). Translation is stopped under pressure, which affects gene expression levels. Hydrostatic pressure in the range of 30–50 MPa usually inhibits the growth of various organisms and the initiation of DNA replication is one of the most sensitive intracellular processes (Abe *et al.*, 1999).

Therefore, cellular functions are perturbed by different degrees of pressure. At higher organization levels (tissues or whole organisms), all of the effects of pressure sum together and may result in pressure sensitivity thresholds that reflect the most sensitive processes. In some cases, compensatory mechanisms can increase the global pressure resistance of an organism. For example, small osmolytes in the cytoplasm may help in stabilizing biomembranes and proteins, thus counteracting pressure effects (Yancey *et al.*, 2002).

Other physico-chemical factors are known to interact with pressure effects. Therefore, depending on the conditions, observed pressure effects may vary. The best-known example is temperature. Temperature increases tend to increase membrane fluidity, counteracting the effects of pressure increases. Conversely, temperature decreases have a similar effect to pressure increases, by increasing the concentration of fatty acids and decreasing membrane fluidity. Osmotic pressure and pH were also shown to interact with pressure. In the case of pH, effects are complicated by the fact that pressure also influences the balance of weak acids, and relatively low pressures are effective in lowering cellular pH. Since the dissociation constants of weak-acids are pressure sensitive, the analysis of pressure effects on pH-dependent processes may be difficult.

14.3 High Pressure Biosphere Collections and Observation Tools

High hydrostatic pressure effects on living systems were initially investigated in 1 atm-adapted creatures submitted to high pressure, instead of using direct observations of high pressure-adapted deep-sea organisms (e.g. Zobell and Johnson, 1949; Marsland, 1950; Kitching, 1954; Zimmerman and Marsland, 1964), due to the lack of appropriate tools to retrieve the latter alive from their habitat. Traditional oceanographic collection equipment, such as deep-sea trawls, traps or nets, usually only recover dead or moribund animals. These methods are useful and provide data on the diversity, ecology, distribution and physiology of freshly collected tissues of deep-sea organisms, but they are clearly insufficient for investigating biological activities *in situ* and live animal behaviour, or when long term observations are needed (e.g. developmental studies). Simulating natural high pressure is therefore essential for maintaining animals in a sufficiently good condition to allow physiological and behavioural studies.

Over the past 30 years, various pressure systems were designed to recover and maintain deep-sea organisms at their ambient natural conditions (Macdonald and Gilchrist, 1978, 1980, 1982; Yayanos, 1978, 1981; Phleger *et al.*, 1979; Yayanos and Dietz, 1983). Today, a wealth of high pressure equipment is available to scientists, ranging from laboratory high pressure generators to systems that are able to invade the high pressure biosphere, such as submersibles capable of sampling piezophilic organisms in the deep-sea.

Laboratory-based or onboard-ship pressure equipment has been designed with different purposes in mind and various characteristics. Marine biologists and microbiologists use isobaric trap systems to sample microorganisms or larger animals without compromising them and, once isolated, to investigate their biology under simulated, realistic, environmental conditions (Macdonald and Gilchrist, 1978; Yayanos, 1978; Quetin and Childress, 1980; Shillito *et al.*, 2001, 2008; Koyama *et al.*, 2002; Drazen *et al.*, 2005; Fig. 14.2). One of the practical limitations in culturing organisms over an extended period at high pressure is the supply of oxygen when aerobes are placed in closed hydrostatic chambers. Therefore, the most up-to-date pressure systems are usually large volume systems equipped with a flow-through facility where oxygen is supplied with the renewed water (Childress *et al.*, 1993; Shillito *et al.*, 2001; Koyama *et al.*, 2002). Such systems permit heat-resistance investigations on live invertebrates endemic to the hottest part of the hydrothermal vent biotope: the wall of active vent chimneys (Shillito *et al.*, 2001, 2006; Ravaux *et al.*, 2003; Cottin *et al.*, 2008, 2010). These studies proved that restoration of pressure conditions was crucial for accurate determination of the thermal limits of the organism studied (Shillito *et al.*, 2006). In addition to studying the physiology and biology of deep-sea organisms within realistic and meaningful conditions, pressure vessels are also used to examine the effects of low pressure or decompression on deep-sea creatures (Yayanos, 1978, 1981; Brauer *et al.*, 1980; Macdonald and Gilchrist, 1980, 1982). Such studies allow us to determine the pressure tolerance boundaries of deep-sea creatures, and recently were used to perform slow and controlled decompression of deep-sea animals in order to achieve the progressive acclimatization of deep-sea creatures to atmospheric pressure (Koyama *et al.*, 2005).

Besides large volume pressure vessels, small pressure chambers dedicated to the study of cells, embryos or larvae are also available (Besch and Hogan, 1996; Koyama *et al.*, 2001; Pradillon *et al.*, 2004; Frey *et al.*, 2006; Oger *et al.*, 2006; Yoshiki *et al.*, 2006; Bao *et al.*, 2010; Fig. 14.2). Most of them offer microscopic imagery facilities, meeting the critical need to observe cellular level processes under pressure because many changes that are induced by moderate pressures are rapidly reversible after decompression (Begg *et al.*, 1983; Bourns *et al.*, 1988). Specific cellular processes such as apoptosis or necrosis, or cellular structure can be visualized with fluorescence (Pagliaro *et al.*, 1995; Frey *et al.*, 2006; Oger *et al.*, 2006). With such equipment, we were able to perform analyses of the pressure and temperature resistance of larval stages of hydrothermal vent organisms (Marsh *et al.*, 2001; Pradillon *et al.*, 2001, 2004, 2005; Brooke and Young, 2009).

In addition to the pressure bioreactors discussed above, another type of approach was recently promoted to observe and monitor the high-pressure biosphere directly, without the need to recreate its habitats in dedicated chambers at the surface. Deep-sea monitoring, using instruments installed on simple long-term mooring systems, or complex networks of measurement tools and

(a)

(b)

Fig. 14.2. Examples of pressure equipment used for the study of deep-sea animals. (a) PIRISM pressure chamber for microscopic observation under pressure. During microscopic observation, the pressurized chamber (pc), equipped with sapphire windows, is disconnected from the pressure line and isolated between pressure valves (v). (b) Periscope, a deep-sea isobaric collection device. The sampling cell (sc) is used to collect organisms (hydrothermal vent shrimps here) using the suction power of the submersible, and is subsequently disconnected from the submersible and introduced into a pressure-retaining device (prd) before ascent to the surface. © Bruce Shillito, Ifremer_Victor6000/campagne Momareto, 2006.

video camera systems are now being used. For example, *in situ* observation with lander systems deployed from a mother ship and equipped with camera and light systems have been used to investigate the feeding behaviour of deep-sea fishes and invertebrates, reproductive characteristics of deep-sea bivalves (Fujiwara *et al.*, 1998), or to monitor the metabolic activity of deep benthic organisms. Yet, major limitations of such systems remain, such as long term energy supply and data storage. The development of cabled offshore observatories are beginning to allow scientists to continuously monitor the deep-sea environment, instead of relying on brief oceanographic cruises or instruments that run out of battery power. These deep-sea observatories provide us with opportunities to observe and measure deep biosphere physiological, metabolic and reproductive functions *in situ*, and are rapidly enhancing our understanding of life in high pressure environments.

14.4 Adaptation to High Hydrostatic Pressure

In aquatic environments, organisms experience pressure variations ranging from a few tenths of an atmosphere (e.g coastal species submitted to depth variations due to tides, or changes in atmospheric conditions) to several hundred atmospheres (deep divers such as sperm whales or krill undergoing vertical migrations down to the bottom of the ocean for feeding; see Section 14.5). Depending on the magnitude of the pressure variation, as well as whether pressure change occurs abruptly or more gradually, the pressure response of the organism will vary and acclimatization mechanisms may be possible to a certain extent.

Piezophilic organisms differ from those that are piezotolerant – the former can only survive under pressure, while the latter usually live at atmospheric pressure but are able to survive pressure increases, although they are not at their optimal physiological functioning under pressure. The term eurybath is used for organisms that can live and function normally over a wide pressure range. Thus, piezotolerant organisms have the capacity to tolerate pressure and acclimatization mechanisms exist that allow them to extend their usual pressure range. Acclimatization potential is probably an evolutionary trait and has a genetic basis. It is, however, different from adaptation. Genetic adaptation implies that natural selection is involved, and usually applies at the population level rather than the individual level. Acclimatization denotes the adjustment of an organism to its environment, either a multi-stress environment, including seasonal and climatic changes, or a single dominant stress. Acclimatization is part of the global adaptation of an organism, but often its genetic basis has to be demonstrated. It refers to responses, and hence to phenotypically plastic, adaptive traits, that is, to those characteristics of an individual that change in response to their environment.

Pressure adaptations may occur through different strategies: (i) evolution of new function/genes through mutations that result in increased stability and/or activity under pressure; (ii) fine tuning of the expression patterns of genes to compensate for the loss of function due to pressure; or (iii) expression of a set of 'pressure-responding' genes.

In addition, non-genetic responses to pressure stress have also been identified. Many deep-sea organisms accumulate small organic molecules within their tissues, called osmolytes, which are usually involved in maintaining cell volume in water-stressed organisms. Some of these osmolytes, such as methylamines (trimethylamine N-oxide TMAO) or trehalose, were more specifically identified as stabilizers under high pressure conditions (Yancey *et al.*, 2002; Yancey, 2005) that would help to counteract pressure-induced perturbations at the membrane and protein folding levels.

14.4.1 Evolution of pressure-resistant functions

Comparisons of congeneric species experiencing similar environmental conditions and having similar life history and ecological characteristics, but having different depth distributions, permit sensitive analyses of depth-related biochemical adaptations.

In many fish and invertebrate marine species, adaptations to life under pressure have been demonstrated in metabolic pathways where affinities and kinetic characteristics of enzymes are maintained under pressure via changes in amino acids that make the enzyme active site become pressure tolerant. In fish, the activity of deshydrogenases (enzymes involved in energy metabolism) is inhibited by high hydrostatic pressure in shallow water species, but not their deep-sea congeneric species (Somero, 1992; Nishiguchi *et al.*, 2008). Pressures of only 5–10MPa are sufficient to disrupt the binding of shallow-water fishes' dehydrogenases to their cofactor, thus inhibiting their normal functioning (Somero, 1990). Deep-living species have enzymes that are more stable under pressure and that do not appear to be affected by low pressure. However, one trade-off is that the efficiency of these enzymes in reduced (perhaps making these species less competitive at shallow depths, if they were metabolically able to live at these depths). Amino acid substitutions were evidenced in deep-sea species, and they are likely to reduce the pressure sensitivity of dehydrogenases by reducing volume increases occurring during enzymatic reactions (Siebenaller and Somero, 1982; Somero, 1992; Nishiguchi *et al.*, 2008). Changes in protein hydration anywhere on the protein surface may also contribute to the sensitivity to pressure and adaptation to pressure can be achieved through amino acid substitutions that influence surface hydration changes. Therefore, shallow living species are prevented from invading deeper regions of the water column by their protein chemistry and pressure adaptations in proteins would appear to play important roles in establishing biogeographic patterns with depth.

Another example of pressure adaptation through the evolution of pressure-resistant variants is given in cytoskeleton proteins (tubulin, actine, myosin). It was shown that the volume change that accompanies subunit self-assembly is reduced for deep-sea species compared to shallow cold water and terrestrial species. The pressure threshold at which adaptation in actin self-assembly appears lies in the range of 20–30MPa in congeneric rattail fishes. A hypothetical model is also proposed to explain the higher pressure tolerance of deep-sea actins that involves changes in amino acids that reduce volume changes due to polymerization through a more compact arrangement of the solvent water with the subunits (Morita, 2010).

14.4.2 Gene expression pattern changes in response to pressure

Most cellular structures are not affected in the pressure range encountered in oceans, but their biological activity might be diminished. Lipid bilayers lose their fluidity (Beney and Gervais, 2001; Winter and Jeworrek, 2009). The activity of enzymes, membrane transporters and membrane based signal-transduction proteins might be perturbed, and there is a general loss of efficiency of the proteome due to high pressure-induced conformational changes.

Compensation of these perturbations might be achieved by increasing levels of expression, thus counteracting activity loss. Such fine tuning of gene expression is expected to play an important role in sub-lethal pressure conditions (around 40MPa) at which most surface organisms are able to survive. Many deep-sea microorganisms preserve membrane functionality at high pressure by increasing the proportion of unsaturated fatty acids in their lipids (DeLong and Yayanos, 1985), which increases membrane fluidity by increasing membrane disorder, thus reducing pressure-dependent packing of lipid bilayers (homeoviscous adaptation). When exposed to a pressure of 200MPa for 30 min, the yeast *Saccharomyces cerevisiae* upregulates the expression of enzymes involved in producing unsaturated fatty acids (Fernandes *et al.*, 2004), as well as genes involved in the response to membrane structure stress (Iwahashi *et al.*, 2003, 2005). Similar observations were made on the piezophilic bacterium *Photobacterium profundum* submitted to high pressure (Vezzi *et al.*, 2005).

Transcriptome-level analyses in bacteria and yeasts identified several genes that are upregulated by high pressure, and in particular chaperone proteins, which assist in maintaining protein folding and function after a pressure shock (Elo *et al.*, 2000; Fernandes *et al.*, 2004; Abe and Minegishi, 2008). In fact, these chaperones are expressed in a variety of stress conditions (temperature extremes, osmotic shock), and are part of a general stress response profile. Other upregulated genes are involved in carbohydrate metabolism while most of the repressed ones were in cell cycle progression and protein synthesis categories (Fernandes *et al.*, 2004; Abe and Minegishi, 2008). At sub-lethal pressures, the stress response observed that involves chaperone proteins also confers resistance to subsequent pressure shock. This might be one of the mechanisms operating in acclimatization. Indeed other stresses that also trigger the synthesis of chaperone proteins can confer subsequent pressure resistance. This provides evidence of an interconnection between stress generated by pressure and other factors (Palhano *et al.*, 2004).

14.4.3 Pressure-induced specific genes

Genes specifically expressed in high pressure conditions, but not at atmospheric pressure, were mostly identified in piezophilic microorganisms for which genome information is available: *Photobacterium profundum, Pyrococcus abyssi* and *Shewanella benthica*. Comparisons between deep and shallow bacterial strains evidenced genes that were present only in the genome of the deep strain or in the genome of the shallow strain. However, none of these genes were clearly related to pressure tolerance, but rather to another parameter that might correlate with depth, e.g. genes involved in photosynthesis exist only in shallow strains, since light only penetrates surface waters (DeLong *et al.*, 2006; Gunbin *et al.*, 2009). This would suggest that adaptation to life under pressure does not require novel functions that do not exist in other organisms.

However, some genes were identified that are not restricted to the genome of piezophilic organisms but that nevertheless appear to be exclusively expressed under pressure. In the piezophilic bacteria *Shewanella benthica* and *S. violacea*, different enzyme complexes involved in the respiratory chain are expressed depending on whether the bacteria are cultivated at atmospheric or high pressure (Kato and Qureshi, 1999). In *P. profundum*, two different porins (outer membrane proteins, OmpH and OmpL) are expressed according to pressure conditions (Bartlett *et al.*, 1989). The expression of these porins is regulated by transmembrane proteins that act as pressure sensors as well as transcriptional regulators of the porins.

14.5 Pressure Variations' Impact on Organisms' Life History Strategies

High pressure is not challenging for organisms that are living exclusively at depth and that have evolved adaptations to withstand such conditions. Therefore, rather than looking at truly piezophilic organisms, this section will deal with animals that usually live at or near the ocean surface, but which are able to face pressure variations that may be stressful for an atmospheric pressure-adapted physiology. These animals are capable of sensing and responding to pressure variations, which range from tenths of an atmosphere to several hundred atmospheres, as part of their life-history strategies.

Many marine organisms naturally face pressure variations over daily to life-span time scales. The larvae of many benthic animals may be exposed to large pressure changes while being entrained at different depth by oceanic currents, before settling on the bottom to begin their adult life (Young *et al.*, 1997; Tyler and Young, 1998; Tyler *et al.*, 2000). Zooplanktonic animals such as krill, small crustaceans, or medusae make daily vertical migrations between the surface and tens to hundreds of metres below it, in order to feed or to escape predators

(Lučić *et al.*, 2009). Eels travelling towards breeding grounds also make daily vertical migrations down to depths of 1000 m, possibly for thermoregulation and gamete maturation control (Vettier *et al.*, 2006; Aarestrup *et al.*, 2009; Jellyman and Tsukamoto, 2010). Deep vertical migrations for thermoregulation or feeding were also reported in tuna (Teo *et al.*, 2007) and white sharks (Boustany *et al.*, 2002). Air-breathing animals such as sea birds, cetaceans and pinnipeds are capable of routine extreme deep diving, for hunting, avoiding surface predators, exploration or social behaviour (Castellini, 2010).

Moreover, organisms living in shallow water areas near coast lines are also subjected to pressure variations, albeit over a much smaller range, due to the waves, tidal cycles, winds or variations in atmospheric pressure. They are capable of sensing and responding to such small pressure variations. Although these small variations are not likely to be stressful and challenging for animals, they are nevertheless felt and response mechanisms have been identified that have consequences for the life-history of these animals.

14.5.1 Deep diving in surface dwellers

Vertical migrations to great depths (more than 500 m), on short time scales (minutes to hours), are observed in some fish, as well as in marine mammals and some penguins (Fig. 14.3). Zooplankton usually only migrate to a depth of a hundred or a few hundred metres. However, recently, Antarctic krill, *Euphausia superba*, which usually lives in the upper 150 m of the Southern Ocean, was observed feeding on the abyssal plain at 3500 m depth (Clarke and Tyler, 2008).

Marine mammals and birds that are able to dive to great depths face a number of challenges relating to prolonged breath-holding and high hydrostatic pressure. Many of these deep-divers can reach 500 m, and some are capable of approaching 2000 m: northern elephant seals (*Mirounga angustirostris*) dive to depths of over 1500 m; sperm whales (*Physeter catadon*) dive to 2000 m (examples in Teloni *et al.*, 2008; Castellini, 2010; Fig. 14.3). They have evolved anatomical, biochemical and physiological adaptations that allow them to freely enter the deep marine environment. Because of prolonged apnoea (suspension of external breathing), they must optimize the size and use of their oxygen stores, and must deal with the accumulation of lactic acid if they rely upon anaerobic metabolism. As hydrostatic pressure compresses gases in the body, there are direct mechanical effects, physical effects such as bubble formation during decompression and also possible physiological effects such as nitrogen narcosis

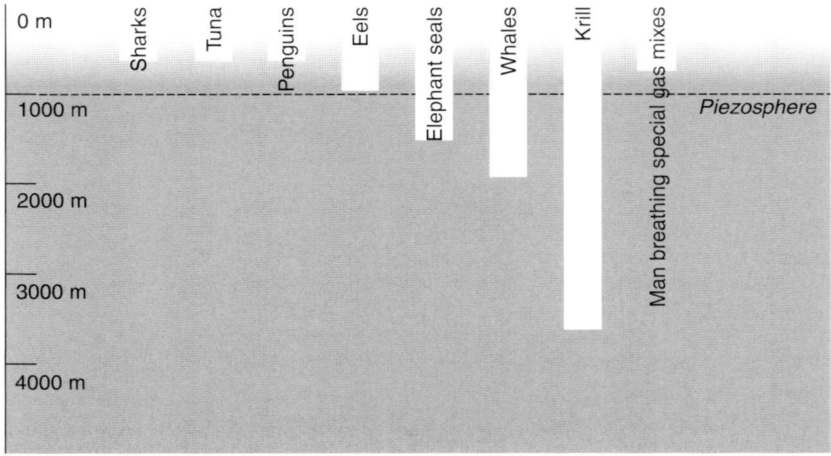

Fig. 14.3. Examples of deep biosphere penetration by 1 atmosphere adapted organisms.

and high pressure nervous syndrome (HPNS). A number of anatomical adaptations have been described that relate to the issue of air spaces and their contraction/re-expansion during the diving cycle (Kooyman, 1989; Castellini *et al.*, 2002; Castellini, 2010). By collapsing their lungs, marine mammals do not breathe compressed air, as is the case with human divers, and avoid mechanical problems with gas exchange at depth. They minimize nitrogen absorption into the blood, thus avoiding the nitrogen narcosis that is observed in human divers, as well as decompression issues due to blood degassing. To avoid the narcotic effects of nitrogen when pressure increases, human deep-divers need to breathe mixtures of gas that have fewer narcotic effects (hydrogen, helium).

But the affects of hydrostatic pressure on diving mammals and birds are broader and may also affect biochemical and cellular level processes. Studies of the glycolytic activity in the red blood cells of marine mammals showed, in general, an increased or unchanged glycolytic activity when submitted to pressures of ~14 MPa for 2 h, whereas the glycolytic activity in terrestrial mammals decreased under pressure (Castellini *et al.*, 2001). Pressure-sensitive mechanisms were proposed to explain this: (i) that glucose transport was affected due to alteration of the red blood cell membranes' properties; (ii) that changes in metabolic pathways occurred as a result of the pressure; and (iii) that there was a direct effect of pressure on glycolytic enzymes in red blood cells. High body temperature does not seem to confer protection against high pressure effects in diving marine mammals (Castellini *et al.*, 2002).

In fishes, daily vertical migrations during breeding periods appear to be a means of regulating sexual maturation using periodic exposure to lower temperatures in deeper areas. Atlantic bluefin tuna (*Thunnus thynnus*) migrate daily down to depths of approximately 600 m while entering the relatively warmer waters of the Gulf of Mexico where they reproduce (Teo *et al.*, 2007). Eels migrate daily down to approximately 1000 m, returning to 200–300 m at night. This clever strategy minimizes predation and the increasing pressure and low temperature helps to regulate gamete maturation, so that spawning coincides with the eels' arrival at their spawning grounds (Aarestrup *et al.*, 2009; Jellyman and Tsukamoto, 2010). Furthermore, the migrating eels have been shown to exhibit higher membrane fluidity and this allows them to enter deep, high pressure areas (Vettier *et al.*, 2006).

14.5.2 Pressure changes in marine embryos and larvae

The planktonic larval stages of marine benthic organisms disperse through water masses before settling on the bottom. During this dispersal phase, they are likely to be exposed to wide variations in environmental conditions. In particular, water currents may carry larvae to different depths and allow them to colonize new habitats, provided that the larvae are able to tolerate pressure changes.

Recent studies of the pressure effects on developing marine embryos and larvae investigated pressure tolerance in these early life-stages and examined whether it would be limiting for their dispersal, and subsequently for the bathymetric distribution of adults. Embryos and larvae can usually tolerate pressure over a range that exceeds that reflecting adult bathymetric distribution (Young *et al.*, 1997; Tyler and Young, 1998; Aquino-Souza *et al.*, 2008). Embryos of the shallow water sea urchin *Echinus acutus* are able to develop at 10 MPa, whereas the larval stage tolerates 20 MPa, which represents deeper areas than those at which adult populations are found (Tyler and Young, 1998). Embryos obtained from adult *E. acutus* collected from depths of 1000 m showed an even higher pressure tolerance and the authors concluded that this particular population was in the process of invading deeper areas.

Pressure tolerance in embryos and larvae might also be modulated by temperature. Indeed, low temperature appears to reduce the pressure tolerance of embryos from shallow-water species (Tyler *et al.*, 2000; Mestre *et al.*, 2009). This might be explained by the fact that low temperatures reduce biomembrane fluidity, thus reinforcing the pressure effect and lowering pressure tolerance thresholds. Invasion of the deep ocean would therefore be easier in regions where the water column is isothermal, such as polar regions, because the larvae would face only pressure increases and temperature would remain constant (Thatje *et al.*, 2005).

Deep-sea hydrothermal vents are exceptional with regard to their very high temperatures. The larval stages of endemic, hydrothermal vent organisms therefore have to face a double challenge during dispersal: they are exposed to both warm vent waters and abyssal cold sea water, and to pressure changes if they travel upwards carried by hydrothermal vent plumes. Embryos and larvae of vent annelids incubated in controlled temperature–pressure conditions in hyperbaric vessels demonstrated a need for high pressure and low temperature (Marsh *et al.*, 2001; Pradillon *et al.*, 2001, 2005; Brooke and Young, 2009). This was particularly significant for the polychaete species *Alvinella pompejana*, which is thermophilic when adult (see also Lutz, Chapter 13, this volume).

Usually, as a consequence of pressure challenges, embryos exhibit abnormal division patterns, membrane breaking and, at sub-lethal pressures, reduced division rates (Pradillon *et al.*, 2005; Mestre *et al.*, 2009). Abnormal division patterns are also observed with alteration of cell relative volume. This reflects the pressure effects on membranes and the cytoskeleton, which cause abnormal cell divisions in developing embryos.

Pressure also exerts mechanical constraints on developing embryos, and could potentially serve as a regulator, inhibiting development when unsuitable conditions are encountered. In fact, it has been shown that growth-generated strains and pressures in developing tissues regulate morphogenesis throughout development (Henderson and Carter, 2002; Farge, 2003; Brouzes and Farge, 2004). Mechanical constraints produced by morphogenetic movements in developing embryos of the fruit fly (*Drosophila*) can trigger the expression of genes involved in the establishment of the axis and segments of the future larvae (Farge, 2003). Specific transduction pathways are activated in cells where mechanical signals are detected leading to the modulation of the expression of genes in the developmental programme.

If mechanical constraints produced internally by the developing embryo can cause a response through the expression of new sets of developmental genes then there is a possibility that external mechanical signals coming from the environment may also trigger a response in the developing embryo. Such responses may be part of an adaptation strategy to avoid development when conditions are not suitable to allow a correct development. Indeed, it has been shown that embryos may respond to hydrostatic pressure changes. In the estuarine mud crab *Rhithropanopeus harrisii*, embryos respond to the cyclical changes in hydrostatic pressure caused by the tide when triggering their release from their mother (Forward and Bourla, 2008). Similarly, in *Drosophila* fruit flies, hydrostatic pressure triggers the activation of oocytes (Horner and Wolfner, 2008). Therefore, hydrostatic pressure may be used by embryonic and larval stages as a signal regulating their development programme.

14.5.3 Sensing pressure changes

Animals appear to be able to sense even small pressure variations of approximately 1 kPa, corresponding to depths of 10 cm of water or less. In the oceans, many animals make use of the vertical dimension and they use pressure information as a proxy for depth in their normal lives (Fraser, 2010). But how do animals and cells sense these pressure variations?

Cellular membranes are one of the most sensitive structures to pressure variations. Therefore, they would be the most obvious candidate for pressure detectors via changes in their fluidity (Beney and

Gervais, 2001). Physical stress such as pressure causes structural modifications to the membrane. Pressure is the input signal and it triggers a reaction, which tends to restore the initial membrane state (homeoviscous adaptation).

The cytoskeleton forms a three-dimensional viscoelastic intracellular network, which affects the way cells sense their extracellular environment and respond to stimuli. It provides a continuous mechanical coupling throughout the cell, which changes as the cytoskeleton remodels. Such mechanical effects can influence ion channel activity at the plasma membrane of cells and may conduct mechanical stresses from the cell membrane to internal organelles. As a result, both rapid responses, such as changes in intracellular Ca^{2+} (Horner and Wolfner, 2008), and slower responses such as gene transcription can be elicited or modulated by mechanical perturbations. In addition to mechanical features, the cytoskeleton also provides a large negatively charged surface on which many signalling molecules, including protein and lipid kinases, phospholipases and GTPases, localize in response to activation of specific transmembrane receptors (Janmey, 1998).

At the level of whole organisms, pressure sensors have been identified that allow animals to use this parameter as a proxy for depth. Fishes with swim-bladders are extremely sensitive to depth change, and the swim-bladder wall is the likely site for pressure-operated stretch receptors. In crabs, the statocyst, an organ normally involved in balance regulation and known as an angular acceleration receptor, also appears to detect hydrostatic pressure variations and therefore can be used as a pressure sensor (Fraser, 2010). Similarly, although anatomically very different, mechanoreceptor systems were also identified in some fish such as sharks, e.g. the use of hair cells.

The mechanisms involved in hydrostatic pressure reception still need to be clarified if we want to understand pressure effects at different levels from cells to whole organisms, as well as the role of pressure in regulating the life history strategies of a wide range of organisms.

14.6 Conclusions

Although most life forms on Earth live under pressures that are much higher than those prevailing on land or in shallow water, the mechanisms and processes involved in pressure adaptation are still not fully understood. High pressure is likely to have prevailed when life emerged on Earth. Indeed, all of the studies performed to date have failed to identify a specifically evolved pressure response; rather we observe pressure adaptations at different organization levels that involve a range of mechanisms. Indeed, virtually all known animal phyla are encountered in high pressure environments, suggesting that life did not colonize deep high pressure environments but is more likely to have evolved from them.

In the present, high pressure effects on molecules or cells are being widely used for a variety of applications: food pasteurization, food organoleptic properties enhancement via pressure-controlled enzymatic activity, preparation of antiviral vaccines, development of antitumoral vaccine, high-pressure-assisted protein refolding, and improvement of the survival rate of cryopreserved gametes and eggs (Aertsen *et al.*, 2009; Rivalain *et al.*, 2010). In addition, piezophilic microorganisms are expected to be a source of high pressure-resistant enzymes and to provide information regarding the structural adaptations that are necessary for mesophilic enzymes to withstand high pressure (Abe and Horikoshi, 2001; Kawano *et al.*, 2004). Indeed, certain enzymatic reactions can be strongly favoured under high pressure. In *Bacillus thermoproteolyticus*, the activity of the enzyme thermolysin, which can generate a precursor of aspartame, an artificial peptide sweetener, is drastically increased at high pressure, resulting in an almost six-fold increase of aspartame precursor formation at 150 MPa (Kunugi and Nomura, 1990). Thus the isolation or the engineering of high pressure-stable enzyme variants could lead to higher biocatalysis yields at higher pressures.

To date most research has focused on pressures exceeding atmospheric pressure rather than lower pressures. However, hypobaric conditions are also challenging for terrestrial life. Low pressure effects on living systems have been examined with regard to the possibility that life forms could exist on other planets with low surface pressures. Nearby planets such as Mars could potentially harbour life forms and research programmes are investigating whether the low pressure atmosphere on Mars would prevent the growth and reproduction of terrestrial organisms (Richards *et al.*, 2006). In this regard, the study of deep-sea organisms, in particular the effect of lower than ambient on them, and how they are able to tolerate decompression, might provide analogues for extraterrestrial environments with much lower pressure conditions than Earth.

References

Aarestrup, K., Okland, F., Hansen, M.M., Righton, D., Gargan, P., Castonguay, M., Bernatchez, L., Howey, P., Sparholt, H., Pedersen, M.I. and McKinley, R.S. (2009) Oceanic spawning migration of the European Eel (*Anguilla anguilla*). *Science* 325, 1660.

Abe, F. (2007) Exploration of the effects of high hydrostatic pressure on microbial growth, physiology and survival: perspectives from piezophysiology. *Bioscience, Biotechnology, and Biochemistry* 71, 2347–2357.

Abe, F. and Horikoshi, K. (1995) Hydrostatic pressure promotes the acidification of vacuoles in *Saccharomyces cerevisiae*. *FEMS Microbiology Letters* 130, 307–312.

Abe, F. and Horikoshi, K. (1998) Analysis of intracellular pH in the yeast *Saccharomyces cerevisiae* under elevated hydrostatic pressure: a study in baro-(piezo-) physiology. *Extremophiles* 2, 223–228.

Abe, F. and Horikoshi, K. (2000) Tryptophan permease gene *TAT2* confers high pressure growth in *Saccharomyces cerevisiae*. *Molecular and Cellular Biology* 20, 8093–8102.

Abe, F. and Horikoshi, K. (2001) The biotechnological potential of piezophiles. *Trends in Biotechnology* 19, 102–108.

Abe, F. and Minegishi, H. (2008) Global screening of genes essential for growth in high-pressure and cold environments: searching for basic adaptive strategies using a yeast deletion library. *Genetics* 178, 851–872.

Abe, F., Kato, C. and Horikoshi, K. (1999) Pressure-regulated metabolism in microorganisms. *Trends in Microbiology* 7, 447–453.

Aertsen, A. and Michiels, C.W. (2005) Mrr instigates the SOS response after high pressure stress in *Escherichia coli*. *Molecular Microbiology* 58, 1381–1391.

Aertsen, A., Meersman, F., Hendrickx, M.E.G., Vogel, R.F. and Michiels, C.W. (2009) Biotechnology under pressure: applications and implications. *Trends in Biotechnology* 27, 434–441.

Aquino-Souza, R., Hawkins, S.J. and Tyler, P.A. (2008) Early development and larval survival of *Psammechinus miliaris* under deep-sea temperature and pressure conditions. *Journal of the Biological Association of the United Kingdom* 88, 453–461.

Arai, S., Kawarai, T., Arai, R., Yoshida, M., Furukawa, S., Ogihara, H. and Yamasaki, M. (2008) Cessation of cytokinesis in *Schizosaccharomyces pombe* during growth after release from high pressure treatment. *Bioscience, Biotechnology, and Biochemistry* 72, 88–93.

Bao, C., Gai, Y., Lou, K., Jiang, C. and Ye, S. (2010) High-hydrostatic-pressure optical chamber system for cultivation and microscopic observation of deep-sea organisms. *Aquatic Biology* 11, 157–162.

Bartlett, D., Wright, M., Yayanos, A.A. and Silverman, M. (1989) Isolation of a gene regulated by hydrostatic pressure in a deep-sea bacterium. *Nature* 342, 572–574.

Bartlett, D.H. (2002) Pressure effects on *in vivo* microbial processes. *Biochimica and Biophysica Acta* 1595, 367–381.

Begg, D.A., Salmon, E.D. and Hyatt, H.A. (1983) The changes in structural organization of actin in the sea urchin egg cortex in response to hydrostatic pressure. *Journal of Cell Biology* 97, 1795–1805.

Beney, L. and Gervais, P. (2001) Influence of the fluidity of the membrane on the response of microorganisms to environmental stresses. *Applied Microbiology and Biotechnology* 57, 34–42.

Besch, S.R. and Hogan, P.M. (1996) A small chamber for making optical measurements on single living cells at elevated hydrostatic pressure. *Undersea and Hyperbaric Medicine* 23, 175–184.

Boonyaratanakornkit, B.B., Park, C.B. and Clark, D.S. (2002) Pressure effects on intra- and intermolecular interactions within proteins. *Biochimica et Biophysica Acta* 1595, 235–249.

Bourns, B., Franklin, S., Cassimeris, L. and Salmon, E.D. (1988) High hydrostatic pressure effects in vivo: changes in cell morphology, microtubules assembly, and actin organization. *Cell Motility and the Cytoskeleton* 10, 380–390.

Boustany, A.M., Davis, S.F., Pyle, P., Anderson, S.D., Le Boeuf, B.J. and Block, B.A. (2002) Expended niche for white sharks. *Nature* 415, 35–36.

Brauer, R.W., Bekman, M.Y., Keyser, J.B., Nesbitt, D.L., Sidelev, G.N. and Wright, S.L. (1980) Adaptation to high hydrostatic pressures of abyssal gammarids from Lake Baikal in eastern Siberia. *Comparative Biochemistry and Physiology* 65A, 119–127.

Bravim, F., de Freitas, J.M., Fernandes, A.A.R. and Fernandes, P.M.B. (2010) High hydrostatic pressure and the cell membrane; stress response of *Saccharomyces cerevisiae*. *Annals of the New York Academy of Sciences* 1189, 127–132.

Brooke, S.D. and Young, C.M. (2009) Where do the embryos of *Riftia pachyptila* develop? Pressure tolerances, temperature tolerances, and buoyancy during prolonged embryonic dispersal. *Deep-Sea Research II* 56, 1599–1606.

Brouzes, E. and Farge, E. (2004) Interplay of mechanical deformation and patterned gene expression in developing embryos. *Current Opinion in Genetics & Development* 14, 367–374.

Castellini, M. (2010) Pressure tolerance in diving mammals and birds. In: Sébert, P. (ed.) *Comparative High Pressure Biology*. Science Publishers, Enfield, New Hampshire, pp. 379–398.

Castellini, M.A., Castellini, J.M. and Rivera, P.M. (2001) Adaptations to pressure in the RBC metabolism of diving mammals. *Comparative Biochemistry and Physiology Part A* 129, 751–757.

Castellini, M.A., Rivera, P.M. and Castellini, J.M. (2002) Biochemical aspects of pressure tolerance in marine mammals. *Comparative Biochemistry and Physiology Part A* 133, 893–899.

Childress, J.J., Lee, R., Sanders, N.K., Felbeck, H., Oros, D., Toulmond, A., Desbruyères, D. and Brooks, J. (1993) Inorganic carbon uptake in hydrothermal vent tubeworms facilitated by high environmental PCO2. *Nature* 362, 147–149.

Clarke, A. and Tyler, P.A. (2008) Adult Antarctic krill feeding at abyssal depths. *Current Biology* 18, 282–285.

Cossins, A.R. and Macdonald, A.G. (1989) The adaptation of biological membranes to temperature and pressure: fish from the deep and cold. *Journal of Bioenergetics and Biomembranes* 21, 115–135.

Cottin, D., Ravaux, J., Léger, N., Halary, S., Toullec, J.-Y., Sarradin, P.-M., Gaill, F. and Shillito, B. (2008) Thermal biology of the deep-sea vent annelid *Paralvinella grasslei*: *in vivo* studies. *Journal of Experimental Biology* 211, 2196–2204.

Cottin, D., Shillito, B., Chertemps, T., Thatje, S., Léger, N. and Ravaux, J. (2010) Comparison of heat-shock responses between the hydrothermal vent shrimp *Rimicaris exoculata* and the related shrimp *Palaemonetes varians*. *Journal of Experimental Marine Biology and Ecology* 393, 9–16.

Daniel, I., Oger, P. and Winter, R. (2006) Origins of life and biochemistry under high-pressure conditions. *Chemical Society Reviews* 35, 858–875.

DeLong, E.F. and Yayanos, A.A. (1985) Adaptation of the membrane lipids of a deep-sea bacterium to changes in hydrostatic pressure. *Science* 228, 1101–1103.

DeLong, E.F., Preston, C.M., Mincer, T., Rich, V., Hallam, S.J., Frigaard, N.U., Martinez, A., Sullivan, M.B., Edwards, R., Brito, B.R., Chisholm, S.W. and Karl, D.M. (2006) Community genomics among stratified microbial assemblages in the ocean's interior. *Science* 311, 496–503.

Domitrovic, T., Fernandes, C.M., Boy-Marcotte, E. and Kurtenbach, E. (2006) High hydrostatic pressure activates gene expression through Msn2/4 stress transcription factors which are involved in the acquired tolerance by mild pressure precondition in *Saccharomyces cerevisiae*. *FEBS Letters* 580, 6033–6038.

Drazen, J.C., Bird, L.E. and Barry, J.P. (2005) Development of a hyperbaric trap-respirometer for the capture and maintenance of live deep-sea organisms. *Limnology and Oceanography: Methods* 3, 488–498.

Elo, M.A., Sironen, R.K., Kaarniranta, K., Auriola, S., Helminen, H.J. and Lammi, M.J. (2000) Differential regulation of stress proteins by high hydrostatic pressure, heat shock, and unbalanced calcium homeostasis in chondrocytic cells. *Journal of Cellular Biochemistry* 79, 610–619.

Erijman, L. and Clegg, R.M. (1995) Heterogeneity of *E. coli* RNA polymerase revealed by high pressure. *Journal of Molecular Biology* 253, 259–265.

Erijman, L. and Clegg, R.M. (1998) Reversible stalling of transcription elongation complexes by high pressure. *Biophysical Journal* 75, 453–462.

Fang, J., Zhang, L. and Bazylinski, D.A. (2010) Deep-sea piezosphere and piezophiles: geomicrobiology and biogeochemistry. *Trends in Microbiology* 18, 413–422.

Farge, E. (2003) Mechanical induction of Twist in the *Drosophila* foregut/stomodeal primordium. *Current Biology* 13, 1365–1377.

Fernandes, P.M.B., Domitrovic, T., Kao, C.M. and Kurtenbach, E. (2004) Genomic expression pattern in *Saccharomyces cerevisiae* cells in response to high hydrostatic pressure. *FEBS Letters* 556, 153–160.

Foguel, D., Suarez, M.C., Ferrão-Gonzales, A.A., Porto, T.C.R., Palmieri, L., Einsiedler, C.M., Andrade, L.R., Lashuel, H.A., Lansbury, P.T., Kelly, J.W. and Silva, J.L. (2003) Dissociation of amyloid fibrils of □-synuclein and transthyretin by pressure reveals their reversible nature and the formation of water-excluded cavities. *Proceedings of the National Academy of Sciences of the United States of America* 100, 9831–9836.

Forward, R.B. and Bourla, M.H. (2008) Entrainment of the larval release rhythm of the crab *Rhithropanopeus harrisii* (Brachyura: Xanthidae) by cycles in hydrostatic pressure. *Journal of Experimental Marine Biology and Ecology* 357, 128–133.

Fraser, P.J. (2010) Pressure sensing: depth sensors and depth. In: Sébert, P. (ed.) *Comparative High Pressure Biology*. Science Publishers, Enfield, New Hampshire, pp. 143–159.

Frey, B., Franz, S., Sheriff, A., Korn, A., Bluemelhuber, G., Gaipl, U., Voll, R., Meyer-Pittroff, R. and Herrmann, M. (2004) Hydrostatic pressure induced death of mammalian cells engages pathways related to apoptosis or necrosis. *Cellular and Molecular Biology* 50, 459–467.

Frey, B., Hartmann, M., Herrmann, M., Meyer-Pittroff, R., Sommer, K. and Bluemelhuber, G. (2006) Microscopy under pressure – an optical chamber system for fluorescence microscopic analysis of living cells under high hydrostatic pressure. *Microscopy Research and Technique* 69, 65–72.

Fujiwara, Y., Tsukahara, J., Hashimoto, J. and Fujikura, K. (1998) *In situ* spawning of a deep-sea vesicomyid clam: evidence for an environmental cue. *Deep-Sea Research I* 45, 1881–1889.

Gibbs, A. and Somero, G.N. (1989) Pressure adaptation of Na+/K+ -ATPase in gills of marine teleosts. *Journal of Experimental Biology* 143, 475–492.

Girard, E., Prangé, T., Dhaussy, A.-C., Migianu-Griffoni, E., Lecouvey, M., Chervin, J.-C., Mezouar, M., Kahn, R. and Fourme, R. (2007) Adaptation of the base-paired double-helix molecular architecture to extreme pressure. *Nucleic Acids Research* 35, 4800–4808.

Goldinger, J.M., Kang, B.S., Chow, Y.E., Paganelli, C.V. and Hong, S.K. (1980) Effect of hydrostatic pressure on ion transport and metabolism in human erythrocytes. *Journal of Applied Physiology* 49, 224–231.

Gross, M. and Jaenicke, R. (1994) Proteins under pressure. The influence of high hydrostatic pressure on structure, function and assembly of proteins and protein complexes. *European Journal of Biochemistry* 221, 617–630.

Gross, M., Lehle, K., Jaenicke, R. and Nierhaus, K.H. (1993) Pressure-induced dissociation of ribosomes and elongation cycle intermediates. *European Journal of Biochemistry* 218, 463–468.

Grossman, Y., Aviner, B. and Mor, A. (2010) Pressure effects on mammalian central nervous system. In: Sébert, P. (ed.) *Comparative High-Pressure Biology*. Science Publishers, Enfield, New Hampshire, pp. 161–186.

Gunbin, K.V., Afonnikov, D.A. and Kolchanov, N.A. (2009) Molecular evolution of the hyperthermophilic archaea of the *Pyrococcus* genus: analysis of adaptation to different environmental conditions. *BMC Genomics* 10, 639.

Hall, A.C., Pickles, D.M. and MacDonald, A.G. (1993) Aspects of eukaryotic cells. In: *Effects of High Pressure on Biological Systems*. Springer-Verlag, Berlin, Germany, pp. 29–85.

Henderson, J.H. and Carter, D.R. (2002) Mechanical induction in limb morphogenesis: the role of growth-generated strains and pressures. *Bone* 31, 645–653.

Horikawa, D.D., Iwata, K.-I., Kawai, K., Koseki, S., Okuda, T. and Yamamoto, K. (2009) High hydrostatic pressure tolerance of four different anhydrobiotic animal species. *Zoological Science* 26, 238–242.

Horner, V.L. and Wolfner, M. (2008) Mechanical stimulation by osmotic and hydrostatic pressure activates *Drosophila* oocytes *in vitro* in a calcium-dependent manner. *Developmental Biology* 316, 100–109.

Ishii, A., Sato, T., Wachi, M., Nagai, K. and Kato, C. (2004) Effects of high hydrostatic pressure on bacterial cytoskeleton FtsZ polymers *in vivo* and *in vitro*. *Microbiology* 150, 1965–1972.

Iwahashi, H., Obuchi, K., Fujii, S. and Komatsu, Y. (1997) Effect of temperature on the role of Hsp104 and trehalose in barotolerence of *Saccharomyces cerevisiae*. *FEBS Letters* 416, 1–5.

Iwahashi, H., Nwaka, S. and Obuchi, K. (2000) Evidence for contribution of neutral trehalase in barotolerance of *Saccharomyces cerevisiae*. *Applied and Environmental Microbiology* 66, 5182–5185.

Iwahashi, H., Nwaka, S. and Obuchi, K. (2001) Contribution of Hsc70 to barotolerance in the yeast *Saccharomyces cerevisiae*. *Extremophiles* 5, 417–421.

Iwahashi, H., Shimizu, H., Odani, M. and Komatsu, Y. (2003) Piezophysiology of genome wide gene expression levels in the yeast *Saccharomyces cerevisiae*. *Extremophiles* 7, 291–298.

Iwahashi, H., Odani, M., Ishidou, E. and Kitagawa, E. (2005) Adaptation of *Saccharomyces cerevisiae* to high hydrostatic pressure causing growth inhibition. *FEBS Letters* 579, 2847–2852.

Janmey, P.A. (1998) The cytoskeleton and cell signaling: component localization and mechanical coupling. *Physiological Reviews* 78, 763–781.

Jellyman, D. and Tsukamoto, K. (2010) Vertical migrations may control maturation in migrating female *Anguilla dieffenbachii. Marine Ecology Progress Series* 404, 241–247.

Kaarniranta, K., Elo, M.A., Lammi, M.J., Goldring, M.B., Eriksson, J., Sistonen, L. and Helminen, H.J. (1998) Hsp 70 accumulation in chondrocytic cells exposed to high continuous hydrostatic pressure coincides with mRNA stabilization rather than transcriptional activation. *Proceedings of National Academy of Sciences of the United States of America* 95, 2319–2324.

Kato, C. and Qureshi, M.H. (1999) Pressure response in deep-sea piezophilic bacteria. *Journal of Molecular Microbiology and Biotechnology* 1, 87–92.

Kawano, H., Nakasone, K., Matsumoto, M., Yoshida, Y., Usami, R., Kato, C. and Abe, F. (2004) Differential pressure resistance in the activity of RNA polymerase isolated from *Shewanella violacea* and *Escherichia coli. Extremophiles* 8, 367–375.

Kitching, J.A. (1954) The effects of high hydrostatic pressures on a suctorian. *Journal of Experimental Biology* 31, 68–75.

Kobori, H., Sato, M., Tameike, K., Hamada, K., Shimada, S. and Osumi, M. (1995) Ultrastructural effects of pressure stress to the nucleus in *Saccharomyces cerevisiae*: a study by immunoelectron microscopy using frozen thin sections. *FEMS Microbiology Letters* 132, 253–258.

Kooyman, G.L. (1989) *Diverse Divers: physiology and behavior*. Springer-Verlag, Berlin, Germany, 200 pp.

Koyama, S., Miwa, T., Sato, T. and Aizawa, M. (2001) Optical chamber system designed for microscopic observation of living cells under extremely high hydrostatic pressure. *Extremophiles* 5, 409–415.

Koyama, S., Miwa, T., Horii, M., Ishikawa, Y., Horikoshi, K. and Aizawa, M. (2002) Pressure-stat aquarium system designed for capturing and maintaining deep-sea organisms. *Deep Sea Research Part I* 49, 2095–2102.

Koyama, S., Nagahama, T., Ootsu, N., Takayama, T., Horii, M., Konishi, S., Miwa, T., Ishikawa, Y. and Aizawa, M. (2005) Survival of deep-sea shrimps (*Alvinocaris* sp.) during decompression and larval hatching at atmospheric pressure. *Marine Biotechnology* 7, 272–278.

Kunugi, S. and Nomura, A. (1990) Pressure as a control factor of enzymatic synthetic reactions. *Annals of the New York Academy of Sciences* 613, 501–505.

Lin, X., Long, L., Shan, X., Zhang, S., Shen, S. and Liu, B. (2006) *In planta* mobilization of mPing and its putative autonomous element Pong in rice by hydrostatic pressurization. *Journal of Experimental Botany* 57, 2313–2323.

Long, L., Lin, X., Zhai, J., Kou, H., Yang, W. and Liu, B. (2006) Heritable alteration in DNA methylation pattern occurred specifically at mobile elements in rice plants following hydrostatic pressurization. *Biochemical and Biophysical Research Communications* 340, 369–376.

Lučić, D., Benović, A., Morović, M., Batistić, M. and Onofri, I. (2009) Diel vertical migration of medusae in the open Southern Adriatic Sea over a short time period (July 2003). *Marine Ecology* 30, 16–32.

Lynch, T.W. and Sligar, S.G. (2002) Experimental and theoretical high pressure strategies for investigating protein-nucleic acid assemblies. *Biochimica et Biophysica Acta* 1595, 277–282.

Macdonald, A.G. (1997) Hydrostatic pressure as an environmental factor in life processes. *Comparative Biochemistry and Physiology Part A* 116, 291–297.

Macdonald, A.G. and Gilchrist, I. (1978) Further studies on the pressure tolerance of deep-sea crustacea, with observations using a new high-pressure trap. *Marine Biology* 45, 9–21.

Macdonald, A.G. and Gilchrist, I. (1980) Effects of hydraulic decompression and compression on deep sea amphipods. *Comparative Biochemistry and Physiology Part A* 67, 149–153.

Macdonald, A.G. and Gilchrist, I. (1982) The pressure tolerance of deep sea amphipods collected at their ambient high pressure. *Comparative Biochemistry and Physiology Part A* 71, 349–352.

Macgregor, R.B.J. (2002) The interactions of nucleic acids at elevated hydrostatic pressure. *Biochimica et Biophysica Acta* 1595, 266–276.

Marsh, A.G., Mullineaux, L.S., Young, C.M. and Manahan, D.T. (2001) Larval dispersal potential of the tubeworm *Riftia pachyptila* at deep-sea hydrothermal vents. *Nature* 411, 77–80.

Marsland, D. (1950) The mechanism of cell division: temperature-pressure experiments on the cleaving eggs of *Arbacia punctulata. Journal of Cellular and Comparative Physiology* 36, 205–227.

Marsland, D. and Landau, J.V. (1954) The mechanisms of cytokinesis: temperature-pressure studies on the cortical gel system in various marine eggs. *Journal of Experimental Zoology* 125, 507–539.

Meersman, F. and Dobson, C.M. (2006) Probing the pressure-temperature stability of amyloid fibrils provides new insights into their molecular properties. *Biochimica et Biophysica Acta* 1764, 452–460.

Mestre, N.C., Thatje, S. and Tyler, P.A. (2009) The ocean is not deep enough: pressure tolerances during early ontogeny of the blue mussel *Mytilus edulis*. *Proceedings of the Royal Society B* 276, 717–726.

Molina-Höppner, A., Sato, T., Kato, C., Gänzle, M.G. and Vogel, R.F. (2003) Effects of pressure on cell morphology and cell division of lactic acid bacteria. *Extremophiles* 7, 511–516.

Morita, T. (2010) High-pressure adaptation of muscle proteins from deep-sea fishes, *Coryphaenoides yaquinae* and *C. armatus*. *Annals of the New York Academy of Sciences* 1189, 91–94.

Nishiguchi, Y., Miwa, T. and Abe, F. (2008) Pressure-adaptive differences in lactate dehydrogenases of three hagfishes: *Eptatretus burgeri*, *Paramyxine atami* and *Eptatretus okinoseanus*. *Extremophiles* 12, 477–480.

Northrop, D.B. (2002) Effects of high pressure on enzymatic activity. *Biochimica et Biophysica Acta* 1595, 71–79.

Oger, P.M. and Jebbar, M. (2010) The many ways of coping with pressure. *Research in Microbiology* 161, 799–809.

Oger, P.M., Daniel, I. and Picard, A. (2006) Development of a low-pressure diamond anvil cell and analytical tools to monitor microbial activities *in situ* under controlled P and T. *Biochimica et Biophysica Acta* 1764, 434–442.

Oliviera, A.C., Gaspar, L.P., Da Poian, A.T. and Silva, J.L. (1994) Arc repressor will not denature under pressure in the absence of water. *Journal of Molecular Biology* 240, 184–187.

Pagán, R. and Mackey, B. (2000) Relationship between membrane damage and cell death in pressure-treated *Escherichia coli* cells: differences between exponential- and stationary-phase cells and variation among strains. *Applied and Environmental Microbiology* 66, 2829–2834.

Pagliaro, L., Reitz, F. and Wang, J. (1995) An optical pressure chamber designed for high numerical aperture studies on adherent living cells. *Undersea and Hyperbaric Medicine* 22, 171–181.

Palhano, F.L., Orlando, M.T.D. and Fernandes, P.M.B. (2004) Induction of baroresistance by hydrogen peroxide, ethanol and cold-shock in *Saccharomyces cerevisiae*. *FEMS Microbiology Letters* 233, 139–145.

Phleger, C.F., McConnaughey, R.R. and Crill, P. (1979) Hyperbaric fish trap operation and deployment in the deep-sea. *Deep-Sea Research A* 26A, 1405–1409.

Pradillon, F., Shillito, B., Young, C.M. and Gaill, F. (2001) Developmental arrest in vent worm embryos. *Nature* 413, 698–699.

Pradillon, F., Shillito, B., Chervin, J.-C., Hamel, G. and Gaill, F. (2004) Pressure vessels for *in vitro* studies of deep-sea fauna. *High Pressure Research* 24, 237–246.

Pradillon, F., Le Bris, N., Shillito, B., Young, C.M. and Gaill, F. (2005) Influence of environmental conditions on early development of the hydrothermal vent polychaete *Alvinella pompejana*. *Journal of Experimental Biology* 208, 1551–1561.

Pribenszky, C., Molnár, M., Cseh, S. and Solti, L. (2005) Improving post-thaw survival of cryopreserved mouse blastocysts by hydrostatic pressure challenge. *Animal Reproduction Science* 87, 143–150.

Pribenszky, C., Du, Y., Molnár, M., Harnos, A. and Vajta, G. (2008) Increased stress tolerance of matured pig oocytes after high hydrostatic pressure treatment. *Animal Reproduction Science* 106, 200–207.

Quetin, L.B. and Childress, J.J. (1980) Observations on the swimming activity of two bathypelagic mysid species maintained at high hydrostatic pressures. *Deep-Sea Research Part A* 27, 383–391.

Ravaux, J., Gaill, F., Le Bris, N., Sarradin, P.-M. and Shillito, B. (2003) Heat-shock response and temperature resistance in the deep-sea vent shrimp *Rimicaris exoculata*. *Journal of Experimental Biology* 206, 2345–2354.

Richards, J.T., Corey, K.A., Paul, A.-L., Ferl, R.J., Wheeler, R.M. and Schuerger, A.C. (2006) Exposure of *Arabidopsis thaliana* to hypobaric environments: implications for low-pressure bioregenerative life support systems for human exploration missions and terraforming on Mars. *Astrobiology* 6, 851–866.

Rivalain, N., Roquain, J. and Demazeau, G. (2010) Development of high pressure in biosciences: Pressure effect on biological structures and potential applications in biotechnologies. *Biotechnology Advances* 28, 659–672.

Robinson, N.J., Thatje, S. and Osseforth, C. (2009) Heartbeat sensors under pressure: a new method for assessing hyperbaric physiology. *High Pressure Research* 29, 422–430.

Roer, T.D. and Shelton, M.G. (1982) Effects of hydrostatic pressure on Na transport in the freshwater crayfish, *Procambarus clarkii*. *Comparative Biochemistry and Physiology* 71A, 271–276.

Rosypal, A.C., Bowman, D.D., Holliman, D., Flick, G.J. and Lindsay, D.S. (2007) Effects of high hydrostatic pressure on embryonation of *Ascaris suum* eggs. *Veterinary Parasitology* 145, 86–89.

Rothschild, L.J. and Mancinelli, R.L. (2001) Life in extreme environments. *Nature* 409, 1092–1101.

Roussel, E.G., Cambon Bonavita, M.-A., Querellou, J., Cragg, B.A., Webster, G., Prieur, D. and Parkes, R.J. (2008) Extending the sub-sea-floor biosphere. *Science* 320, 1046.

Salmon, E.D. (1975) Pressure-induced depolymerization of brain microtubules *in vitro*. *Science* 189, 884–886.

Sato, M., Kobori, H., Ishijima, S.A., Feng, Z.H., Hamada, K., Shimada, S. and Osumi, M. (1996) *Schizosaccharomyces pombe* is more sensitive to pressure stress than *Saccharomyces cerevisiae*. *Cell Structure and Function* 21, 167–174.

Sato, M., Hasegawa, K., Shimada, S. and Osumi, M. (1999) Effects of pressure stress on the fission yeast *Schizosaccharomyces pombe* cold-sensitive mutant nda3. *FEMS Microbiology Letters* 176, 31–38.

Seki, K. and Toyoshima, M. (1998) Preserving tardigrades under pressure. *Nature* 395, 853.

Shillito, B., Jollivet, D., Sarradin, P.-M., Rodier, P., Lallier, F.H., Desbruyères, D. and Gaill, F. (2001) Temperature Resistance of *Hesiolyra bergi*, a polychaetous annelid living on deep-sea vent smoker walls. *Marine Ecology Progress Series* 216, 141–149.

Shillito, B., Le Bris, N., Hourdez, S., Ravaux, J., Cottin, D., Caprais, J.-C., Jollivet, D. and Gaill, F. (2006) Temperature resistance studies on the deep-sea vent shrimp *Mirocaris fortunata*. *Journal of Experimental Biology* 209, 945–955.

Shillito, B., Hamel, G., Duchi, C., Cottin, D., Sarrazin, J., Sarradin, P.-M., Ravaux, J. and Gaill, F. (2008) Live capture of megafauna from 2300 m depth, using a newly-designed pressurised recovery device. *Deep Sea Research Part I* 55, 881–889.

Shimada, S., Andou, M., Naito, N., Yamada, N., Osumi, M. and Hayashi, R. (1993) Effects of hydrostatic pressure on the ultrastructure and leakage of internal substances in the yeast *Saccharomyces cerevisiae*. *Applied Microbiology and Biotechnology* 40, 123–131.

Siebenaller, J.F. and Somero, G.N. (1982) The maintenance of different enzyme activity levels in congeneric fishes living at different depths. *Physiological Zoology* 55, 171–179.

Smeller, L. (2002) Pressure-temperature phase diagrams of biomolecules. *Biochimica et Biophysica Acta* 1595, 11–29.

Somero, G.N. (1990) Life at low volume change: hydrostatic pressure as a selective factor in the aquatic environment. *American Zoologist* 30, 123–135.

Somero, G.N. (1992) Adaptations to high hydrostatic pressure. *Annual Reviews of Physiology* 54, 557–577.

Swezey, R.R. and Somero, G.N. (1985) Pressure effects on actin self-assembly: interspecific differences in the equilibrium and kinetics of the G to F transformation. *Biochemistry* 24, 852–860.

Takai, K., Nakamura, K., Toki, T., Tsunogai, U., Miyazaki, M., Miyazaki, J., Hirayama, H., Nakagawa, S., Nunoura, T. and Horikoshi, K. (2008) Cell proliferation at 122°C and isotopically heavy CH4 production by a hyperthermophilic methanogen under high-pressure cultivation. *Proceedings of the National Academy of Sciences of the United States of America* 105, 10949–10954.

Takano, K.J., Takano, T., Yamanouchi, Y. and Satou, T. (1997) Pressure-induced apoptosis in human lymphoblasts. *Experimental Cell Research* 235, 155–160.

Teloni, V., Mark, J.P., Patrick, M.J.O. and Peter, M.T. (2008) Shallow food for deep divers: dynamic foraging behavior of male sperm whales in a high latitude habitat. *Journal of Experimental Marine Biology and Ecology* 354, 119–131.

Teo, S.L.H., Boustany, A., Dewar, H., Stokesbury, M.J.W., Weng, K.C., Beemer, S., Seitz, A.C., Farwell, C.J., Prince, E.D. and Block, B.A. (2007) Annual migrations, diving behavior, and thermal biology of Atlantic bluefin tuna, *Thunnus thynnus*, on their Gulf of Mexico breeding grounds. *Marine Biology* 151, 1–18.

Thatje, S., Hillenbrand, C.-D. and Larter, R. (2005) On the origin of Antarctic marine benthic community structure. *Trends in Ecology and Evolution* 20, 534–540.

Thatje, S., Casburn, L. and Calcagno, J.A. (2010) Behavioural and respiratory response of the shallow-water hermit crab *Pagurus cuanensis* to hydrostatic pressure and temperature. *Journal of Experimental Marine Biology and Ecology* 390, 22–30.

Todo, Y., Kitazato, H., Hashimoto, J. and Gooday, A.J. (2005) Simple foraminifera flourish at the ocean's deepest point. *Science* 307, 689.

Treude, T., Janßen, F., Queisser, W. and Witte, U. (2002) Metabolism and decompression tolerance of scavenging lysianassoid deep-sea amphipods. *Deep Sea Research Part I* 49, 1281–1289.

Tyler, P.A. and Dixon, D.R. (2000) Temperature/pressure tolerance of the first larval stage of *Mirocaris fortunata* from Lucky Strike hydrothermal vent field. *Journal of the Marine Biological Association of the United Kingdom* 80, 739–740.

Tyler, P.A. and Young, C.M. (1998) Temperature and pressure tolerances in dispersal stages of the genus *Echinus* (Echinodermata: Echinoidea): prerequisites for deep-sea invasion and speciation. *Deep-Sea Research II* 45, 253–277.

Tyler, P.A., Young, C.M. and Clarke, A. (2000) Temperature and pressure tolerance of embryos and larvae of the Antarctic sea urchin *Sterechinus neumayeri* (Echinodermata: Echinoidea): potential for deep-sea invasion from high latitudes. *Marine Ecology Progress Series* 192, 173–180.

Vettier, A., Labbe, C., Amerand, A., Da Costa, G., Le Rumeur, E., Moisan, C. and Sebert, P. (2006) Hydrostatic pressure effects on eel mitochondrial functioning and membrane fluidity. *Undersea and Hyperbaric Medicine* 33, 149–156.

Vezzi, A., Campanaro, S., D'Angelo, M., Simonato, F., Vitulo, N., Lauro, F.M., Cestaro, A., Malacrida, G., Simionati, B., Cannata, N., Romualdi, C., Bartlett, D.H. and Valle, G. (2005) Life at depth: *Photobacterium profundum* genome sequence and expression analysis. *Science* 307, 1459–1461.

Welch, T.J., Farewell, A., Neidhardt, F.C. and Bartlett, D.H. (1993) Stress response of *Escherichia coli* to elevated hydrostatic pressure. *Journal of Bacteriology* 175, 7170–7177.

Whitman, W.B., Coleman, D.C. and Wiebe, W.J. (1998) Prokaryotes: the unseen majority. *Proceedings of the National Academy of Sciences of the United States of America* 95, 6578–6583.

Winter, R. and Dzwolak, W. (2005) Exploring the temperature-pressure configurational landscape of biomolecules: from lipid membranes to proteins. *Philosophical Transactions of the Royal Society A* 363, 537–563.

Winter, R. and Jeworrek, C. (2009) Effect of pressure on membranes. *Soft Matter* 5, 3157–3173.

Yancey, P.H. (2005) Organic osmolytes as compatible, metabolic and counteracting cytoprotectants in high osmolarity and other stresses. *Journal of Experimental Biology* 208, 2819–2830.

Yancey, P.H., Blake, W.R. and Conley, J. (2002) Unusual organic osmolytes in deep-sea animals: adaptations to hydrostatic pressure and other perturbants. *Comparative Biochemistry and Physiology Part A* 133, 667–676.

Yayanos, A.A. (1978) Recovery and maintenance of live amphipods at a pressure of 580 bars from an ocean depth of 5700 meters. *Science* 200, 1056–1059.

Yayanos, A.A. (1981) Reversible inactivation of deep-sea amphipods (*Paralicella capresca*) by a decompression from 601 bars to atmospheric pressure. *Comparative Biochemistry and Physiology* 69(A), 563–565.

Yayanos, A.A. and Dietz, A.S. (1983) Death of a hadal deep-sea bacterium after decompression. *Science* 220, 497–498.

Yayanos, A.A. and Pollard, E.C. (1969) A study of the effects of hydrostatic pressure on macromolecular synthesis in *Escherichia coli*. *Biophysical Journal* 9, 1464–1482.

Yoshiki, T., Toda, T., Yoshida, T. and Shimizu, A. (2006) A new hydrostatic pressure apparatus for studies of marine zooplankton. *Journal of Plankton Research* 28, 563–570.

Young, C.M., Tyler, P.A. and Fenaux, L. (1997) Potential for deep sea invasion by Mediterranean shallow water echinoids: pressure and temperature as stage-specific dispersal barriers. *Marine Ecology Progress Series* 154, 197–209.

Zeng, X., Birrien, J.-L., Fouquet, Y., Cherkashov, G., Jebbar, M., Querellou, J., Oger, P., Cambon Bonavita, M.-A., Xiao, X. and Prieur, D. (2009) *Pyrococcus* CH1, an obligate piezophilic hyperthermophile: extending the upper pressure-temperature limits for life. *ISME Journal* 3, 873–876.

Zimmerman, A.M. and Marsland, D. (1964) Cell division: effects of pressure on the mitotic mechanisms of marine eggs (*Arbacia punctulata*). *Experimental Cell Research* 35, 293–302.

Zobell, C.E. and Cobet, A.B. (1964) Filament formation by *Escherichia coli* at increased hydrostatic pressure. *Journal of Bacteriology* 87, 710–719.

Zobell, C.E. and Johnson, F.H. (1949) The influence of hydrostatic pressure on the growth and viability of terrestrial and marine bacteria. *Journal of Bacteriology* 57, 179–189.

15 Deep Sea

David Hughes
Scottish Association for Marine Science, Argyll, UK

15.1 Introduction

For marine biologists the 'deep sea' begins roughly 200 m below the ocean surface at the level where a distinct topographic boundary – the continental shelf edge – coincides with the ecologically important transition from sunlit surface waters to the dark ocean. By this definition the deep sea covers approximately 65% of the Earth's surface and accounts for over 90% of the ocean's volume. A visiting alien might well regard this immense living space as the typical environment of Planet Earth while viewing the terrestrial biosphere as 'extreme'. However, from our perspective the deep sea is an extreme environment because its defining conditions – darkness, cold and high pressure – are so hostile to human existence. Once viewed as monotonous, unchanging and biologically impoverished, the deep sea is now known to contain a remarkable variety of habitats hosting a significant fraction of the ocean's biodiversity (Koslow, 2007). Deep-sea ecosystems are also far more sensitive to processes occurring at the ocean surface than was believed even a decade ago, a linkage of growing importance in this age of concern about a changing climate (Smith *et al.*, 2009). The deep sea may be 'extreme' but it is not in any sense marginal to the functioning or biodiversity of the global marine ecosystem.

15.2 The Deep-Sea Environment

The deep sea spans the globe and extends vertically from the continental shelf edge to the floors of the deepest trenches. Physical and chemical properties of the water mass filling this vast space change primarily along the gradient of water depth, and depth is also used to define the major physiographic features of the deep-sea floor (Fig. 15.1).

15.2.1 Physical and chemical background

The deep sea is a dark environment. Sunlight intensity declines exponentially through the water column, beginning at the red end of the spectrum. The blue wavelengths penetrate furthest and are still faintly detectable down to approximately 1000 m. This absorbance pattern creates a steep gradient in natural illumination through the mesopelagic (or 'twilight') zone (200–1000 m) with permanent darkness below this. Hydrostatic pressure increases linearly with depth, adding the equivalent of 1 atmosphere (atm) for each 10 m increment from the ocean surface to the seabed. Across most of the globe deep-water temperatures are low (approximate range −1 to +4°C, depending on depth and location) but extremely stable.

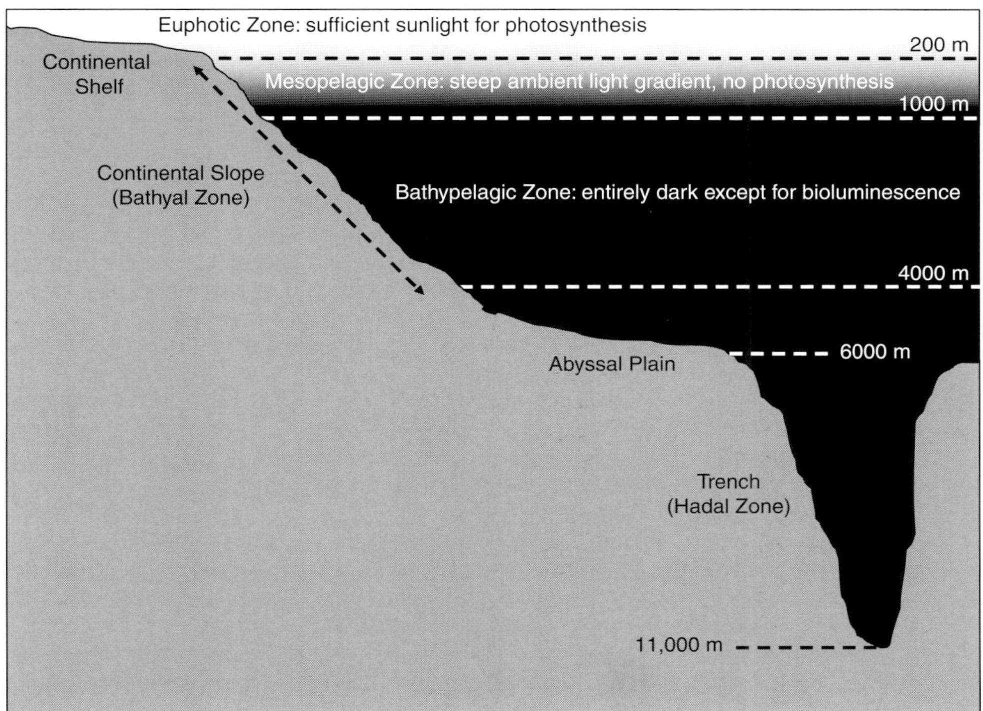

Fig. 15.1. Diagram showing the zonation of the deep-sea environment in relation to water depth and seabed topography. The seabed is shown in an idealized profile extending from the continental shelf to the floor of an oceanic trench. Note that the vertical ranges of the successive bathymetric zones are not shown to the same scale.

Bottom temperatures in the deep Mediterranean and Red Sea are much higher (13°C and 21°C, respectively) but still very constant. Most deep-sea organisms therefore never experience large or rapid changes in environmental temperature. The main exceptions occur at hydrothermal vents, where superheated fluids mix with the ambient, low-temperature bottom water (see Lutz, Chapter 13, this volume). Constant thermal conditions across most of the deep ocean are matched by a stable salinity and pH.

Most of the deep-sea water mass originates as sinking cold surface water in subpolar oceans and as a result is well oxygenated. However, in highly productive upwelling areas bacterial decomposition of sinking organic matter consumes much of the oxygen in a mid-water layer from approximately 100–1000 m depth. These oxygen minimum zones (OMZs) are most intense in the eastern Pacific, the northern Indian Ocean and off south-west Africa. Where an OMZ impinges on the continental margin, benthic communities show distinct depth zonation associated with oxygen gradients, and abundance and diversity may be very low at the most severely hypoxic depths (Levin, 2003; see also Hourdez, Chapter 21, this volume).

Hydrodynamic conditions in the deep sea range from relative quiescence on the abyssal plains to high-energy habitats on continental slopes swept by strong boundary currents or subject to internal tides. Localized current acceleration can also occur over elevations such as seamounts and ridges. The deep-sea bed shows as much physical habitat diversity as the land. Continental margin topography ranges from steep rock to sediment-covered slopes, often incised by deep canyons. The vast expanses of abyssal plain are covered either by biogenic sediments or by fine clays (Fig. 15.2). Elevated rocky habitats exist along the global mid-ocean ridge system and in the form of seamounts, volcanic peaks rising over 1000 m above the surrounding seafloor. Many thousands of seamounts punctuate the abyssal plains, while hills of lower elevation probably number in the millions. Subduction zones around the Pacific Rim are the sites of most of the world's ocean trenches, which in the most extreme cases reach depths in excess of 10,000 m (Jamieson *et al.*, 2009b).

15.2.2 Food for the deep sea

The deep sea is too dark to support photosynthesis. Primary production by green plants – the basis of virtually all terrestrial, freshwater and shallow marine ecosystems – is therefore absent, and deep-sea food webs are fuelled almost entirely by organic particles sinking from surface waters. Particle size ranges from dead phytoplankton cells to the carcasses of whales. Along continental margins, canyons may funnel large quantities of seagrass, macroalgal and terrestrial plant detritus to the deep-sea bed (Vetter and Dayton, 1999).

Particles may be consumed by mid-water animals or degraded by bacteria as they sink through the water column, and the resulting steep decline in available food with increasing depth is perhaps the most important factor structuring deep-sea ecosystems (Rex *et al.*, 2006). Food supply may also be highly variable over time. In regions without marked seasonality the deep sea receives a constant drizzle of small particles ('marine snow'). By contrast, at high latitudes much of the annual flux arrives in concentrated pulses following the spring/summer peak of phytoplankton production. Mucus-bound aggregates of dead cells and zooplankton faecal pellets sink rapidly and accumulate on the deep-sea bed as visible deposits of 'phytodetritus' (Beaulieu, 2002).

Fig. 15.2. The typical landscape of the deep-sea bed: vast areas on the lower continental slope and abyssal plains are covered with fine, biogenic sediments. This photograph, taken at a depth of 2060 m in the Rockall Trough, north-east Atlantic, shows a seabed marked by holes, pits and small mounds created by burrowing fauna. In the centre of the frame are a small sea urchin (*Gracilechinus affinis*) and a large brittlestar (*Ophiomusium lymani*), two representatives of the Phylum Echinodermata which often dominates the deposit-feeding megafauna at bathyal and abyssal depths. Distance across lower edge of frame is approximately 1 m.

Phytodetrital deposition can be very patchy over small spatial scales and the quantities arriving are also extremely variable from year to year (Lampitt *et al.*, 2010). As will be discussed below, the episodic, unpredictable nature of phytodetrital flux exerts a significant influence on the behaviour, life histories and population dynamics of benthic organisms (Gooday, 2002).

Chemosynthetic primary production (chemoautotrophy) using energy derived from redox reactions is carried out by bacteria and Archaea in the water column, in bottom sediments and in localized seafloor habitats such as hydrothermal vents (see Lutz, Chapter 13, this volume) and hydrocarbon seeps. Decomposition of large whale carcasses also creates reducing conditions suitable for microbial chemosynthesis. All these habitats are biomass 'hotspots' with specialized fauna relying on symbiotic partnerships with chemoautotrophic bacteria.

15.2.3 How 'extreme' is the deep sea?

From this brief summary we can see that not all deep-sea conditions are extreme. Most of the deep sea is cold, but high-latitude and montane environments on land experience much lower temperatures. In physiological terms, stable temperature, salinity and generally high oxygen content make the deep sea a rather benign place in comparison with many shallow-water habitats. The label 'extreme' can justifiably be used with respect to hydrostatic pressure, absence of sunlight and scarcity of food. Deep-sea organisms show biochemical adaptations to counter the effects of high pressure on enzyme and cell membrane function (see Pradillon, Chapter 14, this volume) but darkness and food shortage are the factors shaping the visible aspects of body form and behaviour. Away from the scattered oases of vents, seeps and whalefalls, deep-sea ecosystems bear the imprint of resource limitation. These effects are clearly seen in the distribution of biomass, animal body size and food-web structure, and will be the main focus of this chapter.

15.3 Life in a Dark World

The effects of darkness on the body form, behaviour and physiology of deep-sea organisms are most apparent in the mid-water environment. The mesopelagic (200–1000 m) and bathypelagic (1000–4000 m) zones (Fig. 15.1) collectively make up over 1 billion km^3 of living space inhabited by actively-swimming fish, cephalopods and crustaceans, together with a diverse array of gelatinous zooplankton (medusae, siphonophores, comb jellies, larvaceans, salps and other groups). Until recently, research in the deep water column was largely reliant on mid-water trawls, which yield a biased sample of dead or fragmentary specimens divorced from their natural environment. This view is now being transformed by precise sampling and *in situ* observation of living animals made possible by the use of remotely operated vehicles (ROVs) (Robison, 2004).

15.3.1 Vision and bioluminescence

Vision and bioluminescence are inseparable in any discussion of mid-water life in the deep sea. Over 90% of animal species in this environment produce light, in some cases by symbiotic partnerships with luminescent bacteria (Robison, 2004). Besides its near ubiquity, several lines of evidence point to the ecological importance of bioluminescence in the deep water column. Many fish and cephalopods possess complex accessory structures (photophores) which reflect, refract or filter the emitted light, and members of both groups retain functional eyes at depths far below the maximum penetration of sunlight (Fig. 15.3). The visual pigments of lanternfish (Myctophidae), one of the most important mesopelagic fish families, have absorption maxima optimized for detecting bioluminescence rather than downwelling sunlight (Turner *et al.*, 2009). A wide range of functions has been postulated for bioluminescence, some of which are now being confirmed by ROV observations. In species-rich groups such as the Myctophidae contrasting photophore patterns may have a role in intraspecific mate recognition (Herring, 2007). A defensive function is also probably widespread (Widder, 2010). The polychaete *Swima* releases bursts of luminescent fluid when disturbed, probably distracting a predator while the worm makes its escape (Osborn *et al.*, 2009). In other cases light emission may act as a 'burglar alarm' that alerts an attacker's own predators. Conversely, bioluminescence can help predators locate or attract their prey. Predatory dragonfish such as *Aristostomias* have large facial photophores positioned to direct a beam of light forwards. The emission spectra and visual sensitivity of most deep-sea organisms peak in the blue wavelengths (ca. 475 nm) (Widder, 2010), but unusually the 'searchlights' of *Aristostomias* emit red light (Widder *et al.*, 1984). The dragonfish can see its prey while the latter are unaware they have been detected. Of the 54 lanternfish species studied by Turner *et al.* (2009), two (*Myctophum nitidulum* and *Bolinichthys longipes*) have visual pigments capable of detecting red light. While conferring some defence against dragonfish this spectral sensitivity might make the bearers more vulnerable to siphonophores such as *Erenna*, which has red fluorescent lures on its tentacle filaments (Haddock *et al.*, 2005).

The demands of finding food in a dark, three-dimensional void have led to some remarkable adaptations in the visual system of mid-water animals. The mesopelagic fish *Macropinna microstoma* has upward-directed tubular eyes covered by a

Fig. 15.3. Two fish from the mesopelagic zone of the deep sea, showing adaptations to an environment with a steep vertical gradient in ambient light. The barreleye *Opisthoproctus grimaldii* (above) has upwardly-directed tubular eyes probably used to locate bioluminescent prey in the water column above. The dragonfish, *Stomias* sp. (below), has more conventional eyes but has a luminescent lure on the lower jaw to attract prey within reach. The dragonfish also shows the capacious mouth typical of many fish inhabiting the deep-water column. Image reproduced from *'From the Surface to the Bottom of the Sea'* by H. Bouree (1912), photography by Sean Linehan, National Oceanic and Atmospheric Administration (NOAA) (www photolib.noaa.gov/htmls/ship3089.htm).

transparent head shield, an arrangement facilitating the search for bioluminescent siphonophores in the water column above (Robison and Reisenbichler, 2008). Once prey is located, the eyes can be swivelled forward through 90°, allowing the fish to browse with its tiny terminal mouth. The head shield probably protects the eyes from stings while the fish is feeding. Tubular eyes occur in several other mesopelagic fish families (Fig. 15.3) but it is not known whether they function in the same way.

The abundance of bioluminescent organisms declines steeply with increasing water depth. In the Atlantic Ocean off Cape Verde, the frequency of luminescent flashes recorded by an autonomous free-fall lander decreased from 26.7 m^{-3} in the 500–999 m depth band to 1.6 m^{-3} at 2000–2499 m, and only 0.5 m^{-3} between 2500 m and the abyssal seafloor at approximately 4000 m (Priede *et al.*, 2006a). This gradient is reflected in the body form of mid-water fish. Important mesopelagic families such as the lanternfish (Myctophidae), hatchetfish (Sternoptychidae) and viperfish (Chauliodontidae) tend to be silvery, with large, sensitive eyes, while typical bathypelagic fish such as the anglers (Ceratiidae and related families), whalefish (Cetomimidae) and gulper eels (Saccopharyngidae) have dark skin and very small eyes. There is some evidence that bathypelagic fish compensate for the reduced importance of vision by a greater reliance on other senses. Regions of the brain associated with vision are reduced, while those linked with hearing and the mechanosensory lateral line system are enlarged (Kotrschal *et al.*, 1988).

In contrast to the profusion of bioluminescence in the deep water column, very few benthic organisms produce light. Some fish groups living near the deep-sea bed have reduced eyes and presumably rely more on other senses. The minute eyes of the tripodfish *Bathypterois* must be of little use, but specialized pectoral fin rays served by enlarged spinal nerves function as a mechanosensory array. Seabed photographs (Fig. 15.4) show the fish propped up on its elongate pelvic and caudal fins with the pectoral fin rays held extended

Fig. 15.4. Close-up of the tripod fish *Bathypterois* sp. in typical posture on a sediment-covered ridge at 1960 m depth in the Northeast Providence Channel, Bahamas. The elongate pectoral fin rays are held upright to detect swimming prey and are more important sensory organs than the almost unnoticeable eyes. Image courtesy of the Bahamas Deep-Sea Coral Expedition Science Party, National Oceanic and Atmospheric Administration-Ocean Explorer (NOAA-OE)(http://oceanexplorer.noaa.gov/explorations/09deepseacorals/logs/summary/media/tripod_600.html).

like fans, probably to detect small prey swimming near the seabed (Carrassón and Matallanas, 2001).

The elongate, tapering body form widespread among benthopelagic fish may be an adaptation to increase the length, and thus the precision and sensitivity, of the lateral line sensory system (Haedrich, 1996). However, one of the most diverse and abundant fish families in this category – the grenadiers (Macrouridae) – also have large, well-developed eyes (Fig. 15.5). Grenadiers are important benthic scavengers and predators (Drazen *et al.*, 2009) but the diets of some species also include a significant pelagic component (Bergstad *et al.*, 2010). Large eyes are presumably useful for detecting bioluminescent prey in off-bottom feeding excursions. Although benthic light production is generally rare, small bioluminescent swimming organisms appear to congregate around seabed features such as carcasses and coral mounds (Gillibrand *et al.*, 2007). The need to locate these productivity 'hotspots' may contribute to the retention of eyes by benthic fish living in an otherwise lightless environment.

15.3.2 Light and the pace of life

Outwardly visible characters such as eyes and photophores are not the only biological features influenced by the ambient light regime. In pelagic fish species, the activity rates of key metabolic enzymes decline significantly with increasing depth (Childress and Somero, 1979). This metabolic gradient correlates with the higher tissue water content, and lower lipid, protein and caloric values, in deeper-living fish (Childress *et al.*, 1990). Overall, these physiological and biochemical changes are expressed in reduced musculature, lower skeletal mass and a less active lifestyle in bathypelagic and lower mesopelagic fish, compared with species living higher in the water column. Parallel depth-related declines in metabolism and locomotory capacity are also seen in pelagic cephalopods (Fig. 15.6). The powerful,

Fig. 15.5. The grenadier *Coryphaenoides armatus* photographed at 3560 m depth in the southern Rockall Trough, north-east Atlantic. Grenadiers (Macrouridae) are the most abundant predatory and scavenging fish of the lower slope and abyssal plains and have large, well-developed eyes despite the absence of sunlight at these depths.

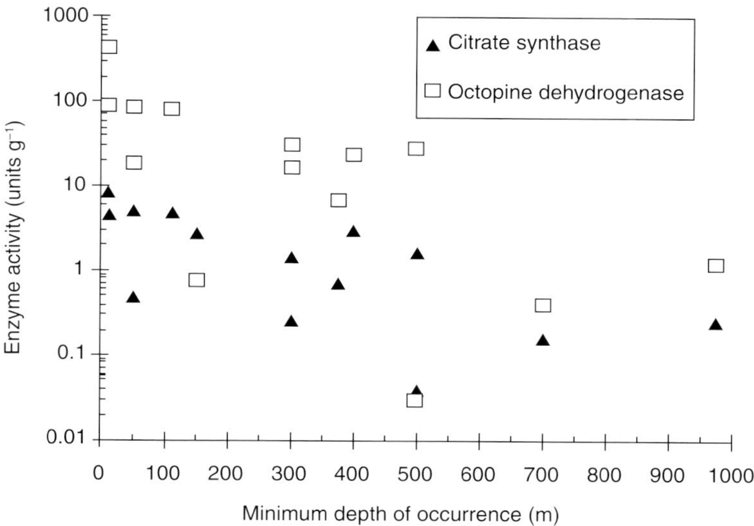

Fig. 15.6. Mass-specific activity of two important metabolic enzymes (citrate synthase and octopine dehydrogenase) in mantle muscle of 14 pelagic cephalopod species from the Pacific Ocean near Hawaii. Enzyme activities are plotted against the minimum depth of occurrence of each species to show the declining trend in metabolic potential from the ocean surface to the base of the mesopelagic zone at 1000 m. B. A. Seibel *et al.* 2000. *Biol Bull.* 198: 284–298. Reprinted with permission from the Marine Biological Laboratory, Woods Hole, MA.

jet-propelled squids occupying the upper few hundred metres of the water column are replaced at greater depths by sluggish, gelatinous forms, which use fins or pulsations of an arm web to move around (Seibel *et al.*, 2000). The fact that both fish and cephalopods have complex, image-forming eyes and largely rely on them to find food and evade predators has led to the formulation of a 'Visual Interaction Hypothesis' (VIH; Childress, 1995) to explain the observed bathymetric trends. As ambient light levels decline with increasing depth, the distances over which predators and prey can see and respond to each other also shrink, relaxing the selection pressure for high-energy metabolism and fast swimming. Crucially, metabolic rates of medusae and other zooplankton show no depth-related declines, a pattern attributed to the fact that these groups do not rely on vision to find food (Thuesen and Childress, 1994). Unlike the mid-water environment, the seabed provides opportunities for concealment. The VIH therefore predicts that metabolism should be less depth-dependent in benthic than in pelagic fish, a pattern broadly consistent with the available data (Drazen and Seibel, 2007).

Metabolic rates of deep-sea fish show the steepest declines across the uppermost 500 m of the water column and level off at around 1000 m (Drazen and Seibel, 2007). This asymptote coincides with the disappearance of the last traces of downwelling sunlight, and is usually taken as further support for the VIH. However, it might also indicate that typical bathypelagic fish have reached the limits of structural and metabolic reduction compatible with the maintenance of a functioning teleost body (Poulson, 2001). Similarly, the limited scope for economy in the sparse metabolically-active tissue component of gelatinous zooplankton may partly explain the lack of any bathymetric trend in energy expenditure.

15.4 Finding Food

In the deep sea the need to find sufficient food to support metabolism, growth and reproduction is a principal factor shaping body form and behaviour. The absence of green plants leaves non-living organic detritus (in various forms) and the bodies of other organisms as the two main forms of available food. Terrestrial food webs based on large, complex vascular plants permit the evolution of many dietary specializations. Insect herbivores, for example, may feed on the leaves of only a single plant species, and may in their turn form the sole prey of specialized carnivores. Extremely restricted diets of this kind are rare or absent in the deep ocean, but it would be wrong to view all deep-sea animals as undiscriminating generalists. Biochemical analyses of diet and *in situ* observations of foraging behaviour are now revealing a range of strategies for making a living in an environment where food is either scarce or of low quality, and sometimes both.

15.4.1 Mid-water filterers and predators

Obtaining food in the water column entails either intercepting some fraction of the downward flux of 'marine snow', or else locating or attracting other mid-water organisms as prey. Many gelatinous zooplankton are small-particle filterers. Appendicularians such as *Bathochordaeus* secrete a delicate 'house' of mucous sheets that traps small particles and channels them to the occupant in the centre (Hamner and Robison, 1992). Salps and doliolids filter food from the water pumped through their barrel-shaped bodies. Gelatinous predators display a variety of feeding strategies (Robison, 2004). ROV observations suggest that some siphonophores, medusae and comb jellies are passive 'trappers', waiting for mobile prey to make contact with trailing, sting-laden tentacles, while others are more active hunters. The narcomedusa *Solmissus marshalli* captures other gelatinous animals with tentacles held outstretched at 180° to the direction of travel, such that prey are not alerted by pulsations of the swimming bell (Raskoff, 2002). Low prey encounter rates in a sparsely populated environment have selected for the capacity to exploit any potential meal. The capacious mouths and distensible stomachs of fish such as the black swallower (*Chiasmodon niger*) are well known, but the ability to ingest very large meals is not confined to fish. By folding elongate food items, the feeding members (gastrozooids) of some siphonophore colonies can also engulf prey much longer than themselves (Pagès and Madin, 2010).

Mid-water animals can partly escape the energetic constraints of the deep sea by foraging higher in the water column where food is more abundant. Diel vertical migration, sometimes covering hundreds of metres, is one of the most ubiquitous behavioural strategies of mesopelagic zooplankton, fish and cephalopods (Robison, 2004). Upward feeding excursions are made at night to minimize exposure to visual predators (Hays, 2003). Distances from the bathypelagic zone to productive surface waters are too great for vertical migration to be a feasible option, and adaptation to these depths emphasizes energy conservation and a 'float-and-wait' predation strategy. With globular, weakly muscled bodies, bathypelagic anglerfish are clearly not active hunters, but rather ambush predators using bioluminescent lures to attract prey within striking distance. Parallel adaptations are seen in bathypelagic cephalopods, including the largest known permanent inhabitants of the deep sea. The Southern Ocean colossal squid (*Mesonychoteuthis hamiltoni*) may reach 500 kg in weight and is often portrayed as a fast, voracious hunter. However, its gelatinous mantle muscles and low metabolic rate are more suggestive of a sluggish predator relying on ambush and stealth rather than active pursuit (Rosa and Seibel, 2010).

15.4.2 Sediment grazers

The deep-sea bed is a kingdom of mud-eaters. Concentrations of sponges, stalked crinoids and other filter feeders occur in areas where water currents increase the flux of suspended particles, but the vast abyssal plains are dominated by deposit-feeders which extract organic matter from the bottom sediments. As a food source, deep-sea sediment has the advantage that it exists in unlimited quantity and requires no effort to find, but on the downside it contains little worth eating. On the Porcupine Abyssal Plain, 4800 m below the productive waters of the north-east Atlantic, total organic carbon content of surface sediment is only 0.37–0.45% by weight (Rabouille *et al.*, 2001), and much of this consists of refractory compounds of little or no nutritive value to animals. In temperate and polar oceans, phytodetritus containing relatively fresh, undegraded algal cells provides a seasonal 'windfall' for the benthic ecosystem (Beaulieu, 2002), but the quantity arriving is unpredictable and its distribution often patchy (Smith *et al.*, 2008). Holothurians (sea cucumbers) – the most abundant and successful large deposit-feeders at abyssal depths – display a range of strategies for the exploitation of this ephemeral food source. Species on the Porcupine Abyssal Plain specialize on different phytodetrital fractions, as shown by the algal pigment profiles of their gut contents (Wigham *et al.*, 2003). *Amperima rosea* has a diet rich in cyanobacteria and lacks pigments characteristic of other algal groups. This small holothurian moves at a relatively rapid pace across the seabed but stops to feed once it has located a favourable patch. Larger species such as *Psychropotes longicauda*, which contain high levels of dinoflagellate and coccolithophore biomarkers, appear to feed continuously while on the move, never stopping for any length of time. Thus, even in a severely resource-limited environment there is room for some degree of specialization and a diet of sediment does not necessarily compel an unselective food intake.

Small holothurians may sometimes gain a free ride to new feeding grounds by drifting in the near-bed currents (Fig. 15.7). The same strategy has been observed in abyssal enteropneusts (acorn worms), which 'touch down' on the seabed, spend 1–2 days

Fig. 15.7. Holothurian (*Benthodytes* sp.) floating above a muddy seabed at 2789 m depth near Davidson Seamount, California. Swimming or drifting in near-bed water currents is a strategy used by some benthic deposit-feeders in the deep sea to locate patches of nutrient-rich sediment. Photo: National Oceanic and Atmospheric Administration/Monterey Bay Aquarium Research Institute (www.photolib.noaa.gov/htmls/expl0790.htm).

grazing the sediment surface in a characteristic spiral pattern, and then drift away, presumably to the next food-rich patch (Smith *et al.*, 2005). Animals living in burrows are far more constrained in their foraging area but are also able to gain a share of the phytodetrital bonanza. Maldanid polychaetes rapidly subduct fresh algal material to depths of 10–13 cm (Levin *et al.*, 1997), while large caches of phytodetritus have also been found in the deep burrows of echiuran worms (Smith *et al.*, 1996). When high-quality food arrives episodically and is only available for short periods, hoarding is an effective strategy for burrowers lacking the mobility of the surface fauna.

Below the mid-ocean gyres and in other oligotrophic (nutrient-poor) seas there is no seasonal flux of phytodetritus. Benthic ecosystems here are sustained by a meagre supply of 'marine snow', already highly degraded by the time it reaches the seafloor. The consequences for benthic biomass and diversity will be considered in Section 15.6, but at the species level not much is known about feeding strategies on these oligotrophic abyssal plains. In the Indian Ocean, echiuran worms which collect surface sediment on a long proboscis show a precise, non-overlapping arrangement of successive proboscis 'strokes' around their burrow opening, affording optimal utilization of the sediment within reach (Ohta, 1984). In such nutrient-starved environments, a sedentary lifestyle may not always be sustainable. Echiurans occupying U-shaped burrows on the Cape Verde Abyssal Plain, North Atlantic, open up a new feeding hole when the surface around the burrow is fully grazed (Bett *et al.*, 1995). By using the previous feeding hole to expel waste material, the animal effectively moves its burrow across the seabed and gains access to ungrazed surface sediment. These examples can be viewed as behavioural responses to the selection pressure exerted by resource scarcity at the abyssal seafloor.

The macrofauna and meiofauna of deep-sea sediments are also predominantly deposit-feeders. Small gut volume precludes bulk ingestion of sediment, and tiny animals are probably obliged to be selective, handling and evaluating individual detrital particles before ingestion (Jumars *et al.*, 1990). The need for particle selectivity in a food-limited environment may be a principal driver of the trend towards miniaturization of body size in the deep-sea benthos, a phenomenon that will be discussed further in Sections 15.5 and 15.6.

15.4.3 Seafloor predators and scavengers

The detritus-based food webs of the deep-sea floor can support carnivores, although at least some of these are probably part-time deposit-feeders. Invertebrate groups such as sea stars (Asteroidea), typically predators or scavengers in shallow water, include many deposit-feeding species in the deep sea. Conversely, some deep-sea representatives of suspension-feeding groups have adopted a facultatively carnivorous lifestyle. Octacnemid ascidians such as *Megalodicopia hians* have widely gaping oral apertures capable of engulfing small crustaceans (Okuyama *et al.*, 2002), while septibranch bivalves suck in prey through a spout-like extension of the shell (Reid and Reid, 1974). Carnivory in these sedentary animals is probably the result of selection pressure to exploit the greatest possible range of food particles, both living and non-living. Another anomalous diet is that of the bivalve *Acesta bullisi*, found among thickets of vestimentiferan tubeworms at Gulf of Mexico hydrocarbon seeps (Järnegren *et al.*, 2005). By clamping its shell around the tube opening of a female worm, *Acesta* ingests lipid-rich eggs directly as they are spawned, a predation mode bordering on parasitism.

Like their deposit-feeding prey, deep-sea benthic predators are not necessarily unselective. Lipid profiles of *Coryphaenoides* grenadiers from the abyssal Pacific indicate a diet of carrion and crustaceans. Echinoderms are apparently not eaten, despite being by far the most abundant benthic megafauna at these depths (Drazen *et al.*, 2009). The high carbonate content and low tissue mass of echinoderms may make them unprofitable as food items. The carrion in grenadier diets comes partly from scavenging of cetacean carcasses at the seabed.

A fresh carcass represents a massive concentration of nutritious food, but such windfalls are sporadic, unpredictable, and likely to attract many competing scavengers. Acoustic tracking data indicate that grenadiers forage actively across the seafloor rather than passively waiting for a carcass to arrive within sensing range (Priede *et al.*, 1990). Carcasses are located by following odour plumes, and possibly also using sound (of a large carcass hitting the seafloor) or light (bioluminescent organisms aggregating around a food-fall). Average time spent at a carcass varies according to season and nutritional status of the habitat – when food is scarce, the fish stay longer (Armstrong *et al.*, 1992). In some cases the main food source may be other scavengers attracted to the carcass. Grenadiers on the Porcupine Abyssal Plain (Jones *et al.*, 1998) and liparid fish in Pacific trenches (Jamieson *et al.*, 2009a) were both observed to eat scavenging amphipods at baited landers, but did not attack the bait itself. A carcass may therefore act as a temporary focus for predator–prey interactions that normally occur widely dispersed across the seafloor.

Carcasses are not the only concentrated food sources at the deep-sea bed. Sunken wood is common enough to support specialized boring bivalves and other invertebrates (Wolff, 1979), while humans have inadvertently added to the list in the form of shipwreck cargos. Decomposing organic substrates in ships' holds can support chemosynthesis-dependent animals normally found at hydrocarbon seeps (Dando *et al.*, 1992). In both cases, dispersive larvae must be able to home-in on 'islands' of suitable habitat surrounded by vast areas of unfavourable terrain, but so far we still have little idea how they achieve this.

15.5 Deep-Sea Life Cycles

Organisms need food not only to sustain basic metabolic functions, but also to support growth and reproduction. It was once thought that resource scarcity and the relative uniformity of the deep sea across time and space would favour a uniform life history in its inhabitants, characterized by slow growth, high longevity and reproduction without a planktonic larval stage. However, as in so many other respects, the deep sea defies simple generalization and it is now clear that a range of life history strategies can prove successful in this environment.

15.5.1 Size, growth and longevity

Some deep-sea animals live life in the slow lane. Off Hawaii, growth rings in the central 'stem' of several species of bushy, deep-water corals indicate radial growth rates of only 10–15 μm year^{-1} (Roark *et al.*, 2006). A colony of the antipatharian coral *Leiopathes glaberrima* was estimated to be 2377 years old (Fig. 15.8). At Gulf of Mexico hydrocarbon seeps, the vestimentiferan tubeworm *Lamellibrachia luymesi* grows around 8 mm year^{-1} and has a lifespan of up to 250 years (Bergquist *et al.*, 2000). The orange roughy (*Hoplostethus atlanticus*), a characteristic seamount fish in both the northern and southern hemispheres, takes 32 years to reach maturity and may live to be 150 (Fenton *et al.*, 1991). Energy budgets may explain the glacial rate of growth in some of these animals. Cold seep vestimentiferans rely on sulfide generated by bacterial metabolism in the underlying sediments, a process occurring at rates that can support only slow growth (Cordes *et al.*, 2005). In contrast, the east Pacific vestimentiferan *Riftia pachyptila*, bathed in sulfide-rich hydrothermal vent fluid, extends its tube at a rate 100 times faster than its resource-starved relative (Lutz *et al.*, 1994). Life on current-swept seamounts entails high metabolic costs and despite a rich food supply, orange roughy and many co-occurring fish species have little surplus energy to support growth (Koslow, 1996). However, instances of very slow growth should be seen in their specific ecological context and not taken as representative of the deep sea in general. Growth and mortality patterns of macrofaunal bivalves from 2900 m in the Rockall Trough, north-east Atlantic, are not very different from those of inshore

Fig. 15.8. Colony of the antipatharian coral *Leiopathes* sp. photographed at 1284 m depth on the rocky slope of the George Bligh Bank, north-east Atlantic. Deep-sea corals such as this have been shown to have a lifespan of over 2000 years. Chirostylid crabs and small brittlestars can be seen amongst the colony branches.

bivalves (Gage, 1991). A gorgonarian coral colony observed for 8 years on the wreck of the RMS *Titanic*, at 3700 m depth off Newfoundland, increased in height by about 1 cm $year^{-1}$ (Vinogradov, 2000), a much higher growth rate than those of the Hawaiian corals at much shallower depths. Thus, a resource-poor environment does not invariably impose a slow rate of growth on its inhabitants.

Two apparently contradictory trends have been noted in the body sizes attained by deep-sea benthic animals. Macro- and meiofauna are generally more diminutive than their shallow-water counterparts and deep-sea community structure is therefore shifted towards the smaller size classes (Kaariainen and Bett, 2006). On the other hand, the largest species of epibenthic groups such as the isopod crustaceans and pycnogonids (sea spiders) occur in the deep sea. Miniaturization in deep-sea sediments may be an adaptation to reduce individual food requirements, allowing sedentary species to maintain viable populations, while gigantism allows mobile scavengers to range widely in search of food (Thiel, 1975). Body size of western Atlantic gastropods seems to follow the 'Island Rule' – the well-known trend towards gigantism in small-bodied groups, and dwarfism in large-bodied groups of terrestrial vertebrates on islands. Consistent with this, gastropod genera that are typically small-bodied in shallow water have significantly larger deep-sea relatives, while the opposite is true for genera that are large-sized in the shallows (McClain *et al.*, 2006). Convergence on an optimal body size for the deep-sea environment may account for the gastropod data, but it remains to be seen whether this explanatory framework is valid for other animal groups.

15.5.2 Reproduction

Like growth rate and longevity, reproduction in the deep sea does not follow a single, uniform pattern. Despite the challenges of finding mates in sparse, widely dispersed populations, there is no evolutionary trend towards asexual reproduction. Animal groups reproducing sexually in shallow water generally do the same in the deep sea. Seafloor topography can provide markers for mate location. Orange roughy and many other fish species aggregate to spawn over seamounts. The Gorda Escarpment off California is a nesting site for two quite different species: a benthic fish, the blob sculpin (*Psychrolutes pictus*), and an octopus (*Graneledone* sp.) both assemble on the crest of this submarine ridge to brood clusters of eggs laid on the rock surface (Drazen *et al.*, 2003). Exploration of the continental margins and mid-ocean ridges will probably uncover additional reproductive 'hotspots' of this kind. Mid-water animals must use other means to find a mate. Permanent attachment of a mated pair, as seen in the deep-sea anglerfish, is the most extreme solution to the problem. The dwarf males of this group are not the only example of extreme sexual dimorphism in mid-water fish. Males of the bathypelagic whalefishes (Cetomimidae) lose their stomach and oesophagus on reaching maturity and rely on a bolus of stored food to sustain them while searching for a mate. Males and females are morphologically so different from each other that they were formerly classified in separate families until their true affinities were revealed by molecular genetics (Johnson *et al.*, 2009).

Many deep-sea benthic animals have no defined breeding season but others display reproductive cycles attuned to the seasonal phytoplankton bloom in surface waters. In the Rockall Trough, north-east Atlantic, the brittlestar, *Ophiura ljungmani*, the seastar, *Plutonaster bifrons*, and the urchin, *Gracilechinus affinis*, all spawn in early spring, and their larvae feed in the plankton before settlement and metamorphosis (Tyler *et al.*, 1982). It was once assumed that the vertical separation of the deep-sea bed from productive surface waters would make brooded or non-feeding larvae obligatory for benthic animals, but it is now clear that this trait is not universal. In prosobranch gastropods of the north-west Atlantic, the proportion of species with planktotrophic larvae actually increases with depth, from less than 25% above 1000 m, to over 50% on the abyssal plain at 4000 m (Rex and Warén, 1982). Sending larvae up to surface waters to feed in the plankton is one way to evade the energetic constraints of life at the deep-sea floor.

In the Rockall Trough seasonally breeding echinoderms with planktotrophic larvae live alongside other species which reproduce year-round and provision their offspring with stored food. This demonstrates once again that contrasting strategies can succeed in the deep sea environment. Perhaps the closest approach to a consistent deep-sea pattern is the near-universal occurrence of non-feeding larvae in the benthic macrofauna and meiofauna. Functioning cells cannot be miniaturized indefinitely, and producing large numbers of small eggs is therefore not an option for the diminutive polychaetes, bivalves and crustaceans of deep-sea sediments. For these tiny animals, brooding eggs packed with stored food gives a better return of surviving offspring than gambling on the lottery of life in the plankton.

15.6 Ecosystem Structure and Functioning

The prevailing conditions of any environment shape not only the behaviour and adaptations of individual organisms but also the structure and dynamics of the ecosystems they inhabit. In the deep ocean, patterns of organic input play a primary role in the distribution of biomass and the flow of matter and energy through food webs. Food supply also influences the species diversity of deep-sea ecosystems, although the relationship is complex and not well understood.

15.6.1 Variation in biomass with depth, geography and time

The exponential decline in total benthic biomass with increasing water depth is probably the best-documented ecological pattern in the deep sea (Rex *et al.*, 2006).

The overall trend is clear but hides some complexity in the responses of different body size classes. Bacterial biomass shows no consistent depth relationship, whereas among metazoans (multicellular animals), larger size classes drop out most rapidly (Fig. 15.9), generating a pattern of declining mean individual body size as depth increases (Thiel, 1975). These differential responses indicate that larger organisms are more sensitive to the progressive reduction in available food. Seabed topography can modify this general pattern. Submarine canyons which trap organic detritus from the coastal zone may support much greater benthic biomass than the adjacent continental slopes (De Leo *et al.*, 2010). Similar processes can generate biomass 'hotspots' even at extreme hadal depths. Beneath the productive upwelling zone off western South America, meiofaunal populations at 7800 m in the Atacama Trench, where organic-rich sediment accumulates, are ten times higher than at nearby bathyal depths (1000–1350 m) (Danovaro *et al.*, 2002). The mid-water ecosystem also exhibits a resource-driven decline in community biomass with increasing depth (Robison, 2004) although animal abundance often rises again near the seabed where resuspended particles provide a richer food supply (Childress *et al.*, 1989).

Over geographic scales, the key role of food input is confirmed by the close correlation of benthic biomass with primary productivity in overlying waters (Wei *et al.*, 2010). Deep-sea life is most abundant at high latitudes and beneath upwelling zones, while the sparsest communities occur below the unproductive central oceanic gyres and in nutrient-poor seas such as the eastern Mediterranean (Kröncke *et al.*, 2003). Biomass contrasts between ocean basins are likely to be highly persistent, but within an individual province some deep-sea ecosystems are much more variable over time than once believed. Time-series observations on the Porcupine

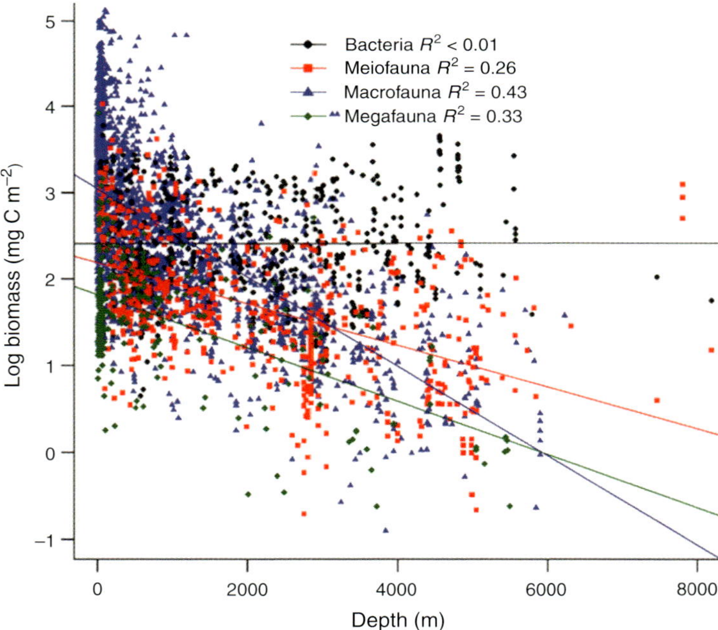

Fig. 15.9. Biomass of four major size classes of benthic organisms as a function of water depth, based on 4872 records from the Census of Marine Life (CoML) Fresh Biomass Database. Biomass is $\log_{10}$ transformed, with the effects of latitude and longitude removed by partial regression. Figure reproduced from Wei *et al.*, 2010.

Abyssal Plain, north-east Atlantic, have revealed striking shifts in benthic communities in the period from 1996 to 1999, and again in 2002 (Billett *et al.*, 2010). Both periods witnessed huge (thousand-fold) increases in populations of the small holothurian *Amperima rosea*. Peaks in *A. rosea* abundance associated with reductions in mean body size point to large-scale recruitment events as the proximate cause. During these '*Amperima* events', populations of sediment-dwelling Foraminifera, metazoan meiofauna and macrofauna also boomed, indicating a response of the entire benthic ecosystem to an environmental change. Comparable, and equally rapid, shifts in benthic community structure have also been documented at a 4100 m deep station in the eastern North Pacific (Ruhl and Smith, 2004). In both localities, there is convincing evidence that the abyssal ecosystem is responding to inter-annual variation in the supply of sinking phytodetritus, in turn linked to changes in surface ocean climate and productivity (Smith *et al.*, 2008). On the Porcupine Abyssal Plain, the *Amperima* event of 2002 followed an exceptionally high input of organic matter to the seabed in 2001 (Lampitt *et al.*, 2010). Our knowledge of temporal change in the deep sea is still drawn from only a handful of locations, but is sufficient to show that benthic community structure is highly sensitive to fluctuating food supply. The abyssal seabed may be far removed from the ocean surface but it is by no means isolated from climate-driven changes taking place there.

15.6.2 Biodiversity in the deep sea

Despite the pervasive food scarcity and low biomass, the deep ocean supports an extraordinary diversity of life. Almost all animal phyla are represented, and at the species level this is one of the richest environments on Earth. Pelagic biodiversity is poorly known, and description of new and distinctive animal species in the deep water column is a regular occurrence (Osborn *et al.*, 2009). Benthic biodiversity has received more attention, as the seafloor is relatively easy to sample quantitatively using low-tech sleds and corers (Fig. 15.10).

Fig. 15.10. A selection of macrofauna from box-core samples taken at approximately 900 m depth in the Porcupine Seabight, north-east Atlantic, representing the major taxonomic groups of animals in this size class. The top row shows amphipod, isopod and tanaid crustaceans, the centre row gastropod and bivalve molluscs and echinoderms, and the lower row polychaete worms. Over the last few decades there has been a growing appreciation of the remarkable species diversity of macrofauna inhabiting deep-sea sediments. Photographs by Lee-Anne Henry, SAMS.

Sampling programmes from the late 1960s onwards have documented high species richness in the meio- and macrofauna of deep-sea sediments (Grassle and Maciolek, 1992). The total number of species in the deep-sea benthos is unknown, but estimates in the range of hundreds of thousands to more than 1 million are widely accepted (Snelgrove and Smith, 2002). Such high levels of diversity were unexpected in a habitat consisting largely of monotonous mud plains lacking the physical structure of other species-rich ecosystems such as tropical rainforests and coral reefs. Much effort has been devoted to considering how small-scale processes such as disturbance and patch dynamics allow large numbers of species to coexist, partly because these processes are important in shallow water, and also because they are amenable to small-scale experimentation *in situ*. However, not surprisingly, there is also evidence for larger-scale trends related to food input (Levin *et al.*, 2001). With respect to water depth, a unimodal pattern with diversity peaking in the mid-bathyal zone (around 2500 m) is common but by no means universal. In different ocean basins this depth-diversity pattern appears to be modified by local factors such as hydrodynamics and the occurrence of Oxygen Minimum Zones (Stuart and Rex, 2009). On a geographic scale, the relationship between overhead primary productivity and deep-sea benthic diversity is also complex and trends for different groups are not always consistent. In the abyssal equatorial Pacific, there is a weak positive correlation between organic flux and species richness for polychaete worms (Glover *et al.*, 2002) but a stronger one for nematodes (Lambshead *et al.*, 2002). One trend that does appear to be consistent is the association of reduced diversity and strongly pulsed organic input, although the causal mechanisms are not clear. Benthic Foraminifera in the deep North Atlantic show lower species richness and diversity in areas receiving seasonal fluxes of phytodetritus (Corliss *et al.*, 2009). Organic-rich trenches also show high benthic standing stocks but relatively low species diversity, perhaps because of frequent disturbance by sediment slides and turbidity flows (Jamieson *et al.*, 2009b).

In abyssal macrofaunal communities there are typically many rare species and only a small number of common ones (Glover *et al.*, 2002). For gastropods and bivalves in the western North Atlantic, Rex *et al.* (2005) suggest that most abyssal species occur at densities too low for successful reproduction, and are instead non-reproductive outliers maintained by a trickle of recruits from centres of abundance on the continental slope. In this perspective, the apparent high species richness of the abyss is deceptive, but it is unlikely that this 'source-sink' model can be applied to all macrofaunal groups. Asellote isopods, for example, show maximum diversity at abyssal depths and appear to have undergone evolutionary radiation there (Raupach *et al.*, 2008). One major group that may be excluded from the abyss by energy limitation is the Chondrichthyes (sharks and rays; Fig. 15.11). In contrast to bony fish, which occur down to hadal depths, sharks and rays appear to have a maximum depth limit of about 3000 m (Priede *et al.*, 2006b). The reason may be that the oil-rich liver that functions as a buoyancy organ in cartilaginous fish is energetically costly to build and maintain. The gas-filled swim bladder or gelatinous tissues of bony fish are more economic options in the food-poor environment of the deep sea.

15.6.3 Food webs and energy flow

Deep-sea food webs, in contrast to those of most other marine and terrestial ecosystems, are principally detritus-based and reliant on external primary production. Important exceptions to this generalization are the various forms of chemoautotrophic ecosystems (vents, seeps, whalefalls and others) at the seafloor. Mid-water food webs are little known, but ROV observations indicate a key role for gelatinous zooplankton, which may constitute up to 25% of pelagic biomass (Robison, 2004). These animals form a complex network of trophic interactions, with gelatinous predators feeding on each other, and on nekton such as fish and

Fig. 15.11. A Portuguese dogfish (*Centroscymnus coelolepis*) and, to the right, a small ray (probably *Raja fyllae*) at 1246 m depth over silt-covered ledges on the George Bligh Bank, north-east Atlantic. Sharks and rays are important predators along continental margins but appear to be absent at abyssal depths.

crustaceans. By packaging particulate material in faecal pellets or mucus-bound aggregates, gelatinous zooplankton also speed up the flux of organic carbon to the benthos (Robison *et al.*, 2005).

At the seabed, some chemosynthetically generated organic carbon enters the wider benthic food web through cropping by mobile predators, and in the form of dispersive larvae released by vent and seep animals. This input may be locally important (MacAvoy *et al.*, 2002), but on a larger scale its contribution is very small in comparison to the flux of organic matter from surface photosynthesis. Stable isotope studies show that detritus-based food webs at the deep-sea floor have up to five trophic levels (Iken *et al.*, 2001; Bergmann *et al.*, 2009), indicating intense recycling of organic carbon. Multiple trophic levels illustrate the complexity of abyssal benthic food webs, despite their reliance on an attenuated supply of organic matter produced thousands of metres above. When fresh phytodetritus arrives at the deep-sea bed, initial consumption and downward mixing is often dominated by metazoan macrofauna, although these animals usually comprise only about 5% of total benthic biomass (Aberle and Witte, 2003). Whatever the precise trophic pathway, evidence suggests that the efficiency of matter and energy flow is positively related to the high biological diversity of deep-sea sediments. Several indices of ecosystem functioning increase exponentially with the species richness and trophic diversity of nematodes, used here as a proxy for total benthic biodiversity (Danovaro *et al.*, 2008). At present we have only a rudimentary understanding of these complex interrelationships, but it is clear that climatic or human-driven reductions in benthic biodiversity could have significant consequences for the deep-sea ecosystem.

15.7 Caves as Deep-Sea Analogues

Darkness and absence of photosynthesis are two of the principal factors shaping the ecosystems of the deep ocean, so it is instructive to consider similarities and contrasts with caves – another 'extreme' environment that shares these characteristics (see Lee *et al.*, Chapter 16, this volume). Coastal marine caves show a marked decline in biomass with distance from the entrance (Gili *et al.*, 1986), correlated with decreasing particle flux and sediment organic content (Fichez, 1990). This horizontal zonation matches the bathymetric resource gradient of the deep sea.

Terrestrial cave faunas include many fish species inhabiting subterranean streams and pools. Living in a dark, resource-poor environment, obligate cave fish might be expected to resemble bathypelagic marine fish in their structure and adaptations but in fact the parallels are not very close (Poulson, 2001). The lack of bioluminescence in freshwater means that cave fish are under no selection pressure to retain vision and most are blind. As we have seen, most deep-sea fish retain functional eyes. Cave fish have no need of the camouflage provided by dark skin, and typically lack body pigmentation. In contrast to the feeble, gelatinous musculature of bathypelagic fish, cave fish need robust, well-muscled bodies to maintain position in often fast-flowing streams. Cave ecosystems resemble those of the deep-sea bed in being dominated by detritus feeders and having few specialist predators (Gibert and Deharveng, 2002).

Cave faunas have long fascinated biologists because of their high levels of endemism, frequency of relict taxa and extreme adaptations to subterranean life. These features point to an important difference between caves and the deep ocean. Early deep-sea exploration, including the pioneering *Challenger* expedition of 1872–1876, was partly inspired by the belief that archaic forms of animal life known from the geological record would be found alive at great depths. These hopes were soon dispelled and we now know that 'living fossils' make up only a small fraction of the deep-sea fauna. In fact, molecular genetic evidence now points in the opposite direction. Stylasterid corals (Lindner *et al.*, 2008), asellote isopods (Raupach *et al.*, 2009) and eels (Inoue *et al.*, 2010) are all groups that have undergone adaptive radiation in the deep ocean, sending representatives into shallow marine and even, in the case of eels, freshwater environments. The classic features of obligate cave animals – loss of eyes and body pigmentation – would be severely maladaptive in most surface environments, and as a result adaptation to subterranean life is probably a one-way ticket. The rarity of cave-to-cave dispersal (Gibert and Deharveng, 2002) suggests that these habitats may be dead ends in both the physical and evolutionary sense. In contrast, the deep-sea environment appears to impose fewer constraints upon the evolutionary potential of its inhabitants.

15.8 Deep-Sea Life Beyond the Earth?

One of the most intriguing results of the *Voyager* and *Galileo* space missions is the discovery that Jupiter's moon Europa probably harbours a liquid water ocean beneath a kilometres-thick crust of ice. This remote world is now a prime target in the search for extraterrestrial life (see Gómez *et al.*, Chapter 26, this volume), and much debate centres on the question of energy sources for a Europan biosphere (Chyba and Phillips, 2002). *Voyager 2*'s encounter with Europa in 1979 coincided with the initial exploration of Earth's deep-sea hydrothermal vents, and as early as 1982 the science-fiction writer Arthur C. Clarke imagined exotic organisms around seafloor vents on Europa in his novel *2010*. Chemosynthesis is still seen as a potential basis for Europan life (McCollom, 1999), but the structure of Earth's deep-sea ecosystems suggests limits to the types of organism we might expect to find there. Here on Earth, no living groups of multicellular organisms are known to have originated at vents. The spectacular tubeworms and other animals that provide the textbook image of vent biology (see Lutz, Chapter 13, this volume) all belong to groups that arose in photosynthesis-based ecosystems and colonized vents in the relatively recent geological past (Little and Vrijenhoek, 2003). The thickness of Europa's ice makes photosynthesis highly unlikely, so if evolution followed a similar course to Earth, we might imagine microbial life clustered around hot vents separated by wide expanses of barren seafloor, with no more than a thin suspension of microorganisms in the water column above. This scenario will remain speculative until we can explore Europa's ocean, perhaps beginning with a search for chemical traces of life entrained in the ice and exposed at the moon's surface (Lipps and Rieboldt, 2005).

At present, Europa's deep ocean can only be visited in the imagination. More than a century after the *Challenger* expedition the largest of Planet Earth's extreme environments, the deep sea, also still keeps many of its secrets. Vast areas remain unexplored and the pace of discovery of exotic species and habitats shows no signs of slowing. We are only beginning to understand how deep-sea ecosystems function and respond to environmental change, and in the face of growing human and climatic pressures this is likely to be the main focus of future research (Glover and Smith, 2003).

References

Aberle, N. and Witte, U. (2003) Deep-sea macrofauna exposed to a simulated sedimentation event in the abyssal NE Atlantic: *in situ* pulse-chase experiments using ^{13}C-labelled phytodetritus. *Marine Ecology Progress Series* 251, 37–47.

Armstrong, J.D., Bagley, P.M. and Priede, I.G. (1992) Photographic and acoustic tracking observations of the behaviour of the grenadier *Coryphaenoides* (*Nematonurus*) *armatus*, the eel *Synaphobranchus bathybius* and other abyssal demersal fish in the North Atlantic Ocean. *Marine Biology* 112, 535–544.

Beaulieu, S. (2002) Accumulation and fate of phytodetritus on the sea floor. *Oceanography and Marine Biology: an Annual Review* 40, 171–232.

Bergmann, M., Dannheim, J., Bauerfeind, E. and Klages, M. (2009) Trophic relationships along a bathymetric gradient at the deep-sea observatory HAUSGARTEN. *Deep-Sea Research I* 56, 408–424.

Bergquist, D.C., Williams, F.M. and Fisher, C.R. (2000) Longevity record for deep-sea invertebrate. *Nature* 403, 499–500.

Bergstad, O.A., Gjelsvik, G., Schander, C. and Høines, Å.S. (2010) Feeding ecology of *Coryphaenoides rupestris* from the Mid-Atlantic Ridge. *PloS ONE* 5, e10453.

Bett, B.J., Rice, A.J. and Thurston, M.H. (1995) A quantitative photographic survey of 'spoke-burrow' type *Lebensspuren* on the Cape Verde Abyssal Plain. *Internationale Revue der Gesamten Hydrobiologie* 80, 153–170.

Billett, D.S.M., Bett, B.J., Reid, W.D.K., Boorman, B. and Priede, I.G. (2010) Long-term change in the abyssal NE Atlantic: the '*Amperima* Event' revisited. *Deep-Sea Research Part II* 57, 1406–1417.

Carrassón, M. and Matallanas, J. (2001) Feeding ecology of the Mediterranean spiderfish *Bathypterois mediterraneus* (Pisces: Chlorophthalmidae) on the western Mediterranean slope. *Fisheries Bulletin* 99, 266–274.

Childress, J.J. (1995) Are there physiological and biochemical adaptations of metabolism in deep-sea animals? *Trends in Ecology and Evolution* 10, 30–36.

Childress, J.J. and Somero, G.N. (1979) Depth-related enzymatic activities in muscle, brain and heart of deep-living pelagic teleosts. *Marine Biology* 52, 273–283.

Childress, J.J., Gluck, D.L., Carney, R.S. and Gowing, M.M. (1989) Benthopelagic biomass distribution and oxygen consumption in a deep-sea benthic boundary layer dominated by gelatinous organisms. *Limnology and Oceanography* 34, 913–930.

Childress, J.J., Price, M.H., Favuzzi, J. and Cowles, D. (1990) Chemical composition of midwater fishes as a function of depth of occurrence off the Hawaiian Islands: food availability as a selective factor? *Marine Biology* 105, 235–246.

Chyba, C.F. and Phillips, C.B. (2002) Europa as an abode of life. *Origins of Life and Evolution of the Biosphere* 32, 47–68.

Cordes, E.E., Arthur, M.A., Shea, K., Arvidson, R.S. and Fisher, C.R. (2005) Modeling the mutualistic interactions between tubeworms and microbial consortia. *PLoS Biology* 3, e77.

Corliss, B.H., Brown, C.W., Sun, X. and Showers, W.J. (2009) Deep-sea benthic diversity linked to seasonality of pelagic productivity. *Deep-Sea Research Part I* 56, 835–841.

Dando, P.R., Southward, A.J., Southward, E.C., Dixon, D.R., Crawford, A. and Crawford, M. (1992) Shipwrecked tube worms. *Nature* 356, 667.

Danovaro, R., Gambi, C. and Della Croce, N. (2002) Meiofauna hotspot in the Atacama Trench, eastern South Pacific Ocean. *Deep-Sea Research Part I* 49, 843–857.

Danovaro, R., Gambi, C., Dell'Anno, A., Corinaldesi, C., Fraschetti, S., Vanreusel, A., Vincx, M. and Gooday, A.J. (2008) Exponential decline of deep-sea ecosystem functioning linked to benthic biodiversity loss. *Current Biology* 18, 1–8.

De Leo, F.C., Smith, C.R., Rowden, A.A., Bowden. D.A. and Clark, M.R. (2010) Submarine canyons: hotspots of benthic biomass and productivity in the deep sea. *Proceedings of the Royal Society B: Biological Sciences* 277, 2783–2792.

Drazen, J.C. and Seibel, B.A. (2007) Depth-related trends in metabolism of benthic and benthopelagic deep-sea fishes. *Limnology and Oceanography* 52, 2306–2316.

Drazen, J.C., Goffredi, S.K., Schlining, B. and Stakes, D.S. (2003) Aggregations of egg-brooding deep-sea fish and cephalopods on the Gorda Escarpment: a reproductive hot spot. *Biological Bulletin* 205, 1–7.

Drazen, J.C., Phleger, C.F., Guest, M.A. and Nichols, P.D. (2009) Lipid composition and diet inferences of abyssal macrourids in the eastern North Pacific. *Marine Ecology Progress Series* 387, 1–14.

Fenton, G.E., Short, S.A. and Ritz, D.A. (1991) Age determination of orange roughy, *Hoplostethus atlanticus* (Pisces: Trachyichthyidae) using ^{210}Pb:^{226}Ra disequilibria. *Marine Biology* 109, 197–202.

Fichez, R. (1990) Decrease in allochthonous organic inputs in dark submarine caves, connection with lowering in benthic community richness. *Hydrobiologia* 207, 61–69.

Gage, J.D. (1991) Biological rates in the deep sea: a perspective from studies on processes in the benthic boundary layer. *Reviews in Aquatic Sciences* 5, 49–100.

Gibert, J. and Deharveng, L. (2002) Subterranean ecosystems: a truncated functional diversity. *Bioscience* 52, 473–481.

Gili, J.M., Riera, T. and Zabala, M. (1986) Physical and biological gradients in a submarine cave on the western Mediterranean coast (north-east Spain). *Marine Biology* 90, 291–297.

Gillibrand, E.J.V., Bagley, P., Jamieson, A., Herring, P.J., Partridge, J.C., Collins, M.A., Milne, R. and Priede, I.G. (2007) Deep sea benthic bioluminescence at artificial food falls, 1000–4800 m depth, in the Porcupine Seabight and Abyssal Plain, North East Atlantic Ocean. *Marine Biology* 150, 1053–1060.

Glover, A.G. and Smith, C.R. (2003) The deep-sea floor ecosystem: current status and prospects of anthropogenic change by the year 2025. *Environmental Conservation* 30, 219–241.

Glover, A.G., Smith, C.R., Paterson, G.L.J., Wilson, G.D.F., Hawkins, L. and Sheader, M. (2002) Polychaete species diversity in the central Pacific abyss: local and regional patterns, and relationships with productivity. *Marine Ecology Progress Series* 240, 157–170.

Gooday, A.J. (2002) Biological responses to seasonally varying fluxes of organic matter to the ocean floor: a review. *Journal of Oceanography* 58, 305–332.

Grassle, J.F. and Maciolek, N.J. (1992) Deep-sea species richness: regional and local diversity estimates from quantitative bottom samples. *American Naturalist* 139, 313–341.

Haddock, S.H.D., Dunn, C.W., Pugh, P.R. and Schnitzler, C.E. (2005) Bioluminescent and red-fluorescent lures in a deep-sea siphonophore. *Science* 309, 263.

Haedrich, R.L. (1996) Deep-water fishes: evolution and adaptation in the Earth's largest living spaces. *Journal of Fish Biology* 49 (Suppl. A), 40–53.

Hamner, W.M. and Robison, B.H. (1992) *In situ* observations of giant appendicularians in Monterey Bay. *Deep-Sea Research* 39, 1299–1313.

Hays, G.C. (2003) A review of the adaptive significance and ecosystem consequences of zooplankton diel vertical migrations. *Hydrobiologia* 503, 163–170.

Herring, P.J. (2007) Sex with the lights on? A review of bioluminescent sexual dimorphism in the sea. *Journal of the Marine Biological Association of the United Kingdom* 87, 829–842.

Iken, K., Brey, T., Wand, U., Voigt, J. and Junghans, P. (2001) Food web structure of the benthic community at the Porcupine Abyssal Plain (NE Atlantic): a stable isotope analysis. *Progress in Oceanography* 50, 383–405.

Inoue, J.G., Miya, M., Miller, M.J., Sado, T., Hanel, R., Hatooka, K., Aoyama, J., Minegishi, Y., Nishida, M. and Tsukamoto, K. (2010) Deep-ocean origin of the freshwater eels. *Biology Letters* 6, 363–366.

Jamieson, A.J., Fujii, T., Mayor, D.J., Solan, M., Matsumoto, A.K., Bagley, P.M. and Priede, I.G. (2009a) Liparid and macrourid fishes of the hadal zone: *in situ* observations of activity and feeding behaviour. *Proceedings of the Royal Society B: Biological Sciences* 276, 1037–1045.

Jamieson, A.J., Fujii, T., Mayor, D.J., Solan, M. and Priede, I.G. (2009b) Hadal trenches: the ecology of the deepest places on Earth. *Trends in Ecology and Evolution* 25, 190–197.

Järnegren, J., Tobias, C.R., Macko, S.A. and Young, C.M. (2005) Egg predation fuels unique species association at deep-sea hydrocarbon seeps. *Biological Bulletin* 209, 87–93.

Johnson, G.D., Paxton, J.R., Sutton, T.T., Satoh, T.P., Sado, T., Nishida, M. and Miya, M. (2009) Deep-sea mystery solved: astonishing larval transformations and extreme sexual dimorphism unite three fish families. *Biology Letters* 5, 235–239.

Jones, E.G., Collins, M.A., Bagley, P.M., Addison, S. and Priede, I.G. (1998) The fate of cetacean carcasses in the deep sea: observations on consumption rates and succession of scavenging species in the abyssal north-east Atlantic Ocean. *Proceedings of the Royal Society B: Biological Sciences* 265, 1119–1127.

Jumars, P.A., Mayer, L.M., Deming, J.W., Baross, J.A. and Wheatcroft, R.A. (1990) Deep-sea deposit-feeding strategies suggested by environmental and feeding constraints. *Philosophical Transactions of the Royal Society of London A* 331, 85–101.

Kaariainen, J.I. and Bett, B.J. (2006) Evidence for benthic body size miniaturization in the deep sea. *Journal of the Marine Biological Association of the United Kingdom* 86, 1339–1345.

Koslow, J.A. (1996) Energetic and life-history patterns of deep-sea benthic, benthopelagic and seamount-associated fish. *Journal of Fish Biology* 49 (Suppl. A), 54–74.

Koslow, J.A. (2007) *The Silent Deep: the Discovery, Ecology and Conservation of the Deep Sea*. University of Chicago Press, Chicago.

Kotrschal, K., Van Staaden, M.J. and Huber, R. (1988) Fish brains: evolution and environmental relationships. *Reviews in Fish Biology and Fisheries* 8, 373–408.

Kröncke, I., Türkay, M. and Fiege, D. (2003) Macrofauna communities in the eastern Mediterranean deep sea. *P.S.Z.N. Marine Ecology* 24, 193–216.

Lambshead, P.J.D., Brown, C.J., Ferreo, T.J., Mitchell, N.J., Smith, C.R., Hawkins, L.E. and Tietjen, J. (2002) Latitudinal diversity patterns of deep-sea marine nematodes and organic fluxes: a test from the central equatorial Pacific. *Marine Ecology Progress Series* 236, 129–135.

Lampitt, R.S., Salter, I., de Cuevas, B.A., Hartman, S., Larkin, K.E. and Pebody, C.A. (2010) Long-term variability of downward particle flux in the deep Northeast Atlantic: causes and trends. *Deep-Sea Research Part II* 57, 1346–1361.

Levin, L.A. (2003) Oxygen Minimum Zone benthos: adaptation and community response to hypoxia. *Oceanography and Marine Biology: an Annual Review* 41, 1–45.

Levin, L.A., Blair, N., DeMaster, D., Plaia, G., Fornes, W., Martin, C. and Thomas, C. (1997) Rapid subduction of organic matter by maldanid polychaetes on the North Carolina slope. *Journal of Marine Research* 55, 595–611.

Levin, L.A., Etter, R.J., Rex, M.A., Gooday, A.J., Smith, C.R., Pineda, J., Stuart, C.T., Hessler, R.R. and Pawson, D. (2001) Environmental influences on regional deep-sea species diversity. *Annual Review of Ecology and Systematics* 32, 51–93.

Lindner, A., Cairns, S.D. and Cunningham, C.W. (2008) From offshore to onshore: multiple origins of shallow-water corals from deep-sea ancestors. *PloS ONE* 3, e2429.

Lipps, J.H. and Rieboldt, S. (2005) Habitats and taphonomy of Europa. *Icarus* 177, 515–527.

Little, C.T.S. and Vrijenhoek, R.C. (2003) Are hydrothermal vent animals living fossils? *Trends in Ecology and Evolution* 18, 582–588.

Lutz, R.A., Shank, T.M., Fornari, D.J., Haymon, R.M., Lilley, M.D., Von Damm, K.L. and Desbruyeres, D. (1994) Rapid growth at deep-sea vents. *Nature* 371, 663–664.

MacAvoy, S.E., Carney, R.S., Fisher, C.R. and Macko, S.A. (2002) Use of chemosynthetic biomass by large, mobile, benthic predators in the Gulf of Mexico. *Marine Ecology Progress Series* 225, 65–78.

McClain, C.R., Boyer, A.G. and Rosenberg, G. (2006) The island rule and the evolution of body size in the deep sea. *Journal of Biogeography* 33, 1578–1584.

McCollom, T.M. (1999) Methanogenesis as a potential source of chemical energy for primary biomass production by autotrophic organisms in hydrothermal systems on Europa. *Journal of Geophysical Research* 104, 30, 729–730, 742.

Ohta, S. (1984) Star-shaped feeding traces produced by echiuran worms on the deep-sea floor of the Bay of Bengal. *Deep-Sea Research* 31, 1415–1432.

Okuyama, M., Saito, Y., Ogawa, M., Takeuchi, A., Jing, Z., Naganuma, T. and Hirose, E. (2002) Morphological studies on the bathyal ascidian *Megalodicopia hians* Oka 1918 (Octacnemidae, Phlebobranchia), with remarks on feeding and tunic morphology. *Zoological Science* 19, 1181–1189.

Osborn, K.J., Haddock, S.H.D., Pleijel, F., Madin, L. and Rouse, G.W. (2009) Deep-sea, swimming worms with luminescent 'bombs'. *Science* 325, 964.

Pagès, F. and Madin, L.P. (2010) Siphonophores eat fish larger than their stomachs. *Deep-Sea Research Part II* 57, 2248–2250.

Poulson, T.L. (2001) Adaptations of cave fishes with some comparisons to deep-sea fishes. *Environmental Biology of Fishes* 62, 345–364.

Priede, I.G., Smith, K.L. Jr and Armstrong, J.D. (1990) Foraging behaviour of abyssal grenadier fish: inferences from acoustic tagging and tracking in the North Pacific Ocean. *Deep-Sea Research Part A* 37, 81–101.

Priede, I.G., Bagley, P.M., Way, S., Herring, P.J. and Partridge, J.C. (2006a) Bioluminescence in the deep sea: free-fall lander observations in the Atlantic Ocean off Cape Verde. *Deep-Sea Research Part I* 53, 1272–1283.

Priede, I.G., Froese, R., Bailey, D.M., Bergstad, O.A., Collins, M.A., Dyb, J.E., Henriques, C., Jones, E.G. and King, N. (2006b) The absence of sharks from abyssal regions of the world's oceans. *Proceedings of the Royal Society B: Biological Sciences* 273, 1435–1441.

Rabouille, C., Witbaard, R. and Duineveld, G.C.A. (2001) Annual and interannual variability of sedimentary recycling studied with a non-steady state model: application to the North Atlantic Ocean (BENGAL site). *Progress in Oceanography* 50, 147–170.

Raskoff, K.A. (2002) Foraging, prey capture and gut contents of the mesopelagic narcomedusa *Solmissus* spp. (Cnidaria; Hydrozoa). *Marine Biology* 141, 1099–1107.

Raupach, M.J., Mayer, C., Malyutina, M. and Wägele, J.-W. (2008) Multiple origins of deep-sea Asellota (Crustacea: Isopoda) from shallow waters revealed by molecular data. *Proceedings of the Royal Society of London B: Biological Sciences* 276, 799–808.

Reid, R.G.B. and Reid, A.M. (1974) The carnivorous habit of members of the septibranch genus *Cuspidaria* (Mollusca: Bivalvia). *Sarsia* 56, 47–56.

Rex, M.A. and Warén, A. (1982) Planktotrophic development in deep-sea prosobranch snails from the western North Atlantic. *Deep-Sea Research Part A* 29, 171–184.

Rex, M.A., McClain, C.R., Johnson, N.A., Etter, R.J., Allen, J.A., Bouchet, P. and Warén, A. (2005) A source-sink hypothesis for abyssal biodiversity. *American Naturalist* 165, 163–178.

Rex, M.A., Etter, R.J., Morris, J.S., Crouse, J., McClain, C.R., Johnson, N.A., Stuart, C.T., Deming, J.W., Thies, R. and Avery, R. (2006) Global bathymetric patterns of standing stock and body size in the deep-sea benthos. *Marine Ecology Progress Series* 317, 1–8.

Roark, E.B., Guilderson, T.P., Dunbar, R.B. and Ingram, B.L. (2006) Radiocarbon-based ages and growth rates of Hawaiian deep-sea corals. *Marine Ecology Progress Series* 327, 1–14.

Robison, B.H. (2004) Deep pelagic biology. *Journal of Experimental Marine Biology and Ecology* 300, 253–272.

Robison, B.H. and Reisenbichler, K.R. (2008) *Macropinna microstoma* and the paradox of its tubular eyes. *Copeia* 4, 780–784.

Robison, B.H., Reisenbichler, K.R. and Sherlock, R.E. (2005) Giant larvacean houses: rapid carbon transport to the deep-sea floor. *Science* 308, 1609–1611.

Rosa, R. and Seibel, B.A. (2010) Slow pace of life of the Antarctic colossal squid. *Journal of the Marine Biological Association of the United Kingdom* 90, 1375–1378.

Ruhl, H.A. and Smith, K.L. Jr (2004) Shifts in deep-sea community structure linked to climate and food supply. *Science* 305, 513–515.

Seibel, B.A., Thuesen, E.V. and Childress, J.J. (2000) Light-limitation on predator–prey interactions: consequences for metabolism and locomotion of deep-sea cephalopods. *Biological Bulletin* 198, 284–298.

Smith, C.R., Hoover, D.J., Doan, S.E., Pope, R.H., DeMaster, D.J., Dobbs, F.C. and Altabet, M.A. (1996) Phytodetritus at the abyssal seafloor across 10° of latitude in the central equatorial Pacific. *Deep-Sea Research Part II* 43, 1309–1338.

Smith, K.L. Jr, Holland, N.D. and Ruhl, H.A. (2005) Enteropneust production of spiral fecal trails on the deep-sea floor observed with time-lapse photography. *Deep-Sea Research Part I* 52, 1228–1240.

Smith, K.L. Jr, Ruhl, H.A., Kaufmann, R.S. and Kahru, M. (2008) Tracing abyssal food supply back to upper-ocean processes over a 17-year time series in the northeast Pacific. *Limnology and Oceanography* 53, 2655–2667.

Smith, K.L. Jr, Ruhl, H.A., Bett, B.J., Billett, D.S.M., Lampitt, R.S. and Kaufmann, R.S. (2009) Climate, carbon cycling and deep-ocean ecosystems. *Proceedings of the National Academy of Sciences of the United States of America* 106, 19211–19218.

Snelgrove, P.V.R. and Smith, C.R. (2002) A riot of species in an environmental calm: the paradox of the species-rich deep-sea floor. *Oceanography and Marine Biology: an Annual Review* 40, 311–342.

Stuart, C.T. and Rex, M.A. (2009) Bathymetric patterns of deep-sea gastropod species diversity in 10 basins of the Atlantic Ocean and Norwegian Sea. *Marine Ecology* 30, 164–180.

Thiel, H. (1975) The size structure of the deep-sea benthos. *Internationale Revue der Gesamten Hydrobiologie* 60, 575–606.

Thuesen, E.V. and Childress, J.J. (1994) Oxygen consumption and metabolic enzyme activities of oceanic California medusae in relation to body size and habitat depth. *Biological Bulletin* 187, 84–98.

Turner, J.R., White, E.M., Collins, M.A., Partridge, J.C. and Douglas, R.H. (2009) Vision in lanternfish (Myctophidae): adapatations for viewing bioluminescence in the deep-sea. *Deep-Sea Research Part I* 56, 1003–1017.

Tyler, P.A., Grant, A., Pain, S.L. and Gage, J.D. (1982) Is annual reproduction in deep-sea echinoderms a response to variability in their environment? *Nature* 300, 747–749.

Vetter, E.W. and Dayton, P.K. (1999) Organic enrichment by macrophyte detritus, and abundance patterns of megafaunal populations in submarine canyons. *Marine Ecology Progress Series* 186, 137–148.

Vinogradov, G.M. (2000) Growth rate of the colony of a deep-water gorgonarian *Chrysogorgia agassizi: in situ* observations. *Ophelia* 53, 101–103.

Wei, C.-L., Rowe, G.T., Escobar-Briones, E., Boetius, A., Soltwedel, T., Caley, M.J., Soliman, Y., Huettman, F., Qu, F., Yu, Z., Pitcher, C.R., Haedrich, R.L., Wicksten, M.K., Rex, M.A., Baguley, J.G., Sharma, J., Danovaro, R., MacDonald, I.R., Nunnally, C.C., Deming, J.W., Montagna, P., Lévesque, M., Weslawski, J.M., Wlodarska-Kowalczuk, M., Ingole, B.S., Bett, B.J., Billett, D.S.M., Yool, A., Bluhm, B.A., Iken, K. and Narayanaswamy, B.E. (2010) Global patterns and predictions of seafloor biomass using Random Forests. *PloS ONE* 5, e15323.

Widder, E.A. (2010) Bioluminescence in the ocean: origins of biological, chemical and ecological diversity. *Science* 328, 704–708.

Widder, E.A., Latz, M.I., Herring, P.J. and Case, J.F. (1984) Far red bioluminescence from two deep-sea fishes. *Science* 225, 512–514.

Wigham, B.D., Hudson, I.R., Billett, D.S.M. and Wolff, G.A. (2003) Is long-term change in the abyssal Northeast Atlantic driven by qualitative changes in export flux? Evidence from selective feeding in deep-sea holothurians. *Progress in Oceanography* 59, 409–441.

Wolff, T. (1979) Macrofaunal utilization of plant remains in the deep sea. *Sarsia* 64, 117–136.

16 Caves and Karst Environments

Natuschka M. Lee,[1] Daniela B. Meisinger,[1] Roman Aubrecht,[2] Lubomir Kovacik,[3] Cesareo Saiz-Jimenez,[4] Sushmitha Baskar,[5] Ramanathan Baskar,[5] Wolfgang Liebl,[1] Megan L. Porter[6] and Annette Summers Engel[7]

[1]Department of Microbiology, Technische Universität München, Freising, Germany; [2]Department of Geology and Palaeontology, Comenius University, Bratislava, Slovakia; [3]Department of Botany, Comenius University, Bratislava, Slovakia; [4]Instituto de Recursos Naturales y Agrobiologia, IRNAS-CSIC, Sevilla, Spain; [5]Department of Environmental Science and Engineering, Guru Jambheshwar University of Science and Technology, Hisar, India; [6]Department of Biological Sciences, University of Maryland, Baltimore, USA; [7]Department of Geology and Geophysics, Louisiana State University, Baton Rouge, USA; now at: Department of Earth and Planetary Sciences, University of Tennessee, Knoxville, USA

16.1 Introduction

Caves have played a fascinating role throughout the history of our planet and of our culture in different ways. Our first associations with caves centre on how they served as primitive dwelling sites for animals and humans, and as settings for mysterious fantasies and myths (Fig. 16.1). Today, scientific exploration has added another aspect to our connection to caves by revealing a plethora of unexpected insights into diverse disciplines, including the natural sciences (geology, palaeontology, climatology, physics, chemistry, biology and cosmology), medical sciences and engineering, as well as social disciplines such as archaeology, theology, and the cultural history of mankind (Fig. 16.2). The foundation for this is the extreme nature of caves, characterized by a lack of light and geographic isolation, but also by nutrient limitation and a range of extreme redox conditions. In recent years, a variety of different types of caves and intriguing cave creatures have been discovered. Therefore, cave-based sciences play an important role in enhancing our understanding of the history of our planet and also form a foundation for exploring novel concepts about the boundaries of life and the evolution of extreme dark life ecosystems on Earth, as well as in other parts of the Universe (Krajick, 2001; NOVA, 2002; SPACE/Malik and Writer, 2005; Forti, 2009).

16.2 Description of Caves: Definition, Distribution and Biogeochemistry

16.2.1 Introduction

Many different definitions have been used to describe a cave or a cavern. The most general way to describe a cave, irrespective of its geological history and location, is to

Fig. 16.1. Entrance zone of a cave: Cascade Caverns, Carter Caves State Resort Park, Kentucky, USA. Depending on the entrance morphology, sunlight only penetrates a limited distance into the cave. © Annette S. Engel.

simply define it as a natural cavity in a rocky environment where at least some part of it is in total darkness. The science of exploration of caves and various karst features is termed speleology (Gunn, 2004). Speleology is a broad interdisciplinary science; a multitude of parameters and disciplines needs to be considered for a thorough exploration of the development of caves and their various impacts on their surroundings worldwide throughout history.

Caves are developed in soluble rocks and constitute a characteristic feature of karst (carbonate rock, such as limestone and dolomite) and pseudokarst (non-carbonate rock) landscapes that covers roughly 15–20% of the Earth's ice-free land surface (Ford and Williams, 2007). Although caves are found in various regions of the Earth and at all latitudes, caves are generally not interconnected over large physiographic provinces, limited by the extent of rock type. Caves come in a wide range of shapes and sizes, from micro-fissures to caverns several thousands of metres deep and high, and hundreds of kilometres in length. Only a few of these are accessible to humans. In fact, many cave openings consist only of microscopic fractures. Despite numerous caving activities all over the world, it is estimated that, even in well explored areas like Europe and North America, so far only 50% of all caves in these regions have been accessed; globally, only ~10% of all caves have been discovered (Eavis, 2009; Engel, 2011). The subsurface can be regarded as one of the least explored environment types on Earth, second only to the deep oceans. Caves often serve as the only available natural entrances and connections to the subsurface, offering fascinating windows into this vast and unexplored habitat.

Because of the increasing interest in cave sciences and all the promising prospects they offer, plus the development of improved caving technology that allows access to even the most difficult caves, many novel types of caves have been discovered during the last decade (Eavis, 2009). Son Doong Cave, Vietnam, found in 2009, contains the largest discovered cavern to date, measuring up to

(a)

(b)

Fig. 16.2. (a) Grotta dei Cervi, Italy (discovered in February 1970), with black Neolithic wall paintings, made with bat guano. © Cesareo Saiz-Jimenez. (b) Castle in front of cave, Predjamski Grad, Slovenia. © Annette S. Engel.

140 by 140 m and 4.5 km in length. The deepest cave, Krubera Cave, Georgia, near the Black Sea, was discovered in 2001 and extends to a depth of –2191 m into the subsurface. Qaqa Mach'ay Cave is the highest cave known, and was discovered in Peru in 2004 at an altitude of nearly 5000 m. Several fascinating underwater caves have also been described, such as those in the Yucatan Peninsula, Mexico, and the Nullarbor

Plain, Australia; each area contains hundreds of kilometres of submerged passageways. However, even caves that have been known about and explored for many years, e.g. Mammoth Cave, Kentucky, USA, also continue to reveal astonishing insights. Due to natural karstification as well as continued exploration, Mammoth Cave is now estimated to cover a distance of around 600 km, and as such represents the longest cave known on Earth (for a summary of spectacular cave types, see NSS GEO2 committee, 2011).

However, it is not only the size and depth of caves that are impressive (Figs 16.3 and 16.4); many of their contents undisputedly continue to capture our imagination and curiosity, e.g. their spectacular speleothems (cave rock and mineral formations), such as stalactites, stalagmites and giant crystals that may be over 70m in length, and fascinating creatures belonging to unique, extreme ecosystems (Taylor, 1999; Krajick, 2001; Culver and Pipan, 2009; Eavis, 2009, http://www.canyonsworldwide.com/crystals/mainframe3.html).

Fig. 16.3. Examples of spectacular constructions in different types of caves. (a) The large room of Škocjan Caves Regional Park, Slovenia, a UNESCO site. This photo highlights the massive size that speleothems can reach (people for scale in the lower left). © Annette S. Engel. (b) Unique formations of unknown nature (possibly biospeleothems) inside a sandstone cave in Venezuela (Aubrecht *et al.*, 2008). © Jan Schloegl and Roman Aubrecht. (c) Entrances to caves can be very small and embedded in water; Cascade Caverns, Carter Caves State Resort Park, Kentucky, USA. © Annette S. Engel. (d) Example of different colour formations due to various microbiological redox processes. Sulfidic spring in a cave with toxic hydrogen sulfidic gases, Lower Kane Cave, Wyoming, USA. © Annette S. Engel.

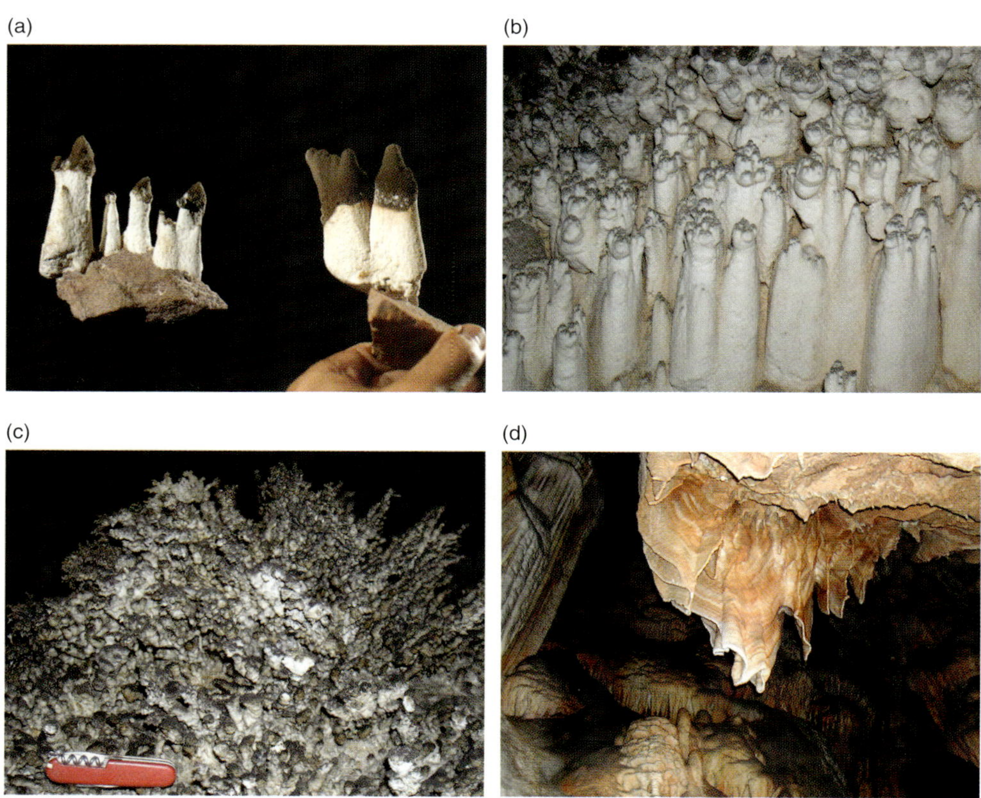

Fig. 16.4. Different types of speleothems and biospeleothems in a sandstone cave in Venezuela. (a–c) sandstone cave in Venezuela, © Roman Aubrecht. (d) Driny Cave in Slovakia, © Lubomir Kovacik.

16.2.2 Classification and formation of caves

Caves are classified using several criteria (Northup and Lavoie, 2001; Gunn, 2004; Engel, 2011): the solid rock/bedrock type, the proximity to the groundwater table, the overall passage morphology and organization (i.e. size, length, depth, routes of fluid flow, etc.) and the speleogenetic history (related to the origin and development of caves). So far, at least 250 different minerals have been described from karst and pseudokarst settings (Hill and Forti, 1997). However, the most common rock types are calcareous rocks (e.g. limestone, which underlies about 15% of Earth's surface) and basaltic rock (e.g. lava tubes). Other less common rock types include other volcanic deposits, gypsum, granite, quartzite, sandstone, salt and even ice.

Although new caves are constantly being formed somewhere on Earth, the majority of the caves so far described and explored have most likely persisted for longer periods, from thousands to millions of years. Irrespective of their speleogenetic histories, all caves are constantly changing over time, either by natural karstification processes or simply due to continued exploration. Depending on the proximity to the groundwater table, two different general modes of cave formation can be discerned: (i) epigenic caves, as the most commonly described caves, are related to surface processes by formation at or proximal to the groundwater table, and are critically dependent on the hydrological conditions of the region; and (ii) hypogenic caves, which are related to subsurface processes because they form by the action of rising fluids, such as water or gases, at or below the water table (Engel, 2011). The most commonly described hypogenic caves to date are sulfidic caves, which are formed by a combination of abiotically and microbially produced sulfuric acid (Engel, 2007). Irrespective of the cave's relation to surface or subsurface processes, the character of the void spaces and the fluid flow patterns are crucial to the continuous evolution of the cave, as these produce the foundation for all flow systems and circulations relevant for the various speleogenetic processes.

Depending on rock type, and geochemical and geophysical (including climatic) conditions, at least six different speleogenetic processes may take place (Northup and Lavoie, 2001; Gunn, 2004; Engel, 2011).

1. Solubilization of the host rock and the precipitation of minerals, which may eventually initiate the formation of speleothems. Speleothems are secondary mineral deposits and may contain a number of different minerals or unconsolidated materials such as clays and organic matter. So far, at least 38 different types of speleothems have been described (e.g. stalactites, stalagmites, helictites, cave pearls, curtains of dripstone, flowstone, rimstone, pool fingers, etc.), based on the formation mechanism and dominant (bio)-geochemical reactions (Hill and Forti, 1997; Boston *et al.*, 2001; Self and Hill, 2003; Melim *et al.*, 2009; Lavoie *et al.*, 2010; see the website of the NOAA Paleoclimatology Speleothem (2011) for information about the value of speleothems for paleoclimatology research; Fig. 16.4). Examples of caves containing a variety of different speleothem types include Mammoth Cave, Kentucky, USA, and Castañar de Ibor Cave, Spain.
2. Solubilization of the host rock from sulfuric acid-driven speleogenesis, which may dissolve calcareous rocks to generate significantly large caverns and often are associated with active microbiological colonization if reactive solutions are still present (Engel, 2007). Examples of caves formed from sulfuric acid speleogenesis include Carlsbad Cavern, New Mexico and Lower Kane Cave, Wyoming, USA, as well as Movile Cave, Romania.
3. Volcanism leading to lava flows that produce lava tube caves (Halliday, 2004), which first may result in sterilized surfaces after a volcanic eruption that may then be colonized through time. Examples of extensive lava tube caves have been described for the Kilauea Volcano, Hawaii, USA.
4. Physical weathering by water, which results in the formation of sea (littoral) caves due to the constant pounding of waves and digging out of seashore cliffs (Bunnell, 2004).
5. Anchialine caves along the coast are formed by the constant solubilization of host rock at a freshwater–saline water interface.
6. Ice (glacier) caves are formed by streams that erode tunnels under and through glaciers (Gunn, 2004; Fig. 16.5).

(a)

(b)

Fig. 16.5. (a) Isfjellelva Cave, Vestre Torellbreen, Svalbard, © Stanislav Rehak. (b) Tone Cave, Tonebreen Glacier, Svalbard, © Stanislav Rehak.

16.2.3 The cave environment

Several parameters contribute to the formation of the cave environment (Northup and Lavoie, 2001; Gunn, 2004; Barton and Northup, 2007; Engel, 2011). Intergranular spaces, pores, joints, fractures and fissures, and dissolutionally enlarged conduits and cave passages form a porosity and permeability continuum that can be colonized by organisms. An opening connected to the surface can have three major zones based on light penetration and intensity: (i) the entrance zone, which is exposed to full sunlight and experiences the daily light cycle; (ii) the twilight zone; and (iii) the dark zone, where no light penetrates (Fig. 16.6). Different types of life, and subsequently adaptations expressed by that life, are generally correlated to these zones because of specific physico-chemical and geochemical conditions related to photic gradients. Entrance and twilight zone conditions are tolerable to a wide variety of organisms, from insects to vertebrates, with little modification to their overall lifestyle.

In contrast to surface habitats, conditions in the deeper parts of caves are generally more stable (i.e. stable temperature throughout the year), but however stable, the conditions can also be considered more extreme simply because it is constantly dark in this zone, thereby making photosynthesis impossible. As such, deep cave environments are generally considered to be extremely oligotrophic (nutrient poor) because many of the resources needed for surface-based ecosystems, such as light and organic matter, are limited. Some systems with direct hydrological connections to the surface may occasionally be subject to catastrophic events, such as floods (Fig. 16.7), whereas caves in deserts may undergo long periods of drought, and lava tube systems may experience renewed volcanic activity. For hypogenic caves, there may also be a range of hazardous conditions, including toxic concentrations of inorganic compounds such as sulfur, heavy metals, lethal gases, or radioactivity (Fig. 16.8). Depending on the rock type, caves can also have extreme pH and redox gradients, such as the interface between

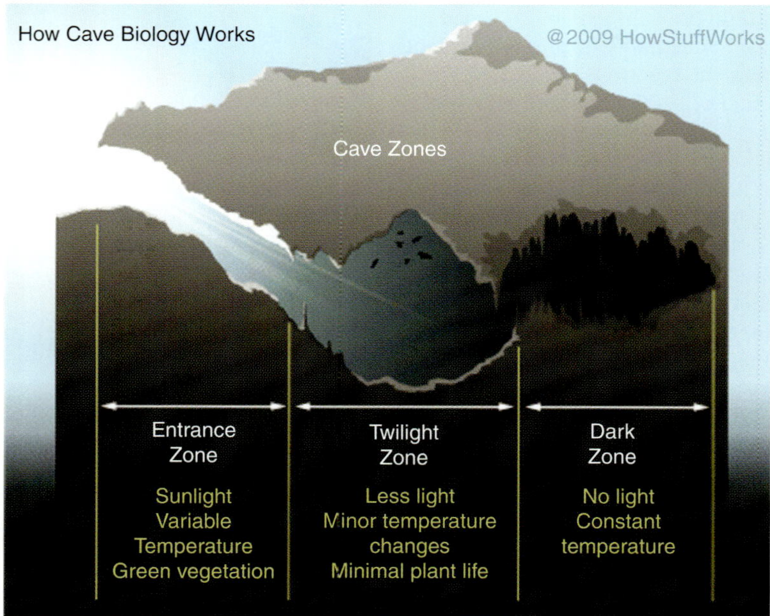

Fig. 16.6. The three zones (entrance, twilight, dark) of a cave (graph from www.howstuffworks.com).

Fig. 16.7. Some cave systems with direct hydrological connections to the surface may occasionally be subject to catastrophic events, such as floods. Here, scientists wade through a flooded cave to reach their sampling site. © Annette S. Engel.

Fig. 16.8. Castañar de Ibor, Spain, is a cave with extremely high radon (^{222}Rn) concentrations reaching up to 50,462 Bq m^{-3}. Cigna (2005) studied 220 caves around the world and calculated an annual average value of 2500 Bq m^{-3}, well below the range of Castañar de Ibor Cave. © Sergio Sanchez-Moral.

the host rock and cave passage atmosphere, or rock and water. The rock itself may contain various reduced compounds that create significantly different redox gradients within short distances. These redox-variable environments play a crucial role in the development of complex micro- and macro-biological communities that each have differing impacts on the speleogenetic history of a cave system (Northup and Lavoie, 2001; Gunn, 2004; Engel, 2011).

16.3 The Biology of Caves

16.3.1 History of biological cave research

The first biological studies in caves were initiated in the Middle Ages in Europe and China. In the centuries that followed, the biology of caves only evoked interest among a limited group of specialists (Culver and Pipan, 2009; Romero, 2009; Engel, 2010). In the late 1970s to early 1980s dark life ecosystems were discovered at hydrothermal vent systems in the Atlantic Ocean (see also Lutz, Chapter 13, this volume), providing insights into the evolution of life adapted to extreme and dark conditions. Shortly after this marine discovery the extreme terrestrial dark life ecosystem in the Movile Cave in Romania was described (Sarbu *et al.*, 1996). Both habitats (vents and caves) revealed astonishingly rich ecosystems of various faunal species, geochemically fuelled by novel species of microorganisms. From this time forward, the modern era of extreme environment research in caves began. With the further development of caving technology and molecular biological tools that enabled the exploration of previously inaccessible caves and of unculturable, novel species, the number of discoveries of extreme ecosystems in different types of caves has consistently increased. Based on the investigations conducted to date, it has become obvious that caves may serve as excellent model systems for several fundamental biological disciplines, in particular geobiology and astrobiology (Taylor, 1999; Boston *et al.*, 2001; Engel and Northup, 2008; see also Gómez *et al.*, Chapter 26, this volume). Therefore, cave research has been used to enhance our understanding of the mechanisms of biological adaptation to extreme conditions, the interactions between organisms and minerals, the role of inorganic matter in different dark environments, the evolution and speciation of biological systems under extreme conditions, and also led to various biotechnological applications (i.e. screening of novel enzymes or pharmaceutical research).

16.3.2 Life in caves

All kinds of life forms (i.e. viruses, bacteria (including cyanobacteria), fungi, algae, protists, plants, animals) have been found in caves (Culver and Pipan, 2009; Romero, 2009), in either an active or fossilized state and in a range of different types of habitats (i.e. rocks, springs, pools, cave walls (Fig. 16.9), or even dispersed in the air). Depending on the environmental conditions, cave organisms can be either motile or sessile, and can be directly or indirectly associated with other organisms in symbiotic associations (e.g. parasitic or mutualistic). In general, ecological interactions profoundly influence the development and maintenance of cave ecosystem dynamics and food webs. Such interactions can include how chemolithoautotrophic microorganisms form the base for some subsurface ecosystems, whereby rich and diverse higher level organisms are sustained, but also can include how bat guano in some caves provides a significant energy and nutrient source for many other organisms (Figs 16.10 and 16.11b).

Depending on the location and the conditions in the cave, organisms may be either non-residents or permanent residents (Culver and Pipan, 2009; Romero, 2009). Non-resident organisms, referred to as accidentals, enter a cave occasionally via wind, water (groundwater, sea spray, rain water), or air, as sediment or spores, or can even be carried into a cave by other animals. Depending on their actions and length of stay in the cave, they may have a profound

Fig. 16.9. Microbial growth on walls in caves: (a) colonies of white and gold actinomycetes and other bacteria on a cave wall, Slovenia, © Annette S. Engel; (b) fungal snottites from Sharps Cave, West Virginia, USA, © Megan Porter; (c) cyanobacterial growth, Cave Bojnice, Slovakia, © Lubomir Kovacik; (d) biovermiculations on the cave wall in the Frasassi Caves, Italy. These features, composed of clays and organic matter, are thought to be formed by microbial and nematode activity, © Annette S. Engel.

influence on the resident populations. The mechanism for the origin of permanent residents is still under debate; the most plausible explanation is that they originally descended from the surface, and entered the subsurface either accidentally or were forced underground by catastrophic events. Once below, they adapted to the conditions.

Depending on the available energy source over time and the survival ability of the species, different types of adaptations in response to the cave environment are likely to have taken place for at least certain species. A good example of this are microbial species that adapted metabolically to a strict chemolithoautotrophic life style, living off inorganic compounds present in the rocks and groundwater, such as metals, sulfur and methane (Engel, 2010). Other examples include higher organisms that adapted to the subterranean environments via various metabolic and morphological adaptations, such as pigment loss, eye loss, wing loss, reduction in size, development of more sensitive sensory organs, limb extension, reduction in metabolic rates and increased longevity (Culver and Pipan, 2009; Romero, 2009; Fig. 16.11a–d). Interestingly, a broad taxonomic range of organisms obligately adapted to the subsurface, in either terrestrial (troglobionts) or aquatic (sytgobionts) habitats, share this suite of characters, referred to as troglomorphy. These obligate subsurface organisms generally also have

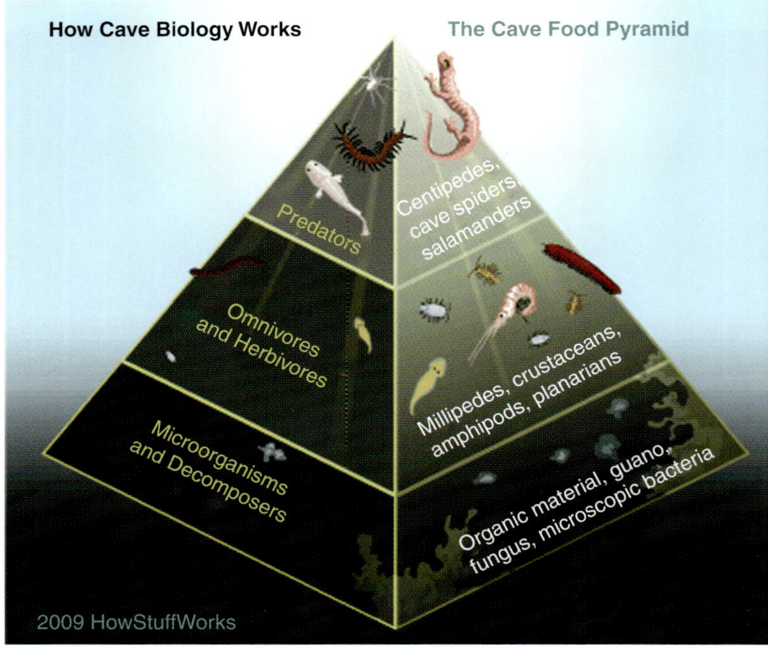

Fig. 16.10. The cave food pyramid (graph from www.howstuffworks.com).

limited possibilities for dispersal, which can further constrain genetic populations to local, and rarely regional, hydrostratigraphic regions.

Viruses

Viruses are the most abundant type of 'biological entity' on Earth, being found wherever there is life, and have probably existed since the first cells evolved. Viruses are capable of infecting all types of organisms, from prokaryotes to plants and animals. It is evident that viruses have had and still have a strong impact on virtually all evolutionary and ecological processes (Abedon, 2008; Forterre, 2010). Unfortunately, there have been no detailed, holistic studies on the ecology of viruses in caves or on the role of viruses or the mechanisms of their impact. There are a few, sporadic reports of high abundances of viruses in various extreme environments or caves, with at least three related categories of case study pointing toward the potentially high significance of viruses in caves: (i) many prokaryotes in various types of extreme ecosystems are in general attacked by viruses (e.g. see examples in Rainey and Oren, 2007); (ii) large numbers of novel viruses have been detected in the subsurface and are thus postulated to play a crucial role (e.g. Kyle *et al.*, 2008); and (iii) several infectious viruses have been reported from caves, in particular from animals like insects or mammals residing in caves. A classic example of this comes from bats: the animals may themselves be attacked by viruses, such as the West Nile virus, or alternatively serve as significant reservoirs of viruses that infect humans and other animals, e.g. emerging zoonotic viruses, such as lyssaviruses, filoviruses and paramyxoviruses (Quan *et al.*, 2010). Furthermore, most of the outbreaks of hemorrhagic fever caused by the lethal Marburg virus in humans and associated with visits to caves and mines in Africa (Kuzmin *et al.*, 2010). The reason for this is unknown because the source of the initial infection is unclear, but a reasonable lesson

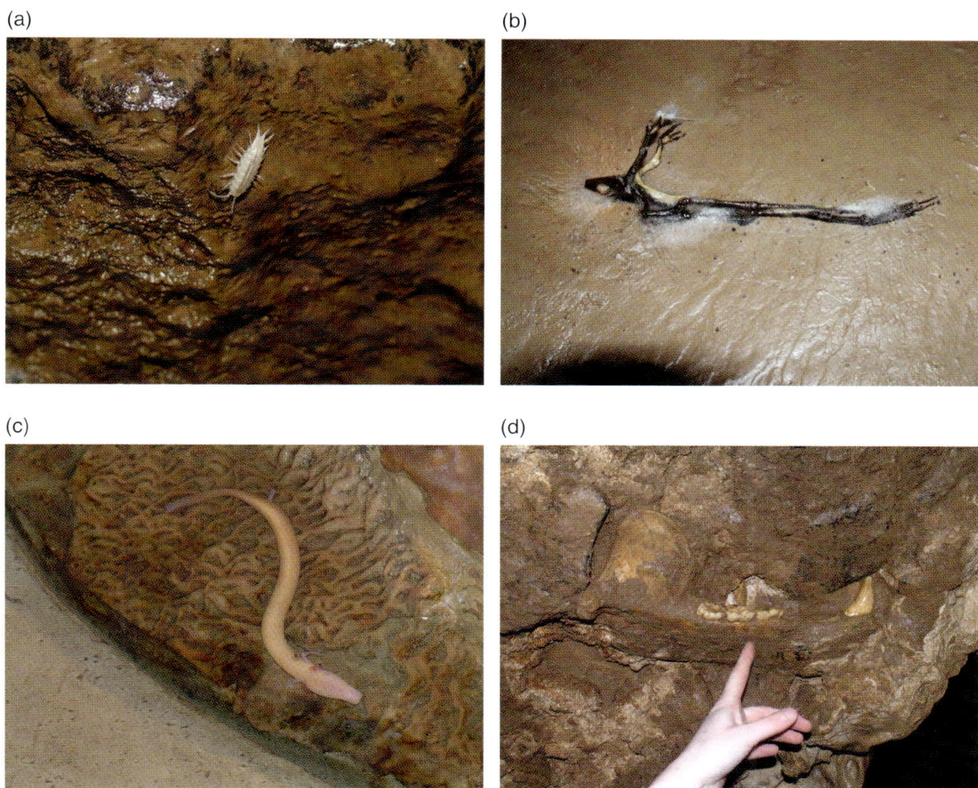

Fig. 16.11. Examples of cave animals: (a) the amphibious non-pigmented isopod *Titanethes albus*, from Planinska Jama, Slovenia, © Megan Porter; (b) dead cave frog overgrown by fungi, Photographer Peter Luptacik, © Elsevier; (c) the blind amphibious, non-pigmented, cave salamander *Proteus anguinus*, Slovenia, © Megan Porter; (d) jaw of the Late Pleistocene cave bear *Ursus spelaeus*, Krizna Jama, Slovenia, © Megan Porter.

to be learned from this case is that the evolution of a rather special gene pool (in this case viral) in isolated portions of caves may have profound impacts on other organisms once they enter these isolated areas. Thus, research into the unique, isolated ecosystems presented by caves may yield many interesting and relevant insights into a number of other biological disciplines.

Bacteria and Archaea

The prokaryotes, which comprise the domains Bacteria and Archaea, are the most abundant group of organisms on our planet. It is well understood that these microbial groups have played a key role in the development of our planet since the early beginnings. The first microbiological studies in caves were performed in the late 1940s using predominately microscopy and enrichment techniques, and these approaches continued for almost three decades. The research revealed few spectacular insights. For example, these studies demonstrated that microbes were prevalent but not as diverse as in surface habitats, and that several geological processes, such as speleothem formation, cave deposits such as saltpetre and moonmilk (Fig. 16.12), iron oxidation (Fig. 16.13) and sulfur oxidation, may be microbially mediated (Northup and Lavoie, 2001; Barton and Northup, 2007; Engel and Northup, 2008; Engel, 2010). Unfortunately, quantitative and undisputable evidence for these hypotheses was missing from these earlier studies.

Fig. 16.12. Moonmilk in Altamira Cave, Spain, most likely of biological origin (Cañaveras *et al.*, 2006).

Furthermore, because standard microbiological approaches are only capable of identifying culturable species and these showed that several identified cave microbes were similar to surface-derived groups, e.g. from soils, it was assumed that the microbes identified in the caves had merely been transported into the cave. The conclusion was, therefore, that no unique cave microbes existed.

Fortunately, with the development of molecular and other analytical techniques since the 1980s, the exploration of non-culturable organisms has become possible. Despite the fact that only ca. 10% of all caves discovered so far have been biologically explored, many different types of bacterial and archaeal groups have now been detected in caves (Engel, 2010). Evidence has mounted that these cave microbes may indeed be unique and genetically divergent from surface groups, which has important implications regarding the role of microbes in distinct geochemical and geobiological processes. Consequently, several hypotheses have been proposed that address questions related to surface–subsurface linkages, biogeography and endemism, as well as microbial adaptation mechanisms to extreme conditions without light. One hypothesis being evaluated suggests that older caves may serve as a long-term reservoir of microbes in the subsurface and thus offer unique possibilities to explore the vast unknown biodiversity of the subsurface. In this way, cave research provides analogues for marine and deep-sea hydrothermal vent systems and possible life on other planetary bodies (Northup and Lavoie, 2001; Barton and Northup, 2007; Engel and Northup, 2008; Engel, 2010).

Many different types of biogeochemical reactions driven by microorganisms have been observed from distinct ecological cave zones (e.g. ammonification, denitrification, nitrification, sulfate reduction, anaerobic sulfide oxidation, metal oxidation, metal reduction, methane cycling, photosynthesis; Northup and Lavoie, 2001; Barton and Northup, 2007; Engel and Northup, 2008; Engel, 2010). Depending on the rock type, the concentration and nature of electron donors and acceptors, availability of oxygen and flux of organic material derived from

(a)

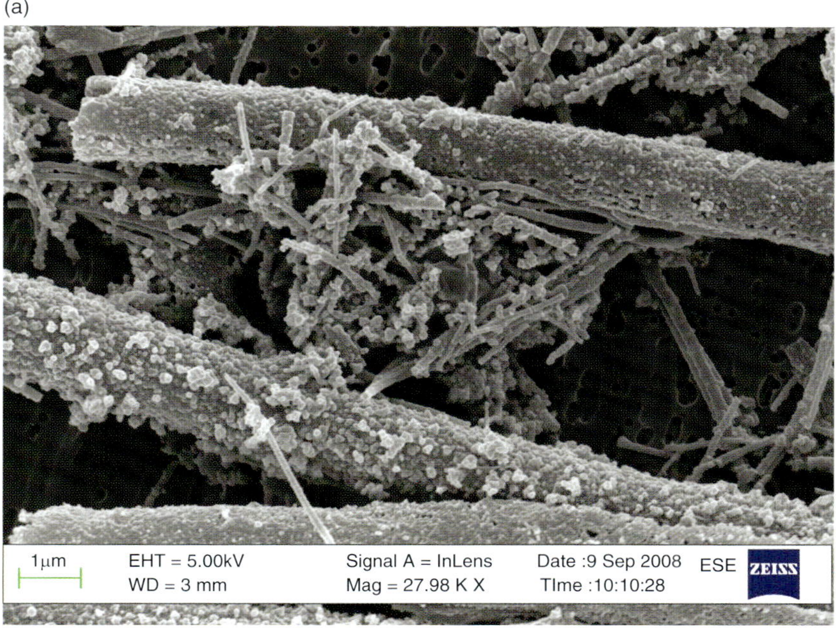

(b)

Fig. 16.13. Precipitates of biological origin produced by various cave bacterial species: (a) biogenic iron precipitated by a dense microbial community on *Leptothrix* sheaths, Borra caves, India; (b) calcium carbonate precipitate. Calcite crystals precipitated *in vitro* by *Bacillus pumilis* isolated from Sahastradhara caves, Dehradun, India (Baskar *et al.*, 2006). © Sushmitha Baskar and Ramanathan Baskar.

the surface, chemolithoautotrophs and heterotrophs play different ecological roles. Extreme environmental conditions also influence the metabolism of oligotrophic, acidophilic, thermophilic and/or sulfidophilic species. Different types of growth patterns may be observed, including single-celled, planktonic life stages to impressive aggregates, such as biofilms, forming either massive microbial mats on cave springs or pools (Fig. 16.14), or microbial draperies ('snottites') on cave wall surfaces (Fig. 16.9).

An important question that is being investigated by current research is what the

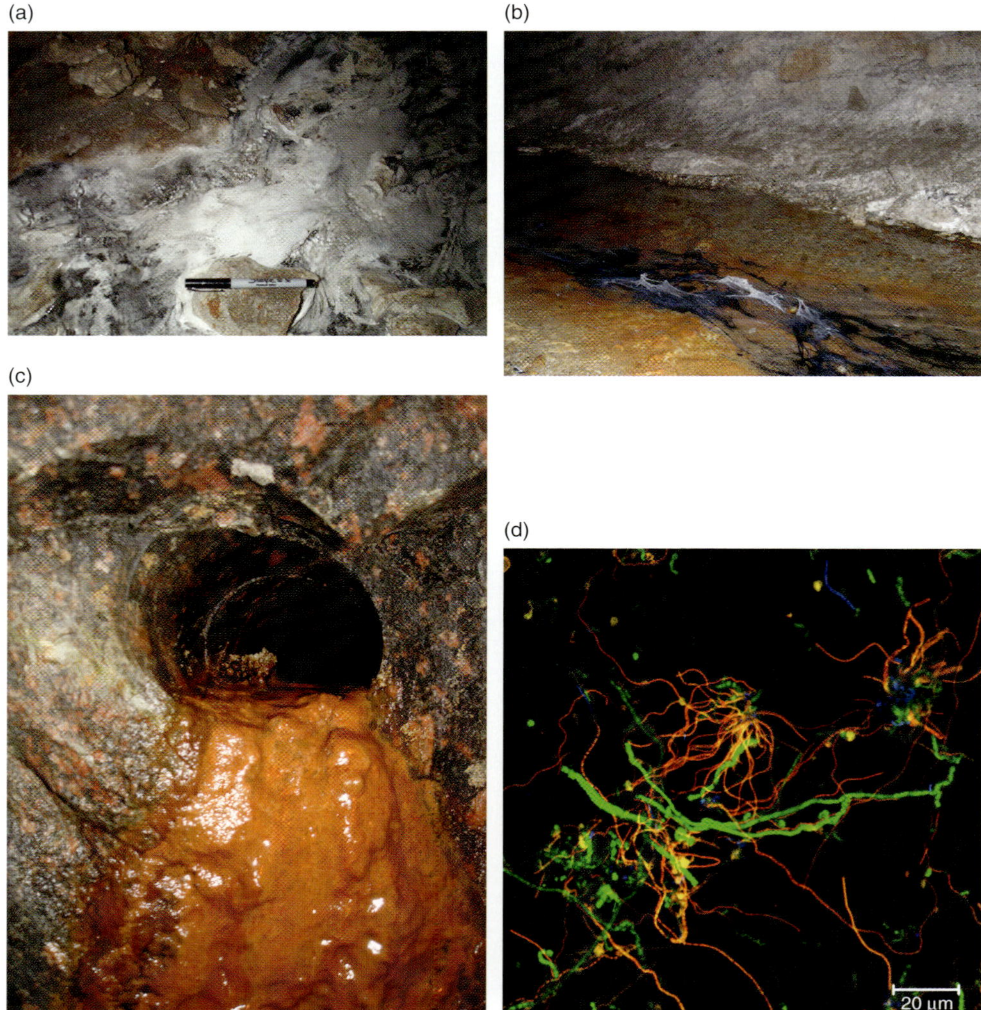

Fig. 16.14. Microbial mats in caves: (a) white filamentous microbial mats in the sulfidic stream of the Pozzo di Cristali in the Frasassi Caves, Italy, © Annette S. Engel; (b) black microbial mats in the sulfidic stream of the Lower Kane Cave, Wyoming, USA, © Natuschka M. Lee; (c) orange microbial mat with iron oxidizing bacteria growing on the flow water dripping out from a drill hole (the subsurface hard rock laboratory, Äspö, Sweden), © Natuschka M. Lee; (d) fluorescence *in situ* hybridization of different types of S-oxidizing filamentous bacteria (novel Epsilonproteobacteria, *Thiothrix* and other unknown species) in a microbial mat in Lower Kane Cave, Wyoming, USA, © Natuschka M. Lee.

source and transportation modes for microorganisms are that bring cells into, distribute cells throughout, and carry cells out of, cave environments. Possible inoculation sources include soil, water (i.e. groundwater, sea spray and rainwater), plants, animals, deep-seated fluids circulating in sedimentary basins, and possibly even the rock itself. Microbial transport is itself linked to various circulation systems, such as vertical and horizontal fluid flow, which affects retention in zones of slow movement and circulation. Slow movement is strongly influenced by sorption onto biofilms that form on nearly every solid and semi-solid surface (e.g. rocks or the shells of macroorganisms; Northup and Lavoie, 2001; Barton and Northup, 2007; Engel and Northup, 2008; Engel, 2010).

As in other ecosystems, biogeochemical processes in caves are controlled by a complex interaction between geochemistry, geophysics and system ecology (for a more detailed review, see Northup and Lavoie, 2001; Barton and Northup, 2007; Engel and Northup, 2008; Engel, 2010). This makes it difficult to distinguish between abiotically and biogenically driven processes in the cave environment. None the less, microbes have been shown to influence many geochemical processes in caves at various stages of speleogenesis, because microbes are considered to be agents of *concentration*, whereby they localize the accumulation of inorganic minerals (e.g. $CaCO_3$ deposits such as moonmilk, or FeS_2 formation from sulfate reduction); *dispersion*, whereby they initiate the solubilization, mobilization and dispersion of insoluble minerals (e.g. Fe(III)-oxide reduction); *fractionation*, whereby they preferentially use one component in a mixture, resulting in the fractionation of elements and isotopes; and *reduction*, whereby they form new compounds due to the use of certain other compounds (e.g. acids from respiration (H_2CO_3), from S_0 oxidation (H_2SO_4), or H_2S production from sulfate reduction) (Ehrlich, 1996).

With this understanding, the focus of previous research has been to distinguish what effects and interactions microbes have on precipitation and dissolution processes in caves. Microbes may promote these processes in either an active (e.g. by enzymes) or passive way. Organisms (live or dead) or their products, such as extracellular substances (EPS), serve as nucleation sites for chemical reactions by sorbing various compounds (e.g. metals to amphoteric functional groups such as carboxyl, phosphoryl and amino constituents, or on to negatively charged cell wall surfaces, sheaths or capsules). These biochemical functional groups provide additional sites for chemical interactions, reduce activation energy barriers, change the system pH, or remove solutes from solution by causing solid phases, like minerals, to precipitate. Precipitation processes result in the formation of different types of products of various sizes, from microscopic structures to large speleothems, and compositions (e.g. calcium carbonate (Fig. 16.13) such as moonmilk (Fig. 16.12) or silicates and clays, iron and manganese oxides, sulfur compounds, or nitrates such as saltpetre). Microbially influenced dissolution and corrosion processes can be mediated by iron-, manganese- and sulfur-oxidizing bacteria, occurring via mechanical attachment and secretion of exoenzymes or from organic or mineral acids (e.g. sulfuric acid) that generate considerable acidity (Northup and Lavoie, 2001; Barton and Northup, 2007; Engel and Northup, 2008; Engel, 2010).

The number of bacterial and archaeal 16S rRNA gene sequences retrieved by standard clone libraries and pyrosequencing from various caves thus far constitute only a small fraction of all 16S rRNA gene sequences retrieved from the environment on a global basis (Engel, 2010). Despite this fact, several interesting insights have been gained. Approximately half of the bacterial phyla, and less than half of the archaeal phyla, are identifiable to a certain extent. The rest of the 16S rRNA gene sequences so far retrieved from cave and karst environments represent novel, so far unculturable species with unknown function. However, certain patterns can be discerned within a number of known phylogentic groups. For instance, some epigenic caves appear to contain *Deltaproteobacteria*, *Acidobacteria*,

Nitrospira and *Betaproteobacteria*, whereas some hypogenic caves, such as sulfidic caves, appear to be dominated either by *Epsilonproteobacteria*, or *Gammaproteobacteria* and *Betaproteobacteria* in microbial mats, or by *Acidimicrobium, Thermoplasmales, Actinobacteria*, bacterial candidate lineages, and some Archaea in 'snottites' on cave wall surfaces (Engel, 2010; Fig. 16.15).

Along with this, several different bacterial genera, such as *Firmicutes, Bacillus, Clostridium* and various enteric bacteria/human indicator bacteria have been discovered in cave systems that are believed to result from some kind of contamination via, for example, local wastewater treatment plants, storm events, or visitors. Fortunately, many of these microbial contaminants have a low persistence in the cave environment, indicating that caves may have the potential to recover from short-lived, occasional contamination events. Nevertheless, there are also a number of caves, especially caves with valuable rock-art paintings, that have had enormous difficulty recovering from invasive species, such as heterotrophic bacteria, phototrophs and fungi, spread by improper cave management, insects, or animals (Section 16.4).

Further research is needed to resolve the questions remaining about the presence of endemic and invasive microbial species in caves and their possible roles. Despite all developments in terms of molecular and microbial analytical tools, only a fraction of the real taxonomic and functional diversity of Bacteria and Archaea in caves has been thoroughly and appropriately described. This is due to several methodological problems, ranging from inadequate sampling technology, insufficient distribution data, and the lack of appropriate methodology to address a research question. Testing specific hypotheses requires modifications to traditional approaches using more holistic, full-cycle methods, in order to more truly identify, quantify and determine the function and activity of culturable, as well as unculturable, species and thus provide correlations and evidence of the role of microbes in different types of speleogenetic processes.

Eukaryotes

Many different types of eukaryotes have been detected in caves worldwide, from single-celled species, such as different types of Protozoa, to a large variety of multicellular species ranging from fungi, invertebrates (flatworms, annelids, millipedes, centipedes, diplurans, insects, collombolans, spiders, mites, crustaceans, scorpions) and vertebrates (amphibians, reptiles, fish, mammals such as bats). For extensive details, see Culver and Pipan (2009). Actively growing plants, algae, or microscopic phototrophs are only found in cave entrances, in streams or other locations where sufficient amounts of light are available, whether natural or artificial due to cave lighting, where as little as 1 μmol photon m^{-2} may support growth (Grobbelaar, 2000).

Depending on the location within a cave zone, adaptation level and residence time in the caves, organisms may be classified as trogloxenes – terrestrial organisms that use the cave merely for occasional shelter, such as bats – or troglophiles – terrestrial organisms that may complete their life cycle in the cave but are still able to survive outside the caves. Many troglophiles therefore maintain at least some of their original senses. Stygoxenes are aquatic trogloxenes and stygophiles are aquatic troglophiles. Trogloxenes and troglophiles are usually found in the entrance or the twilight zone (Howarth, 1980; Culver and Pipan, 2009).

Troglobites and stygobites, obligate terrestrial and aquatic subsurface-dwellers, respectively, have developed astonishing adaptation mechanisms to life in caves (Culver and Pipan, 2009). Troglobites are generally exclusively found in the deeper parts of the caves where it is permanently dark and the humidity often high (up to 95–100%). Aquatic stygobites are generally geographically more wide-ranging than troglobites, and stygobites are more commonly found in caves with tropical and subtropical climates than troglobites (Lamoreux, 2004). Both groups have developed impressive morphological adaptations, as well as physiological mechanisms, to survive e.g. darkness, high humidity and limited food supplies. Typical food

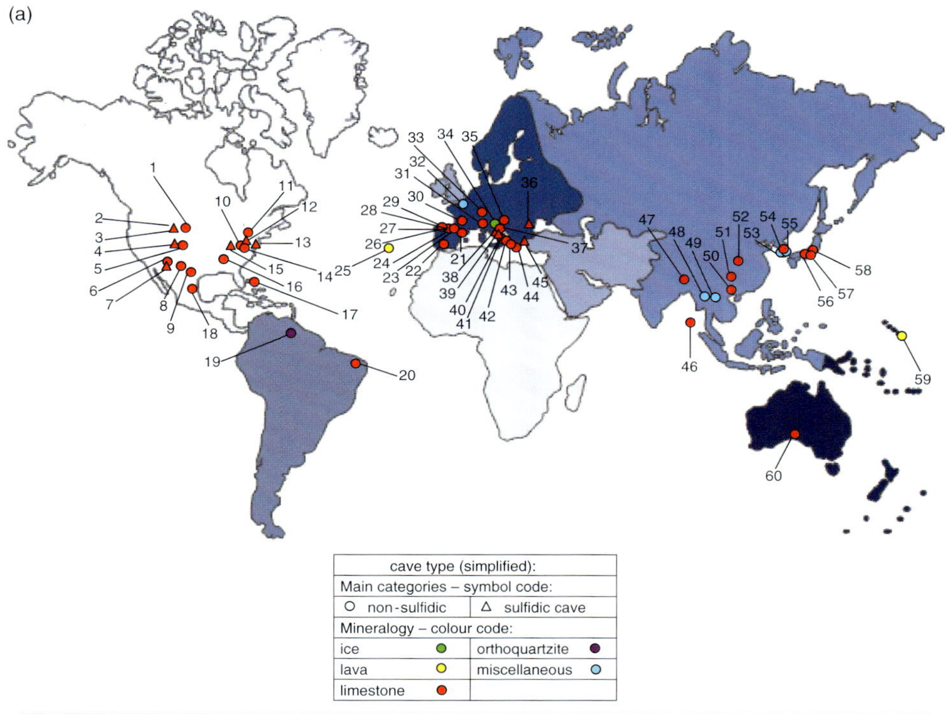

number	cave name, country	seq./diversity[a]	number	cave name, country	seq./diversity[a]
1	Wind Cave, SD, USA	83/14	31	Swallow hole, Yverdon-les-Bains, Switzerland[b]	134/>10
2	Hellespont Cave, WY, USA	11/1	32	Scladina case (cave layer), Belgium	63/10
3	Lower Kane Cave, WY, USA[b, c] (→ Fig. 16.3d; 16.14b, d)	>1500/>17	33	Herrenberg Cave, Germany[c]	9/3
4	Glenwood Springs Cave, CO, USA	334/14	34	Alpine Ice Caves, Werfen, Austria[b, c]	2/1
5	Fairy Cave, CO, USA	37/6	35	Domice Cave/Domica Cave, Slovakia	52/3
6	Kartchner Cave, AZ, USA[b, c]	164/10	36	Movile Cave, Romania[b]	372/14
7	Millipede Cave, AZ, USA	>12/1	37	Pajsarjeva jama Cave, Slovenia[b]	50/11
8	Carlsbad Lechuguilla Cave and Spider Cave, NM, USA[b]	46/9	38	Frasassi Cave system, Italy (→ Fig. 16.9d; 16.14a)	1039/>18
9	Hinds Cave, TX, USA	50/4	39	Cave of Acquasanta Therme, Italy[b]	173/11
10	Mammoth Cave, KY, USA[b]	254/10	40	Saint Callixtus Catacombs, Rome, Italy[b]	8/1
11	Tytoona Cave, PA, USA	13/7	41	Grotta Azzurra of Palinuro Cape, Italy[b]	39/10
12	Parker Cave, KY, USA	17/4	42	Grotta dei Cervi, Italy[b, c] (→ Fig. 16.2a)	2/1
13	Cesspool Cave, VA, USA[b]	129/>6	43	Cave Koutouki, Attika, Greece[c]	1/1
14	Limestone Cave, KY, USA	4/2	44	Cave Kastria, Peloponnese, Greece	
15	Big Sulfur Cave, KY, USA	>30/1	45	Shallow-water Cave, Paxos, Greece	148/9
16	Blowing Spring Cave, AL, USA	9/4	46	Lime Cave, Baratang, India[b, c]	1/1
17	South Andros Black Hole Cave, Bahamas[b, c]	3/1	47	Meghalaya Caves, India[c]	3/2
18	El Zacaton sinkhole, Mexico[b]	1589/27	48	Thai Cave, Thailand[b, c]	9/1
19	Roraima Sur Cave, Venezuela[b]	42/9	49	Pha Tup Cave Forest Park, Thailand	2/1
20	Gruta da Caridade, Rio Grande do Norte, Brazil	2/1	50	Reed Flute Cave, Guilin, Guangxi, China[b, c]	2/1
21	Cave Papellona, Barcelona, Spain[b, c]	1/1	51	Niu-Cave, China	142/14
22	Ardales Cave, Spain (→ Fig. 16.16)	19/7	52	Heshang Cave, China	111/13
23	Santimamiñe Cave, Spain	10/10	53	Cave Apatite DepositsSouth Korea	65/7
24	Covalanas, Cantabria and La Haza Caves, Spain	325/3	54	Pseudo-limestone Cave, South Korea	2/1
25	Lava Tubes, Azores, Terceira, Portugal	515/>10	55	Gold Mine Caves, Kongju, South Korea[b, c]	4/2

Fig. 16.15. (a) Worldmap showing all caves which have been explored with molecular biological methods. (b) 16s rRNA phylogenetic tree showing nearly full sequences retrieved from various caves all over the world. Abbreviations: BPB/GPB = Betaproteobacteria/Gammaproteobacteria; DPB = Deltaproteobacteria; APB = Alphaproteobacteria; EPS = Epsilonproteobacteria; WS3 and OP 10 = novel candidate divisions.

26	Altamira Cave, Spain[b, c] (→ Fig. 16.12)	>581/12	56	Ryugashi Cave, Shizuoka, Japan	1/1
27	Monedas and Chufin Caves, Spain	125/10	57	Cave Sabichi, Ishigakijima, Japan[b]	1/1
28	Llonin Cave and La Garma Cave, Spain	85/10	58	Jomon and Joumon Limestone Caves, Gifu, Japan[b, c]	9/2
29	Tito Bustillo Cave, Spain (→ Fig. 16.17)	41/10	59	Lava Caves and Tubes, HI, USA[b]	302/>18
30	Lascaux Cave, France	681/>4	60	Nullarbor Caves, Australia	52/8

a: The ratio demonstrates the total amount of 16S rRNA gene sequences from the domains Bacteria and Archaea *versus* the amount of identifiable phyla. This information is based on ~ 8,000 partial 16S rRNA gene sequences retrieved from the Gene bank (http://www.ncbi.nlm.nih.gov/) in February 2011. A detailed list of the sequences and the references can be downloaded from the website http://www.microbial-systems-ecology.de/Lee_et_al_2011_cave.html

b: A representative selection of these sequences (> 1,400 nucleotides) were used to calculate an overall 16S rRNA phylogenetic tree (see figure 16.15b and http://www.microbial-systems-ecology.de/Lee_et_al_2011_cave.html) showing all so far retrieved 16S rRNA gene sequences from different parts of the world.

c: This study also describes 16S rRNA gene sequences from isolates.

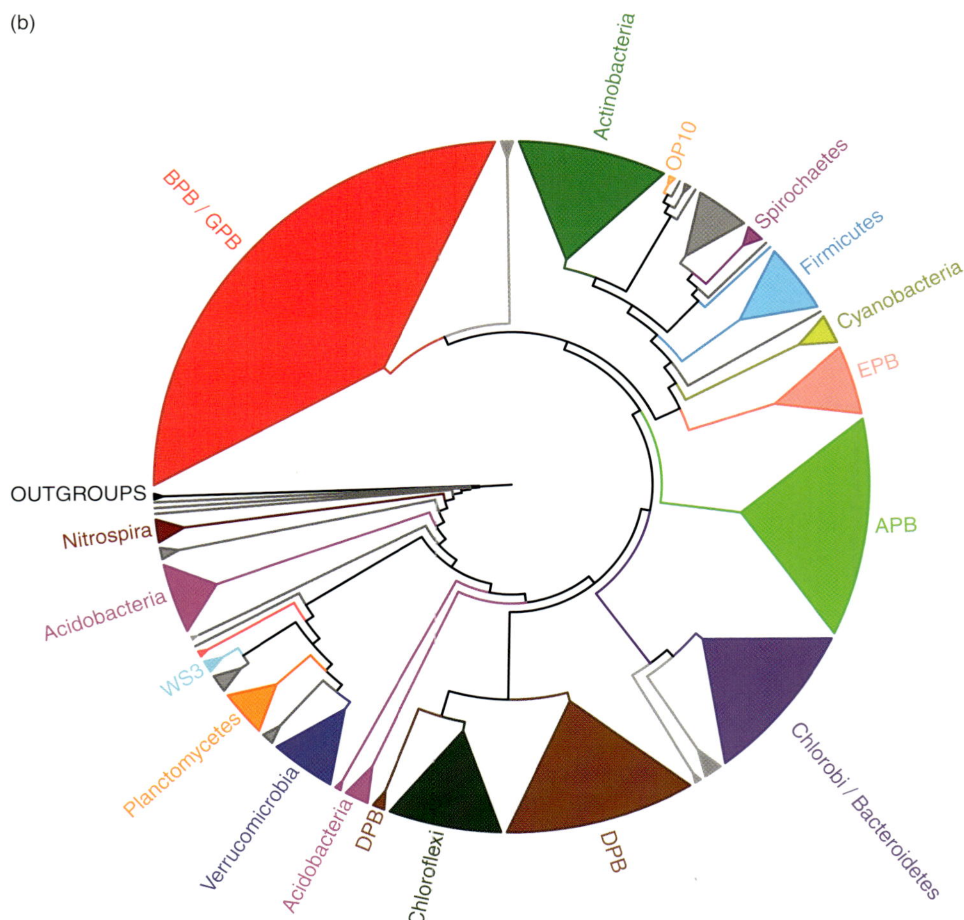

Fig. 16.15. Continued.

sources may include microorganisms, other animals, faeces (e.g. bat guano), carcasses from trogloxenes, or other matter, such as twigs and plant residues delivered into the cave by other animals or aquatic streams. Because of their low metabolic rates, sometimes sedentary lifestyle and infrequent reproduction, trogoblites and stygobites usually live longer than many non-cave species. Several animal categories, such as carabid beetles, gastropods, collembolans and spiders are found in many caves worldwide, while some animal species have so far only been detected in certain caves, e.g. scorpions are mostly found in Mexican caves. The most common troglobite so far described is the carabid beetle of the subfamily Trechinae, which measures approximately 5–7 mm (Barr and Holsinger, 1985). The largest troglobite found so far is the cave salamander (*Proteus anguinus*), commonly referred to as the 'Olm', that can measure up to 30 cm in length (Krajick, 2007; Fig. 16.11c).

To date, nearly 8000 species of troglobites have been described (Krajick, 2007). However, since approximately only 10% of the caves thought to exist worldwide have been discovered, it should be anticipated that many more species and interesting ecological interactions remain to be discovered. A striking example for this is the unknown diversity of micro-eukaryotes in caves. Currently, 18S rRNA gene sequence data suggest that most of the micro-eukaryotes so far are not identifiable (Engel, 2010). Their role is unknown, but the limited knowledge retrieved suggests that at least some of the micro-eukaryotes may have a severe impact on speleothems and deterioration of valuable cave walls and paintings. Different types of other ecological functions can be ascribed, depending on the species and activity level. Some eukaryotic structures, such as fungal hyphae, may serve passively as nuclei for crystallization or as sites for attachment of crystals and thus, indirectly, contribute to a speleological process (Figs 16.11b and 16.16).

Fig. 16.16. Photo of fungi (*Beauveria felina*) on rodent excrement, Ardales Cave, Spain. © Cesareo Saiz-Jimenez.

16.4 Natural and Anthropogenically Disturbed Caves and the Future of Cave Preservation

Unfortunately, our chance of realizing the true biodiversity of caves is endangered by many factors, including the environmental problems caused by mining, drilling, pumping of aquifers, contamination, or invasions by other organisms, including humans via tourism and even research activities. Many cave environments appear to be endangered by several parameters, ranging from climatic catastrophes, through different types of environmental pollution. Even an ecosystem impacted by a minor disturbance, such as a natural flood (Fig. 16.7), can become susceptible to additional, more severe disturbance, like disease or being out competed for nutrients and resources by invading species.

In general, microbial colonization of a cave is a natural process that has occurred a long time before a cave's discovery. However, as soon as a cave, with its already established ecosystem that is finely balanced in terms of ecological interactions and environmental conditions (such as nutrient input) is opened and connected with the exterior environment, the ecosystem becomes subjected to an unaccustomed input of abundant organic matter, or may be impacted by invading communities coming from the surface. This can lead to significant food web changes because the newcomers may exert an enormous pressure on the original inhabitants. In the worst-case scenario, the newcomers may even displace the original populations and communities. There are several dramatic examples of this (Saiz-Jimenez, 2010).

The strongest disturbance recorded so far has been commercial cave mass tourism, which commenced during the second half of the 19th century. At that time, minimal consideration was given to conservation. As a consequence, practices adopted for visits to caves often resulted in irreversible damage caused by thousands of visitors, for example, in the form of lint, litter, and even increased carbon dioxide (from exhaled breath) and altered temperature levels (due to human body heat) in passages with low circulation. In addition, destructive construction works have removed tonnes of rocks and other materials from the topsoil and cave entrances to allow access for tourists to many caves worldwide. Lastly, the introduction of artificial lighting in some tourist caves, sometimes left illuminated all day, is sufficient to turn the darkness into a terrarium in some caves and support the growth of various phototrophs and plants that would not otherwise be able to survive (Fig. 16.17). One example of this is the green alga, *Bracteacoccus minor*, that occurred on the wall paintings in the Lascaux Cave, France. For the most part, the ancient rock-art paintings in the Lascaux Cave and others in Europe are not only threatened by algae, but also by fungal species introduced and spread by human activities. Due to this, a number of caves had to be closed for several years to treat invading organisms with various biocides (e.g. the Lascaux Cave, France; for more detailed information see review by Saiz-Jimenez, 2010; Fig. 16.17).

Another striking example of the impact of disturbance on cave ecosystems is the unexpected outcome of the occurrence of the fungal species *Geomyces destructans*, in some caves in the USA. Although still under investigation, it is highly probable the *G. destructans* is a natural fungal species to humid, cool subsurface environments. In North America, its link to the 'white nose syndrome', which has been associated with the sudden deaths of more than a million bats in the USA, is mysterious (Blehert *et al.*, 2009; Fig. 16.18). Strikingly, the first discovery was made in 2006 in a heavily visited, commercial cave in Schoharie County, New York, USA. After this, the white-nose syndrome fungal infection spread astonishingly fast to over 100 other caves throughout the North American continent – including caves and mines that are not generally accessible to humans. The event, and more significantly the death of so many different valuable bat species, has had several serious consequences: it has led to

Fig. 16.17. Example of an invasive species on the ground in a cave, caused by artificial lighting. This promotes the growth of calcifying cyanobacteria (*Scytonema julianum*) and algae in tourist caves, as shown in Tito Bustillo Cave, Spain (Saiz-Jimenez, 1999). © Cesareo Saiz-Jimenez.

(a)

(b)

Fig. 16.18. Cave bats infected by the 'white nose syndrome', caused by the fungus *Geomyces destructans*. (a) Close-up of little brown bat's nose with fungus, New York, USA. Photo courtesy Ryan von Linden, New York Department of Environmental Conservation (www.fws.gov/whitenosesyndrome/images/3842close-upofnosewithfungus.jpg). (b) Fungus on wing membrane of little brown bat, New York, USA. Photograph courtesy Ryan von Linden, New York Department of Environmental Conservation (www.fws.gov/whitenosesyndrome/images/3845Fungusonwingmembrane.jpg).

the near extinction of several species and has had a tremendous negative ecological impact on agriculture owing to the fact that bats consume enormous numbers of insects. The United States Forest Service has estimated that about 1.1 million kg of insects will go uneaten in the most heavily impacted regions, and it is likely that insect infestation will become a great financial burden on the agricultural sector. Without a natural regulator of insect population, the use of insecticides is likely to increase, which will in turn increase existing environmental problems (for more information see the USDA website on the white nose syndrome). Clearly, research into the biology of caves is not only a matter of exploring unique and extreme ecosystems, but is also fundamental to our understanding of the delicate ecological balances on Earth.

Unfortunately, no tourist cave impacted by severe disturbances has ever been completely restored to its former ecological state (Elliot, 2006). This is particularly the case where the wrong decisions were made in selecting a biocide to employ in Lascaux Cave, since it caused severe irreversible population shifts in the natural cave flora. For instance, in the treatment of the Lascaux Cave, benzalkonium chloride use resulted in the selection of Gram-negative bacterial species that were adapted to this biocide (e.g. *Ralstonia, Pseudomonas* and other pathogenic bacteria). The lesson to be learned from this case is that biocides that are generally considered acceptable for combating microorganisms in surface environments are not necessarily appropriate for use in sensitive ecosystems such as caves. Interestingly, recently discovered rock-art caves such as Chauvet, France, and La Garma, Spain, that have been protected against mass tourism, have so far shown no sign of deterioration. Based on this, it is evident that it is important to protect cave ecosystems from the outset following discovery and perhaps the best protection is to limit the amount of tourism (Saiz-Jimenez, 2010).

Principally, every visitor to a cave, from the professional speleologist to tourists, has the potential to exert a negative impact on the cave ecosystem, especially if the rules for basic safe caving, as outlined by McClurg (1996), are not followed: 'Don't take anything, don't leave anything; don't break or remove cave formations; don't handle or collect cave life; or, in other words: take nothing but pictures; leave nothing but foot prints and kill nothing but time'. It goes without saying, however, that for some regions of the world, caves serve as the only viable commodity and tourism is the only source of revenue. For example, in the Mammoth Cave region, USA, over US$400 million were brought to the area by tourists in 2009, up from an estimate of

US$52 million in 1993 (Stynes, 1999). The natural beauty of caves can also serve as a mechanism for science education to the public. Local to national strategies can be, and should be, enacted to manage and protect vulnerable or at-risk cave ecosystems while also balancing the use of caves for tourism and education.

16.5 Summary and Future Visions of Cave Life Sciences

During the last decades, caves have emerged as fascinating model systems for a number of scientific disciplines. As only a minor fraction of all caves on our planet have been found and explored, much still remains to be discovered. To accomplish this, a number of further developments in caving technology, sampling technology and analytical tools are needed and these must be applied in a systematic way to allow clear correlations to be made. Fortunately, this is being accomplished via increased research in various scientific disciplines, as well as in the emergence of various professional networking organizations (e.g. websites for CRF, the Speleogenesis Network and the UIS). While each cave may pose its own specific research questions, some general examples for future fields of research include:

- Cave life sciences, in general, because our knowledge about cave biodiversity, biogeography, endemism, function and activity status, geo-ecological interactions, population dynamics, nutrient cycling, adaptation mechanisms and life cycles of various cave adapted species is still very limited.
- Biology of different types of extremophilic organisms in oligotrophic, dark life environments.
- Different applications in geobiology, such as interactions between life and minerals and subsurface sciences, as caves may provide a unique opportunity for *in situ* exploration of overall subsurface habitat biodiversity, function and its mutual interactions with the upper spheres.
- Astrobiology whereby caves may serve as interesting model systems (analogues) for hypotheses on the origin and development of life on Earth, as well as in outer space.
- Geotechnological applications, such as drilling, caving and mining.
- Environmental sciences addressing topics such as climatology, global warming, pollution and the screening of potentially novel species with interesting degradation traits (with the potential to allow water purification and/or degradation of contaminants).
- Cultural history such as archaeology and development of conservation techniques.
- Medical sciences and different types of biotechnological applications whereby caves could be explored for, e.g. useful enzymes, pathogenic organisms, screening of antibiotic-producing organisms, to name but a few.

Clearly, cave science is an exciting and ever-expanding field with enormous future potential.

Acknowledgements

We thank all enthusiastic members in our research teams for valuable contributions over the years. Financial support was provided to NL, DM and WL from the Technische Universität München, the Helmholtz Foundation for the 'virtual institute for isotope biogeochemistry–biologically mediated processes at geochemical gradients and interfaces in soil–water systems' and the DFG project FOR 571; from the United States National Science Foundation (DEB-0640835) and the Louisiana Board of Regents (NSF(2010)-PFUND-174) for AES, for RA and LK from the project APVV 0251-07, Slovakia, for CSJ from the project TCP CSD2007-00058, Spain, for SB from the Scientists Pool Scheme CSIR, New Delhi, India, and for RB University Grants Commission (UGC), New Delhi, India.

Websites

Canyons Worldwide: www.canyonsworldwide.com/crystals/mainframe3.html
CRF Cave Research Foundation: www.cave-research.org (accessed 3 May 2011).
Speleogenesis Network: www.network.speleogenesis.info/index.php (accessed 3 May 2011).
UIS International Union of Speleology: www.uis-speleo.org (accessed 2 May 2011).
USDA report on the white nose syndrome: www.invasivespeciesinfo.gov/microbes/wns.shtml (accessed 3 May 2011).
US Show Caves Directory: www.goodearthgraphics.com/showcave/menu.html (accessed 3 May 2011).

References

Abedon, S.T. (2008) *Bacteriophage Ecology: Population Growth, Evolution, and Impact of Bacterial Viruses* (Advances in Molecular and Cellular Microbiology). Cambridge University Press, Cambridge, UK, 526 pp.

Aubrecht, R., Brewer-Carías, Ch., Šmída, B., Audy, M. and Kováčik, L. (2008) Anatomy of biologically mediated opal speleothems in the World's largest sandstone cave: Cueva Charles Brewer, Chimantá Plateau, Venezuela. *Sedimentary Geology* 203, 181–195.

Barr, T.C.J. and Holsinger, J.R. (1985) Speciation in cave faunas. *Annual Review of Ecology and Systematics* 16, 313–337.

Barton, H.A. and Northup, D.E. (2007) Geomicrobiology in cave environments: past, current and future perspectives. *Journal of Cave and Karst Studies* 67, 27–38.

Baskar, S., Baskar, R., Mauclaire, L. and McKenzie, J.A. (2006) Microbially induced calcite precipitation by culture experiments – possible origin for stalactites in Sahastradhara, Dehradun, India. *Current Science* 90, 58–64.

Blehert, D.S., Hicks, A.C., Behr, M., Meteyer, C.U., Berlowski-Zier, B.M., Buckles, E.L., Coleman, J.T.H., Darling, S.R., Gargas, A., Niver, R., Okoniewski, J.C., Rudd, R.J. and Stone, W.B. (2009) Bat white-nose syndrome: an emerging fungal pathogen? *Science* 323, 227.

Boston, P.J., Spilde, M.N., Northup, D.E., Melim, L.A., Soroka, D.A., Kleina, L.G., Lavoie, K.H., Hose, L.D., Mallory, L.M., Dahm, C.N., Crossey, L.J. and Scheble, R.T. (2001) Cave biosignature suites: microbes, minerals and Mars. *Astrobiology Journal* 1, 25–55.

Bunnell, D. (2004) Littoral caves. In: Gunn, J. (ed.) *Encyclopedia of Caves and Karst*. Routledge, New York, pp. 491–492.

Cañaveras, J.C., Cuezva, S., Sanchez-Moral, S., Lario, J., Laiz, L., Gonzalez, J.M. and Saiz-Jimenez, C. (2006) On the origin of fiber calcite crystals in moonmilk deposits. *Naturwissenschaften* 93, 27–32.

Cigna, A.A. (2005) Radon in caves. *International Journal of Speleology* 34, 1–18.

Culver, D.C. and Pipan, T. (2009) *The Biology of Caves and Other Subterranean Habitats*. Oxford University Press, Oxford, UK, 255 pp.

Eavis, A. (2009) An up to date report of cave exploration around the world. In: White, B. (ed.) *Proceedings of 15th International Congress of Speleology*, Kerrville, Texas, pp. 21–25.

Ehrlich, H.L. (1996) *Geomicrobiology*. Marcel Dekker Inc., New York, 719 pp.

Engel, A.S. (2007) On the biodiversity of sulfidic karst habitats. *Journal of Cave and Karst Studies* 69, 187–206.

Engel, A.S. (2010) Microbial diversity of cave ecosystems. In: Barton, L.L., Mandl, M. and Loy, A. (eds) *Geomicrobiology: Molecular and Environmental Perspective*. Springer, the Netherlands, pp. 219–238.

Engel, A.S. (2011) Karst microbial ecosystems. In: Reitner, J. and Thiel, V. (eds) *Encyclopedia of Geobiology*. Springer Encyclopedia of Earth Sciences Series (EESS, formerly Kluwer Edition), Berlin, Germany, pp. 521–531.

Engel, A.S. and Northup, D.E. (2008) Caves and karst as model systems for advancing the microbial sciences. In: Martin, J. and White, W.B. (eds) *Frontiers in Karst Research*. Karst Waters Institute Special Publication 13, Leesburg, Virginia, pp. 37–48.

Elliot, W.R. (2006) Biological dos and don'ts for cave restoration and conservation. In: Hildreth-Werker, V. and Werker, J. (eds) *Cave Conservation and Restoration*. National Speleological Society, Huntsville, Alabama, pp. 33–46.

Ford, D.C. and Williams, P. (2007) *Karst Hydrogeology and Geomorphology*. Wiley, Chichester, UK, 576 pp.

Forterre, P. (2010) Defining life: the virus viewpoint. *Origins of Life and Evolution of the Biosphere* 40, 151–160.

Forti, P. (2009) State of the art in the speleological sciences. In: White, B. (ed.) *Proceedings of 15th International Congress of Speleology*, Kerrville, Texas, pp. 26–31.

Grobbelaar, J.U. (2000) Lithophytic algae: a major threat to the karst formation of show caves. *Journal of Applied Phycology* 12, 309–315.

Gunn, J. (2004) *Encyclopedia of Caves and Karst Science*. Taylor and Francis, New York, 1940 pp.

Halliday, W.R. (2004) Volcanic caves. In: Gunn, J. (ed.) *Encyclopedia of Caves and Karst Science*. Dearborn, London, USA, pp. 760–764.

Hill, C.A. and Forti, P. (1997) *Cave Minerals of the World*, 2nd edn. National Speleological Society, Huntsville, Alabama, 463 pp.

Howarth, F.G. (1980) The zoogeography of specialized cave animals: a bioclimatic model. *Evolution* 34, 394–406.

Krajick, K. (2001) Cave biologists unearth buried treasure. *Science* 293, 2378–2381.

Krajick, K. (2007) Discoveries in the dark. *National Geographic*, September 2007.

Kuzmin, I.V., Niezgoda, M., Franka, R., Agwanda, B., Markotter, W., Breiman, R.F., Shieh, W.-J., Zaki, S.R. and Rupprecht, C.E. (2010) Marburg Virus in Fruit Bat, Kenya. *Emerging Infectious Diseases Journal* 16, 352–354.

Kyle, J.E., Eydal, H.S., Ferris, F.G. and Pedersen, K. (2008) Viruses in granitic groundwater from 69 to 450m depth of the Aspö hard rock laboratory, Sweden. *Journal of the International Society for Microbial Ecology* 2, 571–574.

Lamoreux, J. (2004) Stygobites are more wide-ranging than troglobites. *Journal of Cave and Karst Studies* 66, 18–19.

Lavoie, K.H., Northup, D.E. and Barton, H.A. (2010) Microbial-mineral interactions; geomicrobiology in caves. In: Jain, S.K., Khan, A.A. and Rai, M.K. (eds) *Geomicrobiology*. Science Publishers/CRC Press, Enfield, New Hampshire, pp. 1–21.

McClurg, D. (1996) *Adventure of Caving. A Beginner's Guide for Exploring Caves Softly and Safely*. D&J Press, Carlsbad, New Mexico, 251 pp.

Melim, L.A., Liescheidt, R., Northup, D.E., Spilde, M.N., Boston, P.J. and Queen, J.M. (2009) A Biosignature suite from cave pool precipitates, Cottonwood Cave, New Mexico. *Astrobiology* 9, 907–917.

Northup, D.E. and Lavoie, K. (2001) Geomicrobiology of caves: a review. *Geomicrobiology Journal* 18, 199–222.

NOAA Paleoclimatology Speleothem (Cave Deposit) Data (2011) Available at: www.ncdc.noaa.gov/paleo/speleothem.html (accessed 3 May 2011).

NOVA (2002) Science Programming on Air and Onlive. The mysterious life of caves. Available at: www.pbs.org/wgbh/nova/caves (accessed 3 May 2011).

NSS GEO2 committee on long and deep caves (2011) Available at: www.caverbob.com (accessed 2 May 2011).

Quan, P.L., Firth, C., Street, C., Henriquez, J.A., Petrosov, A., Tashmukhamedova, A., Hutchison, S.K., Egholm, M., Osinubi, M.O., Niezgoda, M., Ogunkoya, A.B., Briese, T., Rupprecht, C.E. and Lipkin, W.I. (2010) Identification of a severe acute respiratory syndrome coronavirus-like virus in a leaf-nosed bat in Nigeria. *mBio* 1, X-Y. pii: e00208-10.

Rainey, F. and Oren, A. (2007) *Extremophiles*. Elsevier Ltd., Oxford, UK, 838 pp.

Romero, A. (2009) *Cave Biology: life in darkness*. Cambridge University Press, Cambridge, UK, 306 pp.

Saiz-Jimenez, C. (1999) Biogeochemistry of weathering processes in monuments. *Geomicrobiology Journal* 16, 27–37.

Saiz-Jimenez, C. (2010) Painted material. In: Mitchell, R. and McNamara, C.J. (eds) *Cultural Heritage Microbiology: Fundamental Studies in Conservation Science*. ASM Press, Washington, DC, pp. 3–13.

Sarbu, S.M., Kane, T.C. and Kinkle, B.K. (1996) A chemoautotrophically based cave ecosystem. *Science* 272, 1953–1955.

Self, C.A. and Hill, C.A. (2003) How speleothems grow: an introduction to the ontogeny of cave minerals. *Journal of Cave and Karst Studies* 65, 130–151.

SPACE/Malik, T. and Writer, S. (2005) Spelunking on Mars: Caves are Hot Spots in Search for Life. Available at: www.space.com/scienceastronomy/scitues_marscaves_050308.html

Stynes, D.J. (1999) *Approaches to Estimating the Economic Impacts of Tourism:* Some *Examples*. Michigan State University, USA, 18 pp.18

Taylor, M.R. (1999) *Dark Life: Martian Nanobacteria, Rock-eating Cave Bugs, and Other Extreme Organisms of Inner Earth and Outer Space*. Scribner, New York, 288 pp.

17 The Deep Biosphere: Deep Subterranean and Subseafloor Habitats

Elanor M. Bell[1] and Verena B. Heuer[2]
[1]*Scottish Institute for Marine Science, Argyll, UK;*
[2]*Organic Geochemistry Group, Department of Geosciences and MARUM Center for Marine Environmental Sciences, University of Bremen, Germany*

17.1 Introduction to the Deep Biosphere

Jules Verne's famous book, *A Journey to the Centre of the Earth*, tells the tale of Professor Lidenbrock, his nephew Axel, their guide Hans and their encounters with a group of weird and wonderful life-forms on their descent into the interior of the Earth (Verne, 1864). When the novel was first published in 1864, the idea of life existing in a deep subsurface realm remote from sunlight and air was genuine science fiction. But towards the end of the 20th century, when scientific drilling and deep mining operations opened windows into environments hundreds of metres under the surface of the Earth, it became clear that Jules Verne was, in a sense, right (Pedersen, 2000). Deep under terrestrial ground and the seafloor, life is very abundant (Whitman *et al.*, 1998; Parkes *et al.*, 2000; Pedersen, 2000), albeit mainly in the form of prokaryotic microorganisms. In fact, the deep subsurface biosphere represents a substantial fraction of Earth's total biomass: first estimates suggested a contribution of 10–30% (Parkes *et al.*, 1994; Whitman *et al.*, 1998). Pedersen (2000) even suggests that the combined biomass of intraterrestrial organisms in all deep subsurface habitats may in fact be equal to the total weight of all marine and terrestrial plants put together.

The discovery of the deep biosphere gives rise to new, fundamental questions about the evolution and distribution of life. Considering the broad range of extreme temperatures and high pressures in the subsurface, and the lack of fresh organic matter as a source of energy and nutrients, the existence and survival of an extensive deep biosphere is remarkable. Geofuels, i.e. gases, that are generated by mineral–water interactions or derived from the deep crust, might provide an alternative energy source in the absence of light and photosynthesis. But, so far, little is known about the individual inhabitants of the deep biosphere, about their genetic diversity, metabolic capabilities and survival strategies. Moreover, the relationships between deep subsurface ecosystems and the surface world also remain to be explored. This chapter will give a synopsis of what is known about the deep biosphere.

17.2 Types of Deep Subsurface Environment

17.2.1 How deep is deep?

The deep biosphere is defined by Pedersen (1993) as being the intraterrestrial life in subterranean (or subsea) environments deeper than 50–100 m. In reality, deep biosphere researchers investigate subsurface life in habitats as diverse as 1 m below the surface of ocean sediments (Jørgensen and Boetius, 2007; Fry *et al.*, 2008), 4 m below the tidal flats of Spiekeroog Island, Germany (Wilms *et al.*, 2006) and hundreds of metres below the sea floor in the basement crust beneath the sediments (Pedersen, 2000). Intraterrestrial life has been found at depths of up to 1626 m in marine sediments (representing up to 111 million years old and ca. 100°C hot sediments on the Newfoundland margin; Roussel *et al.*, 2008), up to 2800 m in continental sedimentary rocks, 5300 m in igneous rock aquifers and in fluid inclusions in ancient salt deposits from salt mines (reviewed in Pedersen, 2000). As far as we know the deep biosphere is dominated by prokaryotic microorganisms. In marine sediments, concentrations of intact prokaryotic cells decrease rapidly with sediment depth, but considerable microbial populations of approximately 10^6 prokaryotic cells are usually still present at the mean oceanic sediment depth of 500 m (Parkes *et al.*, 1994, 2000).

With depth, temperature increases due to the geothermal gradient and, ultimately, it is likely that temperature limits the depth of the deep biosphere. Microbial activity is known to occur at temperatures of up to ca. 110°C but microbial abundance rapidly declines if the temperatures increase further and the upper temperature limit for microbial life as we know it is reported to be approximately 121°C (Stetter *et al.*, 1990; Furnes and Staudigel, 1999; Kashefi and Lovley, 2003).

17.2.2 Continental sedimentary rocks

Investigations carried out on continental sedimentary rocks show that living microbes are present and active down to the deepest levels studied to date (approximately 3000 m) and it is likely that in the future microbes will continue to be found as deep as the temperature allows (Pedersen, 2000). Large phylogenetically and culturally diverse communities of Archaea and Bacteria have been described in North American, African and Scandinavian deep subsurface continental sediments (Pedersen *et al.*, 1996; Chandler *et al.*, 1997; Colwell *et al.*, 1997; Crozier *et al.*, 1999). As in other oligotrophic (nutrient-poor) habitats, metabolic and physiological activity takes place in deep sedimentary rocks but at very slow rates. Interestingly, Fredrickson *et al.* (1997) showed that there was usually a significant correlation between the type of geological strata, for instance in the pore size and organic content, and the level of activity.

Finding life deep in continental sedimentary rocks has important implications for the long-term maintenance of anoxic conditions and the impact of anaerobic biochemical processes on groundwater chemistry. It also implies that natural attenuation of contaminated groundwater by indigenous microbes can occur at depth, but further work is needed to unravel details about these significant processes (Pedersen, 2000) and how they might be applied to the remediation of contaminated groundwaters at shallower depths.

17.2.3 Ancient salt deposits

Extensive, ancient deposits of salt exist deep in subterranean environments worldwide, for example, the 250 million year old deposits located deep under Denmark, Germany and the North Sea (Pedersen, 2000). Such crystalline salt deposits form when saline water evaporates (for instance, following a geological period of climatic warming). During this process small pockets of saline fluid (brine) become trapped between the crystals creating fluid inclusions. Halophilic (salt-loving) microorganisms that were present in the original water are trapped as well and, even in ancient deposits cut off from the surface for millions of years, a wide diversity of culturable, halophilic species can still be isolated and grown. For example, Vreeland *et al.* (1999) isolated a diverse community of microbes from the geologically stable, Permian age Salado salt formation, located 650 m below the New Mexico desert, USA. Population abundances in some areas of the deposit reached as high as 1.0×10^4 colony-forming units per gram of NaCl. These findings have exciting implications for the longevity of halophilic microorganisms and their ability to survive in isolated brine inclusions for millions of years (Pedersen, 2000), as well as the possibility of life on other extraterrestrial planets and bodies (HiRISE, 2007; Section 17.7).

17.2.4 Aquifers in igneous terrestrial rocks

Igneous rocks are the predominant solid constituents of the Earth, formed through cooling of molten or partly molten material at or beneath the Earth's surface. Humans access deep igneous rocks on a regular basis during activities such as mining, tunnelling and drilling, deposition and waste disposal, to name but a few. These activities have led to the discovery of previously unknown microbial ecosystems at depths exceeding 1000 m (Pedersen and Ekendahl, 1990; Pedersen, 1997; Stroes-Gascoyne and Sargent, 1998) in most, if not all, deep aquifers (Pedersen and Ekendahl, 1990). It is now known that a remarkable diversity of species thrive (including many new to science) and a wide range of metabolic activities take place in these deep, extreme environments (e.g. Ekendahl *et al.*, 1994; Pedersen *et al.*, 1996, 1997; Motamedi and Pedersen, 1998; Takai *et al.*, 2001).

The repeated observations of autotrophic, hydrogen-dependent microorganisms in the deep aquifers imply that hydrogen may be an important electron and energy source and carbon dioxide an important carbon source in deep subsurface ecosystems (Pedersen, 2000, 2001), potentially together with abiogenic methane and higher hydrocarbon gases (Sherwood Lollar *et al.*, 2002, 2006). Hydrogen, methane and carbon dioxide have been found at all sites that have been tested for these gases. It has therefore been proposed that the deep biosphere in deep igneous rock aquifers is hydrogen-driven, and likely inhabited by autotrophic methanogens and acetogens that use hydrogen and carbon dioxide to produce methane and acetate, respectively, which in turn serve as substrates for other microorganisms (Stevens and McKinley, 1995; Pedersen, 1997, 2000, 2001). Hydrogen might result from radiolysis of water (Lin *et al.*, 2005). Moreover, production of sulfate from pyrite/barite can support sulfate reducers as observed in ca. 3000 m deep fracture zones in South African goldmines, where nutrient concentrations are higher than in shallower crustal environments and thermophilic sulfate reducers are sustained by geologically produced sulfate, hydrogen and hydrocarbon gases (Kieft *et al.*, 2005; Lin *et al.*, 2006).

17.2.5 Caves

An extensive review of life in caves is provided in Lee *et al.* (Chapter 16, this volume), thus only brief mention will be made of these 'deep' environments here. Caves potentially provide the link between surface and subsurface environments. Their environments vary widely but some larger, deeper examples offer excellent opportunities to study intraterrestrial life. Cunningham *et al.* (1995) described extensive bacterial and fungal colonization in the deep, gypsum and sulfur-bearing hypogenic Lechuguilla Cave, located in the Carlsbad Caverns National Park, New Mexico, USA. At least 90% of the cave lies at a depth of >300 m, thus, external (allochthonous) inputs of organic material are limited. Instead, autotrophic bacteria present in ceiling-bound residues are thought to function as primary producers within the cave food web, providing organic matter for the heterotrophic bacterial and fungal communities also known to exist.

17.2.6 Subseafloor sediments and basement rock

Last but by no means least, deep subsurface environments hosting extensive life have been found deep beneath seafloor sediments and in basement rock (Wellsbury *et al.*, 1997; Fisk *et al.*, 1998; for a review see also Schrenk *et al.*, 2010). The deep subseafloor is potentially the largest ecosystem on Earth, but its existence was only recognized during the 1990s.

Due to the extreme conditions at the seafloor, chiefly high pressure combined with the very low near-surface sediment temperatures, decreasing porosity with increasing depth, and the fact that a large part of the seabed (95%) is at water depths where light intensity is too low to sustain photosynthesis, the seabed was long thought to be flat and biologically inactive, with only insignificant microbial activity (Jannasch *et al.*, 1971; Jannasch and Wirsen, 1973; Fry *et al.*, 2008). However, following the discovery of hydrothermal vent systems and cold seeps in the mid-1970s and mid-1980s, respectively, it became evident that the seafloor environment is a dynamic geo- and biosphere that provides a diverse range of living conditions that are host to rich microbial communities (Jørgensen and Boetius, 2007; Fry *et al.*, 2008). Even plain sediments are densely populated compared to ocean water. They contain 10–10,000-fold more microbial cells per unit volume than productive ocean-surface waters (Jørgensen and Boetius, 2007). On average, 10^8 prokaryotic cells are present in every cubic centimetre of sediment in the top metre of the seafloor (Jørgensen and Boetius, 2007).

During the 1990s, this new perception of the seabed was further extended to include subseafloor life. In order to investigate life in the deep subseafloor, access to unadulterated samples and the detection of life and activity at the limit of technological sensitivity are key challenges that need to be overcome. Subseafloor samples for scientific research started to become available with the Deep Sea Drilling Project (DSDP) in 1968. Microbiological investigations, however, were only initiated in the subsequent Ocean Drilling Program (ODP) towards the end of the 1980s when the acridine orange dye method (Smithwick and David, 1971) and rigorous contamination tests enabled reliable direct counts of total cell numbers in drill cores (Parkes *et al.*, 1994). Following the discovery of abundant microorganisms in deeply buried sediments, in 2002 ODP Leg 201 was devoted to the exploration of life in the deep subseafloor. The deep biosphere became a major research initiative in 2004 when ODP was succeeded by the Integrated Ocean Drilling Program (IODP), an international effort involving 24 countries led by implementing organizations in the USA, Japan and Europe. Within IODP, coordination among engineers, geologists, geochemists and biologists is essential for the development of new tools and drilling technologies and will allow scientists to reach greater sediment depths and to install subseafloor observatories (Fig. 17.1). The latter are particularly important for the investigation of the deep crustal biosphere, as this is barely accessible using conventional drilling techniques because basement recovery is usually low and there are inherently high risks of contamination from drilling fluids.

Due to the concentrated and intense research activities in ODP and IODP, the deep subseafloor is the most extensively studied deep biosphere habitat and will, therefore, form the basis for discussion in the remainder of this chapter.

17.3 Case Study: Deep Subseafloor

17.3.1 The seafloor

The deep-ocean seafloor extends over an area of 320 million km^2 and is best viewed like a slow-moving, giant conveyor belt. The hard rock basement of the seafloor originates at spreading mid-ocean ridges on the bottom of the sea. Lava and mantle rock convect up through these ridges toward the ocean floor and hot fluids are expelled at temperatures of up to 350–400°C (Jørgensen and Boetius, 2007),

Fig. 17.1. The drilling vessel DV *JOIDES Resolution*. © John Beck, IODP/TAMU.

forming hydrothermal features such as vents in the cracks and fissures (see also Lutz, Chapter 13, this volume). Subsequently, plate tectonic forces mean that the newly formed ocean basement inexorably migrates towards subduction zones where it is pushed back under continental shelves, returned to the Earth's hot interior and recycled in a process estimated to take ca. 170 million years (Pedersen, 2000). During the intervening years between formation and subduction, this basement layer accumulates layers of fine grained sediments. On average, the sediments form a layer up to 500 m thick, but they can be as much a 10 km deep at continental margins where the input of terrigenous sediments (sediments derived from erosion on land) is high. Figures 17.2 to 17.6 illustrate the extremely long sediment cores that are collected to study the subseafloor biosphere and their processing.

In the central ocean, aeolian dust or minute carbonate skeletons of planktonic organisms constitute an important fraction of the sediment, whereas terrigenous weathering products predominate in sediments that are closer to the continents. Sedimentation rates vary in different regions of the world's oceans, but in general a sediment cover develops on hard rocks within 5–100 million years of their formation (Schrenk *et al.*, 2010). While the barren ocean crust on mid-ocean ridges and ridge flanks are active hydrological systems with advective water flow, transport of chemicals is predominantly controlled by diffusion in marine sediments. However, when deposited sediments become compressed and geothermally altered in subduction zones, subsurface pore fluids and gases are expelled and form gas-hydrate deposits, gas chimneys, mud volcanoes and diverse seep settings (Jørgensen and Boetius, 2007). These sediments, some of which are millions of years old, host active microbial populations, notably acetogens and methanogens (Wellsbury *et al.*, 1997), the latter of which might be responsible for the enormous

Fig. 17.2. A core is brought from the rig floor to the catwalk during IODP Expedition 301. © Jonathan Rice.

Fig. 17.3. Scientists cut a core into sections. © Carlos Alvarez Zarikian, IODP-USIO/TAMU.

Fig. 17.4. Cores laid out for processing in the core refer. © John Beck, IODP/TAMU.

reservoirs of methane in the aforementioned gas hydrates found globally in subseafloor sediments (Kvenvolden, 1995). There are also strong indications that microbial life is widespread at depth in the crustal rocks below the sediments (Fisk *et al.*, 1998; Bach and Edwards, 2003; Schrenk *et al.*, 2010).

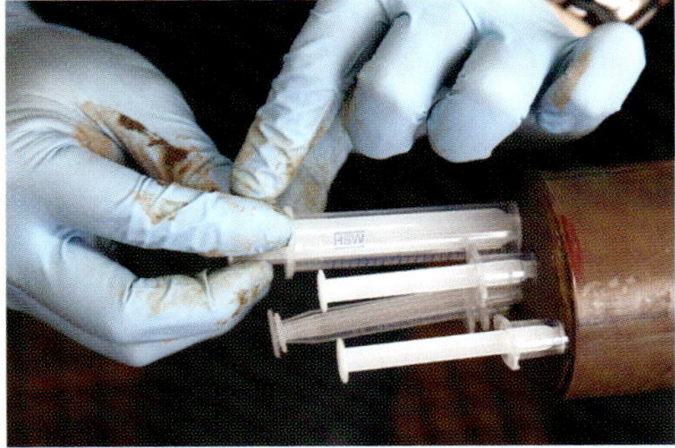

Fig. 17.5. Microbiology samples being taken from the core on the catwalk. © John Beck, IODP/TAMU,

17.3.2 Definition

The actual boundary between seafloor and subseafloor is difficult to identify. Jørgensen and Boetius (2007) have pragmatically defined the seafloor as the top, 1 m layer of the seabed that is bioturbated by animals and porous ocean crust that is penetrated by seawater, whereas the deep subsurface, which harbours the deep biosphere, comprises the rock and sediment that is deeper than 1 m beneath the seafloor. Similarly, Fry *et al.* (2008) chose 1 m beneath seafloor as upper limit for the subseafloor biosphere. In contrast, Teske and Sørensen (2008) define the deep subsurface based on ecological criteria as 'sediment layers with distinct microbial communities that lack a microbial imprint of water column communities' and allow the position of this boundary to vary locally. The findings of Lipp *et al.* (2008) show that these definitions, in fact, overlap. In a wide range of oceanographic settings, the microbial community changes distinctly within the upper metre

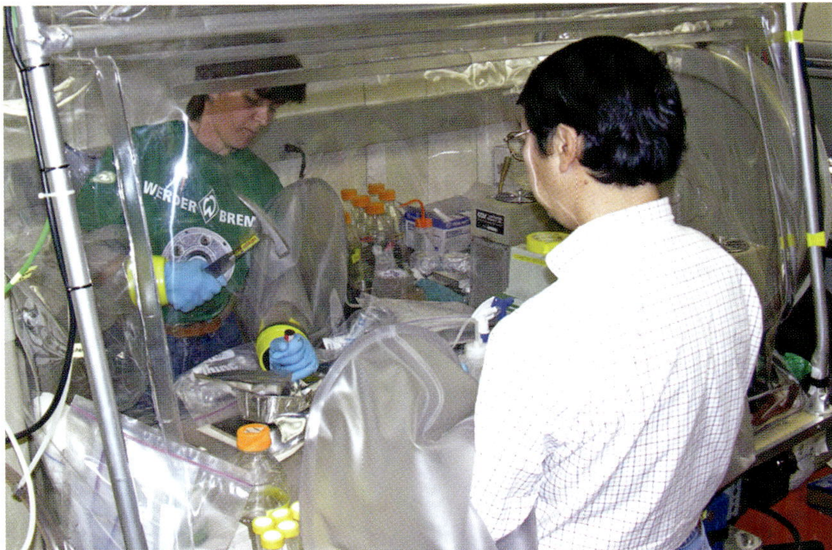

Fig. 17.6. Core samples being prepared in a glove box flushed with nitrogen gas to protect any anaerobic microorganisms from toxic oxygen. Microbiology is not always a delicate science; a hammer and chisel have to be used to break the hard basalt to collect samples for culturing the microorganisms that are thought to live inside the rock (the hammer and chisel are sterilized to kill any bacteria on them before being put in the glove box). © Jonathan Rice.

of sediment. While Bacteria dominate in surface sediments, Archaea constitute a major fraction of the biomass in subsurface sediments buried deeper than 1 m below the seafloor.

17.3.3 The physical and chemical extremes of the deep subseafloor

In the deepsea, the seafloor at or close to the sediment/water interface has important extreme features in common with the deep subseafloor. Both are remote from sunlight, air and photosynthetic ecosystems, and both experience high hydrostatic pressure. Yet, there are also distinct differences between the two with respect to porosity, temperature and the availability of organic matter and electron acceptors.

Pressure

Marine sediments have an average depth of around 500 m, but can be up to 10 km deep with an average overlying water depth of 3800 m. Such water depths produce an average hydrostatic pressure at the seafloor of 380 bar (375 atm) and at maximum water depths (11 km, Mariana Trench, western Pacific Ocean) this can be up to 1100 bar (1085 atm) (Fry *et al.*, 2008). Below the seafloor, hydrostatic pressure is increased further by the lithostatic pressure of the accumulated sediments. For example, in the Bering Sea the pressure 360 m below the seafloor is approximately 400 bar (395 atm; Scholl *et al.*, 2007).

Porosity

While porosity is high in freshly deposited sediments at the seafloor, and can also be sufficient in the upper 200–300 m of deep ocean crusts to provide space and conduits for water and therefore life (Furnes and Staudigel, 1999), deeply buried, compacted sediments and hard rocks provide only limited habitable pore-space in the subseafloor.

Temperature

Temperatures are in general very low at the seafloor (ca. 2–4°C) but increase with depth in the subseafloor at a rate of 25–50°C km^{-1}. In fact, temperature might be the ultimate limit to life in both subseafloor and subterranean environments (Stetter *et al.*, 1990; Furnes and Staudigel, 1999). The upper temperature limit of microbial life ever recorded is ca. 121°C (Kashefi and Lovley, 2003). If this holds true for the deep biosphere, it limits life to the upper 2–4 km of the subseafloor in most oceanic provinces (Whitman *et al.*, 1998). At mid-ocean ridges where hot fluids can be expelled at temperatures of up to 350–400°C (Jørgensen and Boetius, 2007), this temperature limit is exceeded at much shallower depths.

Organic matter and energy

Temperature might set the limits to where life can exist but it is the limited availability of energy in deep environments that make them 'extreme' for the organisms that do live in them. Between 5 and 10 billion t of particulate organic matter is constantly sinking through the oceans' water columns and accumulating as marine sediments. As a result, particulate organic matter (POC) is concentrated 10,000–100,000-fold higher in the sediments than in the overlying seawaters. The vast majority of this organic matter is removed by near-surface microbial activity, but over geological time the remainder accumulates and results in the largest global reservoir of organic carbon (15,000 × 10^{18} g C; Fry *et al.*, 2008).

Despite the vast quantities of accumulated organic matter not all of it is bioavailable, i.e. easily used by organisms. Surface sediments receive relatively fresh and easily degradable organic matter from the overlying water column, but organic matter concentrations are low in the deep subseafloor (0.1–1.0% by dry weight) and the aged, degraded, detrital material is not easily accessible for microbial degradation (Jørgensen and Boetius, 2007). However, it is thought that low-temperature heating during sediment burial might increase the availability of recalcitrant organic matter for microbial degradation in the deep biosphere, just as kerogen is known to break down at

100–150°C to form oil and gas (Wellsbury *et al.*, 1997; Parkes *et al.*, 2007).

Typically, the microbial degradation of organic matter occurs along a cascade of redox reactions with different electron acceptors that yield successively lower energy (Froelich *et al.*, 1979). In shallow sediments this cascade creates a vertical redox zonation that is sustained as electron acceptors such as oxygen, nitrate and sulfate enter the sediment from the overlying ocean. Once all of the sulfate has been reduced, methanogenesis and fermentation become the principal remaining avenues of metabolic activity, primarily because their electron acceptors can be replenished by organic matter degradation. Deep subseafloor sediments deviate from this standard zonation (D'Hondt *et al.*, 2004): on the one hand, the zonation is upset because electron acceptors are introduced into the deep subseafloor not only from the overlying ocean but also from water circulating through the underlying basaltic aquifer, in addition to upward diffusion from ancient brines within the sediments (D'Hondt *et al.*, 2004). On the other hand, the zonation is less distinct and redox reactions with different energy yields co-occur. For example, methanogenesis has been observed in the sulfate-replete, deeply buried sediments of the eastern Pacific even though sulfate-reducers are generally thought to out-compete methanogens in the competition for available substrates (D'Hondt *et al.*, 2004). Moreover, stable isotope data suggest that acetogenesis and methanogenesis, using carbon dioxide and hydrogen to produce acetate and methane, respectively, co-occur in continental margin sediments of the north-east Pacific, despite the fact that methanogens are thought to out-compete acetogens for hydrogen (Heuer *et al.*, 2009). Stable carbon isotope analysis also provides evidence for the uptake of methane carbon into the membrane lipids of methanotrophic Archaea in shallow sediments at cold seeps (Elvert *et al.*, 1999; Hinrichs *et al.*, 1999), but there is no indication of the uptake of methane carbon into archaeal cells in the deep sulfate–methane transition zones at the continental margins studied to date, e.g. off the coast of Peru, even though methanotrophy accounts for a major fraction of carbon cycled in these ecosystems (Biddle *et al.*, 2006).

Hydrogen might provide an alternative to organic matter as an energy source for subseafloor inhabitants. Hydrogen is produced in the subsurface when anoxic seawater reacts with igneous rocks and is a suitable energy source for lithotrophic organisms in fresh basalt along mid-ocean ridges. However, in old and cold ocean crust the rates of hydrogen formation are probably too low to contribute significantly to the energy that is derived from sedimentary organic matter (Jørgensen and Boetius, 2007). The natural radioactivity of sediments might form hydrogen via the radiolysis of water, but this energy source is probably only important in the most energy-depleted systems, such as the South Pacific Gyre (Jørgensen and Boetius, 2007).

Because of the distinct differences between the shallow seafloor and deep subseafloor, metabolic processes in the deep biosphere cannot be predicted by simple extrapolation of findings from marine surface sediments. Therefore, it is likely that the energy-yielding activities of microbial communities are much more diverse in deep subseafloor sedimentary ecosystems than in shallow marine sediments (D'Hondt *et al.*, 2004; Section 17.4.2).

17.4 Life in the Deep Subseafloor

17.4.1 Diversity and community composition

Deep subseafloor communities are dominated by prokaryotic microorganisms in the domains Bacteria and Archaea. Despite the extreme conditions these microbial communities live under, estimates suggest that the seabed hosts the 'hidden majority' of all microbial cells, which comprises between half and five-sixths of the Earth's total microbial biomass and between one-tenth and one-third of the Earth's total living biomass (Parkes *et al.*, 1994; Whitman *et al.*, 1998; Pearson, 2006; Lipp *et al.*, 2008).

As would be expected, the diversity of prokaryotes identified in the subseafloor using cultivation-independent molecular methods is much greater than the number of taxa obtained using cultivation methods (for reviews see Fry *et al.*, 2008 and Teske and Sørensen, 2008). Analyses of 16S rRNA gene libraries reveal a substantial diversity of Bacteria and Archaea in deep sub-seafloor sediments as well as a great variability in the composition of microbial communities in different oceanographic settings. Gammaproteobacteria, Chloroflexi and members of the candidate division JS1 appear to be the most abundant bacterial groups. In a review of 34 different 16S rRNA gene libraries from a wide range of sediments (Fry *et al.*, 2008) these three groups were present in 62–70% of the clone libraries, where they individually represented 17–26% of the clones.

Alpha-, Beta-, Delta- and Epsilon-Proteobacteria were the next most abundant groups. They were found in 3–50% of the libraries, where they accounted for 2–8% of the clones. Direct comparison between surface and subseafloor sediments suggests that the upper layers are dominated by Gamma- and Delta-Proteobacteria while Chloroflexi and candidate division JS1 are more abundant in the deeper layers. Chloroflexi and the JS1 group can also co-occur with one or the other dominating in each case.

The same study reviewed 47 different 16S rRNA libraries with respect to the community composition of Archaea (Fry *et al.*, 2008). Among the Archaea, Crenarchaeota seem to dominate. They represented 73% of the clones, while only 24% of the clones belong to Euryarchaeota (Fry *et al.*, 2008). The most abundant archaeal groupings were the crenarchaeotal groups Miscellanous Crenarchaeotic Group (MCG, 33%) and Marine Benthic Group B (MBG-B, 26%), followed by Marine Group 1 (8%). The South African Gold Mine Groups (SAGMEG, 7%) and the thermophilic Euryarchaeota (7%) were the most abundant euryarchaeotal groups (Fry *et al.*, 2008).

The taxonomic association of phylotypes in 16S rRNA gene libraries can be identified with the help of phylogenetic trees. For the dominant bacterial groups, phylogenetic trees show that many of the Proteobacteria are related to cultured species while there are only few cultured members of Chloroflexi and no cultured members of JS1 (Fry *et al.*, 2008). Chloroflexi is in general an extremely diverse but poorly characterized extremophilic phylum that is not only found in the deep subseafloor but also in hot springs, hydrothermal sediments, soils, waste water and polluted sites (Hugenholtz *et al.*, 1998; Wilms *et al.*, 2006; Fry *et al*, 2008). In contrast, JS1 clones all belong to one monophyletic group that seems to be limited to anoxic sedimentary habitats (Fry *et al.*, 2008). Most of the Archaea represent uncultured lineages, an exception being some euryarchaeotal methanogens, thermophiles and hyperthermophiles with related cultured species. However, these groups accounted for <8% of the archaeal clones (Fry *et al.*, 2008).

17.4.2 Metabolic activity in the deep subseafloor

Considering the extremely low energy flow that is available to individual cells, one would expect deep subseafloor microorganisms to be either adapted for extraordinarily low metabolic activity, inactive or dead. However, several lines of evidence show that the deep subseafloor biosphere is indeed alive and active: (i) geochemical pore-water point to biologically catalysed reactions; (ii) intact cell components have been detected that are known to rapidly degrade after cell death; such components include ribosomes (Schippers *et al.*, 2005) and intact polar lipids from cell membranes (Biddle *et al.*, 2006; Lipp *et al.*, 2008; Lipp and Hinrichs, 2009); and (iii) cell concentrations correspond to the availability of energy, i.e. cell concentrations are highest where concentrations of metabolic products and net rates of sulfate- and iron-reduction are highest, and cell concentrations are lowest where these rates and metabolic product concentrations are lowest. Cell concentrations are particularly high in sulfate–methane transition zones that represent an interface of high biogeochemical energy supply (D'Hondt *et al.*, 2004).

Given this evidence of widespread, active, microbial life in the deep subseafloor, it is likely that microbes are involved in many geochemical processes, such as diagenesis, weathering, precipitation, and in the oxidation or reduction reactions of metals, carbon, nitrogen and sulfur, and a similar role can be assumed for subterranean environments (Pedersen, 1993; Fig. 17.7). However, the physiology and ecological role of even the most dominant bacterial and archaeal groups in the subseafloor is difficult to predict based on currently available cultivation-independent microbiological methods.

While novel and uncultured lineages of bacterial and archaeal 16S rRNA genes dominate in clone libraries from the deep subseafloor, classical sulfate-reducing bacteria or methanogenic and methane-oxidizing Archaea that are known from surface sediments do not seem to constitute a significant fraction of the subseafloor populations (Jørgensen and Boetius, 2007). This observation is surprising, since geochemical gradients in pore-waters suggest that sulfate-reduction, methanogenesis and anaerobic methane oxidation are important metabolic processes in the deep subseafloor (D'Hondt *et al.*, 2002). There might be several reasons for this observation: (i) the processes might be carried out by novel phylogenetic lineages with divergent gene sequences; (ii) sulfate-reducing bacteria, methanogenic and methane-oxidizing Archaea might occur in low population densities that are still sufficient to generate

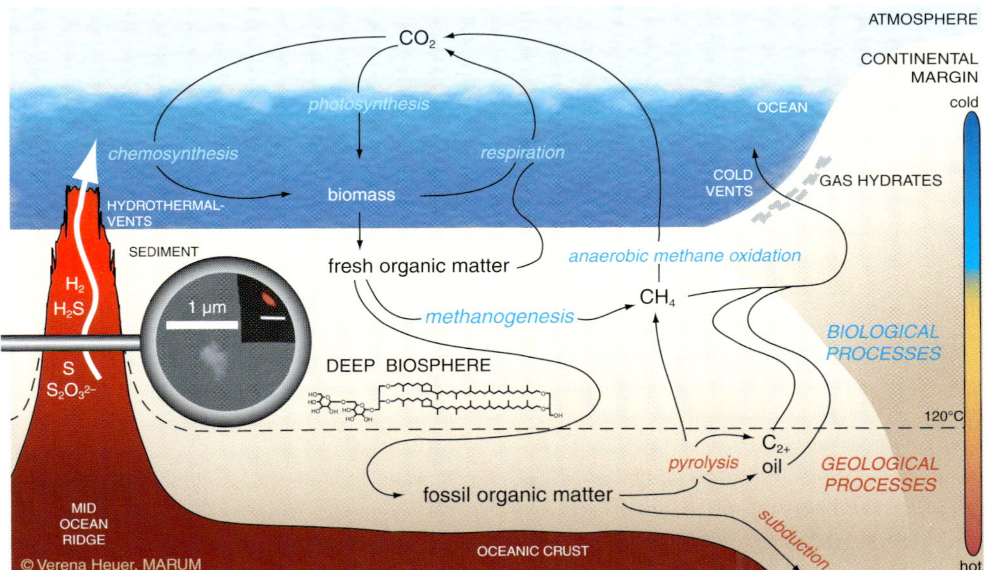

Fig. 17.7. The deep marine biosphere: microscopic images and geochemical analysis of labile organic molecules which rapidly decay after cell depth show that microorganisms are intact and able to live in deeply buried, old sediments. The illustration shows the genetic information carrier ribonucleic acid which shows up red under the microscope following fluorescence *in situ* hybridization, and intact membrane lipids which are identified by mass-spectrometric methods. Following the geothermal gradient, temperature increases with depth in the subsurface and life is probably limited to depth in which temperatures are <120°C. Together with the deep sea, the deep subseafloor is probably the largest continuous ecosystem on Earth. The discovery of the deep biosphere gives rise to numerous new scientific questions, e.g. which role does the deep biosphere have in the global carbon cycle? © Verena Heuer.

the observed sulfate and methane profiles over geological time; and (iii) difficulties in the quantitative extraction of DNA from deep sediments, which contain relatively low microbial biomass, might lead to bias in the amplified genes (Biddle *et al.*, 2006; Teske, 2006).

To date, the investigation of the metabolism of novel bacterial and archaeal lineages suffers from the extremely low cultivability of deep subseafloor prokaryotes. Studies have involved laboratory cultivation since the late 1980s, but in general the cultured members constitute less than 0.1% of the microbial community and do not have close sequence similarities to clones in deep subseafloor clone libraries (D'Hondt *et al.*, 2004; Fry *et al.*, 2008). Among the successfully cultured taxa are the barophilic sulfate-reducing bacterium *Desulfovibrio profundus*, the methanogen *Methanoculleus submarines*, and thermophilic *Firmicutes* in the genus *Thermosediminibacter* (Fry *et al.*, 2008).

For activity estimates intact sediment cores have been spiked with radiotracer substrates, incubated at *in situ* temperature and monitored for accumulation of radiolabelled reaction products over time. Estimates of sulfate reduction use $^{35}SO_4^{2-}$ as radiotracer and measure ^{35}S in the sulfide produced, while estimates of methanogenesis use ^{14}C-labelled substrates such as carbon dioxide or acetate and monitor ^{14}C in the produced methane. In the upper layers of the subseafloor, rates of sulfate reduction and methanogenesis are comparable to those found in anaerobic near-surface sediments (Parkes *et al.*, 2000). But they generally decrease with depth, although specific environmental conditions can enhance activity in deeper layers (Fry *et al.*, 2008). For example, metabolic activities are higher than average in sulfate–methane transition zones which interface zones of high biochemical energy supply (Parkes *et al.*, 2005).

Radiotracer experiments confirm the potential for *in situ* activity that can also be inferred from geochemical pore-water data. While the radiotracer experiments only extend over days to weeks, the geochemical pore-water data integrate processes over many years. Pore-water data strongly suggest that biologically catalysed reactions do consume and release metabolites in the deep subseafloor. They show that dissolved electron acceptors such as sulfate and nitrate exhibit subsurface depletion, whereas metabolic products such as dissolved inorganic carbon, ammonia, sulfide, methane, manganese and reduced, water-soluble iron consistently exhibit concentration maxima deep in the subseafloor (D'Hondt *et al.*, 2002, 2004). The underlying microbial processes include: organic carbon oxidation, ammonification, methanogenesis, methanotrophy, sulfate reduction, manganese reduction, iron reduction, and the production and consumption of formate, acetate, lactate, hydrogen, ethane and propane (D'Hondt *et al.*, 2004). More information on metabolic processes that occur *in situ* can also be obtained from the natural abundance of stable carbon isotopes in water-soluble metabolites. For example, stable carbon isotopic compositions of hydrocarbon gases point to a predominantly biogenic origin for methane in deeply buried continental margin sediments (e.g. Pohlman *et al.*, 2009) and suggest biogenic formation of ethane and propane in subseafloor sediments along novel pathways (Hinrichs *et al.*, 2006), while stable carbon isotopic compositions of volatile fatty acids suggest that the flow of carbon through the pore-water acetate pool changes systematically with depth in deeply buried sediments (Heuer *et al.*, 2009).

In the diffusion-controlled regime of the sedimentary subseafloor, the depth-integrated rate of sulfate reduction is a direct measure of total dissimilatory activity (D'Hondt *et al.*, 2002). At sites where subsurface microbial activity is consistently low, sulfate concentrations are relatively high throughout the sediment column and the activity of subseafloor microorganisms can be approximated by the flux of dissolved sulfate into the sediment. At sites where subsurface activity is relatively high, all dissolved sulfate is consumed in the upper sediment column and microorganisms in deeper sediment produce methane.

The latter diffuses up toward the sulfate-rich sediments where it is consumed by sulfate-reducing methane oxidation. Assuming that in the sulfate–methane transition zone all sulfate is reduced by methane oxidation, the upward diffusive flux of methane, and thus the total rate of microbial activity within the methanogenic zone, can be estimated from the downward flux of dissolved sulfate into the sediment. Using this approach, D'Hondt *et al.* (2002) inferred that the annual metabolic activity of the sedimentary subseafloor is extremely low relative to the annual metabolic activity in the overlying ocean.

Furthermore, although the biomass in subseafloor sediments is two orders of magnitude greater than in ocean waters, the annual rate of subseafloor respiration is two or more orders of magnitude lower than the annual rate of biomass production in the sunlit water of the ocean surface. By combining the estimate of annual respiration rates with cell numbers in deep subseafloor sediments, the mean metabolic activity of individual cells can be estimated. For deep subseafloor sediments in the eastern Pacific, D'Hondt *et al.* (2004) estimated mean sulfate respiration to be around 10^{-18} mol cell^{-1} year^{-1}. Considering the minimum amount of energy needed for reproduction, the theoretical mean generation time of deep subseafloor microorganisms can be expected to exceed 1000 years (Jørgensen and D'Hondt, 2006). Nevertheless, the *in situ* metabolic activity of subseafloor prokaryotes is spectacularly visible in the accumulation of biogenic methane in deeply buried, massive gas hydrates along continental margins (e.g. Kvenvolden, 1995). Moreover, the slow activity of the subseafloor biosphere may directly affect the surface of Earth, e.g. by consuming the greenhouse gas methane in sulfate–methane transition zones.

17.5 Adaptations to Life in Deep Subseafloor Environments

To date, very little is known about the specific adaptations that allow microorganisms living in the deep biosphere to survive under the extreme conditions that exist (Section 17.3.3), in particular the lack of energy. We do, however, know that Archaea and Bacteria are the dominant life forms in the deep biosphere and Archaea constitute a major fraction of the biomass in deep subseafloor sediments.

The first Archaea ever described thrived in conditions that were characterized by high salinity, high temperature and acidity, or strict anoxia, and this led to the assumption that Archaea require extreme conditions. However, this assumption was abandoned as Archaea began to be discovered in a wide range of non-extreme environments as well (Chaban *et al.*, 2006). Many bacteria are also known to thrive under extreme environmental conditions (see other chapters in this volume) and conditions of low energy flux (e.g. Jackson and McInerney, 2002; Adams *et al.*, 2006), and have evolved many adaptations to do so. Nevertheless, diverse evidence now indicates that Archaea are better adapted to chronic energy stress than Bacteria and this may explain their dominance in the deep biosphere. Indeed, Valentine (2007) proposes that chronic energy stress is the primary selective pressure governing archaeal evolution.

Whereas Bacteria seem to focus less on adapting to energy stress, and more on exploiting new or variable resources, Archaea have evolved the ability to grow in extreme conditions in which Bacteria and eukaryotes cannot. Various biochemical adaptations are thought to be responsible (reviewed by Valentine, 2007) including lipid membrane composition, catabolic specificity and energy conservation.

A major distinction between the Archaea and the Bacteria is the chemical structure of lipids composing the cytoplasmic membrane. Bacterial lipids typically consist of fatty acids esterified to a glycerol moiety, whereas archaeal lipids typically consist of isoprenoidal alcohols that are ether-linked to glycerol (Valentine, 2007). As a result, archaeal membranes are less permeable to ions than bacterial membranes and reduce the amount of futile ion cycling *in vivo*. Archaea therefore lose less energy during the maintenance of a chemiosmotic potential than Bacteria (Valentine, 2007) making them competitively superior in energy-limited environments.

Known catabolic functions among Archaea include aerobiosis, fermentation, nitrification, methanogenesis, phototrophy and numerous forms of lithotrophy (Chaban *et al.*, 2006). In extreme conditions, Archaea also rely on these unique catabolic pathways to exclude or out-compete Bacteria.

Last but not least, distinctive mechanisms of energy conservation are characteristic of many Archaea. Such mechanisms include methanogenesis, anaerobic methane oxidation, proton reduction coupled directly to proton translocation, H_2-dependent sulfur reduction and phototrophy (Sapra *et al.*, 2003; Chaban *et al.*, 2006; Valentine, 2007). Valentine (2007) proposes that this variability is an evolutionary tactic that Archaea employ to balance the energetic needs of the cell with the availability of energy in the local environment and, once again, it confers a competitive advantage over Bacteria and other organisms under a wide range of environmental conditions.

However, our knowledge is still very limited and much remains unknown about the specific adaptations of the inhabitants of the deep biosphere. The field of research lies wide open for future investigation and discovery.

17.6 Was the Origin of Life 'Deep'?

The idea that life originated on the surface of our planet, where it was strongly dependent on a hypothetical primordial soup, has recently come up against strong competition (Pedersen, 2000). There are now several suggestions that life originated in the form of a thermophilic lithotroph (Huber and Wächtershäuser, 1997) and that the birthplace was intraterrestrial, perhaps a hydrothermal vent area (Russell and Hall, 1997). In experiments modelling the reactions of the reductive acetyl-coenzyme A pathway at hydrothermal temperatures, Huber and Wächtershäuser (1997) showed that an aqueous slurry of co-precipitated nickel- and iron sulfides (NiS and FeS) converted carbon monoxide (CO) and methanethiol (CH_3SH) into the activated thioester CH_3-CO-SCH_3, which hydrolysed to acetic acid. In the presence of aniline, acetanilide was formed. When NiS-FeS was modified with catalytic amounts of selenium, acetic acid and CH_3SH were formed from CO and H_2S alone. The reaction could be considered the primordial initiation reaction for a chemoautotrophic origin of life (Huber and Wächtershäuser, 1997). Russell and Hall (1997) went on to argue that the synthesis of such 'organic primers', the precursors for primordial life, potentially occurred ca. 4.2 billion years ago at an oceanic redox and pH front when hot (ca. 150°C), extremely reduced, alkaline, bisulfide-bearing, submarine seepage waters interfaced with the acidic, warm (ca. 90°C), iron-bearing Hadean ocean. This would have led to the spontaneous precipitation of a colloidal iron sulfide (FeS) membrane laced with nickel. They suggest that this precipitate acted as a semipermeable catalytic boundary between the seep fluid and ocean waters, and encouraged the synthesis of organic anions by hydrogenation and carboxylation of hydrothermal organic primers. Thus, the role of the deep biosphere and subsurface environments has come under increased scrutiny from evolutionary scientists and may hold a key to our understanding of how life on Earth evolved.

17.7 Search for Extraterrestrial Life in Subsurface Samples from Other Planets

Subsurface life may also prove to be widespread among the planetary bodies of our solar system since, while having inhospitable surfaces, many have subsurface conditions that have been demonstrated on Earth to be suitable for life. For example, temperature and pressure regimes in the interior of many planets might allow liquid water to exist and this, combined with the presence of hydrocarbon compounds, would appear to be the universal prerequisite for life (see Gómez *et al.*, Chapter 26, this volume).

Furthermore, niches that harbour 'deep' life on Earth are also present on other planets; for instance ancient deposits of chloride salts have been discovered scattered across the surface of Mars (HiRISE, 2007) and these could plausibly support life just as ancient salt deposits on Earth have been shown to (Vreeland *et al.*, 1999; Section 17.2.3).

Gold (1992) went further, speculating that subsurface life may in fact be widely distributed throughout the Universe and beyond on planetary-type bodies with suitable subsurface conditions existing as solitary objects in space. If this were the case, Panspermia, the theory of the distribution of life throughout the Universe (McKay *et al.*, 1996) in a protected subsurface environment on meteoroids, asteroids and planetoids over interplanetary and interstellar distances, would become an exciting and plausible possibility (see also Gómez *et al.*, Chapter 26, this volume).

17.8 Conclusions

To date, relatively little is known about the individual inhabitants of the deep biosphere, about their genetic diversity, metabolic capabilities and survival strategies. Moreover, the relationships between, for example, deep subseafloor ecosystems and both the overlying photosynthetic surface world and the underlying oceanic crust, remain to be explored. While electron donors might be supplied in the form of buried organic matter from the surface world, as well as in the form of reduced minerals and thermogenic methane from deep within the Earth, prokaryotes in the subseafloor ultimately rely on electron acceptors from the photosynthetically-oxidized surface world (oxygen, nitrate, sulfate) that enter these sediments by diffusion down past the seafloor and by transport upward from seawater flowing through the underlying basalts. The existence of practically non-growing populations of microorganisms in the deep subseafloor also challenges our understanding of life and evolution on Earth and other planets.

Our increasing scientific interest in the deep biosphere goes well beyond our desire to simply understand different environments on Earth. The continuing contamination of groundwater through the creation of toxic waste dumps and municipal landfills, industrial chemicals, pesticides and fertilizers, and the mobilization of toxic metals by acid rains have created serious environmental problems that demand the attention of the scientific community (Pedersen, 1993, 2000). Moreover, many countries are using, or are seriously considering using, subsurface environments as a repository for low to high level nuclear waste and carbon sequestration. Enhancing our knowledge of deep biosphere ecosystems will enhance our chances of mitigating such environmental problems, for instance by restoring contaminated groundwater environments via bacterial bioremediation (Bouwer, 1992), and could allow us to pursue subsurface dumping and storage of environmental pollutants more safely (e.g. Stroes-Gascoyne and Sargent, 1998).

Moreover, 55 million years ago a global warming catastrophe occurred as a result of rapid melting of methane gas hydrates (Norris and Röhl, 1999), demonstrating the delicate balance of our global climate. This should be taken as a serious warning (Pedersen, 2000). Continued rapid melting of Arctic (e.g. Vinnikov *et al.*, 1999; Holland *et al.*, 2006) and Antarctic sea ice could potentially trigger a similar, modern catastrophe: a sobering example of how microbes in the deep biosphere may exert an enormous influence on global processes (Pedersen, 2000).

References

Adams, C.J., Redmond, M.C. and Valentine, D.L. (2006) Pure-culture growth of fermentative bacteria, facilitated by H_2 removal: bioenergetics and H_2 production. *Applied and Environmental Microbiology* 72, 1079–1085.

Bach, W. and Edwards, K.J. (2003) Iron and sulfide oxidation within the basaltic ocean crust: implications for chemolithoautotrophic microbial biomass production. *Geochimica et Cosmochimica Acta* 67, 3871–3887.

Biddle, J.F., Lipp, J.S., Lever, M.A., Lloyd, K.G., Sørensen, K.B., Anderson, R., Fredricks, H.F., Elvert, M., Kelly, T.J., Schrag, D.P., Sogin, M.L., Brenchley, J.E., Teske, A., House, C.H. and Hinrichs, K.-U. (2006) Heterotrophic Archaea dominate sedimentary subsurface ecosystems off Peru. *Proceedings of the National Academy of Sciences of the United States of America* 103, 3846–3851.

Bouwer, E.J. (1992) Bioremediation of organic contaminants in the subsurface. In: Mitchell, R.E. (ed.) *Environmental Microbiology*. Wiley-Liss, New York, pp. 287–318.

Chaban, B., Ng, S.Y.M. and Jarrell, K.F. (2006) Archaeal habitats – from the extreme to the ordinary. *Canadian Journal of Microbiology* 52, 73–116.

Chandler, D.P., Li, S.M., Spadoni, C.M., Drake, G.R., Balkwill, D.L., Fredrickson, J.K. and Brockman, F.J. (1997) A molecular comparison of culturable aerobic heterotrophic bacteria and 16S rDNA clones derived from a deep subsurface sediment. *FEMS Microbiology Ecology* 23, 131–144.

Colwell, F.S., Onstott, T.C., Delwiche, M.E., Chandler, D., Fredrickson, J.K., Yao, Q.-J., McKinley, J.P., Boone, D.R., Griffiths, R. And Phelps, T.J. (1997) Microorganisms from deep, high temperature sandstones: constraints on microbial colonization. *FEMS Microbiology Reviews* 20, 425–435.

Crozier, R.H., Agapov, P.-M. and Pedersen, K. (1999) Towards complete biodiversity assessment: an evaluation of the subterranean bacterial communities in the Oklo region of the sole surviving natural nuclear reactor. *FEMS Microbiology Ecology* 28, 325–334.

Cunningham, K.I., Northup, D.E., Pollastro, R.M., Wright, W.G. and LaRock, E.J. (1995) Bacteria, fungi and biokarst in lechuguilla caves, Carlsbad Caverns National Park, New Mexico. *Environmental Geology* 25, 2–8.

D'Hondt, S., Rutherford, S. and Spivack, A.J. (2002) Metabolic activity of subsurface life in deep-sea sediments. *Science* 295, 2067–2070.

D'Hondt, S., Jørgensen, B.B., Miller, D.J., Batzke, A., Blake, R., Cragg, B.A., Cypionka, H., Dickens, G.R., Ferdelman, T., Hinrichs, K.-U., Holm, N.G., Mitterer, R., Spivack, A., Wang, G., Bekins, B., Engelen, B., Ford, K., Gettemy, G., Rutherford, S.D., Sass, H., Skilbeck, C.G., Aiello, I.W., Guèrin, G., House, C.H., Inagaki, F., Meister, P., Naehr, T., Niitsuma, S., Parkes, R.J., Schippers, A., Smith, D.C., Teske, A., Wiegel, J., Padilla, C.N. and Acosta, J.L.S. (2004) Distr butions of microbial activities in deep subseafloor sediments. *Science* 306, 2216–2221.

Ekendahl, S., Arlinger, J., Ståhl, F. and Pedersen, K. (1994) Characterization of attached bacterial populations in deep granitic groundwater from the Stripa research mine with 16S-rRNA gene sequencing technique and scanning electron microscopy. *Microbiology* 140, 1575–1583.

Elvert, M., Suess, E. and Whiticar, M.J. (1999) Anaerobic methane oxidation associated with marine gas hydrates: superlight C-isotopes from saturated and unsaturated C-20 and C-25 irregular isoprenoids. *Naturwissenschaften* 86, 295–300.

Fisk, M.R., Giovannoni, S.J. and Thorseth, I.H. (1998) Alteration of oceanic volcanic glass: textural evidence of microbial activity. *Science* 281, 978–980.

Fredrickson, J.K., McKinley, J.P., Bjornstad, B.N., Long, P.E., Ringelberg, D.B., White, D.C., Krumholz, L.R., Suflita, J.M., Colwell, F.S., Lehman, R.M. and Phelps, T.J. (1997) Pore-size constraints on the activity and survival of subsurface bacteria in a late Cretaceous shale-sandstone sequence, northwestern New Mexico. *Geomicrobiology Journal* 14, 182–202.

Froelich, P.N., Klinkhammer, G.P., Bender, M.L., Luedtke, N.A., Heath, G.R., Cullen, D., Dauphin, P., Hammond, D., Hartman, B. and Maynard, V. (1979) Early oxidation of organic matter in pelagic sediments of the eastern equatorial Atlantic: suboxic diagenesis. *Geochimica et Cosmochimica Acta* 43, 1075–1090.

Fry, J.C., Parkes, R.J., Cragg, B.A., Weightman, A.J. and Webster, G. (2008) Prokaryotic biodiversity and activity in the deep subseafloor biosphere. *FEMS Microbiology Ecology* 66, 181–196.

Furnes, H. and Staudigel, H. (1999) Biological mediation in ocean crust alteration: how deep is the deep biosphere? *Earth and Planetary Science Letters* 166, 97–103.

Gold, T. (1992) The deep, hot biosphere. *Proceedings of the National Academy of Sciences of the United States of America* 89, 6045–6049.

Heuer, V.B., Pohlman, J.W., Torres, M.E., Elvert, M. and Hinrichs, K.-U. (2009) The stable carbon isotope biogeochemistry of acetate and other dissolved carbon species in deep subseafloor sediments at the northern Cascadia Margin. *Geochimica et Cosmochimica Acta* 73, 3323–3336.

Hinrichs, K.-U., Hayes, J.M., Sylva, S.P., Brewer, P.G. and DeLong, E.F. (1999) Methane-consuming Archaebacteria in marine sediments. *Nature* 398, 802–805.

Hinrichs, K.-U., Hayes, J.M., Bach, W., Spivack, A.J., Hmelo, L.R., Holm, N.G., Johnson, C.G. and Sylva, S.P. (2006) Biological formation of ethane and propane in the deep marine subsurface. *Proceedings of the National Academy of Sciences of the United States of America* 103, 14684–14689.

HiRISE (2007) Possible ancient salt deposits within unnamed crater in Terra Cimmeria. Available at: http://hirise.lpl.arizona.edu/PSP_005680_1525 (accessed 13 May 2011).

Holland, M.M., Bitz, C.M. and Tremblay, B. (2006) Future abrupt reductions in the summer Arctic sea ice. *Geophysical Research Letters* 33, L23503, doi:10.1029/2006GL028024.

Huber, C. and Wächtershäuser, G. (1997) Activated acetic acid by carbon oxidation on (Fe, Ni)S under primordial conditions. *Science* 276, 245–247.

Hugenholtz, P., Goebel, B.M. and Pace, N.R. (1998) Impact of culture-independent studies on the emerging phylogenetic view of bacterial diversity. *Journal of Bacteriology* 180, 4765–4774.

Jackson, B.E. and McInerney, M.J. (2002) Anaerobic microbial metabolism can proceed close to thermodynamic limits. *Nature* 415, 454–456.

Jannasch, H.W. and Wirsen, C.O. (1973) Deep-sea microorganisms – *in-situ* response to nutrient enrichment. *Science* 180, 641–643.

Jannasch, H.W., Eimhjell, K., Wirsen, C.O. and Farmanfa, A. (1971) Microbial degradation of organic matter in deep sea. *Science* 171, 672–675.

Jørgensen, B.B. and Boetius, A. (2007) Feast and famine – microbial life in the deep-sea bed. *Nature Reviews Microbiology* 5, 770–781.

Jørgensen, B.B. and D'Hondt, S. (2006) A starving majority deep beneath the seafloor. *Science* 314, 932–943.

Kashefi, K. and Lovley, D.R. (2003) Extending the upper temperature limit for life. *Science* 301, 934.

Kieft, T.L., McCuddy, S.M., Onstott, T.C., Davidson, M., Lin, L.H., Mislowack, B., Pratt, L., Boice, E., Lollar, B.S., Lippmann-Pipke, J., Pfiffner, S.M., Phelps, T.J., Gihring, T., Moser, D. and van Heerden, A. (2005) Geochemically generated, energy-rich substrates and indigenous microorganisms in deep, ancient groundwater. *Geomicrobiology Journal* 22, 325–335.

Kvenvolden, K. (1995) Natural gas hydrate occurrence and issues. *Sea Technology* 36, 69–74.

Lin, L.H., Slater, G.F., Lollar, B.S., Lacrampe-Couloume, G. and Onstott, T.C. (2005) The yield and isotopic composition of radiolytic H_2, a potential energy source for the deep subsurface biosphere. *Geochimica et Cosmochimica Acta* 69, 893–903.

Lin, L.H., Wang, P.L., Rumble, D., Lippmann-Pipke, J., Boice, E., Pratt, L.M., Lollar, B.S., Brodie, E.L., Hazen, T.C., Andersen, G.L., DeSantis, T.Z., Moser, D.P., Kershaw, D. and Onstott, T.C. (2006) Long-term sustainability of a high-energy, low-diversity crustal biome. *Science* 314, 479–482.

Lipp, J.S. and Hinrichs, K.-U. (2009) Structural diversity and fate of intact polar lipids in marine sediments. *Geochimica et Cosmochimica Acta* 73, 6816–6833.

Lipp, J.S., Morono, Y., Inagaki, F. and Hinrichs, K.-U. (2008) Significant contribution of Archaea to extant biomass in marine subsurface sediments. *Nature* 454, 991–994.

McKay, D.S., Gibson Jr, E.K., Thomas-Keprta, K.L., Vali, H., Romaneck, C.S., Clemett, S.J., Chillier, X.D.F., Maechling, C.R. and Zare, R.N. (1996) Search for past life on Mars: possible relic biogenic activity in Martian meteorite ALH84001. *Science* 273, 924–930.

Motamedi, M. and Pedersen, K. (1998) *Desulfovibrio aespoeensis* sp. nov., a mesophilic sulfate-reducing bacterium from deep groundwater at Äspö Hard Rock Laboratory, Sweden. *International Journal of Systematic Bacteriology* 48, 311–315.

Norris, R.D. and Röhl, U. (1999) Carbon cycling and chronology of climate warming during the Palaeocene/Eocene transition. *Nature* 401, 775–778.

Parkes, R.J., Cragg, B.A., Bale, S.J., Getliff, J.M., Goodman, K., Rochelle, P.A., Fry, J.C., Weightman, A.J. and Harvey, S.M. (1994) Deep bacterial biosphere in Pacific-Ocean sediments. *Nature* 371, 410–413.

Parkes, R.J., Cragg, B.A. and Wellsbury, P. (2000) Recent studies on bacterial populations and processes in subseafloor sediments: a review. *Hydrogeology Journal* 8, 11–28.

Parkes, R.J., Webster, G., Cragg, B.A., Weightman, A.J., Newberry, C.J., Ferdelman, T.G., Kallmeyer, J., Jorgensen, B.B., Aiello, I.W. and Fry, J.C. (2005) Deep subseafloor prokaryotes stimulated at interfaces over geological time. *Nature* 436, 390–394.

Parkes, R.J., Wellsbury, P., Mather, I.D., Cobb, S.J., Cragg, B.A., Hornibrook, E.R.C. and Horsfield, B. (2007) Temperature activation of organic matter and minerals during burial has the potential to sustain the deep biosphere over geological timescales. *Organic Geochemistry* 38, 845–852.

Pearson, A. (2008) Biogeochemistry – who lives in the sea floor? *Nature* 454, 952–953.

Pedersen, K. (1993) The deep subterranean biosphere. *Earth-Science Reviews* 34, 243–260.

Pedersen, K. (1997) Microbial life in granitic rock. *FEMS Microbiology Reviews* 20, 399–414.

Pedersen, K. (2000) Exploration of deep intraterrestrial microbial life: current perspectives. *FEMS Microbiology Letters* 185, 9–16.

Pedersen, K. (2001) Diversity and activity of microorganisms in deep igneous rock aquifers of the Baltic Shield. In: Fredrickson, J. and Fletcher, M. (eds) *Subsurface Microbiology and Biogeochemistry*. John Wiley, New York, 352 pp.

Pedersen, K. and Ekendahl, S. (1990) Distribution and activity of bacteria in deep granitic groundwaters of southeastern Sweden. *Microbial Ecology* 20, 37–52.

Pedersen, K., Arlinger, J., Ekendahl, S. and Hallbeck. L. (1996) 16S rRNA gene diversity of attached and unattached groundwater bacteria along the access tunnel to the Äspö Hard Rock Laboratory, Sweden. *FEMS Microbiology Ecology* 19, 249–262.

Pedersen, K., Hallbeck, L., Arlinger, J., Erlandson, A.-C. and Jahromi, N. (1997) Investigation of the potential for microbial contamination of deep granitic aquifers during drilling using 16S rRNA gene sequencing and culturing methods. *Journal of Microbiological Methods* 30, 179–192.

Pohlman, J.W., Kaneko, M., Heuer, V.B., Coffin, R.B. and Whiticar, M. (2009) Methane sources and production in the northern Cascadia margin gas hydrate system. *Earth and Planetary Science Letters* 287, 504–512.

Roussel, E.G., Cambon Bonavita, M.-A., Querellou, J., Cragg, B.A., Webster, G., Prieur, D. and Parkes, R.J. (2008) Extending the sub-sea-floor biosphere. *Science* 320, 1046.

Russell, M.J. and Hall, A.J. (1997) The emergence of life from iron monosulphide bubbles at a submarine hydrothermal redox and pH front. *Journal of the Geological Society of London* 154, 377–402.

Sapra, R., Bagramyan, K. and Adams, M.W.W. (2003) A simple energy-conserving system: proton reduction coupled to proton translocation. *Proceedings of the National Academy of Sciences of the United States of America* 100, 7545–7550.

Schippers, A., Neretin, L.N., Kallmeyer, J., Ferdelman, T.G., Cragg, B.A., Parkes, R.J. and Jorgensen, B.B. (2005) Prokaryotic cells of the deep subseafloor biosphere identified as living bacteria. *Nature* 433, 861–864.

Scholl, D., Barth, G., Childs, J. and Gibbons, H. (2007) Sub-Sea-Floor Methane in the Bering Sea – USGS Emeritus Describes Possible Gas-Hydrate Accumulations to the Geophysical Society of Alaska. Available at: http://soundwaves.usgs.gov/2007/04/research3.html (accessed 13 May 2011).

Schrenk, M.O., Huber, J.A. and Edwards, K.J. (2010) Microbial provinces in the subseafloor. *Annual Review of Marine Science* 2, 279–304.

Sherwood Lollar, B., Westgate, T.D., Ward, J.A., Slater, G.F. and Lacrampe-Couloume, G. (2002) Abiogenic formation of alkanes in the Earth's crust as a minor source for global hydrocarbon reservoirs. *Nature* 416, 522–524.

Sherwood Lollar, B., Lacrampe-Couloume, G., Slater, G.F., Ward, J., Moser, D.P., Ghiring, T.M., Lin, L.H. and Onstott, T.C. (2006) Unravelling abiogenic and biogenic sources of methane in the Earth's deep subsurface. *Chemical Geology* 226, 328–339.

Smithwick, R.W. and David, H.L. (1971) Acridine orange as a fluorescent counterstain with the auramine acid-fast stain. *Tubercle* 52, 226–231.

Stetter, K.O., Fiala, G., Huber, G. and Segerer, A. (1990) Hypothermophilic microorganisms. *FEMS Microbiology Reviews* 75, 117–124.

Stevens, T.O. and McKinley, J.P. (1995) Lithoautotrophic microbial ecosystem in deep basalt aquifers. *Science* 270, 450–453.

Stroes-Gascoyne, S. and Sargent, F.P. (1998) The Canadian approach to microbial studies in nuclear waste management and disposal. *Journal of Contaminant Hydrology* 35, 175–190.

Takai, K., Moser, D.P., DeFlaun, M., Onstott, T.C. and Fredrickson, J.K. (2001) Archaeal diversity in waters from deep South African gold mines. *Applied and Environmental Microbiology* 67, 5750–5760.

Teske, A.P. (2006) Microbial communities of deep marine subsurface sediments: molecular and cultivation surveys. *Geomicrobiology Journal* 23, 357–368.

Teske, A. and Sørensen, K.B. (2008) Uncultured Archaea in deep marine subsurface sediments: have we caught them all? *ISME Journal* 2, 3–18.

Valentine, D.L. (2007) Adaptations to energy stress dictate the ecology and evolution of the Archaea. *Nature Reviews Microbiology* 5, 316–323.

Verne, J. (1864) *Journey to the Centre of the Earth*, 1st edn. Pierre-Jules Hetzel.

Vinnikov, K.Y., Robock, A., Stouier, R.J., Walsh, J.E., Parkinson, C.L., Cavalieri, D.J., Mitchell, J.F.B., Garret, D. and Zakharov, V.F. (1999) Global warming and northern hemisphere sea ice extent. *Nature* 286, 1934–1937.

Vreeland, R.H., Piselli Jr, A.F., McDonnough, S. and Meyers, S.S. (1999) Distribution and diversity of halophilic bacteria in a subsurface salt formation. *Extremophiles* 2, 321–331.

Wellsbury, P., Goodman, K., Barth, T., Cragg, B.A., Barnes, S.P. and Parkes, R.J. (1997) Deep marine biosphere fuelled by increasing organic matter availability during burial and heating. *Nature* 388, 573–576.

Whitman, W.B., Coleman, D.C. and Wiebe, W.J. (1998) Prokaryotes: the unseen majority. *Proceedings of the National Academy of Sciences of the United States of America* 95, 6578–6583.

Wilms, R., Köpke, B., Sass, H., Chang, T.S., Cypionka, H. and Engelen, B. (2006) Deep biosphere-related bacteria within the subsurface of tidal flat sediments. *Environmental Microbiology* 8, 709–719.

18 Acidic Environments

Elly Spijkerman and Guntram Weithoff
Department of Ecology and Ecosystem Modelling, Institute for Biochemistry and Biology, Potsdam University, Germany

18.1 Introduction

Extremely acidic environments are defined as those that have a pH well below 4.0 (Rothschild and Mancinelli, 2001). In such environments a great many neutrophiles (species loving neutral pH habitats) cannot persist. For instance, Ntengwe (2005) documented the death of many organisms (e.g. fish and water hyacinth) exposed to acid mining drainage with pH 2.0. Many enzymes involved in basic metabolic processes cannot function at such a low pH and, usually, organisms maintain their intracellular pH close to neutral. As can be expected, only specialized organisms can survive, or indeed thrive, at low pH: the acidophiles (acid-loving) and to some extent the acidotolerant (acid-tolerating) species. Acidophiles include prokaryotes and Archaea, unicellular eukaryotes (Rothschild and Mancinelli, 2001). To withstand such high external concentrations of protons and maintain a neutral intracellular climate, acidophiles have evolved specialized features. However, pH is not their only challenge, in most acidic habitats several additional stresses affect species composition. Acidophilic organisms are often subjected to high temperatures (e.g. Yellowstone National Park, USA, with its geothermal features including geysers that have a temperature of 30–83°C and a pH of 2.7–3.7; Johnson *et al.*, 2003). They very often also experience high concentrations of heavy metals (e.g. lead and zinc), as is the case for e.g. mine tailings (Tan *et al.*, 2008), and in Lake Brown, Western Australia, low pH is combined with high salinity levels of 13.0–23.0% (Mormile *et al.*, 2007). Of course the term 'extreme' is relative; for example, for fish living at a pH of 7 or 8 in a freshwater lake or the ocean, a water body with a pH of 5 will be extremely acidic, whereas acidophilic or acidotolerant organisms would find this benign.

This chapter summarizes the different types of acidic environments that exist and their occurrence worldwide, and goes on to discuss the diversity of life present in them. We restrict ourselves to ecology or ecologically relevant traits, i.e. organisms living in 'natural habitats', summarizing food-web ecology, metabolic and physiological characteristics. Gastrointestinal life-forms and consortia are excluded from the general discussion, although the stomach typically has a pH between 1 and 3 (Burget, 1920) and was long considered too acidic to support a diverse community. Nevertheless, 128 different bacteria from eight phyla were found in the stomach of only 23 individuals and most of them were considered true inhabitants (Bik *et al.*, 2006). One of the phylotypes was a *Deinococcus*-related organism, relatives of which survive extremely high radiation doses (see Bell and Callaghan, Chapter 1, this volume). Interestingly, the stomach bacterium that causes gastric ulcers, *Helicobacter pylori*, is actually an acidotolerant species preferring a neutral environment.

18.2 Types of Acidic Environment

The acidophilic environments included in this chapter cover aquatic habitats, such as rivers and lakes. Most knowledge has been gathered from organisms living in acidic mining lakes (Fig. 18.1) or acidic drainage (e.g. Rio Tinto). In addition, life in both very acidic volcanic lakes and geysers (which are also hot) has been found; in the latter case eukaryotes mainly inhabit the benthos (surface of sediments) or are endolithotrophic (live within stone).

But how does the water in these environments become so acidic? Acidic lakes and rivers have a low pH because sulfur- (and often also iron-) oxidizing bacteria have weathered minerals rich in sulfur (and often also rich in iron, e.g. pyrite and marcasite; Fig. 18.2). Although the process also occurs abiotically, bacterial metabolism greatly increases its speed (Ingledew, 1990). The result of such weathering processes is sulfuric acid and high concentrations of metal ions buffering and thus, stabilizing, a low pH (Friese *et al.*, 1998).

(a)

(b)

Fig. 18.1. Mining Lake 107, Germany, which has a pH of 2.3. (a) The remains of trees present before the flooding of the mine pit are covered with a thick iron crust; (b) the water is a striking red colour due to the high iron concentrations. © Elly Spijkerman.

18.3 Biodiversity in Acidic Environments

There are a great number of studies that describe the species composition of extremely acidic environments. In general, the species diversity of acidic environments is lower than that of neutral habitats (Brown and Wolfe, 2006; Mendez *et al.*, 2008), supporting the hypothesis that only a limited number of species can thrive in these hostile locations. Most of the studies cover the diversity of bacteria and in a great many of them, no species names could be provided as DNA/RNA sequences did not fit with known species from the database (Bond *et al.*, 2000), or exact identity varied with the technique used (Bryan *et al.*, 2006). Despite this, it appears that different habitats harbour either the same and/or similar, i.e. closely related, species (Table 18.1). For example, similar prokaryotes and eukaryotes were identified in Burma, Chile, Germany and Australia (Hawkes *et al.*, 2006; Weithoff *et al.*, 2010).

An excellent overview of the prokaryotic diversity in very acidic habitats is provided in Johnson (1998). Recently, methanotrophs (methane consumers) have been added to the prokaryotic diversity and in 2007, two papers described the isolation of a methanotroph thriving in fumaroles (steam and gas emissions from cracks or fissures often near volcanoes) at temperatures

Fig. 18.2. Pyrite (FeS mineral), also known as fool's gold. © Elly Spijkerman.

of up to 70°C and a pH as low as 1.8 (Dunfield *et al.*, 2007; Pol *et al.*, 2007).

In the case of eukaryotic algae, the dominant classes of phytoplankton in acidic lakes are the Chlorophyceae genera *Chlorella, Chlamydomonas, Scourfeldia*, Chrysophyceae (*Ochromonas, Chromulina, Dinobryon*), Bacillariophyceae (*Eunotia, Fragilaria, Pinnularia*), Euglenophyta (*Lepocinclis teres*) and Dinophyta (*Peridinium, Gymnodinium*) (Nixdorf *et al.*, 2001). Filamentous algae belonging to the Xanthophyceae (*Bumilleria klebsiana, Heterococcus* sp.) and Chlorophyta (*Zygogonium ericetorum, Klebsormidium subtile*) inhabit the littoral zones. In addition to the filamentous algae, the benthos consists of *Euglena mutabilis, Eunotia exigua, Eunotia* cf. *denticulate* and *Nitzschia paleaeformis*. The whole acidic diatom flora, irrespective of temperature, has been revised by DeNicola (2000) who listed over ten genera living in habitats with pH <2.5. Incredibly, several diatom species have a lower pH limit of 1.5: *Pinnularia appendiculata, Pinnularia brebissoni, Pinnularia gentilis, Pinnularia biceps, Pinnularia interrupta* f. *minutissima, Pinnularia mesolepta* and *Pinnularia obscura* (Ciniglia *et al.*, 2007).

The most common macrophyte species in very acidic waters are submerged/floating water reeds, *Juncus bulbosus*, and shoreline stands of emergent *Phragmites australis, Typha latifolia* and *Juncus effusus* (Nixdorf *et al.*, 2001).

Heterotrophic Protozoa are an important component of the microzooplankton in acidic water bodies and they may reach high abundances (Woelfl, 2000). However, detailed taxonomic investigations are rare. Among ciliates, species of the genera *Chilodonella, Euplotes, Oxytricha, Stylonichia, Urotricha* and *Vorticella* have been documented (McConathy and Stahl, 1982; Packroff, 2000). These ciliates may exert a considerable grazing pressure on bacteria (Johnson and Rang, 1993). Another heterotrophic protist often encountered in acidic environments is the sarcodine

Table 18.1. Overview of species diversity in very acidic habitats.

Species name	Location	pH	Reference
Archaea			
Ferroplasma acidiphilum	Lechang Pb/Zn mine tailings (China)	1.9	Tan *et al.*, 2008
Sulfolobus metallicus, Acidianus brierleyi, Thermoplasma volcanium, Acidianus infernos	Acidic hydrothermal vents, Vulcano Island (Aeolians, Italy)	2–3.5	Simmons and Norris, 2002
Bacteria			
Leptospirillum ferriphilum, Acidithiobacillus ferrooxidans, Sulfobacillus thermotolerans, Acidiphilium cryptum	Lechang Pb/Zn mine tailings (China)	1.9	Tan *et al.*, 2008
Acidithiobacillus thiooxidans*	Mine tailing (Spain)	2.5–2.75	Rowe *et al.*, 2007
Fungi			
27 species (*Candida fluviatilis, Rhodosporidium toruloides, Williopsis californica, Cryptococcus* sp., *Lecytophora* sp.)	Iberian Pyrite Belt (Sao Domingos (Portugal) and Rio Tinto (Spain))	1.7–3	Gadanho *et al.*, 2006
350 strains (*Rhodotorula, Tremella, Holtermannia, Leucosporidium, Mrakia, Candida, Penicillium, Scytalidium, Bahusakala, Phoma, Heteroconium, Acremonium, Mortierella*)	Rio Tinto (Spain)	~2	Lopez-Archilla *et al.*, 2004
Eukaryotic algae			
Ochromonas sp., *Chlamydomonas* sp., *Gymnodinium* sp.	Lake 117 (Germany)	3	Beulker *et al.*, 2003
Euglena mutabilis, Pinnularia acoricola	Lake 107 (Germany)	2.3	Lessmann *et al.*, 1999
Cyanidium caldarium	Yellowstone's Norris Geyser Basin (USA)	1.0	Walker *et al.*, 2005
Zygogonium sp.	Acidic soils	1–4	Kristjansson and Hreggvidsson, 1995
Chlamydomonas acidophila, Pinnularia acidojaponica	Lake Katanuma (Japan)	1.8–2.2	Nishikawa and Tominaga, 2001; Doi *et al.*, 2003
Chlamydomonas, Dunaliella, Chlorella, Euglena mutabilis; *Zygnemopsis, Klebsormidium, Pinnularia, Bodo, Ochromonas*	Rio Tinto (Spain)	2.0	Aguilera *et al.*, 2007
Galdieria sp., *Cyanidium* sp., *Cyanidioschyzon* sp., *Chlamydomonas pitchmannii, Pinnularia* sp.	Piciarelli, Campi Flegrei (Italy)	1.0–1.8	Pinto *et al.*, 2007
Eunotia exigua, Pinnularia acoricola, Euglena mutabilis, Ulotrix tenerrima, Chlamydomonas sp.	Sao Domingos (Portugal)	2.25	Wolowski *et al.*, 2008
Keratococcus raphidioides	Lake Caviahue (Argentina)	2.56	Pedrozo *et al.*, 2001

Continued

Table 18.1. Continued.

Species name	Location	pH	Reference
Angiosperms			
Juncus bulbosus, Juncus effusus, Carex rostrata, Eriophorum angustifolium, Phragmites australis, Schoenoplectus lacustris, Typha latifolia, Typha angustifolia	Lusatian mining lakes (Germany) and English mine stream pool	2.1–3.0	Reviewed in Nixdorf *et al.*, 2001
Consumers			
Actinophrys sol (Heliozoa)	Lake 111 (Germany)	2.65	Bell *et al.*, 2006
Oxytrichia granulifera, Euplotes sp. (ciliates)	Rio Tinto (Spain)	2.0	Aguilera *et al.*, 2007
Oxytricha sp., *Vorticella* sp. (ciliates)	Lake Niemegk (Germany)	2.8	Woelfl, 2000
Cephalodella sp., *Elosa worallii, Brachionus sericus* (rotifers, Fig. 18.3)	Lake 130 (Germany)	2.6	Weithoff *et al.*, 2010
Rotatoria rotatoria, Lecane lunaris, Colurella sp., *Trichocerca similis, Lepadella* sp., *Philodina* sp., *Mytilina crassipes* (rotifers)	Different acidic water bodies	1.8–3.6	Deneke, 2000
Chidorus sphaericus, Simocephalus vetulus (Crustaceae)	Lake Osoresan-ko (Japan)	3.0	Deneke, 2000
Artocorisa germari, Cymatia rogenhoferi, Glaenocorisa sp., *Callicorixa praeusta, Corixa dentipes, Ilyocoris cimicoides, Sigara* spp. (Heteroptera)	Lusatian mining lakes (Germany)	<3.0	Wollmann, 2000
Chironomus acerbiphilus (Diptera)	Lake Katanuma (Japan)	2.2	Doi *et al.*, 2001
Galaxias maculates (fish)	Several streams in Westland (New Zealand)	3.9	Collier *et al.*, 1990

[a] *Acidothiobacillus* was formerly called *Thiobacillus*

(Heliozoa) *Actinophrys* (Bell *et al.*, 2006; Weithoff *et al.*, 2010).

The metazoan plankton community is mainly composed of rotifers and very few crustaceans. A number of rotifer species occur at very low pH and represent both major groups of rotifers (i.e. bdelloids and monogononts): *Philodina* sp., *Rotaria rotaria, Cephalodella* spp., *Elosa woralli* (Fig. 18.3), *Brachionus sericus* as well as some rarely encountered species (Pedrozo *et al.*, 2001). Crustaceans are very rare in extremely acidic water bodies, probably due the low availability of calcium carbonate. Only at pH levels ≥3 does the water flea, *Chydorus sphaericus*, occur (Deneke, 2000). Copepods are typically absent in extremely acidic lakes and rivers.

Insects may occur either as imagos or as larvae. Common predatory insects are corixids of the family Corixidae and Naucoridae (Wollmann, 2000). Bright red sediment-dwelling chironomid larvae are also often observed (e.g. *Chironomus acerbiphilus*; Doi *et al.*, 2001).

Higher invertebrates such as molluscs, or vertebrates such as fish, are absent in extremely acidic water bodies, although there are some references to fish encountered in acidic lakes with pH values slightly below 4 (e.g. *Galaxias maculatus*; Collier *et al.*, 1990) and considerably more fish species in waters ranging between pH 4.1 and 4.5 (Olsson *et al.*, 2006).

Table 18.1 provides an overview of the most common species found in extremely acidic environments worldwide; the list is not exhaustive.

18.4 Food Webs and Ecology

The lower species diversity and the absence of higher predators, such as fish, inhabiting acidic environments result in different pelagic (water column) food-web structures compared with those in neutral environments (Gaedke and Kamjunke, 2006). Although the ratio of heterotrophic to autotrophic biomass is often similar in acidic and neutral lakes, food webs in acidic lakes are largely restricted to two trophic levels, in contrast to the four levels found in neutral lakes (Gaedke and Kamjunke, 2006).

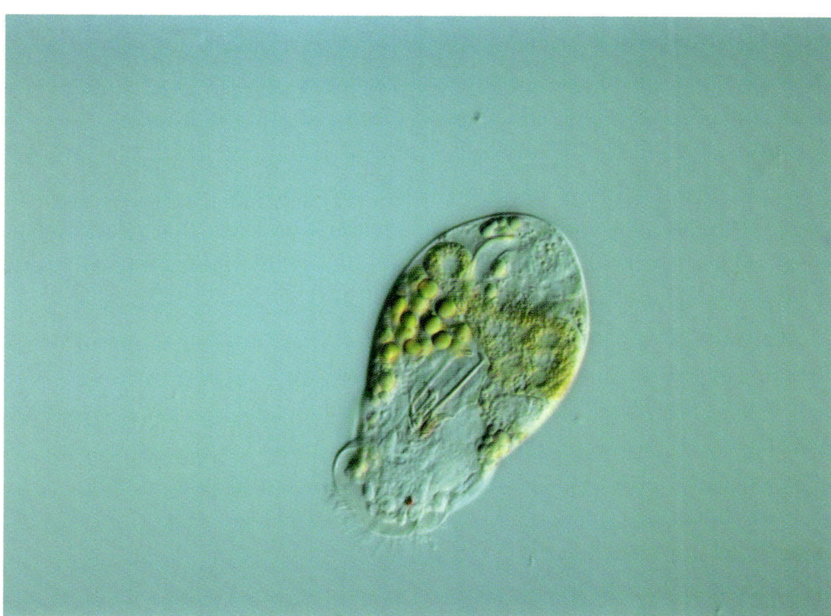

Fig. 18.3. *Elosa worallii* (Rotatoria) isolated from acidic Mining Lake 111, Germany. © Christina Schirmer.

Whilst in neutral lakes fish are the top predators, in the pelagic food web of, for example, the acidic mining Lake 111, Lausitz, Germany (pH 2.65), the heliozoan *Actinophrys sol* (protozoan) was the top predator (Kamjunke *et al.*, 2004; Bell *et al.*, 2006). In addition, benthic-pelagic coupling was very important for recruitment of organisms into the pelagic food web (Bell and Weithoff, 2003); this is not always the case in neutral environments.

Interestingly, it is difficult to group acidophiles into producers and consumers as many species exhibit mixotrophic traits. Mixotrophic organisms combine photosynthesis and the uptake of organic resources as modes of nutrition (Jones, 1994). Two mixotrophic strategies can be distinguished: osmo-mixotrophy and phago-mixotrophy. Osmo-mixotrophic organisms can take up dissolved organic carbon compounds of both autochtonous (*in situ*) and allochthonous (*ex situ*) origin. Phago-mixotrophics are able to ingest particles, e.g. bacteria or small algae. In algae, both types supplement their photosynthetic carbon fixation by the uptake of organic carbon when photosynthesis is limited. An example of an osmo-mixotroph is the green alga *Chlamydomonas acidophila*. In Lake 111, *C. acidophila* is an important primary producer (Kamjunke *et al.*, 2004), limited in its growth mainly by phosphorus (Spijkerman, 2008). The algae can use the natural lake organic carbon for growth in the light and in the dark and in the presence of bacterial competitors (Tittel *et al.*, 2009). The red alga *Galdieria sulphuraria* is also an osmo-mixotrophic alga which favours heterotrophic growth over autotrophic growth if sufficient organic carbon is available (Oesterhelt *et al.*, 2007). In general, many algal taxa growing in an extremely acidic environment can grow heterotrophically on organic compounds, which can be beneficial for survival in such a hostile chemical environment (e.g. Wolowski *et al.*, 2008).

An example of a phago-mixotroph alga is the crysophyte *Ochromonas* sp. In Lake 111, it was shown to graze 68–88% of daily bacterial production (Schmidtke *et al.*, 2006). Since bacteria typically have a lower C:P ratio (i.e. they are relatively rich in phosphorus) compared to algae, phago-mixotrophs benefit from bacterivory via the uptake of phosphorus rather than carbon (Gaedke *et al.*, 2002). This might be one reason for the high abundance of *Ochromonas* sp. in acidic mining lakes.

18.5 Metabolism and Physiology

Do all acidophiles use the same strategy to survive low pH? Acidophiles seem to share distinctive structural and functional characteristics including a reversed membrane potential, highly proton impermeable cell membranes and a predominance of secondary transporters. Furthermore, once protons enter the cytoplasm, methods are required to alleviate the effects of a lowered internal pH (see Baker-Austin and Dopson, 2007 for an excellent review on prokaryotic adaptation). Here we consider a few of the above-mentioned strategies.

18.5.1 Proton exclusion and extrusion (internal neutral pH)

Acidophiles maintain a near neutral intracellular pH, thereby realizing a pH gradient of several pH units across the cellular membrane. Intuitively, it would seem that acidophiles would be capable of using the ΔpH across the membrane to generate large amounts of adenosine triphosphate (ATP), but this would in fact result in rapid acidification of the cytoplasm. Therefore, any protons that enter the cell (through the F_0F_1 ATPase) need to be balanced by extrusion (Baker-Austin and Dopson, 2007). It is therefore believed that acidophiles have an enhanced ATP demand compared to neutrophiles. In acidophilic *Chlamydomonas acidophila*, the rate of ATP consumption was found to increase by 7% in a medium at pH 2 as compared to pH 7 (Messerli *et al.*, 2005). This enhanced ATP demand was not, however, reflected in an enhanced minimum cell quota for phosphorus (Q_0; 2.0–2.1 mg P g C^{-1}; Spijkerman *et al.*, 2007b); instead

Q_0 was similar to that of the neutrophile *Chlamydomonas reinhardtii* (2 mg P g C^{-1}; Lürling and van Donk, 1997).

All bacterial acidophiles contain a range of cytoplasmic buffer molecules that have basic amino acids (e.g. lysine, histidine and arginine) capable of sequestering protons (Baker-Austin and Dopson, 2007). Other buffering molecules present within the cell include the dihydrogen phosphate ion ($H_2PO_4^-$), which has a pKa of 7.2 and whose near-neutral pH is not affected by the addition or removal of protons. Alternatively, potassium can also be used as a buffering substance (Brock *et al.*, 1994) and in the alga *C. acidophila* potassium was accumulated 1000-fold under both K^+-replete and -deplete conditions (Spijkerman *et al.*, 2007b).

Several studies have confirmed a neutral intracellular pH of acidophiles (e.g. Gerloff-Elias *et al.*, 2006). An exception might be the Archaeon, *Ferroplasma acidiphilum*, with a mean intracellular pH of 5.6 (at extracellular pH between 0 and 3; Golyshina *et al.*, 2006). In addition, at temperatures of 70°C and pH 2.0 the extremely thermoacidophilic Archaeon *Metallosphaera sedula* was shown to have an intracellular pH of only 5.4, which decreased further to ca. pH 4.0 when the temperature was increased to 80–85°C (Peeples and Kelly, 1995).

For the freshwater crayfish (*Paranephrops planifrons* (Parastacidae)) low pH tolerance in environments below pH 4 is connected with the inhibition of sodium influx (unpublished results from O.J. Ball cited in Collier *et al.*, 1990).

18.5.2 Proton-induced damage to proteins

The uptake of protons will result in temporary decreases of intracellular pH. The generally higher accumulations of heat shock proteins (Hsp) or chaperones in extremophiles, including acidophiles, has been explained as an adaptation to stressful conditions, enabling fast and efficient repair of damage to protein structure and function (Laksanalamai and Robb, 2004). Moreover, on the gene-expression level, high chaperone expression was demonstrated in acidophiles (Baker-Austin and Dopson, 2007) and Gerloff-Elias *et al.* (2006) confirmed that in the acidophilic, eukaryotic alga, *Chlamydomonas acidophila*, higher concentrations of the heat shock proteins Hsp70, Hsp60 and small Hsps were accumulated than in the neutrophile *C. reinhardtii*. Moreover, Hsp accumulation increased further with decreasing pH (i.e. from pH 3 to 1.5). There are, however, no general rules in protein adaptation (as reviewed in Jaenicke, 2000).

Recently it was found that the proteome of the Archaeon *Ferroplama acidiphilum* contains a uniquely high proportion of iron proteins that might contribute to the pH stability of its enzymes at low pH (Ferrer *et al.*, 2007). Removal of the iron from six purified *F. acidiphilum* proteins resulted in loss of secondary structure and, consequently, protein activity. This led the authors to suggest that the iron is crucial in maintaining the protein's 3D structure and functions as an 'iron rivet' – potentially an evolutionarily ancient property that stabilizes proteins and has been retained in *F. acidiphilum* (Ferrer *et al.*, 2007). Interestingly, all of the *F. acidiphilum* enzymes analysed functioned and were stable *in vitro* in the pH range 1.7–4.0, and had pH optima much lower than the mean intracellular pH of 5.6 (Golyshina *et al.*, 2006). In addition, heat and acid stable α-amylase, cyclomaltodextrinase (neopullulanase), maltose binding protein and endoglucanase have been purified and characterized from the thermo-acidophilic bacterium, *Alicyclobacillus acidocaldarius*, suggesting that, extraordinarily, biotransformation reactions can be performed even at a pH close to zero and temperatures of 100°C (Bertoldo *et al.*, 2004).

18.5.3 Lipid and fatty acid composition

The cell wall is often the primary defence for acidophilic organisms and their constituents act as passive barriers between external and internal pH or buffering substances. Cell wall constituents, accumulation of buffering substances, and fatty acid and lipid composition, in acidophiles all largely differ from those in neutrophiles, and are adaptations observed in Archaea, Bacteria and Eukaryotes.

For example, a *Chlamydomonas* sp. isolated from Lake Katanuma, an acidic (pH 2.2) crater lake in northern Japan that formed following an eruption in the Nasu Volcanic group, had a relatively high accumulation of the lipid triacylglycerol (Yamamoto *et al.*, 1998). The relative percentage of triacylglycerol to the total lipid content in *Chlamydomonas* sp. grown in medium at pH 1 was higher than that in the same species grown at a higher pH (Tatsuzawa *et al.*, 1996). Triacylglycerol is known to decrease membrane lipid fluidity and as a storage lipid. Its accumulation can prevent the osmotic imbalance caused by high concentrations of H_2SO_4. Similarly, the fatty acids in polar lipids from the acidophilic *Chlamydomonas* species were also more saturated than those of neutrophilic species (Tatsuzawa *et al.*, 1996).

A further example is the specific, novel lipid, caldarchaetidylglycerol tetraether that was identified in cell wall-lacking Archaea of the genus *Ferroplasma*. This lipid rendered the cell membrane highly impermeable to protons effectively protecting internal pH (Golyshina and Timmis, 2005).

The extreme chemical conditions present in extremely acidic environments have also resulted in a decreased bio-availability of phosphorus for the growth of algae and bacteria. As a result, biofilm consortia of periphyton and bacteria in the Rio Tinto (Sabater *et al.*, 2003), as well as phytoplankton in various acidic mining lakes (Spijkerman, 2008), are phosphorus limited and exhibit high phosphatase activity independent of external phosphorus concentrations. For instance, laboratory cultures of acidophilic *Chlamydomonas acidophila* incubated in filter-sterilized lake water showed several signs of phosphorus limitation (Fig. 18.4) and the cells were shown to have strongly fragmented chloroplasts potentially as a result of the large number of starch globules they contained. Starch is a storage product for sugars produced via photosynthesis in green algae. These storage products accumulate in the chloroplasts of acidophiles because phosphorus limitation results in an overall decrease in RNA and DNA production (and thus lower cell division rates) whilst photosynthesis is still taking place.

18.5.4 Uptake of nutrients

Many species of bacteria isolated from acidic habitats grow on yeast extract or on glucose. Growth of the thermo-acidophilic Archaeon *Sulfolobus shibatae*, was observed on maltose and one (the low-affinity) uptake system was proton-dependent (Yallop and Charalambous, 1996). However, incorporated protons need extrusion and it is therefore not surprising that numerous proton-driven secondary transporters are present in acidophilic microorganisms (the overall ratio of secondary to primary transporters present on the *Picrophilus torridus* and *Thermoplasma acidophilum* genomes are 10:1 and 5.6:1, respectively; Futterer *et al.*, 2004). These transporters pump H^+ outside the cell and necessary solutes inside and are considered an adaptation to acidic conditions (Baker-Austin and Dopson, 2007).

18.5.5 High metal concentrations

Acidophilic microorganisms inhabit some of the most metal-rich environments known, both natural and man-made, and as such are ideal model systems for the study of microbial metal resistance. It has been suggested that metal tolerance is attributed to active secretion of ions (Weber *et al.*, 2007). The acidophilic Gram-negative bacterium *Acidithiobacillus caldus* was found to be resistant to the arsenic ions arsenate, arsenite, and antimony due to an inducible, chromosomally encoded resistance mechanism (Dopson *et al.*, 2001). In bacteria, the five metal resistance mechanisms identified in neutrophiles were also present in acidophiles, in some cases utilizing homologous proteins, but in many cases the degree of resistance was found to be greater in acidophiles (Dopson *et al.*, 2003).

In *Chlamydomonas acidophila*, a high tolerance for zinc, cadmium, copper and

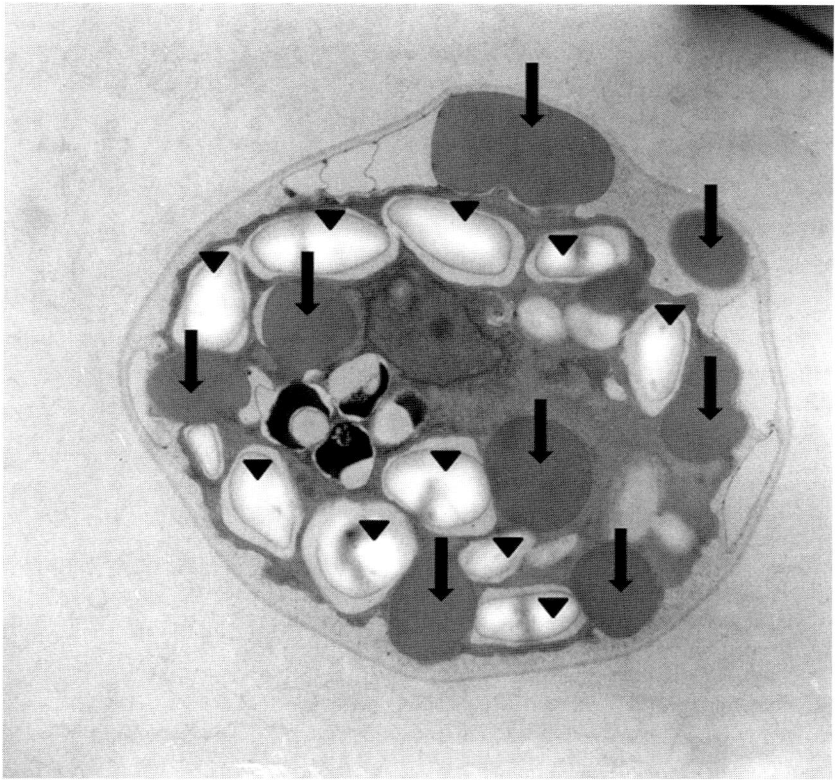

Fig. 18.4. Transmission electron micrograph of the acidophile *Chlamydomonas acidophila* grown in water from Lake 107, Germany (pH 2.3), for 2 weeks. Notice the strongly fragmented chloroplast (arrows) and the starch globules (arrow heads). © Klaus Hausmann.

cobalt was also described (Nishikawa and Tominaga, 2001) and, in another isolate of this alga, a high tolerance for zinc, manganese and aluminium was observed, although a reduction in the growth and the maximum quantum yield for photosynthesis was measured after culturing the species in iron-rich medium (Spijkerman *et al.*, 2007a). Furthermore, the periphytic green alga *Stigeoclonium tenue*, isolated from ditches containing mining drainage water, was zinc tolerant. It was able to grow and reproduce at external zinc concentrations ≥1 mg Zn l^{-1} (Pawlik-Skowronska, 2003). The acidophilic protozoan euglenid, *Euglena mutabilis*, can grow at extremely high iron concentrations (0.7–1.7 g Fe l^{-1}; Casiot *et al.*, 2004) and has the ability to bio-accumulate iron (Mann *et al.*, 1987). *Euglena mutabilis* also contributes to the formation of iron-rich stromatolites by releasing intracellularly stored iron compounds after death, contributing to the solid material of stromatolites and acting as a nucleation site for precipitation of authigenic (generated *in situ*) iron minerals (Brake *et al.*, 2002). In addition, the acidophilic red alga *Cyanidium caldarium* also accumulated iron in acidic mining pits (Nagasaka *et al.*, 2003).

Little is known on the combined effect of metal ions and low pH on zooplankton. A study conducted using several littoral crustaceans revealed that the harmful effect of low pH can be enhanced by high concentrations of aluminium (Havens, 1991). However, the total cation concentrations in these experiments were much lower than the typical concentrations found in acidic mining lakes. It may be that high concentrations of less toxic cations counteract the harmful effects of toxic cations, leading to an enhanced pH-tolerance.

18.6 The Application of Acidophiles

As discussed in Arora and Bell (Chapter 25, this volume) extremophiles have many potential and realized industrial and biotechnological applications. Some examples follow.

18.6.1 Biodegradation of complex molecules

Acidophiles have been successfully employed in biodegradation processes. For example, they are sometimes isolated to grow with (and degrade) aromatic compounds, or complex molecules, such as aliphatic acids, for use as biological treatment agents in acidic waste waters containing both organic and inorganic pollutants, a simpler, cheaper process than traditionally used (Hallberg *et al.*, 1999; Gemmell and Knowles, 2000). In addition, an indigenous, acidophilic microbial community (pH 2.0) isolated from surface water and soil samples from a coal runoff basin at the Westinghouse Savannah River Laboratory site in Aiken, USA, was able to oxidize more than 40% of the hydrocarbon contaminants present (naphthalene and toluene) into carbon dioxide and water (Stapleton *et al.*, 1998), opening up opportunities for hydrocarbon spill remediation.

18.6.2 Bioremediation of iron- and sulfate-rich lakes

Microorganisms indigenous to highly acidic, metalliferous environments are highly diverse, and include bacteria that catalyse the reduction as well as the oxidation of iron and sulfur. Since these reduction reactions can generate alkalinity, they have potential to be employed in the biological remediation of acid mine drainage. For example, *Acidocella facilis, Acidobacterium capsulatum* and all of the *Acidiphilium, Acidocella* and (moderate acidophiles) *Acidobacterium*-like isolates, grown in liquid cultures were able to reduce iron (Coupland and Johnson, 2008). In addition, *Ferroplasma acidarmanus* couples chemoorganotrophic growth on yeast extract to the reduction of ferric iron, between 35 and 42°C and pH 1.0–1.7 (Dopson *et al.*, 2004). In acidic mining-impacted lake sediments, the reduction of ferric (Fe^{3+}) predominated over the reduction of sulfate as long as the sediment remained acidic and carbon-limited (Kusel and Dorsch, 2000). High sulfate surface water is, however, undesirable as it is unsuitable for drinking water purposes. Therefore, the alkalinization of extremely acidic mining lakes probably relies on organic carbon inputs to provide electron donors for microbially induced sulfate reduction. It has been suggested that plant material such as that derived from the pioneer plant, the Bulbous Rush, *Juncus bulbosus*, is an important source of organic matter in lake sediments (Chabbi and Rumpel, 2004). Another study emphasized the general importance of benthic organic carbon production in providing electron donors for alkalization (Kamjunke *et al.*, 2006).

In the future, it would be better to prevent sulfur and iron oxidation altogether during mining activities. To this end, a novel technique termed 'bioshrouding' to safeguard highly reactive, sulfidic mineral tailing deposits has been proposed. In this process, freshly milled wastes are colonized with ferric iron-reducing, heterotrophic, acidophilic bacteria that form biofilms on reactive mineral surfaces, thereby preventing or minimizing colonization by iron sulfide-oxidizing chemolithotrophs, such as *Acidithiobacillus ferrooxidans* and *Leptospirillum* spp. Data from initial experiments showed that the dissolution of pyrite could be reduced by between 57 and 75% by 'bioshrouding' the mineral with three different species of heterotrophic acidophiles (*Acidiphilium, Acidocella* and *Acidobacterium* spp.), under conditions that were conducive to microbial oxidative dissolution of the iron sulfide (Johnson *et al.*, 2008), offering enormous promise for reducing the generation of acidity at mining sites.

18.6.3 Bioleaching of metals

Several isolates of Archaea and bacteria from metal cores have been tested for use in the bioleaching of metals, i.e. metal extraction from ores (e.g. Plumb *et al.*, 2002). However, reports on the success of metal oxidation and leaching have been contrary. For example, thermophilic, acidophilic archaeal isolates were found to be capable of rapidly leaching a chalcopyrite concentrate (up to 91% copper release in 108 h) and are considered useful in bioleaching applications (Plumb *et al.*, 2002). In contrast, moderately thermophilic, iron-oxidizing acidophiles exhibited slower Fe-oxidation rates than mesophilic species (Kinnunen *et al.*, 2003). When oxidizing ferrous iron (Fe^{2+}) in flow-through reactors, different species of iron-oxidizing acidophiles also have varying capacities (Rowe and Johnson, 2008). Overall, it appears that microbial consortia are more robust than pure cultures of mineral-oxidizing acidophiles and also tend to be more effective at bioleaching and bio-oxidizing ores and concentrates.

18.7 Acidic Environments as Extraterrestrial Analogues

As has been demonstrated for extremophiles from other extreme environments on Earth, there is huge potential for using them as analogues for discovering the potential for life on other planets and extraterrestrial bodies. For example, findings by the Mars Exploration Rover (MER) recently confirmed the presence of iron minerals that can only be formed in the presence of water, and are thus indicative of potential life. Thus, chemolithoautotrophic communities living in acidic, iron-rich habitats on Earth can be used as analogues for life on Mars. Indeed, studies on acidophiles suggest that iron-oxidizing bacteria, such as *Leptospirillum ferrooxidans* and *Acidithiobacillus ferrooxidans*, use survival strategies for which it is feasible to identify a past or present hypothetical niche on Mars (Parro *et al.*, 2005; see also Gómez *et al.*, Chapter 26, this volume).

References

Aguilera, A., Amaral-Zettler, L.A., Souza-Egipsy, V., Zettler, E. and Amils, R. (2007) *Eukaryotic community structure from Río Tinto (SW, Spain), a highly acidic river.* In: Seckbach, J. (ed.) *Algae and Cyanobacteria in Extreme Environments.* Springer, Dordrecht, the Netherlands, pp. 465–485.

Baker-Austin, C. and Dopson, M. (2007) Life in acid: pH homeostasis in acidophiles. *Trends in Microbiology* 15, 165–171.

Bell, E.M. and Weithoff, G. (2003) Benthic recruitment of zooplankton in an acidic lake. *Journal of Experimental Marine Biology and Ecology* 285, 205–219.

Bell, E.M., Weithoff, G. and Gaedke, U. (2006) Temporal dynamics and growth of *Actinophrys sol* (Sarcodina: Heliozoa), the top predator in an extremely acidic lake. *Freshwater Biology* 51, 1149–1161.

Bertoldo, C., Dock, C. and Antranikian, G. (2004) Thermoacidophilic microorganisms and their novel biocatalysts. *Engineering in Life Sciences* 4, 521–532.

Beulker, C., Lessmann, D. and Nixdorf, B. (2003) Aspects of phytoplankton succession and spatial distribution in an acidic mining lake (Plessa 117, Germany). *Acta Oecologica - International Journal of Ecology* 24, S25–S31.

Bik, E.M., Eckburg, P.B., Gill, S.R., Nelson, K.E., Purdom, E.A., Francois, F., Perez-Perez, G., Blaser, M.J. and Relman, D.A. (2006) Molecular analysis of the bacterial microbiota in the human stomach. *Proceedings of the National Academy of Sciences of the United States of America* 103, 732–737.

Bond, P.L., Smriga, S.P. and Banfield, J.F. (2000) Phylogeny of microorganisms populating a thick, subaerial, predominantly lithotrophic biofilm at an extreme acid mine drainage site. *Applied and Environmental Microbiology* 66, 3842–3849.

Brake, S.S., Hasiotis, S.T., Dannelly, H.K. and Connors, K.A. (2002) Eukaryotic stromatolite builders in acid mine drainage: implications for Precambrian iron formations and oxygenation of the atmosphere? *Geology* 30, 599–602.

Brock, T.D., Madigan, M.T., Martinko, J.M. and Parker, J. (1994) *Biology of Microorganisms*. Prentice-Hall International, London, UK, 909 pp.

Brown, P.B. and Wolfe, G.V. (2006) Protist genetic diversity in the acidic hydrothermal environments of Lassen Volcanic National Park, USA. *Journal of Eukaryotic Microbiology* 53, 420–431.

Bryan, C.G., Hallberg, K.B. and Johnson, D.B. (2006) Mobilisation of metals in mineral tailings at the abandoned Sao Domingos copper mine (Portugal) by indigenous acidophilic bacteria. *Hydrometallurgy* 83, 184–194.

Burget, G.E. (1920) Note on the flora of the stomach. *Journal of Bacteriology* 5, 299–303.

Casiot, C., Bruneel, O., Personne, J.C., Leblanc, M. and Elbaz-Poulichet, F. (2004) Arsenic oxidation and bioaccumulation by the acidophilic Protozoan, *Euglena mutabilis*, in acid mine drainage (Carnoules, France). *Science of the Total Environment* 320, 259–267.

Chabbi, A. and Rumpel, C. (2004) Decomposition of plant tissue submerged in an extremely acidic mining lake sediment: phenolic CuO-oxidation products and solid-state(13)C NMR spectroscopy. *Soil Biology & Biochemistry* 36, 1161–1169.

Ciniglia, C., Cennamo, P., De Stefano, M., Pinto, G., Caputo, P. and Pollio, A. (2007) *Pinnularia obscura* Krasske (Bacillariophyceae, Bacillariophyta) from acidic environments: characterization and comparison with other acid-tolerant *Pinnularia* species. *Fundamental and Applied Limnology* 170, 29–47.

Collier, K.J., Ball, O.J., Graesser, A.K., Main, M.R. and Winterbourn, M.J. (1990) Do organic and anthropogenic acidity have similar effects on aquatic fauna. *Oikos* 59, 33–38.

Coupland, K. and Johnson, D.B. (2008) Evidence that the potential for dissimilatory ferric iron reduction is widespread among acidophilic heterotrophic bacteria. *FEMS Microbiology Letters* 279, 30–35.

Deneke, R. (2000) Review of rotifers and crustaceans in highly acidic environments of pH values <= 3. *Hydrobiologia* 433, 167–172.

DeNicola, D.M. (2000) A review of diatoms found in highly acidic environments. *Hydrobiologia* 433, 111–122.

Doi, H., Kikuchi, E. and Shikano, S. (2001) Carbon and nitrogen stable isotope ratios analysis of food sources for *Chironomus acerbiphilus* larvae (Diptera Chironomidae) in strongly acidic lake Katanuma. *Radioisotopes* 50, 601–611.

Doi, H., Kikuchi, E., Hino, S., Itoh, T., Takagi, S. and Shikano, S. (2003) Seasonal dynamics of carbon stable isotope ratios of particulate organic matter and benthic diatoms in strongly acidic Lake Katanuma. *Aquatic Microbial Ecology* 33, 87–94.

Dopson, M., Lindstrom, E.B. and Hallberg, K.B. (2001) Chromosomally encoded arsenical resistance of the moderately thermophilic acidophile *Acidithiobacillus caldus*. *Extremophiles* 5, 247–255.

Dopson, M., Baker-Austin, C., Koppineedi, P.R. and Bond, P.L. (2003) Growth in sulfidic mineral environments: metal resistance mechanisms in acidophilic micro-organisms. *Microbiology-Sgm* 149, 1959–1970.

Dopson, M., Baker-Austin, C., Hind, A., Bowman, J.P. and Bond, P.L. (2004) Characterization of *Ferroplasma* isolates and *Ferroplasma acidarmanus* sp nov., extreme acidophiles from acid mine drainage and industrial bioleaching environments. *Applied and Environmental Microbiology* 70, 2079–2088.

Dunfield, P.F., Yuryev, A., Senin, P., Smirnova, A.V., Stott, M.B., Hou, S.B., Ly, B., Saw, J.H., Zhou, Z.M., Ren, Y., Wang, J.M., Mountain, B.W., Crowe, M.A., Weatherby, T.M., Bodelier, P.L.E., Liesack, W., Feng, L., Wang, L. and Alam, M. (2007) Methane oxidation by an extremely acidophilic bacterium of the phylum Verrucomicrobia. *Nature* 450, 879–882.

Ferrer, M., Golyshina, O.V., Beloqui, A., Golyshin, P.N. and Timmis, K.N. (2007) The cellular machinery of *Ferroplasma acidiphilum* is iron-protein-dominated. *Nature* 445, 91–94.

Friese, K., Hupfer, M. and Schultze, M. (1998) Chemical characteristics of water and sediment in acid mining lakes of the Lusatian lignite district. In: Geller, W., Klapper, H. and Salomons, W. (eds) *Acidic Mining Lakes: Acid mine drainage, limnology and reclamation*. Springer, Berlin, Germany, pp. 25–45.

Futterer, O., Angelov, A., Liesegang, H., Gottschalk, G., Schleper, C., Schepers, B., Dock, C., Antranikian, G. and Liebl, W. (2004) Genome sequence of *Picrophilus torridus* and its implications for life around pH 0. *Proceedings of the National Academy of Sciences of the United States of America* 101, 9091–9096.

Gadanho, M., Libkind, D. and Sampaio, J.P. (2006) Yeast diversity in the extreme acidic environments of the Iberian Pyrite Belt. *Microbial Ecology* 52, 552–563.

Gaedke, U. and Kamjunke, N. (2006) Structural and functional properties of low- and high-diversity planktonic foodwebs. *Journal of Plankton Research* 28, 707–718.

Gaedke, U., Hochstadter, S. and Straile, D. (2002) Interplay between energy limitation and nutritional deficiency: Empirical data and foodweb models. *Ecological Monographs* 72, 251–270.

Gemmell, R.T. and Knowles, C.J. (2000) Utilisation of aliphatic compounds by acidophilic heterotrophic bacteria. The potential for bioremediation of acidic wastewaters contaminated with toxic organic compounds and heavy metals. *FEMS Microbiology Letters* 192, 185–190.

Gerloff-Elias, A., Barua, D., Mölich, A. and Spijkerman, E. (2006) Temperature- and pH-dependent accumulation of heat-shock proteins in the acidophilic green alga *Chlamydomonas acidophila*. *FEMS Microbiology Ecology* 56, 345–354.

Golyshina, O.V. and Timmis, K.N. (2005) Ferroplasma and relatives, recently discovered cell wall-lacking archaea making a living in extremely acid, heavy metal-rich environments. *Environmental Microbiology* 7, 1277–1288.

Golyshina, O.V., Golyshin, P.N., Timmis, K.N. and Ferrer, M. (2006) The 'pH optimum anomaly' of intracellular enzymes of *Ferroplasma acidiphilum*. *Environmental Microbiology* 8, 416–425.

Hallberg, K.B., Kolmert, A.K., Johnson, D.B. and Williams, P.A. (1999) A novel metabolic phenotype among acidophilic bacteria: aromatic degradation and the potential use of these organisms for the treatment of wastewater containing organic and inorganic pollutants. *Biohydrometallurgy and the Environment Toward the Mining of the 21St Century, Part A* 9, 719–728.

Havens, K.E. (1991) Littoral zooplankton responses to acid and aluminum stress during short-term laboratory bioassays. *Environmental Pollution* 73, 71–84.

Hawkes, R.B., Franzmann, P.D. and Plumb, J.J. (2006) Moderate thermophiles including '*Ferroplasma cupricumulans*' sp nov dominate an industrial-scale chalcocite heap bioleaching operation. *Hydrometallurgy* 83, 229–236.

Ingledew, W.J. (1990) Acidophiles. In: Edwards, C. (ed.) *Microbiology of Extreme Environments*. Open University Press, Milton Keynes, UK, pp. 33–54.

Jaenicke, R. (2000) Stability and stabilization of globular proteins in solution. *Journal of Biotechnology* 79, 193–203.

Johnson, D.B. (1998) Biodiversity and ecology of acidophilic microorganisms. *FEMS Microbiology Ecology* 27, 307–317.

Johnson, D.B. and Rang, L. (1993) Effects of acidophilic Protozoa on populations of metal-mobilizing bacteria during the leaching of pyritic coal. *Journal of General Microbiology* 139, 1417–1423.

Johnson, D.B., Okibe, N. and Roberto, F.F. (2003) Novel thermo-acidophilic bacteria isolated from geothermal sites in Yellowstone National Park: physiological and phylogenetic characteristics. *Archives of Microbiology* 180, 60–68.

Johnson, D.B., Yajie, L. and Okibe, N. (2008) 'Bioshrouding' – a novel approach for securing reactive mineral tailings. *Biotechnology Letters* 30, 445–449.

Jones, R.I. (1994) Mixotrophy in planktonic protists as a spectrum of nutritional strategies. *Marine Microbial Foodwebs* 8, 287–296.

Kamjunke, N., Gaedke, U., Tittel, J., Weithoff, G. and Bell, E.M. (2004) Strong vertical differences in the plankton composition of an extremely acidic lake. *Archiv für Hydrobiologie* 161, 289–306.

Kamjunke, N., Bohn, C. and Grey, J. (2006) Utilisation of dissolved organic carbon from different sources by pelagic bacteria in an acidic mining lake. *Archiv für Hydrobiologie* 165, 355–364.

Kinnunen, P.H.M., Robertson, W.J., Plumb, J.J., Gibson, J.A.E., Nichols, P.D., Franzmann, P.D. and Pubakka, J.A. (2003) The isolation and use of iron-oxidizing, moderately thermophilic acidophiles from the Collie coal mine for the generation of ferric iron leaching solution. *Applied Microbiology and Biotechnology* 60, 748–753.

Kristjansson, J.K. and Hreggvidsson, G.O. (1995) Ecology and habitats of extremophiles. *World Journal of Microbiology and Biotechnology* 11, 17–25.

Kusel, K. and Dorsch, T. (2000) Effect of supplemental electron donors on the microbial reduction of Fe(III), sulfate, and CO_2 in coal mining-impacted freshwater lake sediments. *Microbial Ecology* 40, 238–249.

Laksanalamai, P. and Robb, F.T. (2004) Small heat shock proteins from extremophiles: a review. *Extremophiles* 8, 1–11.

Lessmann, D., Deneke, R., Ender, R., Hemm, M., Kapfer, M., Krumbeck, H., Wollmann, K. and Nixdorf, B. (1999) Lake Plessa 107 (Lusatia, Germany) – an extremely acidic shallow mining lake. *Hydrobiologia* 408, 293–299.

Lopez-Archilla, A.I., Gonzalez, A.E., Terron, M.C. and Amils, R. (2004) Ecological study of the fungal populations of the acidic Tinto River in southwestern Spain. *Canadian Journal of Microbiology* 50, 923–934.

Lürling, M. and van Donk, E. (1997) Life history consequences for *Daphnia pulex* feeding on nutrient-limited phytoplankton. *Freshwater Biology* 38, 693–709.

Mann, H., Tazaki, K., Fyfe, W.S., Beveridge, T.J. and Humphrey, R. (1987) Cellular lepidocrocite precipitation and heavy metal sorption in *Euglena* sp (unicellular alga): implications for biomineralization. *Chemical Geology* 63, 39–43.

McConathy, J.R. and Stahl, J.B. (1982) Rotifera in the plankton and among filamentous algal clumps in 16 acid strip-mine lakes. *Transactions of the Illinois Academy of Science* 75, 85–90.

Mendez, M.O., Neilson, J.W. and Maier, R.M. (2008) Characterization of a bacterial community in an abandoned semiarid lead-zinc mine tailing site. *Applied and Environmental Microbiology* 74, 3899–3907.

Messerli, M.A., Amaral-Zettler, L.A., Zettler, E., Jung, S.K., Smith, P.J.S. and Sogin, M.L. (2005) Life at acidic pH imposes an increased energetic cost for a eukaryotic acidophile. *Journal of Experimental Biology* 208, 2569–2579.

Mormile, M.R., Hong, B.Y., Adams, N.T., Benison, K.C. and Oboh-Ikuenobe, F. (2007) Characterization of a moderately halo-acidophilic bacterium isolated from Lake Brown, Western Australia. *Instruments, Methods, and Missions for Astrobiology* X, 1–8.

Nagasaka, S., Nishizawa, N.K., Watanabe, T., Mori, S. and Yoshimura, E. (2003) Evidence that electron-dense bodies in *Cyanidium caldarium* have an iron-storage role. *Biometals* 16, 465–470.

Nishikawa, K. and Tominaga, N. (2001) Isolation, growth, ultrastructure, and metal tolerance of the green alga, *Chlamydomonas acidophila* (Chlorophyta). *Bioscience Biotechnology and Biochemistry* 65, 2650–2656.

Nixdorf, B., Fyson, A. and Krumbeck, H. (2001) Review: plant life in extremely acidic waters. *Environmental and Experimental Botany* 46, 203–211.

Ntengwe, F.W. (2005) An overview of industrial wastewater treatment and analysis as means of preventing pollution of surface and underground water bodies – the case of Nkana Mine in Zambia. *Physics and Chemistry of the Earth* 30, 726–734.

Oesterhelt, C., Schmalzlin, E., Schmitt, J.M. and Lokstein, H. (2007) Regulation of photosynthesis in the unicellular acidophilic red alga *Galdieria sulphuraria*. *Plant Journal* 51, 500–511.

Olsson, K., Stenroth, P., Nystrom, P., Holmqvist, N., McIntosh, A.R. and Winterbourn, M.J. (2006) Does natural acidity mediate interactions between introduced brown trout, native fish, crayfish and other invertebrates in West Coast New Zealand streams? *Biological Conservation* 130, 255–267.

Packroff, G. (2000) Protozooplankton in acidic mining lakes with special respect to ciliates. *Hydrobiologia* 433, 157–166.

Parro, V., Rodriguez-Manfredi, J.A., Briones, C., Compostizo, C., Herrero, P.L., Vez, E., Sebastian, E., Moreno-Paz, M., Garcia-Villadangos, M., Fernandez-Calvo, P., Gonzalez-Toril, E., Perez-Mercader, J., Fernandez-Remolar, D. and Gomez-Elvira, J. (2005) Instrument development to search for biomarkers on Mars: terrestrial acidophile, iron-powered chemolithoautotrophic communities as model systems. *Planetary and Space Science* 53, 729–737.

Pawlik-Skowronska, B. (2003) Resistance, accumulation and allocation of zinc in two ecotypes of the green alga *Stigeoclonium tenue* Kutz. coming from habitats of different heavy metal concentrations. *Aquatic Botany* 75, 189–198.

Pedrozo, F., Kelly, L., Diaz, M., Temporetti, P., Baffico, G., Kringel, R., Friese, K., Mages, M., Geller, W. and Woelfl, S. (2001) First results on the water chemistry, algae and trophic status of an Andean acidic lake system of volcanic origin in Patagonia (Lake Caviahue). *Hydrobiologia* 452, 129–137.

Peeples, T.L. and Kelly, R.M. (1995) Bioenergetic response of the extreme thermoacidophile *Metallosphaera sedula* to thermal and nutritional stresses. *Applied and Environmental Microbiology* 61, 2314–2321.

Pinto, G., Ciniglia, C., Cascone, C. and Pollio, A. (2007) Species composition of cyanidiales assemblages in Pisciarelli (Campu flegrei, Italy) and description of *Galdieria phlegrea* sp. nov. In: Seckbach, J. (ed.) *Algae and Cyanobacteria in Extreme Environments*. Springer, Dordrecht, the Netherlands, pp. 489–502.

Plumb, J.J., Gibbs, B., Stott, M.B., Robertson, W.J., Gibson, J.A.E., Nichols, P.D., Watling, H.R. and Franzmann, P.D. (2002) Enrichment and characterisation of thermophilic acidophiles for the bioleaching of mineral sulphides. *Minerals Engineering* 15, II.

Pol, A., Heijmans, K., Harhangi, H.R., Tedesco, D., Jetten, M.S.M. and den Camp, H.J.M.O. (2007) Methanotrophy below pH1 by a new *Verrucomicrobia* species. *Nature* 450, 874–878.

Rothschild, L.J. and Mancinelli, R.L. (2001) Life in extreme environments. *Nature* 409, 1092–1101.

Rowe, O.F. and Johnson, D.B. (2008) Comparison of ferric iron generation by different species of acidophilic bacteria immobilized in packed-bed reactors. *Systematic and Applied Microbiology* 31, 68–77.

Rowe, O.F., Sanchez-Espana, J., Hallberg, K.B. and Johnson, D.B. (2007) Microbial communities and geochemical dynamics in an extremely acidic, metal-rich stream at an abandoned sulfide mine (Huelva, Spain) underpinned by two functional primary production systems. *Environmental Microbiology* 9, 1761–1771.

Sabater, S., Buchaca, T., Cambra, J., Catalan, J., Guasch, H., Ivorra, N., Munoz, I., Navarro, E., Real, M. and Romani, A. (2003) Structure and function of benthic algal communities in an extremely acid river. *Journal of Phycology* 39, 481–489.

Schmidtke, A., Bell, E.M. and Weithoff, G. (2006) Potential grazing impact of the mixotrophic flagellate *Ochromonas* sp (Chrysophyceae) on bacteria in an extremely acidic lake. *Journal of Plankton Research* 28, 991–1001.

Simmons, S. and Norris, P.R. (2002) Acidophiles of saline water at thermal vents of Vulcano, Italy. *Extremophiles* 6, 201–207.

Spijkerman, E. (2008) Phosphorus limitation of algae living in iron-rich, acidic lakes. *Aquatic Microbial Ecology* 53, 201–210.

Spijkerman, E., Barua, D., Gerloff-Elias, A., Kern, J., Gaedke, U. and Heckathorn, S.A. (2007a) Stress responses and metal tolerance of *Chlamydomonas acidophila* in metal-enriched lake water and artificial medium. *Extremophiles* 11, 551–562.

Spijkerman, E., Bissinger, V., Meister, A. and Gaedke, U. (2007b) Low potassium and inorganic carbon concentrations influence a possible phosphorus limitation in *Chlamydomonas acidophila* (Chlorophyceae). *European Journal of Phycology* 42, 327–339.

Stapleton, R.D., Savage, D.C., Sayler, G.S. and Stacey, G. (1998) Biodegradation of aromatic hydrocarbons in an extremely acidic environment. *Applied and Environmental Microbiology* 64, 4180–4184.

Tan, G.L., Shu, W.S., Hallberg, K.B., Li, F., Lan, C.Y., Zhou, W.H. and Huang, L.N. (2008) Culturable and molecular phylogenetic diversity of microorganisms in an open-dumped, extremely acidic Pb/Zn mine tailings. *Extremophiles* 12, 657–664.

Tatsuzawa, H., Takizawa, E., Wada, M. and Yamamoto, Y. (1996) Fatty acid and lipid composition of the acidophilic green alga *Chlamydomonas* sp. *Journal of Phycology* 32, 598–601.

Tittel, J., Wiehle, I., Wannicke, N., Kampe, H., Poerschmann, J., Meier, J. and Kamjunke, N. (2009) Utilisation of terrestrial carbon by osmotrophic algae. *Aquatic Sciences* 71, 46–54.

Walker, J.J., Spear, J.R. and Pace, N.R. (2005) Geobiology of a microbial endolithic community in the Yellowstone geothermal environment. *Nature* 434, 1011–1014.

Weber, A.P.M., Horst, R.J., Barbier, G.G. and Oesterhelt, C. (2007) *Metabolism and Metabolomics of Eukaryotes Living under extreme conditions*. Elsevier Academic Press, San Diego, California, pp. 1–34.

Weithoff, G., Moser, M., Kamjunke, N., Gaedke, U. and Weisse, T. (2010) Lake morphometry and wind exposure may shape the plankton community structure in acidic mining lakes. *Limnologica* 40, 161–166.

Woelfl, S. (2000) Limnology of sulphur-acidic lignite mining lakes. Biological properties: plankton structure of an extreme habitat. *Verhandlungen der Internationale Verein der Limnologie* 27, 1–4.

Wollmann, K. (2000) Corixidae (Hemiptera, Heteroptera) in acidic mining lakes with pH <= 3 in Lusatia, Germany. *Hydrobiologia* 433, 181–183.

Wolowski, K., Turnau, K. and Henriques, F.S. (2008) The algal flora of an extremely acidic, metal-rich drainage pond of Sao Domingos pyrite mine (Portugal). *Cryptogamie Algologie* 29, 313–324.

Yallop, C.A. and Charalambous, B.M. (1996) Nutrient utilization and transport in the thermoacidophilic archaeon *Sulfolobus shibatae*. *Microbiology* 142, 3373–3380.

Yamamoto, Y., Tatsuzawa, H. and Wada, M. (1998) Effect of environmental conditions on the composition of lipids and fatty acids in *Chlamydomonas* isolated from an acidic lake. *Verhandlungen Internationale Vereinigung für Theoretische und Angewandte Limnologie* 26, 1788–1790.

19 Alkaline Environments

Elanor M. Bell
Scottish Institute for Marine Science, Argyll, UK

19.1 Alkaline Environments

Alkalinity is extremely common in inland waterways worldwide, indeed, 80% are on the alkaline side of neutrality (Grant, 2006). However, extremely alkaline environments both manmade and naturally occurring are not restricted to aquatic systems and can be found in a wide range of geographic locations (Table 19.1; Figs 19.1 and 19.2).

19.1.1 Industrial locations

Human industrial processes such as cement manufacture ($Ca(OH)_2$) and casting, mining operations, paper and pulp production (NaOH), food processing (e.g. the lye treatment of potatoes to remove their skins; NaOH and 60–90°C), electroplating and hide treatment, all generate highly alkaline environments (i.e. in excess of pH 11; Grant *et al.*, 1990; Jones *et al.*, 1998). A restricted number of alkali-loving organisms (alkaliphiles) can survive in such transient and often unstable environments. Examples include species in the bacterial genus *Bacillus*, which thrive in alkaline environments but also produce spores capable of surviving between alkaline episodes (Zhilina *et al.*, 2001; Grant, 2006) and the Gram-positive bacterium *Exiguobacterium aurantiacum*, which colonizes the waste from industrial potato peeling processes (Gee *et al.*, 1980; Collins *et al.*, 1983; Grant *et al.*, 1990).

19.1.2 Transient alkaline environments

Transient alkalinity arising naturally through biological activity (e.g. ammonification, sulfate reduction, or photosynthesis) in soils is widespread allowing alkaliphiles to thrive in environments with neutral or acidic bulk pHs (Grant, 2006). Once again, organisms such as *Bacillus* bacteria and the alkaliphilic mycelial prokaryotes *Streptomyces* and *Nocardiopsis*, grow well in these transient environments (Zhilina *et al.*, 2001; Selyanin *et al.*, 2005).

19.1.3 Calcium-dominated groundwaters

Highly alkaline calcium (Ca^{2+})-dominated groundwaters are also present around the world. Their chemistry is determined by the CO_2-mediated weathering of the calcium and magnesium (Mg^{2+}) minerals olivine and pyroxene, which break down releasing the OH^- ions responsible for alkalinity (Grant, 2006). Magnesium is then removed by precipitation as the mineral serpentine (magnesium iron

Table 19.1. Location of alkaline environments. Modified from Grant, 2006.

Continent	Country	Location
Africa	Chad	Lake Munyanyange, Lake Murumuli, Lake Nunyampaka, Lake Bodo, Lake Rombou, Lake Dijikare, Lake Monboio, Lake Yoan
	Egypt	Wadi Natrun
	Ethiopia	Lake Aranguardi, Lake Kilotes, Lake Abiata, Lake Shala, Lake Chilu, Lake Hertale, Lake Metahara
	Kenya	Lake Bogoria, Lake Nakuru, Lake Elmenteita, Lake Magadi, Lake Simbi, Crater Lake (Lake Sonachi), Lake Oloidien, Lake Turkana
	Libya	Lake Fezzan
	Sudan	Dariba Lakes
	Tanzania	Lake Natron, Lake Eyasi, Lake Magad, Lake Manyara, Lake Balangida, Bosotu Crater Lake, Lake Kusare, Lake Tulusia, El Kekhooito, Momela Lakes, Lake Lekandiro, Lake Reshitani, Lake Lgarya, Lake Ndutu
	Uganda	Lake Rukwa North, Lake Katwe, Lake Mahenga, Lake Kikorongo, Lake Nyamunuka
Asia	Armenia	Araxes Plain Lakes
	China	Inner Mongolia, Qaganlimen Nur, Baiyan Nur, Hatong Qagan Nur, Xijuyanhai (Gaxun Nur); Outer Mongolia, various 'nurs'; Sui-Yuan, Cha-Han-Nor and Na-Lin-Nor; Heilungkiang, Hailar and Tsitsihar; Kirin, Fu-U-Hsein and Taboos-Nor; Liao-Ning, Tao-Nan Hsein; Jehol, various soda lakes; Tobet, alkaline deserts; Chahar, Ling-Chai, Shansi, U-Tsu-Hsein; Shensi, She-Hsia-Hsein; Kansu, Ning-Hsia-Hsein, Qinhgai Hu
	India	Lake Looner, Lake Sambhar
	Siberia	Kulunda Steppe, Tanatar Lakes, Bittter Lakes, Cock Lake, Stamp Lake, Krarkul, Chita, Barnaul, Slavgerod, Lake Baikal region, Lake Khatyn
	Turkey	Lake Van, Lake Salda
Australia	Australia	Lake Corangamite, Red Rock Lake, Lake Werowrap, Lake Chidnup, Lake Eyre
Central America	Mexico	Lake Texcoco
Europe	Former Yugoslavia	Pecena Slatina
	Hungary	Lake Feher
North America	Canada	Manito
	USA	Alkali Valley, Albert Lake Lenore, Soap Lake, Big Soda Lake, Owens Lake, Borax Lake, Mono Lake, Searles Lakes, Deep Springs, Rhodes Marsh, Harney Lake, Summer Lake, Surprise Valley, Pyramid Lake, Walker Lake, Union Pacific Lakes (Green River), Ragtown Soda Lakes, Octopus Springs
South America	Chile	Antofagasta
	Venezuela	Langunilla Valley

silicate hydroxide; $(Mg,Fe)_3Si_2O_5(OH)_4$) resulting in $Ca(OH)_2$-dominated environments that can have pH >12 and are highly reducing as a result of ferrous iron (Fe^{2+}) and hydrogen (H_2) release (Grant, 2006). Diverse microbial communities, including many species new to science, inhabit such environments (Kotelnikova *et al.*, 1998; Roadcap *et al.*, 2006).

19.1.4 Soda lakes and deserts

This chapter will focus mainly on soda lakes and soda deserts, which are the major types of naturally occurring, alkaline environments. Soda lakes represent the most stable, high-pH environments on Earth. They are geographically widely

Fig. 19.1. The shore of alkaline Lake Nakuru, Kenya, African Rift Valley. © 2011 iStockphoto LP.

distributed, often in the inner parts of continents and subject to semi-arid climates. They generally form in closed drainage basins exposed to high evaporation rates. The geochemistry of a region directly affects the ionic composition of the ions entering a lake system. Typically, there is a scarcity of Mg^{2+} and Ca^{2+} in the strata surrounding soda lakes, and hence in the water. These cations normally buffer aquatic environments via the precipitation of insoluble carbonates, but their lack means that there is a shift in the carbon dioxide/bicarbonate/carbonate equilibrium of soda lake waters in favour of carbonate (CO_3^{2-}). This, combined with evaporative concentration, means that soda lakes contain an alkaline brine enriched in CO_3^{2-} and chloride (Cl^-) and their pH can range from 8 to >12, akin to the pH of many household cleaning products (Fig. 19.3). An alternative method for the generation of alkalinity involving bacterial sulfate reduction (see Section 19.2.4) has been proposed for the lakes of the Wadi Natrun depression in Egypt (Abd-el-Malek and Rizk, 1963), but this is unlikely to be a general mechanism (Baumgarte, 2003).

Fig. 19.2. An example of an alkaline lake: Lake Eyre, Australia. © 2011 iStockphoto LP.

19.2 The Diversity, Community Composition and Biochemical Groups of Alkaliphiles in Soda Lakes

Despite their extreme chemistry, soda lakes, in particular the more dilute systems, support a rich diversity of life and are some of the most productive aquatic environments on Earth (Grant *et al.*, 1990). One of the most striking things about soda lakes is their colour. Depending on water chemistry they are likely to be green, pink, red or orange, due to the massive permanent or seasonal blooms of microorganisms that they host (Grant *et al.*, 1990; Fig. 19.4).

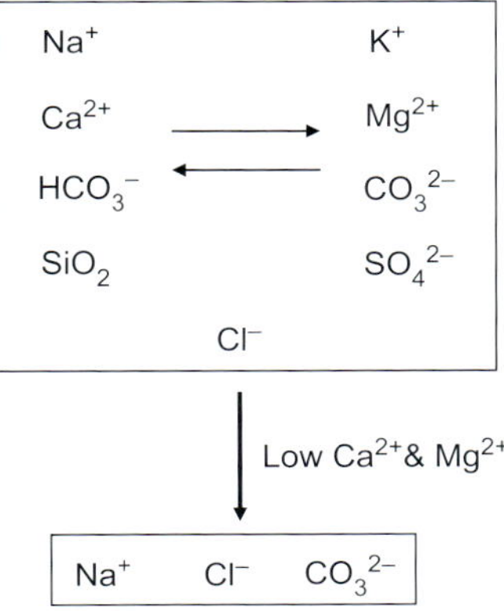

Fig. 19.3. Schematic representation of the creation of alkaline lakes. Soda lake formation is dependent on low levels of Mg^{2+} and Ca^{2+}. The high concentrations of sodium, chloride and carbonate ions can generate alkaline brines with pHs >12. Adapted from Grant, 2004, Figure 2, pp. 21.

The majority of these microorganisms are alkaliphiles (alkali-lovers). Alkaliphiles are organisms which grow very well at greater than pH 8 usually with optima between pH 9 and 10, and sometimes as high as pH 12. Obligate alkaliphiles grow very slowly or are incapable of growth when pH values are near neutral (pH 7; Grant *et al.*, 1990; Horikoshi, 1999). Alkaliphilic organisms, in particular prokaryotes, are ubiquitous and can be found in almost all environments, even in those in which the bulk pH is not particularly alkaline, e.g. soils with transient alkalinity (Grant *et al.*, 1990).

Another major physiological group found in alkaline environments are the haloalkaliphiles, which require both alkaline conditions (>pH 9) and high concentrations of salt (up to 33% wt/vol NaCl; Horikoshi, 1999) for growth. Other polyextremophiles include the thermoalkaliphiles, which also grow at elevated temperatures (>60°C) and ≥pH 9, e.g. *Thermus brockianus* (Thompson *et al.*, 2003).

A wide diversity of types and biochemical groups of alkaliphiles and haloalkaliphiles are encountered in soda lakes and other alkaline environments include the following groups. These will be discussed in the following sections.

Fig. 19.4. The coloured waters of alkaline Laguna Colorada, Bolivia. © 2011 iStockphoto LP.

19.2.1 Primary producers

Extremely high primary productivity is associated with some soda lakes with rates exceeding 10 g m^{-2} day^{-1}. This is likely due to the relatively high ambient temperatures and light intensities, available phosphate and unlimited CO_2 in the carbonate-rich waters (Melack and Kilham, 1974; Grant *et al.*, 1990). Lower rates of up to 1.32 g m^{-2} day^{-1} were recorded in soda lake communities of the Kulunda Steppe, western Siberia (Kompantseva *et al.*, 2009). Dense blooms of cyanobacteria tend to be responsible for the primary production in less alkaline lakes around the world, whilst blooms of both cyanobacteria and alkaliphilic anaerobic phototrophs are responsible in more alkaline environments (Grant *et al.*, 1990; Jones *et al.*, 1998). At higher salinities (approximately 7% NaCl and 10% net soda) unicellular forms of cyanobacteria are common whereas at lower salinities and alkalinities trichomic forms are more prevalent (Zavarzin *et al.*, 1999). In the East African Rift lakes these productive cyanobacterial blooms (principally of *Spirulina* spp.; Zavarzin *et al.*, 1999) are the principal food source for the immense flocks of lesser flamingo (*Phoeniconaias minor*) that inhabit the rift valley (Fig. 19.5). As well as providing a rich source of organic matter and food, the cyanobacteria are also essential for nitrogen fixation in the soda lakes and as oxygen producers. *Spirulina* spp. in Lake Reshitani, East Africa, can produce up to 2 g O_2 m^{-2} h^{-1} during the day (Melack and Kilham, 1974). In Mono Lake, USA, cyanobacteria inhabit the spectacular calcium carbonate, or so-called tufa, columns that rise out of the lake surface. They reside just below the surface of the columns shaded from the intense, desiccating sunlight and in a perfect position to optimize their growth (Nealson, 1999). The same is true of the tufa columns in Lake Van, Turkey, which are covered with coccoid cyanobacteria, as well as hosting a diversity of Proteobacteria, members of the *Cytophaga-Flexibacter-Bacteriodes* group, Actinobacteria and low G+C Gram-positive Firmicutes (López-García *et al.*, 2005).

More alkaline soda lakes typically contain the anaerobic phototrophic purple bacteria genera *Ectothiorhodospira* and *Halorhodospira*, which oxidize hydrogen sulfide with intermediate extra-cellular sulfur deposition (e.g. Imhoff *et al.*, 1979;

Fig. 19.5. Flock of lesser flamingos (*Phoeniconaias minor*) on Lake Nakuru, Kenya. © 2011 iStockphoto LP.

Tindall, 1988). Recently, two new species were isolated from low salinity soda lakes in south-east Siberia, *Thiorhodospira sibirica* (Bryantseva *et al.*, 1999) and *Thioalkalicoccus limnaeus* (Bryantseva *et al.*, 2000a), which use hydrogen sulfide and elemental sulfur as electron donors under anoxic conditions. A few purple non-sulfur bacteria have also been found in soda lakes with low mineralization: *Rhodobaca bogoriensis* was isolated from Lake Bogoria, Kenya, and is capable of both phototrophic and chemotrophic growth (Milford *et al.*, 2000). Other phototrophs including heliobacteria and aerobic phototrophs have been identified in Siberian soda lakes (e.g. Bryantseva *et al.*, 2000b; Sorokin *et al.*, 2000b).

19.2.2 Chemo-organotrophs

Many chemo-organotrophs have been isolated from alkaline environments, both aerobic and anaerobic. All play important roles in the utilization and recycling of the products of photosynthesis. Aerobic chemo-organotrophic bacteria such as Gram-negative bacteria of the *Proteobacteria* lineage, including many in the Halomonadaceae family (the halomonads), have been isolated from soda lakes, as well as alkaline soils and alkaline wastewaters from the alkaline aerobic oxidation of fresh olives.

Aerobic Gram-positive isolates, in both the high and low DNA guanosine/cytosine content (G+C) divisions, are abundant. Alkaliphilic members of the low G+C *Bacillus* group are especially prevalent in soda lakes, likely due to the fact that they have a dormant life stage that can survive drying in the marginal areas around the lakes (Grant, 2006). Examples of high G+C Gram-positive bacteria that have been identified include novel streptomycetes, *Dietzia* spp., *Bogoriella* spp. and uncharacterized relatives of *Nestetenkonia* (summarized by Grant, 2006).

Nevertheless, some extremely alkaline environments host quite different aerobic chemo-organotrophic prokaryotes. The waters of extremely alkaline (pH >11.5), hypersaline (7000 mM Na^+) Lake Magadi, Kenya, are dominated by bright pink or red blooms of haloalkaliphilic Archaea. Representatives of six genera have been identified thus far: *Natronococcus, Natronobacterium, Natrialba, Halorubrum, Natronorubrum* and *Natronomonas*. Some reach abundances of up to 10^8 ml^{-1} in the brines (McGenity and Grant, 1993; Horikoshi, 1999). Representatives of these genera have also been isolated from extremely saline alkaline lakes in other parts of the world, e.g. Bange Lake, Tibet (Xu *et al.*, 1999). Haloalkaliphilic *Bacillus* spp. with a minimum growing requirement of 15% mass per volume NaCl are the only other aerobes that have been cultivated from these environments (Aono and Hirokoshi, 1983; Krulwich *et al.*, 1997, 1998; Horikoshi, 1999; Grant, 2006).

Anaerobic, alkaliphilic, chemo-organotrophs also inhabit soda lakes, in particular anoxic soda lake sediments. Members of the Haloanaerobiales carry out acetogenesis and ammonification (Grant, 2006) and thermophilic prokaryotes capable of tolerating 50°C to boiling can be found in the hot spring areas that often surround soda lakes; for example, the lypolitic, anaerobic bacterium *Thermosyntropha lipoytica*, isolated from the alkaline hot springs of Lake Bogoria, Kenya (Svetlitshnyi *et al.*, 1996). Sacchrolytic and clostridal types such as spirochetes (Zhilina *et al.*, 1996) and *Clostridium* spp. are also common (Zhilina *et al.*, 2005a; Grant, 2006).

19.2.3 Methanogens

Methanotrophy is an important part of the carbon cycle in alkaline environments. Methanotrophs recycle methane carbon into the common pool of organic matter. Obligate alkaliphilic methanogens have been isolated from a number of soda lakes, for example a strain of methanogenic Archaea, *Methanosalsus zhilinaeae*, was isolated from Lake Magadi, Kenya (Kevbrin *et al.*, 1997). This strain is obligately dependent on Na^+ and HCO_3^- and utilizes a variety of one-carbon compounds (C1), likely derived from the anaerobic digestion of cyanobacterial mats common in soda lakes (Grant, 2006). Hydrogen-utilizing bacterial methanogens (*Methanobacterium* spp.) have also been identified but these are probably alkalitolerant rather than alkaliphilic. Furthermore, methane-oxidizing bacteria in the genera *Methylobacter* and *Methylomicrobium* that require aerobic or microaerophilic conditions consume methane in soda lakes (Grant, 2006).

19.2.4 Sulfur-reducing and -oxidizing alkaliphiles

Sulfate reduction by bacteria has been suggested to generate the alkalinity in the lakes of the Wadi Natrun depression in Egypt (Abd-el-Malek and Rizk, 1963). Although this is unlikely to be a general alkalization mechanism (Baumgarte, 2003), rates of sulfate reduction in soda lakes comparable to those in marine sediments have been measured in many parts of the world (Grant, 2006). Moreover, most soda lakes contain black muds due to sulfidogenesis, despite the fact that the high pH in soda lakes prohibits the release of H_2S and sulfide is retained as the odour-free dianion S^{2-}, which is only available in strongly alkaline aqueous solutions.

Alkaliphilic, sulfate-reducing bacteria play an important role as hydrogen sinks in soda lakes as they utilize hydrogen as an electron donor. The bacterium *Desulfonatronovibrio hydrogenovorans* was isolated from Lake Magadi, Kenya, and this species is thought to be responsible for the disappearance of sulfate in the lake. Black sediments interbedded with trona ($NaHCO_3.Na_2CO_3.12H_2O$) in Lake Magadi indicate the production of hydrogen sulfide (H_2S) and subsequent binding of sulfide by iron in the surrounding lava rocks (Zhilina *et al.*, 1997).

19.2.5 Acetogenic and acetoclastic alkaliphiles

Thermophilic acetogens that utilize CO_2 as a terminal electron acceptor and produce acetate as a product of anaerobic respiration, have been isolated from a number of soda lakes around the world (Zavarzin and Zhilina, 2000). For example, the anaerobic bacterium *Thermosyntropha lipoytica* is well known in the hot springs commonly surrounding soda lakes. Similarly, bacteria in the Orders Haloanaerobiales, such as *Natrionella* spp, and Clostridiales, such as *Tindallia* and *Natronoincola* spp., carry out acetogenesis and ammonification under alkaline conditions (Grant, 2006).

In contrast to the hydrogen-utilizing prokaryotes discussed in previous sections, far less is known about those using substrates such as acetate. What is known is that acetate is easily consumed by a number of aerobic alkaliphiles and acetate-oxidizing aerobes have been isolated growing at pH 10 (Zavarzin and Zhilina, 2000; Zhilina *et al.*, 2005b). Moreover, anaerobic alkaliphiles such as purple bacteria and sulfate-reducers, common in soda lakes, can use acetate as an anabolic substrate (Zavarzin and Zhilina, 2000). Aceticlastic methanogens have also been isolated but are only present in low-salt systems (Zavarzin and Zhilina, 2000).

The production and use of acetate promotes syntrophy (cross-feeding) in alkaliphilic microbial communities. For example, *Desulfonatronum cooperativum* is an alkaliphilic, aerobic, hydrogenotrophic, sulfate-reducing bacterium that was isolated from soda lake Khadin, Russia, and was shown to grow lithoheterotrophically on acetate produced by the syntrophic acetate-decomposing community in which it lives (Zhilina *et al.*, 2005b).

19.2.6 Nitrogen-fixing, ammonia- and nitrite-oxidizing, and nitrate-reducing alkaliphiles and nitrification

An active nitrogen cycle exists in soda lakes. Nitrogen fixation is known to be carried out by heterocystous cyanobacteria (e.g. *Cyanospira rippkae*; Florenzano *et al.*, 1985) and many of the aforementioned phototrophic bacteria, and probably chemo-organotrophs, have a nitrogen-fixing capacity (Grant, 2006). Alkaliphilic ammonia and nitrite oxidizers have also been isolated from Siberian and Kenyan soda lakes. These play an important role in biological nitrogen cycling by converting reduced inorganic nitrogen compounds to nitrate (Sorokin *et al.*, 1998). Nitrate-reduction and denitrification occur widely. Lithotrophic bacteria, such as *Thioalkalivibrio denitrificans*, and chemo-organotrophic halomonads are often responsible (Grant, 2006). Ammonium-oxidizing strains tend also to be methane oxidizers; examples include the alkaliphilic methanotroph, *Methylomicrobium* sp. AMO1, isolated from the mixed sediments of five Kenyan soda lakes (Sorokin *et al.*, 2000a). This species is able to oxidize ammonia to nitrite at pH 10–10.5 and is also capable of oxidizing organic sulfur compounds at high pH.

19.2.7 Fungi

Fungi generally germinate and grow well in weakly acidic to neutral pH ranges. However, very early work began to suggest that fungi could tolerate high pHs as well: *Fusarium* spp. and *Penicillium variables* grew at pH 11 (Johnson, 1923), although these species were subsequently shown to neutralize the pH of their environment and may not therefore represent true alkaliphiles (Zak and Wildman, 2004). Various species of fungi also grow well in alkaline soils and alkaline caves with pH 9 and above: the genera *Acremonium* and *Fusarium* are most frequently detected in these environments (Nagai *et al.*, 1995, 1998). In addition, a strain of *Fusarium solani* N4-2 was isolated from an extremely alkaline soda lake in China, lipases from which are currently being used in detergent formulation (Liu *et al.*, 2009; see also Arora and Bell, Chapter 25, this volume).

19.2.8 Protozoa

Alkaline environments can support high numbers of Protozoa, although the highest abundances are generally observed in less alkaline systems. The littoral zone of hyposaline lakes Lakes Alchichica and Atexcac, Mexico (pH 8.4–9.1), support rich protozoan assemblages dominated by ciliates and flagellates: *Cyclidium* spp. and *Stylonychia notophora* dominate the ciliates; *Bodo caudatus*, *Spumella termo* and *Cryptomonas ovata* the flagellates. With the exception of *S. notophora*, which consumes algae, these Protozoa are all bacterivorous (Lugo *et al.*, 1998). Bacterivorous scuticociliates such as *Cyclidium citrullus*, *C. glaucoma*, *Cinetochilum margaritaceum* and *Uronema nigricans*, were also identified in the slightly alkaline waters of a geothermal spring in northern Italy (Madoni and Uluhogian, 1997). Given the huge diversity and abundance of bacteria and other prokaryotes in alkaline environments, the presence of bacterial grazers is likely to influence community structure and exert considerable grazing pressure in these lower alkalinity systems.

19.2.9 Viruses

Viruses are an integral part of aquatic microbial communities and can be another significant source of bacterial mortality (Wommack and Colwell, 2000). Little work has been done on viruses in extremely alkaline environments. However, Jiang *et al.* (2003) observed up to 1×10^9 virus particles ml^{-1} in alkaline, hypersaline Mono Lake, California, USA, higher than the abundance measured in other marine and freshwater systems, but comparable to concentrations observed in hypersaline evaporator ponds used to produce salt from seawater (Guixa-Boixareu *et al.*, 1996).

19.2.10 Animalia

Alkaline environments support very few species of invertebrate, although the densities of those that they do support can be extremely high. For example, the only permanent Metazoa inhabiting alkaline, hypersaline Mono Lake, USA, are the brine shrimp, *Artemia monica*, and the so-called alkali fly, *Ephydra hians* (Thorp and Covich, 2001). High densities of *A. monica* (15–17 individuals l^{-1} in nearshore regions; 6–8 individuals l^{-1} in the pelagic; Conte *et al.*, 1988) are crucial to the survival of migratory birds such as eared grebes, *Podiceps nigricollis*, which depend on them as their sole source of food whilst they moult at the lake each year (Cooper *et al.*, 1984). Unfortunately, removal of freshwater from Mono Lake for human consumption is increasing its salinity and threatens the survival of *A. monica* and hence this simple but essential food chain (Thorp and Covich, 2001).

Mono Lake represents the largest habitat for the alkali fly *E. hians*, and its population density in the lake is the highest of any saline aquatic ecosystem known (Herbst, 1988). The numerical densities and biomass of larvae and pupae are greatest at depths of 0.5–1 m in the lake and on hard substrate; up to 100 g dry weight m^{-2} have been recorded (Herbst and Bradley, 1993). The pupal cases of the fly form extensive, dark deposits around the margins of the lake (Fig. 19.6). Consumers of benthic algae and detritus, the flies' main food sources are the diatom *Nitzschia frustulum*; the filamentous green alga *Ctenocladus circinnatus*; and cyanobacteria, *Oscillatoria* spp.; plus some Protozoa (Herbst, 1986). The volume of algae is directly related to the *E. hians* population size. As with *A. monica* (above), anthropogenically induced salinity increases in Mono Lake threaten to reduce the abundance and diversity of the benthic algal community with knock-on effects on *E. hians* (Herbst and Blinn, 1998). The flies themselves are food for many birds in the area (Herbst, 1986) and the pupae's high fat content meant that Native Americans used to collect them to eat. The alkali fly life cycle is typical of many insects, developing from an egg to larva before pupating and metamorphosing. Female alkali flies walk down a substrate and lay their eggs on algae mats at depths of approximately 3 m. Once the larvae have had time to develop and are ready to pupate they will attach to a hard substrate, such as the characteristic Mono Lake tufa columns. The flies possess two pairs of Malpighian tubules, the anterior pair of which connect to a 'lime gland' containing concretions of pure calcium carbonate. This gland is thought to regulate the high concentrations of carbonate and bicarbonate in their surroundings (Herbst and Bradley, 1989) and contribute to their survival in this extreme chemical environment.

Fig. 19.6. Adult alkali flies, *Ephydra hians*, on surface of alkaline Mono Lake, USA. © David Herbst.

Other alkaline lakes support a higher diversity of metazoans. For example, the large diaptomid copepod *Lovenula africana* inhabits the lower alkalinity crater lakes of Ethiopia. Rotifers such as *Hexarthra jenkinae, Brachionus plicatilis* and *B. dimidiatus* are also very abundant in some of these lakes (Green, 1986).

19.3 Mechanisms of Adaptation to High Alkalinity

Cells face many challenges in an alkaline environment and alkaliphilic organisms have developed a range of strategies to survive. Of greatest significance is the ability of cells to maintain internal pH more than two units lower (i.e. more acidic) than the external environment (Krulwich *et al.*, 1997), a process termed pH homeostasis. Bacteria in the genera *Bacillus* are the most well studied alkaliphiles; they achieve pH homeostasis via both passive and active regulation mechanisms. These are discussed below.

19.3.1 Cell walls

The cell surface is key in discriminating the internal from external environment and maintaining the intracellular neutral environment. Maintaining a negatively charged cell wall is a very effective way of achieving this (Krulwich *et al.*, 1998). The components of the cell walls of several alkaliphilic *Bacillus* spp. have been shown to contain acidic polymers, such as galacturonic acid, gluconic acid, glutamic acid, aspartic acid and phosphoric acid, in addition to the peptidogycans common to neutrophilic (neutral-loving) *Bacillus subtilis*. These acidic wall macromolecules provide a passive barrier to ion flux and their negative charges may give the cell surface its ability to adsorb sodium and hydronium ions and repulse the alkaline hydroxide ions from the external environment (Aono and Horikoshi, 1983; Krulwich *et al.*, 1997; Horikoshi, 1999). For cells growing on fermentable carbon substrates (a carbon source that can be metabolized by anaerobic fermentation), this and other passive mechanisms play an important role in pH homeostasis (Krulwich *et al.*, 1997). However, cells growing on both fermentable and non-fermentable carbon substrates require active mechanisms.

19.3.2 The cell membrane and cytoplasmic pH regulation

After the cell wall, cells have a second barrier between external and internal pH: the cell (plasma) membrane (Fig. 19.7). The plasma membrane actively helps maintain pH homeostasis by first establishing an electrochemical gradient via proton extrusion. This is achieved by respiration or using a proton-transporting ATP-synthase that catalyses the production of ATP under conditions in which the external proton concentration and the bulk chemiosmotic driving forces are low, a process termed ATP-driven H^+ expulsion (e.g. Krulwich *et al.*, 1998). The electrochemical gradient generated by H^+ expulsion powers antiporter proteins that create net acidification of the cytoplasm relative to the outside pH. Na^+/H^+ antiporter proteins (e.g. Krulwich *et al.*, 1998; Horikoshi, 1999; Padan *et al.*, 2005) and K^+/H^+ antiporter proteins (e.g. Horikoshi, 1999) have been identified in alkaliphiles, although there are indications that in some alkaliphiles the process is Na^+-dependent (Krulwich *et al.*, 1997). Lastly, pH homeostasis requires mechanisms to allow Na^+ re-entry to the cell to sustain rapid recycling of Na^+. It is likely that both Na^+-solute symporters and pH-gated Na^+ channels are involved (Sugiyama, 1995; Krulwich *et al.*, 1997). Such cytoplasmic pH regulation can be so effective that alkaliphiles living in an environment of pH 10 or greater can maintain near neutral internal pHs. For example, a strain of the bacterium *Micrococcus* sp. was shown to have an optimal catalytic pH of 7.5 despite living in a pH 10 environment (Horikoshi, 1999).

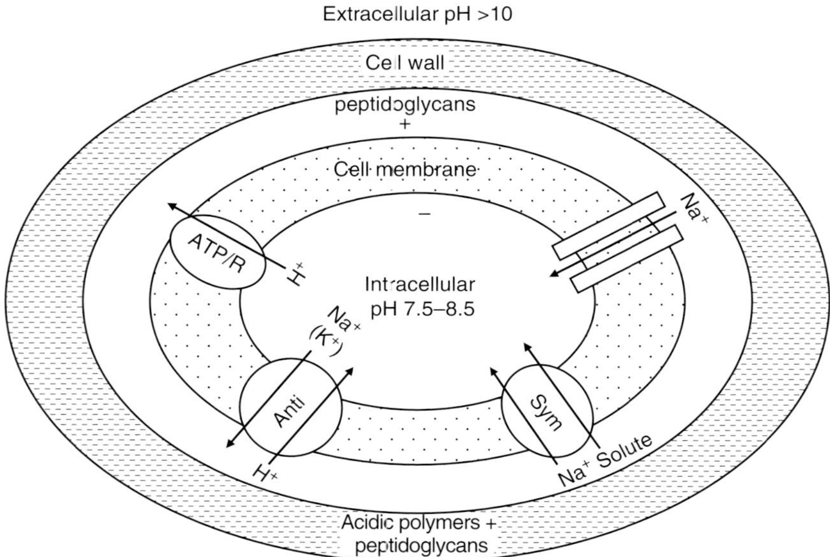

Fig. 19.7. Schematic representation of the cell wall and plasma membrane associated components of pH homeostasis in alkaliphilic bacteria. Negatively charged cell wall (dashed) in which acidic polymers and peptidoglycans are situated. Cell membrane (stipled): ATP/R – electrochemical gradient generated by respiration or ATP-driven H^+ expulsion; Anti – Na^+ or K^+/H^+ antiporter protein; Sym – Na^+/solute symporter. Adapted from Krulwich *et al.*, 1997 and Horikoshi, 1999.

19.3.3 Enzyme stability

There is evidence that alkaliphiles have evolved pH-stable enzymes. Both excreted and surface-located enzymes must be resistant to the effects of extreme pH and able to remain active in the virtual absence of the cations Mg^{2+} and Ca^{2+} (Grant, 2004). Proteases, amylases, glucanotransferases, cellulases, lipases, xylanases, pectinases and chitinases isolated from alkaliphiles have all come under intense scrutiny from scientists and biotechnologists alike and are now employed in a whole suite of industrial processes (Horikoshi, 1999; see Arora and Bell, Chapter 25, this volume).

19.3.4 Alkaliphiles versus neutrophiles

It is still not clear to researchers whether or not the cell wall components, cytoplasmic buffers or antiporters used by alkaliphiles are essentially the same as those used in neutrophiles but present in higher amounts, or whether there are special features and molecules that are required to meet the challenges faced by alkaliphiles (Krulwich *et al.*, 1997). As mentioned above, the cell walls of the alkaliphilic bacteria *Bacillus* spp., often contain peptidogycans common to neutrophilic *Bacillus subtilis* (Horikoshi, 1999). However, neutrophilic *B. subtilis* can utilize either cytoplasmic K^+ or Na^+ antiporters to support pH homeostasis whereas alkaliphilic *Bacillus* spp. appear to be Na^+-specific (summarized in Krulwich *et al.*, 1997). As ever in extremophile research, the more we discover the more we need to find out.

19.4 Case Studies

The diversity and biogeochemistry of some alkaline environments has been covered in depth but presented here are four of the best studied examples that highlight their remarkable nature.

Studied since the 1930s, the best known soda lakes are those of the East African Rift

Valley (e.g. Jenkin, 1936; Grant *et al.*, 1990; Talling, 2006). These systems represent the most naturally occurring alkaline lakes on Earth. Tectonic activity has created a series of shallow basins where groundwater and seasonal streams form permanent standing bodies of water with no significant outflows. Surface evaporation rates exceed dilution in this arid region, thus, Na^+, CO_3^{2-} and Cl^- minerals concentrate creating a suite of lakes ranging from pH 8.5 to pH 12 (Grant, 2006).

19.4.1 Lake Turkana, Kenya

Lake Turkana (formerly Lake Rudolf), Kenya, is the world's largest permanent desert lake and the world's largest alkaline lake. The lake is 260 km long, on average 30 km wide, and has a maximum depth of 114 m. It has an area of approximately 7560 km^2 (Coulter *et al.*, 1986). By volume (ca. 237 km^3) it is also the world's fourth largest salt lake. The majority of the lake is located in Kenya, but the northern end resides in Ethiopia. Lake Turkana is located in an arid, hot region where the mean annual rainfall is less than 250 mm and mean daily temperature ca. 29°C. Three rivers flow into the lake, the Omo, Turkwel and Kerio, but it has no outflows. Thus, water is only lost from the lake via evaporation at rates of ca. 2335 mm $year^{-1}$ (Källqvist *et al.*, 1988). It is the balance of these inflows and evaporation that determine the lake's water level.

The surrounding bedrock is predominantly volcanic and Central Island, located in the middle of the lake, is an active volcano. Consequently, the water chemistry is dominated by sodium and bicarbonate generating an alkalinity >pH 9. The lake is also very saline with a conductivity of 3500 $\mu S\ cm^{-1}$ (Ferguson and Harbott, 1982).

Despite its extreme chemistry, the lake and its surroundings support diverse flora and fauna. Emergent, submerged and floating plants are abundant. Nile crocodiles (*Crocodylus niloticus*; Fig. 19.8), hippopotamus (*Hippopotamus amphibius*) and the vulnerable Turkana mud turtle (*Pelusios broadleyi*) that is endemic to Kenya, inhabit the lake. The lake also supports 50 fish species, 11 of which are endemic (Hopson, 1982) and 84 species of bird, many of which are migratory (Bennun and Njoroge, 1999). Moreover, three species of frog are endemic to the ecoregion: *Poyntonophrynus lughensis*, *Phrynobatrachus natalensis* and *Amietophrynus turkanae*, the latter only

Fig. 19.8. A Nile crocodile (*Crocodylus niloticus*) basks in the shallows. © 2011 iStockphoto LP.

being found on the south-eastern shores of Lake Turkana and in the Uaso Nyiro River, north-central Kenya (IUCN, 2010).

Anthropologists consider the Lake Turkana area to be the cradle of humankind. In 1967, Richard Leakey discovered hominid fossils on its shores at the famous Koobi Fora fossil site (Leakey, 1976). Today, the lake is still home to a number of tribes, including the Turkana (from whom the lake derives its name), the Rendille, Gabra and El Molo (Fig. 19.9). The whole area is now part of the Mount Kulal Biosphere Reserve set up by UNESCO.

19.4.2 Lake Magadi, Kenya

Lake Magadi, Kenya, is the southernmost and lowest lying lake in the Rift Valley, located in a catchment of faulted volcanic rocks (Baker, 1958). The lake has an area of ca. 100 km^2, is 32 km long and 3 km wide, and has a maximum depth of between 1 and 5 m.

Lake Magadi represents the final stage of maximum evaporative concentration. The lake water is a dense sodium carbonate (Na_2CO_3) and sodium chloride (NaCl) brine with a pH of 11.5–12.00. The lake is primarily recharged by saline hot springs (up to 86°C) lying along its north-western and southern shores. These discharge into alkaline lagoons around the lake margins. During the rainy season a thin (<1 m) layer of brine covers much of the saline pan, but since Lake Magadi has no outflows and is located in an arid zone, during the dry season in particular, the brine evaporates depositing vast quantities of minerals, notably trona ($NaHCO_3.Na_2CO_3.12H_2O$; Grant, 2006). It is likely that complex geochemical, hydrological and microbial processes are involved in this mineral deposition, because if it were only a matter of precipitation, thermonatrite ($Na_2CO_3.H_2O$) or natron ($Na_2CO_3.10H_2O$) would precipitate preferentially (Grant, 2006). In places the resultant salt deposits can be 40 m thick and trona is mined on site as a source of soda ash for glass manufacture (Fig. 19.10); the Magadi Soda Factory (established in 1911) lies on the eastern shore of the lake near the Magadi Township. Lake Magadi is also well known for its extensive deposits of siliceous chert, in particular 'Magadi-type chert', which forms from a hydrous sodium silicate mineral precursor, Magadiite ($NaSi_7O_{13}(OH)_3{\cdot}4H_2O$), that was

Fig. 19.9. African village huts, El Molo tribe, Lake Turkana, north Kenya, Great Rift Valley. © 2011 iStockphoto LP.

Fig. 19.10. (a) Thick deposits of salt form at Salar de Uyuni, Bolivia. (b) An example of harvesting salt deposits, Bolivia. © Isaac P. Forster.

discovered at Lake Magadi during the 1960s (Behr, 2002).

In addition to the rich microbial life already discussed in previous sections, a single species of cichlid fish, *Alcolapia grahami*, inhabits the hot, highly alkaline waters of Lake Magadi and some of the hot spring pools at the shoreline where the water temperatures are <45°C (Seegers and Tichy, 1999). *Alcolapia grahami*'s range is restricted to Lakes Magadi and Nakuru, Kenya, and Lake Natron, Tanzania. Lake Magadi was not always saline; during the late Pleistocene to mid-Holocene the lake was freshwater and inhabited by many more fish species, the fossils of which can be found in the sediments around the present shoreline. Lake Magadi is also well known for its rich birdlife, including the charismatic lesser and greater flamingos, *Phoenicopterus minor* and *P. roseus*, respectively, which feed on, but only occasionally breed at, the lake.

19.4.3 Mono Lake, USA

Naturally alkaline and hypersaline Mono Lake, is a closed basin located in California's East Sierra, USA, in a geologically active area at the north end of the Mono-Inyo Craters volcanic chain. It is considered to be the oldest lake in North America. The lake has an area of ca. 182 km^2, is 15 km long, 21 km wide and has a maximum depth of 48 m. Mono Lake resides in the rain shadow of the Sierra Nevada; consequently, the climate can be classified as semi-arid to arid (Johannesson and Lyons, 1994). Inflows to the lake come from groundwater, streams and hot springs, but the lake has no outflows and losses are via evaporation. Annual evaporation for the lake surface has been estimated between 154 and 165 mm, compared with 28–118 mm annual precipitation (Loeffler, 1977).

The author Mark Twain referred to Mono Lake as 'nearly pure lye' (Twain, 1872); indeed the lake's highly concentrated sodium carbonate (Na_2CO_3) brine has a pH of approximately 10 and is rich in chloride (Cl^-), sulfate (SO_4^{2-}), as well as rare earth elements (REE; Johannesson and Lyons, 1994). The salinity of the lake varies depending on the amount of water it contains at any one time. When the lake reached its lowest water levels in 1982 as a result of water being diverted to the City of Los

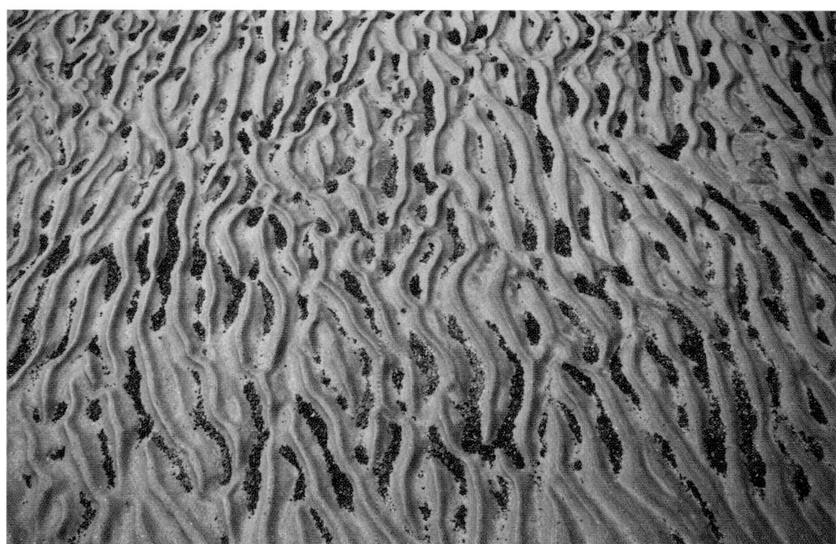

Fig. 19.11. Millions of dead brine shrimp (*Artemia monica*) on the dried lake bed of Mono Lake, USA. © 2011 iStockphoto LP.

Angeles over 480 km away, it reached a salinity of 99 g l^{-1} (compared to an average ocean salinity of ca. 32 g l^{-1}; Fig. 19.11). A legal battle has since reduced the amount of water taken from the basin and the lake is slowly recovering; currently its salinity is approximately 81 g l^{-1} (Mono Lake Committee, 2011). In recent years, freshwater inputs to the lake have meant that density differences across the pycnocline periodically become large enough for meromixis (vertical stratification of the water column) to develop (MacIntyre *et al.*, 1999) and these episodes can have a significant impact on the chemistry and ecology of the lake.

Mono Lake is best known for its spectacular calcium carbonate or tufa columns, which can extend >3 m high and rise above the surface of the water (Fig. 19.12). These are created when calcium-rich groundwater seeps into the alkaline lake water, producing localized calcium precipitation. Similar columns are also formed in alkaline (pH 9.7–9.8) Lake Van, Turkey (Grant, 2006); these can reach a phenomenal 40 m in height (Kempe *et al.*, 1991). The rapid precipitation of minerals traps microorganisms within the columns, leaving behind a microbial fossil record. These offer clues as to how and where life may have originated and are of interest to astrobiologists searching for evidence of extraterrestrial life (Nealson, 1999). For example, meteorites from Mars have been found to contain carbonate globules similar to those of the Mono Lake and Lake Van tufa and raise questions as to whether life could therefore survive passage through space (Hoover, 2005; López-García *et al.*, 2005).

Fig. 19.12. Tufa columns in alkaline Mono Lake, USA. Above the water they no longer grow and are susceptible to erosion. © 2011 iStockphoto LP.

Mark Twain also wrote that Mono Lake contained 'nothing, in fact, that goes to make life desirable' (Twain, 1872). However, more than a century later researchers have demonstrated that the lake is teaming with an abundant diversity of microbial life, both halophilic and alkaliphilic (see Section 19.2). Moreover, the endemic brine shrimp, *Artemia monica*, and the alkali fly, *Ephydra hians*, are abundant, important and charismatic inhabitants (see Section 19.2.10), which in turn support a rich diversity of birds that use Mono Lake to rest and eat for at least part of each year. These include American avocets (*Recurvirostra americana*), killdeer (*Charadrius vociferus*), several species of sandpiper, tens of thousands of migratory eared grebes (*Podiceps nigricollis*; Fig. 19.13) and Wilson's phalaropes (*Phalaropus tricolor*), and the second largest nesting population of California gulls (*Larus californicus*) in the USA (Mono Lake Committee, 2011). Historically, *E. hians* was also consumed by humans. The lake does not, however, support any fish.

19.4.4 Octopus Spring, USA

Octopus Spring, USA, is an alkaline hot spring in the Lower Geyser Basin of

Fig. 19.13. Eared grebe (*Podiceps nigricollis*). © 2011 iStockphoto LP.

Yellowstone National Park. Its outflow channels radiate like the arms of an octopus, hence its name. Water flows from the geothermal source at up to 95°C and into outflow channels where it cools to approximately 83°C. The steady discharge is estimated at 5.7–7.6 l s^{-1} (Hinman and Lindstrom, 1996). There is a lack of hydrogen sulfide in the spring water, resulting in an alkaline system with a pH of 8.4 at the source.

Octopus Spring is rich in microbial life, a colourful succession of which can be observed from its source and radiating along its outflows (Fig. 19.14). Noteworthy examples include the pinkish-purple chemotrophic bacterium *Aquifex pyrophilus* and its close relative *Hydrogenobacter thermophilus*, which form large, pink accumulations of filaments in the spring's outflows close to its source, and grow at temperatures of between 84°C and 88°C. *Aquifex pyrophilus* uses oxygen from the air to oxidize hydrogen gas that is present in the spring water (Reysenbach *et al.*, 1994).

The water cools as it flows away from the source, allowing an increasing number of species to survive and grow in dense microbial mats. A unicellular photosynthetic cyanobacterium resembling the vivid green *Synechococcus* cf. *lividus*, grows well in the outflow channels at temperatures of between 74°C and 42°C. *Synechococcus lividus* co-occurs with a filamentous green non-sulfur bacterium, *Chloroflexus aurantiacus*, which is thought to provide physical cohesiveness to the mats (Ward *et al.*, 1998).

As water temperatures decrease to ca. 57°C, the filamentous cyanobacterium

Fig. 19.14. Alkaline Morning Glory Prismatic Hot Spring, Yellowstone National Park, USA, exemplifying the colourful succession of microbial life present in hot, alkaline environments. © 2011 iStockphoto LP.

Phormidium spp. (also known as *Leptolyngbya*) forms coniform mats that are bright orange in the summer (Ward and Castenholz, 2000; Jahnke *et al.*, 2004). As the temperatures cool to below 43°C, the larvae and adults of ephyrid flies start to graze upon the mats. Then as the waters cools from ca. 40°C to ambient, the nitrogen-fixing cyanobacteria *Calothrix* spp. form dark brown scytonemin-containing mats, which adhere to the siliceous substrates in the channels and around their moist edges (Ward and Castenholz, 2000).

Once again, the incredible microbial communities of alkaline Octopus Spring are considered analogues for life on other planets. In 2007, the NASA Spirit Rover uncovered a white rock layer under the dust whilst travelling across the Gusev Crater on Mars. X-ray spectroscopy revealed that the rock was ca. 90% silica. The science team believe that the silica could have been deposited by a volcanic vent or a hot spring (NASA, 2007), similar to the low sulfur area of Octopus Spring. Further investigation may provide indirect evidence of microbial life in these areas.

19.5 Alkaliphiles in Biotechnology

As discussed in Section 19.3.3 and in more detail in Arora and Bell (Chapter 25, this volume), alkaliphiles have evolved pH-stable enzymes which are resistant to the effects of extreme pH and able to remain active in the virtual absence of the cations Mg^{2+} and Ca^{2+} (Grant, 2004). Because of this, a whole suite of enzymes isolated from alkaliphiles have come under intense scientific scrutiny and many now have a wide range of industrial applications including, but not limited to, detergent additives, hide de-hairing, food processing, recovery of silver from X-ray films, the production of cyclodextrins for foodstuffs, chemicals and pharmaceuticals, biological bleaching of wood pulp, and waste treatments. Alkaliphiles are also a source of metabolites with a range of uses, plus antibiotics and enzyme inhibitors with potential pharmaceutical and biomedical applications (reviewed by Horikoshi, 1999).

19.6 Alkaliphiles as Analogues for Early Life on Earth and on Other Planets

Alkaline environments, in particular soda lakes, are often considered analogues of the terrestrial communities that were thought to be present on Earth's ancient continents during the early Proterozoic period; due to their high salt contents, these communities were only inhabited by prokaryotes. There is further debate as to whether soda lake communities might represent refuges for relic microbial communities (Zavarzin, 1993; Zavarzin and Zhilina, 2000). Akaliphilic communities potentially represent the beginning of the evolutionary line between cyanobacteria and modern-day green algae and higher plants. Alternatively, these ancient saline, epicontinental water bodies may have spurred the development of both ancient and modern saline marine environments with stromatolitic belts and cyanobacterial mats bridging the terrestrial and marine environments (Zavarzin and Zhilina, 2000), and contributed to the evolutionary development of Earth's early atmosphere via photosynthesis and oxygen production.

As with other extreme environments, alkaline environments are also considered analogues of life on other planets in our solar system (see Gómez *et al.*, Chapter 26, this volume, and the alkaline environment case studies discussed in Section 19.4). The processes responsible for the production of ancient, saline, alkaline waters on Earth and the sediments produced in soda lakes, are thought to be analogous to the formation of, and provide evidence for, an ancient, highly alkaline hydrosphere on Mars, traces of which have been detected via remote sensing and during landing missions (Kempe and Kazmierczak, 1997).

References

Abd-el-Malek, Y. and Rizk, S.G. (1963) Bacterial sulphate reduction and the development of alkalinity. II. Laboratory experiments with soils. *Journal of Applied Microbiology* 26, 14–19.

Aono, R. and Horikoshi, K. (1983) Chemical composition of cell walls of alkalophilic strains of *Bacillus*. *Journal of General Microbiology* 129, 1083–1087.

Baker, B.H. (1958) Geology of the Magadi area. *Report of the Geological Survey of Kenya* 42, 81 pp.

Baumgarte, S. (2003) Microbial diversity of soda lake habitats. PhD thesis. The Technical University Carolo-Wilhelmina, Braunschweig, Germany.

Behr, H.J. (2002) Magadiite and Magadi chert: a critical analysis of the silica sediments in the Lake Magadi Basin, Kenya. *Society for Sedimentary Geology (SEPM) Special Publication* 73, pp. 257–273.

Bennun, L. and Njoroge, P. (1999) *Important Bird Areas in Kenya*. Nature Kenya, East African Natural History Society, Nairobi, Kenya.

Bryantseva, I., Gorlenko, V.M., Kompantseva, E.I., Imhoff, J.F., Süling, J. and Mityushina, L. (1999) *Thiorhodospira sibirica* gen. nov., sp. nov., a new alkaliphilic purple sulphur bacterium from a Siberian soda lake. *International Journal of Systematic Bacteriology* 49, 697–703.

Bryantseva, I., Gorlenko, V.M., Kompantseva, E.I. and Imhoff, J.F. (2000a) *Thioalkalicoccus limnaeus* gen. nov., sp. nov., a new alkaliphilic purple sulphur bacterium with bacteriochlorophyll *b*. *International Journal of Systematic and Evolutionary Microbiology* 50, 2157–2163.

Bryantseva, I., Gorlenko, V.M., Kompantseva, E.I., Tourova, T.P., Kuznetsov, B.B. and Osipov, G.A. (2000b) Alkaliphilic heliobacterium *Heliorestis baculata* sp. nov. and emended description of the genus *Heliorestis*. *Archives of Microbiology* 174, 283–291.

Collins, M.D., Lund, B.M., Farrow, J.A.E. and Schleifer, K.H. (1983) Chemotaxonomic study of an alkalophilic bacterium *Exiguobacterium aurantiacum* gen. nov., sp. nov. *Journal of General Microbiology* 129, 2037–2042.

Conte, F.P., Jellison, R.S. and Starrett, G.L. (1988) Nearshore and pelagic abundances of *Artemia monica* in Mono Lake, California. *Hydrobiologia* 158, 173–181.

Cooper, S.D., Winkler, D.W. and Lenz, P.H. (1984) The effects of grebe predation on a brine shrimp population. *Journal of Animal Ecology* 53, 51–64.

Coulter, G.W., Allanson, B.R., Bruton, M.N., Greenwood, P.H., Hart, R.C., Jackson, P.B.N. and Ribbink, A.J. (1986) Unique qualities and special problems of the African Great Lakes. *Environmental Biology of Fishes* 17, 161–183.

Ferguson, A.J.D. and Harbott, B.J. (1982) Geographical, physical and chemical aspects of Lake Turkana. In: Hopson, A.J. (ed.) *Lake Turkana: A report of the findings of the Lake Turkana Project 1972–1975*. Overseas Development Administration, London, Uk, pp. 1–107.

Florenzano, G., Sili, C., Pelosi, E. and Vincenzini, M. (1985) *Cyanospira rippkae* and *Cyanospira capsulata* (gen.nov. and spp. nov.): new filamentous heterocystous cyanobacteria from Mgadi lake (Kenya). *Archives of Microbiology* 140, 301–306.

Gee, J.M., Lund, B.M., Metacalf, G. and Peel, J.L. (1980) Properties of a new group of alkaliphilic bacteria. *Journal of General Microbiology* 117, 9–17.

Grant, W.D. (2004) Introductory chapter: half a lifetime in soda lakes. In: Ventosa, A. (ed.) *Halophilic Microorganisms*. Springer-Verlag, Berlin, Heidelberg, Germany, pp. 17–32.

Grant, W.D. (2006) Alkaline environments and biodiversity. In: Gerday, C. and Glansdorff, N. (eds) *Encyclopedia of Life Support Systems* (EOLSS). Developed under the auspices of the UNESCO, Eolss Publishers, Oxford, UK (www.eolss.net).

Grant, W.D., Mwatha, W.E. and Jones, B.E. (1990) Alkaliphiles: ecology, diversity and applications. *FEMS Microbiology Letters* 75, 255–269.

Green, J. (1986) Zooplankton associations in some Ethiopian crater lakes. *Freshwater Biology* 16, 495–499.

Guixa-Boixareu, N., Calderon-Paz, J.I., Heldal, M., Bratbak, G. and Pedros-Alio, C. (1996) Viral lysis and bacterivory as prokaryotic loss factors along a salinity gradient. *Aquatic Microbial Ecology* 11, 215–227.

Herbst, D.B. (1986) Comparative studies of the population ecology and life history patterns of an alkaline salt lake insect: *Ephydra hians* Say. PhD thesis. Oregon State University, Corvallis, USA, 206 pp.

Herbst, D.B. (1988) Comparative population ecology of *Ephydra hians* Say (Diptera: Ephydridae) at Mono Lake (California) and Albert Lake (Oregon). *Hydrobiologia* 158, 145–166.

Herbst, D.B. and Blinn, D.W. (1998) Experimental mesocosm studies of salinity effects on the benthic algal community of a saline lake. *Journal of Phycology* 34, 772–778.

Herbst, D.B. and Bradley, T.J. (1989) A Malpighian tubule lime gland from an insect inhabiting alkaline salt lakes. *Journal of Experimental Biology* 145, 63–67.

Herbst, D.B. and Bradley, T.J. (1993) A population model for the alkali fly at Mono Lake: depth distribution and changing habitat availability. *Hydrobiologia* 267, 191–201.

Hinman, N.W. and Lindstrom, R.F. (1996) Seasonal changes in silica deposition in hot spring systems. *Chemical Geology* 132, 237–246.

Hoover, R.B. (2005) Microfossils, biominerals, and chemical biomarkers in meteorites. In: Hoover, R.B., Razanov, A.Y. and Paepe, R. (eds) *Perspectives in Astrobiology*. IOS Press, NATO Science Series, Amsterdam, the Netherlands, pp. 43–65.

Hopson, A.J. (1982) *Lake Turkana: a report of the findings of the Lake Turkana Project 1972–1975*. Overseas Development Administration, London, UK, pp. 1–6.

Horikoshi, K. (1999) Alkaliphiles: some applications of their products for biotechnology. *Microbiology and Molecular Biology Reviews* 63, 735–750.

Imhoff, J.F., Sahl, H.G., Soliman, G.S.H. and Trüper, H.G. (1979) The Wadi Natrun: chemical composition and microbial mass developments in alkaline brines of eutrophic desert lakes. *Geomicrobiology Journal* 1, 219–234.

IUCN (2010) IUCN Red List of Threatened Species. Available at: www.iucnredlist.org/apps/redlist/details/54784/0 (accessed 9 February 2011).

Jahnke, L.L., Embaye, T., Hope, J., Turk, K.A., Van Zuilen, M., Des Marais, D.J., Farmer, J.D. and Summons, R.E. (2004) Lipid biomarker and carbon isotopic signatures for stromatolite-forming, microbial mat communities and *Phormidium* cultures from Yellowstone National Park. *Geobiology* 2, 31–47.

Jenkin, P.M. (1936) Reports of the Percy Sladen Expedition to some Rift Valley lakes in Kenya in 1929. VII. Summary of the ecological results with special reference to the alkaline lakes. *Annals and Magazine of Natural History* 18, 133–181.

Jiang, S., Steward, G., Jellison, R., Chu, W. and Choi, S. (2003) Abundance, distribution, and diversity of viruses in alkaline, hypersaline Mono Lake, California. *Microbial Ecology* 47, 9–17.

Johannesson, K.H. and Lyons, W.B. (1994) The rare earth element geochemistry of Mono Lake water and the importance of carbonate complexing. *Limnology and Oceanography* 39, 1141–1154.

Johnson, H.W. (1923) Relationships between hydrogen ion, hydroxyl ion and salt concentrations and the growth of seven soil molds. *Iowa Agriculture Experiment Station Research Bulletin* 76, 307–344.

Jones, B.E., Grant, W.D., Duckworth, A.W. and Owenson, G.G. (1998) Microbial diversity of soda lakes. *Extremophiles* 2, 191–200.

Källqvist, T., Lien, L. and Liti, D. (1988) *Lake Turkana. Limnological Study 1985–1988*. Norwegian Institute for Water Research, Report O-85313, 98 pp.

Kempe, S. and Kazmierczak, J. (1997) A terrestrial model for an alkaline Martian hydrosphere. *Planetary and Space Science* 45, 1493–1495.

Kempe, S., Kazmierczak, J., Landmann, G., Konuk, T., Reimer, A. and Lipp, A. (1991) Largest known microbialites discovered in Lake Van, Turkey. *Nature* 349, 605–608.

Kevbrin, V.V., Lysenko, A.M. and Zhilina, T.N. (1997) Physiology of the alkaliphilic methanogen Z-7936, a new strain of *Methanosalsus zhilinaeae* isolated from Lake Magadi. *Microbiology* 66, 261–266.

Kompantseva, E.I., Komova, A.V., Rusanov, I.I., Pimenov, N.V. and Sorokin, D.Yu. (2009) Primary production of organic matter and phototrophic communities in the soda lakes of the Kulunda steppe (Altai krai). *Microbiology* 78, 643–649.

Kotelnikova, S., Macario, A.J.L. and Pedersen, K. (1998) *Methanobacterium subterraneum* sp. nov., a new alkaliphilic, eurythermic and halotolerant methanogen isolated from deep granitic groundwater. *International Journal of Systematic Bacteriology* 48, 357–367.

Krulwich, T.A., Ito, M., Gilmour, R. and Guffanti, A.A. (1997) Mechanisms of cytoplasmic pH regulation in alkaliphilic strains of *Bacillus*. *Extremophiles* 1, 163–169.

Krulwich, T.A., Ito, M., Hick, D.B., Gilmour, R. and Guffanti, A.A. (1998) pH homeostasis and ATP synthesis: studies of two processes that necessitate inward proton translocation in extremely alkaliphilic *Bacillus* species. *Extremophiles* 2, 217–222.

Leakey, R.E.F. (1976) New hominid fossils from the Koobi Fora formation in Northern Kenya. *Nature* 261, 574 – 576.

Liu, R., Jiang, X., Mou, H., Guan, H., Hwang, H.M. and Li, X. (2009) A novel low-temperature resistant alkaline lipase from a soda lake fungus strain *Fusarium solani* N4-2 for detergent formulation. *Biochemical Engineering Journal* 46, 265–270.

Loeffler, R.M. (1977) Geology and hydrology. In: Winkler, D.W. (ed.) *An Ecological Study of Mono Lake, California*. Institute of Ecology, Publication 12, University of California, Davis, California, pp. 6–38.

López-García, P., Kazmierczak, J., Benzerara, K., Kempe, S., Guyot, F. and Moreira, D. (2005) Bacterial diversity and carbonate precipitation in the giant microbialites from the highly alkaline Lake Van, Turkey. *Extremophiles* 9, 263–274.

Lugo, A., Alcocer, J., del Rosario-Sanchez, Ma. and Escobar, E. (1998) Littoral protozoan assemblages from two Mexican hyposaline lakes. *Hydrobiologia* 381, 9–13.

MacIntyre, S., Flynn, K.M., Jellison, R. and Romero, J.R. (1999) Boundary mixing and nutrient fluxes in Mono Lake, California. *Limnology and Oceanography* 44, 512–529.

Madoni, P. and Uluhogian, F. (1997) Ciliated protozoa of a geothermal sulphur spring. *Hydrobiologia* 353, 161–170.

McGenity, T.J. and Grant, W.D. (1993) The haloalkaliphilic archaeon (archaebacterium) *Natronococcus occultus* represents a distinct lineage within the *Halobacteriales*, most closely related to the other haloalkaliphilic lineage (*Natronobacterium*). *Systematic and Applied Microbiology* 16, 239–243.

Melack, J.M. and Kilham, P. (1974) Photosynthetic rates of phytoplankton in East African alkaline, saline lakes. *Limnology and Oceanography* 19, 743–755.

Milford, A.D., Achenbach, L.A., Jung, D.O. and Madigan, M.T. (2000) *Rhodobaca bogoriensis* gen. nov. and sp. nov., an alkaliphilic purple non-sulfur bacterium from African Rift Valley soda lakes. *Archives of Microbiology* 174, 18–27.

Mono Lake Committee (2011) Available at: www.monolake.org/about/stats (accessed 14 February 2011).

Nagai, K., Sakai, T., Rantiatmodjo, R.M., Suzuki, K., Gams, W. and Okada, G. (1995) Studies on the distribution of alkalophilic and alkali-tolerant soil fungi I. *Mycoscience* 36, 247–258.

Nagai, K., Suzuki, K. and Okada, G. (1998) Studies on the distribution of alkalophilic and alkali-tolerant soil fungi II. Fungal flora in two limestone caves in Japan. *Mycoscience* 39, 293–298.

NASA (2007) Available at: www1.nasa.gov/mission_pages/mer/images/pia09403.html (accessed 16 February 2011).

Nealson, K.H. (1999) The search for extraterrestrial life. *Engineering and Science* 1/2, 30–39.

Padan, E., Bibi, E. and Krulwich, T.A. (2005) Alkaline pH homeostasis in bacteria: new insights. *Biochimica et Biophysica Acta – Biomembranes* 1717, 67–88.

Reysenbach, A.L., Wickham, G.S. and Pace, N.R. (1994) Phylogenetic analysis of the hyperthermophilic pink filament community in Octopus Spring, Yellowstone National Park. *Applied and Environmental Microbiology* 60, 2113–2119.

Roadcap, G.S., Sanford, R.A., Jin, Q., Pardinas, J.R. and Bethke, C.M. (2006) Extremely alkaline (pH>12) groundwater hosts diverse microbial community. *Groundwater* 44, 511–517.

Seegers, L. and Tichy, H. (1999) The *Oreochromis alcalicus* flock (Teleostei: Cichlidae) from lakes Natron and Magadi, Tanzania and Kenya, with descriptions of two new species. *Ichthyological Exploration of Freshwaters* 10, 97–146.

Selyanin, V.V., Oborotov, G.E., Zenova, G.M. and Zvyagintsev, D.G. (2005) Alkaliphilic soil actinomycetes. *Microbiology* 74, 729–734.

Sorokin, D.Y., Muyzer, G., Brinkhoff, T., Kuenen, J.G. and Jetten, M.S. (1998) Isolation and characterization of a novel facultatively alkaliphilic *Nitrobacter* species, *N. alkalicus* sp. nov. *Archives of Microbiology* 170, 345–352.

Sorokin, D.Y., Jones, B.E. and Kuenen, J.G. (2000a) An obligate methylotrophic, methane-oxidizing *Methylomicrobium* species from a highly alkaline environment. *Extremophiles* 4, 145–155.

Sorokin, D.Y., Tourova, T.P., Kusnetsov, B.B., Bryantseva, I.A. and Gorlenko, V.M. (2000b) *Roseinatronobacter thiooxidans* gen. nov., sp. nov., a new alkaliphilic aerobic bacteriochlorophyll-alpha-containing bacteria from a soda lake. *Microbiology* 69, 89–97.

Sugiyama, S. (1995) Na^+-driven flagellar motors as a likely Na^+ re-entry pathway in alkaliphilic bacteria. *Molecular Microbiology* 15, 592.

Svetlitshnyi, V., Rainey, F. and Wiegel, J. (1996) *Thermosyntropha lipolytica* gen. nov., sp. nov., a lipolytic, anaerobic, alkalitolerant, thermophilic bacterium utilizing short- and long-chain fatty acids in syntrophic coculture with a methanogenic archaeum. *International Journal of Systematic Bacteriology* 46, 1131–1137.

Talling, J. (2006) A brief history of the scientific study of tropical African inland waters. *Freshwater Forum* 26, 3037.

Thompson, V.S., Schaller, K.D. and Apel, W.A. (2003) Purification and characterization of a novel thermo-alkali-stable catalase from *Thermus brockianus*. *Biotechnology Progress* 19, 1292–1299.

Thorp, J.H. and Covich, A.P. (2001) An overview of freshwater habitats. In: Thorp, J.H. and Covich, A.P. (eds) *Ecology and Classification of North American Freshwater Invertebrates*. Academic Press, San Diego, California, pp. 19–42.

Tindall, B.J. (1988) Prokaryotic life in alkaline, saline, athalassic environment. In: Rodriguez-Valera, F. (ed.) *Halophilic Bacteria*. CRC Press, Boca Raton, Florida, pp. 31–67.

Twain, M. (1872) Roughing It, Chapter 38. University of Virginia Library: Electronic Text Center, ISBN 0195159799. Available at: http://etext.lib.virginia.edu/etcbin/toccer-new2?id=TwaRoug.sgm&images=images/modeng&data=/texts/english/modeng/parsed&tag=public&part=38&division=div1 (accessed 15 February 2011).

Ward, D.M. and Castenholz, R.W. (2000) Cyanobacteria in geothermal habitats. In: Whitton, B.A. and Potts, M. (eds) *The Ecology of Cyanobacteria: their diversity in time and space*. Kluwer Academic Publishers, Dordrecht, the Netherlands, pp. 37–59.

Ward, D.M., Ferris, M.J., Nold, S.C. and Bateson, M.M. (1998) A natural view of microbial biodiversity within hot spring cyanobacterial mat communities. *Microbiology and Molecular Biology Reviews* 62, 1353–1370.

Wommack, K.E. and Colwell, R.R. (2000) Virioplankton: viruses in aquatic ecosystems. *Microbiology and Molecular Biology Reviews* 64, 69–114.

Xu, Y., Zhou, P. and Tian, X. (1999) Characterization of two novel haloalkaliphilic archaea *Natronorubrum bangense* gen. nov., sp. nov. and *Natronorubrum tibetense* gen. nov., sp. nov. *International Journal of Systematic Bacteriology* 49, 261–269.

Zak, J.C. and Wildman, H.G. (2004) Fungi in stressful environments. In: Mueller, G.M., Bills, G.F. and Foster, M.S. (eds) *Biodiversity of Fungi: inventory and monitoring methods*. Elsevier Academic Press, London, UK, pp. 303–316.

Zavarzin, G.A. (1993) Epicontinental soda lakes as probable relict biotopes of terrestrial biota formation. *Microbiology* 62, 473–479.

Zavarzin, G.A. and Zhilina, T.N. (2000) Anaerobic chemotrophic alkaliphiles. In: Seckbach, J. (ed.) *Journey to Diverse Microbial Worlds*. Kluwer Academic Publishers, the Netherlands, pp. 191–208.

Zavarzin, G.A., Zhilina, T.N. and Kevbrin, V.V. (1999) The alkaliphilic microbial community and its functional diversity. *Microbiology* 68, 503–521.

Zhilina, T.N., Zavarzin, G.A., Rainey, F., Kevbrin, V.V., Kostrikina, N.A. and Lysenko, A.M. (1996) *Spirochaeta alkalica* sp. nov., *Spirochaeta africana* sp. nov., and *Spirochaeta asiatica* sp. nov., alkaliphilic anaerobes from the continental soda lakes in central Asia and the East African Rift. *International Journal of Systematic Bacteriology* 46, 305–312.

Zhilina, T.N., Zavarzin, G.A., Rainey, F.A., Pikuta, E.N., Osipov, G.A. and Kostrikina, N.A. (1997) *Desulfonatronovibrio hydrogenovorans* gen. nov., sp. nov., an alkaliphilic, sulfate-reducing bacterium. *International Journal of Systematic Bacteriology* 47, 144–149.

Zhilina, T.N., Garnova, E.S., Tourova, T.P., Kostrikina, N.A. and Zavarzin, G.A. (2001) *Amphibacillus fermentum* sp. nov. and *Amphibacillus tropicus* sp. nov., new alkaliphilic, facultatively anaerobic, saccharolytic bacilli from Lake Magadi. *Microbiology* 70, 711–722.

Zhilina, T.N., Kevbrin, V.V., Tourova, T.P., Lysenko, A.M., Kostrikina, N.A. and Zavarzin, G.A. (2005a) *Clostridium alkalicellum* sp. nov., an obligately alkaliphilic cellulolytic bacterium from a soda lake in the Baikal region. *Microbiology* 74, 557–566.

Zhilina, T.N., Zavarzina, D.G., Kuever, J., Lysenko, A.M. and Zavarzin, G.A. (2005b) *Desulfonatronum cooperativum* sp. nov., a novel hydrogenotrophic, alkaliphilic, sulfate-reducing bacterium, from a syntrophic culture growing on acetate. *International Journal of Systematic and Evolutionary Microbiology* 55, 1001–1006.

20 Hypersaline Environments

Terry J. McGenity[1] and Aharon Oren[2]
[1]*Department of Biological Sciences, University of Essex, Colchester, UK;*
[2]*Department of Plant and Environmental Sciences, The Institute of Life Sciences, and the Moshe Shilo Minerva Center for Marine Biogeochemistry, The Hebrew University of Jerusalem, Israel*

20.1 Introduction

Hypersaline environments, i.e. those environments that are more salty than seawater, are extremely variable in terms of their overall salinity and ion composition, and differ in many other aspects, such as temperature, pressure and nutrient status. They range from ephemeral drops on the surface of leaves to salt pans as vast as Jamaica; and from sun-baked salinas to dark buried evaporites. Not surprisingly, therefore, there is an immense diversity of organisms that can be called halophiles (salt lovers). In this chapter we try to capture some of that diversity, focusing largely on those organisms that grow optimally at a salinity of more than 100g l^{-1} (approximately three times more saline than seawater), which is mainly, but not exclusively, the realm of microorganisms, and concentrating primarily on their ecology and physiology. Salinity is one of the main factors that influence community composition, and some hypersaline environments are among the most diverse ecosystems on Earth. High salt concentrations attenuate some microbial processes while enhancing others, and so it is imperative that they are considered when modelling global biogeochemical cycles. Furthermore, the tenacity of extreme halophiles, coupled with the preserving properties of salt, makes them ideal candidates for investigating long-term survival.

20.2 Halophiles – the Main Groups

More than 30 years ago, Donn Kushner (1978) proposed a classification of microorganisms on the basis of their salt requirement and tolerance. He distinguished between extreme halophiles (growing best at 2.5–5.2 M NaCl), borderline extreme halophiles (having their optimum at 1.5–4.0 M NaCl), moderate halophiles (preferring media with 0.5–2.5 M NaCl) and halotolerant microorganisms that do not require salt for growth but tolerate salt often in high concentrations. This chapter mostly deals with the extreme and moderate halophiles, growing in environments with salt concentrations above 100g l^{-1} (equivalent to 1.7 M NaCl) up to salt saturation (around 5.2 M or 300 g l^{-1} NaCl; see Box 20.1).

Such halophiles can be found in each of the three domains of life: Archaea, Bacteria and Eukarya (Box 20.2). The best known extreme halophiles are found within the domain Archaea (phylum Euryarchaeota) in the class Halobacteria, which contains the single order Halobacteriales, with a single family the Halobacteriaceae

Box 20.1. Minerals from salt and measurements of salinity

Table 20.1 shows the composition of seawater and how it changes upon evaporation. The most insoluble ions come out of solution first, especially calcium carbonate minerals ($CaCO_3$), followed by gypsum ($CaSO_4.2H_2O$). A balanced inflow and evaporation can result in thick gypsum crusts that often house multicoloured, layered microbial communities (Fig. 20.2b; Canfield *et al.*, 2004). If evaporation exceeds inflow, then crystals of halite (NaCl) will form, perhaps enhanced by extreme halophiles. In order for the highly soluble K^+ and Mg^{2+} salts to precipitate, the salinity must increase considerably, and the resulting brines, called bitterns on account of their bitter taste, create an extremely stressful milieu for life.

Throughout this text we have used the unit g l^{-1} as a measure of salinity. However, when reading the literature a wide range of direct and indirect measures of salt concentration as well as different units are employed. Therefore, the following explanation may help when comparing different studies.

In the halophile literature, mass of dissolved salts per final volume is the most common way of describing salinity, e.g. as g l^{-1} or parts per thousand w/v (= ppt = ‰). Some also use mass of dissolved salts in a given mass of solution (e.g. g kg^{-1} or ppt w/w). This can be determined by slowly drying then weighing a known mass of filtered brine. Oceanographers tend to more commonly use salinity without units (although some use the unit 'practical salinity unit' or 'psu') based on a relative scale, defining a seawater sample with a conductivity at 15°C equivalent with a KCl solution containing a mass of 32.4356g in a total mass of 1 kg of solution as having a salinity of 35.000. This system replaced another measure 'chlorinity' (see UNESCO, 1985).

Measures of conductivity without calibration are often reported, and this is necessary when salinity exceeds 40g l^{-1}, i.e. in hypersaline environments. An increase in the number of ions in water will increase its ability to conduct electricity, and this can be readily measured with calibrated equipment, taking care to recall that conductivity is also temperature dependent. Seawater has a conductivity of 53,088 micro Siemens per cm ($\mu S\ cm^{-1}$) at 25°C. Conversion from conductivity to ‰, for example, requires that the ionic composition of the solution is known, and for hypersaline environments it cannot be assumed that this is going to be like seawater. Refractive index and density can also be used as proxies for salinity.

An extremely useful value, commonly used in the food industry, is 'water activity' (a_w), which can be defined as the vapour pressure of water above a sample divided by that above pure water at the same temperature, and it is related to the mole fraction of water (Brown, 1990). It is unitless and the value at 25°C is usually given. It can be considered as a measure of water that would be available to a cell, i.e. not bound to solutes or in a matrix. Typical a_w values include: pure water, 1; seawater, 0.98; saturated NaCl 0.75.

For hypersaline brines it is preferable to determine the ionic composition, using a range of methods. Care must be taken with units: a molar (M) solution contains 1 mole of solute per litre of solvent, whereas a molal (*m*) solution contains 1 mole of solute per kg of solvent. In very dilute solutions molar and molal concentrations are similar, but diverge as the concentration increases, whereby a solution of *x* M is more concentrated than one of *x m* (Brown, 1990).

(Oren, 2006b). These organisms are often referred to colloquially as 'haloarchaea' to reflect the fact that they belong to the Archaea. In the Euryarchaeota there also are halophilic or highly halotolerant methanogens (order Methanosarcinales, genera *Methanohalobium* and *Methanohalophilus*). We know of no truly halophilic members of the phylum Crenarchaeota. Recently 16S rRNA sequences related to the Nanoarchaeota were retrieved from salt ponds in South Africa and Inner Mongolia (Casanueva *et al.*, 2008), but the nature of the organisms that harbour these sequences is still unknown.

In the domain Bacteria, halophily is widespread (Fig. 20.1): halophiles are found in the phyla Cyanobacteria, Proteobacteria, Firmicutes, Actinobacteria, Spirochaetes and Bacteroidetes. Species with highly different salt requirements are often classified within a single genus or family. Only a few taxonomic groups consist entirely of halophiles (Fig. 20.1): the above-mentioned archaeal order Halobacteriales (Oren, 2006b), the bacterial order Halanaerobiales (Firmicutes) (Oren, 2006c) and the family Halomonadaceae (Gammaproteobacteria) (Arahal and Ventosa, 2006). Considering the

Box 20.2. Application of gene sequence analysis in the taxonomy and ecology of the Halobacteriales and other halophiles

Analysis of the small subunit ribosomal RNA (ssu rRNA; 16S rRNA in prokaryotes and 18S rRNA in eukaryotes) illuminated the natural relations between all life forms (Woese, 1987), resulting in life being divided into three domains, Archaea, Bacteria (Fig. 20.1) and Eukarya, and in the recognition that the Halobacteria belong to the domain Archaea. Most organisms have several very similar copies of ssu rRNA genes, but the Halobacteriaceae are unusual in that species in several genera, such as *Haloarcula* and *Halomicrobium*, have highly heterogeneous copies of 16S rRNA genes, up to 9% in *Halomicrobium mukohataei* (Cui *et al.*, 2009). Such distinct copies of 16S rRNA genes may have arisen recently by lateral gene transfer, and have not yet undergone sufficient recombination to homogenize the sequences. Alternatively, they may represent paralogous genes that have diverged to function under different environmental conditions. Evidence for the latter, so-called ecoparalog hypothesis (Sanchez-Perez *et al.*, 2008), is emerging, as temperature-dependent differential expression has been demonstrated (López-López *et al.*, 2007). This within-species gene heterogeneity complicates the taxonomy of the Halobacteriaceae, however divergent sequences from species of the same genus are not randomly distributed throughout the halobacterial tree, but form distinct clusters. Also, the use of other gene sequences, such as that encoding DNA-dependent RNA polymerase subunit B′ (*rpo*B′), has helped to resolve some taxonomic issues (Minegishi *et al.*, 2010), and analysis of multiple gene sets as well as whole-genome analysis will be valuable in the future.

DNA can be readily extracted from the environment and ssu rRNA genes from diverse organisms amplified, separated and sequenced to obtain a picture of the main organisms present. The heterogeneity of 16S rRNA gene sequences in many halobacterial species, as discussed above, is likely to give an exaggerated picture of their diversity in the environment. However, this approach, coupled with a range of other techniques, has allowed insight into halophilic communities, and even provided focus for the cultivation of important halophiles (e.g. Boxes 20.3 and 20.5). At the same time there are many clades that are often abundant in, and unique to, anoxic hypersaline environments in particular that have so far evaded cultivation, such as the archaeal MSLB-1 group (van der Wielen *et al.*, 2005). Many other genes coding for a specific activity have been used in hypersaline environments, such as the the *dsrAB* gene coding for dissimilatory sulfite reductase and the *mcrA* gene coding for methyl coenzyme-M reductase subunit-A, to identify sulfate-reducing bacteria and methanogens, respectively (e.g. Hallsworth *et al.*, 2007).

Eukarya (eukaryotes), there are halophiles among the green algae (including species of the unicellular flagellate alga *Dunaliella*, which is the main primary producer in most hypersaline environments), the diatoms, the fungi, the flagellate and amoeboid Protozoa, and even among the crustaceans: the brine shrimp *Artemia* spp. (Oren, 2002, 2006a, 2008).

20.3 An Overview of Hypersaline Environments

Typical perceptions of hypersaline environments include large inland lakes, such as the Dead Sea, that are so rich in salts that the high-density brine allows people to float with ease on the surface; alternatively some may picture rufous coastal salterns, rectangular ponds in which sea salt forms by evapotranspiration (Fig. 20.2a). The term ‘hypersaline environment’ is defined as an environment with a higher salinity than seawater (~35 g l^{-1}; Table 20.1). There are, however, so many types of hypersaline environments (Boetius and Joye, 2009) that one chapter cannot begin to cover all of them, and so we will focus largely on environments with salinities from 100 g l^{-1} (approximately three times saltier than seawater) up to the point at which halite precipitates (see Box 20.1). The ecology of environments at such extreme salinities is driven mainly by microbes, specifically halophilic microbes. Monographs by Javor (1989) and Oren (2002) should be consulted for a detailed analysis of hypersaline environments and their biogeochemistry, but the following provides a taster of their diversity in terms

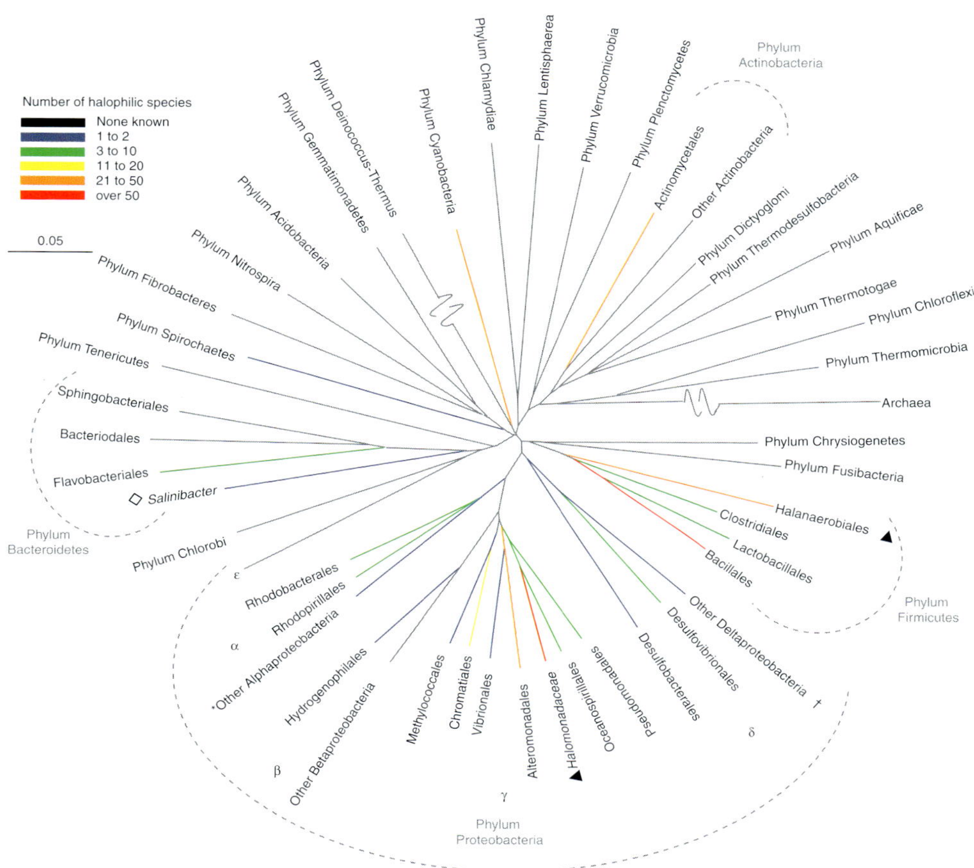

Fig. 20.1. Distribution of halophilic microorganisms in the domain Bacteria. Only organisms that can grow above 100 g l^{-1} NaCl were considered as halophilic, and only organisms described before June 2009 were included. For each group, one representative 16S rRNA gene sequence was selected. The tree was constructed using Jukes-Cantor distance and Neighbour Joining methods. The archaeal outgroup was *Halobacterium noricense*. The bar represents 0.05 nucleotide substitutions per site. ▲ This group (order or family) exclusively contains halophilic organisms. ◊ *Salinibacter ruber* is the only extremely halophilic bacterium known to carry out the salt-in stategy. * One halophilic organism in the order Rhizobiales. † One halophilic organism in the order Bdellovibrionales. Figure prepared with assistance from Audrey Gramain.

of scale, provenance, permanence and ionic composition, and Fig. 20.3 identifies their location.

Hypersaline environments can be havens for wildlife, and frequently have cultural, archaeological, touristic and economic importance (Kurlansky, 2002). They can also be vast, for example the Makgadikgadi salt pans in Botswana and the Salar de Uyuni in Bolivia are both similar in area to Jamaica. Despite these factors, the ecology of many large-scale inland hypersaline environments remains mostly unexplored, and recent 16S rRNA analyses (Box 20.2) of uncultivated archaeal communities in inland salt pans reveal that they are home to many new genera of Halobacteriaceae (Caton *et al.*, 2009). Salt-tolerant and salt-requiring bacteria have also been isolated from saline soils worldwide.

The meeting of sea and land provides an opportunity for evaporative concentration of seawater. In tropical and subtropical regions the relevant combination of tidal

Table 20.1. Concentration of the major cations (Na^+, Mg^{2+}, Ca^{2+}, K^+) and anions (Cl^-, SO_4^{2-}) in various hypersaline brines and seawater (g l^{-1}).

	Na^+	Mg^{2+}	Ca^{2+}	K^+	Cl^-	SO_4^{2-}	Salinity	Notes*	Ref†
Seawater	10.8	1.3	0.4	0.4	19.4	2.7	35		1
Seawater at onset of gypsum precipitation	49.5	6.8	1.7	2.0	91.5	12.5	164		1
Seawater at onset of halite precipitation	98.4	14.5	0.4	4.9	187.0	19.3	324		1
Seawater at onset of potash precipitation	61.4	39.3	0.2	12.8	189.0	51.2	354		1
Great Salt Lake (North Arm), USA	105.0	11.1	0.3	6.7	181.0	27.0	333	North Arm, 1977	1
Dead Sea, Israel-Jordan	39.7	42.4	17.2	7.6	219.0	0.4	327	Lower water mass, 1975	1
Don Juan Pond, Antarctica	11.5	1.2	114.0	0.2	212.0	0.01	339	$CaCl_2$-rich brine, 1962	1
Lake Magadi, Kenya	161.0	0	0	2.3	111.8	16.8	315	$CO_3^{2-}/HCO_3^- = 23.4$; pH = 11	2
Bannock deep-sea brine lake, Mediterranean	97.4	15.8	0.7	5.0	190.0	13.2	322		3
Urania deep-sea brine lake, Mediterranean	80.6	7.7	1.3	4.8	132.2	10.3	237	HS^- concentrations up to 16 mM	3
Discovery deep-sea brine lake, Mediterranean	1.6	121.4	0.1	0.8	336.5	9.2	470	$MgCl_2$-rich brine	3

*Dates of original publications are indicated where changes in salinity have occurred subsequently. The salinity of the Dead Sea, for example, increased to 347 g l^{-1} in 2007, with an increase in Mg^{2+} and a decrease in Na^+ (Bodaker *et al.*, 2010).

†References: 1, Javor (1989) and references therein; 2, Grant (2004); 3, van der Wielen *et al.* (2005).

(a)

(b)

Fig. 20.2. Saltern habitats: (a) Salinas de S'Avall, Mallorca, Spain. The red pond on the left is saturated with NaCl, and has a crust of halite crystals on the surface. The mound of harvested salt is about 2.5 m high. In the lower-salinity pond at the back of the picture wading birds can just be seen, © Rafael Bosch. (b) Gypsum crust from the bottom of a shallow saltern pond (200 g l^{-1} salinity) in Eilat, Israel, showing layered communities of phototrophic microbes. The orange-brown upper layer is dominated by unicellular cyanobacteria; the green layer is dominated by filamentous cyanobacteria; and the red-purple layer contains a dense community of purple sulfur bacteria (e.g. *Halochromatium* spp.) that oxidize sulfide produced by sulfate-reducing bacteria present in the lower grey layer, © Andreas Thywißen.

extent, coastal topography, sediment structure and climate may result in long-lasting sabkhas (flat hypersaline coastal sediments). Even in temperate salt marshes salinities can reach 180 g l^{-1} within several days and then return to seawater salinity in minutes. Therefore, many microbes inhabiting such environments have diverse adaptation mechanisms to tolerate periodic desiccation and rewetting (McKew *et al.*, 2011), and halophiles have been isolated from such environments (see Box 20.4). Hypersaline lagoons are widespread in subtropical coasts, and many have been manufactured to create saltern ponds in which seawater is evaporated until sea salt can be harvested

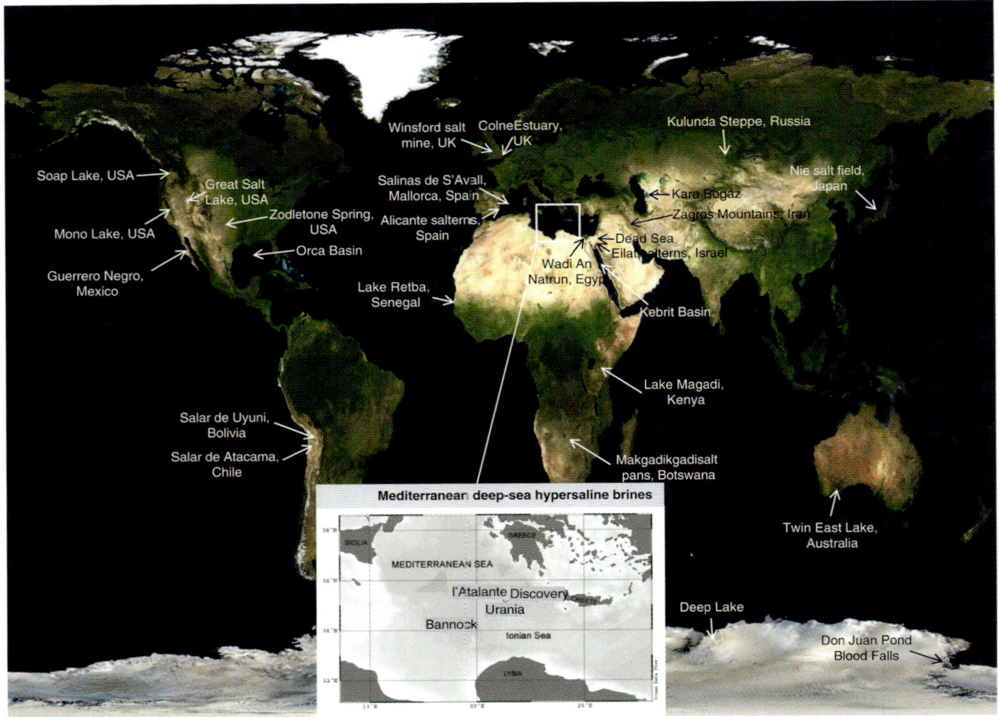

Fig. 20.3. World map showing the location of hypersaline environments discussed in this chapter.

(Fig. 20.2a). The gradient of salinity in the series of ponds has enabled numerous investigations into the effects of salinity on microbial activity and community composition.

We also find halophiles in some unexpected places. Plants that grow on very salty soils, such as the tamarisk tree (*Tamarix*) and *Atriplex* excrete salts from glands on their leaves, which become populated by a diverse community of halophiles (Simon *et al.*, 1994; Qvit-Raz *et al.*, 2008). *Halococcus* spp. were found in the nostril salt glands of a seabird (Brito-Echeverriá *et al.*, 2009), and the nasal cavities of a desert iguana harbour the highly salt-tolerant *Bacillus dipsosauri* (Deutch, 1994). Another interesting niche for halophiles is found in deteriorating archaeological monuments and historical wall paintings (Piñar *et al.*, 2001). Halobacteria are commonly found in salt crystals (Minegishi *et al.*, 2008; Oxley *et al.*, 2010), and the proteolytic enzymes from such halophiles can spoil salted fish and hides. However, halophiles are essential for the production of many nutritious traditionally fermented foods in the Far East.

The vital component of sea ice, which can cover 10% of the earth, is the vein-like network of hypersaline brines. Salts expelled from freezing seawater depress the freezing point of water and thus maintain it in a liquid form that allows halophilic and pyschrophilic life to flourish (see Thomas, Chapter 4, this volume). A subterranean hypersaline brine, intriguingly named 'Blood Falls' (Fig. 20.4a), breaks to the surface in the McMurdo Dry Valleys of Antarctica. The subterranean brine is anoxic and rich in sulfate and iron, and a novel process has been proposed whereby sulfur intermediates, such as sulfite and thiosulfate, drive respiratory ferric iron reduction (Mikucki *et al.*, 2009). Thick deposits of rock salt are found beneath a quarter of the earth's landmass and, as discussed in Section 20.7.6, provide an unusual subterranean habitat and/or repository for halophilic microbes (Fig. 20.4b; Fig. 20.5c). Owing to

Fig. 20.4. Examples of buried hypersaline environments that have been dissolved and/or transferred to the surface. (a) Blood Falls, Antarctica, showing brine from a subterranean hypersaline lake spilling from the Taylor Glacier. The ferrous iron in the brine oxidizes on contact with air, © Elanor Bell. (b) Winsford salt mine, UK; sampling from the area of the mine where moist air condenses on the cool mine walls, forming efflorescences of NaCl, which contain thousands of halophilic Bacteria and Archaea per gram, © Terry J. McGenity. (c) ASTER image of flowing 'glaciers' of NaCl (black, tongue-shaped plumes; about 5 km long) extruded from subterranean salt plugs in the Zagros Mountains in southern Iran, © NASA/GSFC/METI/ERSDAC/JAROS, and US/Japan ASTER Science Team; http://earthobservatory.nasa.gov/IOTD/view.php?id=4168, accessed 9 March 2010.

its high solubility, salt is readily mobilised. For example, deep-sea hypersaline anoxic brine lakes, such as those presently sitting in depressions on the floor of the Mediterranean Sea, derive from dissolved rock salt. Also, rock salt flows; nowhere is this more evident than in the Zagros Mountains in Iran (Fig. 20.4c) where salt is extruded from the top of mountains and flows like a glacier through the valleys. Many inland brine lakes result from solubilized rock salt that has found a route to the surface, but leaching of ions from surrounding rocks by rivers also contributes greatly to the ionic make-up of many salt lakes. In order for the salinity of such lakes to increase, water loss by evaporation must exceed inflow, helped by certain environmental conditions, such as high temperatures, low humidity and rainfall and often high wind speeds, all of which must be coupled with restricted outflow.

20.4 Factors Influencing Diversity, Community Composition and Distribution of Halophilic Microbes

20.4.1 The effect of the ionic composition of brines on halophilic microbes

Hypersaline environments differ greatly in their ionic composition, i.e. the relative

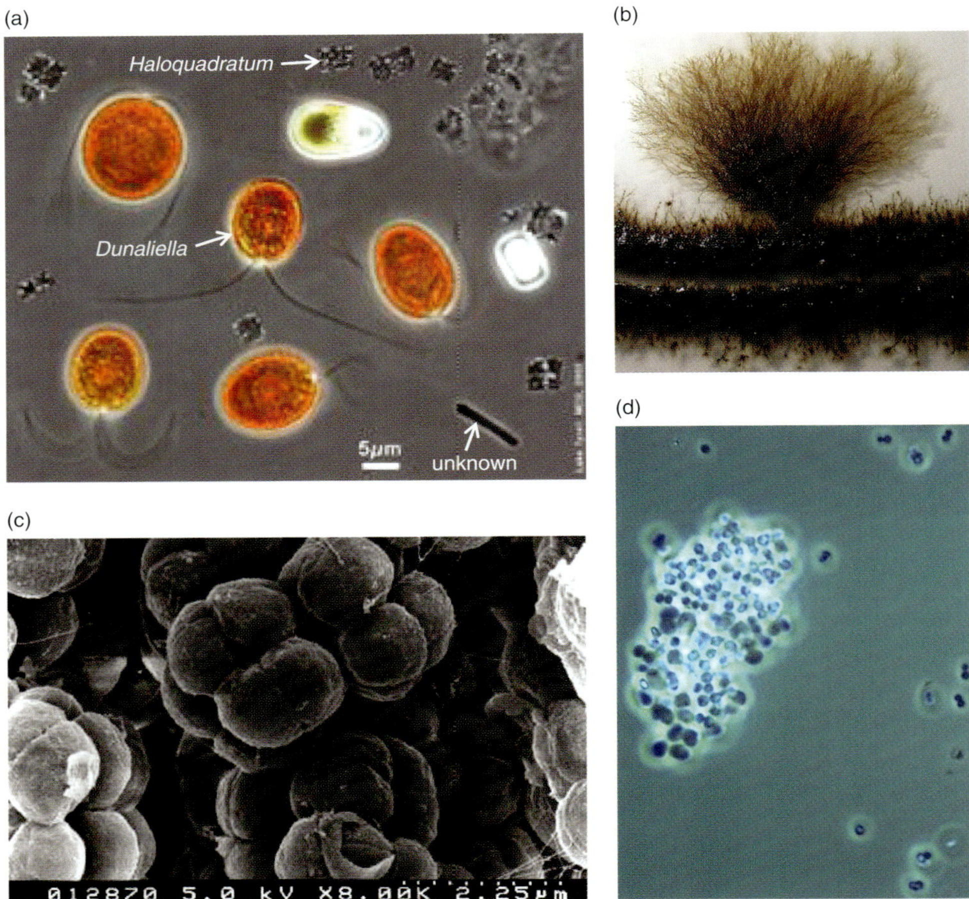

Fig. 20.5. Images of various halophiles. (a) Microscopic image from a natural hypersaline brine (salinity >200 g l^{-1}). Based on distinctive morphologies we can tentatively identify the eukaryotic green alga *Dunaliella salina* living alongside the archaeon *Haloquadratum walsbyi* (flat square with gas vesicles; many cells are dividing like a sheet of postage stamps). A rod-shaped microbe can also be seen, © Mike Dyall-Smith. (b) Colony of the halophilic fungus *Hortaea werneckii* (field of view is approximately 1 cm^2), © Polona Zalar. (c) Scanning electron micrograph of *Halococcus salifodinae* strain Blp, isolated from a salt mine in Austria, © Gerhard Wanner. (d) Phase-contrast mirograph of a low-salt-adapted halobacterial strain isolated from a United Kingdom salt marsh (strain W1 related to *Haladaptatus paucihalophilus*) growing in medium with a salinity of ca. 150 g l^{-1}. Note the large clumps of cells with irregular shapes. Individual cells are ca. 1.5 µm in diameter, © Nils Neuenkirchen.

concentrations of their component salts (see Table 20.1), which in turn can have a major influence on the biology of the environment as well as the limits of life. Hypersaline brines formed by evaporation of seawater have a near-neutral pH and an ionic composition similar to that of seawater (thalassohaline brines with NaCl as the main salt; Table 20.1). In contrast, the ionic composition of athalassohaline brine lakes is determined by the surrounding geology. Some are very rich in divalent cations, such as the Dead Sea (total dissolved salts around 347 g l^{-1}), which is dominated by magnesium (nearly 2 M) and calcium (0.47 M) cations (see Table 20.1). At such high concentrations these ions can be extremely stressful to organisms, as they destabilize and impair the function of essential biological macromolecules such as proteins and membrane lipids. In the absence of more stabilizing ions such as sodium, a concentration of 2.3 M magnesium chloride appears to be the upper limit for life, as shown in studies of Discovery Basin, a $MgCl_2$-saturated lake located at a depth of 3.5 km on the Mediterranean Sea floor (Hallsworth *et al.*, 2007). Members of the Halobacteriaceae isolated from the Dead Sea have a relatively low sodium requirement and a high magnesium tolerance. In contrast, in athalassohaline alkaline salt lakes, in which NaCl, $NaHCO_3$ and Na_2CO_3 are the main salts (Table 20.1), most isolates have a very low magnesium requirement, reflecting the low conentration of divalent cations in these environments.

To date, there is no conclusive evidence of microbes able to grow in $CaCl_2$-dominated Don Juan pond (Table 20.1), presumably because concentrated $CaCl_2$ brines, in addition to restricting water availability to cells, destabilize biological macromolecules in the same way as $MgCl_2$. Extremely saline brines in the Salar de Atacama (700 g l^{-1}) in which lithium salts start to precipitate similarly seem to be devoid of life (Pedrós-Alió, 2004). Determining the frontiers of life in such diverse hypersaline environments and discovering novel types of halophiles provide exciting challenges.

20.4.2 Diversity, evolution and biogeography of halophilic microbes

Salinity was found to be the main factor affecting microbial community composition in an analysis of culture-independent 16S rRNA sequence datasets (Box 20.2) from 111 diverse environments (Lozupone and Knight, 2007). It is generally thought, based on studies from salterns, that in homogeneous water bodies, microbial diversity decreases with increasing salinity. However, the niche heterogeneity imposed by steep redox gradients, especially when salinity is only two or three times that of seawater, as found in Guerrero Negro microbial mats (Mexico), can lead to unprecedented biodiversity (Ley *et al.*, 2006).

Communities of Halobacteriaceae in salt lakes can be very dense: numbers of 10^7–10^8 cells ml^{-1} brine, and even higher, are no exception (Oren, 1994). This is true not only for thalassohaline environments at near-neutral pH, but also for alkaline soda lakes, and at times even for the Dead Sea. Culture-independent, 16S rRNA sequence-based studies of the community composition of halophilic Archaea in saltern and salt lake environments generally show that the dominant species are different from those recovered as colonies on agar plates. Still, with skill and patience it is possible to grow most groups of halophilic Archaea present in saltern crystallizer ponds (Burns *et al.*, 2004b). An important addition to the list of cultured types is the square gas-vacuolate species *Haloquadratum walsbyi* (Box 20.5; Fig. 20.5a).

Recent investigations of continental athalassohaline environments are revealing a wealth of new diversity of both Archaea and Bacteria (Demergasso *et al.*, 2008; Pagaling *et al.*, 2009). For example, *Salinibacter ruber* (Box 20.3) is the main bacterial species at salinities above 250 g l^{-1} in coastal salterns, but a much broader diversity of related organisms in the phylum Bacteroidetes is found in athalassohaline environments (Antón *et al.*, 2008; Demergasso *et al.*, 2008; Pagaling *et al.*, 2009), including the recently isolated *Salisaeta longa* (Vaisman and Oren, 2009). Such analyses suggest that athalassohaline communities are more distinct from each other than thalassohaline communities. The causes of this phenomenon include a more diverse ionic composition as well as the altitude and/or remoteness of such environments, and thus the limitations to microbial transfer leading to divergent evolution (see Pagaling *et al.*, 2009).

In contrast, the ponds in thalassohaline coastal salterns that are saturated with salt (crystallizer ponds; Fig. 20.2a) all have a similar ionic composition. It is intriguing to learn how the extreme halophiles in geochemically similar but geographically remote environments compare. By analysing strains of the archaeal genus *Halorubrum* in Spanish and Algerian salterns, Papke *et al.* (2007) conclude that mutation rates exceed migration rates leading to distinct species being present in the different salterns. However, it will be interesting to learn how these factors differ in other species that may have a different dominant mode of mutation or means of dispersal.

Box 20.3. *Salinibacter ruber*, an extremely halophilic member of the Bacteria

Until about a decade ago, the extremely halophilic Archaea of the family Halobacteriaceae were considered to be the only group of aerobic heterotrophic microbes inhabiting NaCl-saturated brines in natural lakes and saltern crystallizer ponds. The discovery of *Salinibacter*, a representative of the Bacteroidetes branch of the Bacteria, therefore came as a surprise. The first evidence for the existence of a member of the Bacteria that contributes significantly to the prokaryote community in saltern crystallizer ponds came from fluorescence *in situ* hybridization studies. While the flat square–rectangular cells (*Haloquadratum walsbyi*, see Box 20.5) reacted with an Archaea-specific probe, an abundant slightly curved, rod-shaped organism was stained by probes designed to detect Bacteria, and specifically the Bacteroidetes (Antón *et al.*, 1999). Experiments showed that the bacterium also multiplies in salt-saturated brines (Antón *et al.*, 2000) and takes up amino acids and acetate (Rosselló-Mora *et al.*, 2003).

The isolation of the organism in pure culture by two independent groups soon followed. After inoculation of saltern brines on high-salt agar plates and development of red colonies, one group performed colony hybridization with a 16S rRNA-directed probe designed to detect the new phylotype; the second group examined the polar lipid content of randomly selected colonies, and found that a number of colonies that were red, like colonies of halobacteria, contained Bacteria-type lipids. Comparison of the isolates showed that they were very similar, and the new organism was described as *Salinibacter ruber* (Antón *et al.*, 2002). Subsequent genomic and phenotypic comparison of *Salinibacter* isolates obtained from hypersaline environment all around the globe shows that they are markedly similar (Antón *et al.*, 2008), and the red colour of many salt-saturated brines (Fig. 20.2a) can in part be attributed to *Salinibacter*.

Salinibacter has a number of unusual properties. Physiologically it is unlike most other halophilic and halotolerant representatives of the Bacteria, which accumulate organic solutes for osmotic balance (Fig. 20.6). *Salinibacter* uses the same strategy as *Halobacterium* and relatives: KCl is the osmotic solute, and the entire intracellular machinery is adapted to the presence of high salt. Analysis of its genome sequence suggests that many of its genes were derived by horizontal gene transfer from the Halobacteriaceae that inhabit the same environments (Mongodin *et al.*, 2005). For example, *Salinibacter* possesses retinal proteins similar to the bacteriorhodopsin, halorhodopsin and sensory rhodopsins of the Halobacteriaceae.

Some extreme halophiles, for example, can be transported by seawater whereas most would lyse in this relatively low-salt milieu (Box 20.4). Alternatively, many extreme halophiles are trapped and maintain viability inside crystals of halite (NaCl), which could be dispersed by wading birds or, if sufficiently small, by wind. The global distribution of extremely halophilic species such as *Salinibacter ruber* (Box 20.3) and *Haloquadratum walsbyi* (Box 20.5) attests to their having such a means of dispersal.

It is also important to consider that each year millions of tonnes of rock salt are dispersed by natural (Fig. 20.4c) and human activities; these activities and the flow of brines derived from evaporites move halophiles between geographically different areas and even geologically different ages (see Section 20.7.6 for evidence of life in ancient salt).

In summary, because of their patchy distribution and the high salt requirement of their inhabitants, hypersaline environments provide natural testing grounds for ecological and biogeographical theories, and opportunities to explore how physical isolation, age, habitat size, fluctuation in salinity and variation in habitat types influence microbial community composition and diversity. However, in order to address these questions it is important to know how extreme halophiles are transported from one hypersaline environment to another.

20.4.3 Multiple extremes

Many halophiles are 'polyextremophiles', organisms adapted not only to high salt but to one or more additional forms of environmental extreme (Rothschild and Mancinelli, 2001; Bowers *et al.*, 2009). For example, growth in shallow hypersaline ponds results in exposure

Box 20.4. Low-salt-requiring halobacteria

The halophilic Archaea of the family Halobacteriaceae are considered as the halophiles par excellence. *Halobacterium salinarum*, the type species of the type genus of the family, grows in saturated NaCl. Moreover, minimal salt concentrations of around 100–150 g l^{-1} are required for structural stability. At lower concentrations the cells' proteins, including the glycoprotein of which the cell wall is built, denature, and the cells lyse. Therefore it was surprising that *Halococcus* sp. was isolated from seawater (35 g l^{-1} salt) (Rodriguez-Valera *et al.*, 1979). Cells of the genus *Halococcus* do not possess a wall built of glycoprotein subunits as most other genera within the family, but instead have a rigid polysaccharide cell wall. This explains why the cells did not lyse, but how the intracellular enzymatic machinery remained intact when supposedly osmotic equilibrium with seawater was established, is still unclear. There is no evidence that *Halococcus* grows in seawater, but it can survive and so disperse in this milieu.

Analysis of DNA from sediments of the Colne Estuary (UK), with salinities around 35 g l^{-1}, revealed the presence of members of the Halobacteriaceae (Munson *et al.*, 1997). These findings stimulated attempts to cultivate halobacteria, and strains from three phylogenetic lineages of the Halobacteriaceae were isolated (Purdy *et al.*, 2004). Remarkably, some of these isolates (HA group-1) could grow at NaCl concentrations of 25 g l^{-1} (salinity ca. 35 g l^{-1}), far below the minimum concentration required for growth of *Halobacterium salinarum*. Another low-salt environment that harbours a varied community of halophilic Archaea is Zodletone Spring, a sulfide-rich spring in Oklahoma, USA (Elshahed *et al.*, 2004). Several species of Halobacteriaceae with a relatively low NaCl requirement for growth were isolated. The strain requiring the lowest NaCl concentration (47 g l^{-1}) belonged to a new genus and was named *Haladaptatus paucihalophilus* (Savage *et al.*, 2007). Furthermore, two strains of Halobacteriaceae from a traditional Japanese salt field survived prolonged suspension in salt concentrations as low as 5 g l^{-1} (Fukushima *et al.*, 2007). Phylogenetic and phenotypic comparison reveals that these two isolates from Japan and the HA group-1 isolates from England belong to the genus *Haladaptatus*, very closely related to *Haladaptatus paucihalophilus* from the USA. Figure 20.5d shows that the UK representatives of this genus form very large clusters of cells when growing at salinities below 125 g l^{-1}.

An unusual niche for allegedly extreme halophilic Archaea was discovered in geothermal vents in diverse locations such as Kamchatka, Hawaii, New Mexico, California and Wyoming. Halobacteria related to *Haloarcula* were recovered from steam that emerges from the fumaroles. Leaching of salts and minerals through the porous volcanic rock may have enabled growth of halophiles in the geothermal areas, but there is no record of extremely high salinities normally associated with development of members of the Halobacteriaceae. Even more intriguing is the observation that the isolates survived exposure to 75 °C for 5–30 min (Ellis *et al.*, 2008). It will be interesting to understand the molecular mechanisms enabling these organisms to retain their viability at low salt concentrations and high temperatures.

An even more unexpected location for Halobacteriales is the human gastro-intestinal (GI) tract; Oxley *et al.* (2010) found a large diversity of halobacterial 16S rRNA gene sequences, albeit in low abundance, in GI-tract biopsies and faeces, and even demonstrated their presence in an enrichment culture from one biopsy in medium containing 25 g l^{-1} NaCl. It thus appears that we regularly ingest halobacterial species with salt and fermented foods, and that some survive in the digestive system.

to very high levels of ultraviolet light, which would lead to DNA damage and cell death in unadapted organisms. *Halobacterium salinarum*, an Archaeon that thrives in such environments, is among the most UV-resistant of any organism (see Section 20.7.6).

Hypersaline environments are found deep in the Earth's crust (Section 20.7.6) and at the bottom of the deep sea (Section 20.7.5), and high pressure is known to significantly influence metabolism (Pradillon, Chapter 14, this volume). Growth at high salinity generally enhances growth or survival at elevated pressures (see Kaye and Baross, 2004). However, the interplay between temperature, salinity, salt type and pressure in different species is an area that is ripe for exploration.

Hypersaline acidic environments are rare, but members of the Halobacteriaceae capable of growing between pH 4 and 6 have been isolated from salt crystals (Minegishi *et al.*, 2008). Twin East Lake in Australia is both highly acidic (pH 3) and hypersaline (160 g l^{-1}) and harbours many bacteria, such as sphingomonads that are not typically found in neutral lakes of

Box 20.5. *Haloquadratum walsbyi*, the square archaeon

In 1980, Tony Walsby, a British microbiologist, spent a sabbatical at the Hebrew University of Jerusalem. During his stay in Israel he visited the Sinai Peninsula. Being interested in halophilic microorganisms, he collected a sample from a near-shore brine pool coloured red by halophilic Archaea. Back in the laboratory he saw that most of the cells were of a peculiar type: perfectly square or rectangular, and extremely thin (Fig. 20.5a). Many others had probably seen such shapes before, but had never recognized those as being living cells. For Walsby, who had been studying gas vesicles of prokaryotes for many years, the presence of the characteristic gas vesicles did not leave any doubt about the prokaryotic nature of the squares. He published his observations in *Nature* under the title 'A square bacterium' (Walsby, 1980).

It was quickly recognized that these flat square Archaea are extremely widespread, and in fact they form the dominant type of cells in the brines of most saltern crystallizer ponds worldwide. Still, only little could be learned about their properties as the organism could not be grown in culture. Thus, the nature of their polar lipids was deduced from the analysis of the lipids extracted from brine dominated by the square cells (Oren *et al.*, 1996).

When in 1995 the result of a cultivation-independent study of the 16S rRNA sequences recovered from a Spanish saltern crystallizer pond was published, the most common sequence, known as the 'SPht' phylotype (or the 'Susanna' phylotype, named after Susanna Benlloch who authored the study; Benlloch *et al.*, 1995), was only remotely related to the recognized genera within the Halobacteriaceae. Fluorescence *in situ* hybridization (FISH) studies using probes designed to react with that phylotype then showed that it belonged to Walsby's square Archaea (Antón *et al.*, 1999). Combining FISH with autoradiography showed that amino acids and acetate were readily used by the cells (Rosselló-Mora *et al.*, 2003).

In 2004, cultivation of the elusive square Archaea was announced simultaneously by a group from Australia (Burns *et al.*, 2004a) and a group of Dutch and Spanish colleagues (Bolhuis *et al.*, 2004). After it had become clear that the isolates were nearly identical, the groups jointly published a formal description of the species *Haloquadratum walsbyi*, 'the square salt organism of Walsby' (Burns *et al.*, 2007). The genome sequence yielded much useful information on the special adaptation of the organism to life at low water activity (Bolhuis *et al.*, 2006), and an environmental genomics study of the square Archaea in a Spanish crystallizer pond showed that, although most of the genes are shared by all, a large number of accessory genes are present only in part of the cells, and thus the total genetic potential of the community is larger than each individual cell in the pond (Legault *et al.*, 2006). Comparative genomic analysis of the Spanish and Australian strains revealed remarkable similarity, raising intriguing questions about dispersal and evolution of *Haloquadratum walsbyi* (Dyall-Smith *et al.*, 2011).

The commentary 'Archaea with square cells' that Tony Walsby wrote 25 years after his first discovery of the squares (Walsby, 2005) provides an excellent overview of the history of the discovery and all the work that has led to our current understanding of the biology of one of the main players in the saltern ecosystem.

equivalent salinity (Mormile *et al.*, 2009). The inhabitants of alkaline hypersaline lakes are discussed in Section 20.7.4.

Microbial activity has been detected in cold, hypersaline environments and moderately halophilic psychrotolerant strains have been cultured, but the only psychrotolerant extreme halophile to our knowledge is *Halorubrum lacusprofundi* from Deep Lake, Antarctica (Franzmann *et al.*, 1988). Although its temperature optimum is 31–37°C, growth is still possible at temperatures as low as 4°C, and so it could grow during the short periods when the temperature of Deep Lake increases from subzero to a maximum of 11.5°C. High-altitude salt pans can have extremely wide daily and seasonal fluctuations in temperature, but there has been little investigation of the effect of this phenomenon on halophilic communities or pure cultures.

While many members of the Halobacteriaceae can grow up to 50°C, the most thermophilic halophiles are Bacteria. For example, *Halothermothrix orenii*, from a salt lake in Tunisia, grows anaerobically up to 68°C (optimum 60°C) and up to 200g l^{-1} salt (Cayol *et al.*, 1994). The genome sequence of this intriguing organism was recently published (Mavromatis *et al.*, 2009). An isolate from Wadi An Natrun, *Natranaerobius thermophilus*, a representative of a new order, the Natranaerobiales, grows optimally at 54°C (up to 56°C) at pH 9.5, and at a total Na^+

concentration between 3.3 and 3.9 M (range: 3.1–4.9 M) (see Bowers *et al.*, 2009). A better understanding of the compounding and in some cases off-setting effects of multiple extremes on organisms is needed if we are going to identify and even extend the frontiers of life.

20.5 Mechanisms of Adaption to High Salinity

20.5.1 'Salt-in' and 'compatible-solute' strategies

Biological membranes are permeable to water, and so when a non-halophile is in a high-salt environment, water will exit the cell, leaving behind a highly concentrated cytoplasm in which proteins, nucleic acids and other macromolecules lose their structure and function (Fig. 20.6a). An organism adapted to living at high salinity, i.e. at a low water activity (see Box 20.1), should have a cytoplasm that is at the same osmotic pressure as the brines of its medium. In fact, in order for the cell to maintain a turgor pressure (all cells do, with the possible exception of some halophilic Archaea of the family Halobacteriaceae), the intracellular osmotic pressure should even exceed that of the extracellular environment (Brown, 1976, 1990).

Halophiles have developed two basically different strategies to achieve osmotic balance (Fig. 20.6). One is based on the accumulation of KCl to molar concentrations; the other keeps the intracellular ionic concentrations low, while using low-molecular-weight neutral organic solutes. The latter are called 'osmolytes' or 'compatible solutes' because even at high concentrations they are compatible with the functioning of

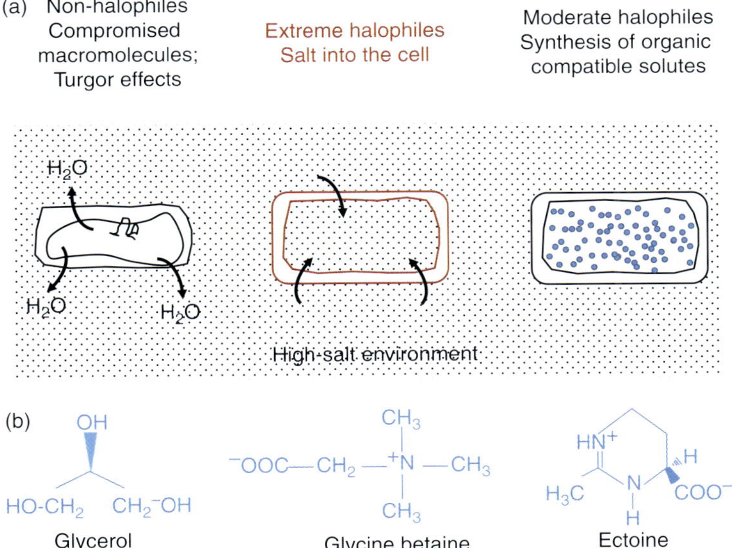

Fig. 20.6. Schematic representation of modes of adaptation to hypersaline environments. (a) Water moves out of the cells of non-adapted microbes resulting in a concentrated cytoplasm and loss of structure and function of biological macromolecules (indicated by the irregular line). Extreme halophiles adopt the salt-in strategy (predominantly with KCl inside the cells), which leads to osmotic balance, but means that their cellular machinery must be specifically adapted to function in a high-salt milieu. Moderate halophiles have a more flexible but energetically expensive strategy of synthesising low-molecular-weight organic solutes that provide osmotic equilibrium and that are compatible with biological macromolecules. Compatible solutes may also be transported into the cell from the environment if they are available. (b) Structures of three important compatible solutes. © Terry J. McGenity.

cellular macromolecules (Galinski, 1995; Grant, 2004; Oren, 2006a).

The first option ('salt-in' to balance 'salt out') is used by few groups of prokaryotes only: the archaeal family Halobacteriaceae, bacterial aerobe *Salinibacter* (Box 20.3) and the anaerobic fermentative Halanaerobiales (Low-G+C branch of the Firmicutes). In the Halobacteriaceae, intracellular KCl above 4.5 M was measured in cells growing in saturating NaCl concentrations. Those microorganisms that use the 'salt-in' strategy are obligate extreme halophiles, i.e. they are unable to adapt to life at low salinity because their complete intracellular enzymatic machinery is active only in the presence of molar concentrations of KCl. To achieve this, the entire proteome has to be modified. Proteins from such extreme halophiles are typically highly acidic, with a great excess of the negatively charged glutamate and aspartate over the positively charged lysine and arginine residues (Lanyi, 1974), creating a negatively charged surface, which facilitates retention of a hydration shell around the protein. Recently, it has been shown that it may not be the negative charge that is important rather the fact that negatively charged amino acids also have a short side chain, and thus decrease the surface area of the protein accessible to ions (Tadeo *et al.*, 2009). To maintain such proteins in their native form, molar concentrations of salts are necessary; without sufficient salt the proteins denature.

The second strategy, i.e. accumulation of organic compatible solutes (often called 'osmolytes'), is energetically more demanding than the salt-in strategy (Oren, 1999b), but allows much more flexibility and adaptability to changing conditions as there is no need for special protein structures and the intracellular concentrations of the solutes can be regulated according to the salinity of the environment. We find this strategy in most halophilic and halotolerant members of the Bacteria (with the exceptions discussed above), in eukaryotic algae, in fungi, and also in methanogenic Archaea. Among the solutes used for osmotic balance are glycerol (in the eukaryotic alga *Dunaliella*), amino acid derivatives such as glycine betaine and ectoine (1,4,5,6-tetrahydro-2-methyl-4-pyrimidine carboxylic acid), sugars such as sucrose and trehalose, and many others (Fig. 20.6b). Some are widely used; other compounds are limited to selected groups of microorganisms. Thus, glucosylglycerol is almost exclusively found in moderately halophilic or highly halotolerant cyanobacteria, and Nε-acetyl-α-lysine and Nδ-acetylornithine have thus far been detected only in aerobic members of the Firmicutes.

A combination of salt-in and compatible-solute strategies is found in some microbes (see Roberts, 2005). Other mechanisms are seen in multicellular organisms: certain higher plants, for example, can exclude salts or sequester them into their vacuoles (Glenn *et al.*, 1999). In addition, some microorganisms can respond to salt concentrations beyond their window of adaptation by forming resistant structures, such as spores, or, as seen with diatoms, by migrating to an area more favourable to growth (McKew *et al.*, 2011). Indeed, the filamentous cyanobacterium Coleofasciculus (*Microcoleus*) *chthonoplastes* migrated to the surface of an intertidal mat (from the United Arab Emirates) when the salinity of the overlying water was less than 150 g l^{-1}, and moved downwards into the sediment when the salinity exceeded 150 g l^{-1} (Kohls *et al.*, 2010). This rapid-response system was called 'halotaxis' (salt-influenced movement) after Kohls and colleagues had ruled out the influence of factors other than salinity, such as light, gravity, a_w, and concentrations of oxygen and hydrogen sulfide.

20.5.2 (Post)-genomic insights

Genomics, coupled with diverse methods for high-throughput analysis of metabolites, proteins and expressed genes, provides a window into adaptation to extreme environments. Since the first genome sequence of an extremely halophilic archaeon, *Halobacterium salinarum* NRC-1 (Ng *et al.*, 2000), several other extreme halophiles have been genome sequenced (11 for Halobacteriaceae and one for *Salinibacter ruber*; as of March 2010), and investigated via diverse 'omics technologies (Boxes 20.3 and 20.5; Soppa *et al.*, 2008). Genome analysis is providing insights into mechanisms of adaptation to fluctuating salinity in *Chromohalobacter salexigens*, which has a broad salt tolerance, and in *Salinibacter ruber*. For example, broad salt tolerance may be enhanced by possession of paralogous proteins that perform the same function with each working optimally at a different salinity (Sanchez-Perez *et al.*, 2008).

Metabolic network reconstruction and comparative analysis of the halophilic Archaea, *Halobacterium salinarum, Haloarcula marismortui, Haloquadratum walsbyi* and *Natronomonas pharaonis*, revealed several analogous metabolic pathways (e.g. for biosynthesis of nucleotides) and some that were divergent, notably folate synthesis (Falb *et al.*, 2008). Several bacteria-like enzymes as well as novel metabolic routes (e.g. ribose-5-phosphate biosynthesis) deserve closer attention (Falb *et al.*, 2008). With improved capability for genetic modification, some intriguing findings have emerged, for example the role of small RNAs in regulating gene expression and thus the capability of *Haloferax volcanii* to cope with low-salt and high-temperature stress (Soppa *et al.*, 2009). Therefore, genome analysis opens a window into the evolution of Archaea, Bacteria and Eukarya, and is providing deep insights into adaptations to high salt (see Ma *et al.*, 2010).

20.6 Ecological Guilds and Biochemical Groups of Halophiles

Most microbial modes of metabolism (Fig. 20.7) known from freshwater and marine environments have also been identified in hypersaline ecosystems. The world of halophiles is highly diverse, and includes aerobes and anaerobes, heterotrophs, phototrophs and chemolithoautotrophs, methanogens, and organisms that live by fermentation or anaerobic respiration. However, there are a number of processes common in low-salt environments that do not function at high salt (Fig. 20.8). As stated above, osmotic adaptation and life at high salt concentrations is energetically costly. The decrease in functional diversity within hypersaline ecosystems with the increase in salt concentration can to a large extent be understood from energy considerations. According to a proposed theory that explains nearly all observations (Oren, 1999b, 2011), the upper salt concentration limit at which a certain dissimilatory process can occur is dictated by energy constraints. Types of metabolism that yield a lot of energy can by found up to the highest salinities; for those metabolic processes that provide little energy, the upper salt limit depends on the mode of osmotic adaptation used (see

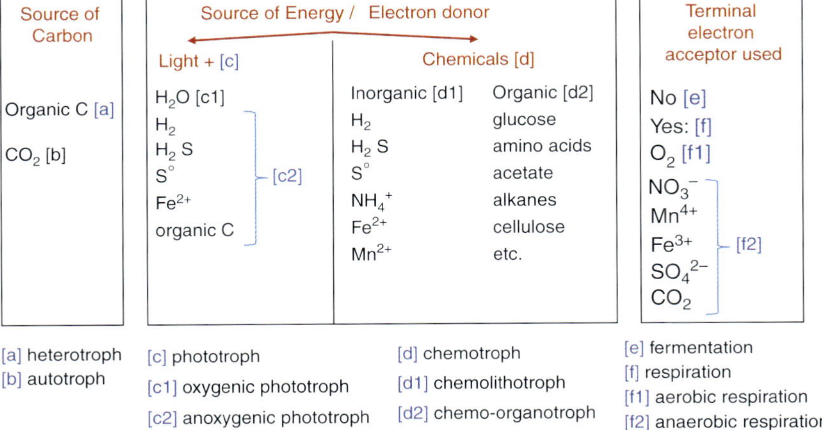

Fig. 20.7. Definitions and examples of different modes of metabolism used by the various ecological guilds described in this chapter. Energy generation by disproportionation is not shown in this scheme, but involves the conversion of a chemical species to oxidized and reduced forms, e.g. thiosulfate ($S_2O_3^{2-}$) reacting to form sulfate (SO_4^{2-}, oxidized) and sulfide (S^{2-}, reduced). © Terry J. McGenity.

Section 20.5.1). For example, because the salt-in strategy is energetically less costly, fermentation by members of the Halanaerobiaceae can proceed nearly up to the highest salinities (Fig. 20.8), while most other Bacteria rely on the synthesis of organic compatible solutes, which, due to the high energetic cost, reduces their capacity to function at extremely high salinities.

20.6.1 Oxygenic phototrophs

The green algal genus *Dunaliella* is found in hypersaline lakes worldwide (Oren, 2005; Ben-Amotz *et al.*, 2009). Blooms of *Dunaliella* species have been reported both from thalassohaline lakes such as Great Salt Lake, Utah (Stephens and Gillespie, 1976) and saltern crystallizer ponds, as well as from athaloassohaline environments such as the Dead Sea (Oren *et al.*, 1995). They are found living alongside members of the Halobacteriaceae as shown in Fig. 20.5a. *Dunaliella* accumulates photosynthetically produced glycerol as its compatible solute (Ben-Amotz and Avron, 1973). Under high-light and nutrient stress, some strains produce massive amounts of β-carotene, which protects against oxidative damage to the cells, and turns cells from green to orange-red. *Dunaliella* species have been exploited for the commercial production of β-carotene, and are under investigation as a source of biofuels.

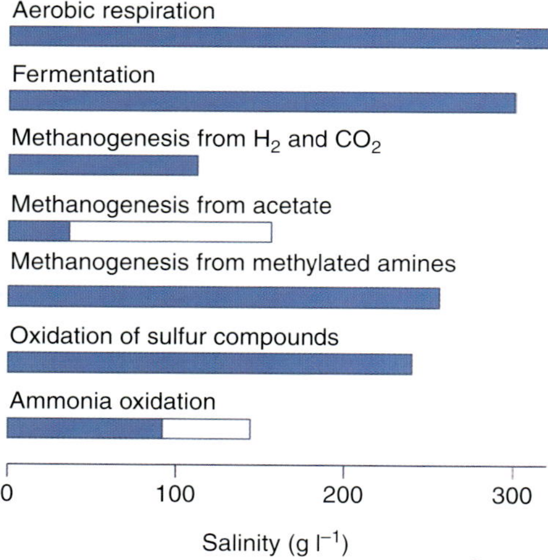

Fig. 20.8. Upper salinity tolerance of selected microbial processes. Closed bars indicate data from pure cultures, and open bars represent data from *in situ* measurements where the values exceed those for pure cultures. The reason why some microbes, Halanaerobiaceae, can grow by fermentation at such high salinity is because they adopt the salt-in strategy rather than the more energetically expensive compatible-solute strategy to cope with high salinities (see text for details). Data from Oren (2011).

Other types of oxygenic phototrophs often encountered in hypersaline environments include diatoms (generally up to 100–150 g l^{-1} salt only) and cyanobacteria. Cyanobacteria are especially prominent in benthic microbial mats in shallow lakes and saltern ponds. In salterns, where gypsum precipitates at salt concentrations between 150 and 200 g l^{-1}, colourful layers of cyanobacteria are often found within the gypsum crusts on the bottom of the ponds: an upper orange-brown layer of unicellular types (*Aphanothece halophytica* and related forms), below which a green layer of filamentous forms (*Phormidium* spp. and others) is seen (Fig. 20.2b; Caumette *et al.*, 1994).

20.6.2 Anoxygenic phototrophs

In the above-mentioned microbial mats and gypsum crusts, a purple layer of anoxygenic phototrophic sulfur bacteria is generally found (Fig. 20.2b). These photoautotrophic bacteria include genera that store elemental sulfur intracellularly (*Halochromatium, Marichromatium, Thiohalocapsa*) and forms that excrete sulfur (*Ectothiorhodospira, Halorhodospira*). *Halorhodospira* species also abound in hypersaline soda lakes (Caumette *et al.*, 1994; Ollivier *et al.*, 1994; Section 20.7.4). Photoheterotropic halophilic bacteria have also been described (*Rhodovibrio salinarum, Rhodovibiro sodomense, Rhodothalassium salexigens*) that can grow in up to 200–240 g l^{-1} salt.

An altogether different mode of phototrophic life is shown by *Halobacterium salinarum*. In addition to growth by aerobic respiration or fermentation, anaerobic growth can be supported by light absorbed by the retinal-containing protein bacteriorhodopsin, which traverses the cell membrane (Fig. 20.9; Hartmann *et al.*, 1980). To what extent *Halobacterium* and other members of its family exploit this option in their natural environment is unknown.

20.6.3 Aerobic chemo-organoheterotrophs: extreme halophiles

In aerobic hypersaline environments at or approaching salt saturation the heterotrophic microbial community is dominated by members of the Halobacteriaceae. Until recently it was assumed that halophilic Bacteria could not compete with the Archaea; however, the discovery of the red extremely halophilic *Salinibacter ruber* has changed our concepts (Box 20.3). Over a hundred species of Halobacteriaceae have been described, classified into 32 genera (Oren, 2006b; see also Papke *et al.*, 2007 about species concepts for the Halobacteriaceae; numbers as of March 2010). Most representatives are pigmented red by 50-carbon carotenoids (α-bacteroruberin and derivatives), and retinal-based pigments (the light-driven proton pump bacteriorhodopsin (Fig. 20.9) and the light-driven chloride pump halorhodopsin) may also contribute to the pigmentation.

Members of the Halobacteriaceae are typically extreme halophiles that need at least 100–150 g l^{-1} salt for growth and for structural stability (Section 20.5.1). However, in recent years members of the group have also been found in relatively low salt environments (Box 20.4). Most species of Halobacteriaceae lead a typical aerobic life style (as far as possible in environments where the availability of oxygen is often limited due to its low solubility in salt-saturated brines). Amino acids and small organic acids are the preferred carbon and energy source for most species; many types also use simple sugars, as well as

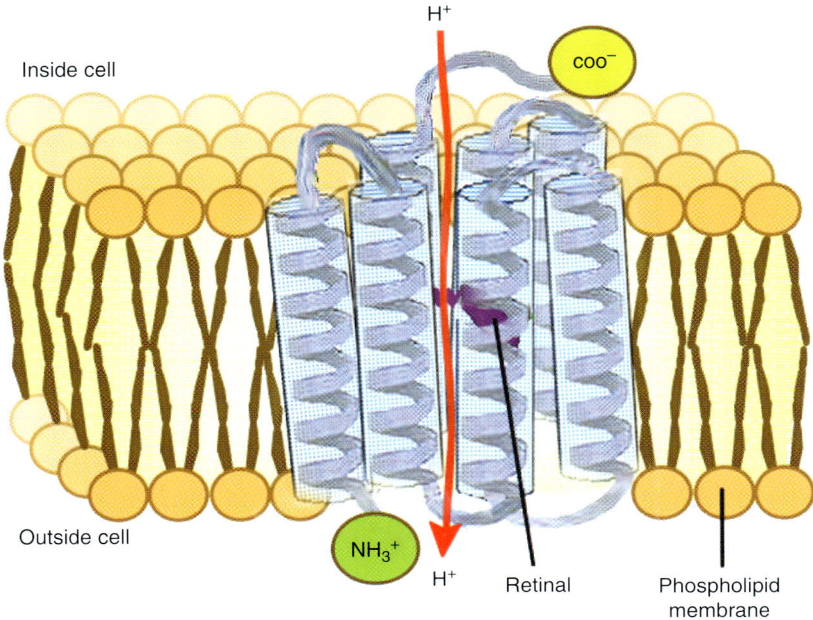

Fig. 20.9. Bacteriorhodopsin, an outwardly directed proton pump driven by light. Light induces a conformational change in the retinal molecule. As this is covalently bound to one of the seven alpha helices it induces conformational change in the protein channel, which intitiates a series of reactions resulting in the transfer of a proton to the outside of the membrane. The extruded protons are transferred to a transmembrane ATPase, through which they can return to the cytoplasm, converting ADP to ATP. Figure prepared by Roshena E. Arevalo.

polymers such as starch, proteins and lipids. Hydrocarbons and aromatic compounds are used by a few species only. Many members of the family also can grow anaerobically by denitrification, by fermentation of arginine and/or by using the light-driven bacteriorhodopsin proton pump (Section 20.6.2).

20.6.4 Aerobic chemo-organoheterotrophs: moderate halophiles

In the domain Bacteria we find a great diversity of moderately halophilic aerobic heterotrophic organisms. These belong to different phyla, notably the Gammaproteobacteria and the Firmicutes (Fig. 20.1; Ventosa *et al.*, 1998). Best known are the many representatives of the family Halomonadaceae, with *Halomonas* and *Chromohalobacter* as the largest genera (Arahal and Ventosa, 2006). These Gammaproteobacteria are highly versatile. Most species can grow over a wide range of salt concentrations, from near-zero up to 150–200 g l^{-1} and sometimes higher. They synthesize ectoine as their main organic compatible solute, but they also can use glycine betaine taken up from the medium. Endospore-forming members of the Firmicutes, e.g. the genus *Halobacillus*, are widespread in hypersaline soils and other high-salt environments.

20.6.5 Aerobic chemolithoautotrophs

In high-salt environments some chemolithoautotrophic processes do occur, while others do not. Thus, nitrification has been detected in the alkaline Mono Lake, California at 80–90 g l^{-1} salt (Joye *et al.*, 1999), but was never demonstrated above 150 g l^{-1}. The most salt-tolerant autotrophic ammonia-oxidizer ever cultured, '*Nitrosococcus halophilus*', grows only to 94 g l^{-1} salt and has its optimum at just 40 g l^{-1} (Koops *et al.*, 1990). On the other hand we know many halophilic chemolithoautotrophs that obtain their energy from the oxidation of reduced sulfur compounds. Some grow in neutral environments (e.g. *Halothiobacillus halophilus* from a salt lake in Western Australia, growing up to 240 g l^{-1} salt, with an optimum at 50–60 g l^{-1}; Wood and Kelly, 1991) and *Thiohalorhabdus denitrificans*, isolated from Siberian salt lakes and Mediterranean salterns, which grows between 11 and 290 g l^{-1} NaCl with an optimum at 175 g l^{-1} (Sorokin *et al.*, 2006, 2008c). Others such as *Thiohalospira halophila*, *Thiohalospira alkaliphila* and *Thioalkalivibrio halophilus* are haloalkaliphiles, and tolerate Na^+ concentrations up to 5.4 M at pH 8–9 (Sorokin and Kuenen, 2005a, b; Sorokin *et al.*, 2008b). The alkaline Mono Lake (approximately 90 g l^{-1} salt) and Sears Lake in California (>300 g l^{-1} salts), both of which have high arsenic concentrations, harbour bacteria that perform chemlithoautotrophic oxidation of As(III) to As(V) (Oremland *et al.*, 2005). *Alkalilimnicola ehrlichii* (Gammaproteobacteria), an isolate from Mono Lake, oxidizes arsenite at salt concentrations up to 190 g l^{-1} using oxygen or nitrate as electron acceptor (Hoeft *et al.*, 2007).

The bioenergetic model described at the beginning of this section (Oren, 1999b, 2011) can explain most of the observed differences in the upper salt limit of the different chemolithoautotrophic processes: ammonia and nitrite are relatively oxidized substrates, and their oxidation therefore yields very little energy; sulfide, elemental sulfur and arsenite are much more reduced, and the energy yield per mole of electrons transferred to oxygen is accordingly higher (note the standard free energy ($\Delta G^{\circ\prime}$) of the reactions below), enabling the cells to obtain sufficient energy for the production of the large amount of organic compatible solutes needed at increased salt concentrations (see Fig. 20.8).

Ammonia oxidation: $NH_4^+ + 1.5O_2 \rightarrow NO_2^- + 2H^+ + H_2O$ ($\Delta G^{\circ\prime} = -274.6$ kJ per reaction) (20.1)

Sulfide oxidation: $HS^- + 2O_2 \rightarrow SO_4^{2-} + H^+$ ($\Delta G^{\circ\prime} = -797$ kJ per reaction) (20.2)

However, the model fails to explain the apparent absence of aerobic methane oxidation in high-salt environments: methane is often formed in anaerobic hypersaline environments, but no microbially mediated oxidation appears to occur in the aerobic layers, in spite of the fact that such an oxidation would yield more than sufficient energy for growth as well as for osmotic adaptation. A few halophilic and halotolerant methanotrophs have been isolated and characterized (Trotsenko and Khmelenina, 2002), and there is an occasional report of methane utilization at high salinity (Sokolov and Trotsenko, 1995), but the activity of methanotrophs in hypersaline ecosystems appears to be limited (Conrad *et al.*, 1995).

20.6.6 Anaerobic: fermentative organisms

Fermentative processes are found in anaerobic sediments of hypersaline lakes up to very high salt concentrations (Oren, 1988b; Ollivier *et al.*, 1994). A variety of halophilic microorganisms living by fermentation have been characterized. Phylogenetically these belong to disparate groups.

While the Archaea of the family Halobacteriaceae normally have an aerobic lifestyle, members of the genus *Halobacterium* can also grow anaerobically by fermentation of arginine with ornithine, ammonia and carbon dioxide as products (Hartmann *et al.*, 1980).

In the domain Bacteria, phylum Firmicutes, the order Halanaerobiales with families Halanaerobiaceae and Halobacteroidaceae consists entirely of anaerobic halophiles, and most obtain their energy by fermentation of simple sugars to ethanol, acetate, hydrogen and carbon dioxide (Oren, 2006c). There are also fermentative Bacteria in other phylogenetic groups, such as *Clostridium halophilum* and the highly unusual *Haloplasma contractile* (Antunes *et al.*, 2009; see also Section 20.7.5).

20.6.7 Anaerobic: denitrifiers and miscellaneous others

Anaerobic respiration, in which organic substrates are oxidized using electron acceptors other than molecular oxygen, is found in the halophile world as well. Some halophiles reduce well known electron acceptors such as nitrate and sulfate, as well as fumarate, dimethylsulfoxide, trimethylamine-*N*-oxide (Oren, 2006b), manganese IV (Yu *et al.*, 2010), and even selenate and arsenate (Oremland *et al.*, 2005).

Denitrification, the dissimilatory reduction of nitrate to the gaseous products nitrogen and nitrous oxide, is performed by a variety of halophiles. Many members of the Halobacteriaceae reduce nitrate to nitrite, and some (e.g. *Haloferax mediterranei* and *Haloarcula marismortui*) are capable of anaerobic growth during which nitrate is converted to molecular nitrogen (Mancinelli and Hochstein, 1986). Nitrate reducers, including true denitrifiers, are also commonly found in the domain Bacteria (Ventosa *et al.*, 1998).

20.6.8 Anaerobic: sulfate reducers

Seawater is rich in sulfate, and sulfate is thus abundantly available as an electron acceptor in thalassohaline hypersaline environments, and it is present in many athalassohaline lakes as well. Thus, sulfate can serve as the terminal electron acceptor in hypersaline anaerobic environments, and sulfate reduction has been reported to occur even at proposed salinities of 422 g l^{-1} (Porter *et al.*, 2007a). Several studies of uncultivated sulfate reducers based on analysis of *dsr*AB gene sequences identify the presence of a phylogenetically broad array of sulfate reducers, with members of the family Desulfohalobiaceae generally predominant (e.g. Kjeldsen *et al.*, 2007).

However, not all types of sulfate-reducing bacteria known from the marine environment have been isolated at high salt concentrations. Sulfate-reducing bacteria are commonly divided in two groups: the 'incomplete oxidizers' that oxidize lactate, and a few additional substrates and excrete acetate, and the 'complete oxidizers' that mainly use acetate as electron donor, oxidizing it to CO_2. Representatives of the first type were reported to grow up to quite high salinities. The most halotolerant reported is probably *Desulfohalobium retbaense* from Lake Retba, Senegal (optimum at 100 g l^{-1} salt, growing up to 240 g l^{-1}; Ollivier *et al.*, 1991, 1994). The haloalkaliphilic *Desulfonatronospira thiodismutans* and *Desulfonatronospira delicata*, isolated from the Kulunda Steppe, Altai, Russia, oxidize lactate and ethanol at total salt concentrations between 1 and 4 mole l^{-1}, and they also can grow as chemolithoautotrophs by disproportionation (Fig. 20.7) of sulfite or thiosulfate (Sorokin *et al.*, 2008a). On the other hand, 'complete oxidizers' appear to be absent at such high salinities. *Desulfobacter halotolerans*, recovered from Great Salt Lake, Utah, is the most salt-tolerant example known, but grows optimally at 10–20 g l^{-1} salt only, with slow growth up to 130 g l^{-1}. A probable explanation is that the small amount of energy gained during the oxidation of acetate with sulfate as electron acceptor is insufficient to offset the high cost of the production of organic compatible solutes (Oren, 1999b, 2011).

20.6.9 Anaerobic: methanogens and acetogens

Methanogenic Archaea can obtain energy from a small number of reactions only. These include the splitting of acetate to CH_4 and CO_2, and the reduction of CO_2 with H_2 as electron donor. These two reactions are responsible for most biological methane production on Earth. In addition, methane can be generated by disproportionation of methanol, methylated amines and dimethylsulfide.

No methanogens have yet been described that grow on acetate at salt concentrations exceeding that of seawater. Also methanogenesis from H_2+CO_2 does not seem to occur in hypersaline sediments above 100–120 g l^{-1} salt. The most halotolerant isolate known that performs this reaction, *Methanocalculus halotolerans* isolated from an oil well, grows up to 120 g l^{-1} NaCl only, with an optimum at 50 g l^{-1} (Ollivier *et al.*, 1998; Fig. 20.8). Leakage of certain compatible solutes from microbial cells to the environment can provide methylated amines and dimethylsulfide which can support methanogenesis at near-saturating salt concentrations, and these reactions appear to be responsible for nearly all methanogenesis in hypersaline sediments (Oremland and King, 1989; McGenity, 2010). Cultured species such as *Methanohalobium evestigatum* and *Methanohalophilus portucalensis* grow up to 240–250 g l^{-1} salt.

The homoacetogenic reaction, in which CO_2 is reduced by H_2 to yield acetate, does function at high salt, as shown by the isolation of *Acetohalobium arabaticum* that grows up to 250 g l^{-1} salt. Bioenergetic constraints may once more explain the observations (Oren, 1999b, 2011; Fig. 20.8): methanogens synthesize organic compatible solutes; accordingly methanogenesis at high salinity is based on reactions that yield relatively more energy such as the disproportionation of methylated amines, but not on reactions with a low energy yield such as the reduction of CO_2 or the splitting of acetate. CO_2 reduction by hydrogen in the homoacetogenic reaction yields even less energy than methanogenesis, but the organisms performing the reaction belong to the Halanaerobiales, a group of organisms that do not need to synthesize organic osmotic solutes but use the salt-in strategy of osmotic adaptation.

20.6.10 Fungi

The existence of halophilic fungi and their possible role in hypersaline ecosystems has been largely ignored until about 10 years ago. However, fungi, and notably black yeasts, inhabit the hypersaline waters of solar salterns (Gunde-Cimerman *et al.*, 2009). Genera such as *Wallemia*, *Debaromyces* and *Hortaea* (Fig. 20.5b) are an integral part of hypersaline ecosystems worldwide (Gunde-Cimerman *et al.*, 2009). To what extent these fungi contribute to the overall heterotrophic activity in hypersaline ecosystems remains to be assessed.

20.6.11 Protozoa

The importance of Protozoa in hypersaline ecosystems was recognized only recently.

Many types of ciliate, flagellate and amoeboid Protozoa were described from salt lakes and salterns, and some of these grow up to NaCl saturation (Park *et al.*, 2003; Cho, 2005; Hauer and Rogerson, 2005; Elloumi *et al.*, 2009). Examples of recently isolated species of heterotrophic flagellates from saltern crystallizer ponds are *Pleurostomum flabellatum* (growing optimally at 300 g l^{-1} salt and requiring at least 150–200 g l^{-1}) (Park *et al.*, 2007) and *Halocafetaria seosinensis* (optimum at 150 g l^{-1} salt, requiring at least 75 g l^{-1}, and still growing at 350 g l^{-1}) (Park *et al.*, 2006). Culture-independent, 18S rRNA sequence-based studies show the existence of many more yet-unknown Protozoa and other eukaryotes, even in anoxic hypersaline environments (Alexander *et al.*, 2009). Analysis of the microbial food web along the salinity gradient in a Spanish saltern showed that the abundance of heterotrophic prokaryotes in ponds of intermediate salinity was mainly controlled by eukaryotic predators, showing that such Protozoa play an important role in hypersaline ecosystems (Pedrós-Alió *et al.*, 2000b).

20.6.12 Animalia

While there is extensive evidence showing that the diversity of Animalia declines with increasing hypersalinity (Elloumi *et al.*, 2009), many species survive in the benthos as cysts that can be revived when more favourable conditions return (Moscatello and Belmonte, 2009). The most famous example of a halophilic animal is the brine shrimp (*Artemia salina* and related species), which can be extremely abundant, especially between salinities of 100 and 250 g l^{-1}, and is thought to feed on *Dunaliella salina* and other microalgae (Javor, 1989; Pedrós-Alió, 2004). The ecological role of *Artemia salina* deserves more attention, not only as a grazer, but also as a habitat for surface-dwelling halophiles and a source of food for extreme halophiles. Brine shrimps have been harvested as a fish feed, and their cysts are so resistant to desiccation that specially bred varieties are exported as the novelty 'pet', sea monkeys. Brine fly larvae of the genus *Ephydra* are frequently abundant in hypersaline environments and, together with brine shrimps, are preyed on by a diversity of wading birds.

A remarkable recent finding was the detection of three species of the animal phylum Loricifera (*Spinoloricus* sp., *Rugiloricus* sp. and *Pliciloricus* sp.) that are new to science and metabolically active in the sediments of the anoxic and hypersaline L'Atalante basin (see Section 20.7.5; Danovaro *et al.*, 2010). It is suggested that these Metazoa may grow in association with endosymbiotic bacteria.

20.6.13 Viruses

Bacterial and archaeal viruses (phages) are an integral part of any ecosystem, and hypersaline ecosystems are no exception. In fact, because of low or no predation at the highest salinities their ecological importance is great, and hypersaline environments thus provide an ideal model for understanding the role of phages in controlling the microbial community. Electron microscopy shows that phage-like particles abound in salt lakes, and their number can exceed the number of prokaryotic cells by 1 to 2 orders of magnitude. This is true not only for saltern ponds (Guixa-Boixareu *et al.*, 1996), but also for athalassohaline lakes such as Mono Lake, USA (Jiang *et al.*, 2004) and the Dead Sea (Oren *et al.*, 1997). In saltern crystallizer ponds viral lysis is far more important than bacterivory by Protozoa in regulating the prokaryotic community densities (Guixa-Boixareu *et al.*, 1996). Viruses of halophilic Archaea are morphologically very diverse and include head-and-tail types, spherical, and spindle-shaped phages, lytic types as well as types propagated in a carrier state, double-stranded DNA and single-stranded DNA viruses with a membrane envelope (Porter *et al.*, 2007b; Jäälinoja *et al.*, 2008; Pietilä *et al.*, 2009). Recent years have seen an explosion in the investigation of the genomes of halophages in hypersaline environments and in culture (see Ma *et al.*, 2010). For example, using a metagenomic approach, Santos *et al.* (2010) demonstrated that the phage community in saltern crystallizer ponds (Section 20.7.1) was very different from that in marine environments, and they were even able to disentangle the genomes of phages infecting *Salinibacter ruber* and *Haloquadratum walsbyi*.

20.7 Case Studies

The microbiology and biogeochemistry of some hypersaline environments has been covered in depth. Here, we focus on six contrasting environments including thalassohaline coastal salterns, three inland salt lakes, one with an ionic composition proportional to seawater (Great Salt Lake, USA), another enriched in magnesium and calcium (Dead Sea), and a collection of alkaline soda lakes. Finally, deep-sea anoxic brine lakes and buried salt deposits are considered.

20.7.1 Salterns

Coastal salterns for the production of salt from seawater are among the best studied hypersaline systems, and much of the information available on the microbial diversity and ecophysiology of halophiles was derived from investigations in salterns (Fig. 20.2a). Much of our knowledge has come from salterns in Alicante (Spain) and Eilat (Israel). The saltern environment is ideal for research: salterns are easily accessible, they present a gradient from seawater salinity to halite saturation and sometimes beyond, and the salinity of each pond is kept almost constant, so that each pond contains its characteristic microbial assemblage which remains stable over time (Casamayor *et al.*, 2002). The ponds are shallow, so that plenty of light reaches the bottom, enabling the development of stratified microbial mats dominated by colourful cyanobacteria and anoxygenic phototrophs (Fig. 20.2b; Caumette *et al.*, 1994; Canfield *et al.*, 2004; Oren *et al.*, 2009). Cultivation-independent 16S rRNA studies of the NaCl-saturated crystallizer pond sediment communities reveal a higher abundance of Bacteria than Archaea, as found in the anoxic deep-sea brine lakes (Section 20.7.5), including a wide range of phylotypes that have so far evaded cultivation (López-López *et al.*, 2010). The crystallizer brines are red because of their dense communities of Halobacteriaceae, particularly *Haloquadratum* (Section 20.4.2; Box 20.5), *Dunaliella* (Fig. 20.5a) and *Salinibacter* (Box 20.3). Healthy microbial communities in the saltern ponds are of great importance in the salt-making process, so that the understanding of the microbiology of the salterns has applied aspects as well (Javor, 2002).

Saltern ecosystems are perfect for the study of microbial processes as well, including measurements of rates of primary production and nutrient conversion and elucidation of the interrelationships (cooperation, competition, predation, etc.) between the different components of the microbial communities in the water and in the sediments (Pedrós-Alió *et al.*, 2000a; Joint *et al.*, 2002; Canfield *et al.*, 2004; Oren *et al.*, 2009).

20.7.2 Great Salt Lake, USA

Great Salt Lake, Utah, USA, has been little studied compared to other hypersaline lakes such as the Dead Sea. This is to be regretted since it not only is one of the largest salt lakes, but its salinity has changed dramatically in the past decades. The division of the lake by a rockfill railroad causeway in the 1950s caused a decrease in salinity in the South Arm, while in the North Arm, where no rivers enter the lake, the salt concentration increased (Table 20.1). Climate fluctuations have caused temporary changes in the water level. Today the salinity in the North Arm approaches saturation, while the South Arm has about 120 g l^{-1} salts.

A number of studies by Fred Post in the 1970s (Post, 1977) provide some information on the microbiology of the lake at the time.

The dynamics of *Dunaliella*, the main primary producer in the water column, were studied (Stephens and Gillespie, 1976), as were the communities of halophilic Archaea and cyanobacteria (*Aphanothece halophytica*). A few new species from Great Salt Lake have been described, such as the halophilic archaeon *Halorhabdus utahensis* and the moderately haltolerant acetate-oxidizing sulfate reducer *Desulfobacter halotolerans*. Recently there has been a renewed interest in the biology of the lake, and the first results of microbial surveys, using both cultivation-dependent and independent approaches, have now been published (Baxter *et al.*, 2005; Kjeldsen *et al.*, 2007; Weimer *et al.*, 2009).

20.7.3 The Dead Sea

The Dead Sea, which is actually an inland lake, is famous for being so saline that people can float with ease on its surface. Its athalassohaline brines, dominated by divalent cations, make it an interesting environment for the study of microbial tolerance limits. Owing to the decrease in water level, the salinity of the waters is steadily increasing. As the lake is saturated with respect to NaCl, halite precipitates to the bottom, resulting in a further increase in the ratio between divalent and monovalent cations, which depresses the water activity to levels that make it inimicable for life. Today the lake is virtually devoid of life, but when, following exceptionally rainy winters, the upper metres of the water column become diluted at least 10–15% by fresh water from the Jordan River and from rain floods from the surrounding mountains, dense microbial blooms can develop. Such phenomena were observed in 1980 and in 1992, when first the alga *Dunaliella* grew (Oren and Shilo, 1982; Oren *et al.*, 1995), followed by halophilic Archaea of the family Halobacteriaceae (Oren, 1988a, 1993, 1999a). Up to 35 million prokaryotes ml^{-1} were counted, colouring the brines red due to their bacterioruberin content. In the 1980 bloom, bacteriorhodopsin was shown to be present as well in the community (Oren and Shilo, 1981). Species isolated from the Dead Sea include the Archaea *Haloferax volcanii, Haloarcula marismortui, Halorubrum sodomense* and *Halobaculum gomorrense*, the anaerobic fermentative Bacteria *Halobacteroides halobius, Sporohalobacter lortetii* and *Orenia marismortui* (Halanaerobiales). Virus-like particles were also observed in the water column at the time of a microbial bloom.

Metagenomic studies have recently been performed to learn more about the composition and the functioning of the microbial community in the Dead Sea (Bodaker *et al.*, 2010). The 1992 bloom, for example, was shown to be dominated by a *Halobacterium*-like species, whereas the low-biomass community of 2007 was much more diverse and even. A striking feature was the abundance of putative CorA magnesium ion channel genes, suggesting that cellular expulsion of Mg^{2+} is important for survival of Dead Sea inhabitants (Bodaker *et al.*, 2010).

20.7.4 Soda lakes

Athalassohaline alkaline salt lakes (or soda lakes), rich in NaCl, $NaHCO_3$ and Na_2CO_3, are usually formed by dissolution of rocks that are low in magnesium and calcium, which would otherwise cause carbonate to precipitate (Grant and Jones, 2000; Grant, 2004). The most studied lakes are those in the East African Rift Valley, and continental Russia and the USA. In addition to being able to tolerate such high pHs and elevated salinities, microbes inhabiting soda lakes have to cope with low availability of NH_4^+, caused by weak dissociation of ammonia at high pH. An accumulation of stressful but volatile NH_3 may occur in enclosed alkaline-saline systems such as sea ice or locally in soda lakes. Despite this, soda lakes are among the most productive ecosystems in the world owing to an excess of CO_2, frequently high temperatures and high light irradiance. They attract wading birds such as flamingos, which often breed there and have a beak that is perfectly adapted to feed on cyanobacteria from the genus *Spirulina* that abound in the less saline lakes. Nutrients are supplied by the birds' guano, which may overcome the deficiency in available nitrogen. Indeed, one of the main inhabitants of soda lakes, the archaeal haloalkaliphile *Natronomonas pharaonis,* is equipped with a system for the uptake and conversion of diverse nitrogen species (Falb *et al.*, 2005). A remarkable feature of a *Halomonas* species from arsenic-rich Mono Lake, USA, made the headlines recently when it was reported to be able to replace the phosphate in its DNA with arsenate, which is normally considered to be highly toxic (Wolfe-Simon *et al.*, 2011).

High pH is a key driver in differentiating microbial community composition in saline lakes (Pagaling *et al.*, 2009). The composition of the main microbes in soda lakes from different continents has been determined by direct extraction of nucleic acids and 16S rRNA gene sequencing, e.g. Wadi An Natrun (Egypt; Mesbah *et al.*, 2007), Kenyan lakes (Rees *et al.*, 2004), Baer Lake (Inner Mongolia, China; Ma *et al.*, 2004) and Soap Lake (USA; Dimitriu *et al.*, 2008). There is a high degree of within-lake spatial and temporal variation, especially along salinity and redox gradients of a meromictic soda lake (Dimitriu *et al.*, 2008). Numerous protein-coding genes have been examined in a similar direct way. For example, Foti *et al.* (2007) demonstrated the presence of diverse and novel sulfate-reducers, coupled with the detection of very high rates of sulfate reduction in several Siberian soda lakes. Additionally, sulfur oxidation (Sorokin and Kuenen, 2005a) and numerous other biogeochemical cycles have been demonstrated in soda lakes (Grant and Jones, 2000). Soda lakes are discussed in more detail by Bell (Chapter 19, this volume).

20.7.5 Deep-sea anoxic hypersaline brine lakes

The deep sea and its sediments are among the last earthly frontiers for discovery. Surveys of the sea floor frequently return unusual reflections indicating the presence of deep-sea basins filled with dense, highly saline waters. Such hypersaline anoxic brine lakes have been particularly studied in the Eastern Mediterranean Sea, the Red Sea and the Gulf of Mexico (Fig. 20.3). The combinations of extremely high salinity and high pressure in such environments have selected for new and diverse assemblages of microorganisms with features that are highly useful to biotechnology industries that will form the basis of sustainable technologies (Ferrer *et al.*, 2005).

A comprehensive multidisciplinary research programme was aimed at elucidating the biology of the deep-sea hypersaline anoxic basins, located at depths of 3.2–3.6 km in the Eastern Mediterranean Sea. These hypersaline brine lakes are situated on the Mediterranean Ridge, where tectonic processes have deformed sediments so that seawater comes into contact with evaporites that were deposited during the Messinian salinity crisis (5.96–5.33 million years ago; Krijgsman *et al.*, 1999). The resultant highly saline brines can accumulate in depressions in the seafloor, and, because there is little mixing with the overlying seawater, anaerobic conditions prevail in the brines, and an extremely steep redox gradient and halocline is formed between seawater and the hypersaline brines with a vertical extension of about 2 m (Daffonchio *et al.*, 2006). This chemocline is inhabited by a highly stratified and diverse world of microorganisms, primarily Bacteria that occur at high densities (Sass *et al.*, 2001; van der Wielen *et al.*, 2005; Daffonchio *et al.*, 2006). Many novel phylogenetic lineages of both Bacteria and Archaea have been found in the deep-sea brines and other anoxic hypersaline environments, such as group MSBL-1, however their function remains unknown (van der Wielen *et al.*, 2005). Diverse and novel eukaryotes from these environments have been described in Sections 20.6.11 and 20.6.12. The chemocline from the four main basins has been studied in detail, and each tells a different story, partly a consequence of the different questions addressed, but also because of variations in chemical composition (Daffonchio *et al.*, 2006; Hallsworth *et al.*, 2007; Yakimov *et al.*, 2007; Borin *et al.*, 2009).

It is particularly interesting to compare the chemoclines of the NaCl-rich brines (e.g. Bannock Basin derived from dissolution of halite) and the unusual $MgCl_2$-saturated brine (Discovery Basin, derived from dissolution of bischofite) (Table 20.1; Fig. 20.3). The convergence of terminal electron acceptors, nutrients, carbon and energy sources (including material that has settled at the interface owing to the high density of the brine) has stimulated microbial activity and biomass in the NaCl-rich chemoclines (Daffonchio *et al.*, 2006). In contrast, life becomes severely inhibited in the chemocline of Discovery Basin at a critical concentration (Hallsworth *et al.*, 2007). Despite the energy-yielding redox couplings in the $MgCl_2$-rich chemocline, the macromolecule-destabilizing effects of $MgCl_2$ above a concentration of 2.3 M were considered to be incompatible with the life-forms being analysed (Hallsworth *et al.*, 2007).

These multiple extreme environments perhaps serve as analogues for the deep-sea brines proposed to occur on Jupiter's moon, Europa. It is therefore intriguing to investigate how the brines and their sediments may preserve cells, spores and nucleic acids for long periods (Danovaro *et al.*, 2005; Borin *et al.*, 2008; Sass *et al.*, 2008), and to test whether they have been in part populated by microbes that were living in Messinian salt (McGenity *et al.*, 2008).

Exploration of similar brines from the Shaban and the Kebrit Deep on the bottom of the Red Sea also demonstrated the presence of a wealth of novel types of halophiles (Eder *et al.*, 2001, 2002), and two novel microorganisms were recently cultivated. The first is *Halorhabdus tiamatea,* a non-pigmented representative of the Halobacteriaceae. In contrast to other members of the family, it preferentially lives anaerobically by fermentation rather than by aerobic respiration (Antunes *et al.*, 2008). The second is *Haloplasma contractile*, a highly unusual contractile microbe belonging to a new lineage of the Bacteria, between the Firmicutes and the Mollicutes, and classified in a new order, the Haloplasmatales. It grows between 15 and 180 g l^{-1} salt, and lives by fermentation or denitrification (Antunes *et al.*, 2009). Antunes *et al.* (2011) provide a nice review of the deep-sea anoxic hypersaline brine lakes, focusing on those in the Red Sea.

The brines at the bottom of the Gulf of Mexico were also the subject of renewed studies (Joye *et al.*, 2009), and certain brine lakes are even surrounded by dense beds of *Bathymodiolus* mussels which house chemoautotrophic bacteria presumably fed by sulfide or methane from the anoxic brine lake (MacDonald *et al.*, 1990). Thus these microbially dominated habitats influence the geochemical cycles of the wider oceans.

20.7.6 Buried salt deposits

Buried salt deposits have intrigued scientists for many years, primarily because of the notion that the halophiles isolated from them are the remnants or descendants of populations that have been trapped over geological time. Several research groups have independently cultivated halophiles from rock salt that has been surface sterilized (mainly Archaea from the family Halobacteriaceae (see McGenity *et al.*, 2000), but also endospore-forming Bacteria of the Bacillaceae (Vreeland *et al.*, 2000)). This does not in itself imply that the halophiles are as old as the rock formation. It is essential to take into account possible alterations, especially dissolution and recrystallization of salt. However, evidence is accumulating to suggest that halobacteria, in particular, are able to remain viable in rock salt over millions of years (see McGenity *et al.*, 2000, 2008). For example, strains of *Halobacterium noricense* were isolated from 121 million-year-old primary halite (Vreeland *et al.*, 2007), and independent studies have detected very similar halobacteria in other subterranean hypersaline environments and rock salt (see McGenity *et al.*, 2000, 2008; Gruber *et al.*, 2004; Gramain *et al.*, 2011).

Halobacteria have many characteristics that predispose them to life in rock salt.

When halite crystallizes, brine inclusions are formed within the crystals, and microorganisms are trapped within these inclusions (Fig. 20.10). Laboratory experiments have shown that such bacteria can remain viable for long periods (Norton and Grant, 1988), and halobacteria are among the best survivors. Intracellular ions, which halobacteria have in abundance, protect biological macromolecules from radiation and free radicals (Nicastro *et al.*, 2002; Tehei *et al.*, 2002; Kish *et al.*, 2009). *Halobacterium salinarum*, which is frequently isolated from salt crystals, tolerates radiation by means of diverse mechanisms (McCready and Marcello, 2003; McGenity *et al.*, 2008), including a high ratio of manganese to iron in the cell, which protects proteins from oxidative damage (Kish *et al.*, 2009). The absence of light, reduced cosmic radiation, high salinity and anoxic conditions inside buried salt crystals all contribute to limit cellular damage. Nevertheless, it is assumed that long-living microbes will need a source of carbon and energy to repair damaged macromolecules. McGenity *et al.* (2008) proposed (with certain provisos) that, by feeding on co-entombed microbes, halobacteria could survive for tens to hundreds of millions of years.

The evidence for halobacterial survival in halite over tens to hundreds of thousands of years is now extremely strong. A salt core from beneath Death Valley has proven the perfect natural laboratory to investigate survival, with three independent research groups isolating halobacteria from similar locations along the core's depth (see Schubert *et al.*, 2010). Schubert *et al.* (2010) observed a heterogeneous distribution of cells in brine inclusions along the core, and isolated several halobacteria from these hot spots, and Schubert *et al.* (2009) showed microscopically that these cells were co-trapped with *Dunaliella* spp., and were particularly adept at growing on glycerol. The green algal cells would have contained up to 7 M glycerol that they use as a compatible solute. Upon the death of *Dunaliella* cells, the glycerol would have become available to halobacteria co-entombed in the fluid inclusions.

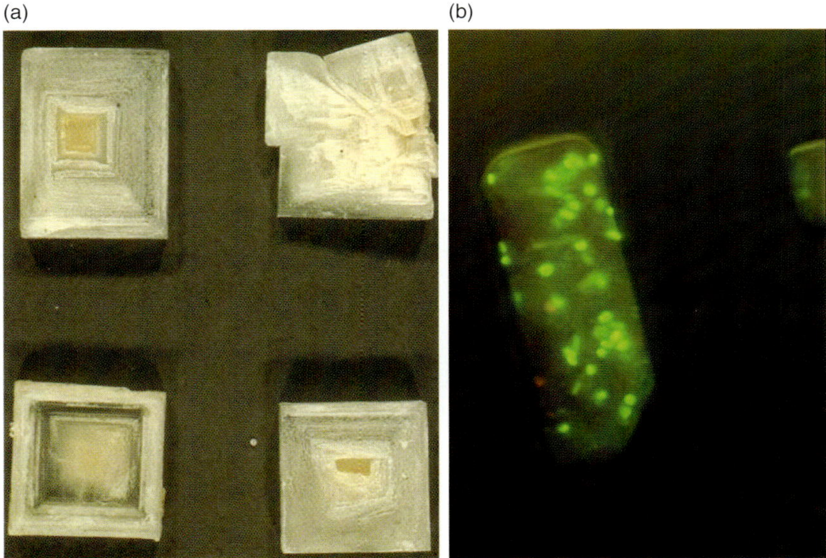

Fig. 20.10. Halobacteria in salt crystals. (a) Laboratory-made crystals of NaCl encasing *Halorubrum saccharovorum*. The cloudiness of the halite (NaCl) crystals is due to the large number of brine inclusions. Each crystal is ca. 1 cm square, © Terry J. McGenity. (b) Fluorescence microscopic image of *Halobacterium salinarum* NRC-1 inside a single brine inclusion (ca. 60 μm long) of a halite crystal. Cells have been stained with the dye SYTO 9 indicating that the entombed cells are alive, © Sergiu Fendrihan.

20.8 Societal Importance of Hypersaline Environments and Halophiles

Many hypersaline environments attract visiting birds, which in turn attracts tourism. Sea-salt manufacture is accelerated by its red, carotenoid-rich microbial inhabitants. However, neither the underlying science of this process, nor the degree to which consumed halophiles trapped in salt crystals affect human nutrition, have been intensively studied. Halophiles have great economic potential, some realized, such as carotenoid production by *Dunaliella* spp., and many requiring optimization, such as bioplastic production by *Haloferax mediterranei* (Oren, 2010). Moreover, the halobacterial light-induced, outwardly directed proton pump, bacteriorhodopsin (Fig. 20.9), is both an important model for transmembrane receptors and has been used in applications ranging from security cards to materials testing. The compatible solute ectoine from *Halomonas elongata* is used in personal-care products. It is produced by sequentially increasing and decreasing the salinity of growth medium to encourage biosynthesis and then export from the cell, and has thus been referred to as 'microbial milking'. Hypersaline and alkaline environments are being investigated using metagenomics, and have yielded some fascinating and novel enzymes of biotechnological interest (Ferrer *et al.*, 2005; see also Arora and Bell, Chapter 25, this volume).

Halobacteria, with their longevity, ability to grow at low water activity and radiation resistance, have received considerable attention from astrobiologists, especially since the discovery of halite on Mars and possible hypersaline oceans in the depths of Europa. The tenacity of halobacteria is illustrated by *Halorubrum chaoviator*, which was dried on to a quartz disc and transported by the Biopan facility into space, where it was exposed to and survived extreme radiation (Mancinelli *et al.*, 2009; see also Gómez *et al.*, Chapter 26, this volume).

References

Alexander, E., Stock, A., Breiner, H.-W., Behnke, A., Bunge, J., Yakimov, M.M. and Stoeck, T. (2009) Microbial eukaryotes in the hypersaline anoxic L'Atalante deep-sea basin. *Environmental Microbiology* 11, 360–381.

Antón, J., Llobet-Brossa, E., Rodríguez-Valera, F. and Amann, R. (1999) Fluorescence *in situ* hybridization analysis of the prokaryotic community inhabiting crystallizer ponds. *Environmental Microbiology* 1, 517–523.

Antón, J., Rosselló-Mora, R., Rodríguez-Valera, R. and Amann, R. (2000) Extremely halophilic bacteria in crystallizer ponds from solar salterns. *Applied and Environmental Microbiology* 66, 3052–3057.

Antón, J., Oren, A., Benlloch, S., Rodríguez-Valera, F., Amann, R. and Rosselló-Mora, R. (2002) *Salinibacter ruber* gen. nov., sp. nov., a novel extreme halophilic member of the Bacteria from saltern crystallizer ponds. *International Journal of Systematic and Evolutionary Microbiology* 52, 485–491.

Antón, J., Peña, A., Santos, F., Martínez-García, M., Schmitt-Kopplin, P. and Rosselló-Móra, R. (2008) Distribution, abundance and diversity of the extremely halophilic bacterium *Salinibacter ruber*. *Saline Systems* 4, 15.

Antunes, A., Taborda, M., Huber, R., Moissl, C., Nobre, M.F. and da Costa, M.S. (2008) *Halorhabdus tiamatea* sp. nov., a non-pigmented, extremely halophilic archaeon from a deep-sea, hypersaline anoxic basin of the Red Sea, and emended description of the genus *Halorhabdus*. *International Journal of Systematic and Evolutionary Microbiology* 58, 215–220.

Antunes, A., Rainey, F.A., Wanner, G., Taborda, M., Pätzold, J., Nobre, M.F., da Costa, M.S. and Huber, R. (2009) A new lineage of halophilic, wall-less, contractile bacteria from a brine-filled deep on the Red Sea. *Journal of Bacteriology* 190, 3580–3587.

Antunes, A., Ngugi, D.K. and Stingl, U. (2011) Microbiology of the Red Sea (and other) deep-sea anoxic brine lakes. *Environmental Microbiology Reports* 3, 416–433.

Arahal, D.R. and Ventosa, A. (2006) The family *Halomonadaceae*. In: Dworkin, M., Falkow, S., Rosenberg, E., Schleifer, K.-H. and Stackebrandt, E. (eds) *The Prokaryotes. A Handbook on the Biology of Bacteria*, 3rd edn, vol. 6. Springer, New York, pp. 811–835.

Baxter, B.K., Litchfield, C.D., Sowers, K., Griffith, J D., Arora DasSarma, P. and DasSarma, S. (2005) Microbial diversity of Great Salt Lake. In: Gunde-Cimerman, N., Oren, A. and Plemenitaš, A. (eds) *Adaptation to Life at High Salt Concentrations in Archaea, Bacteria, and Eukarya*. Springer, Dordrecht, the Netherlands, pp. 11–25.

Ben-Amotz, A. and Avron, M. (1973) The role of glycerol in the osmotic regulation of the halophilic alga *Dunaliella parva*. *Plant Physiology* 51, 875–878.

Ben-Amotz, A., Polle, J.E.W. and Subba Rao, D.V. (eds) (2009) *The Alga* Dunaliella*: Biodiversity, Physiology, Genomics and Biotechnology*. Science Publishers, Enfield, New Hampshire.

Benlloch, S., Martínez-Murcia, A.J. and Rodríguez-Valera, F. (1995) Sequencing of bacterial and archaeal 16S rRNA genes directly amplified from a hypersaline environment. *Systematic and Applied Microbiology* 18, 574–581.

Bodaker, I., Sharon, I., Suzuki, M.T., Reingersch, R., Shmoish, M., Andreishcheva, E., Sogin, M.L., Rosenberg, M., Belkin, S., Oren, A. and Béjà, O. (2010) The dying Dead Sea: comparative community genomics in an increasingly extreme environment. *ISME Journal* 4, 399–407.

Boetius, A. and Joye, S. (2009) Thriving in salt. *Science* 324, 1523–1525.

Bolhuis, H., te Poele, E.M. and Rodríguez-Valera, F. (2004) Isolation and cultivation of Walsby's square archaeon. *Environmental Microbiology* 6, 1287–1291.

Bolhuis, H., Palm, P., Wende, A., Farb, M., Rampp, M., Rodriguez-Valera, F., Pfeiffer, F. and Oesterhelt, D. (2006) The genome of the square archaeon 'Haloquadratum walsbyi': life at the limits of water activity. *BMC Genomics* 7, 169.

Borin, S., Crotti, E., Mapelli, F., Tamagnini, I., Corselli, C. and Daffonchio, D. (2008) DNA is preserved and maintains transforming potential after contact with brines of the deep anoxic hypersaline lakes of the Eastern Mediterranean Sea. *Saline Systems* 4, 10.

Borin, S., Brusetti, L., Mapelli, F., D'Auria, G., Brusa, T., Marzorati, M., Rizzi, A., Yakimov, M., Marty, D., DeLange, G., van der Wielen, P., Bolhuis, H., McGenity, T.J., Polymenakou, P., Malinverno, E., Giuliano, L., Corselli, C. and Daffonchio, D. (2009) Sulfur cycling and methanogenesis primarily drive microbial colonization of the highly sulfidic Urania deep hypersaline basin. *Proceedings of the National Academy of Sciences of the United States of America* 106, 9151–9156.

Bowers, K.J., Mesbah, N.M. and Wiegel, J. (2009) Biodiversity of polyextremophilic *Bacteria*: does combining the extremes of high salt, alkaline pH and elevated temperature approach a physico-chemcial boundary for life? *Saline Systems* 5, 9.

Brito-Echeverría, J., López-López, A., Yarza, P., Antón, J. and Rosselló-Móra, R. (2009) Occurrence of *Halococcus* spp. in the nostrils salt glands of the seabird *Calonextris diomedea*. *Extremophiles* 13, 557–565.

Brown, A.D. (1976) Microbial water stress. *Bacteriological Reviews* 40, 803–846.

Brown, A.D. (1990) *Microbial Water Stress Physiology. Principles and Perspectives*. John Wiley and Sons, Chichester, UK, 328 pp.

Burns, D.G., Camakaris, H.M., Janssen, P.H. and Dyall-Smith, M.L. (2004a) Cultivation of Walsby's square haloarchaeon. *FEMS Microbiology Letters* 238, 469–473.

Burns, D.G., Camakaris, H.M., Janssen, P.H. and Dyall-Smith, M.L. (2004b) Combined use of cultivation-dependent and cultivation-independent methods indicates that members of most haloarchaeal groups in an Australian crystallizer pond are cultivable. *Applied and Environmental Microbiology* 70, 5258–5265.

Burns, D.G., Janssen, P.H., Itoh, T., Kamekura, M., Li, Z., Jensen, G., Rodríguez-Valera, F., Bolhuis, H. and Dyall-Smith, M.L. (2007) *Haloquadratum walsbyi* gen. nov., sp. nov., the square haloarchaeon of Walsby, isolated from saltern crystallizers in Australia and Spain. *International Journal of Systematic and Evolutionary Microbiology* 57, 387–392.

Canfield, D.E., Sørensen, K.B. and Oren, A. (2004) Biogeochemistry of a gypsum-encrusted microbial ecosystem. *Geobiology* 2, 133–150.

Casamayor, E.O., Massana, R., Benlloch, S., Øvreås, L., Díez, B., Goddard, V.J., Gasol, J.M., Joint, I., Rodríguez-Valera, F. and Pedrós-Alió, C. (2002) Changes in archaeal, bacterial and eukaryal assemblages along a salinity gradient by comparison of genetic fingerprinting methods in a multipond solar saltern. *Environmental Microbiology* 4, 338–348.

Casanueva, A., Galada, N., Baker, G.C., Grant, W.D., Heaphy, S., Jones, B., Ma, Y., Ventosa, A., Blamey, J. and Cowan, D.A. (2008) Nanoarchaeal 16S rRNA gene sequences are widely dispersed in hyperthermophilic and mesophilic halophilic environments. *Extremophiles* 12, 651–656.

Caton, T.M., Caton, I.R., Witte, L.R. and Schneegurt, M.A. (2009) Archaeal diversity at the Great Salt Plains of Oklahoma described by cultivation and molecular analysis. *Microbial Ecology* 58, 519–528.

Caumette, P., Matheron, R., Raymond, N. and Relexans, J.-C. (1994) Microbial mats in the hypersaline ponds of Mediterranean salterns (Salins-de-Giraud, France). *FEMS Microbiology Ecology* 13, 273–286.

Cayol, J.-L., Ollivier, B., Patel, B.K.C., Prensier, G., Guezennec, J. and Garcia, J.-L. (1994) Isolation and characterization of *Halothermothrix orenii* gen. nov., sp. nov., a halophilic, thermophilic, fermentative, strictly anaerobic bacterium. *International Journal of Systematic Bacteriology* 44, 534–540.

Cho, B.C. (2005) Heterotrophic flagellates in hypersaline waters. In: Gunde-Cimerman, N., Oren, A. and Plemenitaš, A. (eds) *Adaptation to Life at High Salt Concentrations in Archaea, Bacteria, and Eukarya*. Springer, Dordrecht, the Netherlands, pp. 543–549.

Conrad, R., Frenzel, P. and Cohen, Y. (1995) Methane emission from hypersaline microbial mats: lack of aerobic methane oxidation activity. *FEMS Microbiology Ecology* 16, 297–305.

Cui, H.-L., Zhou, P.-J., Oren, A. and Liu, S.-J. (2009) Intraspecific polymorphism of 16S rRNA genes in two halophilic archaeal genera, *Haloarcula* and *Halomicrobium*. *Extremophiles* 13, 31–37.

Daffonchio, D., Borin, S., Brusa, T., Brusetti, L., van der Wielen, P.W.J.J., Bolhuis, H., Yakimov, M.M., D'Auria, G., Giuliano, L., Marty, D., Tamburini, C., McGenity, T.J., Hallsworth, J.E., Sass, A.M., Timmis, K.N., Tselepides, A., de Lange, G.J., Hübner, A., Thomson, J., Varnavas, S.P., Gasparoni, F., Gerber H.W., Malinverno, E., Corselli, C. and Biodeep Scientific Party (2006) Stratified prokaryote network in the oxic-anoxic transition of a deep-sea halocline. *Nature* 440, 203–207.

Danovaro, R., Corinaldesi, C., Dell'Anno, A., Fabiano, M. and Corselli, C. (2005) Viruses, prokaryotes and DNA in the sediments of a deep-hypersaline anoxic basin (DHAB) of the Mediterranean Sea. *Environmental Microbiology* 7, 586–592.

Danovaro, R., Dell'Anno, A., Pusceddu, A., Gambi, C., Heiner, I. and Kristensen, R.M. (2010) The first Metazoa living in permanently anoxic conditions. *PLoS ONE* 8, 30.

Demergasso, C., Escuardo, L., Casamayor, E.O., Chong, G., Balagué, V. and Pedrós-Alió, C. (2008) Novelty and spatio-temporal heterogeneity in the bacterial diversity of hypersaline Lake Tebenquiche (Salar de Atacama). *Extremophiles* 12, 491–504.

Deutch, C.E. (1994) Characterization of a novel salt-tolerant *Bacillus* sp. from the nasal cavities of desert iguanas. *FEMS Microbiology Letters* 121, 55–60.

Dimitriu, P.A., Pinkart, H.C., Peyton, B.M. and Mormile, M.R. (2008) Spatial and temporal patterns in the microbial diversity of a meromictic soda lake in Washington State. *Applied and Environmental Microbiology* 74, 4877–4888.

Dyall-Smith, M.L., Pfeiffer, F., Klee, K., Palm, P., Gross, K., Schuster, S.C., Rampp, M. and Oesterhelt, D. (2011) *Haloquadratum walsbyi*: limited diversity in a global pond. *PLoS ONE* 6, e20968.

Eder, W., Jahnke, L.L., Schmidt, M. and Huber, R. (2001) Microbial diversity of the brine-seawater interface of the Kebrit Deep, Red Sea, studied via 16S rRNA gene sequences and cultivation methods. *Applied and Environmental Microbiology* 67, 3077–3085.

Eder, W., Schmidt, M., Koch, M., Garbe-Schönberg, D. and Huber, R. (2002) Prokaryotic phylogenetic diversity and corresponding geochemical data of the brine-seawater interface of the Shaban Deep, Red Sea. *Environmental Microbiology* 4, 758–763.

Ellis, D.G., Bizzoco, R.W. and Kelley, S.T. (2008) Halophilic *Archaea* determined from geothermal steam vent aerosols. *Environmental Microbiology* 10, 1582–1590.

Elloumi, J., Carrias, J.-F., Ayadi, H., Sime-Ngando, T. and Bouaïn, A. (2009) Communities structure of the planktonic halophiles in the solar saltern of Sfax, Tunisia. *Estuarine, Coastal and Shelf Science* 81, 19–26.

Elshahed, M.S., Najar, F.Z., Roe, B.A., Oren, A., Dewers, T.A. and Krumholz, L.R. (2004) Survey of archaeal diversity reveals abundance of halophilic *Archaea* in a low-salt, sulfide- and sulfur-rich spring. *Applied and Environmental Microbiology* 70, 2230–2239.

Falb, M., Pfeiffer, F., Palm, P., Rodewald, K., Hickmann, V., Tittor, J. and Oesterhelt, D. (2005) Living with two extremes: conclusions from the genome sequence of *Natronomonas pharaonis*. *Genome Research* 15, 1336–1343.

Falb, M., Müller, K., Königsmaier, L., Oberwinkler, T., Horn, P., von Gronau, S., Gonzalez, O., Pfeiffer, F., Bornberg-Bauer, E. and Oesterhelt, D. (2008) Metabolism of halophilic archaea. *Extremophiles* 12, 177–196.

Ferrer, M., Golyshina, O.V., Chemikova, T.N., Khachane, A.N., dos Santos, V.A.P.M., Yakimov, M.M., Timmis, K.N. and Golyshin, P.N. (2005) Microbial enzymes mined from the Urania deep-sea hypersaline anoxic basin. *Chemistry & Biology* 12, 895–904.

Foti, M., Sorokin, D.Y., Lomans, B., Mussman, M., Zacharova, E.E., Pimenov, N.V., Kuenen, J.G. and Muyzer, G. (2007) Diversity, activity and abundance of sulfate-reducing bacteria in saline and hypersaline soda lakes. *Applied and Environmental Microbiology* 73, 2093–2100.

Franzmann, P.D., Stackebrandt, E., Sanderson, K., Volkman, J.K., Cameron, D.E., Stevenson, P.L., McMeekin, T.A. and Burton, H.R. (1988) *Halobacterium lacusprofundii* sp. nov., a halophilic bacterium isolated from Deep Lake, Antarctica. *Systematic and Applied Microbiology* 11, 20–27.

Fukushima, T., Usami, R. and Kamekura, M. (2007) A traditional Japanese-style salt field is a niche for haloarchaeal strains that can survive in 0.5% salt solution. *Saline Systems* 3, 2.

Galinski, E.A. (1995) Osmoadaptation in bacteria. *Advances in Microbial Physiology* 37, 273–328.

Glenn, E.P., Brown, J.J. and Blumwald, E. (1999) Salt tolerance and crop potential of halophytes. *Critical Reviews in Plant Sciences* 18, 227–255.

Gramain, A., Chong Díaz, G., Demergasso, C., Lowenstein, T.K. and McGenity, T.J. (2011) Archaeal diversity along a subterranean salt core from the Salar Grande (Chile). *Environmental Microbiology* 13, 2105–2121.

Grant, W.D. (2004) Life at low water activity. *Philosophical Transactions of the Royal Society London B: Biological Sciences* 359, 1249–1267.

Grant, W.D. and Jones, B.E. (2000) Microbial diversity and ecology of alkaline environments. In: Seckbach, J. (ed.) *Journey to Diverse Microbial Worlds*. Kluwer Academic Publishers, the Netherlands, pp. 177–190.

Gruber, C., Legat, A., Pfaffenhuemer, M., Radax, C., Weidler, G., Busse, H.J. and Stan-Lotter, H. (2004) *Halobacterium noricense* sp nov., an archaeal isolate from a bore core of an alpine Permian salt deposit, classification of *Halobacterium* sp. NRC-1 as a strain of *H. salinarum* and emended description of *H. salinarum. Extremophiles* 8, 431–439.

Guixa-Boixareu, N., Calděron-Paz, J.I., Heldal, M., Bratbak, G. and Pedrós-Alió, C. (1996) Viral lysis and bacterivory as prokaryotic loss factors along a salinity gradient. *Aquatic Microbial Ecology* 11, 213–227.

Gunde-Cimerman, N., Ramos, J. and Plemenitaš, A. (2009) Halotolerant and halophilic fungi. *Mycological Research* 113, 1231–1241.

Hallsworth, J.E., Yakimov, M.M., Golyshin, P.N., Gillion, J.L.M., D'Auria, G., de Lima Alves, F., La Cono, V., Genovese, M., McKew, B.A., Hayes, S.L., Harris, G., Giuliano, L., Timmis, K.N. and McGenity, T.J. (2007) Limits of life in $MgCl_2$-containing environments: chaotropicity defines the window. *Environmental Microbiology* 9, 801–813.

Hartmann, R., Sickinger, H.-D. and Oesterhelt, D. (1980) Anaerobic growth of halobacteria. *Proceedings of the National Academy of Sciences of the United States of America* 77, 3821–3825.

Hauer, G. and Rogerson, A. (2005) Heterotrophic Protozoa from hypersaline environments. In: Gunde-Cimerman, N., Oren, A. and Plemenitaš, A. (eds) *Adaptation to Life at High Salt Concentrations in Archaea, Bacteria, and Eukarya*. Springer, Dordrecht, the Netherlands, pp. 522–539.

Hoeft, S.E., Switzer Blum, J., Stolz, J.F., Tabita, F.R., Witte, B., King, G.M., Santini, J.M. and Oremland, R.S. (2007) *Alkalilimnicola ehrlichii* sp nov., a novel, arsenite-oxidizing haloalkaliphilic gammaproteobacterium capable of chemoautotrophic or heterotrophic growth with nitrate or oxygen as the electron acceptor. *International Journal of Systematic and Evolutionary Microbiology* 57, 504–512.

Jäälinoja, H.T., Roine, E., Laurinmäki, P., Kivelä, H.M., Bamford, D.H. and Butcher, S.J. (2008) Structure and host-cell interaction of SH1, a membrane-containing, halophilic euryarchaeal virus. *Proceedings of the National Acadamy of Sciences of the United States of America* 105, 8008–8013.

Javor, B. (1989) *Hypersaline Environments. Microbiology and Biogeochemistry*. Springer, Berlin, Germany, 384 pp.

Javor, B.J. (2002) Industrial microbiology of solar salt production. *Journal of Industrial Microbiology and Biotechnology* 28, 42–47.

Jiang, S., Steward, G., Jellison, R., Chu, W. and Choi, S. (2004) Abundance, distribution, and diversity of viruses in alkaline, hypersaline Mono Lake, California. *Microbial Ecology* 47, 9–17.

Joint, I., Henriksen, P., Garde, K. and Riemann, B. (2002) Primary production, nutrient assimilation and microzooplankton grazing along a hypersaline gradient. *FEMS Microbiology Ecology* 39, 245–257.

Joye, S.B., Connell, T.L., Miller, L.G., Oremland, R.S. and Jellison, R.S. (1999) Oxidation of ammonia and methane in an alkaline, saline lake. *Limnology and Oceanography* 44, 178–188.

Joye, S.B., Samarkin, V.A., Orcutt, B.M., MacDonald, I.R., Hinrichs, K.-U., Elvert, M., Teske, A.P., Lloyd, K.G., Lever, M.A., Montoya, J.P. and Meile, C.D. (2009) Metabolic variability in seafloor brines revealed by carbon and sulphur dynamics. *Nature Geosciences* 2, 349–354.

Kaye, J.Z. and Baross, J.A. (2004) Synchronous effect of temperature, hydrostatic pressure, and salinity on growth, phospholipid profiles, and protein patterns of four *Halomonas* species isolated from deep-sea hydrothermal-vent and sea surface environments. *Applied and Environmental Microbiology* 70, 6220–6229.

Kish, A., Kirkali, G., Robinson, C., Rosenblatt, R., Jaruga, P., Dizdaroglu, M. and DiRuggiero, J. (2009) Salt shield: intracellular salts provide cellular protection against ionizing radiation in the halophilic archaeon, *Halobacterium salinarum* NRC-1. *Environmental Microbiology* 11, 1066–1078.

Kjeldsen, K.U., Loy, A., Jakobsen, T.F., Thomsen, T.R., Wagner, M. and Ingvorsen, K. (2007) Diversity of sulphate-reducing bacteria from an extreme hypersaline sediment, Great Salt Lake. *FEMS Microbiology Ecology* 60, 287–298.

Kohls, K., Abed, R.M.M., Polerecky, L., Weber, M. and de Beer, D. (2010) Halotaxis of cyanobacterial in an intertidal hypersaline mat. *Environmental Microbiology* 12, 567–575.

Koops, H.-P., Böttcher, B., Möller, U., Pommerening-Röser, A. and Stehr, G. (1990) Description of a new species of *Nitrosococcus*. *Archives of Microbiology* 154, 244–248.

Krijgsman, W., Hilgen, F.J., Raff, I., Sierro, F.J. and Wilson, D.S. (1999) Chronology, causes and progression of the Messinian salinity crisis. *Nature* 400, 652–655.

Kurlansky, M. (2002) *Salt: a world history*. Jonathan Cape, London, UK, 498 pp.

Kushner, D.J. (1978) Life in high salt and solute concentrations. In: Kushner, D.J. (ed.) *Microbial Life in Extreme Environments*. Academic Press, London, UK, pp. 317–368.

Lanyi, J.K. (1974) Salt-dependent properties of proteins from extremely halophilic bacteria. *Bacteriological Reviews* 38, 272–290.

Legault, B.A., López-López, A., Alba-Casado, J.C., Doolittle, W.F., Bolhuis, H., Rodríguez-Valera, F. and Papke, R.T. (2006) Environmental genomics of '*Haloquadratum walsbyi*' in a saltern crystallizer indicates a large pool of accessory genes in an otherwise coherent species. *BMC Genomics* 7, 171.

Ley, R.E., Harris, J.K., Wilcox, J., Spear, J.R., Miller, S.R., Bebout, B.M., Maresca, J.A., Bryant, D.A., Sogin, M.L. and Pace, N.R. (2006) Unexpected diversity and complexity of the Guerrero Negro hypersaline microbial mat. *Applied and Environmental Microbiology* 72, 3685–3695.

López-López, A., Benlloch, S., Bonfa, M., Rodriguez-Valera, F. and Mira, A. (2007) Intragenomic 16S rDNA divergence in *Haloarcula marismortui* is an adaptation to different temperatures. *Journal of Molecular Evolution* 65, 687–696.

López-López, A., Yarza, P., Richter, M., Suárez-Suárez, A., Antón, J., Niemann, H. and Rosselló-Móra, R. (2010) Extremely halophilic microbial communities in anaerobic sediments from a solar saltern. *Environmental Microbiology Reports* 2, 258–271.

Lozupone, C.A. and Knight, R. (2007) Global patterns in bacterial diversity. *Proceedings of the National Acadamy of Sciences of the United States of America* 104, 11436–11440.

Ma, Y., Zhang, W., Xue, Y., Zhou, P., Ventosa, A. and Grant, W.D. (2004) Bacterial diversity of the inner Mongolian Baer Soda Lake as revealed by 16S rRNA gene sequence analyses. *Extremophiles* 8, 45–51.

Ma, Y., Galinski, E.A., Grant, W.D., Oren, A. and Ventosa, A. (2010) Halophiles 2010: life in saline environments. *Applied and Environmental Microbiology* 76, 6971–6981.

MacDonald, I.R., Reilly, J.F., Guinasso, N.L., Brooks, J.M., Carney, R.S., Bryant, W.A. and Bright, T.J. (1990) Chemosynthetic mussels at a brine-filled pockmark in the northern Gulf of Mexico. *Science* 248, 1096–1099.

Mancinelli, R.L. and Hochstein, L.I. (1986) The occurrence of denitrification in extremely halophilic bacteria. *FEMS Microbiology Letters* 35, 55–58.

Mancinelli, R.L., Landheim, R., Sánchez-Porro, C., Dornmayr-Pfaffenhuemer, M., Gruber, C., Legat, A., Ventosa, A., Radax, C., Ihara, K., White, M.R. and Stan-Lotter, H. (2009) *Halorubrum chaoviator* sp. nov., a haloarchaeon isolated from sea salt in Baja California, Mexico, Western Australia and Naxos, Greece. *International Journal of Systematic and Evolutionary Microbiology* 59, 1908–1913.

Mavromatis, K., Ivanova, N., Anderson, I., Lykidis, A., Hooper, S.D., Sun, H., Kunin, V., Lapidus, A., Hugenholtz, P., Patel, B. and Kyrpides, N.C. (2009) Genome analysis of the anaerobic thermohalophilic bacterium *Halothermothrix orenii*. *PloS ONE* 4, e4192.

McCready, S. and Marcello, L. (2003) Repair of UV damage in *Halobacterium salinarum*. *Biochemistry Society Transactions* 31, 694–698.

McGenity, T.J. (2010) Methanogens and methanonogenesis in hypersaline environments. In: Timmis, K.N. (ed.) *Handbook of Hydrocarbon and Lipid Microbiology*. Springer, Berlin, Germany, pp. 666–680.

McGenity, T.J., Gemmell, R.T., Grant, W.D. and Stan-Lotter, H. (2000) Origins of halophilic microorganisms in ancient salt deposits. *Environmental Microbiology* 2, 243–250.

McGenity, T.J., Hallsworth, J.E. and Timmis, K.N. (2008) Connectivity between 'ancient' and 'modern' hypersaline environments, and the salinity limits of life. *CIESM Workshop Monographs no.33: The Messinian Salinity Crisis from Mega-deposits to Microbiology – A Consensus Report*. Almeria (Spain), 7–10 November 2007, pp. 115–120.

McKew, B.A., Taylor, J.D., McGenity, T.J. and Underwood, G.J.C. (2011) Resistance and resilience of benthic biofilm communities from a temperate saltmarsh to desiccation and rewetting. *ISME Journal* 5, 30–41.

Mesbah, N.M., Abou-El-Ela, S.H. and Wiegel, J. (2007) Novel and unexpected prokaryotic diversity in water and sediments of the alkaline, hypersaline lakes of the Wadi An Natrun, Egypt. *Microbial Ecology* 54, 598–617.

Mikucki, J.A., Pearson, A., Johnston, D.T., Turchyn, A.V., Farquhar, J., Schrag, D.P., Anbar, A.D., Priscu, J.C. and Lee, P.A. (2009) A contemporary microbially maintained subglaceal ferrous 'ocean'. *Science* 324, 397–400.

Minegishi, H., Mizuki, T., Echigo, A., Fukushima, T., Kamekura, M. and Usami, R. (2008) Acidophilic haloarchaeal strains are isolated from various solar salts. *Saline Systems* 4, 16.

Minegishi, H., Kamekura, M., Itoh, T., Echigo, A., Usami, R. and Hashimoto, T. (2010) Further refinement of *Halobacteriaceae* phylogeny based on full-length RNA polymerase subunit B′ (*rpoB′*) gene. *International Journal of Systematic and Evolutionary Microbiology* 60, 2398–2408.

Mongodin, M.E.F., Nelson, K.E., Duagherty, S., DeBoy, R.T., Wister, J., Khouri, H., Weidman, J., Balsh, D.A., Papke, R.T., Sanchez Perez, G., Sharma, A.K., Nesbø, C.L., MacLeod, D., Bapteste, E., Doolittle, W.F., Charlebois, R.L., Legault, B. and Rodríguez-Valera, F. (2005) The genome of *Salinibacter ruber*: convergence and gene exchange among hyperhalophilic bacteria and archaea. *Proceedings of the National Academy of Sciences of the United States of America* 102, 18147–18152.

Mormile, M.R., Hong, B.-Y. and Benison, K.C. (2009) Molecular analysis of the microbial communities of Mars analog lakes in Western Australia. *Astrobiology* 9, 919–930.

Moscatello, S. and Belmonte, G. (2009) Egg banks in hypersaline lakes of the South-East Europe. *Saline Systems* 5, 3.

Munson, M.A., Nedwell, D.B. and Embley, T.M. (1997) Phylogenetic diversity of Archaea in sediment samples from a coastal salt marsh. *Applied and Environmental Microbiology* 63, 4729–4733.

Ng, W.V., Kennedy, S.P., Mahairas, G.G., Berquist, B., Pan, M., Shukla, H.D., Lasky, S.R., Baliga, N.S., Thorsson, V., Sbrogna, J., Swartzell, S., Weir, D., Hall, J., Dahl, T.A., Welti, R., Goo, Y.A., Leithauser, B., Keller, K., Cruz, R., Danson, M.J., Hough, D.W., Maddocks, D.G., Jablonski, P.E., Krebs, M.P., Angevine, C.M., Dale, H., Isenbarger, T.A., Peck, R.F., Pohlschröder, M., Spudich, J.L., Jung, K.H., Alam, M., Freitas, T., Hou, S.B., Daniels, C.J., Dennis, P.P., Omer, A.D., Ebhardt, H., Lowe, T.M., Liang, R., Riley, M., Hood, L. and DasSarma, S. (2000) Genome sequence of *Halobacterium* species NRC-1. *Proceedings of the National Academy of Sciences of the United States of America* 97, 12176–12181.

Nicastro, A.J., Vreeland, R.H. and Rosenzweig, W.D. (2002) Limits imposed by ionizing radiation on the long-term survival of trapped bacterial spores: beta radiation. *Journal of Radiation Biology* 78, 891–901.

Norton, C.F. and Grant, W.D. (1988) Survival of halobacteria within fluid inclusions in salt crystals. *Journal of General Microbiology* 134, 1365–1373.

Ollivier, B., Hatchikian, C.E., Prensier, G., Guezennec, J. and Garcia, J.-L. (1991) *Desulfohalobium retbaense* gen. nov. sp. nov., a halophilic sulphate-reducing bacterium from sediments of a hypersaline lake in Senegal. *International Journal of Systematic Bacteriology* 41, 74–81.

Ollivier, B., Caumette, P., Garcia, J.-L. and Mah, R.A. (1994) Anaerobic bacteria from hypersaline environments. *Microbiological Reviews* 58, 27–38.

Ollivier, B., Fardeau, M.-L., Cayol, J.-L., Magot, M., Patel, B.K.C., Prensier, G. and Garcia, J.-L. (1998) *Methanocalculus halotolerans* gen. nov., sp. nov., isolated from an oil-producing well. *International Journal of Systematic Bacteriology* 48, 821–828.

Oremland, R.S. and King, G.M. (1989) Methanogenesis in hypersaline environments. In: Cohen, Y. and Rosenberg, E. (eds) *Microbial Mats. Physiological Ecology of Benthic Microbial Communities*. American Society for Microbiology, Washington, DC, pp. 180–190.

Oremland, R.S., Kulp, T.R., Switzer Blum, J., Hoeft, S.E., Baesman, S., Miller, L.G. and Stolz, J.F. (2005) A microbial arsenic cycle in a salt-saturated extreme environment. *Science* 308, 1305–1308.

Oren, A. (1988a) The microbial ecology of the Dead Sea. In: Marshall, K.C. (ed.) *Advances in Microbial Ecology*, Vol. 10. Plenum Publishing Company, New York, pp. 193–229.

Oren, A. (1988b) Anaerobic degradation of organic compounds at high salt concentrations. *Antonie van Leeuwenhoek* 54, 267–277.

Oren, A. (1993) The Dead Sea – alive again. *Experientia* 49, 518–522.

Oren, A. (1994) The ecology of the extremely halophilic archaea. *FEMS Microbiology Reviews* 13, 415–440.

Oren, A. (1999a) Microbiological studies in the Dead Sea: future challenges toward the understanding of life at the limit of salt concentrations. *Hydrobiologia* 405, 1–9.

Oren, A. (1999b) Bioenergetic aspects of halophilism. *Microbiology and Molecular Biology Reviews* 63, 334–348.

Oren, A. (2002) *Halophilic Microorganisms and their Environments*. Kluwer Scientific Publishers, Dordrecht, the Netherlands, 600 pp.

Oren, A. (2005) A hundred years of *Dunaliella* research – 1905–2005. *Saline Systems* 1, 2.

Oren, A. (2006a) Life at high salt concentrations. In: Dworkin, M., Falkow, S., Rosenberg, E., Schleifer, K.-H. and Stackebrandt, E. (eds) *The Prokaryotes. A Handbook on the Biology of Bacteria*, 3rd edn, vol. 2. Springer, New York, pp. 263–282.

Oren, A. (2006b) The order *Halobacteriales*. In: Dworkin, M., Falkow, S., Rosenberg, E., Schleifer, K.-H. and Stackebrandt, E. (eds) *The Prokaryotes. A Handbook on the Biology of Bacteria*, 3rd edn, vol. 3. Springer, New York, pp. 113–164.

Oren, A. (2006c) The order *Haloanaerobiales*. In: Dworkin, M., Falkow, S., Rosenberg, E., Schleifer, K.-H. and Stackebrandt, E. (eds) *The Prokaryotes. A Handbook on the Biology of Bacteria*, 3rd edn, vol. 4. Springer, New York, pp. 804–817.

Oren, A. (2008) Microbial life at high salt concentrations: phylogenetic and metabolic diversity. *Saline Systems* 4, 2.

Oren, A. (2010) Industrial and environmental applications of halophilic microorganisms. *Environmental Technology* 31, 825–834.

Oren, A. (2011) Thermodynamic limits to microbial life at high salt concentrations. *Environmental Microbiology* 13, 1908–1923.

Oren, A. and Shilo, M. (1981) Bacteriorhodopsin in a bloom of halobacteria in the Dead Sea. *Archives of Microbiology* 130, 185–187.

Oren, A. and Shilo, M. (1982) Population dynamics of *Dunaliella parva* in the Dead Sea. *Limnology and Oceanography* 27, 201–211.

Oren, A., Gurevich, P., Anati, D.A., Barkan, E. and Luz, B. (1995) A bloom of *Dunaliella parva* in the Dead Sea in 1992: biological and biogeochemical aspects. *Hydrobiologia* 297, 173–185.

Oren, A., Duker, S. and Ritter, S. (1996) The polar lipid composition of Walsby's square bacterium. *FEMS Microbiology Letters* 138, 135–140.

Oren, A., Bratbak, G. and Heldal, M. (1997) Occurrence of virus-like particles in the Dead Sea. *Extremophiles* 1, 143–149.

Oren, A., Sørensen, K.B., Canfield, D.E., Teske, A.P., Ionescu, D., Lipski, A. and Altendorf, K. (2009) Microbial communities and processes within a hypersaline gypsum crust in a saltern evaporation pond (Eilat, Israel). *Hydrobiologia* 626, 15–26.

Oxley, A.P.A., Lanfranconi, M.P., Würdemann, D., Ott, S., Schreiber, S., McGenity, T.J., Timmis, K.N. and Nogales, B. (2010) Halophilic Archaea in the human intestinal mucosa. *Environmental Microbiology* 9, 2398–2410.

Pagaling, E., Wang, H., Venables, M., Wallace, A., Grant, W.D., Cowan, D.A., Jones, B.E., Ma, Y., Ventosa, A. and Heaphy, S. (2009) Microbial biogeography of six salt lakes in Inner Mongolia, China, and a salt lake in Argentina. *Applied and Environmental Microbiology* 75, 5750–5760.

Papke, R.T., Zhaxybayeva, O., Feil, E.J., Sommerfeld, K., Muise, D. and Doolittle, W.F. (2007) Searching for species in haloarchaea. *Proceedings of the National Acadamy of Sciences of the USA* 104, 14092–14097.

Park, J.S., Kim, H.J., Choi, D.H. and Cho, B.C. (2003) Active flagellates grazing on prokaryotes in high salinity waters of a solar saltern. *Aquatic Microbial Ecology* 33, 173–179.

Park, J.S., Cho, B.C. and Simpson, A.G.B. (2006) *Halocafetaria seosinensis* gen. et sp. nov. (Bicosoecida), a halophilic bacterivorous nanoflagellate isolated from a solar saltern. *Extremophiles* 10, 493–504.

Park, J.S., Simpson, A.G.B., Lee, W.J. and Cho, B.C. (2007) Ultrastructure and phylogenetic placement within Heterolobosea of the previously unclassified, extremely halophilic heterotrophic flagellate *Pleurostomum flabellatum* (Ruinen 1938). *Protist* 158, 397–413.

Pedrós-Alió, C. (2004) Trophic ecology of solar salterns. In: Ventosa, A. (ed.) *Halophilic Microorganisms*. Springer, Berlin, Germany, pp. 33–61.

Pedrós-Alió, C., Calderón-Paz, J.I., MacLean, M.H., Medina, G., Marassé, C., Gasol, J.M. and Guixa-Boixereu, N. (2000a) The microbial food web along salinity gradients. *FEMS Microbiology Ecology* 32, 143–155.

Pedrós-Alió, C., Calderón-Paz, J.I. and Gasol, J.M. (2000b) Comparative analysis shows that bacterivory, not viral lysis, controls the abundance of heterotrophic prokaryotic plankton. *FEMS Microbiology Ecology* 32, 157–165.

Pietilä, M.K., Roine, E., Paulin, L., Kalkkinen, N. and Bamford, D.H. (2009) An ssDNA virus infecting archaea: a new lineage of viruses with a membrane envelope. *Molecular Microbiology* 72, 307–319.

Piñar, C., Saiz-Jimenez, C., Schabereiter-Gurtner, C., Blanco-Varela, M.T., Lubitz, W. and Rölleke, S. (2001) Archaeal communities in two disparate deteriorated ancient wall paintings: detection, identification and temporal monitoring by denaturing gradient gel electrophoresis. *FEMS Microbiology Ecology* 37, 45–54.

Porter, D., Roychoudhury, A.N. and Cowan, D. (2007a) Dissimilatory sulphate reduction in hypersaline coastal pans: activity across a salinity gradient. *Geochimica et Cosmochimica Acta* 71, 5102–5116.

Porter, K., Russ, B.E. and Dyall-Smith, M.L. (2007b) Virus–host interactions in salt lakes. *Current Opinion in Microbiology* 10, 418–424.

Post, F.J. (1977) The microbial ecology of the Great Salt Lake. *Microbial Ecology* 3, 143–165.

Purdy, K.J., Cresswell-Maynard, T.D., Nedwell, D.B., McGenity, T.J., Grant, W.D., Timmis, K.N. and Embley, T.M. (2004) Isolation of haloarchaea that grow at low salinities. *Environmental Microbiology* 6, 591–595.

Qvit-Raz, N., Jurkevitch, E. and Belkin, S. (2008) Drop-size soda lakes: transient microbial habitats on a salt-secreting desert tree. *Genetics* 178, 1615–1622.

Rees, H.C., Grant, W.D., Jones, B.E. and Heaphy, S. (2004) Diversity of Kenyan soda lake alkaliphiles. *Extremophiles* 8, 63–71.

Roberts, M.F. (2005) Organic compatible solutes of halotolerant and halophilic microorganisms. *Saline Systems* 1, 5.

Rodriguez-Valera, F., Ruiz-Berraquero, F. and Ramos-Cormenzana, A. (1979) Isolation of extreme halophiles from seawater. *Applied and Environmental Microbiology* 38, 164–165.

Rosselló-Mora, R., Lee, N., Antón, J. and Wagner, M. (2003) Substrate uptake in extremely halophilic microbial communities revealed by microautoradiography and fluorescence *in situ* hybridization. *Extremophiles* 7, 409–413.

Rothschild, L.J. and Mancinelli, R.L. (2001) Life in extreme environments. *Nature* 409, 1092–1101.

Sanchez-Perez, G., Mira, A., Nyirő, G., Pašic, L. and Rodriguez-Valera, F. (2008) Adapting to environmental changes using specialized paralogs. *Trends in Genetics* 24, 154–158.

Santos, F., Yarza, P., Parro, V., Briones, C. and Antón, J. (2010) The metavirome of a hypersaline environment. *Environmental Microbiology* 12, 2965–2976.

Sass, A.M., Sass, H., Coolen, M.J.L., Cypionka, H. and Overmann, J. (2001) Microbial communities in the chemocline of a hypersaline deep-sea basin (Urania Basin, Mediterranean Sea). *Applied and Environmental Microbiology* 67, 5392–5402.

Sass, A.M., McKew, B.A., Sass, H., Fichtel, J., Timmis, K.N. and McGenity, T.J. (2008) Diversity of *Bacillus*-like organisms isolated from deep-sea hypersaline anoxic sediments. *Saline Systems* 4, 8.

Savage, K.N., Krumholz, L.R., Oren, A. and Elshahed, M.S. (2007) *Haladaptatus paucihalophilus* gen. nov., sp. nov., a halophilic archaeon isolated from a low-salt, high-sulfide spring. *International Journal of Systematic and Evolutionary Microbiology* 57, 19–24.

Schubert, B.A., Lowenstein, T.K., Timofeeff, M.N. and Parker, M.A. (2009) How do prokaryotes survive in fluid inclusions in halite for 30 k.y.? *Geology* 37, 1059–1062.

Schubert, B.A., Lowenstein, T.K., Timofeeff, M.N. and Parker, M.A. (2010) Halophilic Archaea cultured from ancient halite, Death Valley, California. *Environmental Microbiology* 12, 440–454.

Simon, R.D., Abeliovich, A. and Belkin, S. (1994) A novel terrestrial halophilic environment: the phylloplane of *Atriplex halimus*, a salt-excreting plant. *FEMS Microbiology Ecology* 14, 99–110.

Sokolov, A.P. and Trotsenko, Y.A. (1995) Methane consumption in (hyper)saline habitats of Crimea (Ukraine). *FEMS Microbiology Ecology* 18, 299–304.

Soppa, J., Baumann, A., Brenneis, M., Dambeck, M., Hering, O. and Lange, C. (2008) Genomics and functional genomics with haloarchaea. *Archives of Microbiology* 190, 197–215.

Soppa, J., Straub, J., Brenneis, M., Jellen-Ritter, A., Heyer, R., Fischer, S., Granzow, M., Voss, B., Hess, W.R., Tjaden, B. and Marchfelder, A. (2009) Small RNAs of the halophilic archaeon *Haloferax volcanii*. *Biochemcial Society Transactions* 37, 133–136.

Sorokin, D.Y. and Kuenen, J.G. (2005a) Haloalkaliphilic sulfur-oxidizing bacteria in soda lakes. *FEMS Microbiology Reviews* 29, 685–702.

Sorokin, D.Y. and Kuenen, J.G. (2005b) Chemolithotrophic haloalkaliphiles from soda lakes. *FEMS Microbiology Ecology* 52, 287–295.

Sorokin, D.Y., Tourova, T.P., Lysenko, A.M. and Muyzer, G. (2006) Diversity of culturable halophilic sulfur-oxidizing bacteria in hypersaline habitats. *Microbiology* 152, 3013–3023.

Sorokin, D.Y., Tourova, T.P., Henstra, A.M., Stams, A.J.M., Galinski, E.A. and Muyzer, G. (2008a) Sulfidogenesis under extremely haloalkaline conditions by *Desulfonatronospira thiodismutans* gen. nov., sp. and *Desulfonatronospira delicate* sp. nov. – a novel lineage of *Deltaproteobacteria* from hypersaline soda lakes. *Microbiology* 154, 1444–1453.

Sorokin, D.Y., Tourova, T.P., Muyzer, G. and Kuenen, J.G. (2008b) *Thiohalospira halophila* gen. nov., sp. nov. and *Thiohalospira alkaliphila* sp. nov., novel obligately chemolithoautotrophic, halophilic, sulfur-oxidizing gammaproteobacteria from hypersaline habitats. *International Journal of Systematic and Evolutionary Microbiology* 58, 1685–1692.

Sorokin, D.Y., Tourova, T.P., Galinski, E.A., Muyzer, G. and Kuenen, J.G. (2008c) *Thiohalorhabdus denitrificans* gen. nov., sp. nov., an extremely halophilic, sulfur-oxidizing, deep-lineage gammaproteobacterium from hypersaline habitats. *International Journal of Systematic and Evolutionary Microbiology* 58, 2890–2897.

Stephens, D.W. and Gillespie, D.M. (1976) Phytoplankton production in the Great Salt Lake, Utah, and a laboratory study of algal response to enrichment. *Limnology and Oceanography* 21, 74–87.

Tadeo, X., López-Méndez, B., Trigueros, T., Laín, A., Castaño, D. and Millet, O. (2009) Structural basis for the amino acid composition of proteins from halophilic Archaea. *PLoS Biology* 7, e1000257.

Tehei, M., Franzetti, B., Maurel, M.C., Vergne, J., Hountondji, C. and Zaccai, G. (2002) The search for traces of life: the protective effect of salt on biological macromolecules. *Extremophiles* 6, 427–430.

Trotsenko, Y.A. and Khmelenina, V.N. (2002) Biology of extremophilic and extremotolerant methanotrophs. *Archives of Microbiology* 177, 123–131.

UNESCO (1985) *The International System of Units (SI) in Oceanography*. UNESCO Technical Papers in Marine Science, 45. International Association for the Physical Sciences of the Ocean. UNESCO, Paris, France, 124 pp.

Vaisman, N. and Oren, A. (2009) *Salisaeta longa* gen. nov., sp. nov., a red halophilic member of the Bacteroidetes. *International Journal of Systematic and Evolutionary Microbiology* 59, 2571–2574.

van der Wielen, P.W.J.J., Bolhuis, H., Borin, S., Daffonchio, D., Corselli, C., Giuliano, L., D'Auria, G., de Lange, G.J., Huebner, A., Varnavas, S.P., Thomson, J., Tamburini, C., Marty, D., McGenity, T.J., Timmis, K.N. and BioDeep Scientific Party (2005) The enigma of prokaryotic life in deep hypersaline anoxic basins. *Science* 307, 121–123.

Ventosa, A., Nieto, J.J. and Oren, A. (1998) Biology of aerobic moderately halophilic bacteria. *Microbiology and Molecular Biology Reviews* 62, 504–544.

Vreeland, R.H., Rosenzweig, W.D. and Powers, D.W. (2000) Isolation of a 250 million-year-old halotolerant bacterium from a primary salt crystal. *Nature* 407, 897–900.

Vreeland, R.H., Jones, J., Monson, A., Rosenzweig, W.D., Lowenstein, T.K., Timofeeff, M., Satterfield, C., Cho, B.C., Park, J.S., Wallace, A. and Grant, W.D. (2007) Isolation of live Cretaceous (121–112 million years old) halophilic Archaea from primary salt crystals. *Geomicrobiology Journal* 24, 275–282.

Walsby, A.E. (1980) A square bacterium. *Nature* 283, 69–71.

Walsby, A.E. (2005) Archaea with square cells. *Trends in Microbiology* 13, 193–195.

Weimer, B.C., Rompato, G., Parnell, J., Gann, R., Ganesan, B., Navas, C., Gonzalez, M., Clavel, M. and Albee-Scott, S. (2009) Microbial diversity of Great Salt Lake, Utah. In: Oren, A., Naftz, D.L., Palacios, P. and Wurtsbaugh, W.A. (eds) *Saline Lakes around the World: Unique Systems with Unique Values*. The S.J. and Jessie E. Quinney Natural Resources Research Library, College of Natural Resources, Utah State University, Utah, pp. 15–22.

Woese, C.R. (1987) Bacterial evolution. *Microbiological Reviews* 51, 221–271.

Wolfe-Simon, F., Switzer Blum, J., Kulp, T.R., Gordon, G.W., Hoeft, S.E., Pett-Ridge, J., Stolz, J.F., Webb, S.M., Weber, P.K., Davies, P.C.W., Anbar, A.D. and Oremland, R.S. (2011) A bacterium that can grow by using arsenic instead of phosphorus. *Science* 332, 1163–1166.

Wood, A.P. and Kelly, D.P. (1991) Isolation and characterisation of *Thiobacillus halophilus* sp. nov., a sulphur-oxidising autotrophic eubacterium from a Western Australian hypersaline lake. *Archives of Microbiology* 156, 277–280.

Yakimov, M.M., La Cono, V., Denaro, R., D'Auria, G., Decembrini, F., Timmis, K.N., Golyshin, P.N. and Giuliano, L. (2007) Primary producing prokaryotic communities of brine, interface and seawater above the halocline of deep anoxic lake L'Atalante, Eastern Mediterranean Sea. *ISME Journal* 1, 743–755.

Yu, D., Sorokin, G. and Muyzer, G. (2010) Bacterial dissimilatory MnO_2 reduction at extremely haloalkaline conditions. *Extremophiles* 14, 41–46.

21 Hypoxic Environments

Stéphane Hourdez
Station Biologique de Roscoff, CNRS – UPMC, France

21.1 Introduction

Oxygen in the Earth's atmosphere and oceans has evolved over geological times from a negligible presence at the origin of our planet to the present level of 21% (Cloud, 1973; Holland, 2006; see also Lomax, Chapter 2, this volume). These vast amounts of oxygen were produced by photosynthetic cyanobacteria and, after all the dissolved iron contained in the oceans was oxidized, diffused into the atmosphere. Holland (2006) distinguishes five stages in the evolution of oxygen levels on Earth. During stage one (3.85–2.45 billion years ago (Ga)), oxygen was mostly absent both in the atmosphere and in the oceans, with the possible exception of small areas in shallow oceans. During the following stage (2.45–1.85 Ga), the levels of atmospheric oxygen rose to values of 3–4%, the shallow oceans were mildly oxic and the deep oceans remained anoxic (devoid of oxygen). Stage three (1.85–0.85 Ga) was a rather stable period for the atmospheric oxygen, and oceans (surface and deep) were mildly oxygenated. Stage four (0.85–0.54 Ga) saw a rise in oxygen values up to about 20%, the shallow oceans followed but the deep oceans remained anoxic. Finally, during stage five (0.54 Ga to present) the atmospheric oxygen values rose up to 32% during the Carboniferous before returning to the current 21%, the shallow oceans were oxygenated but the oxygenation level of the deep oceans fluctuated greatly.

All these changes had a profound effect on organisms and their evolution. At first a waste product and a toxic molecule, some prokaryotic cells found a way to use oxygen to oxidize molecules and produce energy (respiration). Later, this allowed the rise of eukaryotic cells 1.6–2.1 Ga (Knoll *et al.*, 2006), whose mitochondria are likely the product of the endosymbiosis established with a heterotrophic aerobic proteobacterium (Sagan, 1967; Margulis, 1975). The first evidence for large colonial organisms also follows the rise of oxygen levels 2.45–2.32 Ga (El Albani *et al.*, 2010). During the Carboniferous, the high level of oxygen is thought to have permitted arthropod, as well as amphibian, gigantism (Dudley, 1998). The late Permian transition to hypoxia (low oxygen levels) brought the downfall of these gigantic life forms. The post-Palaeozoic fossil record indicates that there has been a succession of dysoxic and anoxic episodes in the deep sea, some of them global (Jacobs and Lindberg, 1998). These anoxic events triggered the extinction of most, if not all, the metazoans in the deep, anoxic basins. As a result, the current deep-sea fauna is typically younger than 60 million years.

At present, although some prokaryotes and eukaryotes are facultative or obligatory anaerobes, nearly all metazoans require oxygen at least during part of their life cycle. One exception among metazoans was recently discovered in the deep anoxic hypersaline L'Atalante Basin in the Mediterranean. Three new species of the animal phylum Loricifera have been found to live and thrive there, in the complete absence of molecular oxygen (Danovaro *et al.*, 2010). Their cells do not contain mitchondria but instead hydrogenosome-like organelles.

On Earth today, hypoxic environments are mostly found at high altitude, in underground burrows and in some localized aquatic environments (marine and fresh water) (Fig. 21.1). Although these environments are challenging and would be considered extreme by many organisms, being able to deal with the conditions can allow the resident organisms to access food sources and/or escape predators. Metazoa offer us an excellent insight into the adaptations that organisms from hypoxic habitats possess (Fig. 21.2). Many unicellular organisms do inhabit hypoxic and anoxic habitats, but their adaptations are usually found at the metabolic level and/or at the organellar level (Ginger *et al.*, 2010). This review does not deal with photosynthetic organisms.

21.2 Aerial Environments

21.2.1 Subterranean burrows

Although the atmosphere at a given altitude is very homogeneous, the structure of subterranean burrows could limit air renewal and hypoxic conditions may develop, especially in heavy soil and soils with a high water content.

Fossorial mammals are adapted to spend the majority of their lives underground. Mole rats of the genus *Spalax* inhabit sealed subterranean burrows. There are several species of *Spalax*, all living in hypoxic conditions, but *S. ehrenbergi* is probably the most extreme: it lives in heavy

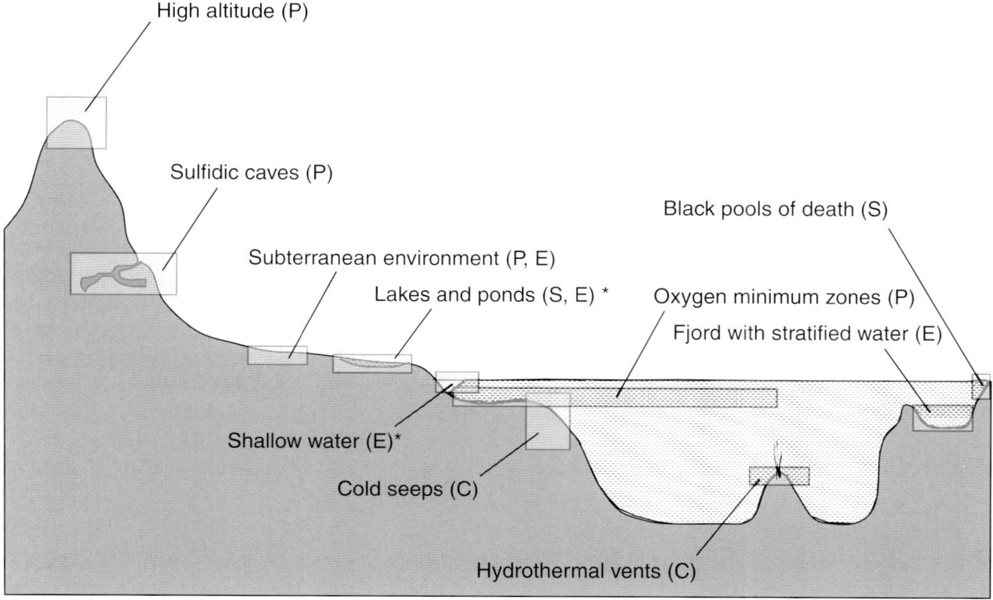

Fig. 21.1. Environments with permanent (P), chronic (C), seasonal (S) and episodic (E) hypoxia. The asterisks denote environments for which human impact has been shown to induce hypoxic events. © Stéphane Hourdez.

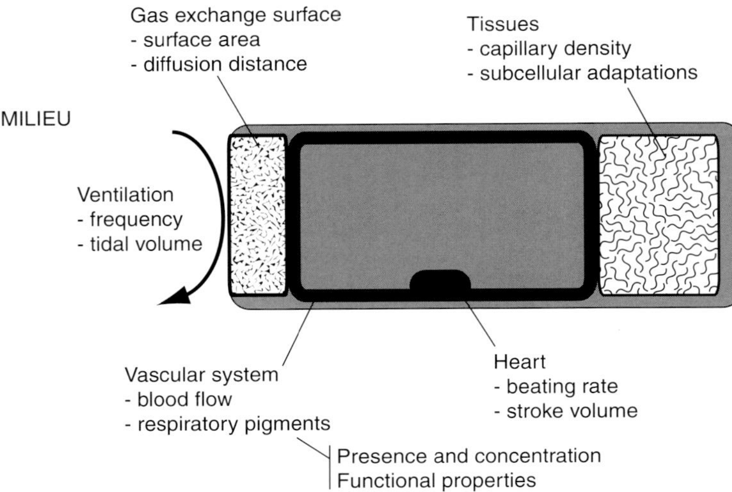

Fig. 21.2. Gas transfer in a theoretical animal from the milieu to the tissues. For each level, parameters that can be modified to improve oxygen delivery are indicated. © Stéphane Hourdez.

clay soils that, when flooded, allow strongly hypoxic and hypercapnic (too much carbon dioxide; CO_2) conditions to develop, with oxygen values as low as 7.2% and CO_2 as high as 6.1% (normal atmospheric levels are 21% O_2 and <0.04% CO_2). In the laboratory, this species can survive 3% O_2 and 15% CO_2 for up to 14 h while the laboratory rat, *Rattus norvegicus*, died after 2–4 h under the same conditions (Avivi *et al.*, 1999). These mole rats possess a greater density of mitochondria than lab rats, and a higher capillary density, resulting in a shorter diffusion distance between the blood and the mitochondria (Widmer *et al.*, 1997). The same study also revealed a significantly increased pulmonary diffusing capacity that likely facilitates oxygen uptake from the hypoxic air the animals inhale. Adaptations found in these rats also involve intracellular globins (such as neuroglobins, cytoglobins and myoglobins; Avivi *et al.*, 2010) that probably help diffusion of oxygen into the cells and help prevent damage due to hypoxia, expression of erythropoietin and inducible factor-1α, as well as a higher haematocrit and a greater oxygen affinity of the blood (see Avivi *et al.*, 2010 and references therein). The higher haematocrit is a direct result of erythropoietin production and inducible factor-1α will induce the production of specific genes in the hypoxia response pathway, including some that will eventually affect the amount of 2,3-diphosphoglycerate (DPG). The European mole and other fossorial mole-rats also possess haemoglobin with a high oxygen affinity compared to other similar-sized terrestrial mammals (van Aardt *et al.*, 2007). This high affinity is the result of a reduced sensitivity to DPG, a molecule that can decrease oxygen affinity in mammalian haemoglobins (Campbell *et al.*, 2010).

Similarly, invertebrates that live in heavy soils and closed burrows can be exposed to environmental hypoxia. Schmitz and Harrison (2004) reviewed the mechanisms of hypoxia tolerance in air-breathing arthropods. The species studied are usually only transiently exposed to low oxygen concentrations and the responses observed only allow the animals to cope with temporary hypoxia. Nevertheless, there is good potential for hypoxia tolerance. Little is known about the long-term effects (e.g. on growth or reproduction) of hypoxia in these invertebrates. A study revealed that exposure to sustained hypoxia during development in insects increases tracheal dimensions, allowing greater volumes of air into the body (Schmitz and Harrison, 2004).

21.2.2 High altitude

Although the proportion of oxygen in the air remains constant, the atmospheric pressure decreases as altitude increases: at 5500 m the pressure is about half of that at ocean level and at 12,000 m it is only 21% of the ocean-level pressure. As a result, the partial pressure of oxygen is reduced and organisms have to deal with this environmental hypoxia. A few animals have been studied with respect to their ability to deal with hypoxia. In mammals, as in other vertebrates, the haemoglobin present in red blood cells seems to play an important role in adaptation to chronic hypoxia (for a review see Storz, 2007). The classic example of the left-shifted oxygen dissociation curves (i.e. high affinity) in llama and vicuña (family Camelidae) has been cited in many textbooks. This increased affinity can be the result of either an increase of the intrinsic oxygen affinity of the protein or of the decrease of sensitivity to effectors that usually decrease oxygen affinity. In high-altitude camelids, a single change on the β-globin chain compared to lowland camelids explains the high oxygen affinity by removing two binding sites for 2,3-DPG (2,3-diphosphoglycerate), a molecule that lowers the oxygen affinity of haemoglobin by maintaining it in its low affinity conformation (Poyart *et al.*, 1992). These high-altitude camelids and many more animals have now been studied and give us a broader view of adaptation to high altitude.

While bar-headed geese, *Anser indicus*, do not always live at very high altitude, they are the record holders for altitude survival as they perform an annual trans-Himalayan migration, commonly flying at 9000 m altitude. They have also been observed flying at nearly 11,000 m altitude where the partial pressure of oxygen is only 39 mm Hg (i.e. 5.2 kPa, about 25% of the ocean-level value). This tremendous accomplishment relies not only on morphological adaptations to support flight in the rarefied air (slightly larger wing area for a bird of this weight; Lee *et al.*, 2008) but also on respiratory adaptations. When experimentally exposed to levels of oxygen similar to those found at 11,000 m altitude, bar-headed geese fared better than their low-altitude relatives. The geese achieve this by increasing their tidal volume (the lung volume representing the normal volume of air displaced between normal inspiration and expiration when extra effort is not applied), thereby improving parabronchial ventilation (Milsom and Scott, 2008). As a consequence, the oxygen loading of bar-headed goose blood is enhanced. Diffusion into the vascular system is also facilitated by the slightly greater oxygen affinity of their haemoglobin when compared to their low altitude relatives (Petschow *et al.*, 1977; Rollema and Bauer, 1979). Only four amino acids differ between the bar-headed goose and its low-altitude relative, the greylag goose (*Anser anser*). The crystal structure of bar-headed goose haemoglobin revealed that two of these amino acid changes reduce the contacts between the α_1 and β_1 subunits, loosening their interface and thereby, potentially, increasing their affinity for oxygen (Liu *et al.*, 2001).

In aquatic species, hypoxia can be more pronounced at night as photosynthesis ceases and only respiration occurs. Andean frogs of the genus *Telmatobius* (Order Anuran) live from 2000 m to over 4000 m altitude. The Titicaca water frog, *T. culeus*, lives in Lake Titicaca at an altitude of 3812 m. It has reduced and poorly vascularized lungs. However, the species compensates for its diminutive lungs by having large and highly vascularized skin folds hanging from the dorsum, sides and hind limbs, which likely increase oxygen uptake (Macedo, 1960). Under hypoxia, the water in contact with the skin is renewed by a bobbing behaviour (Hutchison *et al.*, 1976). Compared to other Anuran amphibians from low-altitude areas, the erythrocytes in the blood of *T. culeus* are small, the haematocrit is high (i.e. the frog has a higher oxygen carrying capacity) and its blood has a high affinity for oxygen (Hutchison *et al.*, 1976). The Peruvian water frog, *T. peruvianus*, is the only non-bird and non-mammal animal in which molecular adaptation of haemoglobin to high altitude has been studied to date (Weber *et al.*, 2002). It lives at an altitude of 3800 m, and compared to the lowland species the

African clawed frog, *Xenopus laevis*, haemoglobins from *T. peruvianus* display a higher affinity for oxygen. This high affinity results from a nearly complete lack of chloride sensitivity in the highland species, a characteristic that correlates with only two amino acid replacements (Weber *et al.*, 2002). Haemoglobins from *T. peruvianus* however do still exhibit sensitivity to ATP (adenosine triphosphate) and 2,3-DPG, allowing a modulation of their properties as a function of the physiological state of the animals.

Ostojic *et al.* (2000) studied the haemoglobins from three subspecies of the Peruvian toad, *Bufo spinulosus*, that occupy different altitudinal niches. *Bufo spinulosus limensis* lives at sea level, in areas that average 20°C during daytime. *Bufo s. trifolium* was collected at 3100 m where the average daytime temperature was 15°C, and *B. s. flavolineatus* was collected at 4100 m where the average daytime temperature was 10°C. While the blood oxygen affinity at 20°C was comparable for the two lower-altitude species, the high-altitude species displayed a marked increase in affinity. At their ecological average temperature, the difference was even more pronounced for the high-altitude species whose blood exhibited very high affinities for oxygen. The Bohr effect (a decrease of affinity with decreasing pH) was also more pronounced in the high-altitude species (Ostojic *et al.*, 2000), likely compensating the high affinity and facilitating the unloading of oxygen near metabolically active organs.

A very interesting, similar altitudinal gradient has been observed in deer mice, *Peromyscus maniculatus* (Storz *et al.*, 2009). These authors studied haemoglobin variants in populations of *P. maniculatus* that inhabited different altitudes. A multi-locus analysis of the globin genes and linkage disequilibrium showed that adaptation to high altitude involves parallel functional differentiation at multiple unlinked gene duplicates (two α-globin paralogs on a chromosome, and two β-globin genes on another chromosome). Several alleles of the α- and the β-globins exist and exhibit variations in frequencies between lowland and highland animals. The β-globin allele shows different sensitivity to 2,3-DPG. The most abundant allele found at higher elevations has a suppressed 2,3-DPG sensitivity, locking the haemoglobin in its high affinity state and ensuring a better oxygen loading in the hypoxic high-altitude air (Storz *et al.*, 2010). Here, again, only a few key amino acid changes are involved.

Although humans did not evolve at high altitude, their physiology allows them to at least temporarily transit through these environments. In response to high altitude, humans from lowland areas first increase their breathing and heart resting rates and, after a few days, typically experience an increase in 2,3-DPG in their red blood cells and increase their haematocrit (production of additional red-blood cells in response to erythropoietin production by the kidneys). Although an increased oxygen carrying capacity associated with the higher haematocrit improves oxygen delivery to the organs, it also comes with an increased workload for the circulatory system and may cause long-term damage (Storz, 2007). The high viscosity of the blood also poses problem in capillary circulation and during pregnancy. In addition, the increased 2,3-DPG concentration lowers the oxygen affinity of the haemoglobin, and as a result, lowers the actual oxygen uptake in the lungs. This seems counter-productive and is probably due to the fact that reaction to hypoxia in humans is thought to have originally evolved in response to anaemia (Storz, 2010). In anaemia, the origin of hypoxia is internal and can be corrected by increasing red blood cell production and lowering the haemoglobin affinity to favour oxygen delivery to the organs. The pulmonary oxygen concentration remains high and, even with a lower affinity, the haemoglobin is still nearly saturated when it comes out of the lungs. However, in the hypobaric hypoxia response, the source of the hypoxia is external and loading in the lungs decreases. Despite these limitations and maladaptation, some human populations in the Andes, in Himalaya and in Ethiopia only live at high altitude. There is evidence that human populations in the Andes and in Himalaya have taken different paths in adaptations to high altitude; research on highland Ethiopians is only just beginning (Beall, 2006). Andeans, typically have a higher haemoglobin concentration in their blood and a lower oxygen saturation that eventually leads to an arterial oxygen content that is 16% higher than sea-level reference values. In Tibetans residing in the Himalayas, there is no increase of haemoglobin concentration (compared to sea-level values) and profoundly reduced oxygen saturation, yielding an overall decrease of arterial oxygen content of 10% compared to sea-level reference values. In Ethiopians, the single study performed to date indicated no significant difference in haemoglobin concentration, oxygen saturation and arterial oxygen content compared to sea-level populations (Beall, 2006). Recent genomic work on Tibetan populations revealed that genes in the hypoxia inducible factor (HIF) pathway have been subject to strong and recent selection in Tibetan highlanders (see Storz, 2010 for a review). It seems that selection has favoured the lack of erythropoietic response, thereby preventing complications due to an increased blood viscosity. The absence of a similar response in Andean people may be due to the fact they colonized these altitudes more recently.

Adaptations to hypobaric (low pressure) hypoxia has long fascinated biologists. This may be due to the relative ease of access to high altitude areas for studying and the feeling that this frontier was readily accessible for thousands of years, whereas underwater environments have only recently been more easily explored. Although in animals that are restricted to high altitude, adaptations seem to affect the sequence of the respiratory pigment(s), adaptation in humans seems rather to affect the regulation of gene expression. These differences are probably due to the time for which the different species have occupied high altitude areas.

21.3 Aquatic Environments

The diffusion of gases is extremely reduced in water when compared to the atmosphere (Altman, 1958) and, in the absence of perturbations, gas pockets with radically different compositions can exist side by side. This allows very sharp gradients in the conditions experienced by aquatic organisms and potentially the very close existence of different communities that are adapted to the specific conditions of each environment. One must consider different settings in the various aquatic environments: some areas, such as oxygen minimum zones (OMZs), are exposed to constantly low oxygen concentrations over at least tens of thousands of years, others can be seasonal or episodic (e.g. fjords, closed bays) and some are exposed to chronic and highly variable levels of hypoxia. Only environments that have been continuously hypoxic will be inhabited by fauna with specific adaptations to this 'extreme', while in areas with episodic or seasonal hypoxia, the fauna will be eliminated and later recolonize the affected areas.

21.3.1 Oxygen minimum zones (OMZs)

Although some very extensive areas of the world's oceans exhibit little planktonic primary production, in areas of intense upwelling there is high phytoplankton production. This organic matter eventually sinks and is decomposed in mid-water, consuming the available dissolved oxygen at these depths. Should this high oxygen demand occur in conjunction with slow circulation (or mixing) and oxygen-poor source waters, some very important mid-water oxygen minima develop, usually between depths of 200 and 1000 m (Wyrtki, 1962; Kamykowski and Zentara, 1990). Such minima are found in large areas of the eastern Pacific Ocean, in the south-east Atlantic off West Africa, and in the northern Indian Ocean. These minima intercept the continental margin at the shelf and continental slope (i.e. exist at bathyal depths), creating extensive seafloor habitats subject to permanent hypoxia that persist over thousands of years (Reichart *et al.*, 1998). OMZs represent over 1 million km^2 of permanently hypoxic shelf and bathyal seafloor, 59% of which occur in the northern Indian Ocean (Helly and Levin, 2004). Although some planktonic Metazoa rely on temporary anaerobic metabolism and migrate to oxygenated water layers to pay this oxygen debt later, numerous planktonic species and most benthic species have instead developed specific adaptations to deal with the permanently low oxygen concentrations and maintain an aerobic metabolism.

Levin (2003) reviewed the adaptations and community responses of benthos, and Childress and Seibel (1998) reviewed the adaptations of the pelagic species that permanently inhabit OMZs. Although the conditions are bound to be extreme and very challenging, protozoan and metazoan assemblages can and do thrive in these environments. In the sediments, the density of meiofauna (protists and Metozoa) is actually greater than in areas at similar depths outside OMZs. This is likely the result of reduced competition for local food supplies (numerous sulfide-oxidizing bacteria often live on the organic rich sediments) and reduced predation. Macro- and mega-fauna are often found in greater densities on the edges of the OMZs but their density and diversity drops sharply in the core of the OMZs. Some taxa seem to be more tolerant of the challenging conditions encountered in the OMZ cores. Childress and Seibel (1998) suggest that the pelagic species present can cope with hypoxia by lowering their metabolic demand, using anaerobic metabolism, or by improving their effectiveness at extracting oxygen from the hypoxic environment. Although the pelagic species considered do exhibit a lower oxygen demand compared to their shallow-water relatives, this oxygen requirement is very similar to that of other pelagic species that live in well-oxygenated areas. Reliance on anaerobic metabolism is not energetically favourable, and only planktonic species with a migratory behaviour will use this strategy. This leaves the strategy of improved oxygen uptake for all the species permanently residing in OMZs (benthic and pelagic). This improved oxygen extraction may involve a combination of a better ventilatory capacity, enhanced percentage removal of oxygen from the water circulating over the gills, large gill surface areas, short diffusion distances, respiratory pigments with a high affinity for oxygen to ensure extraction from the hypoxic water, and a strong Bohr effect that allows the release of oxygen from these high affinity pigments to metabolically active tissues.

Species that inhabit OMZs have been shown to possess enlarged gills compared to close relatives of similar sizes, for example, ampeliscid amphipods (Levin, 2003) and various families of Polychaetes collected off Oman (Lamont and Gage, 2000). The nematode *Glochinema bathyperuvensis* possesses numerous body spines and cuticular protrusions that are thought to aid oxygen uptake (Neira *et al.*, 2001). The comparatively reduced size of benthic organisms from OMZs could also facilitate oxygen diffusion into their bodies although other explanations have also been suggested (Levin, 2003). Numerous species also possess respiratory pigments with high affinities for oxygen that allow them to better extract oxygen from the hypoxic surrounding waters (Childress and Seibel, 1998; Levin, 2003) and facilitate diffusion by maintaining optimal oxygen partial pressure differences between the water and the inside of the body. This characteristic has been observed in the benthic shortspine thornyhead fish, *Sebastolobus alascanus*, in pelagic fish, crustacea and the giant red mysid, *Gnathophausia ingens*. The presence of respiratory pigments in itself is unusual for some invertebrate taxa. It allows a greater capacitance of the body fluids and increases the efficiency of oxygen transport. In particular, haemoglobin has been reported in the mytilid mussel *Amydalum anoxicolum* (Oliver, 2001) from the Arabian Sea and in other bivalve taxa from the Oman margin OMZ (P.G. Oliver, Cardiff, 2003, personal communication in Levin, 2003). The occurrence of such molecules in a great diversity of bivalve taxa in which they are not usually found (Veneroidea, Crassatelloidea, Lucinoidea and Pectinoidea) most likely corresponds to an adaptive convergence rather than a plesiomorphic character and calls for further studies of these groups. A high ventilatory and blood circulation capacity has also been documented in pelagic crustaceans, but no similar study is available for benthic OMZ species (Childress and Seibel, 1998).

21.3.2 Temporary/seasonal hypoxia in marine systems

In an enclosed bay or fjord, dense brine can form and produce a stable pycnocline (layering of water masses due to density differences), which may resist mixing for months or years and lead to anoxic bottom waters in these deep water basins (Gallagher and Burton, 1988). Josefson and Widbom (1988) followed the evolution of the benthic community in the Gullmar Fjord, Sweden, after a severe hypoxic event. The macrofauna completely disappeared during the hypoxic event and later recolonized the fjord. Opportunistic capitellid polychaetes were the initial colonizers and the community slowly developed. A year and a half after the event, the community had still not completely recovered. In contrast, the meiofaunal community was not clearly affected by the hypoxic event. These observations confirm previous studies and indicate that macrofauna is in general more sensitive to low oxygen values than meiofauna.

Similarly, during winter ice formation, sinking brine can pool in small depressions at the seafloor. When the ice melts and retreats, the brine pools are left in these depressions and, should the conditions remain stable, oxygen will quickly become depleted. The benthos in these locations will either move or die there. Vagrant species may get trapped and quickly killed in this anoxic and toxic (high H_2S) brine. The areas are relatively small (on average 9±7 m^2) and have been termed 'black pools of death'. They can remain on the seafloor until weather conditions trigger the mixing of the water column down to these depths (Kvitek *et al.*, 1998). Once reoxygenated, the sediment patch will be recolonized by the surrounding benthic fauna.

In both cases (fjords and 'black pools of death'), there is usually no specialized fauna able to withstand the hypoxic and anoxic conditions. Instead, the megafauna is eliminated and the areas are recolonized when the conditions become favourable again. Under some circumstances, the sulfide horizon that is usually found in the sediment can also reach the water column and will make survival more difficult.

21.3.3 Chronically hypoxic and highly variable zones

The deep-sea fauna usually obtains its nutrition from sinking organic matter formed by photosynthesis in the photic zone. This organic matter is slowly degraded during the descent and, as a result, these great depths typically harbour little life. Deep-sea hydrothermal vents (Corliss and Balard, 1977; see also Lutz, Chapter 13, this volume) and cold seep areas represent oases of life the deep sea. The very high biomass found in these environments contrasts sharply with the desert-like appearance of the surrounding deep sea at similar depths. At hydrothermal vents and cold seeps, the primary production is local, based on chemosynthesis by bacteria, either free-living or symbiotic (Fisher, 1995). This discovery revolutionized our view and understanding of oceans at great depths. Although the local production and the biomass are very high, the diversity of animals is small, with only a limited number of species at a given site, and about 95% of these species are usually endemic (Tunnicliffe, 1991). This very high endemicity and low number of species is probably the result of the extreme conditions that also characterize hydrothermal vents and cold seeps. The fluids that are emitted in both types of chemosynthesis-based ecosystems are reduced (containing sulfide, methane, and sometimes hydrogen at some hydrothermal vents), and the reduced chemicals are used by the chemosynthetic bacteria as a source of energy. The very same fluids that are essential for local primary production are also responsible for the chronic hypoxia that characterizes both types of environments. At hydrothermal vents, the hydrothermal fluid is hot (up to 400°C) and anoxic, with high concentrations of CO_2 and sulfide. Other toxic compounds, including heavy metals, are also present. Although metazoan life is not encountered in pure vent fluids, mixing of these fluids with the surrounding deep-sea water is very chaotic and the resulting oxygen concentrations are very variable (Johnson *et al.*, 1988). At cold seeps, large amounts of sulfide come out of the sediment and will react with oxygen, decreasing its concentration in the overlying water. Although exposed to different levels of hypoxia, animals from these environments require as much oxygen as their close relatives from well-oxygenated environments (Hourdez and Weber, 2005) and this fact has stimulated interest from different research groups studying adaptations to low levels of oxygen.

Hourdez and Lallier (2007) reviewed the adaptations of metazoans from hydrothermal vents and cold seeps to chronic hypoxia, and Hourdez and Weber (2005) reviewed the adaptation in haemoglobins from hydrothermal vent and cold seep animals. Most of the endemic hydrothermal vent and cold seep species are invertebrates. The zoarcid fish *Thermarces cerberus* is a noteworthy exception. All other species reported to date are thought to only visit these environments temporarily. *Thermarces cerberus* inhabits thickets of the giant tubeworm *Riftia pachyptila* on the East Pacific Rise. When compared with shallow water species in the same family, *T. cerberus* exhibits much higher oxygen affinities. One of the major haemoglobin components in *T. cerberus* also exhibits a decreased chloride affinity, maintaining it in a constantly high affinity state (Weber *et al.*, 2003). On the other hand, the eel *Symenchelis parasitica* is not endemic to hydrothermal vents and only casually visits this environment. In this species, the haemoglobin system is not very different from that of other similar fish, indicating that it has less need for a high affinity haemoglobin system (Weber *et al.*, 2003; Hourdez and Weber, 2005).

Invertebrates, with their simpler body plan, may offer more possibilities for adaptations compared to the more rigid framework of vertebrates. As a result, adaptations to hypoxia could affect other organizational levels in invertebrates. Similar to OMZ invertebrates, all species from hydrothermal vents and cold seeps studied to date possess respiratory pigments (haemoglobins and haemocyanins) with very high intrinsic oxygen affinities (see Hourdez and Lallier, 2007). This high affinity for oxygen allows oxygen uptake even from waters with very low partial pressures. These high affinities are compensated for by a strong Bohr effect that allows the release of oxygen near metabolically-active organs (local decrease of pH due to CO_2 production by respiring cells). The polychaete worm *Branchipolynoe symmytilida* possesses two types of haemoglobins in its coelomic fluid. For one of them, there is also a specific effect of CO_2 (independent of pH) that further decreases its affinity for oxygen (Hourdez *et al.*, 1999a): this may represent an evolutionary adaptation. Respiratory pigments are found in most (but, interestingly, not all) species from these chemosynthetic communities (Hourdez and Lallier, 2007). This is particularly meaningful for species that belong to taxonomic groups otherwise renowned for lacking respiratory pigments. Species of the scaleworm genus *Branchipolynoe* have been particularly well studied. Shallow-water relatives are devoid of circulating haemoglobins while *Branchipolynoe* species possess large amounts of extracellular haemoglobin in their coelomic fluid (Hourdez *et al.*, 1999b). In the ancestor of these species, a myoglobin-like gene was likely duplicated and, after molecular tinkering, was expressed as an extracellular haemoglobin (Projecto-Garcia *et al.*, 2010). In some species, the presence of respiratory pigments could also provide an oxygen storage system that buffers the periods of extreme hypoxia or anoxia (Hourdez and Lallier, 2007). In *Branchipolynoe* species, this storage could represent up to 1.5 h of aerobic metabolism, one order of magnitude more than in shallow water species.

Morphological adaptations have also been observed in various invertebrate groups from hydrothermal vents and cold seeps. In particular, gill surface areas are greatly enlarged in polychaetes (Hourdez and Lallier, 2007) but usually not in crustaceans (Decelle *et al.*, 2010). Instead, in crustaceans the scaphognathite (a mouth part appendage responsible for renewing the water in the gill chamber) is enlarged in species from hydrothermal vents and cold seeps. As a result, one beat of the scaphognathite in these species moves more water over the gills than those of their relatives from well-oxygenated areas. Ventilation (and adaptation at this level) is usually more difficult to evaluate in annelid worms but scanning electron microscopy observations of gill surfaces has revealed the presence of numerous cilia that most likely help renew the water diffusion layer (Hourdez and Lallier, 2007). The presence of gills in some scale-worm species is very unusual as shallow water species of the same family are usually devoid of such structures. In *Branchipolynoe*, these gills are mere expansions of the body-wall, communicating with the general coelomic cavity, and without the blood vessels that are usually found in other annelids' gills (Hourdez and Jouin-Toulmond, 1998). In other species of annelids, the diffusion distances through the gills are also reduced compared to their shallow water relatives, thereby improving gas diffusion (Hourdez and Lallier, 2007). In shrimp, however, there is no such shortening of diffusion distances, possibly because crustacean gills have already reached a physical limit in all species (Decelle *et al.*, 2010).

In the hydrothermal vent-endemic polychaete family Alvinellidae, an internal gas exchange organ is greatly enhanced compare to the closely related family Terebellidae. This organ allows gas exchange between the haemoglobins contained in the vascular and coelomic compartments (Hourdez and Lallier, 2007). The very close affinities of both types of respiratory pigments allows bidirectional exchange, likely allowing storage of oxygen in the coelomic system and buffering oxygen variations, in particular in the cerebral ganglion located directly downstream of the gas exchange system.

21.3.4 Freshwater systems

In addition to the sometimes extensive areas of marine hypoxic waters, various freshwater environments periodically or chronically experience hypoxia. These include lakes that undergo euthrophication, small lakes and ponds with winter ice-covers, and sulfidic caves.

Some freshwater systems have recently been exposed to eutrophication as a result of, for example, increased inorganic fertilizer loadings that stimulate rapid and enhanced primary production. Extremely eutrophic conditions can lead to hypoxia and anoxia in aquatic environments as a result of the combined bacterial degradation of sinking organic matter, and stratified water masses in summer when the upper layers become very warm and buoyant. As in the previously discussed marine systems, such eutrophic events do not allow animals to develop adaptations for the often rapid hypoxia they experience.

Lakes, in particular, are increasingly eutrophic during summer time. Lake Erie, USA, for instance experiences seasonal hypoxia due to the decomposition of phytoplanktonic blooms triggered by high phosphorus levels as a result of human activity (Arend *et al.*, 2010). Not all species of fish have the same sensitivity to hypoxia but the functioning of the ecosystem is nevertheless affected as some species migrate toward more oxygenated waters while others remain (Arend *et al.*, 2010). In addition to fish that are of direct interest for fishing, some studies have focused on invertebrates that form an important link in the food chain. In particular, benthic cyclopoid copepods seem to avoid anoxic water layers and swim towards oxygenated layers (where they are more exposed to predation), while some planktonic species enter a resting stage to survive the anoxic periods, only becoming active once more favourable oxygen conditions return (Tinson and Laybourn-Parry, 1985).

Although the eutrophication of some lakes is relatively recent, some other freshwater environments experience seasonal hypoxia and some species have evolved to cope with extended periods of hypoxia. The crucian carp, *Carassius carassius*, that inhabits ponds covered with ice in winter, has been extensively studied. The ice all but stops oxygen diffusion into the water and respiration of aquatic organisms will result in hypoxia that will remain until the ice melts. Although the carp's metabolism decreases in response to the colder temperature, its metabolic demand remains important, and carp are able to regulate their oxygen uptake down to low oxygen values (Johnston and Bernard, 1984). Individual carp previously acclimated to low partial pressures of oxygen fared better under hypoxic conditions. Their muscle capillary density was unchanged after acclimation to low environmental values but their mitochondrial volume value greatly increased (+67%) in slow-muscle fibres after acclimation. This could represent an adaptation to increase the utilization of circulating oxygen stores at low oxygen partial pressures. This in turn, would favour oxygen uptake in the gills by decreasing the veinous oxygen partial pressure. The crucian carp is also known for its ability to modify the morphology of its gills (Sollid *et al.*, 2003). Under normal oxygen (normoxic) conditions, the carp's gills lack lamellae (the primary site of oxygen uptake in fish), leading to unusually small respiratory surface areas. The secondary lamellae are actually embedded in cell masses which, when exposed to hypoxia, experience apoptosis and reduced cell proliferation (Sollid *et al.*, 2003). As a result, the lamellae protrude, and the respiratory surface area is increased by approximately 7.5-fold. This change is reversible and the cell masses will proliferate again under normoxic conditions, minimizing osmoregulatory costs (Sollid *et al.*, 2003).

Besides the seasonally and episodically hypoxic environments of ponds and lakes, some freshwater environments are chronically hypoxic. Sulfidic caves are such environments. They were discovered in the 1980s and their inhabiting unique ecosystems have quickly sparked interest. Engel (2007) reviewed the biology of sulfidic caves around the world. Numerous taxa have been found inhabiting sulfidic caves but few studies have focused on their response and adaptation to hypoxia. In the cave mollies *Poecilia mexicana* and *P. sulphuraria* (Poeciliidae, Teleostei), the presence of sulfide in the water forces the fish to breathe the water's surface (Tobler *et al.*, 2009). As a result, the fish spent less time foraging and had less food in their stomachs than fish under normoxic conditions. At the same time, they were potentially more exposed to predation. Although predation by the giant water bug *Belostoma* sp. was not significantly different between sulfidic and non-sulfidic sites, for some unclear reason males were more often taken by predators. This may eventually affect the genetics of the species, as less males will contribute to the gene pool. Numerous other hypoxia-adapted species have been found in sulfidic caves but are as yet poorly understood and call for further study. The amphipod *Niphargus ictus* revealed the first occurrence of a chemoautotrophic sulfide-oxidizing symbiosis that is not marine (Dattagupta *et al.*, 2009). This species and others (closely related) in the same cave system are likely exposed to different degrees of hypoxia (Flot *et al.*, 2010) and could be of great interest for comparative approaches to study adaptations in cave invertebrates that have evolved for hundreds of thousand years under these conditions (Dattagupta *et al.*, 2009).

Aquatic settings with their limited diffusion rates and possible further advection limitations by density differences of water masses allow the establishment of episodic, seasonal, or chronic hypoxic conditions. Depending on the extent of area affected by hypoxia and the time for which this hypoxia has been occurring, organisms are either temporarily eliminated or possess specific adaptations to cope with the conditions and reap the benefits of living in these areas. The great diversity of phyla inhabiting aquatic hypoxic environments offers the possibility to study adaptations and possible convergence in these adaptations. Although the marine hypoxic habitats and ice-covered ponds have now been studied in some detail, work on the sulfidic caves has only begun.

21.4 Hypoxia in a Changing World

The occurrence of hypoxic events in the world has increased in the recent years, in particular in shallow water where human influences are the strongest (Diaz and Rosenberg, 2008; Vaquer-Sunyer and Duarte, 2008). Although not all aquatic organisms exhibit the same sensitivity to hypoxia, the impact of removing some or most of the species in an area is bound to affect the whole ecosystem (Vaquer-Sunyer and Duarte, 2008). Eutrophication due to human activities is on the rise, the increased phytoplankton and benthic algae produced are then decomposed by bacteria and, in combination with density stratification, strong hypoxic conditions can develop (Diaz and Rosenberg, 2008). As a result, mass death events of fish and benthic organisms have been recorded that significantly affect fisheries plus local and wider economies (Levin *et al.*, 2009). Where these hypoxic events are nearly annual events (e.g. the Gulf of Mexico shelf), the benthic communities recuperate relatively fast because they are constantly kept at an early successional stage (Boesch and Rabalais, 1991). When these events are more exceptional, recovery is slow; in the New York Bight, for instance, it took more than 2 years after the 1976 catastrophic hypoxia for the communities to return to their normal state (Sindermann and Swanson, 1979; Boesch and Rabalais, 1991).

The organism response here is very similar to that of episodic and seasonal hypoxic events. Depending on the severity of the hypoxia, only some of the metazoans may die and new animals will recolonize the area when the conditions are no longer detrimental. As these conditions are only transitory, no species has evolved long enough under hypoxic conditions to possess specific adaptations.

21.5 Conclusions and Future Directions of Research

In the different types of environments exposed to chronic hypoxia, modification of the properties of respiratory pigments represents a common adaptation in vertebrates and invertebrates alike. These adaptations often require only a few amino acid changes. Invertebrates are usually more structurally plastic than vertebrates and display morphological adaptations as well, including increased ventilatory capacity, increased gill surface areas and reduced diffusion distances. On the other hand, vertebrate morphological adaptations seem to be less common and less pronounced. A few exceptions do however exist: in the Andean frog *Telmatobius culeus* the skin forms folds to increase the area available for gas exchange; in mole rats capillaries are more numerous in the muscles compared to other (lowland) rat species; and in the crucian carp, reversible respiratory surface area enlargement occurs under hypoxic conditions. Some cellular and subcellular modifications also exist to improve oxygen diffusion.

The tremendous amount of data brought forth by modern sequencing technologies now make it possible to compare species over a great number of genes and regulatory motifs in the species' genomes. These comparisons will be most meaningful when comparing closely related species (or populations of a species) that inhabit contrasting habitats. This approach is not based on *a priori* and could allow the discovery of pathways not initially suspected in the adaptation to hypoxia.

Work on organisms from extreme environments has revolutionized our understanding of life and the limits of this life. The discovery of hydrothermal vents came as a surprise and forced researchers to revise their view of the deep-sea. Life so close to the limits can stimulate metabolic and morphological innovations if the drive is strong enough (e.g. access to food sources, protection from predators). Not so long ago, researchers believed that metazoan life absolutely required the presence of molecular oxygen in large amounts. The discovery of the three species of *Loricifera* that live and thrive in the complete absence of molecular oxygen in the deep L'Atalante Basin (Danovaro *et al.*, 2010) has challenged this view. The phrase 'life as we know it' is often used to talk about the possible presence of organisms in some environments or other worlds. As we study more and more extreme environments, our knowledge expands the limits of life as we know it. We could now imagine complex (i.e. at least multicellular) life forms even under anoxic conditions, opening up fascinating possibilities for life in places such as Antarctic subglacial Lake Vostok, or even the subglacial oceans of the Jovian moon Europa.

References

Altman, P.L. (1958) *Handbook of Respiration*. W.B. Saunders Co., Philadelphia, 403 pp.

Arend, K.K., Beletsky, D., De Pinto, J., Ludsin, S.A., Roberts, J.J., Rucinski, D.K., Scavia, D., Schwab, D.J. and Hook, T.O. (2010) Seasonal and interannual effects of hypoxia on fish habitat quality in central Lake Erie. *Freshwater Biology* 56, 366–383.

Avivi, A., Resnik, M.B., Joel, A., Nevo, E. and Levi, A.P. (1999) Adaptive hypoxic tolerance in the subterranean mole rat *Spalax ehrenbergi*: the role of vascular endothelial growth factor. *FEBS Letters* 452, 133–140.

Avivi, A., Gerlach, F., Joel, A., Reuss, S., Burmester, T., Nevo, E. and Hankeln, T. (2010) Neuroglobin, cytoglobin, and myoglobin contribute to hypoxia adaptation of the subterranean mole rat *Spalax*. *Proceedings of the National Academy of Sciences of the United States of America*, early edition doi/10.1073/pnas.1015379107.

Beall, C.M. (2006) Andean, Tibetan, and Ethiopian patterns of adaptation to high-altitude hypoxia. *Integrative and Comparative Biology* 46, 18–24.

Boesch, D.F. and Rabalais, N.N. (1991) Effects of hypoxia on continental shelf benthos: comparisons between the New York Bight and the northern Gulf of Mexico. *Geological Society, Special Publication* 58, 27–34.

Campbell, K.L., Storz, J.F., Signore, A.V., Moriyama, H., Catania, K.C., Payson, A.P., Bonaventura, J., Stetefeld, J. and Weber, R.E. (2010) Molecular basis of a novel adaptation to hypoxic-hypercapnia in a strictly fossorial mole. *BMC Evolutionary Biology* 10, 214.

Childress, J.J. and Seibel, B.A. (1998) Life at stable low oxygen levels: adaptations of animals to oceanic oxygen minimum layers. *Journal of Experimental Biology* 201, 1223–1232.

Cloud, P. (1973) Paleoecological significance of banded iron-formation. *Economic Geology* 68, 1135–1143.

Corliss, J.B. and Ballard, R.D. (1977) Oases of life in the cold abyss. *National Geographic* 152, 441–454.

Danovaro, R., Dell'Anno, A., Pusceddu, A., Gambi, C., Heiner, I. and Kristensen, R.M. (2010) The first Metazoa living in permanently anoxic conditions. *BMC Biology* 8, 30.

Dattagupta, S., Schaperdoth, I., Montanari, A., Mariani, S., Kita, N., Valley, J.W. and Macalady, J.L. (2009) A novel symbiosis between chemoautotrophic bacteria and a freshwater cave amphipod. *ISME Journal* 3, 935–943.

Decelle, J., Andersen, A.C. and Hourdez, S. (2010) Morphological adaptations to chronic hypoxia in deep-sea decapod crustaceans from hydrothermal vents and cold-seeps. *Marine Biology* 156, 1259–1269.

Diaz, R.J. and Rosenberg, R. (2008) Spreading of dead zones and consequences for marine ecosystems. *Science* 321, 926–929.

Dudley, R. (1998) Atmospheric oxygen, giant Paleozoic insects and the evolution of aerial locomotor performance. *Journal of Experimental Biology* 201, 1043–1050.

El Albani, A., Bengtson, S., Canfield, D.E., Bekker, A., Macchiarelli, R., Mazurier, A., Hammarlund, E.U., Boulvais, P., Dupuy, J.-J., Fontaine, C., Fürsich, F.T., Gauthier-Lafaye, F., Janvier, P., Javaux, E., Ossa, F.O., Pierson-Wickmann, A.-C., Riboulleau, A., Sardini, P., Vachard, D., Whitehouse, M. and Meunier, A. (2010) Large colonial organisms with coordinated growth in oxygenated environments 2.1 Gyr ago. *Nature* 466, 100–104.

Engel, A.S. (2007) Observations on the biodiversity of sulfidic karst habitats. *Journal of Cave and Karst Studies* 69, 187–206.

Fisher, C.R. (1995) Toward an appreciation of hydrothermal-vent animals: their environment, physiological ecology, and tissue stable isotope values. In: Humphris, S.E., Zierenberg, R.A., Mullineaux, L.S. and Thomson, R.E. (eds) *Seafloor Hydrothermal Systems: Physical, Chemical, Biological, and Geological Interactions*. Geophysical Monograph 91. American Geophysical Union, Washington, DC, pp. 297–316.

Flot, J.-F., Wörheide, G. and Dattagupta, S. (2010) Unsuspected diversity of *Niphargus* amphipods in the chemoautotrophic cave ecosystem of Frasassi, central Italy. *BMC Evolutionary Biology* 10, 171–184.

Gallagher, J.B. and Burton, H.R. (1988) Seasonal mixing of Ellis Fjord, Vestfold Hills, East Antarctica. *Estuarine and Coastal Shelf Science* 27, 363–380.

Ginger, M.L., McFadden, G.I. and Michels, P.A.M. (2010) The evolution of organellar metabolism in unicellular eukaryotes. *Philosophical Transactions of the Royal Society of London, Part B* 365, 693–698.

Helly, J.J. and Levin, L. (2004) Global distribution of naturally occurring marine hypoxia on continental margins. *Deep-Sea Research I* 51, 1159–1168.

Holland, H. (2006) The oxygenation of the atmosphere and oceans. *Philosophical Transactions of the Royal Society of London, Part B* 361, 903–915.

Hourdez, S. and Jouin-Toulmond, C. (1998) Functional anatomy of the respiratory system of Branchipolynoe (Annelida; Polychaeta), commensal with mussels from deep-sea hydrothermal vents. *Zoomorphology* 118, 225–233.

Hourdez, S. and Lallier, F.H. (2007) Adaptations to hypoxia in hydrothermal vent and cold-seep invertebrates. *Reviews in Environmental Science and Biotechnology* 6, 143–159.

Hourdez, S. and Weber, R.E. (2005) Molecular and functional adaptations in deep-sea hemoglobins. *Journal of Inorganic Biochemistry* 99, 130–141.

Hourdez, S., Martin-Jézéquel, V., Lallier, F.H., Weber, R.E. and Toulmond, A. (1999a) Characterization and functional properties of the extracellular coelomic hemoglobins from the deep-sea, hydrothermal vent scaleworm *Branchipolynoe symmytilida*. *Proteins* 34, 435–442.

Hourdez, S., Lallier, F.H., Green, B.N. and Toulmond, A. (1999b) Hemoglobins from deep-sea scale-worms of the genus *Branchipolynoe* (Polychaeta, Polynoidae): a new type of quaternary structure. *Proteins* 34, 427–434.

Hutchison, V.H., Haines, H.B. and Engbretson, G. (1976) Aquatic life at high altitude: respiratory adaptations in the Lake Titicaca frog, *Telmatobius culeus*. *Respiratory Physiology* 27, 115–129.

Jacobs, D. and Lindberg, D.R. (1998) Oxygen and evolutionary patterns in the sea: onshore/offshore trends and recent recruitment of deep-sea faunas. *Proceedings of the National Academy of Sciences of the United States of America* 95, 9396–9401.

Johnson, K.S., Childress, J.J. and Beehler, C.L. (1988) Short-term temperature variability in the Rose Garden hydrothermal vent field: an unstable deep-sea environment. *Deep-Sea Research* 35, 1711–1721.

Johnston, I.A. and Bernard, L.M. (1984) Quantitative study of capillary supply to the skeletal muscles of crucian carp *Carassius carrassius* L.: effect of hypoxic acclimation. *Physiological Zoology* 57, 9–18.

Josefson, A.B. and Widbom, B. (1988) Differential response of benthic macrofauna and meiofauna to hypoxia in the Gullmar Fjord basin. *Marine Biology* 100, 31–40.

Kamykowski, D. and Zentara, S.J. (1990) Hypoxia in the world ocean as recorded in the historical data set. *Deep-Sea Research* 37, 1861–1874.

Knoll, A.H., Javaux, E.J., Hewitt, D. and Cohen, P. (2006) Eukaryotic organisms in Proterozoic oceans. *Philosophical Transactions of the Royal Society of London, Part B* 361, 1023–1038.

Kvitek, R.G., Conlan, K.E. and Iampietro, P.J. (1998) Black pools of death: hypoxic, brine-filled ice gouge depressions become lethal traps for benthic organisms in a shallow Arctic embayment. *Marine Ecology Progress Series* 162, 1–10.

Lamont, P.A. and Gage, J.D. (2000) Morphological responses of macrobenthic polychaetes to low oxygen on the Oman continental slope, NW Arabian Sea. *Deep-Sea Research Part II* 47, 9–24.

Lee, S.Y., Scott, G.R. and Milsom, W.K. (2008) Have wing morphology or flight kinematics evolved for extreme high altitude migration in the bar-headed goose? *Comparative Biochemistry and Physiology – C: Toxicology and Pharmacology* 148, 324–331.

Levin, L.A. (2003) Oxygen minimum zone benthos: adaptation and community response to hypoxia. *Oceanography and Marine Biology: An Annual Review* 41, 1–45.

Levin, L.A., Ekau, W., Gooday, A.J., Jorissen, F., Middelburg, J.J., Naqvi, S.W.A., Neira, C., Rabalais, N.N. and Zhang, J. (2009) Effect of natural and human-induced hypoxia on coastal benthos. *Biogeosciences* 6, 2063–2098.

Liu, X.-Z., Li, S.-L., Jing, H., Liang, Y.-H., Hua, Z.-Q. and Lu, G.-Y. (2001) Avian haemoglobins and structural basis of high affinity for oxygen: structure of bar-headed goose aquomet haemoglobin. *Acta Crystallographica Section D: Biological Crystallography* 57, 775–783.

Macedo, H. (1960) Vergleichende Untersuchungen an Arten der Gattung *Telmatobius*. *Zeitschrift für Wissenschaftliche Zoologie* 163, 355–396.

Margulis, L. (1975) Symbiotic theory of the origin of eukaryotic organelles; criteria for proof. *Symposium of the Society for Experimental Biology* 29, 21–38.

Milsom, W.K. and Scott, G. (2008) Respiratory adaptations in the high flying bar-headed goose. *Comparative Biochemistry and Physiology Part C: Toxicology and Pharmacology* 148, 460.

Neira, C., Gad, G., Arroyo, N.L. and Decraemer, W. (2001) *Glochinema bathyperuvensis* sp. n. (Nematoda, Epsilonematidae): a new species from Peruvian bathyal sediments, SE Pacific Ocean. *Contributions to Zoology* 70, 147–159.

Oliver, P.G. (2001) Functional morphology and description of a new species of *Amygdalum* (Mytilidoidea) from the oxygen minimum zone of the Arabian Sea. *Journal of Molluscan Studies* 67, 225–241.

Ostojic, H., Monge, C. and Cifuentes, V. (2000) Hemoglobin affinity for oxygen in three subspecies of toads (*Bufo* sp.) living at different altitudes. *Biological Research* 33, 5–10.

Petschow, D., Würdinger, I., Baumann, R., Duhm, J., Braunitzer, G. and Bauer, C. (1977) Causes of high blood O_2 affinity of animals living at high altitude. *Journal of Applied Physiology* 42, 139–143.

Poyart, C., Wjacman, H. and Kister, J. (1992) Molecular adaptation of hemoglobin function in mammals. *Frontiers in Respiratory Physiology* 90, 3–17.

Projecto-Garcia, J., Zorn, N., Jollivet, D., Schaeffer, S.W., Lallier, F.H. and Hourdez, S. (2010) Origin and evolution of the unique tetra-domain hemoglobin from the hydrothermal vent scale-worm *Branchipolynoe*. *Molecular Biology and Evolution* 27, 143–152.

Reichart, G.L., Lourens, L.J. and Zachariasse, W.J. (1998) Temporal variability in the northern Arabian Sea oxygen minimum zone (OMZ) during the last 225,000 years. *Paleoceanography* 13, 607–621.

Rollema, H.S. and Bauer, C. (1979) The interaction of inositol pentaphosphate with the hemoglobins of high-land and low-land geese. *Journal of Biological Chemistry* 254, 12038–12043.

Sagan, L. (1967) On the origin of mitosing cells. *Journal of Theoretical Biology* 14, 255–274.

Schmitz, A. and Harrison, J.F. (2004) Hypoxic tolerance in air-breathing invertebrates. *Respiratory Physiology and Neurobiology* 141, 229–242.

Sindermann, C.J. and Swanson, R.L. (1979) Oxygen depletion and associated benthic mortalities in New York Bight, 1976, historical and regional perspective. *NOAA Professional Paper* 11, 1–16.

Sollid, J., De Angelis, P., Gundersen, K. and Nilsson, G.E. (2003) Hypoxia induces adaptive and reversible gross morphological changes in crucian carp gills. *Journal of Experimental Biology* 206, 3667–3673.

Storz, J.F. (2007) Hemoglobin function and physiological adaptation to hypoxia in high-altitude mammals. *Journal of Mammalogy* 88, 24–31.

Storz, J.F. (2010) Genes for high altitudes. *Science* 329, 40–41.

Storz, J.F., Runck, A.M., Sabatino, S.J., Kelly, J.K., Ferrand, N., Moriyama, H., Weber, R.E. and Fago, A. (2009) Evolutionary and functional insights into the mechanism underlying high-altitude adaptation of deer mouse hemoglobin. *Proceedings of the National Academy of Sciences of the United States of America* 106, 14450–14455.

Storz, J.F., Runck, A.M., Moriyama, H., Weber, R.E. and Fago, A. (2010) Genetic differences in hemoglobin function between highland and lowland deer mice. *Journal of Experimental Biology* 213, 2565–2574.

Tinson, S. and Laybourn-Parry, J. (1985) The behavioural responses and tolerance of freshwater benthic cyclopoid copepods to hypoxia and anoxia. *Hydrobiologia* 127, 257–263.

Tobler, M., Riesch, R.W., Tobler, C.M. and Plath, M. (2009) Compensatory behaviour in response to sulphide-induced hypoxia affects time budgets, feeding efficiency, and predation risk. *Evolution Ecology Research* 11, 935–948.

Tunnicliffe, V. (1991) The biology of hydrothermal vents: ecology and evolution. *Oceanography and Marine Biology: An Annual Review* 29, 319–407.

Van Aardt, W.J., Bronner, G. and Buffenstein, R. (2007) Hemoglobin-oxygen-affinity and acid-base properties of blood from the fossorial mole-rat, *Cryptomys hottentotus pretoriae*. *Comparative Biochemistry and Physiology Part A* 147, 50–56.

Vaquer-Sunyer, R. and Duarte, C.M. (2008) Threshold of hypoxia for marine biodiversity. *Proceedings of the National Academy of Sciences of the United States of America* 105, 15452–15457.

Weber, R.E., Ostojic, H., Fago, A., Dewilde, S., Hauwaert, M.-L., Moens, L. and Monge, C. (2002) Novel mechanism for high-altitude adaptation in hemoglobin of the Andean frog *Telmatobius peruvianus*.

American Journal of Physiology – Regulatory, Integrative and Comparative Physiology 283, R1052–R1060.

Weber, R.E., Hourdez, S., Knowles, F. and Lallier, F.H. (2003) Hemoglobin function in deep-sea and hydrothermal vent fish: *Symenchelis parasitica* (Anguillidae) and *Thermarces cerberus* (Zoarcidae). *Journal of Experimental Biology* 206, 2693–2702.

Widmer, H.R., Hoppeler, H., Nevo, E., Taylor, C.R. and Weibel, E.R. (1997) Working underground: respiratory adaptations in the blind mole rat. *Proceedings of the National Academy of Sciences of the United States of America* 94, 2062–2067.

Wyrtki, K. (1962) The oxygen minima in relation to ocean circulation. *Deep-Sea Research* 9, 11–23.

22 High Ultraviolet Radiation Environments

Kevin K. Newsham[1] and Andrew T. Davidson[2]
[1]*British Antarctic Survey, Natural Environment Research Council, Cambridge, UK;*
[2]*Australian Antarctic Division, Kingston, Tasmania, Australia*

22.1 Introduction

This chapter deals with the effects of solar ultraviolet (UV) radiation on organisms in terrestrial and aquatic habitats. It is not meant to be an exhaustive review, but rather an overview of the main processes leading to UV radiation reaching ecosystems, the broad effects of UV radiation on microbes, plants and animals in terrestrial and aquatic ecosystems, and how these organisms cope with UV exposure.

UV radiation is the portion of the electromagnetic spectrum that is shorter than visible radiation but longer than x-rays. It has wavelengths (i.e. the distance between the consecutive crests of its sinusoidal waves) of between 10 nm and 400 nm. The shortest wavelengths of solar UV radiation, termed vacuum UV (10–280 nm), are not encountered in the natural environment, since they are strongly absorbed by water and oxygen in the atmosphere long before they reach the Earth's surface and are thus disregarded here. However, significant amounts of radiation with wavelengths above 280 nm do penetrate the Earth's atmosphere to the ground. This radiation is divided into two regions, the UV-B (280–315 nm) and UV-A (315–400 nm) wavelengths.

Defining what constitutes 'extreme' in terms of UV exposure is very difficult because it depends on the past history of exposure of an organism. If the organism in question originates from a high UV environment, the effects of a given dose of UV are likely to be much less than if it comes from a low or zero UV environment, because organisms living in high UV environments are likely to be pre-adapted (Section 22.4).

22.2 Biological Effects of UV Radiation

Photons – the elementary particles that form electromagnetic radiation – carry different amounts of energy depending on their wavelength. Although UV-B represents <1% of the total solar radiation reaching the Earth's surface, the biological effects of this radiation are significant. This is because photons of short wavelengths carry more energy than those of longer wavelengths, with UV-B radiation carrying 3.9–4.4 eV per photon, compared with 3.1–3.9 eV per photon of UV-A radiation. It should be recognized that UV radiation does have positive health benefits, since it stimulates the synthesis of vitamin D_3 in skin, a lack of which causes diseases such as rickets, but over-exposure to UV radiation, and UV-B in particular, is harmful to a wide range of organisms. This is principally because DNA, the molecule that codes for most forms of life on Earth, strongly absorbs UV-B radiation. Typically, exposure to photons of UV-B causes dimers to form in the DNA helix, in which bonds are formed between adjacent, rather than opposite, cytosine bases, producing mutations in the DNA code that interfere with its replication. Alternatively, when photons of UV enter a cell, they can cause direct damage to membranes and organelles, or can cause indirect damage by forming reactive oxygen species and free radicals. These ions or very small molecules are highly reactive, owing to the presence of unpaired electrons in their structures, and lead to oxidative stress and damage to the cell. Thus, because photons of UV-B radiation carry more energy per photon than those of longer wavelengths of radiation reaching the Earth's surface, the absorption of these photons by molecules in cells, typically nucleic acids, is harmful to most forms of life.

22.3 Factors Affecting the Receipt of UV Radiation

The receipt of UV radiation by ecosystems is influenced by many different factors. UV radiation entering the Earth's atmosphere is subject to absorption by ozone and water, scattering by air molecules, particulates and clouds, and reflection back into space. Since ozone is the principal gas in the atmosphere that absorbs UV-B radiation from the Sun, it has an important influence on the receipt of UV-B at the Earth's surface. The gas absorbs little UV-A radiation but, instead, preferentially absorbs UV-B wavelengths, and the loss of ozone from the atmosphere, discussed in more detail below, leads to increased irradiances of UV-B radiation at the Earth's surface. Clouds also have an important influence over the irradiance of UV received at the Earth's surface since they trap radiation by internal scattering, increasing the path length of radiation before it reaches the ground. Their transparency to UV is determined by their height and type, with, for example, low-level stratus clouds reducing UV radiation dose by up to 75%, but high-level cirrus only reducing the UV dose slightly, relative to clear skies. The presence or absence of clouds therefore has profound effects on the dose of UV radiation received at ground level.

A further factor affecting the receipt of UV radiation is solar elevation. At solar noon, there is less atmosphere for UV radiation to travel through before it reaches the ground and, assuming cloudless skies, UV dose is highest at this time of day. Time of year has the same effect, with daily doses of UV being highest close to the summer solstice. Altitude also has a strong effect on the amount of UV radiation received at the Earth's surface, with an approximate 6% increase in skin burning UV irradiance for every kilometre increase in altitude. This is because there is less scattering and absorption of radiation in thin atmospheres, and also because there are fewer tropospheric pollutants at higher altitudes, which reduce the amount of UV radiation reaching the ground.

Exposure to UV is further enhanced in alpine and polar habitats because of the higher albedo of ice and snow surfaces in these regions. Albedo, the ratio of reflected to incident radiation, which has a range between 0 and 1, is higher for snow and ice than for other surfaces. Snow-covered ice is highly reflective, having an albedo of about 0.95, with fresh snow having an albedo of 0.80. These values are higher than those for soil, vegetation and water (0.02–0.20). Exposure to UV reflected from snow and ice is thus greater in polar and alpine regions than at lower latitudes and altitudes.

Another factor that influences UV receipt, particularly in polar habitats, is stratospheric ozone depletion. The physical processes that lead to ozone depletion are caused to a large extent by manmade substances released to the Earth's atmosphere during the 1960s and 1970s. These substances, known as chlorofluorocarbons (or CFCs), were formerly used as refrigerants and aerosol propellants. The presence in the atmosphere of these long-lived substances has had a dramatic effect on the concentrations of ozone in the Earth's stratosphere over the past four decades. Between the 1970s and 1980s, the amount of springtime ozone in the stratosphere – the region of the atmosphere located between 8 km and 50 km above the Earth's surface – fell from approximately 320 Dobson units (DU) to about 200 DU (Farman *et al.*, 1985). This loss was due to chemical reactions catalysed by chlorine monoxide, derived from CFCs, which photocatalyses ozone to oxygen. Owing to the long-lived nature of CFCs, ozone depletion in the Earth's atmosphere is expected to continue until at least the middle of the 21st century (World Meteorological Organization, 2007; see also Glover and Neal, Chapter 24, this volume).

The chemical reactions that lead to ozone depletion only occur in extremely cold atmospheres. The reactions take place on the surfaces of nacreous clouds, which form each winter in cold (less than −78°C) air masses over polar regions. Because the atmosphere over the Antarctic is colder than that over the Arctic, frequently falling below −78°C in temperature during winter, ozone depletion is typically more intense over the South Pole than it is over the North Pole, with a mean 15% loss of ozone over the Arctic in the last two decades, but a 30% springtime loss of ozone from over the Antarctic during the same period (World Meteorological Organization, 2007). Ozone depletion can lead to substantial areas of the Earth's surface, including inhabited areas of Patagonia and Australasia, lacking a full protective ozone layer, with consequent increases in UV-B radiation at ground level.

The main processes affecting the amount of UV radiation reaching the Earth's surface are thus the concentration of ozone in the atmosphere, which is seasonally affected by the depletion of the gas in the stratosphere each spring over the poles, and cloud cover, altitude, reflectance from snow and ice, tropospheric pollution and solar elevation.

22.4 Responses of Terrestrial Organisms to UV-B Radiation

UV-B radiation is known to have widespread effects on organisms in terrestrial ecosystems. There are two principal routes by which UV-B exposure can affect terrestrial organisms, termed direct and indirect effects. Direct effects refer to the influence of UV-B radiation on an organism, such as the DNA damage sustained by many organisms on exposure to UV-B; indirect effects refer to secondary effects of UV-B on an organism that have been mediated by others.

22.4.1 Plants

Plants have important roles as the dominant autotrophs of terrestrial ecosystems and as crops. They are sessile photosynthetic organisms that are unable to avoid exposure to solar radiation, and have therefore been the focus for much UV-B research in recent decades. Early experiments into the effects of UV-B radiation on plants typically took place in growth chambers or glasshouses under artificial light sources, at minimal levels of visible radiation and extreme doses of UV-B radiation. These experiments indicated substantial negative effects of UV-B radiation exposure on photosynthetic rates and plant growth (Fiscus and Booker, 1995). These early findings raised concern within the research community that the exposure of plants to projected doses of UV-B radiation under depleted ozone columns might have detrimental effects on plant growth. However, it was soon realized that these effects of UV-B exposure on plants were primarily owing to the low doses of UV-A and visible radiation applied in growth chambers. Because UV-A and visible radiation are necessary for the photorepair of DNA, the low doses of these portions of the spectrum applied in growth chambers tended to exaggerate the effects of UV-B radiation on plants.

Subsequent UV-B studies were based outdoors. In outdoor experiments, UV-B treatments are often applied from UV-emitting fluorescent lamps against a background of full solar radiation, avoiding the problems associated with early UV-B studies. The lamps provide an above-ambient UV-B treatment, usually simulating a level of ozone depletion of between 15% and 30%. Outdoor lamp experiments have indicated much more subtle effects of UV-B radiation on plant growth and photosynthesis than those observed in earlier growth chamber experiments.

In recent years, the view that plant photosynthesis and growth are not substantially affected by elevated UV-B exposure in outdoor experiments has been supported by many studies. Lamp experiments in temperate, tropical, Arctic and even Antarctic habitats have each confirmed this view. Much of the information from Arctic habitats has been derived from a long-term lamp experiment at Abisko in northern Sweden, which has exposed dwarf shrub species to elevated UV-B radiation corresponding to a 15% depletion in ozone column since 1993 (Björn *et al.*, 1999). The UV-B treatment applied each year at Abisko has had few effects on the growth of species such as *Empetrum hermaphroditum, Vaccinium uliginosum* and *V. vitis-idaea*, although the stem elongation of *V. myrtillus* has been shown to be reduced by UV-B exposure (Phoenix *et al.*, 2001). The UV-B treatment has, however, led to significant increases in the concentrations of sunscreen pigments in the upper mesophyll layers of leaves (Semerdjieva *et al.*, 2003). These pigments are phenolic compounds synthesized by the phenylpropanoid biosynthetic pathway. The expression of key genes in this pathway, such as that encoding for phenylalanine ammonium lyase, is up-regulated by exposure to UV-B radiation. The resulting pigments synthesized by the pathway, collectively known as flavonoids, contain carbon ring structures in their skeletons, which are efficient UV absorbers, helping to protect plant tissues from the harmful effects of UV-B radiation.

An alternative approach used in outdoor studies is to place screens over plants to reduce the dose of UV-B radiation that they receive. The screens are usually constructed from two types of plastic, one of which absorbs, and the other of which transmits, UV-B radiation. This approach is technologically much less demanding than the use of lamps and is suitable for deployment in remote locations where power sources cannot be accessed. For example, screens have been used to study the influence of UV-B exposure on the two native Antarctic flowering plant species, the Antarctic pearlwort (*Colobanthus quitensis*) and Antarctic hair grass (*Deschampsia antarctica*), growing on the Danco Coast on the Antarctic Peninsula (Day *et al.*, 2001). Typically, the concentrations of sunscreen pigments are increased in plants exposed to near-ambient UV-B radiation, and the biomass of plants is reduced by up to 40%, with corresponding reductions in photosynthetic parameters, relative to plants exposed to reduced UV-B radiation. In the Arctic, recent screen experiments based at Zackenberg in north-eastern Greenland have similarly shown that UV-B exposure lowers the net photosynthesis of the bog or northern bilberry, *Vaccinium uliginosum*, by approximately 25%, and increases the concentration of sunscreen pigments in plant tissues by approximately the same amount (Albert *et al.*, 2008). It is likely that the substantial reductions in UV-B exposure experienced by plants under screens, compared with the more modest increases in UV-B exposure experienced by those under lamps, account for the different results of the experiments on *Vaccinium* species at Abisko and Zackenberg.

Plant responses to UV-B radiation can also be assessed using a non-manipulative approach, in which concentrations of sunscreen pigments are measured periodically and then correlated with natural fluctuations in solar UV-B radiation dose. Analyses of data from such experiments have shown increased concentrations of these pigments in the tissues of plants during periods of high UV-B exposure (e.g. Newsham *et al.*, 2002). These studies have much in common with so-called hindcasting studies, in which past UV-B exposures are predicted by measuring the concentrations of sunscreen pigments present in herbarium specimens of lower plants. Such studies have recently been fine-tuned by analysing the ratio of two flavonoids, luteolin and apigenin, in the tissues of herbarium specimens of the moss *Bryum argenteum* (Ryan *et al.*, 2009). This study recorded a significant negative relationship between the ratio of luteolin to apigenin in *B. argenteum* tissues and total ozone in the atmosphere at close to the time of collection of the specimens. Luteolin, which has an additional hydroxyl group attached to its skeleton, is a more efficient scavenger of free radicals than apigenin, accounting for the increase in the concentration of the former flavonoid when plants are exposed to UV-B radiation.

A wide range of effects of UV-B radiation exposure on plants have been recorded in numerous scientific studies and it is beyond the scope of this chapter to review all of these responses. However, in order to arrive at a consensus for how plants respond to UV-B exposure, meta-analyses have been made on the data arising from them. A meta-analysis typically consists of taking data from primary research publications, recording the response ratio (the logarithm of the treatment mean divided by the control mean), and then analysing the data to determine if the mean response ratio for a parameter is less than or greater than zero, indicating a reduction or an increase in a parameter in response to UV-B exposure, respectively. Only parameters that exhibit consistent increases or decreases to treatments across different research studies are identified by meta-analyses as being responsive to UV-B exposure. Three such studies on plant response to UV-B exposure have been published over the past decade, focused on temperate, tropical and polar plant species. In broad terms, each of these studies has identified similar responses of plants to UV-B exposure. That of Searles *et al.* (2001), which analysed data from lamp experiments simulating between 10% and >20% losses of ozone column, indicated reductions of plant height and shoot biomass of 3% and 14%, respectively (Fig. 22.1a). In addition, the concentrations of sunscreen pigments in foliage were increased by 10%, relative to controls, after plants had been exposed to UV-B radiation (Fig. 22.1a). In an analysis of plant responses to UV-B radiation in lamp, screen and non-manipulative experiments in polar regions (Newsham and Robinson, 2009), UV-B exposure was found to increase the concentration of sunscreen pigments per unit of leaf weight by 7% (Fig. 22.1b). Shoot biomass and plant height were decreased by 15% and 10%, respectively (Fig. 22.1b). Neither this analysis nor that of Searles *et al.* (2001) found effects of UV-B exposure on photosynthetic rate or the concentrations of chlorophylls or carotenoids in plant tissues (Figs 22.1a, b). Li *et al.* (2010) found that relatively moderate supplements of UV-B radiation from fluorescent lamps (corresponding to 18–40% of that present in solar radiation, arising from 9–20% losses of ozone column) led to 5% reductions in the height of herbaceous plant species, with no effect on woody species (Fig. 22.1c). UV-B exposure was also found to result in 11–16% increases in the concentrations of sunscreen pigments in the foliage of both plant groups (Fig. 22.1c) but, in contrast with the other two meta-analyses, also led to 18% reductions in carotenoid concentrations in herbaceous species, and reductions in photosynthetic rate of 11% and 22% in herbaceous and woody species, respectively (Fig. 22.1c).

In summary, early studies into the effects of UV-B exposure on plants predicted widespread deleterious effects on many species. Subsequent outdoor experiments under natural light conditions indicated that UV-B exposure was unlikely to have dramatic effects on plant growth and photosynthesis, but would likely result in 3–15% reductions in plant height and shoot biomass, and 7–16% increases in the concentrations of sunscreen pigments in aboveground tissues.

22.4.2 Microbes and soils

In comparison with plants, microbes are particularly sensitive to UV-B exposure because of the short path lengths through their cells. For example, single microbial cells of less than 20 µm diameter are unable to synthesize sufficient sunscreen pigments to absorb UV-B radiation (Garcia-Pichel, 1994). Microbes have pivotal roles in the functioning of terrestrial ecosystems, particularly because of their involvement in the decomposition of soil organic matter, a process that returns otherwise unavailable organic nutrients to the soil in inorganic

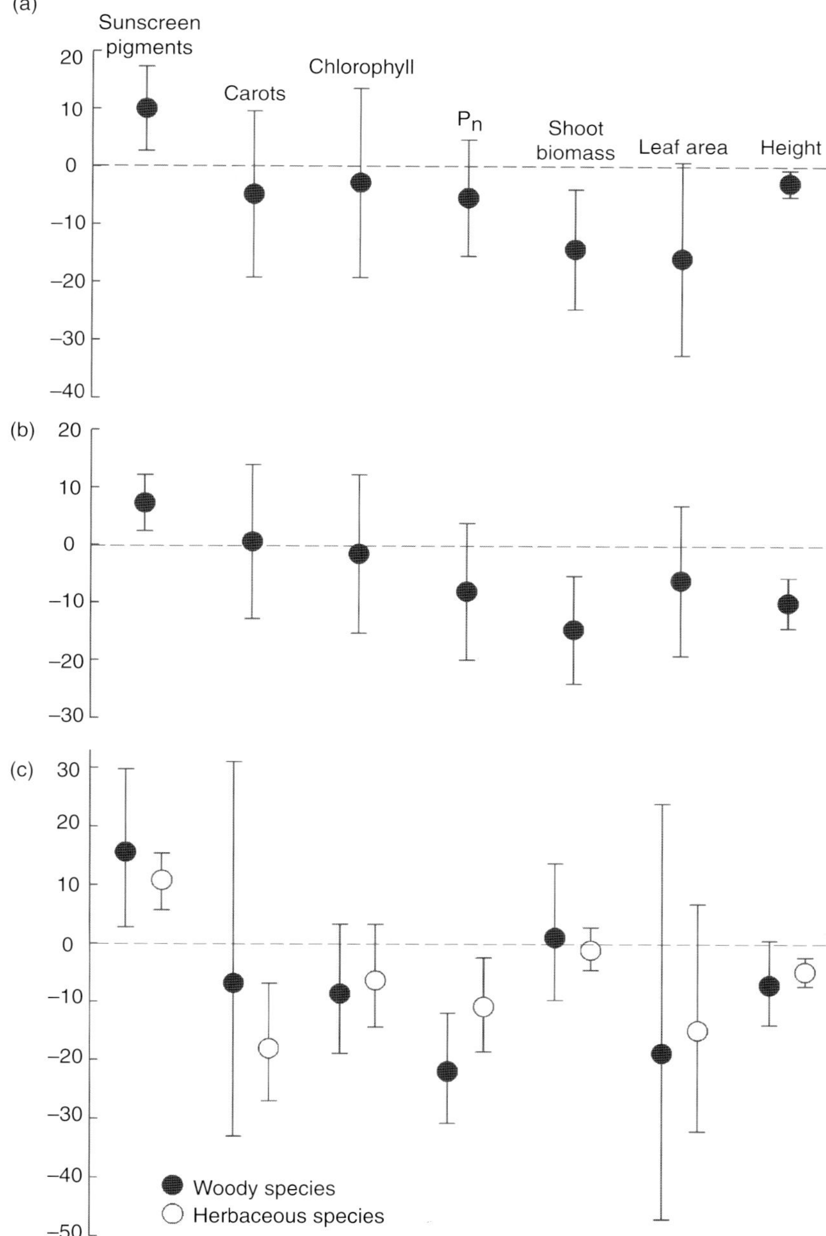

Fig. 22.1. Percentage changes in plant pigmentation, photosynthetic rate and morphology in response to UV-B radiation exposure. Data are redrawn from the meta-analyses of (a) Searles *et al.* (2001), (b) Newsham and Robinson (2009) and (c) Li *et al.* (2010). Circles represent means. The filled and open circles in (c) represent means for woody and herbaceous plant species, respectively. The bars attached to means represent 95% confidence intervals (a, b) and 95% bootstrap confidence intervals (c). Where the bars do not cross the horizontal dashed lines, there is a significant effect of UV-B exposure on a parameter at $P<0.05$. Abbreviations: carots, carotenoids; P_n, net photosynthesis.

form that can be assimilated by plant roots. UV-B radiation has been shown to have widespread effects on this process. In their pioneering work in this area, Gehrke *et al.* (1995) found that the litter of *Vaccinium uliginosum* was less frequently colonized by two species of soil fungi in the Subarctic when exposed directly to UV-B radiation. Decomposition of the litter, measured both as mass loss of litter and CO_2 release, was inhibited by the UV-B treatment. Subsequent studies showed similar negative direct effects of UV-B exposure on the litters of tree species, which were attributed to effects on decomposer microbes. However, it is apparent that a process called photodegradation, caused largely by the exposure of lignin in litter to UV-B radiation, may lead to enhanced decomposition rates of litters, particularly in arid and semi-arid environments.

Direct effects of UV-B radiation in vegetated habitats are only likely to take place in glades, since the majority of UV-B will be attenuated by the leaf canopy. Studies have thus assessed the indirect effects of UV-B exposure on plant litter decomposition. These studies have exposed plants to UV-B radiation during growth, and then decomposed the litter in mesh bags, usually on forest floors, in environments with much less UV-B exposure than at canopy level. Such experiments have revealed remarkably different effects of UV-B exposure during growth on the decomposition of different plant species' litters, with some studies showing reductions in decomposition rates of litters exposed to UV-B radiation during growth and others showing increases in decomposition rates (Rozema *et al.*, 1997; Newsham *et al.*, 2001).

Other below ground processes are known to be affected by UV-B exposure. Although mycorrhizas, which are widespread symbioses between plant roots and soil fungi, exist below ground, it has been suggested that their functioning is altered by the exposure of above ground plant parts to UV-B radiation, which is thought to change patterns of carbon allocation to the soil. This is of importance because mycorrhizas, and not roots, are the principal organs through which the majority of plants absorb nutrients from soil. As with the indirect effects of UV-B radiation on plant litter decomposition, a clear consensus for how the mycorrhizal symbiosis responds to UV-B exposure has yet to be reached, with both reductions and increases in the frequencies of mycorrhizal fungi having been recorded in roots (Rinnan *et al.*, 2005).

Cyanobacteria are also known to be affected by exposure to UV-B radiation. These microbes, which are present in vegetation and at soil surfaces, are of importance because they fix a significant amount of the nitrogen that enters terrestrial ecosystems from the atmosphere. Exposure to UV-B radiation in Arctic ecosystems has been shown to depress the amount of nitrogen fixed by free-living and lichen-associated cyanobacteria (Solheim *et al.*, 2006). Similarly, in hot deserts, the nitrogen fixation and photosynthesis of biological soil crusts, which are assemblages of lichens, cyanobacteria and mosses, is also inhibited by exposure to UV-B radiation (Belnap *et al.*, 2008).

Overall, direct UV-B exposure thus reduces the growth of microbes, often because the cells of these organisms are too small to produce sufficient pigments to absorb the radiation. Indirect effects of UV-B radiation in terrestrial ecosystems, for example on mycorrhizas and the decomposition of plant litter in soil, are apparently widespread, with both positive and negative effects of UV-B exposure on the symbiosis and decomposition having been shown.

22.4.3 Animals

There is little information on the effects of UV-B radiation on mammals and birds in terrestrial ecosystems. It is reasonable to anticipate few effects of UV-B exposure on these animals, given their body hair or plumage, their frequently nocturnal habits, and their usually short life spans, which allow any cancers or damage to eyes to remain asymptomatic. Nevertheless, as in humans, sunburn and squamous cell carcinomas have been shown to form in domestic animals with sparsely haired and lightly pigmented areas of skin that are exposed to sunlight. Cancers of the eye have similarly been shown to occur in livestock and domestic animals exposed to UV-B radiation (de Gruijl *et al.*, 2003).

Research on lower vertebrates, and particularly amphibians, has indicated that UV-B radiation has an important role in the survival of these animals in the natural environment. A meta-analysis of the responses of amphibians to UV-B exposure indicates a twofold decline in the survival of amphibians relative to controls shielded from UV-B radiation, with embryos and larvae, but not juveniles, being susceptible to UV-B exposure, and salamanders being more susceptible than frogs and toads (Bancroft *et al.*, 2008).

Invertebrates are also known to be affected by UV-B radiation exposure. Some insects, such as thrips, avoid direct exposure to UV-B radiation, whilst others, such as honeybees and hornets, use UV-B and UV-A radiation as navigation aids. In broad terms, the level of herbivory by insects on plants is increased when UV-B is removed from solar radiation by screens. For example, thrips have been shown to prefer the leaves of soybean that have been shielded from UV-B radiation over those exposed to UV-B, with fewer animals on, and less damage to, leaves of the latter plants (Mazza *et al.*, 1999). This observation is consistent with the view that UV-B exposure is likely to deter feeding by insects on plants because of the increased synthesis of flavonoids, which are also thought to have a role in plant defence against herbivores. Effects of UV-B radiation on invertebrate behaviour have also been observed in polar ecosystems. For example, in screen experiments on the Danco Coast of the western Antarctic Peninsula, Convey *et al.* (2002) found reductions in the numbers of the springtails in soil beneath plants that had been exposed to UV-B radiation, relative to those protected from UV-B.

Thus, exposure to UV-B radiation is unlikely to influence higher animals in the natural environment, but is known to affect lower animals such as amphibians and insects. UV-B exposure has deleterious effects on the embryos and larvae of amphibians, and is known to reduce the feeding of insects on plants, indirectly as well as having effects on insect behaviour.

22.5 The Influence of UV-B on Aquatic Ecosystems

Aquatic environments are inhabited by organisms ranging from viruses and bacteria to large vertebrates such as whales, encompassing many taxonomic kingdoms that have little in common, and yet they are united in the nature of the damage caused by UV-B and the mechanisms that their inhabitants employ to withstand this radiation. Like the terrestrial environment, the amount of UV-B-induced damage sustained by aquatic organisms is determined by the amount of UV-B radiation that reaches the organism and its capacity to withstand UV-B exposure.

22.6 Aquatic UV-B Irradiances

We have seen above that the amount of UV-B reaching the surface of a water body is increased by ozone depletion, altitude and high albedo (Section 22.3). Unlike the terrestrial system, which is effectively two dimensional, aquatic organisms occupy a turbulent, three-dimensional environment in which UV-B irradiance changes greatly with depth and water clarity. There are many factors that attenuate UV reaching the sea surface (Section 22.3), resulting in UV-B comprising only 0.8% of the overall surface solar irradiance (Fig. 22.2). The spectral irradiances beneath the water are further reduced by absorption and scattering by the water itself and the particles and dissolved organic matter that it contains. Water best transmits green and blue light (~500 nm wavelength) and increasingly absorbs light at shorter and longer wavelengths (Davidson, 2006; Fig. 22.2). Until Jerlov (1950), UV radiation was thought to be inconsequential for aquatic environments. It is now recognized that absorption by water alone allows measurable amounts of UV-B radiation to penetrate up to 60 m depth in the clearest of waters (Gieskes and Kraay, 1990). This penetration is reduced by scattering, mutual shading among suspended particles and absorption by coloured compounds, including plant pigments and dissolved organic matter. Together, the water and its contents cause a rapid decline in UV-B irradiance with depth, resulting in a gradient from high to negligible UV-B that is compressed into near surface waters. UV-B irradiances are only high enough to cause biological effects in the upper 20–30 m of transparent ocean waters (Fig. 22.2), but this declines to as little as 1.5 m in highly productive environments and/or those containing high concentrations of dissolved organics (Zagarese and Williamson, 2000). For a given UV-B irradiance at the water surface, the amount of UV-B-induced damage sustained by an organism is determined by the water clarity and the speed and depth of vertical movement (Neale *et al.*, 1998, 2003). Any organism that swims or is carried close to the surface experiences rapid changes in UV-B irradiance, making it difficult or impossible to measure the amount of UV-B radiation that it has received. Our inability to measure and simulate natural sunlight exposures has largely confounded attempts over 15 years to determine the impact of UV-B on natural communities of planktonic organisms (see Section 22.9).

Vertical mixing in aquatic systems is controlled by the competing influences of solar warming and sea ice melt, which reduce the mixed depth, and wind that deepens and accelerates mixing (Neale *et al.*, 2003). Diurnal and seasonal changes in solar warming of the ocean surface cause large variations in the vertical movement of planktonic particles. Warming produces a lens of surface water that is buoyant and can trap cells, exposing them to high UV-B irradiances (Neale *et al.*, 2003). Models suggest that global warming may intensify this process, especially at higher latitudes where warming is expected to be greatest, exposing cells to much higher UV-B irradiances than ozone depletion (Sarmiento and LeQuéré, 1996; Matear and Hirst, 1999).

In polar marine waters, melting sea ice releases buoyant fresher water, which can result in mixed layers of 20 m or less for up to 6 days (Davidson, 2006). This traps plankton in a high nutrient, high light environment as the pack ice margin retreats in spring and summer. These conditions generate blooms that supply 25–67% of all the phytoplankton production in Antarctic waters and support the wealth of life for which Antarctica is renowned. Yet this also traps plankton blooms in shallow waters that are penetrated by UV-B radiation at a time of ozone depletion.

22.7 Tolerance Mechanisms

The sensitivity of aquatic organisms to UV-B damage differs greatly among species due to differences in the extent to which they can prevent damage or repair the damage that they sustain. Four mechanisms can be used to enhance an organism's tolerance to UV-B exposure; avoidance, screening, quenching and repair (Fig. 22.3). The widespread occurrence of these tolerance mechanisms indicates that UV exposure is an important factor in aquatic environments. Avoidance and screening by pigments are protective mechanisms that reduce the incoming UV-B radiation. Avoiding a high UV-B climate is beyond the capability of planktonic organisms that are largely unable to control their depth. Organisms that are able to control their depth may avoid high UV-B irradiance by migrating downward or retreating to shade (negative phototaxis). But UV-B can damage phototaxic organisms, making them unable to avoid high light (Häder, 1986) and many organisms studied can detect UV-A wavelengths but none have proven able to detect UV-B (Zagarese and Williamson, 2000). Consequently, any light-induced behavioural responses controlled by UV-A and/or visible light may not be appropriate where depletion of stratospheric ozone has enhanced only UV-B wavelengths.

Like terrestrial organisms (Section 22.4), those in aquatic systems can protect themselves from UV-B by producing UV-absorbing

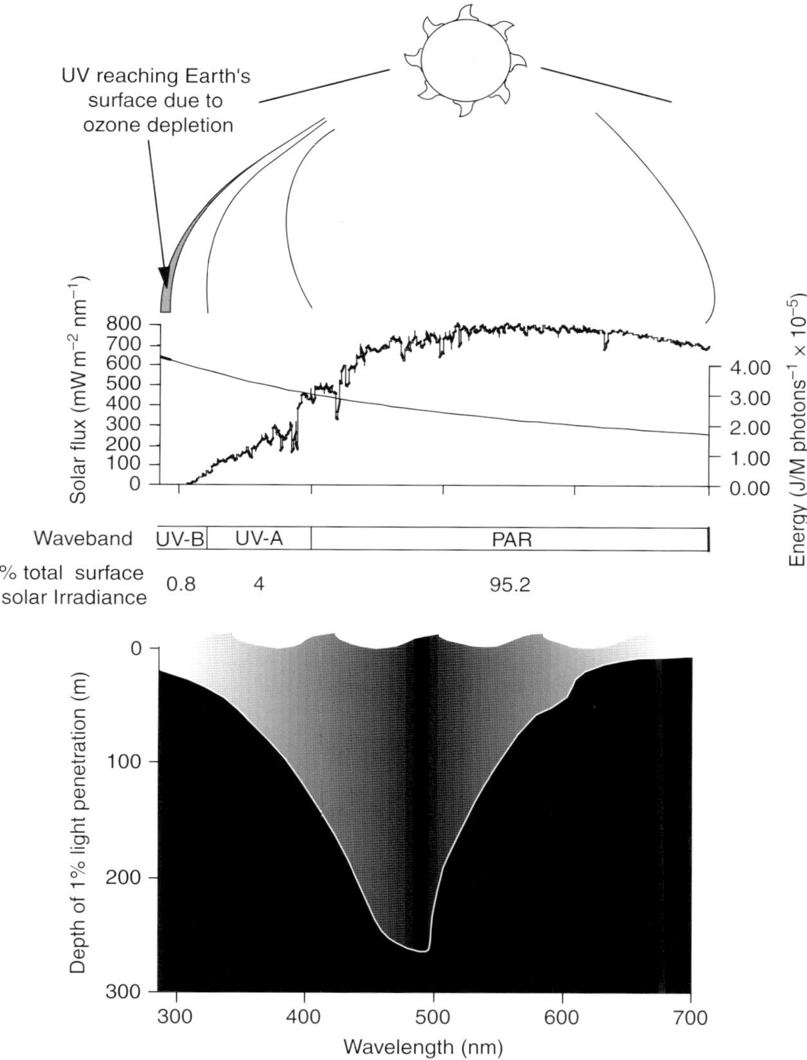

Fig. 22.2. The solar spectral flux, molar photon energy, wavelengths enhanced by ozone depletion and spectral penetration of light in the marine environment. Solar spectral flux was calculated from the UVSpec model for noon on the summer solstice for Davis Station, Antarctica, an albedo of 0.5 and column ozone of 300 Dobson units. Photon energy was calculated after Kirk (1994).

sunscreens, including mycosporine-like amino acids (MAAs), carotenoids and melanin. MAAs are produced by algae (phytoplankton and macrophytes) but can be transferred to higher organisms via the food chain, especially subtidal and intertidal invertebrates and fish, which concentrate and locate the MAAs to protect UV-B sensitive organelles and life stages (Hessen, 2003; Fig. 22.3). Other external features such as mucilage, membranes, the silica shell of diatoms (frustules), the theca of dinoflagellates and the scales of naked flagellates may protect unicellular organisms from UV-B exposure. The extracellular mucilage of colonies of the alga *Phaeocystis antarctica*

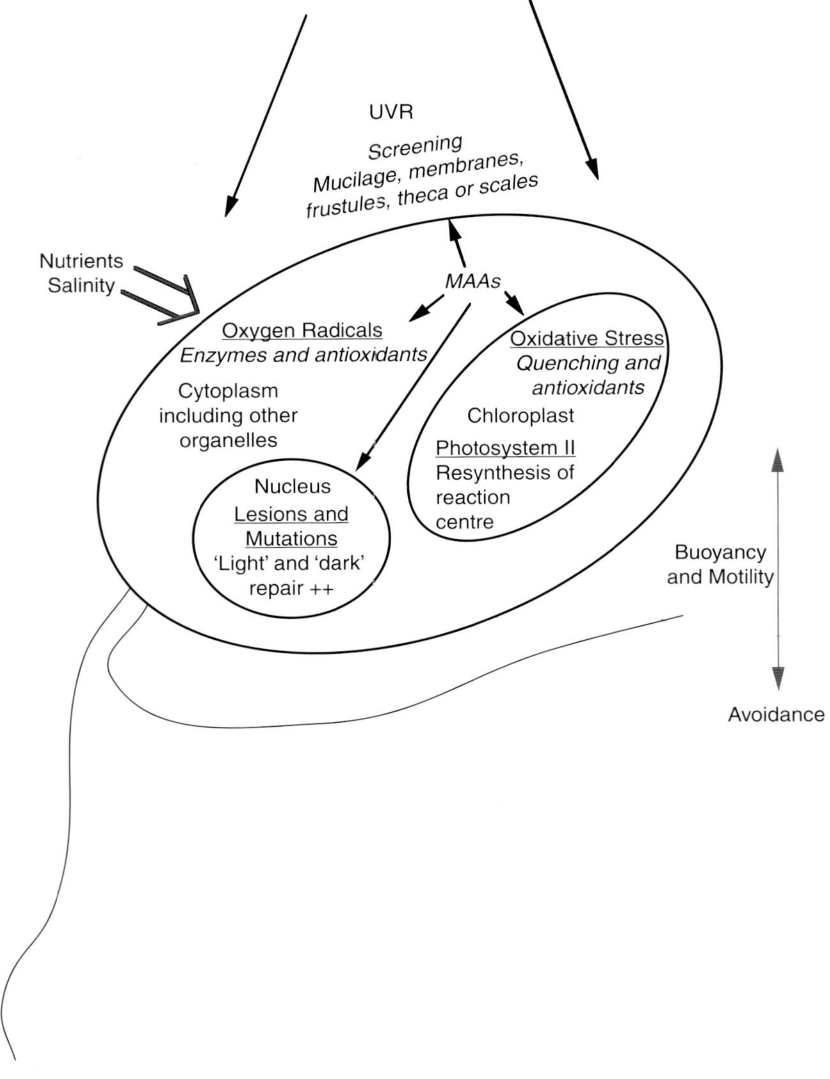

Fig. 22.3. Factors that influence the tolerance of microalgae to UV radiation (UVR) exposure. Underlined text indicates UV impacts on cells. Text in italics indicate tolerance strategies involved in acclimation. Mycosporine-like amino acids (MAAs) may act as screening agents and antioxidants. The nucleus appears to be the main site of DNA damage but extra-nuclear damage may also occur in chloroplasts and mitochrondria. Nutrients and salinity mediate rates of damage and repair, while buoyancy and motility may allow avoidance of UVR in some taxa and/or environments.

reportedly provides substantial protection against UV-B radiation (Marchant *et al.*, 1991), but protection of diatoms by their frustule is minimal (Davidson *et al.*, 1994). Carotenoids also absorb UV radiation but are primarily thought to protect against UV-B by acting as antioxidants (see below). Melanins occur across the animal kingdom from zooplankton to insects and vertebrates (including humans) and are used for both UV absorption and camouflage. Concentrations of these screening compounds are higher in organisms occupying high UV environments and commonly increase on exposure to UV-B radiation. While this suggests that these sunscreens play an important role in protecting against UV-B damage, the evidence for this is largely circumstantial.

The remaining two strategies are recovery mechanisms, which mitigate chemical damage in cells by quenching (neutralizing oxygen radicals), or by repairing damage to the DNA or photosynthetic systems (Fig. 22.3). A range of biomolecules that are critical to cell function are damaged by absorption of UV-B (e.g. DNA, RNA, enzymes, photosynthetic pigments, membranes and proteins including hormones and histones; Tevini, 1993). Excess energy from absorption of UV-B can generate damaging reactive oxygen species and free radicals (O_2^-, H_2O_2, $^1O_2$ and OH; Section 22.2; Kieber *et al.*, 2003). These radicals can be neutralized by antioxidants, including carotenoids and xanthophyll pigments, sulfur compounds (dimethylsulfoniopropionate, DMSP) and a range of enzymes (Sunda *et al.*, 2002; Hessen, 2003; Kieber *et al.*, 2003). Antioxidants have largely been studied in phytoplankton, but have also been reported from symbiotic dinoflagellates, corals and sea anemones (Banaszak, 2003). Like sunscreens, UV-B-induced increases in the concentrations of antioxidants suggest that they have a protective role, but the evidence for this is again circumstantial (Davidson, 2006).

Repair of UV-B-induced DNA damage has received much attention and exhaustive reviews of the processes involved are readily available elsewhere (e.g. Buma *et al.*, 2003). Briefly, studies show large differences among species in the amount of DNA damage sustained and the ability to repair this damage at a given UV-B irradiance. Damage to DNA may be repaired in three ways. The two main methods are 'light' and 'dark' repair (Fig. 22.3), but the relative importance of these in rectifying UV-B-induced DNA damage is unknown. Light repair (photoreactivation) occurs in organisms from bacteria to vertebrates and uses wavelengths of between 350 and 450 nm to repair UV-B-induced lesions that block DNA transcription. Dark repair uses a number of enzymes to remove and replace damaged portions of DNA. Post-replication repair may also be used, though its role in repairing DNA damage is unknown. Finally, studies, primarily on cyanobacteria, have shown that proteins at the core of the photosynthetic process are damaged by UV-B exposure, but that this is at least partly compensated for by UV-B-induced formation of oxygen radicals (above) that enhance the rate at which these proteins are resynthesized (Davidson, 2006).

Differences in the effectiveness of tolerance mechanisms cause large intra- and inter-specific variations in UV-B sensitivity that are mediated by environmental factors such as temperature, nutrients and salinity (Hessen, 2003). Karentz *et al.* (1991) showed DNA damage varied 100-fold among species and over 10-fold among members of the same genus, while Neale *et al.* (1998) measured differences in the sensitivity of phytoplankton production to UV exposure of ±46%. Exposure to UV-B radiation often causes an increase in the synthesis of protective pigments and enzymes that scavenge oxygen radicals, indicating that an organism's sensitivity to UV-B exposure is mediated by its light history. Thus, any rapid increase in UV-B exposure, such as the sudden disappearance of sea ice or the removal of a protective canopy of seaweeds by storms, is likely to be more damaging to organisms than a gradual increase in exposure. It should be considered that having the ability to adjust the level of tolerance to suit the UV irradiance precludes an organism having to spend energy maintaining an excessive level of protection or repair.

22.8 Responses of Aquatic Organisms to UV-B Radiation

The discovery of the ozone hole (Section 22.3) led to a proliferation of research into the effects of UV-B radiation on marine organisms, especially in Antarctic waters where ozone depletion is greatest.

These studies typically showed that high UV-B irradiances can damage marine organisms, but there are also reports showing that ambient or moderate UV-B irradiances have little effect, or may even benefit some organisms (Gustavson *et al.*, 2000; Thomson *et al.*, 2008).

22.8.1 Phytoplankton

The effect of UV-B on rates of primary production by phytoplankton, the base of the marine food web, has received the greatest scientific attention. This radiation can inactivate or destroy photosynthetic pigments and proteins, inhibiting phytoplankton production in near-surface waters by 6.4–60% (e.g. Karentz and Lutze, 1990) and can even inhibit production by 5–23% beneath sea-ice (Ryan and Beaglehole, 1994; Schofield *et al.*, 1995). Studies by Smith *et al.* (1992) and Helbling *et al.* (1994) indicated that ozone depletion of 33–50% reduced photosynthesis in the Antarctic marginal ice zone by 4–12%. Over the entire Southern Ocean, the estimated inhibition fell to <0.15% because of the rapid attenuation of UV-B radiation in the water column and strong attenuation of UV radiation by sea ice (Helbling *et al.*, 1994; Arrigo *et al.*, 2003). Thus, even severe ozone depletion is predicted to have a relatively small overall effect on phytoplankton production in Antarctic waters. Arctic phytoplankton are at least as sensitive as elsewhere (Helbling *et al.*, 1996). Recent climate warming has caused sea ice to become thinner and to melt earlier, potentially exposing them to higher UV-B irradiances (Häder *et al.*, 2006). However, there are apparently no published estimates of the inhibitory effect of ozone depletion on phytoplankton production in Arctic waters, or the compounding effect of climate change on this process. Some studies from other marine and freshwater systems have also reported little inhibition of photosynthesis by UV-B radiation, perhaps due to light history, high nutrients or a prevalence of UV-tolerant species (Hobson and Hartley, 1983; Gala and Giesy, 1991; Neale, 2000). Thus the responses of phytoplankton to UV-B radiation are not simply a function of the amount or rate of UV-B exposure. Counter-intuitively, some studies have even shown promotion of phytoplankton by moderate levels of UV-B (Thomson *et al.*, 2008). The mechanism(s) of this promotion are uncertain, but may involve tolerance mechanisms being enhanced by UV-B exposure (Section 22.7) or shortwave radiation being used in photosynthesis (Gao *et al.*, 2007).

Exposure of phytoplankton to UV-B radiation also reportedly reduces cell size and enzyme activities, concentrations of ATP, lipids and fatty acids, and the rates of production of amino acids, nutrient uptake and nitrogen use, cell division, motility, phototaxis, growth and survival (Davidson, 2006). Consequently, cells exposed to UV-B are usually less vigorous and competitive compared with other members of the community. Given the observed variation in sensitivity to UV-B among species (Section 22.7), it is unsurprising that studies show UV-B to have the potential to change the composition of phytoplankton communities (Davidson, 2006). Studies of individual phytoplankton taxa from Antarctic waters demonstrate that they differ in sensitivity to UV-B radiation, but caution needs to be exercised when interpreting such studies. Davidson and Marchant (1994) showed *Phaeocystis antarctica*, which is one of the most abundant phytoplankton taxa in Antarctic waters, contains high concentrations of MAAs and that cell size, growth and cell-specific rates of primary production are enhanced by exposure to UV-B radiation. In contrast, they found that *P. antarctica* dies at UV-B irradiances around half that of a range of Antarctic diatom species, which largely lack UV-absorbing compounds. Despite this, competition experiments showed that exposing mixed cultures to near-surface UV-B radiation for ≤2 days favours *P. antarctica* relative to diatoms (Davidson *et al.*, 1996). Thus, the apparent sensitivity of a species can vary according to the tolerance mechanism that is studied.

22.8.2 Benthic plants

Sessile aquatic plants such as macroalgae and seagrasses can suffer the same range of damage as phytoplankton and have similar tolerance and repair mechanisms. However, being attached in one location brings other challenges to escape UV-B exposure. Like phytoplankton, macroalgae and seagrasses can suffer DNA damage, oxidative stress and reduced growth and production when exposed to UV-B radiation. They also have similar mechanisms that allow them to tolerate this exposure, such as light avoidance, production of screening compounds, quenching of oxygen radicals and DNA repair, the effectiveness of which differs among species. Being anchored in one location, benthic plants cannot escape exposure to high UV-B irradiances. However, some benthic plants have thick epidermal layers, thick leaves or the capacity to sacrifice cells nearest to the sun to shield the cells beneath (Trocine *et al.*, 1981; Dawson and Dennison, 1996; Bischof *et al.*, 2002). Others can be protected by a covering of epiphytes or detritus on their surface and/or other aquatic plants growing nearby. Whatever the tolerance mechanism employed, studies show that UV-B tolerance differs among species and that acclimation to the ambient light climate results in increasing tolerance as the depth declines (Bischof *et al.*, 1998; van der Poll *et al.*, 2002). Thus UV-B helps define the depth zonation of sessile plants in aquatic systems, with their tolerance levels limiting the water depth in which a species can effectively compete for light and space with its neighbours.

22.8.3 Bacteria

Aquatic bacteria are also sensitive to UV-B exposure, which jeopardizes their vital roles in decomposing organic matter and recycling carbon and nutrients. As for phytoplankton, UV-B damages bacterial DNA and reduces rates of production, metabolism, growth and survival. It also reduces the production of ecto-enzymes, which are used by bacteria to break down complex food into simple molecules that they can absorb (Jeffrey *et al.*, 2000). Studies show that bacterial species differ greatly in sensitivity to UV-B radiation. Arrieta *et al.* (2000) showed differences in the sensitivity of different bacterial communities causing inhibition of bacterial production of between 21 and 92% for leucine uptake and 14 and 84% for thymidine uptake. But bacterial sensitivity to UV-B is mediated by the availability of substrates, with nutrient replete cells being less sensitive than cells that are starved (Aas *et al.*, 1996). UV-B-induced inhibition of phytoplankton could reduce the availability of resources for bacteria, making them more sensitive to damage. In contrast, other studies suggest that UV-B radiation may have little effect or be beneficial to aquatic bacteria as UV-B can cleave complex dissolved organic matter, making it more available to bacteria (Herndl *et al.*, 1997). Furthermore, UV-B-induced mortality of phytoplankton may increase the availability of substrate for bacterial growth (Davidson and van der Heijden, 2000). Bacteria also have light and dark repair mechanisms that enable them to recover rapidly from UV-B-induced damage to DNA (e.g. Kaiser and Herndl, 1997) and it is likely that the differences in their effectiveness are reflected in the large variation in UV-B sensitivity observed among bacterial species. Thus exposure to UV-B favours tolerant taxa, and this tolerance reduces the sensitivity of the bacterial community to damage (Piquet *et al.*, 2010). In addition, bacterial mortality could decline if UV-B is more damaging to viruses and bacterivorous Protozoa than to bacteria themselves.

22.8.4 Viruses

Viruses are an important cause of bacterial and phytoplankton mortality that act to maintain species and genetic diversity in plankton communities, but they too are inhibited or destroyed by UV-B radiation. UV-B exposure reduces viral abundance, viability and infectivity. Solar UV-B increases rates of viral decay by between 0.1 h^{-1} and 1.0 h^{-1} (Jeffrey *et al.*, 2000) and reduces their ability to infect host cells by between 30 and

92% (Cheng *et al.*, 2007). Like phytoplankton and bacteria (above), the depth and rate of mixing greatly affects the UV-B-induced DNA damage suffered by viruses. High rates of UV-B damage occur in shallow waters under calm conditions but no damage accumulates when mixing is significant (Weinbauer *et al.*, 1999). The light and dark DNA repair mechanisms of viral hosts also benefit viruses, restoring infectivity to between 40 and 80% of particles damaged by sunlight (Jeffrey *et al.*, 2000). They also appear to be more resistant to UV-B damage in summer than in winter, perhaps due to the greater effectiveness of the tolerance mechanisms of their hosts at this time of year. Other studies show that UV-B radiation can cause an increase in viral abundance, which has been attributed to UV-B-induced changes in viral infections that favour lysogeny (Jeffrey *et al.*, 2000). Thus it would appear that lysogeny and host-mediated repair maintain high rates of viral infectivity in aquatic systems despite exposure to UV-B radiation. This may in part be due to the surprising finding that viral infection can confer extra protection against UV-B damage to the hosts by infecting them with genes that encode for the repair of UV-B-induced DNA damage (Jacquet and Bratbak, 2003).

22.8.5 Animals

The effects of UV-B radiation on aquatic animals from Protozoa to vertebrates have received less attention than for other organisms, but animals also show adaptations to cope with UV radiation. The mechanisms that they use to tolerate UV-B are similar to those employed by single-celled organisms. Unicellular animals (protozoans) can be damaged by UV-B, reducing their growth, motility, production, grazing rates and survival (Sommaruga, 2003). Conversely, studies using entire microbial communities show UV-B radiation can benefit some Protozoa. UV-B-induced mortality of phytoplankton can increase the abundance of their food (dissolved organic carbon and/or bacterial cells), or kill grazers that eat small heterotrophs (Mostajir *et al.*, 2000; Davidson and Belbin, 2002). In higher animals, UV-B can damage DNA, suppress the immune system and reduce life span, growth, reproduction and survival. For example, in the estuary and Gulf of St Lawrence in Canada, natural UV-B irradiances may be responsible for 32% of the mortality of eggs produced by the copepod *Calanus finmarchicus* (Hessen, 2003). Natural irradiances are also high enough to cause reproductive problems, developmental abnormalities in nauplii, skin lesions, cataracts, and to reduce the spawning success of fish (Zagarese and Williamson, 2000; Hessen, 2003).

Multicellular aquatic animals, while using the tolerance strategies discussed above, have added extra abilities to their arsenal of UV-B tolerance mechanisms. Planktonic crustaceans and at least some fish larvae swim away from light, achieving vertical migrations that carry them a few to hundreds of metres deeper during the day. It is likely that these vertical migrations are primarily to avoid being eaten by visual predators such as fish (Hessen, 2003) but other studies show that UV radiation can be the primary driver of the vertical distributions of zooplankton such as *Daphnia* (water fleas) in high UV environments (Rautio and Tartarotti, 2010). Eggs and embryos of animals are especially sensitive to UV-B irradiation and it is possible that, while tolerating UV-B exposure as larvae and/or adults, their populations could be reduced by exposure to UV-B during early developmental stages. These early life stages are sometimes protected by pigments such as MAAs (Section 22.7). Alternatively, exposure to UV-B radiation can be avoided by timing reproduction to coincide with low UV irradiances, or by sensitive life stages occupying deep water (Hessen, 2003). Multicellular animals are also able to produce melanins (Section 22.7), which provide substantial protection against UV-B radiation. Melanin production is more prevalent in environments with higher UV irradiances but this pigment also causes dark coloration of organisms such as Arctic and alpine species of *Daphnia*, making them more visible to grazers than unpigmented (hyaline) strains of *Daphnia* of lower

UV-tolerance (Hessen, 2003). Thus tolerance mechanisms can come at a cost and, like maintaining an appropriate level of UV-B tolerance to avoid wasting energy (Section 22.7), the use of melanin can become a trade-off between reduced mortality due to UV-B and increased mortality owing to predation.

22.9 Effects of UV-B on Aquatic Communities

Aquatic organisms are linked by the food web, and the effects of UV-B radiation on the growth, production and survival of organisms at one trophic level are likely to have consequences for the organisms that directly or indirectly depend on them as food (Mostajir *et al.*, 2000; Sommaruga, 2003). Controlled experiments and field observations have identified the targets for UV-B within cells, the consequences of damage to cell metabolism and the potential for increased UV-B to favour UV-tolerant species. The potential effects of interactions among trophic levels were first demonstrated in experiments conducted by Bothwell *et al.* (1994). In their study, grazers were more inhibited by UV-B radiation than the phytoplankton that they consume. Consequently, exposure of the community to UV-B radiation caused a counter-intuitive increase in phytoplankton abundance due to the decline in grazing. Experiments show that UV-B radiation can change the composition and abundance of aquatic communities, and alter the relative biomass of different trophic levels (Mostajir *et al.*, 2000; Davidson and Belbin, 2002; Sommaruga, 2003; Thomson *et al.*, 2008). To date there is little evidence from natural environments of UV-B-induced changes to aquatic communities (Sommaruga, 2003). This is partly due to the inherent spatial and temporal variability of aquatic systems, making it difficult or impossible to detect UV-B-induced changes against a noisy background. It is also due to a range of methodological problems that limit or preclude extrapolation of findings to the natural environment (reviewed by Davidson, 2006). Determining the significance of increased UV-B radiation for natural aquatic systems is extremely difficult and the crucial question, 'What are the effects of enhanced solar UV-B on natural aquatic communities?' remains largely unanswered (Vernet, 2000; Davidson, 2006).

22.10 Conclusions

UV radiation is known to have effects on the survival and growth of a wide range of both terrestrial and aquatic organisms, and often those inhabiting extreme environments in polar regions. Multicellular organisms, such as animals and plants, are typically able to cope with exposure to UV radiation by either avoiding sunlight or using protective mechanisms, but microbes on Earth are particularly susceptible to the deleterious effects of UV owing to their frequent inability to synthesize sufficient amounts of protective pigments to absorb shortwave radiation entering their cells. Nevertheless, evidence in recent years indicates that lichenized fungal and algal cells survive extreme doses of UV radiation under space conditions on orbiting satellites (Sancho *et al.*, 2007). These remarkable findings suggest that viable microbes could be transferred between planets, thus supporting the panspermia hypothesis of Arrhenius (1903; see also Gómez *et al.*, Chapter 26, this volume).

References

Aas, P., Lyons, M.M., Pledger, R., Mitchell, D.L. and Jeffrey, W.H. (1996) Inhibition of bacterial activities by solar radiation in nearshore waters and the Gulf of Mexico. *Aquatic Microbial Ecology* 11, 229–238.

Albert, K.R., Mikkelsen, T.N. and Ro-Poulsen, H. (2008) Ambient UV-B radiation decreases photosynthesis in high arctic *Vaccinium uliginosum*. *Physiologia Plantarum* 133, 199–210.

Arrhenius, S. (1903) Die verbreitung des lebens im weltenraum. *Die Umschau* 7, 481–485.

Arrieta, J.M., Weinbauer, M.G. and Herndl, G.J. (2000) Interspecific variability in the sensitivity to UV radiation and subsequent recovery in selected isolates of marine bacteria. *Applied and Environmental Microbiology* 66, 1468–1473.

Arrigo, K.R., Lubin, D., van Dijken, G.L., Holm-Hansen, O. and Morrow, E. (2003) The impact of a deep ozone hole on Southern Ocean primary production. *Journal of Geophysical Research* 108, C5, 10.1029/2001JC001226, 23-1–19.

Banaszak, A.T. (2003) Photoprotective physiological and biochemical responses of aquatic organisms. In: Helbling, E.W. and Zagarese, H. (eds) *UV Effects in Aquatic Organisms and Ecosystems*. Comprehensive Series in Photochemistry and Photobiology Vol. 1. The Royal Society of Chemistry, Cambridge, UK, pp. 329–356.

Bancroft, B.A., Baker, N.J. and Blaustein, A.R. (2008) A meta-analysis of the effects of ultraviolet-B radiation and its synergistic interactions with pH, contaminants, and disease on amphibian survival. *Conservation Biology* 22, 987–996.

Belnap, J., Phillips, S.L., Flint, S.D., Money, J. and Caldwell, M.M. (2008) Global change and biological soil crusts: effects of ultraviolet augmentation under altered precipitation regimes and nitrogen additions. *Global Change Biology* 14, 670–686.

Bischof, K., Hanlet, D., Tüg, H., Karsten, U., Brouwer, P.E.M. and Weincke, C. (1998) Acclimation of brown algal photosynthesis to ultraviolet radiation in Arctic coastal waters (Spitzbergen, Norway). *Polar Biology* 20, 388–395.

Bischof, K., Peralta, G., Kräbs, G., van der Poll, W., Péréz-Lloréns, J.L. and Breeman, A.M. (2002) Effects of UVB radiation on canopy structure of *Ulva* communities from southern Spain. *Journal of Experimental Botany* 53, 1–11.

Björn, L.O., Callaghan, T.V., Gehrke, C., Gwynn-Jones, D., Lee, J.A., Johanson, U., Sonesson, M. and Buck, N.D. (1999) Effects of ozone depletion and increased ultraviolet-B radiation on northern vegetation. *Polar Research* 18, 331–337.

Bothwell, M.L., Sherbot, D. and Pollock, C.M. (1994) Ecosystem response to solar ultraviolet-B radiation: influence of trophic level interactions. *Science* 265, 97–100.

Buma, A.G.J., Boelen, P. and Jeffrey, W.H. (2003) UVR-induced damage in aquatic organisms. In: Helbling, E.W. and Zagarese, H. (eds) *UV Effects in Aquatic Organisms and Ecosystems*. Comprehensive Series in Photochemistry and Photobiology Vol. 1. The Royal Society of Chemistry, Cambridge, UK, pp. 291–327.

Cheng, K., Zhao, Y., Du, X., Zhang, Y., Lan, S. and Sho, Z. (2007) Solar radiation-driven decay of cyanophage infectivity and photoreactivation of cyanophage and host cyanobacteria. *Aquatic Microbial Ecology* 48, 13–18.

Convey, P., Pugh, P.J.A., Jackson, C., Murray, A.W.A., Ruhland, C.T., Xiong, F.S. and Day, T.A. (2002) Response of Antarctic terrestrial microarthropods to long-term climate manipulations. *Ecology* 83, 3130–3140.

Davidson, A.T. (2006) Effects of ultraviolet radiation on microalgal growth, survival and production. In: Rao, S.D.V. (ed.) *Algal Cultures, Analogues of Blooms and Application*. Science Publishers Inc., New Hampshire, pp. 715–767.

Davidson, A.T. and Belbin, L. (2002) Exposure of Antarctic marine microbial communities to ambient UV radiation: effects on protistans. *Aquatic Microbial Ecology* 27, 159–174.

Davidson, A.T. and Marchant, H.J. (1994) Impact of ultraviolet radiation on *Phaeocystis* and selected species of Antarctic marine diatoms. In: Weiler, C.S. and Penhale, P.A. (eds) *Ultraviolet Radiation in Antarctica: Measurement and Biological Effects*. Antarctic Research Series 62, American Geophysical Union, Washington, DC, pp. 160–187.

Davidson, A.T. and van der Heijden, A. (2000) Exposure of Antarctic marine microbial communities to ambient UV radiation: effects on bacterioplankton. *Aquatic Microbial Ecology* 21, 257–264.

Davidson, A.T., Bramich, D., Marchant, H.J. and McMinn, A. (1994) Effects of UV-B radiation on growth and survival of Antarctic marine diatoms. *Marine Biology* 119, 507–515.

Davidson, A.T., Marchant, H.J. and de la Mare, W.K. (1996) Natural UVB exposure changes the species composition of Antarctic phytoplankton in mixed culture. *Aquatic Microbial Ecology* 10, 299–305.

Dawson, S.P. and Dennison, W.C. (1996) Effects of ultraviolet and photosynthetically active radiation on five seagrass species. *Marine Biology* 125, 629–638.

Day, T.A., Ruhland, C.T. and Xiong, F. (2001) Influence of solar ultraviolet-B radiation on Antarctic terrestrial plants: results from a 4-year field study. *Journal of Photochemistry and Photobiology B: Biology* 62, 78–87.

de Gruijl, F.R., Longstreth, J., Norval, M., Cullen, A.P., Slaper, H., Kripke, M.L., Takizawa, Y. and van der Leun, J.C. (2003) Health effects from stratospheric ozone depletion and interactions with climate change. *Photochemical and Photobiological Sciences* 2, 16–28.

Farman, J.C., Gardiner, B.G. and Shanklin, J.D. (1985) Large losses of total ozone in Antarctica reveal seasonal ClOx/NOx interaction. *Nature* 315, 207–210.

Fiscus, E.L. and Booker, F.L. (1995) Is increased UV-B a threat to crop photosynthesis and productivity? *Photosynthesis Research* 43, 81–92.

Gala, W.R. and Giesy, J.P. (1991) Effects of ultraviolet radiation on the primary production of natural phytoplankton assemblages in Lake Michigan. *Ecotoxicology and Environmental Safety* 22, 345–361.

Gao, K., Wu, Y., Li, G., Wu, H., Villafane, V.E. and Helbling, E.W. (2007) Solar UV radiation drives CO_2 fixation in marine phytoplankton: a double edged sword. *Plant Physiology* 144, 54–59.

Garcia-Pichel, F. (1994) A model for self-shading in planktonic organisms and its implications for the usefulness of ultraviolet sunscreens. *Limnology and Oceanography* 39, 1704–1717.

Gehrke, C., Johanson, U., Callaghan, T.V., Chadwick, D. and Robinson, C.H. (1995) The impact of enhanced ultraviolet-B radiation on litter quality and decomposition processes in *Vaccinium* leaves from the subarctic. *Oikos* 72, 213–222.

Gieskes, W.W.C. and Kraay, G.W. (1990) Transmission of ultraviolet light in the Weddell Sea: report of the first measurements made in the Antarctic. *BIOMASS Newsletter* (College Station, Texas) 12, 12–14.

Gustavson, K., Garde, K., Wängberg, S.-Å. and Selmer, J.-S. (2000) Influence of UV-B radiation on bacterial activity in coastal waters. *Journal of Plankton Research* 22, 1501–1511.

Häder, D.-P. (1986) Effects of solar and artificial UV radiation on motility and phototaxis in the flagellate *Euglena gracilis. Journal of Photochemistry and Photobiology B: Biology* 44, 651–656.

Häder, D.-P., Kumar, H.D., Smith, R.C. and Worrest, R.C. (2006) Effects of UV radiation on aquatic ecosystems and interactions with climate change. *The Environmental Effects Assessment Panel Report for 2006*, UNEP, Nairobi, Kenya, pp. 95–134.

Helbling, E.W., Villafañe, V.E. and Holm-Hansen, O. (1994) Effects of ultraviolet on Antarctic marine photosynthesis with particular attention to the influence of mixing. In: Weiler, C.S. and Penhale, P.A. (eds) *Ultraviolet Radiation in Antarctica: Measurement and Biological Effects*. Antarctic Research Series 62, American Geophysical Union, Washington, DC, pp. 207–228.

Helbling, E.W., Eilertsen, H.C., Villafañe, V.E. and Holm-Hansen, O. (1996) Effects of UV radiation on post-bloom phytoplankton populations in Kvalsund, North Norway. *Journal of Photochemistry and Photobiology B: Biology* 33, 255–259.

Herndl, G.J., Brugger, A., Hager, S., Kaiser, D., Obernosterer, I., Reitner, B. and Slezak, D. (1997) Role of ultraviolet-B radiation on bacterioplankton and the availability of dissolved organic matter. *Plant Ecology* 128, 42–51.

Hessen, D.O. (2003) UVR and pelagic metazoans. In: Helbling, E.W. and Zagarese, H. (eds) *UV Effects in Aquatic Organisms and Ecosystems*. Comprehensive Series in Photochemistry and Photobiology, Vol. 1. The Royal Society of Chemistry, Cambridge, UK, pp. 399–430.

Hobson, L.A. and Hartley, F.A. (1983) Ultraviolet radiation and primary production in a Vancouver Island fjord, British Columbia, Canada. *Journal of Plankton Research* 5, 325–331.

Jacquet, S. and Bratbak, G. (2003) Effects of ultraviolet radiation on marine virus-phytoplankton interaction. *FEMS Microbiology Ecology* 44, 279–289.

Jeffrey, W.H., Kase, J.P. and Wilhelm, S.W. (2000) UV radiation effects on heterotrophic bacterioplankton and viruses in marine ecosystems. In: de Mora, S., Demers, S. and Vernet, M. (eds) *The Effects of UV Radiation in the Marine Environment*. Cambridge University Press, Cambridge, UK, pp. 237–278.

Jerlov, N.G. (1950) Ultraviolet radiation in the sea. *Nature* 116, 111–112.

Kaiser, E. and Herndl, G.J. (1997) Rapid recovery of marine bacterioplankton activity after inhibition by UV radiation in coastal waters. *Applied and Environmental Microbiology* 63, 4026–4031.

Karentz, D. and Lutze, L.H. (1990) Evaluation of biologically harmful ultraviolet radiation in Antarctica with a biological dosimeter designed for aquatic environments. *Limnology and Oceanography* 35, 549–561.

Karentz, D., Cleaver, J.E. and Mitchell, D.L. (1991) Cell survival characteristics and molecular responses of Antarctic phytoplankton to ultraviolet-B radiation. *Journal of Phycology* 27, 326–341.

Kieber, D.J., Peake, B.M. and Scully, N.M (2003) Reactive oxygen species in aquatic ecosystems. In: Helbling, E.W. and Zagarese, H. (eds) *UV Effects in Aquatic Organisms and Ecosystems*. Comprehensive Series in Photochemistry and Photobiology, Vol. 1. The Royal Society of Chemistry, Cambridge, UK, pp. 251–290.

Kirk, J.T.O. (1994) *Light and Photosynthesis in Aquatic Ecosystems*. Cambridge University Press, Cambridge, UK, 509 pp.

Li, F.-R., Peng, S.-L., Chen, B.-M. and Hou, Y.-P. (2010) A meta-analysis of the responses of woody and herbaceous plants to elevated ultraviolet-B radiation. *Acta Oecologia* 36, 1–9.

Marchant, H.J., Davidson, A.T. and Kelly, G.J. (1991) UV-B protecting compounds in the marine alga *Phaeocystis pouchetii* from Antarctica. *Marine Biology* 109, 391–395.

Matear, R. and Hirst, A.C. (1999) Climate change feedback on the future oceanic CO_2 uptake. *Tellus* 51, 722–733.

Mazza, C.A., Zavala, J., Scopel, A.L. and Ballaré, C.L. (1999) Perception of solar UVB radiation by phytophagous insects: behavioral responses and ecosystem implications. *Proceedings of the National Academy of Sciences of the United States of America* 96. 980–985.

Mostajir, B., Demers, S., de Mora, S.J., Bukata, R.P. and Jerome, J.H. (2000) Implications of UV radiation for the food web structure and consequences on the carbon flow. In: de Mora, S., Demers, S. and Vernet, M. (eds) *The Effects of UV Radiation in the Marine Environment*. Cambridge Environmental Chemistry Series, Vol. 10. Cambridge University Press, Cambridge, UK, pp. 310–320.

Neale, P.J. (2000) Spectral weighting functions for quantifying effects of UV radiation in marine ecosystems. In: de Mora, S., Demers, S. and Vernet, M. (eds) *The Effects of UV Radiation in the Marine Environment*. Cambridge Environmental Chemistry Series, Vol. 10. Cambridge University Press, Cambridge, UK, pp. 72–100.

Neale, P.J., Davis, R.F. and Cullen, J.J. (1998) Interactive effects of ozone depletion and vertical mixing on photosynthesis of Antarctic phytoplankton. *Nature* 392, 585–589.

Neale, P.J., Helbling, E.W. and Zagarese, H.E. (2003) Modulation of UVR exposure and effects by vertical mixing and advection. In: Helbling, E.W. and Zagarese, H. (eds) *UV Effects in Aquatic Organisms and Ecosystems*. Comprehensive Series in Photochemistry and Photobiology, Vol. 1. The Royal Society of Chemistry, Cambridge, UK, pp. 107–136.

Newsham, K.K. and Robinson, S.A. (2009) Responses of plants in polar regions to UV-B exposure: a meta-analysis. *Global Change Biology* 15, 2574–2589.

Newsham, K.K., Anderson, J.M., Sparks, T.H., Splatt, P., Woods, C. and McLeod, A.R. (2001) UV-B effect on *Quercus robur* leaf litter decomposition persists over four years. *Global Change Biology* 7, 479–483.

Newsham, K.K., Hodgson, D.A., Murray, A.W.A., Peat, H.J. and Smith, R.I. (2002) Response of two Antarctic bryophytes to stratospheric ozone depletion. *Global Change Biology* 8, 972–983.

Phoenix, G.K., Gwynn-Jones, D., Callaghan, T.V., Sleep, D. and Lee, J.A. (2001) Effects of global change on a sub-Arctic heath: effects of enhanced UV-B radiation and increased summer precipitation. *Journal of Ecology* 89, 256–267.

Piquet, A.M.-T., Bolhuis, H., Davidson, A.T. and Buma, A.G.J. (2010) Seasonal succession and UV sensitivity of marine bacterioplankton at an Antarctic coastal site. *FEMS Microbiology Ecology* 73, 68–82.

Rautio, M. and Tartarotti, B. (2010) UV radiation and freshwater zooplankton: damage, protection and recovery. *Freshwater Reviews* 3, 105–131.

Rinnan, R., Keinänen, M.M., Kasurinen, A., Asikainen, J., Kekki, T.K., Holopainen, T., Ro-Poulsen, H., Mikkelsen, T.N. and Michelsen, A. (2005) Ambient ultraviolet radiation in the Arctic reduces root biomass and alters microbial community composition but has no effects on microbial biomass. *Global Change Biology* 11, 564–574.

Rozema, J., Tosserams, M., Nelissen, H.J.M., van Heerwaarden, L., Broekman, R.A. and Flierman, N. (1997) Stratospheric ozone reduction and ecosystem processes: enhanced UV-B radiation affects chemical quality and decomposition of leaves of the dune grassland species *Calamagrostis epigeios*. *Plant Ecology* 128, 284–294.

Ryan, K.G. and Beaglehole, D. (1994) Ultraviolet radiation and bottom-ice algae: laboratory and field studies from McMurdo Sound, Antarctica. In: Weiler, C.S. and Penhale, P.A. (eds) *Ultraviolet Radiation in Antarctica: Measurements and Biological Effects*. American Geophysical Union, Antarctic Research Series, Vol. 62, Washington, DC, pp. 229–242.

Ryan, K.G., Burne, A. and Seppelt, R.D. (2009) Historical ozone concentrations and flavonoid levels in herbarium specimens of the Antarctic moss *Bryum argenteum*. *Global Change Biology* 15, 1694–1702.

Sancho, L.G., de la Torre, R., Horneck, G., Ascaso, C., de los Rios, A., Pintado, A., Wierzchos, J. and Schuster, M. (2007) Lichens survive in space: results from the 2005 LICHENS experiment. *Astrobiology* 7, 443–454.

Sarmiento, J.L. and LeQuéré, C. (1996) Oceanic carbon dioxide uptake in a model of century-scale global warming. *Science* 274, 1346–1350.

Schofield, O., Kroon, B.M.A. and Prézelin, B.B. (1995) Impact of ultraviolet-B radiation on photosystem II activity and its relationship to the inhibition of carbon fixation rates for Antarctic ice algal communities. *Journal of Phycology* 31, 703–715.

Searles, P.S., Flint, S.D. and Caldwell, M.M. (2001) A meta-analysis of plant field studies simulating stratospheric ozone depletion. *Oecologia* 127, 1–10.

Semerdjieva, S.I., Sheffield, E., Phoenix, G.K., Gwynn-Jones, D., Callaghan, T.V. and Johnson, G.N. (2003) Contrasting strategies for UV-B screening in sub-Arctic dwarf shrubs. *Plant, Cell and Environment* 26, 957–964.

Smith, R.C., Prézelin, B.B., Baker, K.S., Bidigare, R.R., Boucher, N.P., Coley, T., Karentz, D., MacIntyre, S., Matlick, H.A., Menzies, D., Ondrusek, M., Wan, Z. and Waters, K.J. (1992) Ozone depletion: ultraviolet radiation and phytoplankton biology in Antarctic waters. *Science* 255, 952–959.

Solheim, B., Zielke, M., Bjerke, J.W. and Rozema, J. (2006) Effects of enhanced UV-B radiation on nitrogen fixation in Arctic ecosystems. *Plant Ecology* 182, 109–118.

Sommaruga, R. (2003) UVR and its effects on species interactions. In: Helbling, E.W. and Zagarese, H. (eds) *UV Effects in Aquatic Organisms and Ecosystems*. Comprehensive Series in Photochemistry and Photobiology, Vol. 1. The Royal Society of Chemistry, Cambridge, UK, pp. 485–508.

Sunda, W., Kieber, D.J., Kiene, R.P. and Huntsman, S. (2002) An antioxidant function for DMSP and DMS in marine algae. *Nature* 418, 317–320.

Tevini, M. (1993) Molecular biological effects of ultraviolet radiation. In: Tevini, M. (ed.) *UV-B Radiation and Ozone Depletion: Effects on Humans, Animals, Plants, Microorganisms and Materials*. Lewis Publishers, Boca Raton, Florida, pp. 1–15.

Thomson, P.G., Davidson, A.T. and Cadman, N. (2008) Seasonal changes in effects of ambient UVR on natural communities of Antarctic marine protists. *Aquatic Microbial Ecology* 52, 131–147.

Trocine, R.P., Rice, J.D. and Wells, G.N. (1981) Inhibition of seagrass photosynthesis by ultraviolet-B radiation. *Plant Physiology* 68, 74–81.

van der Poll, W.H., Eggert, A., Buma, A.G.J. and Breeman, A.M. (2002) Temperature dependence of UV effects in Arctic and temperate isolates of three red macrophytes. *Journal of Phycology* 37, 59–68.

Vernet, M. (2000) Effects of UV radiation on the physiology and ecology of marine phytoplankton. In: de Mora, S., Demers, S. and Vernet, M. (eds) *The Effects of UV Radiation in the Marine Environment*. Cambridge Environmental Chemistry Series, Vol. 10. Cambridge University Press, Cambridge, UK, pp. 237–278.

Weinbauer, M.G., Wilhelm, S.W., Suttle, C.A., Pledger, R.J. and Mitchell, D.L. (1999) Sunlight-induced DNA damage and resistance in natural viral communities. *Aquatic Microbial Ecology* 17, 111–120.

World Meteorological Organization (2007) *Scientific Assessment of Ozone Depletion: 2006*. Global Ozone Research and Monitoring Project – Report No. 50, Geneva, Switzerland.

Zagarese, H.E. and Williamson, C.E. (2000) Impact of solar UV on zooplankton and fish. In: de Mora, S., Demers, S. and Vernet, M. (eds) *The Effects of UV Radiation in the Marine Environment*. Cambridge Environmental Chemistry Series, Vol. 10. Cambridge University Press, Cambridge, UK, pp. 279–309.

23 Life in a Changing Climate

Roland Psenner and Birgit Sattler
University of Innsbruck, Institute of Ecology, Austria

23.1 Introduction

The question of life in a changing climate is a matter of scales and perspectives. Depending on temporal and spatial dimensions the answer can be quite different. What is, for instance, the level of biodiversity we look at: shall we consider genes, species, habitats, or ecosystems? Which extreme ecosystems – certainly neither the deep biosphere nor hydrothermal vents – will particularly be affected by climate change? Are we interested in microorganisms with a wide distribution range, or in plants and animals with restricted ranges? What is the relevant time period: recent warming, the Holocene, the sequence of glaciations and interstadials, or do we consider even longer periods in the evolution of life? Shall we focus on air temperature, precipitation and deposition of nutrients, metals and organic pollutants, or is it more relevant to consider changes in marine and air currents? How important are invasive species, inclusively that of humans? Are we concerned with the abiotic or the biotic extremes potentially influenced?

23.2 Will Extreme Species, Communities, Habitats and Ecosystems be Affected by Climate Change?

The debate about how biodiversity may influence stability, resilience and performance of ecosystems (Chapin *et al.*, 2000) is far from being settled, but it is generally assumed that biodiversity is pivotal to the functioning of ecosystems after strong perturbations. So the central question of this chapter is: will extreme species, communities, habitats and ecosystems affected by climate change increase in numbers, size and productivity, or must we expect negative consequences, or even extinction? Up to now, we have been used to raising hypotheses first and then to substantiate or refute them by experiments, but in the era of the so-called data intensive science, these steps are inverted and theories are created by a sophisticated treatment of immense datasets (Ozdemir *et al.*, 2011). Scientifically more embarrassing, however, are requests to anticipate future developments, as suggested by the title of this chapter; and confronted with the complexity of climate and ecosystems, we may never know enough to take action, but we feel that we must act in time to prevent problems of global dimensions.

It is sensible, therefore, to look at the recent past and the most conspicuous species. Parmesan (2006) studied the effects of recent climate change, with a focus on plant and animal phenology, and states in his review that species with a restricted range, such as polar and mountaintop species, seem to be extremely vulnerable to climate warming because of severe range contractions, consequently entire species have gone extinct due to recent climate change. Based on his observations, the spatial focus on alpine and polar ecosystems makes sense, not only because there the effects of climate change appear to be most severe, but also because the alpine and nival (of, relating to, or growing in or under snow) zone extends all over the world, from 80 °N to 70 °S, with elevations from close to sea level at the poles to >4000 m at the equator. Sattler *et al.* (Chapter 8, this volume) argue that it is these particular environments that will vanish with rising temperatures and associated abiotic factors, but increasing availability of liquid water may, for some time, also enhance density and productivity of adapted, mostly prokaryotic, populations, enhancing the biotic pressures, such as competition for resources, experienced by organisms in the short term. Gornitz (2009) lists seven important cryospheric elements with a surface of >140 million km^2 at their maximum extension, so the consideration of cold ecosystems is – not only because of their feedback effects on climate – of global importance (Anesio *et al.*, 2009; see also Chapters 3 through 10, this volume). In the long run, snow and ice areas will shrink considerably and may be reduced to retreats or refuges for species and communities unable to adapt to warmer conditions. Recent and ongoing changes in cold environments, however, may also depend on other factors related to global change, such as high depositions of nitrogen, particularly in the Arctic and large regions of the northern hemisphere (Hodson *et al.*, 2005; Phoenix *et al.*, 2006).

In many articles and in most of the chapters cited above, the temporal scale is very short – mainly the last decades – and the focus is on air temperature warming. From an aquatic perspective, temperature affects a plethora of physical and chemical constants. Viscosity, for instance, changes twofold if water warms or cools by 20°C, and the ion product of water is ten times smaller at 0°C compared to its value at 25°C. So even minute changes in water temperature will influence conditions crucial for microorganisms, for instance the flux of nutrients to the surface of plankton cells (where viscosity dominates molecular diffusion), and virtually all chemical equilibria depending on pH. Thus, the influence of even minor temperature changes on biogeochemical processes can hardly be overestimated. The metabolic theory of ecology (Brown *et al.*, 2004), based on the known Arrhenius equation ($k = A\ e^{-E/RT}$), aims at explaining not only productivity and abundance of organisms, but also species diversity by two principles, i.e. temperature and body mass. In their formula, $\ln (P\ e^{E/kT}) = ¾ \ln (M) + C$, the logarithm of the temperature-corrected biomass production rate P scales with ~0.75 of the logarithm of the body mass M over 40 orders of magnitude (in this equation, k is the Boltzmann constant). This simple law states that temperature rules everything alive, from animal behaviour to biodiversity, and consequently ecosystem processes and services on which we unconditionally depend. Interestingly, in aquatic ecosystems global warming seems to favour the smaller species, from bacteria to fish (Daufresne *et al.*, 2009). If this also applies to other ecosystems and species, we may see a general productivity boost since, according to the formula of Brown *et al.* (2004), both body mass and temperature are the relevant factors for productivity – unless other factors become limiting.

Today, several hundred articles describe the short-term effects of recent global warming on species and populations (Parmesan, 2006 and references therein) and potential mismatches between prey and predators (Edwards and Richardson, 2004; Lewandowska and Sommer, 2010), but little is known about the long-term effects of climate change. Therefore, we will discuss two aspects of climatic impacts on cold ecosystems: (i) the effects of Holocene climate oscillations on aquatic invertebrate species; and (ii) the still enigmatic release of enormous amounts of heavy metals, particularly nickel, from melting permafrost.

23.3 The Effects of Holocene Climate Oscillations on Aquatic Invertebrate Species

Ilyashuk *et al.* (2011) reconstructed summer temperatures and productivity of a small Alpine lake situated at nearly 2800 m altitude in the Eastern Alps, by studying the remains of chironomids (midges) deposited in the sediment since the end of the last ice age. The authors showed that July temperatures during the early Holocene were more than 4°C warmer than today and that a rapid cooling began between 5000 to 4500 years before present. The warm-loving *Micropsectra radialis*-type dominated the community of benthic chironomids soon after the origin of the lake more than 10,000 years ago (Fig. 23.1). It did so with some interruptions, for instance during the well-known 8.2 Ka cold event, until ~4500 years before present when the cold-loving species (*Pseudodiamesa nivosa*-type) re-appeared and remained in the lake until present times. The ups and downs of species adapted to warmer/colder or nutrient-rich/oligotrophic conditions show at least five things: (i) a dominant species, such as *P. nivosa*, disappears from the sediment record for millennia, but reappears when its environmental conditions improve; (ii) *M. radialis* never disappears completely, but is reduced to very low numbers during cold periods; (iii) other species occur for some time, disappear, and never come back; (iv) temperature oscillations during the Holocene were large if compared to recent climate change, but the input of organic matter and, as a consequence, lake productivity also showed strong fluctuations – perhaps an interesting parallel to recent warming with its concomitant increase in nutrient deposition; and (v) the driving forces of climate change may have changed since the beginning of the Anthropocene *sensu* Crutzen (Zalasiewicz *et al.*, 2010), but the sheer dimension of natural variations in the past seems to be at least equivalent to, or larger than, recent climate change.

23.4 The Release of Heavy Metals from Melting Permafrost

Thies *et al.* (2007) detected a phenomenon that still needs explanation, i.e. the release of ions such as sulfate, calcium and magnesium, from melting permafrost that may occur in the Alps at elevations above 2500 m. In a small lake in the Eastern Alps, situated at 2682 m altitude at the feet of a rock glacier, sulfate and magnesium concentrations have increased more than 20-fold in the course of a decade. More discomforting, however, is the fact that the concentration of nickel has also increased tremendously, exceeding drinking water standards by more than an order of magnitude, while a small pond nearby, not influenced by permafrost, has metal concentrations below detection limit. It is obvious that sulfate, base cations and metals are released from melting permafrost, but the mechanisms of accumulation are hitherto unclear. A survey of other lakes with permafrost or rock glaciers in their catchment showed elevated concentrations of nickel (R. Psenner, 2011, unpublished results), but the analysis of ice cores from rock glaciers is still under way. In the meantime, what we know is that benthic invertebrates (chironomids) show deformations of their mouth parts, a clear signal of metal pollution (E. Ilyashuk, 2011, unpublished results).

23.5 Conclusion

In our opinion, these two examples are an indication that climate change can have large and unpredictable effects on cold environments. Ice ecosystems in particular (as discussed for instance in Sattler *et al.*, Chapter 8, this volume) are supposed to undergo the most severe transformations since even a slight temperature increase may change the aggregate state of water in permafrost soils, glaciers and snow. Thus, in the near future we may see more ecological 'surprises' (Lindenmayer *et al.*, 2010) as climate change continues to alter, both positively and negatively, the 'extremes' that organisms from a wide variety of environments face.

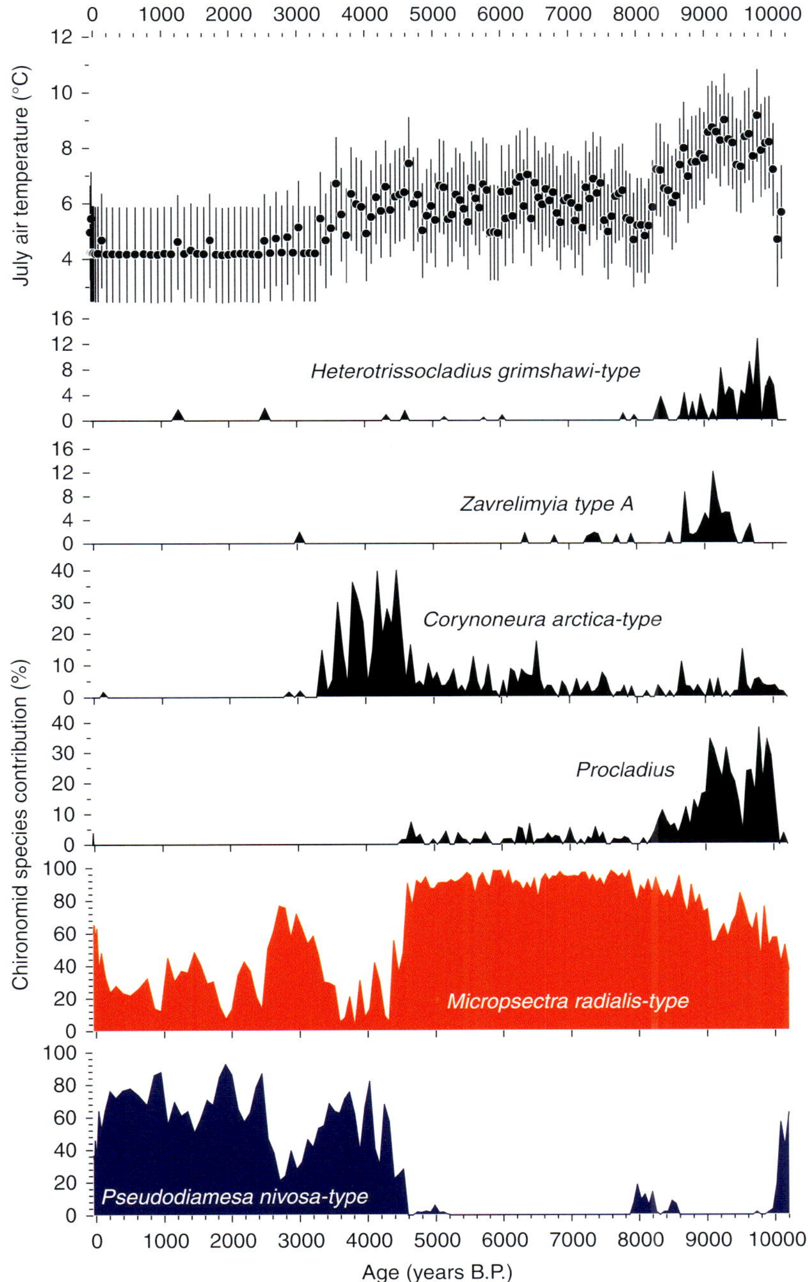

Fig. 23.1. Percentage abundance of selected chironomid taxa in the sediment record of Schwarzsee ob Sölden, Austria, and inferred July air temperature with standard error. Only species that reach >10% abundance in at least one sample are shown. The cold-loving species *Pseudodiamesa nivosa*-type is absent for 5 millennia, except during the 8.2 Ka cold spell, but is almost continuously dominant since the middle of the Holocene. The warm-loving *Micropsectra radialis*-type dominates the assemblage until ~4500 BP when the climate deteriorates significantly. Modified from Ilyashuk *et al.*, 2011.

References

Anesio, A.M., Sattler, B., Hodson, A.J., Fritz, A. and Psenner, R. (2009) High microbial activities on glaciers: importance to the global cycle. *Global Change Biology*, doi: 10.1111/j.1365-2486.2008.01758.x.

Brown, J.H., Gillooly, J.F., Allen, A.P., Savage, V.M. and West, G.B. (2004) Towards a metabolic theory of ecology. *Ecology* 85, 1771–1789.

Chapin, S.F., Zavaleta, E.S., Eviner, V.T., Naylor, R.L., Vitousek, P.M., Reynolds, H.L., Hooper, D.U., Lavorel, S., Sala, O.E., Hobbie, S.E., Mack, M.C. and Díaz, S. (2000) Consequences of changing biodiversity. *Nature* 405, 234–242.

Daufresne, M., Lengfellner, K. and Sommer, U. (2009) Global warming benefits the small in aquatic ecosystems. *Proceedings of the National Academy of Sciences of the United States of America* 106, 12788–12793.

Edwards, M. and Richardson, A.J. (2004) Impact of climate change on marine pelagic phenology and trophic mismatch. *Nature* 430, 881–884.

Gornitz, V. (2009) *Encyclopedia of Paleoclimatology and Ancient Environments*. Springer-Verlag, Heidelberg, Germany, 1049 pp.

Hodson, A.J., Mumford, P.N., Kohler, J. and Wynn, P.M. (2005) The High Arctic glacial ecosystem: new insights from nutrient budgets. *Biogeochemistry* 72, 233–256.

Ilyashuk, E., Koinig, K.A., Heiri, O., Ilyashuk, B.P. and Psenner, R. (2011) Holocene temperature variations at a high-altitude site in the Eastern Alps: a chironomid record from Schwarzsee ob Sölden, Austria. *Quarternary Science Reviews* 30, 176–191.

Lewandowska, A. and Sommer, U. (2010) Climate change and the spring bloom: a mesocosm study on the influence of light and temperature on phytoplankton and mesozooplankton. *Marine Ecology Progress Series* 405, 101–111.

Lindenmayer, D.B., Likens, G.E., Krebs, C.J. and Hobbs, R.J. (2010) Improved probability of detection of ecological 'surprises'. *Proceedings of the National Academy of Sciences of the United States of America* 107, 21957–21962.

Ozdemir, V., Smith, C., Bongiovanni, K., Cullen, D., Knoppers, B.M., Lowe, A., Peters, M., Robbins, R., Stewart, E., Yee, G., Yi-Kuo, Yu. and Kolker, E. (2011) Policy and data-intensive scientific discovery in the beginning of the 21st century. *OMICS: A Journal of Integrative Biology* 15, 221–225. doi:10.1089/omi.2011.0007.

Parmesan, C. (2006) Ecological and evolutionary responses to recent climate change. *Annual Review of Ecology, Evolution, and Systematics* 37, 637–669.

Phoenix, G.K., Hicks, W.K., Cinderby, S., Kuylenstierna, J.C.I., Stocks, W.D., Dentener, F.J., Giller, K.E., Austin, A.T., Lefroy, R.D.B., Gimeno, B.S., Ashmore, M.R. and Ineson, P. (2006) Atmospheric nitrogen deposition in world biodiversity hotspots: the need for a greater global perspective in assessing N deposition impacts. *Global Change Biology* 12, 470–476.

Thies, H., Nickus, U., Tessadri, R. and Psenner, R. (2007) Unexpected response of high Alpine lake waters to climate warming. *Environmental Science and Technology* 41, 7424–7429.

Zalasiewicz, J., Williams, M., Steffen, W. and Crutzen, P. (2010) The new world of the Anthropocene. *Environmental Science and Technology* 44, 2228–2231.

24 Anthropogenic Extreme Environments

Adrian G. Glover and Lenka Neal
Zoology Department, The Natural History Museum, London, UK

24.1 Introduction

Humans would appear to have evolved a fairly successful strategy to deal with extremes: mollify the local living space at the expense of that outside it. Using seal fat to keep the body warm. Constructing a hut from whale bones and skins. Removing forests and growing monocultures to concentrate food supply. Burning hydrocarbon reservoirs for energy. Humans deal with extremes by creating extremes. Over time we have increased our mollified, 'extreme-free' bubble of living space to an extent where there is little 'outside' left. We now shunt the extremes from one part of the planet to another and those on the receiving end must quickly move them on, or move on themselves.

The ability of humans to modify their local, or even regional, environment is not unique to our species. Indeed it is obvious that our planet has undergone a long history of extreme environmental perturbations long before humans ever existed. It was biology after all that created the highly oxygenated world we live in (Holland, 1984; Lomax, Chapter 2, this volume). However, human-induced or anthropogenic extremes are important and interesting for several reasons. First, these extremes are potentially damaging to the impacted people, societies and even the entire planet. Second, these extremes can teach us about biological processes, especially ecosystem responses to perturbations. Finally, having learned about them, we may be able to learn how to modify our activities to reduce any potential impacts on people and the ecosystems they live in.

If the preceding chapters of this book have shown one thing, it is that life has an extraordinary ability to persist in extreme environments. As scientists have poked, prodded and sampled the world's ecosystems, there has been a consistent trend of locating life where it is least expected: worms living in the Arctic ice; Bacteria in subsea, subterranean rocks; photosynthetic microbes in the sulfurous pools of the Yellowstone National Park, USA. That is one perspective. Another is that these ecosystems are in fact almost barren: life has found a way, but only just. They may teach us much about biology, but are not perhaps models for the Earth we may want to inhabit. The purpose of this chapter is to highlight rather less of the 'success against all the odds' and rather more of the dangers. Extremes may provide the thrills, but the home we want to live in is the bit in the middle.

Anthropogenic extremes are in most cases quite different to natural extremes.

In general, they arise over much shorter time-scales; fauna and flora that is not either pre-adapted for the extreme, or able to evolve very rapidly, will usually disappear. In contrast, natural extremes such as high-latitude ecosystems or deep-sea ecosystems may have been relatively stable over thousands or even millions of years. This has allowed fauna to disperse and evolve unique adaptations to these extreme ecosystems. Humans have evolved some fairly unique abilities to modify large areas in very short periods of time. These may be dramatic – deforestation and agriculture for example – or more subtle, such as influencing the degree of ultraviolet radiation on the planet. It is an interesting, and difficult, question to decide whether or not some of the ecosystems we have created this way are in fact 'extreme'. The coppiced woodlands of southern England, teeming with wildlife, are probably not considered extreme, whereas a palm oil plantation in a formerly-rainforested part of Borneo may well be. Thus our vision of 'extreme' is most likely influenced to a great extent by our various biases in conservation.

We have arranged this chapter around seven types of potential anthropogenic extremes, driven by chemical or physical processes (Table 24.1). Within each section we survey how these extremes may be generated and what life can persist within them. These extremes may occur in aquatic or terrestrial ecosystems although some of them are only relevant to one of these. Further, for some extremes, such as those induced by climate change, or changes in UV radiation, we have been deliberately brief as these subjects are covered in detail in earlier chapters.

24.2 Chemical Extremes: Inorganic

Inorganic chemical extremes represent some of the more dramatic and potentially toxic anthropogenic impacts on the environment. They include some of the more headline-grabbing environmental stories, such as the acidification of lakes and rivers or pollution by heavy metals, as well as more subtle impacts such as salinization. The spatial impacts of inorganic chemical extremes are generally local to regional, with the temporal impact lasting from months to decades.

24.2.1 Acidification

Acidification refers to a reduction in pH, or elevated levels of hydrogen ions. In their extreme form, acids have highly corrosive properties. In milder form, they can still have dramatic impacts on natural ecosystems. We consider here two forms of anthropogenic acidification which may have 'extreme' consequences. The first is the formation of acidic rainwater caused by pollutants. The second is the acidification of the oceans caused by elevated atmospheric carbon dioxide. The type of impacts, spatial and temporal scales are rather different in both cases.

The relationship between the reduced pH of rainwater ('acid rain') and atmospheric pollution from sulfur dioxide (SO_2) and nitrogen oxide (NO_x) has long been known. These chemicals are released into the atmosphere by human activities such as power generation and transport. They react in the atmosphere with local moisture, creating sulfuric and nitric acid, respectively, which can fall to the earth as precipitation (Mason, 2002). The input of acids into freshwater lakes, rivers and soils can cause a loss of the natural buffering of pH in these ecosystems (Bouwman *et al.*, 2002). In particular, freshwater systems that are already naturally acidic (e.g. in areas of granite or peat-based soil) are most affected. Typical impacts in freshwater include the loss of fish, reduced diversity and fundamental shifts in community structure (Mason, 2002). In terrestrial systems such as forests, the impacts are more complex where the acidification renders the forests more vulnerable to a range of other variables, such as frost, drought and pests (Bouwman *et al.*, 2002).

The passage of national and international legislation in the 1980s and early 1990s has

Table 24.1. A classification of anthropogenically-induced extremes, with principal drivers, examples and references

Extreme	Principal drivers	Key examples and references
Chemical: inorganic	Acidification	Soil acidification driven by SO_2 and NO emissions (Bouwman *et al.*, 2002) Freshwater acidification driven by SO_2 and NO emissions (Mason, 2002) Oceanic acidification driven by elevated atmospheric CO_2 (Orr *et al.*, 2005)
	Salinization	Soil and freshwater salinization driven by irrigation (Ghassemi *et al.*, 1995)
	Metal pollution	Lead, nickel, cadmium, zinc, copper, mercury and aluminium (Hopkin, 1989; Luoma and Rainbow, 2008)
	Other inorganics	Gases (chlorine, ammonia, methane) and Anions (cyanides, fluorides, sulfides and sulfites) (Mason, 2002)
Chemical: organic	Oil pollution	Oceanic oil spills (e.g. *Exxon Valdez*, Deepwater Horizon) (Peterson *et al.*, 2003; Sindermann, 2006) Terrestrial fuel/oil spills (e.g. hydrocarbon-contaminated soils) (Aislabie *et al.*, 2001; Ollivier and Magot, 2005)
	Eutrophication and hypoxia	Coastal marine hypoxia (Gray *et al.*, 2002), freshwater hypoxia (Smith *et al.*, 2006) and terrestrial eutrophication (Smith *et al.*, 1999)
	Other organics	Organochlorine pesticide release, polychlorinated biphenyls (PCBs) (Mason, 2002)
Physical: radiation	Nuclear weapons use and testing	Marshall Islands radionuclide contamination (Robison and Noshkin, 1999)
	Nuclear accidents	Chernobyl radionuclide contamination (Møller and Mousseau, 2006)
	Nuclear waste	Disposal of unshielded nuclear waste at sea (Smith *et al.*, 1998)
	Ozone hole damage	Ultraviolet (UV-B) radiation increases in high-latitudes (Karentz and Bosch, 2001)
Physical: temperature	Climate change	Warming such that near-extreme ecosystems become extreme (Parmesan, 2006)
	Thermal pollution	Thermal shifts in freshwater driven by pollution and changes in flow (Caissie, 2006)
Physical: pressure	Sudden pressure increase	Nuclear blasts, explosions causing temporary pressure increase (Simenstad, 1973)
Physical: light	Chronic or periodic illumination	Change in foraging habits, navigation, disorientation, mortality, entrapment (Longcore and Rich, 2004)
Physical: habitat	Urbanization, agriculture, water resources	In general terms, the conversion of natural habitats to man-made cities or agricultural areas and use of water resources

limited (at least in Europe and the USA) the emission of acid-generating chemicals into the atmosphere. A corresponding 'chemical recovery' in the form of rising pH, alkalinity and acid-neutralizing capacity has been seen in a number of well-studied ecosystems, such as the freshwater lakes of boreal Canada, but the biological recovery has been more complex and difficult to measure (Vinebrooke *et al.*, 2003; Monteith *et al.*, 2005).

While the acidification of boreal lakes, rivers and forests has in general been a regional issue, dependent on sources of pollutants and natural buffers, the acidification of the oceans in response to elevated CO_2 is now thought to be a more pressing global problem (Orr *et al.*, 2005). In terms of media and public interest, the 'acid rain' story has been replaced by the 'acid ocean' story. Is this another example of an anthropogenic extreme ecosystem?

Surface ocean pH is already 0.1 unit lower than pre-industrial values, caused apparently by the steady rise of CO_2 in the atmosphere (Fig. 24.1) (Orr *et al.*, 2005). Under 'business-as-usual' IS92a emission scenarios from the IPCC, ocean pH will be another 0.3–0.4 units lower by the end of this century (Orr *et al.*, 2005). Marine ecosystems at risk of major change include coral reefs and deep-sea ecosystems. Reefs are created by calcifying organisms and under lower pH scenarios, reef-building becomes energetically more costly (Kleypas and Yates, 2009). While ocean acidification is a potentially huge problem over the next century, it probably has not yet resulted in ecosystems we humans would define as extreme, but may in the future generate extremes for other organisms, especially the aforementioned calcifying species.

24.2.2 Salinization

Human-induced salinization, or alkalization, is a well-studied phenomenon in soils. It is typically associated with semi-arid areas where irrigation is widely used in agriculture. As water (containing salts) irrigates the land, the salts are left behind when the plants consume the water. Over time, salinity builds up in the soils, and more water has to be applied as the plants themselves struggle to take up water from the higher-salinity soils around their roots. Salinization has detrimental effects on plant growth and can ultimately lead to soil erosion and desertification. As the soil ecosystem is pushed towards this extreme, more salt-loving or salt-tolerant plants must be grown.

Australia is particularly impacted by this human-induced extreme. Approximately

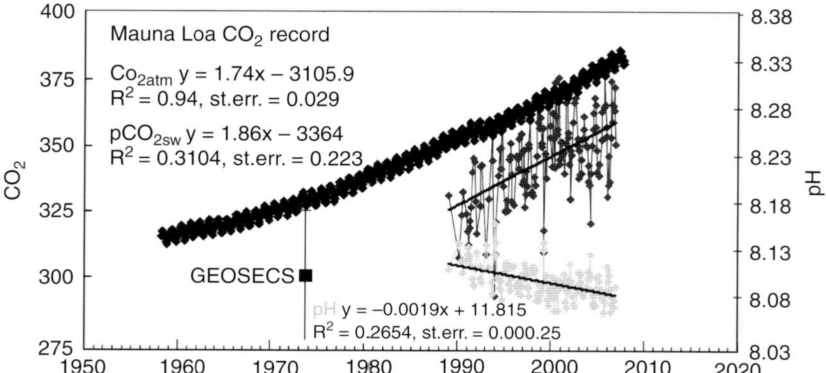

Fig. 24.1. A time-series of atmospheric measurements of CO_2 taken at Mauna Loa volcano in parts per million volume (black triangles), pCO_2 (dark grey triangles) at the oceanic station ALOHA and surface ocean pH (light grey triangles) at the oceanic station ALOHA in the subtropical North Pacific Ocean. A decrease in oceanic pH (increase in acidity) is observed between the measured years 1989 and 2007. Adapted from Doney *et al.*, 2009.

30% of the land area of Australia is affected by salinization and major research programmes are underway to try and understand better the processes involved and potential solutions in a world where there is ever-increasing demand for food (Rengasmy, 2006).

24.2.3 Metal pollution

Incidents of metal pollution and pollution attributed to various issues surrounding metal mines are one of the most dramatic types of human-induced extreme. Mines are effectively sites where metals are removed from stable ore-based reservoirs, concentrated and redistributed for various purposes. When accidents happen, the results can have impacts over spatial scales of perhaps thousands of hectares and temporal scales of years to decades.

A recent example is the Hungarian Ajka alumina plant accident which occurred on 4 October 2010 (Fig. 24.2). Over 1 million m^3 of bright red liquid waste was released from an alumina plant; the mud was highly alkaline (pH 12) and contained toxic trace metals above soil limits as well as being slightly radioactive (Ruyters *et al.*, 2011). However, although this spill was spectacular visually (owing to the high concentrations of red iron oxides), the toxicity was primarily a result of the alkalinity rather than the metals.

A more dramatic incident of metal pollution, since relatively well studied, occurred at the Los Frailes pyrite mine in 1998, with major impacts on the Doñana National Park, Spain. Approximately 5 million m^3 of highly acidic (pH 2), metal-rich waste was accidentally released from the mine tailing ponds causing widespread ecological impact over a wetland ecosystem (Pain *et al.*, 1998). Heavy metals including zinc, lead, copper, arsenic and cadmium were distributed across the floodplain and shown to contaminate both soils and accumulate in wildlife (Simon *et al.*, 1999).

Fig. 24.2. The toxic sludge spill in Ajka, Hungary overtaking the villages of Kolontar and Devescsar in October 2010. The spill is indicated by the white shaded areas. © Digtalglobe.

24.2.4 Other inorganics

Other inorganics may enter natural ecosystems from human sources. These include highly toxic anions such as cyanides, fluorides, sulfides and sulfites, as well as gases such as chlorine, ammonia and methane. Cyanide, for example, is extensively used in industries and may enter ecosystems as a result of metal finishing and mining (Dash *et al.*, 2009). Microbial species are able to degrade cyanides, using it as a nitrogen and carbon source, converting it to ammonia and carbonate under both aerobic and anaerobic conditions (Dash *et al.*, 2009). Other species, including humans, are poisoned by it.

24.3 Chemical Extremes: Organic

In the extreme south-eastern corner of Texas, USA, a small hill named Spindletop was the location of what was probably the world's first oil spill, named the Lucas Gusher in 1901. It was not considered a spill at the time; the discovery was hailed as a major success in oil production, and heralded the 20th-century oil boom and the rise of the USA as a major industrial power. Only 500 km from that site, 110 years later, the Macondo oil well below the Deepwater Horizon oceanic drilling rig blew out, resulting in the largest oceanic oil spill to date (Fig. 24.3). Oil spills normally attract great public interest, but are not the only form of organically induced extreme ecosystem, or probably the most significant. Organic inputs from human activities can also lead to eutrophication and oxygen depletion, which are potentially far more damaging. Other organic extremes may also be generated from chronic inputs of chemicals such as organochlorines and polychlorinated biphenyls (PCBs).

24.3.1 Oil pollution

The 20th-century oil boom has resulted in a situation where humans have concentrated

Fig. 24.3. The extent of the Deepwater Horizon oceanic oil spill in the Gulf of Mexico on 24 May 2010. The spill is indicated by the white areas on the ocean. The location of the source (the Macondo Well) is indicated by the dark spot in the lower portion of the image. © NASA.

petrogenic hydrocarbons in vast quantities on drilling rigs, storage ships, pipelines, refineries and vehicles. These anthropogenic concentrations dwarf natural concentrations of both biogenic or petrogenic hydrocarbons, but it must be remembered that these are natural compounds that are always present in natural ecosystems. In studies of the impacts of oil pollution, much controversy seems to arise from what constitutes a return to a pre-impacted state. This is exacerbated by an often poor knowledge of baseline conditions, in particular the very typical natural variations in ecosystems over annual to decadal scales, a quite typical scale for post-impact studies.

Extreme oil pollution events usually end up with the oil being spilled into the sea. Even in scenarios where it is spilled on land (e.g. the Lucas Gusher, or more dramatically the Kuwaiti oil spills after the 1991 Gulf War) the principal impacts have been marine. Oil in the marine environment will usually spread over the sea surface, from where it evaporates, dissolves, emulsifies (forming a 'mousse' that may wash ashore), sediments to the seafloor and biodegrades. The various impacts of oil spills are well reviewed elsewhere as are some of the more famous case studies, such as the spill of the *Exxon Valdez* in 1989 and the *Braer* in 1993. A common theme among these studies is the surprisingly rapid recovery of oiled littoral zones; rather less is known about the impacts in deep water. At some famous oil spills, where a vast 'clean-up' operation has been demanded by media, politicians and the public, there have been dramatic and long-term environmental impacts from the clean-up itself. After the *Torrey Canyon* spill of 1967, highly toxic oil-dispersants were used to clean rocky shores, impacting these sites for as much as 15 years after the event. In comparison, the untreated oily shores recovered rapidly (Hawkins and Southward, 1992).

Perhaps of greater relevance to this chapter are the studies of oil spill bioremediation, in particular those involving obligate hydrocarbon-degrading bacteria. The most recent review lists over 200 bacterial, cyanobacterial, algal and fungal genera of microbes, representing more than 500 species and strains that are known to be obligate oil-degraders (Yakimov *et al.*, 2007) and worthy of consideration as extremophiles. These are species which have evolved to exploit natural hydrocarbon sources, cold seeps along continental margins being one obvious example. Less obvious are species which have evolved remarkable adaptations to consume the oils inside sunken whale-carcasses on the seafloor (Verna *et al.*, 2010). Following spills, blooms of these types of microbes are responsible for the degradation of many oil constituents, and a large, if difficult to measure, percentage of the total spill. These microbes are understandably the focus of much biotechnological research.

24.3.2 Eutrophication and hypoxia

It has been known since the early 20th century that plant growth may be limited by nutrients such as nitrogen (N) and phosphorous (P). However, it is only since the 1960s and 1970s that an understanding of the impacts of humans mobilizing N and P would have on ecosystems, in particular coastal areas. It is now understood that humans have approximately doubled the rate of N input into the terrestrial N-cycle, and that these rates are still increasing (Smith *et al.*, 1999). N, or P, added to the land may accumulate in the soils, become entrained in surface waters, migrate to groundwaters or enter the atmosphere. Further, additional N is added to the atmosphere from the burning of fossil fuels, which subsequently is returned via wet or dry deposition.

One of the main impacts of eutrophication is a general intensification of all biological activity and dramatic changes in community composition and the structure of food webs. In aquatic systems, two common trends are a shift in algal species composition, and an increase in the frequency and intensity of algal blooms (Fig. 24.4) (Smith *et al.*, 2006). The most serious impacts lead to a decrease in water transparency, the accumulation of unused

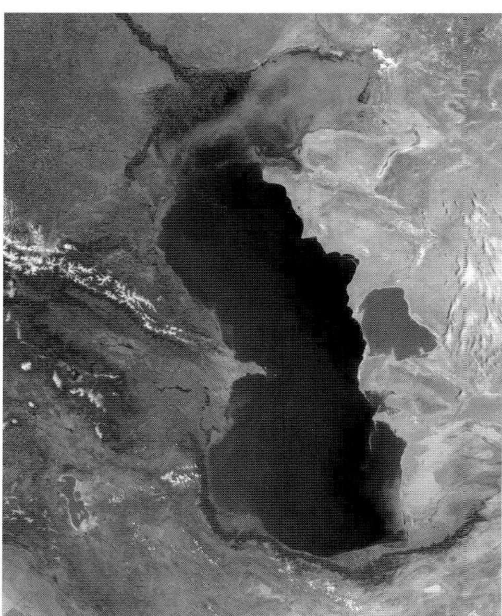

Fig. 24.4. A view of the Caspian Sea from the MODIS sensor on the Terra satellite taken on 11 June 2003. The image shows a large area of eutrophication (the pale grey water) in the water in the upper section of the sea near the mouth of the Volga River indicating the presence of algae. © NASA.

organic matter and harmful algal blooms. The final stage of eutrophication leads to hypoxia (low oxygen) or even anoxia (no oxygen), so called 'dead zones'.

Observations from coastal ecosystems have shown how phytoplankton communities may shift from diatom dominance to a progressively increasing dominance by other protozoan taxa such as dinoflagellates and phytoflagellates (Cloern, 2001). The so-called 'red tide' events are normally the result of major dinoflagellate blooms, the red colour during daytime accompanied by high levels of bioluminescence at night. These algal blooms are of great significance to the food industry as their presence can lead to toxicity in shellfish in particular.

A more serious impact on ecosystems in general comes from a decrease in dissolved oxygen (DO) levels in bottom waters, which can occur as blooming planktonic algae die, sink to the seafloor and fuel microbial respiration (Diaz and Rosenberg, 2008). Hypoxia occurs at DO levels below 2 ml O_2 l^{-1}, causing behavioural changes in fauna; below 0.5 ml l^{-1}, mass mortality occurs. Major hypoxic or anoxic events have been recorded in 'dead zones' in seas such as the Baltic, Kattegat, Black Sea, Gulf of Mexico and East China Sea (Diaz and Rosenberg, 2008). A number of studies over the last few decades have documented the progression of hypoxia in marine ecosystems (Diaz and Rosenberg, 2008). Initially, eutrophication enhances deposition of organic matter and hence microbial growth, creating oxygen demand. DO levels can become more depleted as the water column stratifies, and hypoxia occurs transiently accompanied by mass die-offs. Further eutrophication and nutrient build-up leads to seasonal, or at least periodic hypoxia with boom–bust population cycles in the fauna. If this stage lasts for many years, and DO continues to fall, anoxia can become established, with microbially generated hydrogen sulfide and bacterial mats forming on the seafloor, as has been documented in the Baltic.

Anthropogenic-induced marine hypoxia has clear 'extreme' impacts on ecosystems. For example, in the Baltic Sea it is estimated that approximately 264,000 t of carbon and 30% of the secondary production is 'missing' as a result of the Baltic anoxia (Diaz and Rosenberg, 2008). Pelagic species such as cod experience habitat compression where the shallow halocline limits their potential spawning success. Similar studies estimate that hypoxia in the Chesapeake Bay, USA, reduces secondary production by approximately 5%. It has been suggested that the Baltic Sea would be a third to a half more productive if dead zones were eliminated (Diaz and Rosenberg, 2008).

In terrestrial ecosystems, numerous classical ecological studies have shown the importance of N as a key element regulating productivity, diversity and community composition (Tilman, 1982). The amount of N input to terrestrial ecosystems has approximately doubled as a result of human activity, principally the use of nitrogenous fertilizers, the combustion of fossil fuels and the farming of legume crops (Smith *et al.*, 1999). The input of N into terrestrial ecosystems can have several biotic 'extreme' effects, mainly a massive increase in primary production and subsequent shifts in the community composition and diversity of ecosystems. For example, in the Netherlands, anthropogenic N has resulted in a shift from heathlands to grasslands and forest. In the UK, up to five-fold decreases in species diversity have been observed in grasslands following N enrichment (Smith *et al.*, 1999). N additions can also have long-term impacts on soil chemistry; at Rothamstead Experimental Research Station (UK) acidification of the soil has been reported following ammonia addition (Smith *et al.*, 1999).

24.3.3 Other organics

Other organic pollutants released into the environment may lead to temporary or even long-term 'extreme' conditions for life. Two examples are organochlorine pesticides (e.g. DDT), widely used in the control of disease vectors such as mosquitoes, and polychlorinated biphenyls (PCBs), used in manufacturing industries. These compounds are fat soluble and biologically very stable, as such they can accumulate in animal body fats and biomagnify along food chains. Concentration factors from water to the top predators can be as much as 10 million times (Mason, 2002).

Pesticides such as DDT commonly enter aquatic ecosystems as runoff from farmlands, sewage and effluent. Atmospheric transport and precipitation can also disperse these chemicals. The initial toxic effects can be quite high, but the main problem arises from a bioaccumulation in the tissues of top predators. The classic example is from Clear Lake, California, USA, where DDD insecticide (related to DDT) used throughout the 1950s resulted in accumulation of 80,000 times normal levels in western grebes (*Aechmorphus occidentalis*) and the almost complete extinction of the grebe population (Mason, 2002). There has been a reduction in the use of organochlorine-based pesticides in the developed world, and some recovery has been noted in previously contaminated freshwater ecosystems, such as the Great Lakes of North America (Mason, 2002).

As with organochlorines, PCBs are highly stable compounds that can persist in ecosystems for many years. Most PCBs were produced as mixtures containing a variety of congeners with differing levels of toxicity, the most dangerous with toxicity levels approaching that of dioxin, one of the most potent of all environmental contaminants (Mason, 2002). There are many examples of PCBs being detected in river systems. For example, two power generators on the Hudson River, New York State, USA, released PCBs into the river eventually leading to the complete closure of the shellfish industry (Mason, 2002). There is a four-fold increase in the PCB concentration of roach in the River Seine below the city of Paris, France (Mason, 2002) and Beluga whales (*Delphinapterus leucas*) in the St Lawrence River estuary, Canada, have shown evidence of PCB-related pathologies (Mason, 2002).

24.4 Physical Extremes: Radiation

The release of ionizing radiation into the environment by human activities has been one of the greatest concerns to people since the development of atomic weapons and nuclear power generation. However, radiation is a natural physical process and in many areas of the world, for example near sources of radon gas, low levels of ionizing radiation are always present. At certain levels, ionizing radiation can start to have a dramatic and deleterious impact on living organisms, in particular damage to cellular DNA. At very high levels, acute radiation sickness can kill organisms very rapidly. Anthropogenic sources of ionizing radiation include nuclear weapons use and testing, and accidents such as Chernobyl. Other forms of natural radiation, such as UV radiation, may be enhanced by human activities such as the release of CFCs, and as such we consider them here briefly within this section, although they are covered in more detail in Newsham and Davidson, Chapter 22, this volume.

24.4.1 Nuclear weapons use and testing

In the mind's eye, nuclear weapons always represent one of the most awesome examples of potential human power over the environment. The development of fission-based bombs during the 1940s and subsequent development of much larger nuclear fusion-based devices during the 1950s presented humans with the potential to destroy much of life on the planet. Only two devices have ever been used in combat, over the Japanese cities of Hiroshima and Nagasaki in 1945. Much has been written about these events, and the impacts on the people and cities affected. Of more relevance to this review are the potential radiological impacts on ecosystems of the many more nuclear tests that were carried out during the 1940s and 1950s.

Nuclear tests have been carried out in the air, underwater and in subterranean rock strata. Airburst and underwater tests continued until the signing of the Limited Test Ban Treaty in 1963, and the main nuclear powers switched to underground tests until the signing of a Comprehensive Test Ban Treaty in 1996. Non-signatories such as India, Pakistan and potentially North Korea have still carried out some further testing. The 1963 treaty is most relevant here as radionuclides are only released in significant quantities by airburst or underwater tests, although a powerful pressure wave can still have impacts from underground testing (see section 24.6). The best studied test area for airburst and underwater tests is the Marshall Islands in the South Pacific. This was the site of the largest test ever carried out by the USA, the 15 Mt 'Bravo' and the unusual underwater test 'Operation Crossroads' (Fig. 24.5).

The principal extreme radiological effect from nuclear weapon use and testing is nuclear fallout, that is, contamination of the surrounding area with radioactive substances. One of the better-studied examples is from Bikini Atoll, where the 'Bravo' test led to significant contamination of the island and surroundings. Owing to miscalculations by the nuclear scientists, the bomb was two and a half times more powerful than expected, and the design of the bomb led to significant fallout over a wide area. The short- and medium-term impacts of this were primarily felt by the Marshallese living on the islands of Rongelap and Rongerik. Long-term studies of the impacts of the Marshall Island tests have revealed the fate of radionuclides in fish species, principally isotopes such as ^{137}Cs, ^{60}Co and ^{207}Bi, which have been detected over the period 1964–1995 (Robison and Noshkin, 1999). Despite the enormous

Fig. 24.5. An underwater nuclear weapon test photographed on 25 July 1946 as part of Operation Crossroads in the Bikini Atoll. The image shows a large body of radioactive water being dispersed across the lagoon, where ships are moored to test the destructive powers of the weapon. © United States Department of Defense.

power of these weapons and the potential for radiological contamination, Bikini Atoll today could not be described as an extreme ecosystem. Indeed, the vast majority of reef species have returned and even the giant crater of the Bravo test itself is well populated with coral and fish species (Richards *et al.*, 2008). The most significant impact has been on the economic, social and cultural well-being of the islanders themselves.

24.4.2 Nuclear accidents

A long list of relatively minor (even if dramatic at the time) nuclear accidents is completely dwarfed by the catastrophe of the Chernobyl disaster in April 1986. The explosion of one of the four nuclear reactors and subsequent fire released between 9.35×10^3 and 1.25×10^4 petabecquerel (PBq) of radionuclides into the atmosphere (Møller and Mousseau, 2006) and could be said to have created a locally 'extreme' ecosystem with many far-field, continent-wide, impacts as well. Humans tackling the fire at Chernobyl were exposed to very high levels of radiation and 55 people were killed from radiation sickness. Although many of the radioisotopes released decayed within days, some such as ^{137}Cs, ^{90}Sr and ^{239}Pu are still common in the 'exclusion zone' that now surrounds the reactor. The small city, Pripyat, that was built to house workers in the Chernobyl complex was evacuated and subsequently abandoned – an extreme urban ecosystem now returning to nature as the result of an extreme anthropogenic impact (Fig. 24.6).

Several investigations have reported the biological impacts of Chernobyl and have been recently reviewed (Møller and Mousseau, 2006). The radiation reduced the levels of antioxidants and natural immunity in exposed organisms such as barn swallows *Hirundo rustica*, which caused increased frequency of partial albinism. Twenty-five published studies showed an increase in mutations or cytogenetic abnormalities associated with the accident, although many of these studies suffered from poor sample sizes. The official United Nations report on the Chernobyl incident estimated human deaths attributable to Chernobyl as 10,000, although it has been suggested that these estimates are premature as longer-term radiation exposure effects are likely to continue to manifest (Møller and Mousseau, 2006). Unfortunately, there have been very few real studies of the ecological impacts of Chernobyl other than the research on albinism in barn swallows and speculations as to the wider effects (Møller and Mousseau, 2006).

Fig. 24.6. The abandoned city of Pripyat in 2009 next to the Chernobyl nuclear power station (in the distance on the far right), with trees and vegetation starting to overwhelm the former city. © Matti Paavonen.

24.4.3 Nuclear waste

Nuclear waste is a by-product of the nuclear power industry and has presented numerous problems over the years for safe storage and disposal. One of the central philosophical problems of nuclear waste disposal is that any site where the waste will be safe and stable is generally difficult to monitor. Efforts have concentrated on deposition in geologically-stable rock strata on land, but have generally been blocked by local opposition (e.g. Yucca Mountain, Nevada). As such, the majority of high-level waste is still kept in containers close to where it is generated, with people paid to look after it. An exception to the idea of terrestrial disposal has been disposal in the ocean. The British government considered the disposal of high-level wastes in sediment-penetrating canisters on the Madeira Abyssal Plain during the 1980s, but later rejected the option. However, between 1949 and 1982 about 220,000 drums of low-level waste were dumped directly on the seafloor in the north-east Atlantic by European countries (Glover and Smith, 2003). Between 1946 and 1970, approximately 75,000 drums of low-level waste were dumped by the USA at three sites in the Atlantic and Pacific. Few data are available on the amounts of waste dumped by the former Union of Soviet Socialist Republics (USSR) at sea. The deep-sea nuclear dumpsites have been monitored periodically. Few changes have been noted in the surrounding ecosystem, although measurable, albeit low, levels of radionuclides have been reported in some benthic invertebrates close to the sites (Smith *et al.*, 1998).

24.4.4 Ultraviolet radiation

Ozone (O_3) molecules present in the upper atmosphere naturally reflect incoming ultraviolet (UV) radiation, creating conditions more beneficial to life on Earth. The release of chlorofluorocarbons (CFCs) and other atmospheric anthropogenic pollutants has been blamed for a decline in atmospheric ozone, most notably in the Antarctic (Fig. 24.7) during the 1970s and 1980s (Karentz and Bosch, 2001). This 'ozone hole' has lead to concomitant increases in UV-B radiation in high latitude ecosystems, creating a form of anthropogenic extreme. Although the Montreal Protocol (limiting CFC use and release) has improved ozone levels, there are still ongoing impacts. One of the key potential impacts of UV-B radiation is a decline in primary productivity in phytoplankton, and there is some evidence suggesting that this has happened in some Antarctic marine ecosystems (Karentz and Bosch, 2001). However, plankton are also resilient and can respond rapidly to changes in UV radiation, for example by adjusting their degree of UV tolerance (Karentz and Bosch, 2001). The full impacts of the ozone hole on Antarctic ecosystems is in reality not yet understood, but the ozone hole has diminished in size from its 2006 maximum as a result of the actions prescribed by the Montreal Protocol.

24.5 Physical Extremes: Temperature

Perhaps the most obvious anthropogenic extreme is, and will be, a function of climatic change driven mainly by the release of carbon dioxide, methane and nitrous oxide into the atmosphere (see Psenner and Sattler, Chapter 23, this volume). These 'greenhouse gases' are now thought to be driving a relatively rapid increase in global temperatures. One possibility is that this will lead to a rise in extreme events (e.g. droughts, floods) and for areas that are on the brink of 'extreme' it may prove a tipping-point. In addition to the obvious problem of climate change, we consider here some more local impacts, such as thermal pollution in rivers.

24.5.1 Climate change

While there is a huge wealth of recent reviews of the impacts of climate change (e.g. see Parmesan, 2006 and Psenner and

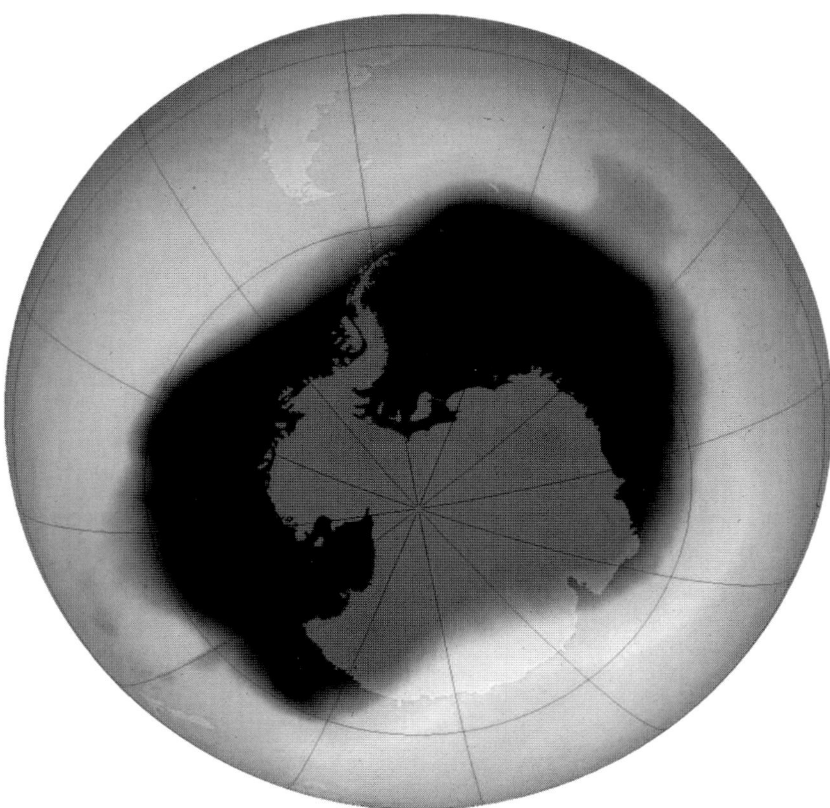

Fig. 24.7. A reconstruction of the ozone hole (here in dark grey and black) over Antarctica between 21 and 30 September 2006 based on satellite data. © NASA.

Sattler, Chapter 23, this volume), it is worth discussing here as to whether these can be considered extreme. It is now accepted that ecological changes are already occurring in the marine, freshwater and terrestrial ecosystems as a direct result of anthropogenic climate change (Parmesan, 2006 and references therein). By far the dominant impact observed has been changes in species phenologies. This is not surprising given humans' biased interest in planting, seasons and agriculture, not to mention cultural fascination with the passing of the seasons, particularly in northern Europe. Species range shifts are also often reported. In particular, recent studies have focused on range shifts in high latitudes, for example the shifts in distributions of penguins associated with sea ice in the Antarctic. It has also been established that elevational shifts in distributions occur in montane regions (Parmesan, 2006).

Phenological changes, geographical range shifts and elevational shifts are highly significant, and have and will lead ultimately to species extinctions. But it is difficult to consider these to be creating, as yet, extreme ecosystems at the edge of the tolerances of life. One potentially very significant impact from climate change is an increase in the frequency of extreme events such as droughts, flooding and storms (IPCC, 2007). The impacts of climate change on extreme ecosystems are covered in greater detail in Psenner and Sattler, Chapter 23, this volume.

24.5.2 Thermal pollution

While less significant than anthropogenic climate change, it is also important to consider more local-scale thermal effects. One example is the thermal pollution of rivers, recently reviewed. Water temperature is one of the most important physical parameters of freshwater ecosystems, and determines the overall health of the ecosystem. It can influence growth rates and distributions of organisms, for example salmonids have quite specific requirements in terms of temperature range. In terms of anthropogenic impacts, of most significance appears to be the relationship between streamside forest removal and river water temperature. Most studies seem to indicate quite significant increases in river water temperatures following forest removal – for example an increase of 7.8°C in the mean monthly maximum water temperatures in Oregon's Alsea River Basin, USA. Thermal effluents, reductions in river flow (e.g. from irrigation or hydroelectric schemes) and water releases to dams can also alter water temperature and significantly alter local and regional ecosystems.

24.6 Physical Extremes: Pressure

It is perhaps difficult to imagine any anthropogenic extreme involving increasing atmospheric pressure, but indeed this is what happened during the 137 underground nuclear tests carried out on Mururoa Atoll in French Polynesia, southern Pacific Ocean, by the French armed forces and the impacts of this have actually been scientifically studied (Planes *et al.*, 2005). A feature of underground nuclear tests is that thermal and radiation effects are contained, but the shock wave yielding peak pressures of 6,000–12,000 kPa reaches the surface some 0.25 and 1.0 s after the explosion killing virtually all fish within 2000 m of the test site (Bablet *et al.*, 1995).

The tests at Muroroa have provided a rather unusual scenario of continuous exposure of a natural ecosystem to high pressure waves during the testing period from 1976–1995. The tests were interesting from an ecological point of view as they removed nearly all the fish but without otherwise destroying the habitat (the coral reef for example). In the scientific investigation of the long-term impacts of this perturbation, it was observed that recovery of fish populations was extremely rapid, on the order of from 1 to 5 years for the resurrection of the fish assemblage structure (Planes *et al.*, 2005). The authors concluded that the rapid recovery was a result of the habitat being still fairly intact, and the possibility of recolonization from nearby relatively healthy reef fish populations. However, in the short term the ecological disruption was extreme.

24.7 Physical Extremes: Light

As every star-gazer knows, artificial light – of which the extent and intensity has increased over the last century – obscures the observation of the night sky. However, there are more serious consequences of light pollution with substantial effects on the biology and ecology of species in the wild, particularly of those adapted to the nocturnal lifestyle. Ecological light pollution includes chronic or periodically increased illumination, unexpected changes in illumination, and direct glare (Longcore and Rich, 2004). Wild animals can experience increased orientation or disorientation from additional illumination and are attracted to or repulsed by glare, which affects foraging, reproduction, communication and other critical behaviours. The more subtle influences of artificial night lighting on the behaviour and community ecology of species include extended foraging under artificial lights in diurnal birds and reptiles as well as behaviours, such as territorial singing in birds (Longcore and Rich, 2004). Constant artificial night lighting may also disorient organisms accustomed to navigating in a dark environment. Among the more serious and well studied consequences of light pollution are disorientation of hatchling sea turtles emerging from nests on sandy beaches, effects on the egg-laying behaviour of female sea turtles, a reduction in visual capability in frogs, mortality in insects and entrapment in buildings or collisions with structures in birds (Longcore and Rich, 2004).

24.8 Physical Extremes: Habitat Modification

From the anthropogenic extremes reviewed so far, it might be arguable that the majority have quite local to at most regional impacts, and over relatively short time scales. Obvious exceptions to this are impacts from climate change and changes in UV-B radiation. But at a landscape level, some quite vast changes to the Earth's habitat have occurred as a result of the presence of humans. The rise of agriculture and urbanization must also be considered in the sense of the creation of extreme ecosystems, albeit with relatively little scientific study. The rapidly expanding city of Shenyang in China, seen from the International Space Station (Fig. 24.8) is an example of the spread of urbanization across a previously natural landscape. Equally dramatic is the spread of agriculture and the impact at the landscape level in countries such as Brazil (Fig. 24.9).

In the 2009 revision of the United Nations World Urbanization Report (United Nations, 2010) it was documented that, for the first time, the population of the world passed the 50% urbanization level. The trend towards urbanization will continue, most rapidly in developing countries with an expected 6.3 billion people living in cities by 2050 (United Nations, 2010). Urban environments are habitats and hence ecosystems as much as any other, with their own unique fauna and flora. Many of these organisms must be adapted to these conditions which, by the standards of what was previously present, could be considered 'extreme'.

During the latter half of the 20th century, global food production doubled and for important grains such as wheat, rice and maize, the rate of production actually

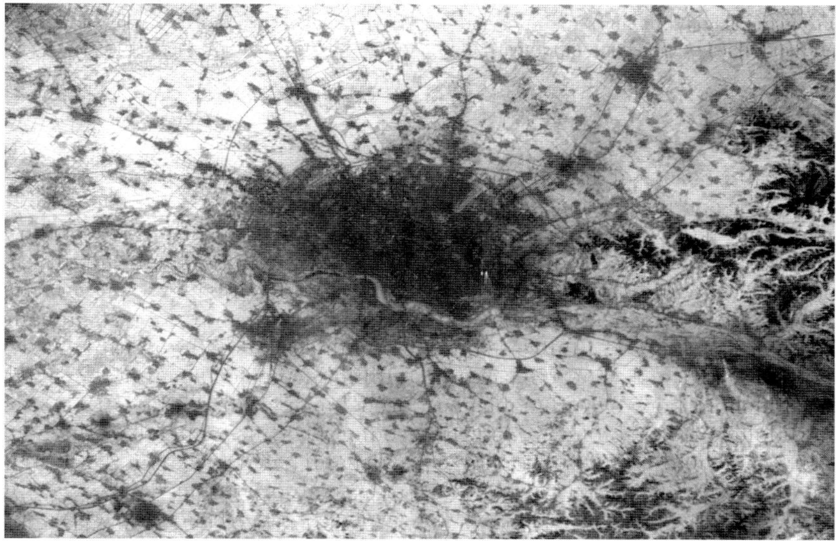

Fig. 24.8. Shenyang, the sixth largest city in China, imaged from the International Space Station in January 2005. The spread of the city limits is indicated by the dark regions against the agricultural areas which are covered in snow. South-east trending plumes are visible from large industrial complexes in the lower portion of the image. © NASA.

Fig. 24.9. Human use of land for agriculture is demonstrated in this photograph of a region of Minas Gerais State, Brazil, taken in February 2011 by Expedition 26 crew of the International Space Station. © NASA.

increased faster than the human population growth rate (Tilman, 1999). This has resulted in a better supply of food, and reduced malnourishment in humans. However, a doubling of the amount of land in cultivation and approximate seven-fold and three-fold increases in nitrogen and phosphorous fertilization respectively could be said to have had extreme consequences in some regions, particularly marine and freshwater ecosystems. Estimates are inherently difficult and controversial, but it is possible that to double world food production once more, to feed an increasing population, will require 18% more arable land (Tilman, 1999), comparable in size to cultivating all the currently forested land in the USA. If we are to consider a monoculture of maize as an 'extreme ecosystem' then we must remember that maize crops now cover some 140 million ha of the Earth (Tilman, 1999). Coupled with this is the extreme pressure such massive cultivation places on our already diminishing freshwater resources and increasingly saline, nutrient-depleted soils. These factors in turn lead to increased chemical fertilizer use further exacerbating environmental problems.

24.9 Conclusions

The persistence of life in extreme environments has been a constant source of surprise to biologists. The regular discoveries of microbial life in ever-more extreme and distant places on our planet have changed our perceptions of the boundaries to where life may persist. The discovery of hydrothermal vent ecosystems in 1977 changed our knowledge of these boundaries even further. Not only were microbes present, but large metazoans including remarkable giant tubeworms, bivalves and associated fauna living on these apparently 'toxic' vent chimneys (Lutz, Chapter 13, this volume). It was these discoveries that re-launched the discipline of astrobiology – the search for life on other planets – at a time when scientists were beginning to think of the solar system as being largely azoic (Gómez *et al.*, Chapter 26, this volume). Where there is an energy source, water and temperatures within the limits of enzyme activity – and sometimes at those limits – life has generally found a way to persist.

This chapter is a brief overview of the many and varied ways that humans have been able to apply powerful forcing factors to ecosystems. Some are quite obvious – the release of toxic sludge from a mine – others more subtle. Some, such as at Bikini Atoll, may teach us much about the resilience of ecosystems, while others, such as eutrophication, illustrate how unresilient ecosystems can be. Many of these forcing factors have been the substance of media headlines, international political incidents, inquiries by learned panels and major environmental regulatory frameworks. But a common theme is that our scientific knowledge of the actual impacts of many of them is often quite poor. For example, it might be expected that the Chernobyl disaster of 1986 would have led to a vast wealth of research into the local and regional ecological impacts of the event. But apart from the quite intensive research into the human tragedy of the disaster, there has been remarkably little done. A major review on the ecological impacts of Chernobyl contains hard, convincing data for only a single species of barn swallow. Numerous other studies are cited but the authors admit that the sample sizes are small and the impacts not very obvious.

The truth is that post-impact long-term studies are costly and difficult to fund once the initial media interest has waned. Quality ecological data collection is always more far more expensive and time-consuming than local industry, politicians and even scientific committees usually realize. The stochastic spatial and temporal locations of these events also create a problem. It is extraordinarily rare for one of these events to occur in a location for which there is a quality set of baseline data. It is even rarer if that baseline data includes a reasonable time-series element – decadal-scale datasets for example. One major anthropogenic incident for which useful baseline data does exist is the Deepwater Horizon oil spill of 2010. At the time of this publication, very little has yet been published, although funded investigations are underway. However, even though a wealth of data on the deep-sea biology of the Gulf has been collected, there are no well-established deep-sea time-series stations in this region (Glover *et al.*, 2010).

The Earth is naturally quite a variable place. As we have seen in this volume, it contains extremes in time and space that go beyond anything humans can easily create. Even our awe at the vast power of nuclear weapons is in reality dwarfed by natural events such as past asteroid impacts or periods of volcanic activity. At those scales, the extremes we list in this chapter seem rather minor. Our extremes are also perhaps not that useful as natural analogues; it is probably more the other way around. The natural extremes are probably where we need to look to learn how to deal with our own problems. But deal with them we must – for the sake of our own health and well-being, and the conservation options that we choose as a society to value.

References

Aislabie, J., Fraser, R., Duncan, S. and Farrell, R.L. (2001) Effects of oil spills on microbial heterotrophs in Antarctic soils. *Polar Biology* 24, 308–313.

Bablet, J.P., Gout, P.B. and Goutière, G. (1995) *Les Expérimentations Nucléaires, III. Le milieu vivant et son évolution*. Direction des Centres d'Expérimentations Nucléaires, Monaco.

Bouwman, A.F., Van Vuuren, D.P., Derwent, R.G. and Posch, M. (2002) A global analysis of acidification and eutrophication of terrestrial ecosystems. *Water, Air and Soil Pollution* 141, 349–382.

Caissie, D. (2006) The thermal regime of rivers: a review. *Freshwater Biology* 51, 1389–1406.

Cloern, J.E. (2001) Our evolving conceptual model of the coastal eutrophication problem. *Marine Ecology Progress Series* 210, 223–253.

Dash, R.R., Gaur, A. and Balomajumder, C. (2009) Cyanide in industrial wastewaters and its removal: a review on biotreatment. *Journal of Hazardous Materials* 163, 1–11.

Diaz, R.J. and Rosenberg, R. (2008) Spreading dead zones and consequences for marine ecosystems. *Science* 321, 926–929.

Doney, S.C., Fabry, V.J., Feely, R.A. and Kleypas, J.A. (2009) Ocean acidification: the other CO_2 problem. *Annual Review of Marine Sciences* 1, 169–192.

Ghassemi, F., Jakeman, A.J. and Nix, H.A. (1995) *Salinisation of Land and Water Resources: human causes, extent, management and case studies*. CAB International, Wallingford, UK, 544 pp.

Glover, A.G. and Smith, C.R. (2003) The deep-sea floor ecosystem: current status and prospects of anthropogenic change by the year 2025. *Environmental Conservation* 30, 219–241.

Glover, A.G., Gooday, A.J., Bailey, D.M., Billett, D.S.M., Chevaldonné, P., Colaço, A., Copley, J., Cuvelier, D., Desbruyères, D., Kalogeropoulou, V., Klages, M., Lampadariou, N., Lejeusne, C., Mestre, N.C., Paterson, G.L.J., Perez, T., Ruhl, H.A., Sarrazin, J., Soltwedel, T., Soto, E.H., Thatje, S., Tselepides, A., Van Gaever, S. and Vanreusel, A. (2010) Temporal change in deep-sea benthic ecosystems: a review of the evidence from recent time-series studies. *Advances in Marine Biology* 58, 1–95.

Gray, J.S., Shiu-sun Wu, R. and Or, Y.Y. (2002) Effects of hypoxia and organic enrichment on the coastal marine environment. *Marine Ecology Progress Series* 238, 249–279.

Hawkins, S.J. and Southward, A.J. (1992) The Torrey Canyon oil spill: recovery of rocky shore communities. In: Thayer, G.W. (ed.) *Restoring the Nation's Environment*. University of Maryland Sea Grant Publications, Maryland, pp. 583–631.

Holland, H.D. (1984) *The Chemical Evolution of the Atmosphere and Oceans*. Princeton University Press, Princeton, New Jersey, 598 pp.

Hopkin, S.P. (1989) *Ecophysiology of Metals in Terrestrial Invertebrates*. Elsevier Applied Science Publishers, London, UK, 366 pp.

IPCC (2007) *Intergovernmental Panel on Climate Change Fourth Assessment Report*. Cambridge University Press, Cambridge, UK.

Karentz, D. and Bosch, I. (2001) Influence of ozone-related increases in ultraviolet radiation on Antarctic marine organisms. *American Zoologist* 41, 3–16.

Kleypas, J.A. and Yates, K.K. (2009) Coral reefs and ocean acidification. *Oceanography* 22, 108–117.

Longcore, T. and Rich, C. (2004) Ecological light pollution. *Frontiers in Ecology and the Environment* 2, 191–198.

Luoma, S.N. and Rainbow, P.S. (2008) *Metal Contamination in Aquatic Environments*. Cambridge University Press, Cambridge, UK, 588 pp.

Mason, C. (2002) *Biology of Freshwater Pollution*. Pearson Education Limited, Harlow, UK, 400 pp.

Møller, A.P. and Mousseau, T.A. (2006) Biological consequences of Chernobyl: 20 years on. *Trends in Ecology and Evolution* 21, 200–207.

Monteith, D.T., Hildrew, A.G., Flower, R.J., Raven, P.J., Beaumont, W.R.B., Collen, P., Kreiser, A.M., Shilland, E.M. and Winterbottom, J.H. (2005) Biological responses to the chemical recovery of acidified fresh waters in the UK. *Environmental Pollution* 137, 83–101.

Ollivier, B. and Magot, M. (2005) *Petroleum Microbiology*. ASM Press, Washington, DC, 365 pp.

Orr, J.C., Fabry, V.J., Aumont, O., Bopp, L., Doney, S.C., Feely, R.A., Gnanadesikan, A., Gruber, N., Ishida, A., Joos, F., Key, R.M., Lindsay, K., Maier-Reimer, E., Matear, R., Monfray, P., Mouchet, A., Najjar, R.G., Plattner, G.-K., Rodgers, K.B., Sabine, C.L., Sarmiento, J.L., Schlitzer, R., Slater, R.D., Totterdell, I.J., Weirig, M.-F., Yamanaka, Y. and Yool, A. (2005) Anthropogenic ocean acidification over the twenty-first century and its impact on calcifying organisms. *Nature* 437, 681–686.

Pain, D.J., Sánchez, A. and Meharg, A.A. (1998) The Doñana ecological disaster: contamination of a world heritage estuarine marsh ecosystem with acidified pyrite mine waste. *Science of the Total Environment* 222, 45–54.

Parmesan, C. (2006) Ecological and evolutionary responses to recent climate change. *Annual Review of Ecology, Evolution and Systematics* 37, 637–669.

Peterson, C.H., Rice, S.D., Short, J.W., Esler, D., Bodkin, J.L., Ballachey, B.E. and Irons, D.B. (2003) Long-term ecosystem response to the Exxon Valdez oil spill. *Science* 302, 2082–2086.

Planes, S., Galzin, R., Bablet, J.P. and Sale, P.F. (2005) Stability of coral reef fish assemblages impacted by nuclear tests. *Ecology* 86, 2578–2585.

Rengasmy, P. (2006) World salinization with emphasis on Australia. *Journal of Experimental Botany* 57, 1017–1023.

Richards, Z.T., Beger, M., Pinca, S. and Wallace, C.C. (2008) Bikini Atoll coral biodiversity resilience five decades after nuclear testing. *Marine Pollution Bulletin* 56, 503–515.

Robison, W.L. and Noshkin, V.E. (1999) Radionuclide characterization and associated dose from long-lived radionuclides in close-in fallout delivered to the marine environment at Bikini and Enewetak Atolls. *Science of the Total Environment* 237–238, 311–327.

Ruyters, S., Mertens, J., Vassilieva, E., Dehandschutter, B., Poffijn, A. and Smolders, E. (2011) The red mud accident in Ajka (Hungary): plant toxicity and trace metal bioavailability in red mud contaminated soil. *Environmental Science and Technology* 45, 1616–1622.

Simenstad, C.A. (1973) Biological effects of underground nuclear testing on marine organisms. I. Review of documented shock effects, discussion of mechanisms of damage, and predictions of Amchitka test effects. *Proceedings of the First Conference on the Environmental Effects of Explosives and Explosions* (30–31 May 1973). Available at: www.stormingmedia.us/30/3047/0304777.html.

Simon, M., Ortiz, I., Garcia, I., Fernández, E., Fernández, J., Dorronsoro, C. and Aguilar, J. (1999) Pollution of soils by the toxic spill of a pyrite mine (Aznalcollar, Spain). *Science of the Total Environment* 242, 105–115.

Sindermann, C.J. (2006) *Coastal Pollution: Effects on Living Resources and Humans*. Taylor & Francis, Boca Raton, Florida, 280 pp.

Smith, C.R., Present, T.M.C. and Jumars, P.A. (1998) *Development of Benthic Biological Monitoring Criteria for Disposal of Low-level Radioactive Waste in the Abyssal Deep Sea*. Environmental Protection Agency, Publisher, Washington, DC.

Smith, V.H., Tilman, G.D. and Nekola, J.C. (1999) Eutrophication: impacts of excess nutrient inputs on freshwater, marine and terrestrial ecosystems. *Environmental Pollution* 100, 179–196.

Smith, V.H., Joye, S.B. and Howarth, R.W. (2006) Eutrophication of freshwater and marine ecosystems. *Limnology and Oceanography* 51, 351–355.

Tilman, D. (1982) *Resource Competition and Community Structure*. Princeton University Press, Princeton, New Jersey, 296 pp.

Tilman, D. (1999) Global environmental impacts of agricultural expansion: the need for sustainable and efficient practices. *Proceedings of the National Academy of Sciences* 96, 5995–6000.

UN (2010) *World Urbanization Prospects: the 2009 Revision*. United Nations, Publisher, New York, 44 pp.

Verna, C., Ramette, A., Wiklund, H., Dahlgren, T.G., Glover, A.G., Gaill, F. and Dubilier, N. (2010) High symbiont diversity in the bone-eating worm *Osedax mucofloris* from shallow whale-falls in the North Atlantic. *Environmental Microbiology* 12, 2355–2370.

Vinebrooke, R.D., Graham, M.D., Findlay, D.L. and Turner, M.A. (2003) Resilience of epilithic algal assemblages in atmospherically and experimentally acidified boreal lakes. *Ambio* 32, 196–202.

Yakimov, M.M., Timmis, K.N. and Golyshin, P.N. (2007) Obligate oil-degrading marine bacteria. *Current Opinion in Biotechnology* 18, 257–266.

25 Biotechnological Applications of Extremophiles: Promise and Prospects

Rajesh Arora[1] and Elanor M. Bell[2]
[1]*Staff Officer, Office of Distinguished Scientist and Chief Controller Research and Development (Life Sciences and International Cooperation), Defence Research and Development Organization (DRDO), New Delhi, India;*
[2]*Scottish Institute for Marine Science, Argyll, UK*

25.1 Introduction

Microbial life flourishes in almost every environment on Earth, and possibly also on comets and on other planets within the Universe. One is often astounded that microbes can survive under conditions that humans would not even dare to venture into under normal circumstances. However, as discussed in the preceding chapters in this volume, research carried out over recent decades has shown that microbial communities can flourish in the most diverse and extreme conditions, including extremes of temperature (>50°C; <4°C), pressure (>500 atm), pH (>12; <1), salinity (>1.0 M NaCl), oxygen tension, high radiation and nutrient depletion. The organisms that can survive these extreme limits of geophysical and geochemical environmental conditions, extremophiles, have developed highly efficient mechanisms. Often these involve the development of novel metabolic pathways, the use of specialized enzymes (extremozymes) and/or the secretion of various biomolecules (extremolytes) and these are coming under increasing scrutiny from scientists and technologists as means to help humans live outside our own 'extreme envelope'.

In the future, when Earth potentially becomes inhospitable, or resources become constrained and shortages begin to hamper human life, space travel and long-duration stays, even colonization of other planets within or outside of our solar system, may become routine. Our understanding of extremophiles and the substances they produce could help us understand mechanisms and develop strategies for human survival under such conditions. Aside from these futuristic possibilities, extremophilic organisms are currently being effectively utilized for a diverse array of contemporary applications and we have entered a new era of biotechnological research and development (Fig. 25.1).

Initial interest in extremophiles was stimulated by the discovery of thermostable enzymes (biological catalysts) that help extremophiles function under extreme conditions. The application of enzymes derived from extremophiles, so-called extremozymes (Gerday and Glansdorff, 2007), has made available a wide range of biomolecules resistant to the extreme conditions present in a wide range of industrial processes, and has led to a huge number of industrial applications (Tables 25.1 and 25.2). Many of these applications could

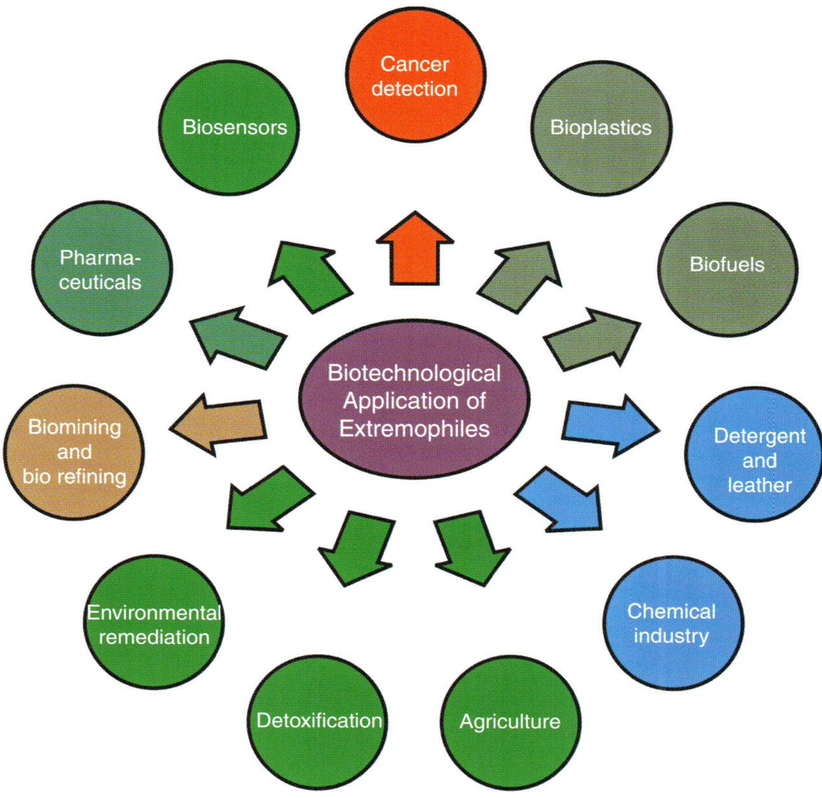

Fig. 25.1. Some applications of extremophiles in biotechnology. © Rajesh Arora.

Table 25.1. Source of some enzymes commonly used in the industry.

Nature of extremophile	Enzymes	Uses
Thermophilic and hyperthermophilic microbes	Alcohol dehydrogenases, amylases, cellulases, chitinases, esterases, glucoisomerases, xylanases, lipases, pectinases, proteases, pullulanases	Several industries
Alkaliphiles	Elastases, keratinases	Cosmetology
	Alginases, alkaline proteases, amylases, α-galactosidase, β-galactosidase, cellulases, catalases, DNases, glucanases, lipases, pectinases, pullulanases, RNases, restriction enzymes, maltose dehydrogenase, penicillinase, glucose dehydrogenase, uricase, polyamine oxidase, β-mannanase, β-mannosidase, β1,3-xylanase	Detergent and leather industry
Thermostable enzymes	DNA polymerases, DNA ligases, restrictases, phosphatises	Molecular biology, medicine, forensics
Psychrophiles	β-glucanases, cellulases, pectinases, proteinases	Food industry and waste treatment

Table 25.2. Extremolytes and their specific applications.

Extremophilic organisms	Extremophile type	Product	Applications	References
Pseudoalteromonas strain 721	Thermophile	Octasaccharide repeating unit with two side chains	Gelling properties	Rougeaux *et al.*, 1999a; Guézennec, 2002
Alteromonas macleodii subsp. *fijiensis*	Thermophile	Sulfated heteropolysaccharide, high uronic acids with pyruvate. The repeating unit is a branched hexasaccharide containing Glc, Man, Gal, GlcA, GalA, pyruvated mannose	Thickening agent in food-processing industry, biotoxification and waste-water treatment, bone healing, treatment of cardiovascular diseases	Raguenes *et al.*, 1996; Loaëc *et al.*, 1997, 1998; Rougeaux *et al.*, 1998; Colliec Jouault *et al.*, 2001; Zanchetta and Guézennec, 2001
Thermococcus litoralis	Thermophile	Mannose	Biofilm formation	Rinker and Kelley, 1996
Geobacillus sp. strain 4004	Thermophile	A pentasaccharide repeating unit: two of them with a gluco-galacto configuration and three with a manno configuration. Gal:Man:GlcN:Arab (1.0:0.8:0.4:02)	Pharmaceutical application	Nicolaus *et al.*, 2002
Bacillus thermodenitrificans strain B3-72	Thermophile	Trisaccharide repeating unit and a manno-pyranosidic configuration. Man:Glc (1:0.2)	Immunomodulatory and antiviral activities	Gugliandolo and Maugeri, 1998; Arena *et al.*, 2009
Bacillus licheniformis strain B3-15	Thermophile	Mannose is the main monosaccharide. Tetrasaccharide repeating unit and a manno-pyranosidic configuration	Antiviral activity	Maugeri *et al.*, 2002; Arena *et al.*, 2006
Pseudoalteromonas strain SM9913	Barophile	Linear arrangement of α-(1→6) linkage of glucose with a high degree of acetylation	Flocculation behaviour and bio-sorption capacity	Qin *et al.*, 2007; Li *et al.*, 2008
Pseudoalteromonas strain CAM025	Barophile	Sulfated heteropolysaccharide, high levels of uronic acids with acetyl groups Glc:GalA:Rha:Gal (1:0.5:0.1:0.08)	Cryoprotection	Mancuso Nichols *et al.*, 2004
Pseudoalteromonas strain CAM306	Barophile	Sulfated heteropolysaccharide, high levels of uronic acids with acetyl and succinyl groups GalA:Glc:Man:GalNAc:Ara (1:0.8:0.84:0.36:0.13)	Trace metal binding	Mancuso Nichols *et al.*, 2004

Colwellia psychrerythraea strain 34H	Barophile	n.r.	Cryoprotection	Marx *et al.*, 2009
Haloferax mediterranei	Halophile	→4)-β-D-GlcpNAcA-(1→6)-ά-D-Manp-(1→4)-β-D-GlcpNAcA-3-O-SO3⁻ -(1→	Candidate in oil recovery, especially in oil deposits with high salinity concentrations	Anton *et al.*, 1988; Parolis *et al.*, 1996
Hahella chejuensis	Halophile	EPS named EPS-R Glc:Gal (0.68:1.0)	Biosurfactant and detoxification of polluted areas from petrochemical oils	Lee *et al.*, 2001
Halomonas alkaliantartica strain CRSS	Halophile	Glc:Fru:GlcN:GalN (1.0:0.7:0.3:trace)	High viscosity	Poli *et al.*, 2004, 2007
Alteromonas macleodii subsp. *fijiensis* strain HYD657	Barophile, thermophile	The repeating unit is an undesaccharide with three side chains. Gal:Glc:Rha:Fuc:Man:GlcA:GalA:3-0-(1 carboxyethyl)-D-GlcA (1:0.42:0.85:0.5:0.42:0.5:0.5:0.5)	Cosmetics (patent PCT 94907582-4)	Cambon-Bonavita *et al.*, 2002
Alteromonas strain 1644	Barophile, thermophile	Main chain of five sugars with a side chain of three sugars including a dicarboxylic acid. Glc:Gal:Glca:3Lac-GlcA:GalA	Heavy metal binding	Bozzi *et al.*, 1996a, b
Alteromonas infernos strain 785	Barophile, thermophile	Glc:Gal:GlcA:GalA (1:1:0.7:0.4)	Anticoagulant activity	Colliec Jouault *et al.*, 2001
Archaeoglobus fulgidus	Acidophile	DGP	Thermostabilization of proteins and rubredoxin	Lamosa *et al.*, 2000, 2003
Artemia franciscana	Halophile	GADD45beta	In medicine, industry and diagnostics	Petrovick *et al.*, 2010
Bacillus thermoleovorans IHI-91	Thermophile	Lipases	In various biotechnological processes such as wastewater treatment	Markossian *et al.*, 2000
Desiccation-tolerant extremophiles	Xerophile	Stabilized mammalian cell-based biosensors	Ability to achieve air-dry stabilization of mammalian (especially human) cells with subsequent recovery following rehydration	Bloom *et al.*, 2001
Dunaliella	Halophile	β-carotene Ectoine	Enzyme stabilizer, also applied in cosmetic products	Oren, 2010
Elyria rufescens/*Bryopsis* sp. (mollusc/green alga)	–	Kahalalide F	Treatment of patients with severe psoriasis	Hamann *et al.*, 1996

Continued

Table 25.2. Continued.

Extremophilic organisms	Extremophile type	Product	Applications	References
Extremophiles	–	Extremolytes, e.g. ectoines	Minimize the denaturation of biopolymers, cell protectants in skin care and as protein-free stabilizers of proteins and cells	Lentzen and Schwarz, 2006
Extremophiles	–	Exopolysaccharides (EPSs)	Provide non-pathogenic products, appropriate for applications in food, pharmaceutical and cosmetics industries as emulsifiers, stabilizers, gel agents, coagulants, thickeners and suspending agents etc.	Nicolaus *et al.*, 2010
Galdieria sulphuraria (Cyanidiales)	Thermoacidophile		Thermo-stable enzymes are useful in biotechnology. Also, tolerates high concentrations of toxic metal ions such as cadmium, mercury, aluminium and nickel, suggesting potential application in bioremediation	Weber *et al.*, 2004
Geobacillus, Anoxybacillus and *Aeribacillus* genera	Thermophile	Lipases, proteases, and amylases, produce exopolymers and acids.	Enhanced oil recovery, show emulsifying activity and biodesulfurization	Pinzón-Martínez *et al.*, 2010
Halichondria okadai (sponge, synthetic)	–	E7389 (halichondrin B derivative)	Treatment for breast cancer	Bai *et al.*, 1991
Halomonas salina DSM 5928	Halophile	Ectoine	Free or immobilized cells for continuous culture to produce ectoine	Zhang *et al.*, 2009
Halorhodospira halochloris	Halophile	Ectoine	Enzyme stabilization against heating, and drying	Lippert and Galinski, 1992
			Protection of LDH against heat and freeze-thawing	Goller and Galinski, 1999
			Inhibition of insulin amyloid formation	Arora *et al.*, 2004
			Stabilization of tobacco cells against hyperosmotic stress	Nakayama *et al.*, 2000
			Block of UVA-induced ceramide release in human keratinocytes	Grether-Beck *et al.*, 2005
			Protection of the skin barrier against water loss and dying out	Bunger and Driller, 2004

			Protection of the skin barrier against UV radiation	Beyer *et al.*, 2000
			Reduction of UV-induced SBSs	Bunger *et al.*, 2001
			Prevention of UVA-induced photo-ageing	Bunger and Driller, 2004
			Cytoprotection of keratinocytes	Bummino *et al.*, 2005
Lyngbya majuscule (cyanobacterium)	–	Curacin A	Potent inhibitor of cell growth and mitosis	Simmons *et al.*, 2005
Rhodothermus marinus	Thermophile	Mannosylglycerate	Stabilization of enzymes against thermal stress and freeze drying	Borges *et al.*, 2002
			Stabilization of recombinant nuclease	Ramos *et al.*, 1997
Streptomyces strain	–	Hydroxyectoine	Protection of oxidative protein damage (LDH)	Andersson *et al.*, 2000
			Reduction of VLS in immunotoxin therapy	Barth, 2000
			Stabilization of retroviral vaccines	Cruz *et al.*, 2006
			Induction of thermotolerance in *E. coli*	Malin and Lapidot, 1996
			Protection of *P. putida* against anhydrobiotic stress	Manazanera *et al.*, 2002
Thermophilic microorganisms	Thermophile	Various enzymes like DNA polymerase	Used mainly in animal genomics. Potential for applications in biosensors, compatible solutes in skin care products, drug excipients, treatment for respiratory disease, radioprotectants, peptide antibiotics, drug delivery and anticancer therapeutics.	Irwin, 2010
Vibrio diabolicus strain HE800	Barophile, thermophile	A linear tetrasaccharide repeating unit. Uronic acid: GlcN:GalA 1:0.5:0.5 $\rightarrow$3)-β-D-GlcpNAc-(1$\rightarrow$4)-β-D-GlcpA-(1$\rightarrow$4)-β-D-GlcpA-(1$\rightarrow$4)-α-D-GalpNAc-(1$\rightarrow$	Bone regeneration and cicatrizing material (patent US 7015206B2)	Raguenes *et al.*, 1997 Rougeaux *et al.*, 1999b; Zanchetta *et al.*, 2003a, b

help humans survive at biotic and abiotic extremes that we would not otherwise cope with. For example, they aid in production and preservation of high quality food stuffs for us to consume in otherwise food-poor environments, process hydrocarbons to produce high performance fabrics that can be worn to survive at temperature extremes. Extremophiles have also proved themselves to be novel, model systems which scientists utilize to perform life-saving medical research, particularly drug development, as well as helping in energy generation and environmental remediation (Arora, 2012).

Extremophiles have also been shown to synthesize a plethora of other compounds, including but not limited to antioxidants, biocatalysts, cytoprotectants, antifreeze proteins, pharmaceutically useful compounds, membrane stabilizers and radiation countermeasure agents. Many of these have been evaluated and are already used medically and industrially; for others there is still a need to find novel applications, and yet more remain to be discovered.

The extremophiles employed and the environments they originate from are wide ranging: thermophiles (heat-lovers) have found widespread industrial and pharmaceutical application. Psychrophiles (cold-lovers) have been used to obtain cold-active enzymes, which find use in the detergent and food industries, in specific biotransformations and environmental bioremediations, specialized uses in contact lens cleaning fluids and reducing the lactose content of milk. Ice-nucleating proteins derived from psychrophiles have potential uses in the manufacture of ice cream or artificial snow; lipids isolated from Antarctic marine psychrophiles are used as dietary supplements in the form of polyunsaturated fatty acids (Russell, 1998). These are but a few examples of how psychrophiles have been employed in industry.

Barophiles (pressure-lovers) have yielded several natural products of potential use in human health and environmental bioremediation, while piezophiles (pressure-lovers) are of interest since they can be sources of novel restriction endonucleases or other DNA-binding proteins.

Alkaliphilic microorganisms have made a significant impact in the application of biotechnology for the manufacture of several consumer and pharmaceutically important products (Fujinami and Fujisawa, 2010), including food processing (potassium hydroxide-mediated removal of potato skins), cement manufacture (or casting), alkaline electroplating, leather tanning, paper and board manufacture, indigo fermentation, rayon manufacture and herbicide manufacture (Takahara and Tanabe, 1962; Gee *et al.*, 1980; Agnew *et al.*, 1995; Mueller *et al.*, 1998).

Among halophiles, the Archaeon *Halobacteria salinarum* produces the photosynthetic protein Bacteriorhodopsin that is used e.g. in optical information recording, and the alga *Dunaliella* sp. is used for the commercial production of β-carotene. Other halophiles are used to produce polymers (polyhydroxyalcanoates and polysaccharides), enzymes, and compatible solutes that have enhanced processes such as oil recovery, cancer detection, drug screening and the biodegradation of residues and toxic compounds (Ventosa and Nieto, 1995).

These potentially profitable discoveries have generated a large amount of interest in extremophile research in recent decades and consequently scientists around the world are increasingly exploring the novel applications of extremophiles and their extremozymes and extremolytes. It is applications such as those briefly introduced above that will be discussed in detail in the remainder of this chapter.

25.2 Industrial Applications of Extremophiles

25.2.1 Extremophilic enzymes in biotechnology

It is well known that a major constraint to the biotechnological application of enzymes derived from mesophiles (organisms that grow best at moderate temperatures, typically between 25 and 40°C) is their low stability to heat, pH, organic solvents and proteolytic degradation. This has now been circumvented via the use of extremozymes. Extremozymes offer new biotechnological opportunities for biocatalysis and biotransformations because they remain stable under extreme stressors, for example, at both high and low temperatures, over a wide range of pHs, ionic strengths, salinities, and possess the ability to function in organic solvents that would denature most other enzymes. In addition, extremozymes often have higher reaction rates, the capability of destroying and/or eliminating xenobiotics (chemical compounds foreign to a given biological system), and the ability to modulate the hyper-accumulation of substances such as heavy metals, pollutants and radionuclides. Moreover, the use of extremozymes allows industrial processes to closely approach the gentle, efficient processes that take place in nature.

One of the most significant applications of an extremozyme and one that revolutionized molecular biology was the discovery and isolation of Taq DNA polymerase from the thermophilic bacterium *Thermus aquaticus*, which lives in neutral and alkaline springs in the Great Fountain region of Yellowstone National Park, USA (Brock and Freeze, 1969). The enzyme, thermostable at up to 80°C, was isolated in 1976 (Chien *et al.*, 1976) but it was only in the 1980s that it was employed in the polymerase chain reaction (PCR) DNA amplification technique, eliminating the need to add enzyme after every cycle of thermal denaturation of the DNA (Saiki *et al.*, 1988). PCR is now a common, indispensable technique widely used for an array of applications, including DNA cloning for sequencing, DNA-based phylogeny, functional analysis of genes, diagnosis of hereditary and infectious diseases, the detection of contaminants in forensics, and for as agricultural uses. Taq polymerase has also been used for numerous other applications in research laboratories and industries. Other thermostable extremozymes from thermophiles find applications in the chemical, food, pharmaceutical, paper and textile industries (Vieille and Zeikus, 2001; Fujiwara, 2002), for instance, amylase for use in the production of glucose and xylanase for use in the paper bleaching process.

At the other end of the temperature scale, enzymes isolated from psychrophiles, such as lipases, proteases and cellulases have been used as additives in the preparation of detergents working at low temperature, or as additives in the frozen food industry. Cold active lipases, in particular, are the enzymes of choice for organic chemists, pharmacists, biophysicists, biochemical and process engineers, biotechnologists, microbiologists and biochemists (Joseph *et al.*, 2008) because they maintain high catalytic activity at low temperatures. They have a broad spectrum of biotechnological applications such as additives in detergents, additives in food industries, environmental bioremediations, biotransformation, molecular biology applications and heterologous gene expression in psychrophilic hosts to prevent formation of inclusion bodies. Examples include the cold active lipases isolated from the fungus *Candida antarctica*, which have been patented for use as industrial biocatalysts.

Examples of industrially important extremozymes isolated from extremophiles living in other types of extreme environment include, but are not limited to, alkaliphiles: Cao *et al.* (1992) isolated four alkaliphilic bacteria, NT-2, NT-6, NT-33 and NT-82, producing pectinase and xylanase. NT-33 has an excellent capacity for degumming ramie (*Boehmeria nivea*) fibres and these are increasingly being used in the textile industry to produce high strength ramie–cotton fabric composites. Similarly, peptic lyase produced by the alkaliphilic *Bacillus* sp. strain GIR 277, has been used to improve the production of a type of Japanese paper (Horikoshi, 2006).

Many other examples of extremozyme use exist, some of which are summarized in Table 25.1. As research continues, inevitably more will be isolated and applications found for them.

25.2.2 Extremozyme use in the detergent industry

The use of microbial enzymes in laundry processes dates back to the middle of the last century. Nowadays, the detergent industry is the largest single market for enzymes (25–30% of total sales) such as proteases, amylases, cellulases, lipases and pullulanases, and the use of small amounts (usually 0.4–0.8% crude enzyme by weight) in detergent formulations is common in most countries. Over half of all detergents presently available contain extremozymes. In particular, the discovery of alkaliphilic- or alkalitolerant-based biomolecules led to the introduction of proteolytic enzymes, classified as serine proteases, which have revolutionized the processes involved (Rai and Mukherjee, 2011).

The main reasons why enzymes from alkaliphiles have become so important in the detergent industry are: (i) their long term stability; (ii) energy and cost-efficiency; (iii) quicker and more reliable product; (iv) reduced effluents; and (v) stability in the presence of detergent additives such as bleach activators, softeners, bleaches and perfumes (Ulukanliand and Diúrak, 2002).

Specific examples include the ore-wash laundry Biotex® detergent that was launched in the early 1960s and contained an alkaline proteinase called Alcalase® (Subtilisin Carlsberg), produced from the bacterium *Bacillus licheniformis* by Nero Industries (Grant *et al.*, 1990). By the late 1960s, enzymes in detergents were quite popular in Europe and the USA. However, subsequent consumer reaction to adverse publicity, coupled with high costs and poor quality performance led to their withdrawal. However, since then scientists have been able to isolate a large number of enzymes that function very efficiently in solution at a pH of between 8 and 10.5. These include: alginases, alkaline proteases, amylases, α-galactosidase, β-galactosidase, cellulases, catalases, DNases, glucanases, pectinases, pullulanases, RNases, restriction enzymes, maltose dehydrogenase, penicillinase, glucosedehydrogenase, uricase, polyamine oxidase, β -mannanase, and β-Mannosidase, β1,3-xylanase (Horikoshi, 1991, 1999).

In recent years there has been a spectacular revival for detergent enzymes due to substantial innovations and technological improvements, changing patterns of consumer demands and governmental regulations fuelled partly by environmental concerns and need for energy-efficient processes. Recent examples of extremozyme use in the industry include a pullulanase isolated from an alkaliphilic *Bacillus* sp. strain KSM-1876 (Ara *et al.*, 1992, 1995, 1996) that is a good candidate for a dishwashing detergent additive, and an alkaline protease isolated from alkaliphilic *Bacillus* sp. strain KSM-K16, suitable for use in both powder and liquid detergents (Kobayashi *et al.*, 1992, 1994, 1995).

25.2.3 Extremozyme use in the leather industry

Traditional leather tanning has relied on the natural enzyme processes taking place during the growth of bacteria naturally present in animal dung for dehairing and bating of hides. The application of extremophilic (in particular alkaliphilic) bacterial enzymes in the manufacture of leather has resulted in quicker and more reliable processes (Rai and Mukherjee, 2011). Proteinases such as Alcalase® and Milezyme® or Reverdase® are used for cleaning and to prevent deterioration of the hides.

25.2.4 Application of extremophiles for biomining

Biomining – the use of extremophilic microorganisms to recover precious and base metals from mineral ores and concentrates – has developed into a successful and expanding area of biotechnology in recent years and has distinctive advantages over traditional mining. Two main processes are used: biooxidation and bioleaching. Biooxidation is a microbially-mediated oxidation process in which the valuable metals remain in the solid phase but become enriched; the solution may then be discarded. This contrasts with bioleaching, where the valuable metal is solubilized in water filtered through mineral ores in a process facilitated by the metabolism of bioleaching microbes. The metals are subsequently recovered from the leachate. Biomining has shown itself to require less capital, reduced operating costs and less skilled operating and maintenance personnel than the traditional pressure oxidation or roasting techniques.

How extremophiles survive in a biooxidizing or bioleaching environment has remained an enigma but it is known that extremophiles adopt adaptive responses to these extremes (Valenzuela *et al.*, 2006) and have the exceptional capability to transform most toxic metals to less volatile forms through: (i) intracellular complexation; (ii) decreased accumulation; (iii) extracellular complexation; or (iv) sequestration in the periplasm.

Biooxidation

The acidophile *Acidithiobacillus ferrooxidans* was the first bacterium discovered that was able to oxidize minerals and it is now routinely used to extract copper, gold and other metals. Now acidophilic, chemolithotrophic microorganisms, capable of oxidizing iron and sulfur are routinely used in industrial processes to recover metals from minerals. The thermoacidophilic archaeon *Sulfolobus metallicus* has also been used in industrial biomining processes to extract copper, gold and other metals.

Major considerations in large-scale gold mining are the reduction of operational cost and making the process environment friendly. Using a combination of the moderately thermophilic, sulfur-oxidizing bacterium *Acidithiobacillus caldus* and *Leptospirillum*, a genus of iron-oxidizing bacteria, in both biooxidation and bioleaching of refractory gold ore has led to the development of low-cost, highly efficient, environmental friendly processes that can be used in large gold mines (Xie *et al.*, 2009). Both extremophiles can resist the relatively high concentrations of arsenic ions produced during the process. Penev and Karamenev (2009) have also advocated the use of *Leptospirillum* spp. in ferrous iron (Fe^{2+}) biooxidation as they are more tolerant of the lower pH and higher redox potentials of the extraction media, and able to tolerate higher cultivation temperatures than mesophilic organisms. Therefore, they are good candidates for novel applications in biomachining, nanoparticle synthesis and microbial fuel cells.

Norris *et al.* (2009) advocated the use of thermotolerant *Thiobacillus prosperus* and similar bacteria isolated from saline, hydrothermal vents in Vulcano, Italy, in pyrite oxidation and copper sulfide ore leaching, since these bacteria can oxidize ferrous iron and sulfur in the presence of salt (NaCl) at concentrations at least twice that of seawater. Indeed, *T. prosperus* is considered to be the most salt tolerant, acidophilic iron (II)-oxidizing bacterium (Zammit *et al.*, 2009).

Bioleaching

In recent decades there has been increased interest in the study of bioleaching of elements such as arsenic from minerals as a result of extremophile research. For example, it was shown that the acidophilic bacterium *Acidithiobacillus ferrooxidans* could degrade arsenopyrite (FeAsS) in gold–arsenic concentrates. Indeed, Ubaldini *et al.* (1997) demonstrated that bioleaching arsenopyrite over a refractory mineral using a mixture of *A. ferrooxidans* and *A. thiooxidans* could increase gold extraction from 55.3% to 96.8%, effectively establishing the economic feasibility of the technology and its potential for alleviating related environmental problems.

Similarly, the moderately thermophilic, ferrous iron-oxidizing acidophilic bacterium *Acidimicrobium ferrooxidans*, isolated from a copper leaching dump (Clark and Norris, 1996), can be used to process nickel sulfide concentrates and also to leach copper and nickel ores (Davis-Belmar and Norris, 2009). Strains of thermoacidophilic, Archaea, *Acidianus brierleyi* and *Metallosphaera sedula*, are also effective solubilizers of copper from chalcopyrite ($CuFeS_2$) concentrates and can facilitate copper recoveries of over 95% in bioleaching reactor systems operated at 70°C (Dinkla *et al.*, 2009).

25.2.5 Application of extremophiles in the biorefining industry

Extremophiles have been extensively used in biorefining industries. For example, glucoamylase from the thermophilic crenarchaeon *Sulfolobus solfataricus*, which is optimally active at 90°C and pH 5.5–6.0, is used in the saccharification step in starch processing that involves hydrolysis of oligosaccharides into glucose/glucose syrups (Turner *et al.*, 2007). Similarly, xylanase (Xyn10A) from the thermophilic, halophilic bacterium *Rhodothermus marinus*, is used to improve brightness in the bleaching sequences of hard- and softwood pulps prepared by Kraft processing (conversion of wood into wood pulp consisting of almost pure cellulose fibres) at 80°C, thereby increasing the process rate and eliminating the use of toxic chlorine for bleaching.

25.2.6 Application of extremophiles in the chemical industry

Extremophiles find applications in the chemical industry for a wide variety of processes. A new aldo-keto reductase enzyme from the hyperthermophilic bacterium *Thermotoga maritima*, originally isolated from geothermally-heated marine sediment, is stable up to 80°C and retains over 60% activity for 5 h at this temperature (Willies *et al.*, 2010). It has therefore proved useful in the production of primary alcohols from substrates such as benzaldehyde, 1,2,3,6-tetrahydrobenzaldehyde and para-anisaldehyde. Furthermore, a novel thymidine kinase (TmTK) isolated from *T. maritime* shows high substrate specificity at the organism's native growth temperature (82°C) but turns promiscuous at 37°C. This TmTK is being used to increase the general understanding of substrate promiscuity among extremozymes and investigate the structural and functional consequences of changing protein dynamics (Lutz *et al.*, 2007).

Wang *et al.* (2010) demonstrated the industrial importance of the thermostable extremozyme deoxy-D-ribose-5-phosphate aldolase, from a hyperthermophilic crenarchaeon, *Hyperthermus butylicus*, that is resistant to high concentrations of acetaldehyde and retains 80% residual activity after heating at 100°C for 10 min. As such it should prove useful in industrial settings where high concentrations of acetaldehyde are necessary.

25.3 Medical and Biomedical Applications of Extremophiles

25.3.1 Pharmaceuticals

Recent reports of the ability of extremophiles to synthesize new chemical entities (NCEs) and their unrivalled ability to synthesize exploitable natural products show immense promise for the pharmaceutical industry (Poli *et al.*, 2009, 2010; Kumar *et al.*, 2010; Dhaker *et al.*, 2011). A number of pharmaceutically active drugs viz., antibacterial, antialgal, antihelminthic, antivirals, antiprotozoals, immunomodulatory, anticancer, cardioprotective compounds, antioxidants, radioprotective compounds, anticoagulants, have already been isolated from diverse extremophiles (Table 25.2; Poli *et al.*, 2009; Kumar *et al.*, 2010). In addition, genetic engineering offers further opportunities for the production of entirely new chemically relevant compounds from naturally occurring extremolytes. A selection of pharmaceutical applications is discussed below:

A number of antioxidants and anticancer drugs have been obtained from extremophilic organisms. When glycine is supplied as a carbon source, it has been reported that the acidophilic green microalga *Chlamydomonas acidophila*, isolated from Tinto River, Spain, accumulates high concentrations of lutein (ca. 10 g kg^{-1} dry weight) and produces large amounts of zeaxanthin, both well-known natural antioxidants (Cuaresma *et al.*, 2011). Both compounds are potentially effective nutritional supplements for patients suffering from macular degeneration.

Ectoine (1,4,5,6-Tetrahydro-2-methyl-4-pyrimidinecarboxylic acid) is a small organic molecule first identified in the halophilic bacterium *Ectothiorhodospira halochloris* (Galinski *et al.*, 1985), but has since been found in a wide range of Gram-negative and Gram-positive bacteria (Steger *et al.*, 2004;

Jebbar *et al.*, 2005; Garcia-Estepa *et al.*, 2006; Vargas *et al.*, 2008). Ectoine protects these extremophiles against the dehydration caused by the high temperatures, high salt concentrations and low water activity in their native environment. In humans, ectoine stabilizes proteins and other cellular structures, and can be used to protect the skin against the deleterious effects of non-ionizing UV-A radiation and dehydration. The anti-ageing effects of ectoine are also well documented.

A psychrophilic bacterial strain, designated XA-1, was isolated from Xi'an hot springs in China and shown to positively influence the immune response and protect the health status of carp (*Cyprinus carpio*) against *Aeromonas hydrophila* infection (Wang *et al.*, 2011). The protective mechanisms conferred have potential applications in humans and are currently being investigated. Similarly, the immunomodulatory and anticancer properties of the thermophilic bacterium *Bacillus thermodenitrificans*, isolated from a marine hydrothermal vent (Manachini *et al.*, 2000), have been established and are being developed for human use, as are the immunostimulatory activities of a semiquinone glucoside derivative (SQGD) isolated from a radioresistant bacterium, *Bacillus* sp. (Kumar *et al.*, 2011).

Very recently, potential anticancer drug discoveries have been made using two ceramide derivatives, bathymodiolamides A and B, isolated from the deep-sea hydrothermal vent invertebrate mussel *Bathymodiolus thermophilus* (Andrianasolo *et al.*, 2011). When screened by a ApopScreen cell-based screen these compounds showed enormous potential for apoptosis induction and potential anticancer activity. There are many more examples of pharmaceutical derived from extremophiles but this chapter cannot hope to describe them all, especially since new discoveries are being made on a regular basis.

25.3.2 Biosensing

Extremophilic enzymes have been employed for the development of simple and highly sensitive biosensors (de Champdoré *et al.*, 2007; De Stefano *et al.*, 2008). For example, thermophilic enzymes have been used for the construction of optical nanosensors, stable and non-consuming analytes. These innovative devices are based on the ability of thermophilic enzymes to bind the substrate at room temperature, without transforming it. The binding of substrate to thermophilic enzyme can be monitored as fluorescence variation of the enzyme. Fluorescence detection, due to simplicity and sensitivity, is now the dominant analytical tool in medical testing, biotechnology and drug discovery. Other biosensing applications include but are not limited to the following:

Cancer detection

The previous section discussed cancer treatments but other extremolytes can be used in cancer detection. Prolidase is a metallopeptidase that is ubiquitous in nature and has been isolated from a number of extremophiles, in particular the hyperthermophilic archaeon *Pyrococcus furiosus* (Theriot *et al.*, 2009). Prolidase specifically hydrolyses dipeptides with a prolyl residue in the carboxy terminus (NH2–X–/–Pro–COOH) and increased prolidase activity in melanoma cell lines has led investigators to create cancer prodrugs targeting this enzyme.

Chemical warfare agents

Prolidase also has the ability to degrade toxic organophosphorus (OP) compounds by cleaving the P-F and P-O bonds in the nerve agents sarin and soman, and conferring protection against chemical warfare agents (Theriot *et al.*, 2009). The applications of prolidase for detoxifying OP nerve agents include its incorporation into fire-fighting foams and as biosensors for OP compound detection.

Glucose sensing

The development of a variety of diabetic health care devices, including continuous, non-invasive, painless glucose monitoring, control of an insulin pump and a warning system for hyper- and hypoglycaemic conditions has been a long-standing research goal. The use of manipulated enzymes as probes for the design of implantable and non-consuming glucose fluorescence biosensors is a topic of recent interest. A thermostable glucokinase from the thermophilic bacterium *Bacillus stearothermophilus* has been tested for use as a reversible glucose sensor (D'Auria *et al.*, 2002) and employed in glucose assays (Tomita *et al.*, 1995). It has also been postulated that glucose dehydrogenase (GD) from the thermoacidophilic archaeon *Thermoplasma acidophilum* can be employed in glucose sensing because it can function in the absence of a coenzyme (D'Auria *et al.*, 2000a). Moreover, De Stefano *et al.* (2008) cloned the gene coding for a thermostable, D-trehalose/D-maltose-binding protein (TMBP) in the hyperthermophilic archaeon *Thermococcus litoralis*, and the recombinant protein was expressed in *Escherichia coli*. The recombinant TMBP exhibited the same functional and structural properties as the native one, exhibited a highly thermostable nature and a high affinity for sugar (maltose, trehalose and glucose) demonstrating yet more potential for the design of advanced optical non-consuming analyte biosensors for glucose detection.

Cation sensing: a thermostable enzyme as a probe for sodium sensing

In clinical blood analyses, the measurement of sodium and potassium concentrations is routine (Smith *et al.*, 1988; Minta and Tsien, 1989; Oguz and Akkaya, 1998). The availability of simple optical methods for rapid point-of-care testing, especially for potassium which is measured during hypertensive screening, would be immensely helpful. A variety of fluorescence probes have been developed that respond to sodium and/or potassium. Most of these responses are based on partially selective binding of these cations to crown ethers (Pedersen, 1988). A sodium sensing method developed using the enzyme pyruvate kinase from the thermoacidophilic bacterium *Bacillus acidocaldarius* has been reported (D'Auria *et al.*, 2000b). This enzyme catalyses the essentially irreversible transphosphorylation step from phosphoenolpyruvate to ADP, the last step of the glycolytic pathway and a reaction that requires magnesium and potassium ions. Substrate binding by the enzyme is an important issue in the manufacture of such biosensors and there is a need to focus more research in this direction with a view to developing highly sensitive methods, reducing costs, and accelerating and simplifying analysis of various compounds.

25.3.3 Antibiotic production and resistance to antibiotics

Bacteria isolated from highly alkaline environments (pH 9–10.5) have exhibited the ability to produce antibiotics (Horikoshi, 1991), and several metabolites produced by alkaliphilic bacteria have been identified (Eltem and Ucar, 1998). An alkaliphilic, aerobic actinomycete, *Nocardiopsis dassonvillei* strain OPC-15, isolated under alkaline conditions, has been shown to produce phenazine antibiotics (Tsujibo *et al.*, 1988).

However, it should be noted that the direct production of antibiotics by alkaliphilic microorganisms may not be straightforward: several antibiotics are unstable under alkaline conditions and there is a possibility that the antibiotics produced are destroyed during alkaliphile cultivation. Nevertheless, some antibiotic-producing alkaliphiles, especially actinomycetes, are able to grow in neutral media and it is hoped that these can be exploited in the future.

25.4 Environmental Applications of Extremophiles

25.4.1 Bioremediation of heavy metals and hydrocarbons

With increasing hydrocarbon pollution and heavy metal contamination of environments worldwide, bioremediation has become an area of great interest and one in which extremophiles hold immense promise.

In the presence of oxygen, microbial reduction of chromium (Cr^{6+}) is commonly catalysed by soluble enzymes. A soluble Cr^{6+} reductase, ChrR, has been purified from the Gram-negative, saprotrophic soil bacterium, *Pseudomonas putida* MK1, and offers promise for Cr^{6+} bioremediation in a wide range of environments (Cheung and Gu, 2007). Similarly, mercury-resistant bacterial strains belonging to the genus *Psychrobacter* have been shown to possess high levels of resistance against $HgCl_2$ and CH_3HgCl and show potential for the bioremediation of mercury-polluted sediments (Pepi *et al.*, 2011).

Two halotolerant and alkaliphilic bacteria, designated as *Halomonas* sp. 19-A and Y2, have recently been isolated from wheat straw black liquor and shown to use guaiacol, vanillin, dibenzo-p-dioxin, biphenyl and fluorene, as carbon and nitrogen sources. The two strains produced carboxymethylcellulase (CMCase), xylanase, lipase, amylase and pullulanase which exhibited high activities even under extreme pH and salinity conditions; they too are promising agents for use in bioremediation and industrial processes (Yang *et al.*, 2010). Additionally, a sulfate-reducing bacterium (isolate M1), similar to the neutrophile *Desulfosporosinus orientis*, and a Gram-negative (non sulfate-reducing) acidophile (isolate PFBC) similar to *Acidocella aromatica*, have been isolated that can be utilized in sulfidogenesis in acidic liquors for the selective recovery of heavy metals from wastewaters (Kimura *et al.*, 2006).

More recently Tapilatu *et al.* (2010) have advocated the role of halophilic Archaea, belonging to the genera *Haloarcula* (strain MSNC 2) and *Haloferax* (strains MSNC 4, MSNC 14, and MSNC 16), and collected from a shallow crystallizer hypersaline pond in Camargue, France, in the degradation of hydrocarbons in both pristine and hydrocarbon-contaminated hypersaline environments.

25.4.2 Bioremediation of radionuclides

The high cost of remediating radioactive waste sites from nuclear weapons production has stimulated the development of bioremediation strategies. Radioresistant microorganisms are known to survive in ionizing and non-ionizing radiation environments. The DNA and protein protective properties of extremophiles can be exploited for biotechnological applications and thus assist humans. Brim *et al.* (2000) developed a radiation resistant bacterium for the treatment of mixed radioactive wastes containing ionic mercury using the thermophilic bacterium *Deinococcus radiodurans*, the most radiation resistant organism described to date. *Deinococcus geothermalis* is another extremely radiation-resistant thermphilic bacterium and is well known for its ability to grow at temperatures as high as 55°C. Both bacteria have been engineered for *in situ* bioremediation of radioactive wastes (Brim *et al.*, 2003, 2006). They are capable of reducing a variety of radionuclides (e.g. Hg(II), Fe(III)-nitrilotriacetic acid, U(VI), Cr(VI)) to less mobile, less toxic compounds at elevated temperatures, and show promise for the bioremediation of contaminated sites, in particular those in which toluene and other fuel hydrocarbons are commonly found in association with radionuclides.

More attention needs to be focused on the bioremediation of radionuclides because several such compounds pose a great risk to the health of humans and the environment. The increasing use of nuclear energy around the globe, despite all of the precautions in place, poses a serious risk of release of such agents into the environment, all too recently witnessed in the Chernobyl reactor accident in the former USSR and the recent Fukushima incident in Japan (Sharma and Arora, 2011). As many as 11 radionuclides, including ^{131}I, and ^{137}Cs, are supposed to have been released in the sea water during the Fukushima incident. Increases in environmental radioactivity pose a danger that may persist for many years, and radionuclides have the potential of entering the food chain. However, bioremediation can be utilized effectively to reduce radionuclides which have a half life of several years.

25.5 Extremophiles as Food Additives

The food industry is one in which extremophiles have played an important role. Extremophilic marine organisms, in particular, have been widely used in the food industry, and new biotechnological tools have been applied for their cultivation and isolation of unique bioactive compounds. Developments and upcoming areas of research that utilize advances in biotechnology in the production of food ingredients from marine sources have been reviewed in a number of other publications (Rasmussen and Morrissey, 2007; Poli *et al.*, 2010).

Enzymes extracted from extremophilic fish and marine microorganisms provide numerous advantages over the traditional enzymes used in food processing due to their ability to function at extremes of temperature and pH. Psychrophilic fish proteins such as collagens and their gelatin derivatives are stable at relatively low temperatures and have been used in heat-sensitive processes such as gelling and clarifying. Similarly, polysaccharides derived from extremophilic algae, including algins, carrageenans and agar, are widely used for their ability to form gels and act as thickeners and stabilizers in a variety of foods. At the other temperature extreme, thermostable polygalacturonase from a thermophilic mould, *Sporotrichum thermophile*, is optimally active at 55 °C and, when added to fruit pulps with xylanase and cellulose, increases the yield of fruit juices (Turner *et al.*, 2007).

Psychrophilic microbes are known to produce hydrolases such as β-glucanases, cellulases, pectinases and proteinases, which find application in the food industry. The halophilic, lactic acid bacterium, *Tetragenococcus halophila*, is used in soy sauce production via fermentation that involves high salt concentration. Other halophilic bacteria employed include *Lactobacillus planetarium, Halobacterium salinarum, Halococcus* spp. and *Bacillus* spp.

Corn Products International Co. in the USA began producing cyclodextrins viz., β-CD by using *Bacillus macerans* CGTase in 1969. β-CD has uses in large quantities in food stuffs, chemicals and pharmaceuticals. Several unmodified cyclodextrins synthesized by extremophiles find use as food additives to stabilize volatile compounds such as docosahexanoic acid and eicosapentaenoic acid; to preserve foods; to maintain the flavours of confectioneries, tea, fruit and vegetable juices, and condiment pastes (ginger, garlic and mustard); to protect natural colours in coloured foods and beverages against the atmospheric oxidation; to improve the taste and odour of soy milk; and to reduce the hygroscopicity of sweets, grated cheeses, soup powders and seasonings etc. Cyclodextrins also find application as deodorants in gargles, chewing gums and edible mouth fresheners (Astray *et al.*, 2009).

Microbial enzymes used in food processing are typically sold as enzyme preparations that contain not only a desired enzyme activity, but also other metabolites of the production strain, plus added materials such as preservatives and stabilizers. These additives must be of food grade and meet the stringent regulatory standards. Well characterized, non-pathogenic, non-toxic extremophilic microbial strains, particularly those with a history of safe use in food enzyme manufacture, have been advocated for genetic modification and strain improvement strategies (Pariza and Johnson, 2001). Such exploration and use of extremophiles in food production is essential if we are to achieve sustainable global food production (Dijkshoorn *et al.*, 2010).

25.6 Extremophile Use in Agriculture

Extremophile use in agriculture is a new and evolving field. McCoy *et al.* (2009) reported the isolation of an enzyme with sarcosine dimethylglycine methyltransferase (SDMT) activity from the acidophilic, thermophilic red microalga *Galdieria sulphuraria*, which has a melting temperature of ~60°C. This enzyme may play a role in maintaining the alga's osmotic homeostasis in its highly saline environment and has generated biotechnological interest in reducing environmental stress in plants and industrial bacteria.

25.7 Sustainably Harnessing Extremophiles

The importance of extremophiles can be gauged from the fact that they have a multitude of applications in various sectors. However, the ever increasing human need for new pharmaceutically and industrially relevant compounds necessitates that extremophiles be harvested in a sustainable manner without irreparable loss of biodiversity. Bioprospection usually leads to increased demand on the target organisms and can result in overharvesting from natural environments, as has been observed with several pharmaceutically important drugs derived from natural sources.

Developing countries and the poor local populations remain at a major disadvantage since they do not usually benefit from the commercial exploitation of their own bioactive and industrially important natural resources due to lack of appropriate legislation, money and in-country expertise. A paucity of national legislation or effective international agreements related to conservation of biodiversity has led to overharvesting of the extremophiles in various parts of the world, including Antarctica. This trend does not augur well for their sustainable use. There is an urgent requirement for a broad financially viable, cost-effective viewpoint, realistic policy framework and effective planning, coordination and implementation strategies. Traditional knowledge about extremophiles, wherever harvested, should include a provision for benefit sharing, biodiversity conservation and best-practice procedures for collection etc. The issue of royalty to the local populations would in turn ensure that the biodiversity is conserved *in situ* to the best of local ability. It is imperative to handle carefully and pragmatically the complex biodiversity issues covered by the Convention on Biological Diversity (CBD, 1992), such as: (i) considering biodiversity in its various forms; (ii) planning for conservation of biological diversity both *in situ* and *ex situ*; (iii) sustainable use of bioactive compounds; (iv) a fair and equitable sharing of the benefits derived; (v) rational use and holistic management of biological diversity; and (vi) ensuring the recognition of the country of origin of the bioresource(s). The IPR rights of the various parties involved should also be adequately protected. In addition, provider countries should consider the time and cost it takes to develop a product, the volumes sold and average profit, and the likelihood that a product will be developed from a given collaboration. For example, industrial enzymes have a much lower profit margin than pharmaceuticals, but they are cheaper to develop compared to pharmaceuticals and can yield commercial products in half or less the time (Laird and ten Kate, 1999; ten Kate, 1999; Ernst and Young, 2005). Following the guidelines of the CBD will go a long way toward ensuring the sustainable harnessing of extremophiles and extremolytes, and the equitable sharing of benefits, whilst conserving Earth's biodiversity.

25.8 The Future of Extremophile Biotechnology

Extremophile and extreme environment research not only offers us an insight into the evolutionary processes governing life on Earth and on other planets, but can also help humans live outside their own 'extreme envelope'. Biotechnology, biomedical research, food development and modification, hydrocarbon refinement and many other fields, expand our human potential to survive biotic and abiotic extremes.

As we have shown, since extremophiles were first discovered, thousands of molecules and drug candidates derived from them have been evaluated for their potential application across a wide range of industries. Even though several microbes have already served humans by synthesizing drugs that save lives, extremophiles have so far not been evaluated to their fullest potential and only a minor fraction of the microorganisms on Earth have so far been exploited. Novel developments in the cultivation and production of extremophiles, and also developments related to the cloning and expression of their genes in heterologous hosts, will increase their potential applications in a multitude of areas. However, as we do discover and exploit extremophiles and extremolytes, there is a pressing need for sustainable harnessing and equitable sharing of the benefits.

At present we are still far away from understanding the intricate mechanisms that render extremophiles more resistant to harsh environments than would ordinarily be the case. Understanding the processes at a molecular level will help develop even wider biotechnological and biomedical applications. Some specific issues that require further research include: the use of rDNA technology for the cloning and expression of genes involved in secondary metabolite biosynthesis for augmenting production, the standardization of growth and production conditions, the use of bioreactors and downstream processing, and the characterization of secondary metabolites utilizing state-of-the-art techniques. Thus, it is clear that the field remains wide open for further exploration, discovery, research and development.

Acknowledgements

Dr Rajesh Arora is immensely grateful to Dr W. Selvamurthy, Chief Controller Research and Development (Life Sciences and International Cooperation), Defence Research and Development Organization, New Delhi, India for whole-hearted support. He would also like to thank his students for their help.

References

Agnew, M.D., Koval, S.F. and Jarrell, K.F. (1995) Isolation and characterisation of novel alkaliphiles from bauxite-processing waste and description of *Bacillus vedderi* sp. nov. *Systematic and Applied Microbiology* 18, 221–230.

Andersson, M.M., Breccia, J.D. and Hatti-Kalu, R. (2000) Stabilizing effect of chemical additives against oxidation of lactate dehydrogenase. *Biotechnology Applied Biochemistry* 32, 145–153.

Andrianasolo, E.H., Haramaty, L., McPhail, K.L., White, E., Vetriani, C., Falkowski, P. and Lutz, R. (2011) Bathymodiolamides A and B, ceramide derivatives from a deep-sea hydrothermal vent invertebrate mussel *Bathymodiolus thermophilus*. *Journal of Natural Products* (in press).

Anton, J., Meseguer, I. and Rodriguez-Valera, F. (1988) Production of an extracellular polysaccharide by *Haloferax mediterranei*. *Applied Environmental Microbiology* 10, 2381–2386.

Ara, K., Igarashi, K., Hagihara, H., Sawada, K., Kobayashi, T. and Ito, S. (1992) Purification and some properties of an alkaline pullulanase from alkaliphilc *Bacillus* sp. KSM-1876. *Bioscience, Biotechnology and Biochemistry* 56, 62–65.

Ara, K., Igarashi, K., Sacki, K. and Ito, S. (1995) An alkaline amylopullulanase from alkaliphilic *Bacillus* sp. KSM-1378: kinetic evidence for two independent active sites for the α-1,4 and α-1,6 hydrolytic reactions. *Bioscience, Biotechnology and Biochemistry* 59, 662–666.

Ara, K., Igarashi, K., Hagihara, H., Sawada, K., Koabayashi, T. and Ito, S. (1996) Separation of functional domains for the α-1,4 and α-1,6 hydrolytic activities of a *Bacillus* amylopullulanase by limited proteolysis with papain. *Bioscience, Biotechnology and Biochemistry* 60, 634–639.

Arena, A., Gugliandolo, C., Stassi, G., Pavone, B., Iannello, D., Gugliandolo, C. and Bisignono, G. (2006) Antiviral and immunoregulatory effect of a novel exopolysaccharide from a marine thermotolerant *Bacillus licheniformis*. *International Immunopharmacology* 6, 8–13.

Arena, A., Gugliandolo, C., Stassi, G., Pavone, B., Iannelo, D., Bisignano, G. and Maugeri, T.L. (2009) An exopolysaccharide produced by *Geobacillus themodenitrificans* strain B3-72: Antiviral activity on immunocompetent cells. *Immunology Letters* 123, 132–137.

Arora, A., Ha, C. and Park, C.B. (2004) Inhibition of insulin amyloid formation by small stress molecules. *FEBS Letters* 564, 121–125.

Arora, R. (2012) Microbial biotechnology: energy and environment. CAB International, Wallingford, United Kingdom, in press.

Astray, G., Barreiro, C.G., Mejuto, J.C., Otero, R.R. and Gandara, J.S. (2009) A review on the use of cyclodextrins in foods. *Food Hydrocolloids* 23, 1631–1640.

Bai, R.L., Paull, K.D., Herald, C.L., Malspeis, L. and Pettiet, G.R. (1991) Halichondrin B and homohalichondrin B, Marine natural products binding in the vinca domain of tubulin. Discovery of tubulin-based mechanism of action by analysis of differential cytotoxicity data. *Journal of Biological Chemistry* 266, 15882–15889.

Barth, S. (2000) Pharamaceutical Preparation European Patent EP000001183047.

Beyer, N., Driller, H. and Bunger, J. (2000) Ectoin: an innovative, multifunctional active substance for the cosmetic industry. *SOFW-Journal* 126, 27–29.

Bloom, F.R., Price, P., Lao, G., Xia, J.L., Crowe, J.H., Battista, J.R., Helm, R.F., Slaughter, S. and Potts, M. (2001) Engineering mammalian cells for solid-state sensor applications. *Biosensors and Bioelectronics* 16, 603–608.

Borges, N., Ramos, A., Raven, N., Sharp, R.J. and Santos, H. (2002) Comparative study of the thermostabilizing properties of mannosylglycerate and other compatible solutes on model enzymes. *Extremophiles* 6, 209–216.

Bozzi, L., Milas, M. and Rinaudo, M. (1996a) Characterization and solution properties of a new exopolysaccharide excreted by the bacterium *Alteromonas* sp. strain 1644. *International Journal of Biological Macromolecules* 18, 9–17.

Bozzi, L., Milas, M. and Rinaudo, M. (1996b) Solution and gel rheology of a new exopolysaccharide excreted by the bacterium *Alteromonas* sp. strain 1644. *International Journal of Biological Macromolecules* 18, 83–91.

Brim, H., McFarlan, S.C., Fredrickson, J.K. and Kenneth, W. (2000) Engineering *Deinococcus radiodurans* for metal remediation in radioactive mixed waste environments. *Nature Biotechnology* 18, 85–90.

Brim, H., Venkateswaran, A., Kostandarithes, H.M., Fredrickson, J.K. and Daly, M.J. (2003) Engineering *Deinococcus geothermalis* for bioremediation of high-temperature radioactive waste environments. *Applied and Environmental Microbiology* 69, 4575–4582.

Brim, H., Osborne, J.P., Kostandarithes, H.M., Fredrickson, J.K., Wackett, L.P. and Daly, M.J. (2006) *Deinococcus radiodurans* engineered for complete toluene degradation facilitates Cr(VI) reduction. *Microbiology* 152, 2469–2477.

Brock, T.D. and Freeze, H. (1969) *Thermus aquaticus*, a nonsporulating extreme thermophile. *Journal of Bacteriology* 98, 289–297.

Bummino, E., Schiraldi, C., Baroni, A., Paoletti, I., Lamberti, M., de Rosa, M. and Tufano, M.A. (2005) Ectoine from halophilic microorganisms induces the expression of hsp70 and hsp70B′ in human keratinocytes modulating the proinflammatory response. *Cell Stress Chaperones* 10, 197–203.

Bunger, J. and Driller, H. (2004) Ection: an effective natural substance to prevent UVA-induced premature photoaging. *Skin Pharmacology and Physiology* 17, 232–237.

Bunger, J., Degwert, J. and Driller, H. (2001) the protective function of compatible solute ectoine on the skin cells and its biomolecules with respect to UV-radiation, immunosuppression and membrane damage. *IFSCC Magazine* 4, 1–6.

Cambon-Bonavita, M.-A., Raguenes, G., Jean, J., Vincent, P. and Guézennec, J. (2002) A novel polymer produced by a bacterium isolated from a deep-sea hydrothermal vent polychaete annelid. *Journal of Applied Microbiology* 93, 310–315.

Cao, J., Zheng, L. and Chen, S. (1992) Screening of pectinase produced from alkalophilic bacteria and study on its potential application in degumming of ramie. *Enzyme and Microbial Technology* 14, 1013–1016.

Cheung, K.H. and Gu, J. (2007) Mechanism of hexavalent chromium detoxification by microorganisms and bioremediation potential: a review. *International Biodeterioration and Biodegradation* 59, 8–15.

Chien, A., Edgar, D.B. and Trela, J.M. (1976) Deoxyribonucleic acid polymerase from the extreme thermophile *Thermus aquaticus*. *Journal of Bacteriology* 127, 1550–1557.

Clark, D.A. and Norris, P.R. (1996) *Acidimicrobium ferrooxidans* gen. nov., sp. nov.: mixed-culture ferrous iron oxidation with *Sulfobacillus* species. *Microbiology* 142, 785–790.

Colliec Jouault, S., Chevolot, L., Helley, D., Ratiskol, J., Bros, A., Singuin, C., Roger, O. and Fischer, A.M. (2001) Characterization, chemical modifications and in vitro anticoagulant properties of an exopolysaccharide produced by *Alteromonas infernos*. *Biochemica et Biophysica Acta* 1528, 141–151.

Convention on Biological Diversity (1992) Available at: www.cbd.int/doc/legal/cbd-en.pdf (accessed 1 May 2011).

Cruz, P.E., Silva, A.C., Roldão, A., Carmo, M., Carronodo, M.J. and Alves, P.M. (2006) Screening of novel excipients for improving the stability of retroviral and adenoviral vectors. *Biotechnology Progress* 22, 568–576.

Cuaresma, M., Casal, C., Forján, E. and Vílchez, C. (2011) Productivity and selective accumulation of carotenoids of the novel extremophile microalga *Chlamydomonas acidophila* grown with different carbon sources in batch systems. *Journal of Industrial Microbiology and Biotechnology* 38, 167–177.

D'Auria, S., Di Cesare, N., Gryczynski, Z., Gryczynski, I., Rossi, M. and Lakowicz, J.R. (2000a) A thermophilic apoglucose dehydrogenase as nonconsuming glucose sensor. *Biochemical and Biophysical Research Communications* 274, 727-731.

D'Auria, S., Rossi, M., Herman, P. and Lakowicz, J.R. (2000b) Pyruvate kinase from the thermophilic eubacterium *Bacillus acidocaldarius* as probe to monitor the sodium concentrations in the blood. *Biophysical Chemistry* 84, 167–176.

D'Auria, S., DiCesare, N., Staiano, M., Gryczynski, Z., Rossi, M. and Lakowicz, J.R. (2002) A novel fluorescence competitive assay for glucose determinations by using a thermostable glucokinase from the thermophilic microorganism *Bacillus stearothermophilus*. *Analytical Biochemistry* 303, 138–144.

Davis-Belmar, C.S. and Norris, P.R. (2009) Ferrous iron and pyrite oxidation by '*Acidithiomicrobium*' species. *Advanced Materials Research* 71–73, 271–274.

de Champdoré, M., Staiano, M., Rossi, M. and D'Auria, S. (2007) Proteins from extremophiles as stable tools for advanced biotechnological applications of high social interest. *Journal of the Royal Society Interface* 4, 183–191.

De Stefano, L., Vitale, A., Rea, I., Staiano, M., Rotiroti, L., Labella, T., Rendina, I., Aurilia, V., Rossi, M. and D'Auria, S. (2008) Enzymes and proteins from extremophiles as hyperstable probes in nanotechnology: the use of D-trehalose/D-maltose-binding protein from the hyperthermophilic archaeon *Thermococcus litoralis* for sugars monitoring. *Extremophiles* 12, 69–73.

Dhaker, A.S., Marwah, R., Damodar, R., Gupta, D., Gautam, H.K., Sultana, S. and Arora, R. (2011) Isolation and *in vitro* evaluation of antioxidant and radioprotective properties of a novel extremophile from mud volcano: implications for management of radiation emergencies. *Molecular and Cellular Biochemistry* (in press).

Dijkshoorn, L., De Vos, P. and Dedeurwaerdere, T. (2010) Understanding patterns of use and scientific opportunities in the emerging global microbial commons. *Research in Microbiology* 161, 407–413.

Dinkla, I.J.T., Gericke, M., Geurkink, B.K. and Hallberg, K.B. (2009) *Acidianus brierleyi* is the dominant thermoacidophile in a bioleaching community processing chalcopyrite containing concentrates at 70°C. *Advanced Materials Research* 71–73, 67–70.

Eltem, R. and Ucar, F. (1998) The determination of antimicrobial activity spectrums of 23 *Bacillus* strains isolated from Denizli-Acigol (Bitter Lake) which is soda lake (Na_2SO_4). *Kukem Dergisi* 21, 57–64.

Ernst and Young (2005) Beyond Borders: A Global Perspective. Global Biotechnology Report. Available at: www.ey.com (accessed 30 April 2011).

Fujinami, S. and Fujisawa, M. (2010) Industrial applications of alkaliphiles and their enzymes-past, present and future. *Environmental Technology* 31, 845–856.

Fujiwara, S. (2002) Extremophiles: development of their special function and potential source. *Journal of Bioscience and Bioengineering* 94, 518–525.

Galinski, E.A., Pfeiffer, H.P. and Trüper, H.G. (1985) 1, 4, 5, 6-Tetrahydro-2-methyl-4-pyrimidinecarboxylic acid A novel cyclic amino acid from halophilic phototrophic bacteria of the genus *Ectothiorhodospira*. *European Journal of Biochemistry* 149, 135–139.

Garcia-Estepa, R., Argandona, M., Reina-Bueno, M., Capote, N., Iglesias-Guerra, F., Nieto, J.J. and Vargas, C. (2006) The ectD gene, which is involved in the synthesis of the compatible solute hydroxyectoine, is essential for thermoprotection of the halophilic bacterium *Chromohalobacter salexigens*. *Journal of Bacteriology* 188, 3774–3784.

Gee, J.M., Lund, B.M., Metcalf, G. and Peel, J.L. (1980) Properties of a new group of alkalophilic bacteria. *Journal of General Microbiology* 117, 9–17.

Gerday, C. and Glansdorff, N. (2007) *Physiology and Biochemistry of Extremophiles*, 1st edn. American Society for Microbiology (ASM) Press, Washington, DC, 429 pp.

Goller, K. and Galinski, E.A. (1999) Protection of a model enzyme (lactate dehydrogenase) against heat, urea and freeze-thaw treatment by compatible solute additives. *Journal of Molecular Catalysis B Enzymatic* 7, 37–45.

Grant, W.D., Mwatha, W.E. and Jones, B.E. (1990) Alkaliphiles: ecology, diversity and applications. *FEMS Microbiology Reviews* 75, 255–270.

Grether-Beck, S., Timmer, A., Felsner, I., Brenden, H., Brammertz, D. and Krutmann, J. (2005) Ultraviolet A-induced signaling involves a ceramide-mediated autocrine loop leading to ceramide de novo synthesis. *Journal of Investigative Dermatology* 125, 545–553.

Guézennec, J.G. (2002) Deep-sea hydrothermal vents: a new source of innovative bacterial exopolysaccharide produced of biotechnological interest. *Journal of Industrial Microbiology and Biotechnology* 29, 204–208.

Gugliandolo, C. and Maugeri, T.L. (1998) Temporal variations of culturable mesophilic heterotrophic bacteria from a marine shallow hydrothermal vent of the island of Vulcuno (Eolian Islands, Italy). *Microbiology Ecology* 36, 13–22.

Hamann, M.T., Otto, C.S., Scheuer, P.J. and Dunbar, D.C. (1996) Kahalaides: bioactive peptides from a marine mollusk *Elysia rufescens* and its algal diet *Bryopsis* sp. *Journal of Organic Chemistry* 61, 6594–6600.

Horikoshi, K. (1991) *Microorganisms in Alkaline Environments*. Kodansha, Tokyo, Japan, 275 pp.

Horikoshi, K. (1999) Alkaliphiles: some applications of their products for biotechnology. *Microbiology and Molecular Biology Reviews* 63, 735–750.

Horikoshi, K. (2006) Alkaliphiles – from an industrial point of view. *FEMS Microbiology Reviews* 18, 259–270.

Irwin, J.A. (2010) Extremophiles and their application to veterinary medicine. *Environmental Technology* 31, 857–869.

Jebbar, M., Sohn-Bosser, L., Bremer, E., Bernard, T. and Blanco, C. (2005) Ectoine-induced proteins in *Sinorhizobium meliloti* include an Ectoine ABC-Type transporter involved in osmoprotection and ectoine catabolism. *Journal of Bacteriology* 187, 1293–1304.

Joseph, B., Ramteke, P.W. and Thomas, G. (2008) Cold active microbial lipases: some hot issues and recent development. *Biotechnology Advance* 26, 457–470.

Kimura, S., Hallberg, K.B. and Johnson, D.B. (2006) Sulfidogenesis in low pH (3.8-4.2) media by a mixed population of acidophilic bacteria. *Biodegradation* 17, 159–167.

Kobayashi, T.H., Kanai, R., Hayashi, T., Akiba, T., Akaboshi, R. and Horikoshi, K. (1992) Haloalkaliphilic maltotriose-forming α-amylase from the archaebacterium *Natronococcus* sp. Strain Ah-36. *Journal of Bacteriology* 174, 3439–3444.

Kobayashi, T.H., Kanai, R., Aono, K., Horikoshi, K. and Kudo, T. (1994) Cloning, expression and nucleotide sequence of the α-amylase gene from haloalkaliphilic archaeon *Natronococcus* sp. Strain Ah-36. *Journal of Bacteriology* 176, 5131–5134.

Kobayashi, T., Hakamada, Y., Adachi, S., Hitomi, J., Yoshimatsu, T., Koike, K., Kawai, S. and Ito, S. (1995) Purification and properties of an alkaline protease from alkaliphilic *Bacillus* sp. KSM-K16. *Applied Microbiology and Biotechnology* 43, 473–481.

Kumar, R., Patel, D.D., Bansal, D.D., Mishra, S., Mohammed, A., Arora, R., Sharma, A., Sharma, R.K. and Tripathi, R.P. (2010) Extremophiles: sustainable resource of natural compounds-extremolytes. In: Singh, O.V. and Harvey, S.P. (eds) *Sustainable Biotechnology: Sources of Renewable Energy*. Springer, Dordrecht, Heidelberg, Germany, pp. 279–294.

Kumar, R., Sharma, R.K., Bansal, D.D., Patel, D.D., Mishra, S., Miteva, L., Dobreva, Z., Gadjeva, V. and Stanilova, S. (2011) Induction of immunostimulatory cytokine genes expression in human PBMCs by a novel semiquinone glucoside derivative (SQGD) isolated from a *Bacillus* sp. INM-1. *Cellular Immunology* 267, 67–75.

Laird, S.A. and ten Kate, K. (1999) Natural products and the pharmaceutical industry. In: ten Kate, K. and Laird, S.A. (eds) *The Commercial Use of Biodiversity: Access to Genetic Resources and Benefit-Sharing*. Earthscan, London, UK, 67 pp.

Lamosa, P., Burke, A., Peist, R., Huber, R., Liu, M.Y., Sliva, G., Rodrigues-Pousada, C., LeGall, J., Maycock, C. and Santos, H. (2000) Thermostabilization of proteins by diglycerol phosphate, a new compatible solute from the hyperthermophile *Archaeoglobus fulgidus*. *Applied and Environmental Microbiology* 66, 1974–1979.

Lamosa, P., Turner, D.L., Ventura, R., Maycook, C. and Santos, H. (2003) Protein stabilization by compatible solutes. Effect of diglycerol phosphate on the dynamics of Desulfovibriogigasrubredoxin studied by NMR. *European Journal of Biochemistry* 270, 4606–4614.

Lee, H.K., Chun, J., Moon, E.J., Ko, S.H., Lee, H.S. and Bae, K.S. (2001) *Hahella chejuensis* gen. nov., an extracellular-polysaccharide-producing marine bacterium. *International Journal of Systematic and Evolutionary Microbiology* 51, 661–666.

Lentzen, G. and Schwarz, T. (2006) Extremolytes: natural compounds from extremophiles for versatile applications. *Applied Microbiology and Biotechnology* 72, 623–634.

Li, W.W., Zhou, W.Z., Zhang, Y.Z., Wang, J. and Zhu, X.B. (2008) Flocculation behaviour and mechanism of an exopolysaccharide from the deep-sea psychrophilic bacterium *Pseudoalteromonas* sp. SM9913. *Bioresource Technology* 99, 6893–6899.

Lippert, K. and Galinski, E.A. (1992) Enzyme stabilization by ectoine-type compatible solutes: protection against heating, freezing and dying. *Applied Microbiology and Biotechnology* 37, 61–65.

Loaëc, M., Olier, R. and Guézennec, J. (1997) Uptake of lead, cadmium and zinc by a novel bacterial exopolysaccharide. *Water Research* 31, 1171–1179.

Loaëc, M., Olier, R. and Guézennec, J. (1998) Chelating properties of bacterial exopolysaccharides from deep-sea hydrothermal vents. *Carbohydrate Polymers* 35, 65–70.

Lutz, S., Lichter, J. and Liu, L. (2007) Exploiting temperature-dependent substrate promiscuity for nucleoside analog activation by thymidine kinase from *Thermotoga maritima*. *Journal of the American Chemical Society* 129, 8714–8715.

Malin, G. and Lapidot, A. (1996) Induction of synthesis of tetra hydro pyrimidine derivatives in *Streptomyces* strain and their effect on *Escherichia coli* in response to osmotic and heat stress. *Journal of Bacteriology* 178, 385–395.

Manachini, P.L., Mora, D., Nicastro, G., Parini, C., Stackebrandt, E., Pukall, R. and Fortina, M.G. (2000) *Bacillus thermodenitrificans* sp. nov., nom. rev. *International Journal of Systematic and Evolutionary Microbiology* 50, 1331–1337.

Manazanera, M., Garcia de Castro, A., Tondervik, A., Rayner-Brandes, M., Strom, A.R. and Tunnacliffe, A. (2002) Hydroxyectoine is superior to trehalose for anhydrobiotic engineering of *Pseudomonas putida* KT2440. *Applied and Environmental Microbiology* 68, 4328–4333.

Mancuso Nichols, C.A., Garon, S., Bowman, J.P., Raguénès, G. and Guézennec, J. (2004) Production of exopolysaccharide by Antarctic marine bacterial isolates. *Journal of Applied Microbiology* 96, 1057–1066.

Markossian, S., Becker, P., Mrkl, H. and Antranikian, G. (2000) Isolation and characterization of lipid-degrading *Bacillus thermoleovorans* IHI-91 from an Icelandic hot spring. *Extremophiles* 4, 365–371.

Marx, J.G., Carpenter, S.D. and Deming, J.W. (2009) Production of cryoprotectant extracellular polysaccharide substances (EPS) by the marine psychrophilic bacterium *Colwellia psychrerythraea* strain 34H under extreme conditions. *Canadian Journal of Microbiology* 55, 63–72.

Maugeri, T.L., Gugliandolo, C., Panico, A., Lama, L., Gambacorta, A. and Nicolaus, B.A. (2002) Halophilic thermotolerant *Bacillus* isolated from a marine hot spring able to produce a new exopolysaccharide. *Biotechnology Letters* 24, 515–519.

McCoy, J.G., Bailey, L.J., Ng, Y.H., Bingman, C.A., Wrobel, R., Weber, A.P.M., Fox, B.G. and Phillips, G.N. Jr (2009) Discovery of sarcosine dimethylglycine methyltransferase from *Galdieria sulphuraria*. *Proteins* 74, 368–377.

Minta, A. and Tsien, R.Y. (1989) Fluorescent indicator for cytosolic sodium. *Journal of Biological Chemistry* 264, 19449–19457.

Mueller, R.H., Jorks, S., Kleinsteuber, S. and Babel, W. (1998) Degradation of various chlorophenols under alkaline conditions by Gram-negative bacteria closely related to *Oschrobactrum anthropi*. *Journal of Microbiology* 38, 269–281.

Nakayama, H., Yoshida, K., Ono, H., Murooka, Y. and Shimyo, A. (2000) Ectoine, the compatible solute of *Halomonas elongata*, confers hyperosmotic tolerance in cultured tobacco cells. *Plant Physiology* 122, 1239–1247.

Nicolaus, B., Lama, L., Panico, A., Schiano Moriello, V., Romano, I. and Gambacorta, A. (2002) Production and characterization of exopolysaccharides excreated by themophilic bacteria from shallow, marine hydrothermal vents of flegrean areas (Italy). *Systematic and Applied Microbiology* 25, 319–325.

Nicolaus, B., Kambourova, M. and Oner, E.T. (2010) Exopolysaccharides from extremophiles: from fundamentals to biotechnology. *Environmental Technology* 31, 1145–1158.

Norris, P.R., Nicolle, J. Le C., Calvo-Bado, L. and Angelatou, V. (2009) Pyrite oxidation by halotolerant, thermotolerant bacteria. *Advanced Materials Research* 71–73, 75–78.

Oguz, U. and Akkaya, E.U. (1998) A squaraine-based sodium selective fluorescent chemosensor. *Tetrahedron Letters* 39, 5857–5860.

Oren, A. (2010) Industrial and environmental applications of halophilic microorganisms. *Environmental Technology* 31, 825–834.

Pariza, M.W. and Johnson, E.A. (2001) Evaluating the safety of microbial enzyme preparation used in food processing: update for a new century. *Regulatory Toxicology and Pharmacology* 33, 173–186.

Parolis, H., Parolis, L.A.S., Boan, I.F., Rodriguez-Valera, F., Widmalm, G., Manca, M.C., Jansson, P.-E. and Sutherland, I.W. (1996) The structure of the exopolysaccharide produced by the halophilic *Archaeon Haloferax mediterranei* strain R4 (ATCC 33500). *Carbohydrate Research* 295, 147–156.

Pedersen, C.J. (1988) The discovery of crown ethers. *Science* 241, 536–540.

Penev, K. and Karamanev, D. (2009) Kinetics of ferrous iron oxidation by *Leptospirillum ferriphilum* at moderate to high total iron concentrations. *Advanced Materials Research* 71–73, 255–258.

Pepi, M., Gaggi, C., Bernardini, E., Focardi, S., Lobianco, A., Ruta, M., Nicolardi, V., Volterrani, M., Gasperini, S., Trinchera, G., Renzi, P., Gabellini, M. and Focardi, S.E. (2011) Mercury resistant bacterial strains *Pseudomonas* and *Psychrobacter* spp., isolated from sediments of Orbetello Lagoon (Italy) and their possible use in bioremediation process. *International Biodeterioration and Biodegradation* 65, 85–91.

Petrovick, M.S., Nargi, F.E., Towle, T., Hogan, K., Bohane, M., Wright, D.J., MacRae, T.H., Potts, M. and Helm, R.F. (2010) Improving the long-term storage of a mammalian biosensor cell line via genetic engineering. *Biotechnology and Bioengineering* 106, 474–481.

Pinzón-Martínez, D.L., Rodríguez-Gómez, C., Miñana-Galbis, D., Carrillo-Chávez, J.A., Valerio-Alfaro, G. and Oliart-Ros, R. (2010) Thermophilic bacteria from Mexican thermal environments: isolation and potential applications. *Environmental Technology* 31, 957–966.

Poli, A., Schiano Moriello, V., Esposito, E., Lama, L., Gambacorta, A. and Nicolaus, B. (2004) Exopolysaccharide production by a new *Halomonas alkaliantarctica* sp. nov., isolated from saline lake Cape Russel in Antarctica, an alkalophilic moderately halophilic, exopolysaccharide-producing bacterium. *Systematic and Applied Microbiology* 30, 31–38.

Poli, A., Esposito, E., Orlando, P., Lama, L., Giordano, A., de Appolonia, F., Nicolaus, B. and Gambacorta, A. (2007) *Halomonas alkaliantarctica* sp. nov., isolated from saline lake Cape Russell in Antarctica, an alkalophilic moderately halophilic, exopolysaccharide-producing bacterium. *Systematic and Applied Microbiology* 30, 31–38.

Poli, A., Kazak, H., Gurleyendag, B., Tommonaro, G., Pieretti, E.T.O. and Nicolaus, B. (2009) High level synthesis of levan by a novel *Halomonas* species growing on defined media. *Carbohydrate Polymers* 78, 651–657.

Poli, A., Anzelmo, G. and Nicolaus, B. (2010) Bacterial exopolysaccharides from extreme marine habitats: production, characterization and biological activities. *Marine Drugs* 8, 1779–1802.

Qin, K., Zhu, L., Chen, L., Wang, P.G. and Zhang, Y. (2007) Structural characterization and ecological roles of a novel exopolysaccharide from the deep-sea psychrotolerant bacterium *Pseudoalteromonas* sp. SM9913. *Microbiology* 153, 1566–1572.

Raguenes, G., Pignet, P., Gautheir, G., Peres, A., Christen, R., Rougeaux, H., Barbier, G. and Guézennec, J. (1996) Description of a new polymer-secreting bacterium from a deep-sea hydrothermal vent, *Alteromonas macleodii* subsps. *fijiensis*, and preliminary characterization of the polymer. *Applied and Environmental Microbiology* 62, 67–73.

Raguenes, G., Christen, R., Guézennec, J., Pignet, P. and Barbier, G. (1997) *Vibrio diabolicus* sp. nov., a new polysaccharide-secreting organism isolated from a deep-sea hydrothermal vent polychaete annelid, *Alvinella pompejana*. *International Journal of Systematic Bacteriology* 47, 989–995.

Rai, S.K. and Mukherjee, A.K. (2011) Optimization of production of an oxidant and detergent-stable alkaline β-keratinase from *Brevibacillus* sp. strain AS-S10-II: application of enzyme in laundry detergent formulations and in leather industry. *Biochemical Engineering Journal* 54, 47–56.

Ramos, A., Raven, N.D.H., Sharp, R.J., Bartolucci, S., Rossi, M., Cannio, R., Lebbink, J., Oost, J., vos, W.M. and Santos, H. (1997) Stabilization of enzymes against thermal stress and freeze-drying by mannosylglycerate. *Applied and Environmental Microbiology* 63, 4020–4025.

Rasmussen, R.S. and Morrissey, M.T. (2007) Marine biotechnology for production of food ingredients. *Advances in Food and Nutrition Research* 52, 237–292.

Rinker, K.D. and Kelly, R.M. (1996) Growth physiology of the hyperthermophillic *Archaeon Thermococcus litoralis*: Development of a surface of a Surface-Free Defined medium, Characterization of an Exopolysaccharide, and Evidence of Biofilm Formation. *Applied and Environmental Microbiology* 12, 4478–4485.

Rougeaux, H., Talaga, P., Carlson, R.W. and Guézennec, J. (1998) Structural studies of an exopolysaccharide produced by *Alteromonas macleodii* subsps. *fijiensis* originating from a deep-sea hydrothermal vent. *Carbohydrate Research* 312, 53–59.

Rougeaux, H., Guézennec, J., Carlson, R.W., Kervarec, N., Pichon, R. and Talaga, P. (1999a) Structural determination of the exopolysaccharide of *Pseudoalteromonas* strain HYD 721 isolated from a deep-sea hydrothermal vent. *Carbohydrate Research* 315, 273–285.

Rougeaux, H., Kervarec, N., Pihon, R. and Guézennec, J. (1999b) Structure of the exopolysaccharide of *Vibrio diabolicus* isolated from a deep-sea hydrothermal vent. *Carbohydrate Research* 322, 40–45.

Russell, N.J. (1998) Molecular adaptations in psychrophilic bacteria: potential for biotechnological applications. *Advances in Biochemical Engineering/Biotechnology* 61, 1–21.

Saiki, R.K., Gelfand, D.H., Stoffel, S., Scharf, S.J., Higuchi, R., Horn, G.T., Mullis, K.B. and Erlich, H.A. (1988) Primer-directed enzymatic amplification of DNA with a thermostable DNA polymerase. *Science* 239, 487–491.

Sharma, R.K. and Arora, R. (2011) Fukushima, Japan – an Apocalypse in the making? *Journal of Pharmacy and Bioallied Sciences* (in press).

Simmons, T.L., Andrianasolo, E., McPhail, K., Flatt, P. and Gerwick, W.H. (2005) Marine natural products as anticancer drugs. *Molecular Cancer Therapeutics* 4, 333–342.

Smith, G.A., Hesketh, T.R. and Metcalfe, J.C. (1988) Design and properties of a fluorescent indicator of intracellular free Na+ concentrations. *Biochemical Journal* 250, 227–232.

Steger, R., Weinand, M., Krämer, R. and Morbach, S. (2004) LcoP, an osmoregulated betaine/ectoine uptake system from *Corynebacterium glutamicum*. *FEBS Letters* 573, 155–160.

Takahara, Y. and Tanabe, O. (1962) Studies on the reduction of indigo in industrial fermentation vat (XIX). Taxonomic characterisation of strain No.S-8 *Journal of Fermentation Technology* 40, 77–80.

Tapilatu, Y.H., Grossi, V., Acquaviva, M., Militon, C., Bertrand, J.C. and Cuny, P. (2010) Isolation of hydrocarbon-degrading extremely halophilic archaea from an uncontaminated hypersaline pond (Camargue, France). *Extremophiles* 14, 225–231.

ten Kate, K. (1999) Biotechnology in fields other than healthcare and agriculture. In: ten Kate, K. and Laird, S.A. (eds) *The Commercial Use of Biodiversity: Access to genetic resources and benefit-sharing*. Earthscan, London, UK, 416 pp.

Theriot, C.M., Tove, S.R. and Grunden, A.M. (2009) Biotechnological applications of recombinant microbial prolidases. *Advances in Applied Microbiology* 68, 99–132.

Tomita, K., Nomura, K., Kondo, H., Nagata, K. and Tsubota, H. (1995) Stabilized enzymatic reagents for measuring glucose, creatine kinase and γ-glutamyltransferase with thermostable enzymes from a thermophile, *Bacillus stearothermophilus*. *Journal of Pharmaceutical and Biomedical Analysis* 13, 477–481.

Tsujibo, H., Sato, T., Inui, M., Yamamoto, H. and Inamori, Y. (1988) Intracellular accumulation of phenazine antibiotics produced by an alkalophilic actinomycete. I. Taxonomy, isolation and identification of the phenazine antibiotics. *Agricultural and Biological Chemistry* 52, 301–306.

Turner, P., Mamo, G. and Nordberg Karlsson, E. (2007) Potential and utilization of thermophiles and thermostable enzymes in biorefining. *Microbial Cell Factories* 6, 9.

Ubaldini, S., Veglio, F., Toro, L. and Abbruzzese, C. (1997) Biooxidation of arsenopyrite to improve gold cyanidation: study of some parameters and comparison with grinding. *International Journal of Mineral Processing* 52, 65–80.

Ulukanliand, Z. and Diúrak, M. (2002) Alkaliphilic microorganisms and habitats. *Turkish Journal of Biology* 26, 181–191.

Valenzuela, L., Chi, An., Beard, S., Orell, A., Guiliani, N., Shabanowitz, J., Hunt, D.F. and Jerez, C.A. (2006) Genomics, metagenomics and proteomics in biomining microorganisms. *Biotechnology Advances* 24, 197–211.

Vargas, C., Argandona, M., Reina-Bueno, M., Rodriguez-Moya, J., Fernandez-Aunion, C. and Nieto, J.J. (2008) Unravelling the adaptation responses to osmotic and temperature stress in *Chromohalobacter salexigens*, a bacterium with broad salinity tolerance. *Saline Systems* 15, 4–14.

Ventosa, A. and Nieto, J.J. (1995) Biotechnological applications and potentialities of halophilic microorganisms. *World Journal of Microbiology and Biotechnology* 11, 85–94.

Vieille, C. and Zeikus, G.J. (2001) Hyperthermophilic enzymes: sources, uses, and molecular mechanisms for thermostability. *Microbiology and Molecular Biology Reviews* 65, 1–43.

Wang, G.X., Liu, Y.T., Li, F.Y., Gao, H.T., Lei, Y. and Liu, X.L. (2011) Immunostimulatory activities of *Bacillus simplex* DR-834 to carp (*Cyprinus carpio*). *Fish and Shellfish Immunology* 29, 378–387.

Wang, Q., Chen, R., Du, P., Wu, H., Pei, X., Yang, B., Yang, L., Huang, L., Liu, J. and Xie, T. (2010) Cloning and characterization of thermostable-deoxy-D-ribose-5-phosphate aldolase from *Hyperthermus butylicus*. *African Journal of Biotechnology* 9, 2898–2905.

Weber, A., Oesterhelt, C., Gross, W., Bräutigam, A., Imboden, L., Krassovskaya, I., Linka, N., Truchina, J., Schneidereit, J. and Voll, H. (2004) EST-analysis of the thermo-acidophilic red microalga *Galdieria sulphuraria* reveals potential for lipid A biosynthesis and unveils the pathway of carbon export from rhodoplasts. *Plant Molecular Biology* 55, 17–32.

Willies, S., Isupov, M. and Littlechild, J. (2010) Thermophilic enzymes and their applications in biocatalysis: a robust aldo-keto reductase. *Environmental Science and Technology* 31, 1159–1167.

Xie, X.H., Xiao, S.M., Wang, H.T. and Liu, J.S. (2009) Biooxidation of refractory gold ores by mixed moderate thermophiles in airlift bioreactor. *Advanced Materials Research* 71–73, 469–472.

Yang, C., Wang, Z., Li, Y., Niu, Y., Du, M., He, X., Ma, C., Tang, H. and Xu, P. (2010) Metabolic versatility of halotolerant and alkaliphilic strains of *Halomonas* isolated from alkaline black liquor. *Bioresource Technology* 101(17), 6778–6784.

Zammit, C.M., Mutch, L.A., Watling, H.R. and Watkin, E.L.J. (2009) The characterization of salt tolerance in biomining microorganisms and the search for novel salt tolerant strains. *Advanced Materials Research* 71–73, 283–286.

Zanchetta, P. and Guézennec, J. (2001) Surface thermodynamics of osteoblasts: relation between hydrophobicity and bone active biomaterials. *Colloids Surface* 22, 301–307.

Zanchetta, P., Lagarde, N. and Guézennec, J. (2003a) A new bone-healing material: a hyaluronic acid-like bacterial exopolysaccharide. *Calcified Tissue International* 72, 74–79.

Zanchetta, P., Lagarde, N. and Guézennec, J. (2003b) Systematic effects on bone healing of a new hyaluronic acid-like bacterial exopolysaccharide. *Calcified Tissue International* 73, 232–236.

Zhang, L.H., Lang, Y.J. and Nagata, S. (2009) Efficient production of ectoine using ectoine-excreting strain. *Extremophiles* 13, 717–724.

26 Extreme Environments on Earth as Analogues for Life on Other Planets: Astrobiology

Felipe Gómez.[1] Miloš Barták[2] and Elanor M. Bell[3]
[1]*Centro de Astrobiología (INTA-CSIC), Madrid, Spain;*
[2]*Section of Plant Physiology and Anatomy, Department of Experimental Biology, Masaryk University, Brno, Czech Republic;*
[3]*Scottish Institute for Marine Science, Argyll, UK*

26.1 Introduction

The possibility of life elsewhere in the Universe has long been a topic for debate, as the only example of life that we know about is life on Earth. We therefore discuss the potential for life outside Earth under the assumption that, should we find it, it would be similar to our planet's. With this in mind, astrobiology could be defined as the study of the origin and evolution of life on Earth and the possibility of life outside it.

There exist a couple of controversial pieces of evidence for extraterrestrial life: first, evidence of life reported in the ALH84001 meteorite that was found in the Allan Hills region of Antarctica in 1984 was believed to be from Mars. The meteorite contained: (i) hydrocarbons that are the same as breakdown products of dead microorganisms on Earth; (ii) mineral phases consistent with by-products of bacterial activity; and (iii) tiny carbonate globules, which may be 'microfossils' of primitive bacteria, all within close proximity to each other (McKay *et al.*, 1996). Second, a methane 'signal' was detected in Mars' atmosphere by telescopes on Earth and instruments on board the European Space Agency's orbiting Mars Express craft (Formisano *et al.*, 2004; Krasnopolsky *et al.*, 2004). Methane is not a stable molecule in the Martian atmosphere and if it were not replenished in some way, it would only last a few hundred years before it disappeared. There are two possible renewable sources: (i) active volcanoes, none of which have yet been found on Mars; or (ii) methanogenic microbes that produce methane from hydrogen and carbon dioxide. These organisms do not require oxygen and are thought to be the sort of microbes that could survive on Mars. Both of the possibilities are interesting because they indicate that some ongoing process is releasing gas to the atmosphere and that the planet is still alive. However, neither of these pieces of evidence is conclusive and our search for life on Mars and other planets continues.

Life as we know it on Earth, requires water as the solvent in which biochemical reactions take place, but also the elements carbon, hydrogen, oxygen, nitrogen, sulfur and phosphorus. These elements appear to be widespread throughout the Universe and they have been identified in many components of the interstellar medium.

Therefore life could indeed be universally carbon-based and water the universal solvent; the *sine qua non* element for life as we know it.

The study of the physical and chemical limits for life on Earth, i.e. life in extreme environments, could help us to understand the potential for life existing on other celestial bodies. Earth's extreme environments often possess many of the characteristics (e.g. physico-chemical parameters, low water activity, geological or mineralogical compositions and processes) present on other planetary bodies and can therefore be used as analogues for these extraterrestrial environments and the study of potential life on them. Two classic examples that will be discussed in detail later in this chapter are the use of the Atacama Desert, Chile, as an analogue for life on Mars because of its low water activity, and subglacial Lake Vostok, Antarctica, as an analogue for potential life under the ice present on Jupiter's frozen moon, Europa.

Looking beyond Earth's extreme environments to the organisms that live in them, the extremophiles, also widens the limits at which we perceive life to be possible and, by extrapolation, increases the potential for extraterrestrial life.

In addition, Earth's extreme environments and the extremophiles that live in them provide ideal test-beds for the development and testing of instruments and techniques for space missions searching for extraterrestrial life. The logistic difficulties, physico-chemical extremes, and the difficulty of detecting and identifying life in Earth's extreme environments, mean that they represent relevant natural field facilities for testing new exploration tools and techniques in the hope that we are poised to conclusively identify extraterrestrial life if and when it is encountered.

26.2 Learning Lessons from Extremophiles

Some groups of extremophilic organisms are of particular interest to astrobiologists. A good example are hyperthermophiles, which are able to grow at very high temperatures (80–100°C), including temperatures above the boiling point of water (Dwyer *et al.*, 1988; Bryant and Adams, 1989). Of special interest are the hyperthermoacidophilic Archaebacteria, which are able to grow under the extreme conditions typically found in submarine hot springs and hydrothermal vents – sulfur-enriched waters at very low pH (pH 1–2), high hydrostatic pressure and high temperature (100°C) (Bouthier de la Tour *et al.*, 1990; Dahl *et al.*, 1990; DiMarco *et al.*, 1990) – using metabolisms based on sulfur reduction and hydrogen oxidation (Belkin *et al.*, 1986; Ahring and Westerman, 1987; Kjems *et al.*, 1990).

Similarly, acidophiles (acid-lovers) (Fig. 26.1) relying on iron–sulfur metabolism and living in, for example, sulfuric acid

Fig. 26.1. Endolithic phototrophs included in salt precipitates from the Rio Tinto, Spain. Microorganisms establish protected micro-niches inside layered salt deposits. © Felipe Gómez.

springs (pH 1–2) that emerge from mines and are associated with pyrite deposits, are good candidates for astrobiological research because the mineralogy associated with their terrestrial habitats resembles that found in some places on Mars' surface (i.e. Terra Meridiani).

In 2008, the Phoenix mission landed on the polygon-patterned northern plains of the Vastitas Borealis, Mars, and uncovered a permafrost-like shallow ice table at a depth of 5 to 18 cm (Smith *et al.*, 2009). Since then, psychrophiles (cold-lovers) living in Earth's permafrost regions have attracted interest as analogues for life at Mars' North Pole, especially since some psychrophiles are reported to actively metabolize at temperatures as low as −20°C in Siberian permafrost (Mazur, 1980). Permafrost-dwelling, chemolithotrophic microorganisms that have the ability to use inorganic compounds, such as reduced minerals, as energy sources for their metabolism have also come under scrutiny (Gómez *et al.*, 2008).

An increase in the water activity level inside hygroscopic salt deposits due to water deliquescence (absorption by minerals with a strong affinity for water) (Davila *et al.*, 2010) in evaporation ponds, where the halophiles (salt-lovers) grow in saturated salt solutions and withstand desiccation for long periods of time, has also widened the possibilities for life on extraterrestrial bodies, and in protected endolithic micro-niches (Clark and Van Hart, 1981; Gómez *et al.*, 2011). The existence of a water reservoir (most likely saline) underneath the icy crust of the Jupiter's Moon, Europa, and the identification of salt deposits on the surface of Mars has also significantly increased astrobiological interest in halophiles and halophilic habitats.

26.3 Extreme Environments as Analogues for Astrobiology

As previous chapters within this volume have demonstrated, a wide range of extreme environments exist on Earth and some bear remarkable similarities to the conditions known or thought to exist on extraterrestrial bodies in our universe. Therefore, they can be employed as analogues for the study of extraterrestrial life. Some examples are discussed in detail below.

26.3.1 The Dry Valleys, Antarctica

The Antarctic hosts numerous environments that can be considered Martian analogues (Fig. 26.2). The Dry Valleys are one of the best examples. The region represents the largest ice-free area in Antarctica and one of the driest places on Earth as a result of extremely low annual precipitation, and the katabatic winds that evaporate the moisture from the soil. The temperature in the valleys rarely gets above 0°C and annual rainfall is less than 10 mm, conditions that closely resemble those that exist in some locations on the surface of Mars. Other similarities between the Dry Valleys and Mars include the basaltic geological formations, extensive permafrost areas with polygonal terrains and other permafrost patterns that bear amazing morphological similarities; these features can be used for geological process studies.

In spite of the extreme conditions experienced in the Dry Valleys, bacteria, algae, fungi and lichen have been identified colonizing, for example, the interior of sandstone deposits, porous rocks and surface pavements, and bacteria have been identified in hot volcanic fumaroles at Mount Erebus, an active volcano located nearby. If life exists on Mars, it is likely that it will be discovered in similar locations.

The natural habitats of the Dry Valleys are near the limits of life on Earth and resemble those hypothesized for early Mars. Therefore, organisms that thrive in the Dry Valleys in endolithic communities, stromatolitic sediments in ice-covered lakes, and transient riverbeds should be considered test models for survival in extraterrestrial environments (Wynn-Williams and Edwards, 2000). Photosynthetic organisms such as free-living cyanobacteria and cryptoendolithic (living in the rock) vegetation represent typical examples.

Fig. 26.2. Some Antarctic ecosystems might be considered Martian analogues due to waste areas of deglaciated rocks and stony surfaces exposed to extremes in temperature: (a) the Northern coast of James Ross Island, Antarctica, with the peak of the Bibby Point; (b) view from the stony plateau of Bibby Point to the northern coast of James Ross Island; (c) volcanic rocks between Berry Hill and the Lachman Crags mesa, James Ross Island; (d) view on the Devil's Rock group and perpendicular walls of the Lachman Crags mesa, James Ross Island. Images © Milos Barták.

Cryptoendolithic microbial communities were discovered in the 1970s (Friedmann and Ocampo, 1976) in the Ross Desert, Antarctica, living sheltered under rock surfaces in micropores, where they find favourable nanoclimates for their survival and growth. Among these microbial communities, the lichen-dominated cryptoendolithic community is the most studied and best understood (Friedmann and Ocampo-Friedmann, 1984). In recent years, intensive research has been carried out to determine how representatives of the Dry Valleys' cryptoendolithic communities can cope with extraterrestrial conditions (Onofri *et al.*, 2008).

At a larger scale, another important feature of the Dry Valleys which makes them of astrobiological interest is the presence of well protected subsurface ecosystems, such as lakes with perennial ice covers, and extensive permafrost areas (Kapitsa *et al.*, 1996). Data from Galileo spacecraft reported the possible existence of a salty ocean underneath the ice crust of one of Jupiter's moons, Europa. The Dry Valley lakes are considered to be good models for developing new technologies for exploration of such extraterrestrial environments. Furthermore, some ice landforms have also been suggested as terrestrial analogues for the sublimation behaviour of Martian icecaps, where surface melting is absent but processes such as sublimation are the key to the formation of unusual ice features like the dark spirals and circular pits observed on Mars.

26.3.2 Lake Vostok, Antarctica

It is not only surface lakes such as those discussed above in the Dry Valleys that are important analogues. Antarctic subglacial lakes might also offer astrobiologists unique insights into the possibility of life on other planets. To date, >300 subglacial lakes have been found under the Antarctic ice sheet, and of these, Lake Vostok is the largest (see Pearce, Chapter 7, this volume).

Subglacial Lake Vostok was first noticed in 1973 by scientists from Scott Polar Research Institute, UK, using ice-penetrating radar (Oswald and Robin, 1973). Tremendous interest was stimulated in the scientific community and the lake has subsequently been studied using several ice-penetrating techniques, for example, airborne radio-echo sounding and penetrating radar imaging, as well as space-borne radar altimetry. Lake Vostok was delimited in 1996 by British and Russian scientists (Siegert *et al.*, 2001). It is located 4 km beneath the surface of the Central Antarctic ice sheet, is 240 × 50 km in size and has 22 cavities of liquid water. Lake Vostok is divided into two basins separated by a ridge and it is speculated that both have a different water composition. The temperature of the water is approximately −3°C but it remains liquid due to the high pressure exerted by the weight of the ice above it.

To probe the waters of Lake Vostok for life without contaminating the environment, plans were developed by NASA's Jet Propulsion Laboratory, California, USA in 2001. The procedure will involve using a melter probe, the so-called 'cryobot', to melt down through the ice over Lake Vostok, unspooling a communications and power cable as it goes. The cryobot will carry with it a small submersible, called a 'hydrobot', which will be deployed when the cryobot has melted as far as the ice–water interface. The hydrobot will then swim off to 'look for life' with a camera and other instruments. Full access to the lake's water column is planned by an international team of scientists for the 2011/12 season (Pearce, Chapter 7, this volume).

Lake Vostok and other Antarctic subglacial lakes will provide the opportunity to test technologies which may in the future be applied to the search for life on other planets, for example, on Mars, Jupiter's moons Europa and Callisto, and Saturn's largest moon, Titan (Abyzov *et al.*, 2001; Price *et al.*, 2002). All of these planetary bodies have confirmed or suspected ice sheets beneath which liquid water could harbour life, protected from the extreme extraterrestrial conditions.

26.3.3 Svaldbard, Norwegian Arctic

Svalbard is a Norwegian archipelago located in the Arctic Circle approximately midway between mainland Norway and the North Pole, and is also the northernmost populated place on Earth. Svalbard consists of a group of islands located between 74 °N and 81 °N, and 10 °E to 35 °E. Spitsbergen is the largest island and its frozen soils with wide permafrost areas and glaciers mean that it is an important location for astrobiological field campaigns studying life under extreme conditions and microbial habitability at very low temperature.

Such field campaigns are organized every year by the astrobiological community and National Space Agencies, collectively termed Arctic Mars Analog Svalbard Expeditions (AMASE). The objective of AMASE is to determine how to search for life on Mars, and how to develop and test the appropriate technology to do so, for example, the technology for future Mars missions onboard the Mars Science Laboratory (NASA) and ExoMars (European Space Agency) rovers.

The conditions are so extreme that only extremophiles can thrive on Svalbard's basaltic, carbonate-rich rocks or in glacial ice, conditions that could prove similar to equatorial Martian basalts or its icy regions. If we want to search for life on Mars we need to develop and test protocols to find past and present habitable environments, and to avoid biological contamination that could be taken to Mars from our planet (or could come back from Mars to Earth in any Mars sample return mission). AMASE is developing such systems to search for life and prevent contamination. The research will also facilitate exploration of other icy, planetary bodies that could potentially harbour microbial life, e.g. Jupiter's moon Europa, and Saturn's moons, Titan and Enceladus.

A number of hydrothermal sites are also located on Spitsbergen, and life detection strategies based on antibody microarrays have been tested in these locations. Such techniques could be widely applicable in astrobiological research in the future.

26.3.4 Rio Tinto, Spain

The Rio Tinto is a 100 km-long river located on the south-west Iberian peninsula, Huelva province (Andalusia, Spain). The river originates in the Sierra Morena (Sierra de Aracena) mountains in the area called Peña de Hierro. Its waters flow south/south-west to the Gulf of Cadiz (Atlantic Ocean). The Rio Tinto represents an extreme environment with a constant acidic pH (mean 2.3) and high heavy metal concentrations (Fe, Cu, Zn, As and Cr) in solution. Iron is the driving element in the ecosystem and is used as an electron donor and an electron acceptor, allowing a full iron cycle to operate. Indeed, oxidized iron is responsible for the dark red colour of the waters (iron concentrations in some locations are as high as 22 g l^{-1}; Fig. 26.3). As a result, several ancient cultures (e.g. the Phoenicians) called the Rio Tinto the 'River of Fire'.

According to the geological record of the area, the first mining activities commenced 5000 years ago. The Rio Tinto continues to be of mining interest because it is located in an area rich in sulfides and iron minerals with a huge pyritic belt beneath. These sulfides and iron-rich deposits originated from an ancient hydrothermal system.

There are numerous chemical and mineralogical similarities between the Rio Tinto and Mars, and these have led to the former being used as a Mars analogue (e.g. Fernández-Remolar *et al.*, 2004). Space missions have reported areas on the surface of Mars that are rich in sulfur and iron minerals. On Earth, sulfide minerals are made in hydrothermal systems associated with basaltic volcanism; basaltic volcanism is also widespread on Mars. The Rio Tinto and Mars share some important minerals as well: jarosite, an iron magnesium sulfate salt which is only precipitated in low pH solutions with high metal concentration, is the most important example. Jarosite precipitates on the microbial mats and river bed of the Rio Tinto system and the NASA Opportunity Rover also located it on the Martian surface.

(a)

(b)

Fig. 26.3. (a) and (b) The Rio Tinto, Spain, an extreme environment driven by iron chemistry. This 100 km river has a mean pH of 2.3 and an extremely high iron concentration in its waters. © Felipe Gómez.

The presence of jarosite on the surface of Mars is indicative of the past presence of acidic, liquid water and hence, the potential for life.

Despite the extreme chemical conditions in the Rio Tinto, its waters host a high biodiversity of organisms including Bacteria, Archaea, photosynthetic and heterotrophic Protozoa, yeast and filamentous fungi. It is the metabolic activity of chemolithotrophic prokaryotes, mainly iron- and sulfur-oxidizers, that are responsible for the extreme chemistry of the water. Recently published studies report that iron can also protect the different organisms thriving in its waters from radiation (Gómez *et al.*, 2007).

Phototrophs sheltering from particular light wavelengths have also been reported in endolithic ('inside rock') niches in the salt deposits in the Rio Tinto region (Gómez *et al.*, 2011). Molecular ecology techniques and microscope observations demonstrated that microbial communities and biofilm formations comprising the algae *Dunaliella* sp., *Cyanidium* sp. and other phototrophs, were present in the deposits. Some prokaryotes have also been identified. Inside the salt deposits, the organisms experience optimal irradiance and humidity, and thermal and pH stability, therefore they avoid having to deal with the extreme conditions outside (Fig. 26.1). Such endolithic niches are also potential havens for past and/or present life on Mars.

As a Mars analogue, the Rio Tinto source area was selected as a field site for a drilling simulation campaign, the MARTE project ('Mars Analog Research and Technology Experiment'). This project was a NASA ASTEP co-funded programme in collaboration with the Spanish government, dedicated to Mars exploration technology development that commenced in 2003. There were three main objectives of MARTE: (i) to search for and characterize subsurface life in the Rio Tinto, along with the physical and chemical properties and sustaining energy sources of its environment; (ii) to perform a high fidelity simulation of a robotic Mars drilling mission to search for life; and (iii) to demonstrate the drilling, sample handling and instrument technologies relevant to searching for life on Mars.

26.3.5 Atacama Desert, Chile

The Atacama Desert, Chile, is considered to be the driest place on Earth (Fig. 26.4; see also Rewald *et al.*, Chapter 11, this volume). It extends in a 1000 km strip along the Pacific coast of South America. The desert is the direct result of a unique climate regime: a strong Pacific anticyclone and the Humboldt Current from the north create a temperature inversion which promotes extreme aridity. Moisture is also blocked by the Chilean Coast Range and Los Andes mountains. Other important features of the Atacama Desert are the nitrate anomaly and salt precipitation which occur due to the 'camanchaca', a fog coming from the Pacific Ocean most mornings which contains dissolved salts that are deposited and precipitated during sunrise. The lack of precipitation and high salt levels result overall in very low water activity in the area. The Atacama Desert also experiences high levels of ultraviolet radiation, huge temperature ranges, high winds, and dust storms capable of breaking down organic matter. Between the coastal range and the medial range there is a broad valley, which is the most extreme zone of the Atacama Desert. This area contains very low amounts of organic matter, and what little is present is refractory and easily degraded by temperature (Navarro-González *et al.*, 2003).

These extreme physico-chemical characteristics mean that the Atacama Desert is a perfect analogue for extraterrestrial life. Indeed, several Atacama field campaigns have demonstrated the presence of extremophilic life forms in even the most extreme arid zones (Gómez-Silva *et al.*, 2007). Light is the main energy source for these ecosystems due to the low level of organic carbon resources available, but in order to avoid the harmful effects of the high ambient UV radiation, many phototrophic organisms (e.g. cyanobacteria) establish themselves in endolithic or

Fig. 26.4. The Atacama Desert is recognized as the driest place on Earth. Low numbers of bacteria are isolated in some areas of this arid desert. © Felipe Gómez.

hypolithic bacterial biofilms, or within translucent rock substrates such as halite, gypsum or quartz. These protected micro-niches are excellent candidate locations for life on Mars' surface (Gómez *et al.*, 2011) and future space missions will target them.

The low microbial life detection level on the desert soil and low organic matter present also makes the Atacama Desert a good natural laboratory for testing the detection limits of astrobiological equipment prior to space missions. Furthermore, several robots and rover vehicles have been tested in the region.

26.3.6 Kamchatka hot springs, Russia

The Kamchatka Peninsula is one of four regions in the world that has extensive geyser activity, the others being Iceland, New Zealand and Yellowstone National Park in the USA. The relevance of such environments in planetary science is not only that they allow a study of the complex inter-relationship between volcanism and landform development, highly relevant to some planetary bodies, but that the hot springs also host diverse microbiology. The microorganisms living in these springs survive in extreme conditions both in terms of temperature and water composition (e.g. highly acidic).

The specific area for study at Kamchatka is the unique Uzon-Geysernaia twin caldera that forms a 9 × 18 km depression, which originated as a result of large explosive eruptions in the late Pleistocene. Present geothermal activity is concentrated in a 0.3 × 5 km zone filled with 30 geysers, boiling springs, gas-steam jets, mudpots, small mud volcanoes, hot lakes and springs. These habitats contain a wide variety of colonies of blue-green algae and thiobacteria. Boiling waters in the axial part of the thermal field are rich in boron, silicon and ammonium chloride and have high concentrations of alkali metals and ore elements. The rest of the hydrotherms are sulfate-chloride-sodium, hydrocarbonate-sulfate-calcium-sodium and hydrocarbonate. Gas discharge includes CO_2, N_2, H_2, H_2S, CH_4 and radon.

Complex methane-naphthene-aromatic type hydrocarbons, possibly of biogenic nature, occur in the thermal fields of the Uzon Caldera. The site is therefore both a possible analogue for prebiotic conditions on Earth and an analogue for some regions on the surface of Jupiter and Saturn's moons (e.g. Europa and Titan, respectively). The Kamchatka region also offers Mars analogue terrains. For example, the landscape of the Tolbachik Volcano was used in the Soviet/ Russian space programme for testing Mars rovers for the Mars-96 mission.

26.4 Extremophilic Organisms as Analogues for Extraterrestrial Life

26.4.1 Lichens as models for astrobiological research

Lichens that thrive in extreme environments on Earth, such as on high mountains and in the polar regions, are considered good organisms for tests in extraterrestrial environments (Fig. 26.5). There are several reasons for this. The first is their ability to withstand long periods in a dry state with either no or limited inhibition of their physiological processes after being rewetted. The second is their ability to cope with extreme cold and freezing temperatures because of numerous protective mechanisms. Indeed it is well documented that lichens can survive in a freezer for a long time and be returned to normal physiological activity in a matter of hours when hydrated and exposed to room temperature. The third is that the majority of lichens growing in sunny habitats are tolerant of high light levels, and in particular high UV radiation levels.

Over the last decades, numerous field- and laboratory-based ecophysiological studies focused on lichen responses to extremes, such as repetitive dehydration (e.g. Lange *et al.*, 2001), freezing temperatures (e.g. Barták *et al.*, 2007) and resistance to high doses of both UV radiation and visible light (e.g. Gauslaa and Solhaug, 2004), have been carried out. The results have shown lichens to be very resistant to the extreme conditions and fluctuations present in the Antarctic, and suggest that they should also be resistant to the similarly extreme conditions present on both extraterrestrial bodies and in outer space.

Fig. 26.5. Dark-pigmented thalli of *Umbilicaria decussata* are typical representatives of the extremophilic lichen flora of James Ross Island, Antarctica. This foliose lichen species forms dish-like structures no larger than 3 cm in diameter. © Milos Barták.

Several protective mechanisms help lichens to survive under extreme conditions. Among them, the various mechanisms that confer cryoresistance are of great importance. Lichens contain relatively high amounts of polyols (sugar alcohols) that prevent the freezing of intracellular water and help them to maintain their cells physiologically active even at freezing temperatures. When exposed to subzero temperatures, lichens also produce antifreeze proteins, i.e. special proteins that allow subcellular structures to continue functioning. Another mechanism is ice nucleation activity, which alters the form of intercellular ice crystals and removes the probability of sharp crystal edges forming that might damage cellular structures, i.e. leads to the creation of ice crystals with rounded edges.

Lichens are well adapted to rapid changes in water supply and can survive for long periods in a dry state. When dehydrated, lichens are physiologically inactive, which helps them to avoid the negative effects of extremely low temperature, as well as excess light. In a wet state, lichens must cope with the negative effects of UV and photosynthetically active radiation that may destroy the chloroplasts of their symbiotic algae and/or cyanobacteria. Numerous UV screening pigments and compounds are present in the upper cortex of lichens (formed by fungal hyphae) and provide them with extremely effective protection from harmful radiation. These pigments account for why many lichens are dark-coloured. In cells of the symbiotic algae located below the upper cortex, numerous carotenoids and antioxidative enzymes help to protect photosynthetic apparatus when lichen thalli are exposed to high light doses. Under such conditions, reactive oxygen species are formed that are harmful to chlorophyll molecules. The carotenoids zeaxanthin, lutein and carotene, in particular, scavenge the reactive oxygen species effectively and thus protect the chloroplasts.

Astrobiological experiments involving lichens

As discussed above, lichens represent an important group that copes well with both high and low temperatures, frequent desiccation and other negative physical factors. Recently, questions were raised as to whether such extremophilic organisms could survive during interplanetary transfer and whether life could be present on other planets in our solar system. Indeed, it is speculated that extremophilic bacteria and fungi, in particular, may thrive on extraterrestrial bodies if water is present, e.g. under the icy surfaces of some of Jupiter's moons or in underground caverns on Mars. To learn more about the survival of lichens in outer space, two major exposure experiments have been carried out within the last decade. Short-term exposure of lichens to extraterrestrial (orbital) conditions was accomplished in 2005 within the FOTON-M2 spacecraft mission using the BIOPAN facility. Based on promising results from this, a long-term (18 months) lichen exposure experiment was performed within the EXPOSE-E experiment onboard the International Space Station (ISS) in 2008/09.

The short-term extraterrestrial experiment and several preliminary ground-based laboratory tests were made in the European Space Agency's BIOPAN research facility. Two aluminium plates with several holes in which experimental lichen species were located were exposed to a variety of conditions. Half of the lichen samples were kept in the dark and the other half exposed to light passing through selective filters. Because the sensitivity of lichens to extraterrestrial doses of UV radiation was the main focus, two Antarctic lichen species thriving on Earth on stone surfaces in sunny locations and therefore tolerant of a wide range of temperatures and hydration/dehydration cycles were chosen for the experiments: *Rhizocarpon geographicum* and *Xanthoria elegans* (Fig. 26.6).

During pre-flight experiments that simulated space conditions, the responses of the two lichens to 40°C, vacuum or high UV radiation as single factors, and exposed to a combination of vacuum and high UV, were evaluated. These experiments proved that *R. geographicum* and *X. elegans* were good model species for an orbital experiment. The BIOPAN facility was subsequently installed onboard spacecrafts with dry lichen samples and sent into orbit. Before and after the

Fig. 26.6. (a) *Rhizocarpon geographicum*, a lichen species that has been tested in orbital experiments EXPOSE-E. The thalli presented here were collected from James Ross Island, Antarctica; (b) *Xanthoria elegans*, another lichen species that has been tested in orbital experiments EXPOSE-E. The characteristically orange thalli grow on stone surfaces in deglaciated areas on James Ross Island. Images © Milos Barták.

mission, the physiological status of the lichen samples was tested using chlorophyll fluorescence techniques that allowed scientists to evaluate the primary photochemical processes of photosynthesis in the lichen's symbiotic algae. Microscopic studies were also made to estimate the proportion of living to dead cells in the lichen thalli. It was concluded that photosynthetic activity was fully restored and the majority of algal cells survived space conditions with no signs of damage to cell wall or cell ultrastructure. The results obtained during the BIOPAN mission led to the expansion of this basic experimental design and in 2008/09, seeds of *Arabidopsis thaliana* (the most common experimental plant species), samples of lichens, fungi and their symbionts, were accommodated in a special case, sent into space and, using the specialized manipulative control mechanisms present onboard the International Space Station's EXPOSE-E facility, exposed to the extremely low pressure and high UV radiation conditions present in outer space. To mimic a Martian environment, some samples were simultaneously exposed to an artificial atmosphere consisting mainly of CO_2. All of the test organisms showed a considerable level of resistance to outer space conditions.

Pre- and post-flight measurements were also made to assess a lichen species' response to a combination of extreme factors. *Aspicilia fruticulosa*, the most resistant lichen species to space conditions, was simultaneously exposed to Martian-like conditions (UV radiation, freezing temperature, simulated Mars atmosphere) in a simulation chamber (de la Torre *et al.*, 2010). Despite the fact that *A. fruticulosa* was collected from central Spain, it exhibited a similarly high resistance to the experimental Martian-like conditions as the two Antarctic species tested in previous experiments, demonstrating the ubiquity of these remarkable tolerance mechanisms.

26.4.2 Fungi

Fungi collected in Antarctic habitats have also been tested at ground facilities to evaluate their tolerance to extraterrestrial factors. Among them, cryptoendolithic meristematic fungi are of great importance because they can survive inside rocks at freezing temperatures for a long time. In a dry state, Antarctic meristematic fungi can also withstand high temperatures due to their ability to produce extracellular polymeric substances (EPS), that help the species to survive even when dry and physiologically inactive for a long time. To test their ability to cope with extraterrestrial conditions, samples of sandstone were collected from the Linnaeus Terrace and other locations in Southern Victoria Land, Antarctica. From these, two species of rock-inhabiting fungi were isolated (*Cryomyces antarcticus* and *Cryomyces minteri*), grown on agars at controlled conditions and, subsequently, experimentally exposed to a variety of factors mimicking the Martian climate (Onofri *et al.*, 2008). The experiments were carried out at the German Aerospace Centre (DLR, Köln, Germany) as verification before the organisms could be sent to space within the EXPOSE-E space mission. The fungal isolates were tested under the following conditions: vacuum, varying temperature (−20/+20°C), monochromatic UV-C radiation and high polychromatic UV radiation. The responses of *C. antarcticus* and *C. minteri* to these conditions were evaluated both by culture (number of formed colonies) and staining methods. For all combinations of experimental factors, very good viability and resistance was found in all tested samples. It was, therefore, concluded that both *C. antarcticus* and *C. minteri* were suitable candidates for space flight and long-term permanence in space.

26.4.3 Animalia

Not only fungi and lichens have been tested for their survival in space. Tardigrades (water bears), small animals approximately 1 mm in length that can be found living in mosses in various Earth ecosystems, enter a dormant state when dehydrated. In this state, they lose almost all water and their bodies contract forming a so-called tun. Nevertheless, dehydrated tardigrades manage to maintain all of their body structures and, when rehydrated, are able to restore their full function. Due to this ability, tardigrades were used for a unique astrobiological experiment. The animals were first dehydrated to induce a state of dormancy and then sent into space onboard the FOTON-M3 spacecraft to orbit 270 km above the Earth (Jönsson *et al.*, 2002). During the space mission, the chamber containing the tardigrades was opened and the animals were exposed to vacuum and cold. Some of the samples were also exposed to UV radiation coming directly from the Sun. After the return of the spacecraft to Earth, the animals were rehydrated and some of them survived. Thus, they proved highly resistant to outer space conditions and it was concluded that they had the potential to withstand the extreme space environment, i.e. the cosmic rays, a near vacuum and freezing cold.

Overall, therefore, the three groups of organisms discussed above, lichen, fungi and tardigrades, suggest that life is entirely possible in extraterrestrial environments.

26.5 Panspermia – Orbital, Extraterrestrial Environments

The theory of Panspermia (Arrhenius, 1908) states that life can be dispersed in space, and even from planet to planet, in the form of micropropagules, e.g. spores or microorganisms. It was proven that meteoric material approximately 1 cm in thickness is sufficient to protect microorganisms against the negative effects of the UV radiation that they would be exposed to during transport, indicating that meteorites are potential vectors for the dispersal of life in and through space (Horneck *et al.*, 2001, 2008).

This concept of meteoric vectors is referred to as Lithopanspermia (Nicholson *et al.*, 2000; Mileikowsky *et al.*, 2000) and has both supporters and critics. Presently, an enormous scientific effort is focused on the question of whether or not microorganisms, such as fungi and bacteria, embedded in meteoric or rock material can survive space conditions for a sufficient period of time to support the Lithopanspermia theory. The theory assumes that microorganisms, embedded in rocks, must cope with and survive three steps: (i) escape from the planet by impact ejection; (ii) a journey through space over extended time periods; and (iii) landing on another planet or astral body. It would appear that steps (i) and (iii) (escape and landing) are the critical phases since they subject the organisms to extremely high velocities and potentially high pressure and temperature. To test whether microbial life is capable of surviving such conditions, bacterial endospores from *Bacillus subtilis* were exposed to simulated meteorite impact conditions for collisionally-produced rock fragments from a medium-sized terrestrial planet, using a special explosive set-up (Horneck *et al.*, 2001). The most extreme conditions that *B. subtilis* had to survive were a peak shock pressure of 32 GPa and a post-shock temperature of approximately 250°C. Results showed that a substantial fraction of spores were able to survive the severe shock pressure and extreme temperatures, lending support to the possibility of Lithopanspermia. Moreover, recently the possibility of life spores being dispersed through space within ice blocks has been suggested in the Ice Impact Ejecta concept (Houtkooper, 2010).

26.6 Recent and Future Directions for Astrobiology

Astrobiology is an emerging field with tremendous potential and influences far beyond its own boundaries. What the origin of life on Earth was and whether life exists on other planetary bodies are key questions waiting to be answered. It is indisputable that in spite of promising results with bacteria, fungi, lichens and other organisms, their ability to survive extraterrestrial conditions must be tested in much more detail. Extreme environments are promising natural field sites in which to test the limits of life and investigate adaptations to harsh terrestrial and extraterrestrial conditions. Space simulation facilities have also been developed in a few countries to allow researchers to expose both biological and physical samples to simulated extraterrestrial conditions. Two groups of experiments are especially important: (i) Experiment Verification Tests (EVT) that help scientists to specify the test parameters for space flights with associated biological programmes; and (ii) Experiment Sequence Tests (EST), which simulate sample assemblies, exposure to selected space parameters, and sample disassembly.

Recent technological improvements and the possibility of upgrading existing experimental designs allow us to evaluate not only the responses of model organisms to a variety of simulated extraterrestrial conditions, but also to exploit more advanced experimental designs on board spacecraft. In the near future we can expect to be able to take an increasing number of biological experiments into space and examine the physical and chemical properties, as well as the potential for survival and biological evolution, of extremophilic organisms beyond Earth. Moreover, testing the ability of terrestrial organisms to survive on Mars or other extraterrestrial bodies will become technologically possible using landing modules and the concept of robotized transplant experiments.

On one hand, taking experiments into space and bringing samples home carries with it risks related to the biological safety of such experiments and, therefore, ethical issues that must be carefully evaluated. On the other hand, the coming decades will be critical for astrobiological missions to Mars and Jupiter's moon, Europa. These missions will be crucial for enhancing our understanding of the limits of life, its potential to exist beyond planet Earth, and for finding important clues to the origin of life in our planet. It is also certain to offer us new opportunities for biotechnological and biomedical development. Overall, any knowledge we gain will impact not only science in general but human thinking as well.

References

Abyzov, S.S., Mitskevich, I.N., Poglazova, M.N., Barkov, I.N., Lipenkov, V.Y., Bobin, N.E., Koudryashov, B.B., Pashkevich, V.M. and Ivanov, M.V. (2001) M croflora in the basal strata at Antarctic ice core above the Vostok lake. *Advances in Space Research* 28 701–706.

Ahring, B.K. and Westerman, P. (1987) Thermophilic anaerobic degradation of butyrate by a butyrate-utilizing bacterium in coculture and triculture with methanogenic bacteria. *Applied and Environmental Microbiology* 53, 429–433.

Arrhenius, S. (1908) *Worlds in the Making: The Evolution of the Universe*. Harper and Row, New York, 264 pp.

Barták, M., Váczi, P., Hájek, J. and Smykla, J. (2007) Low temperature limitation of primary photosynthetic processes in Antarctic lichens *Umbilicaria antarctica* and *Xanthoria elegans*. *Polar Biology* 31, 47–51.

Belkin, S., Wirsen, C.O. and Jannasch, H.W. (1986) A new sulfur-reducing, extremely thermophilic eubacterium from a submarine thermal vent. *Applied and Environmental Microbiology* 51, 1180–1185.

Bouthier de la Tour, C., Portener, C., Nadal, M., Stetter, K.O., Forterre, P. and Duguet, M. (1990) Reverse gyrase, a hallmark of the hypothermophilic. *Journal of Bacteriology* 172, 6803–6808.

Bryant, F.O. and Adams, M.W.W. (1989) Characterization of hydrogenase fron the hyperthermophilic archaebacterium, *Pyrococcus furiosus*. *Journal of Biological Chemistry* 264, 5070–5079.

Clark, B.C. and Van Hart, D.C. (1981) The salts of Mars. *Icarus* 45, 370–378.

Dahl, C., Koch, H.G., Keuken, O. and Truper, H.G. (1990) Purification and chelation and characterization of ATP sulfurylase from the extremely thermophilic archaebacterial sulfate-reducer, *Archaeoglobus fulgidus*. *FEMS Microbiology Letters* 67, 27–32.

Davila, A.F., Duport, L.G., Melchiorri, R., Jänchen, J., Valea, S., de Los Rios, A., Fairén, A.G., Möhlmann, D., McKay, C.P., Ascaso, C. and Wierzchos, J. (2010) Hygroscopic salts and the potential for life on Mars. *Astrobiology* 10, 617–628.

de la Torre, R., Sancho, L.G., Horneck, G., de los Ríos, A., Wierzchos, J., Olsson-Francis, K., Cockell, C.S., Rettberg, P., Berger, T., de Vera, J.-P.P., Ott, S., Frías, J.M., Melendi, P.G., Lucas, M.M., Reina, M., Pintado, A. and Demets, R. (2010) Survival of lichens and bacteria exposed to outer space conditions – results of the Lithopanspermia experiments. *Icarus* 208, 735–748.

DiMarco, A.A., Smet, K.A., Konisky, J. and Wolfe, R.S. (1990) The formylmethanofuran-tetrahydromethanopterin formyltransferase from *Methanobacterium thermoautotrophicum*. *Journal of Biological Chemistry* 265, 472–476.

Dwyer, D.F., Weeg-Aerssens, E., Shelton, D.R. and Tiedje, J.M. (1988) Bioenergetic conditions of butyrate metabolism by a syntrophic, anaerobic bacterium in coculture with hydrogen-oxidizing methanogenic and sulfidogenic bacteria. *Applied and Environmental Microbiology* 54, 1354–1359.

Fernández-Remolar, D., Gómez-Elvira, J., Gómez, F., Sebastian, E., Martín, J., Rodriguez, J.A., Torres, J., González-Kesler, C. and Amils, R. (2004) The Tinto River, an extreme acidic environment under control of iron, as an analog of the *Terra Meridiani* hematite site of Mars. *Planetary and Space Science* 52, 239–248.

Formisano, V., Atreya, S., Encrenaz, T., Ignatiev, N. and Giuranna, M. (2004) Detection of methane in the atmosphere of Mars. *Science* 306, 1758–1761.

Friedmann, E.I. and Ocampo, R. (1976) Endolithic blue-green algae in the Dry Valleys: primary producers in the Antarctic desert ecosystem. *Science* 193, 1247–1249.

Friedmann, E.I. and Ocampo-Friedmann, R. (1984) The Antarctic cryptoendolithic ecosystem: relevance to exobiology. *Origins of Life and Evolution of Biospheres* 14, 771–776.

Gauslaa, Y. and Solhaug, K.-A. (2004) Photoinhibition in lichens depends on cortical characteristics and hydration. *The Lichenologist* 36, 133–143.

Gómez, F., Aguilera, A. and Amils, R. (2007) Soluble ferric iron as an effective protective agent against UV radiation: Implications for early life. *Icarus* 191, 352–359.

Gómez, F., Prieto-Ballesteros, O., Fernández-Remolar, D., Rodríguez-Manfredi, J.A., Fernández-Sampedro, M., Postigo Cacho, M., Torres Redondo, J., Gómez-Elvira, J. and Amils, R. (2008) Microbial diversity in a permafrost environment of a volcanic-sedimentary Mars analog: Imuruk Lake, Alaska. In: Kane, D.L. and Hinkel, K.M. (eds) *Ninth International Conference on Permafrost*, Institute of Northern Engineering, University of Alaska Fairbanks, Alaska, pp. 503–528. ISBN: 978-0-9800179-2-2.

Gómez, F., Rodríguez-Manfredi, J.A. and Amils, R. (2011) Protected endolithic niches on Earth as models for habitability on Mars. *Geophysical Research Abstracts*, Vol. 13, EGU General Assembly.

Gómez-Silva, B., Rainey, F.A., Warren-Rhodes, K.A., McKay, C.P. and Navarro-González, R. (2007) Atacama Desert soil microbiology. In: Dion, P. and Nautiyal, C.S. (eds) *Microbiology of Extreme Soils*. Springer-Verlag, Berlin, Heidelberg, Germany, 388 pp.

Horneck, G., Rettberg, P., Reitz, G., Wehner, J., Eschweiler, U., Strauch, K., Panitz, C., Starke, V. and Baumstark-Khan, C. (2001) Protection of bacterial spores in space, a contribution to the discussion on Panspermia. *Origins of Life and Evolution of the Biospheres* 31, 527–547.

Horneck, G., Stoeffler, D., Ott, S., Hornemann, U., Cockell, C.S., Moeller, R., Meyer, C., de Vera, J.P., Fritz, J., Schade, S. and Artemieva, N. (2008) Microbial rock inhabitants survive hypervelocity impacts on Mars-like host planets: first phase of lithopanspermia experimentally tested. *Astrobiology* 8, 17–44.

Houtkooper, J.M. (2010) Glaciopanspermia: seeding the terrestrial planets with life? *Planetary and Space Science*, doi:1001016/j.pss.2010.09.003.

Jönsson, K.I., Rabbow, E., Schill, R.O., Harms-Ringdahl, M. and Petra Rettberg, P. (2002) Tardigrades survive exposure to space Earth orbit. *Current Biology* 17, 729–731.

Kapitsa, A., Ridley, J.K., Robin, G. de Q., Siegert, M.J. and Zotikov, I. (1996) Large deep freshwater lake beneath the ice of central East Antarctica. *Nature* 381, 684–688.

Kjems, J., Leffers, H., Olesen, T., Holz, I. and Garrett, R.A. (1990) Sequence, organization and transcription of the ribosomal RNA operon and the downstream TRNA and protein genes in the archaebacterium *Thermoflum pendens*. *Systematic and Applied Microbiology* 13, 117–127.

Krasnopolsky, V.A., Maillard, J.P. and Owen, T.C. (2004) Detection of methane in the Martian atmosphere: evidence for life? *Icarus* 172, 537–547.

Lange, O.L., Green, T.G.A. and Heber, U. (2001) Hydration-dependent photosynthetic production of lichens: what do laboratory studies tell us about field performance? *Journal of Experimental Botany* 52, 2033–2042.

Mazur, P. (1980) Limits to life at low temperatures and at reduced water contents and water activities. *Origins of Life* 10, 137–159.

McKay, D.S., Gibson, E.K., Jr, Thomas-Keprta, K.L., Vali, H., Romanek, C.S., Clemett, S.J., Chillier, X.D.F., Maechling, C.R. and Zare, R.N. (1996) Search for past life on Mars: possible relic biogenic activity in Martian meteorite ALH84001. *Science* 273, 924–930.

Mileikowsky, C., Cucinotta, F.A., Wilson, J.W., Gladman, B., Horneck, G., Lindegren, L., Melosh, H.J., Rickman, H., Valtonen, M. and Zheng, J.Q. (2000) Natural transfer of viable microbes in space. Part 1: From Mars to Earth and Earth to Mars. *Icarus* 145, 391–427.

Navarro-González, R., Rainey, F.A., Molina, P., Bagaley, D.R., Hollen, B.J., De la Rosa, J., Small, A.M., Quinn, R.C., Grunthaner, F.J., Cáceres, L., Gómez-Silva, B. and McKay, C.P. (2003) Mars-like soils in the Atacama Desert, Chile, and the dry limit of microbial life. *Science* 302, 1018–1021.

Nicholson, W.L., Munakata, N., Horneck, G., Melosh, H.J. and Setlow, P. (2000) Resistance of *Bacillus* endospores to extreme terrestrial and extraterrestrial environments. *Microbial and Molecular Biology Reviews* 64, 548–572.

Onofri, S., Barreca, D., Selbmann, L., Isola, D., Rabbow, E., Horneck, G., de Vera, J.-P.P., Hatton, J. and Zucconi, L. (2008) Resistance of Antarctic black fungi and cryptoendolithic communities to simulated space and Martian conditions. *Studies in Mycology* 61, 99–109.

Oswald, G.K.A. and Robin, G. de Q. (1973) Lakes beneath the Antarctic Ice Sheet. *Nature* 245 (5423): 251–2544. doi:10.1038/245251a0.

Price, P.B., Nagornov, O.V., Bay, R., Chirkin, D., He, Y., Miocinovic, P., Richards, A., Woschnagg, K., Koci, B. and Zagorodnov, V. (2002) Temperature profile for glacial ice at the South Pole: implications for life in a nearby subglacial lake. *Proceedings of the National Academy of Sciences of the United States of America* 99, 7844–7847.

Siegert, M.J., Cynan Ellis-Evans, J., Tranter, M., Mayer, C., Petit, J.-R., Salamatin, A. and Priscu, J.C. (2001) Physical, chemical and biological processes in Lake Vostok and other Antarctic subglacial lakes. *Nature* 414 (6864): 603–609.

Smith, P.H., Tamppari, L.K., Arvidson, R.E., Bass, D., Blaney, D., Boynton, W.V., Carswell, A., Catling, D.C., Clark, B.C., Duck, T., DeJong, E., Fisher, D., Goetz, W., Gunnlaugsson, H.P., Hecht, M.H., Hipkin, V., Hoffman, J., Hviid, S.F., Keller, H.U., Kounaves, S.P., Lange, C.F., Lemmon, M.T., Madsen, M.B., Markiewicz, W.J., Marshall, J., McKay, C.P., Mellon, M.T., Ming, D.W., Morris, R.V., Pike, W.T., Renno, N., Staufer, U., Stoker, C., Taylor, P., Whiteway, J.A. and Zent, A.P. (2009) H_2O at the Phoenix Landing Site. *Science* 325, 58–61.

Wynn-Williams, D.D. and Edwards, H.G.M. (2000) Antarctic ecosystems as models for extraterrestrial surface habitats. *Planetary and Space Science* 48, 1065–1075.

27 Concluding Remarks

Elanor M. Bell
Scottish Institute for Marine Science, Argyll, UK

27.1 Extremes and Extremophiles

There are a multitude of environments on planet Earth and beyond defined as extreme, and organisms are continually being discovered that are capable not only of surviving but also thriving in them. These organisms are the so-called extremophiles. Some extremophiles merely tolerate an extreme and become dominant over other organisms, whilst other extremophiles are supremely well adapted to the conditions and thrive without release of competition. In many extreme environments the inhabitants are in fact poly-extremophiles that are adapted to more than one extreme condition at the same time.

The terms extreme and extremophile are, however, relatively anthropocentric and we humans judge habitats and their inhabitants based on what would be considered extreme for our existence. In fact, from an extremophile perspective, the way *we* live is *extreme*. Similarly, we define extremes of the physical (abiotic) environment (extreme heat, cold, wet, dry, anoxic, toxic, etc.) around a baseline that is acceptable for human life. This construct does not take into account either the temporal dynamics of an environment, extreme events such as natural disasters, or the developmental cycle of an organism in which the tolerance of the same environment differs at different stages in the life cycle (e.g. newborn versus adult). The construct fails to consider environments in which biotic interactions are so extreme that they determine community structure and the survival of an individual. It also omits transition zones – ecotones – that are continually shifting and dynamic regions, often exhibiting a wider range of abiotic and biotic conditions over an annual period than the adjacent environments. These ecotones can be considered extreme in that they require organisms to continually adapt their behaviour, phenology and interactions with other species.

As scientific, biotechnological and biomedical interest in extreme environments and extremophiles increases, and technological advances are made, investigation of them will begin to address these omissions. However, at present most of what we know about extreme environments comes from the study of more traditionally defined habitats: extremely cold or hot, extremely deep and subject to high hydrostatic pressure, acidic or alkaline, extremely saline, hypoxic and anoxic, or subject to high doses of radiation. All these environments play host to diverse communities of extremophilic organisms, principally prokaryotic Archaea and

Bacteria, which are adapted in some way to their extreme habitat. In all instances, the extremophiles exhibit measurable rates of growth, reproduction, metabolism and productivity, some of which can exceed those measured in equivalent non-extremophiles.

27.2 Lessons from Past Extremes

Extreme environments offer us a window into the past. The idea that life originated on the surface of our planet, where it was strongly dependent on a hypothetical primordial soup, has recently come up against strong competition (Pedersen, 2000). There are now several suggestions that life originated in the form of a thermophilic lithotroph (Huber and Wächtershäuser, 1997) and that the birthplace was, perhaps, a hydrothermal vent area (Russell and Hall, 1997). Moreover, a wealth of information on ancient climates, extreme environments and life is retained in the rock record, the reading of which is becoming ever more refined as technology and scientific techniques advance. Past changes in the UV-B flux, temperature and changes in the atmospheric composition of O_2 and CO_2 created past extreme environments, and have left behind clues in the geological record that we can use to predict what might happen in the future. On shorter time scales, the two largest mass extinction events at the end of the Permian and the end of the Cretaceous can both be regarded as periods in which extreme environments existed, albeit for different reasons. Our increasing knowledge of these has led many commentators to suggest that we are currently in the grip of the sixth mass extinction event as a result of anthropogenically-driven changes. In a world where we are increasingly faced with anthropogenically-induced extremes, such as prolonged climate change, it would be good to remember 'that the past may be the key to the future' (Beerling, 1998). Therefore it may be time for us to start learning some lessons from the past.

27.3 Influence on Global Processes and Anthropogenically Induced Extremes

From the holistic viewpoint, modern extreme environments cannot be considered remote and separate from the rest of the biosphere. At a basic level, many extremophiles are the principal food source for higher organisms, e.g. the brine shrimp, *Artemia monica*, and alkali fly, *Ephydra hians*, abundant in extremely alkaline Mono Lake, USA, are important food sources for migratory birds (Cooper *et al.*, 1984). At a more complex level, many extreme environments and extremophiles potentially exert profound influences on global processes. For instance, permafrost regions currently act as carbon sinks but may become net producers of greenhouse gases if global warming continues to cause their disappearance, thus providing a positive feedback to the warming (Thomas *et al.*, 2008). There is also the sobering possibility of rapid melting of methane gas hydrates, also as a result of continued global warming, that demonstrates how microbes in the deep biosphere could exert an enormous influence on global processes (Norris and Röhl, 1999; Pedersen, 2000).

Thus, our increasing scientific interest in extreme environments and extremophiles goes well beyond simple curiosity about different environments on Earth. The continuing contamination of groundwater through the creation of toxic waste dumps and municipal landfills, industrial chemicals, pesticides and fertilizers, and the mobilization of toxic metals by acid rains have created serious environmental problems that demand the attention of the scientific community (Pedersen, 1993, 2000). Moreover, many countries are using, or are seriously considering using extreme environments, such as the deep sea and/or deep biosphere, as repositories for low to high level nuclear waste and carbon sequestration. Hopefully, by enhancing our knowledge of extreme environments and extremophiles we will increase our chances of mitigating environmental problems, e.g. via bioremediation of contaminated groundwater, and be able to pursue dumping and storage of environmental pollutants more safely.

27.4 Use of Extremophiles in Allowing Humans to Live Outside our Own 'Extreme Envelope'

The discovery of extreme environments and flourishing communities of organisms living in them on Earth has opened up a huge range of biotechnological and biomedical opportunities and, as a result, this area of extremophile research has expanded rapidly in the last 20 years. Now genomes are regularly sequenced, patents filed and concerted national and international programmes of research are launched on a regular basis. The extremophiles that can live at the extreme limits of geophysical and geochemical environmental conditions have developed highly efficient mechanisms or survival strategies to enable them to do so. Often these involve the development of novel metabolic pathways, the use of specialized enzymes (extremozymes) and/or the secretion of various biomolecules (extremolytes). These extremozymes and extremolytes are coming under increasing scrutiny from scientists and technologists for their realized, and yet to be discovered, potential to help us live outside our own, human, 'extreme envelope'. Many of these products have been evaluated and are already used medically and industrially, for others there is still a need to find novel applications, and yet more remain to be discovered.

27.5 Outlook for the Discovery of Life on Other Planets

Last but not least, our continued discovery of extreme environments and understanding of the extreme conditions that life is capable of enduring has also, literally, opened up new worlds of possibilities for the discovery of life on other planets and extraterrestrial bodies with similarly extreme environments as those described on Earth. This also lends support to the theory of Panspermia, the transport of life from one planet to another via, for example, meteorites, bringing us full circle in the debate about how and where life on Earth originated.

27.6 Conclusion

In conclusion, the study of extreme environments and extremophiles is a relatively new scientific field, but one that is expanding rapidly and regularly leads to exciting new discoveries. Continued technological advancements allow us to delve further into the past to investigate the evolution of life as we know it, probe more logistically difficult realms, look ever more closely at cellular ultrastructures and metabolic processes, and extend our search for life deeper into space. They may also help us to make more accurate predictions about the future of life on Earth and how we humans might mitigate some of the new global environmental extremes we are creating.

References

Beerling, D.J. (1998) The future as the key to the past for palaeobotany? *Trends in Ecology and Evolution* 13, 311–316.

Cooper, S.D., Winkler, D.W. and Lenz, P.H. (1984) The effects of grebe predation on a brine shrimp population. *Journal of Animal Ecology* 53, 51–64.

Huber, C. and Wächtershäuser, G. (1997) Activated acetic acid by carbon oxidation on (Fe, Ni)S under primordial conditions. *Science* 276, 245–247.

Norris, R.D. and Röhl, U. (1999) Carbon cycling and chronology of climate warming during the Palaeocene/Eocene transition. *Nature* 401, 775–778.

Pedersen, K. (1993) The deep subterranean biosphere. *Earth-Science Reviews* 34, 243–260.

Pedersen, K. (2000) Exploration of deep intraterrestrial microbial life: current perspectives. *FEMS Microbiology Letters* 185, 9–16.

Russell, M.J. and Hall, A.J. (1997) The emergence of life from iron monosulphide bubbles at a submarine hydrothermal redox and pH front. *Journal of the Geological Society of London* 154, 377–402.

Thomas, D.N., Fogg, G.E., Convey, P., Fritsen, C.H., Gili, J.-M., Gradinger, R., Laybourn-Parry, J., Reid, K. and Walton, D.W.H. (2008) *The Biology of Polar Regions*, 2nd edn. Biology of Habitats Series. Oxford University Press, Oxford, UK, 416 pp.

Index